ATOM-MOLECULE COLLISION THEORY

A Guide for the Experimentalist

PHYSICS OF ATOMS AND MOLECULES

1976:
ELECTRON AND PHOTON INTERACTIONS WITH ATOMS
Edited by H. Kleinpoppen and M.R.C. McDowell

1978:
PROGRESS IN ATOMIC SPECTROSCOPY, Parts A and B
Edited by W. Hanle and H. Kleinpoppen

1979:
ATOM–MOLECULE COLLISION THEORY: A Guide for the Experimentalist
Edited by Richard B. Bernstein

In preparation:
COHERENCE AND CORRELATION IN ATOMIC COLLISIONS
Edited by H. Kleinpoppen and J.F. Williams

THEORY OF ELECTRON–ATOM COLLISIONS
By P.G. Burke and C.J. Joachain

ATOM-MOLECULE COLLISION THEORY

A Guide for the Experimentalist

Edited by

Richard B. Bernstein

Columbia University
New York, New York

PLENUM PRESS · NEW YORK AND LONDON

Library of Congress Cataloging in Publication Data

Main entry under title:

Atom–molecule collision theory.

(Physics of atoms and molecules)
Includes index.
1. Collisions (Nuclear physics) I. Bernstein, Richard Barry, 1923-
QC794.6.C6A82 539.7'54 78-27380
ISBN-13: 978-1-4613-2915-2 e-ISBN-13: 978-1-4613-2913-8
DOI: 10.1007/ 978-1-4613-2913-8

First Printing – June 1979
Second Printing – April 1984

Softcover reprint of the hardcover 1st edition 1979
A Division of Plenum Publishing Corporation
233 Spring Street, New York, N.Y. 10013

Contributors

R. B. Bernstein, Department of Chemistry, Columbia University, New York, New York 10027

M. S. Child, Theoretical Chemistry Department, University of Oxford, Oxford OX1 3TG, England

D. J. Diestler, Department of Chemistry, Purdue University, West Lafayette, Indiana 49707

D. A. Dixon, Department of Chemistry, University of Minnesota, Minneapolis, Minnesota 55455

W. R. Gentry, Department of Chemistry, University of Minnesota, Minneapolis, Minnesota 55455

J. L. Kinsey, Department of Chemistry, Massachusetts Institute of Technology, Cambridge, Massachusetts 02139

D. J. Kouri, Departments of Chemistry and Physics, University of Houston, Houston, Texas 77004

P. J. Kuntz, Hahn-Meitner-Institut für Kernforschung, 1000 Berlin 39, West Germany

R. D. Levine, Department of Physical Chemistry, The Hebrew University, Jerusalem, Israel

J. C. Light, The James Franck Institute and The Department of Chemistry, University of Chicago, Chicago, Illinois 60637

J. T. Muckerman, Chemistry Department, Brookhaven National Laboratory, Upton, New York 11973

M. D. Pattengill, Chemistry Department, University of Kentucky, Lexington, Kentucky 40506

H. Pauly, Max-Planck-Institut für Strömungsforschung, 3400 Göttingen, West Germany

J. Reuss, Fysisch Laboratorium, Katholieke Universiteit, Nijmegen, The Netherlands

H. F. Schaefer III, Department of Chemistry, University of California, Berkeley, California 94720

D. H. Secrest, School of Chemical Sciences, University of Illinois, Urbana, Illinois 61801

S. Stolte, Fysisch Laboratorium, Katholieke Universiteit, Nijmegen, The Netherlands

D. G. Truhlar, Department of Chemistry, University of Minnesota, Minneapolis, Minnesota 55455

R. E. Wyatt, Department of Chemistry, The University of Texas, Austin, Texas 78712

Preface

The broad field of *molecular collisions* is one of considerable current interest, one in which there is a great deal of research activity, both experimental and theoretical. This is probably because elastic, inelastic, and reactive intermolecular collisions are of central importance in many of the fundamental processes of chemistry and physics.

One small area of this field, namely atom–molecule collisions, is now beginning to be "understood" from first principles. Although the more general subject of the collisions of polyatomic molecules is of great importance and intrinsic interest, it is still too complex from the viewpoint of theoretical understanding. However, for atoms and simple molecules the essential theory is well developed, and computational methods are sufficiently advanced that calculations can now be favorably compared with experimental results.

This "coming together" of the subject (and, incidentally, of physicists and chemists!), though still in an early stage, signals that the time is ripe for an appraisal and review of the theoretical basis of atom–molecule collisions.

It is especially important for the experimentalist in the field to have a working knowledge of the theory and computational methods required to describe the experimentally observable behavior of the system. By now many of the alternative theoretical approaches and computational procedures have been tested and intercompared. More-or-less optimal methods for dealing with each aspect are emerging. In many cases working equations, even schematic algorithms, have been developed, with assumptions and caveats delineated.

Thus a book on atom–molecule collision theory would encourage the experimentalist in the field to make full use of the best available theoretical–computational methods in the interpretation and analysis of the data and also in the design of new experiments.

The purpose, then, of this compendium is to serve as a state-of-the-art "handbook," a timely, practical reference work for the *user* of atom–molecule collision theory. It is indeed intended as *A Guide for the Experimentalist.*

Richard B. Bernstein

Department of Chemistry
Columbia University
New York, New York

Contents

Chap. 1. Introduction to Atom–Molecule Collisions: The Interdependency of Theory and Experiment

Richard B. Bernstein

Chap. 2. Interaction Potentials I: Atom–Molecule Potentials

Henry F. Schaefer III

Chap. 3. Interaction Potentials II: Semiempirical Atom–Molecule Potentials for Collision Theory

P. J. Kuntz

Chap. 4. Elastic Scattering Cross Sections I: Spherical Potentials

H. Pauly

Chap. 5. Elastic Scattering Cross Sections II: Noncentral Potentials
S. Stolte and J. Reuss

Chap. 6. Inelastic Scattering Cross Sections I: Theory
John C. Light

Chap. 7. Inelastic Scattering Cross Sections II: Approximation Methods
Donald Jack Kouri

Chap. 8. Rotational Excitation I: The Quantal Treatment
Don Secrest

Chap. 9. Rotational Excitation II: Approximation Methods
Donald Jack Kouri

Chap. 10. Rotational Excitation III: Classical Trajectory Methods
M. D. Pattengill

Chap. 11. Vibrational Excitation I: The Quantal Treatment
Don Secrest

Chap. 12. Vibrational Excitation II: Classical and Semiclassical Methods
W. Ronald Gentry

Chap. 13. Electronic Excitation: Nonadiabatic Transitions

M. S. Child

Chap. 14. Reactive Scattering Cross Sections I: General Quantal Theory

John C. Light

Chap. 15. Reactive Scattering Cross Sections II: Approximate Quantal Treatments

Robert E. Wyatt

Chap. 16. Reactive Scattering Cross Sections III. Quasiclassical and Semiclassical Methods

Donald G. Truhlar and James T. Muckerman

Chap. 17. Direct-Mode Chemical Reactions I: Methodology for Accurate Quantal Calculations

Robert E. Wyatt

Chap. 18. Direct-Mode Chemical Reactions II: Classical Theories

Donald G. Truhlar and David A. Dixon

Chap. 19. Complex-Mode Chemical Reactions: Statistical Theories of Bimolecular Reactions

John C. Light

Chap. 20. Collision-Induced Dissociation I: Quantal Treatment

Dennis J. Diestler

Chap. 21. Collision-Induced Dissociation II: Trajectories and Models

P. J. Kuntz

Chap. 22. Information-Theoretic Approach: Application to Molecular Collisions

R. D. Levine and J. L. Kinsey

1

Introduction to Atom–Molecule Collisions: The Interdependency of Theory and Experiment

Richard B. Bernstein

1. General Introduction

For a century, since Boltzmann, it has been recognized that intermolecular collisions are primarily responsible for most observable rate phenomena, both physical and chemical, in fluid systems. The goal of research in the field of heavy-particle collisions is to gain a detailed, quantitative description (and thus an "understanding") of these rate processes. starting from first principles.[1]

Landmarks in our progress toward the quantitative understanding of physical and chemical rate processes can often be associated with a few key books, which now stand out sharply from a distance. Among these are the 1939 monograph by Chapman and Cowling, *Mathematical Theory of Non-Uniform Gases*[2]; the 1941 text by Glasstone, Laidler, and Eyring, *Theory of Rate Processes*[3]; the seminal 1949 edition of the monograph of Mott and Massey, *The Theory of Atomic Collisions*[4]; and the "green bible" of 1954 by Hirschfelder, Curtiss, and Bird, *Molecular Theory of Gases and Liquids*.[5]

Indeed, the conceptual approach to the theory of molecular collisions has been well known for decades. Starting with the Hamiltonian of the system (needing only the masses and Coulomb's law) the electronic potential energy surfaces governing the motion of the nuclei can be computed (e.g., by variational solution of the Schrödinger equation). Then the various

Richard B. Bernstein • Department of Chemistry, Columbia University, New York, New York 10027.

quantum state-to-state scattering cross sections can be obtained from the S matrix generated by the (usually numerical) solution of the wave equation for the nuclear motion.[2,6,7]

Unfortunately, only for few-electron systems are *ab initio* potential functions obtainable with sufficient accuracy for use in the scattering calculations.[8] And then, only for simple atom–atom or atom–diatom systems can one execute accurately these scattering calculations. Of course, with the ever-increasing efficiency of modern computers it should soon be possible to follow this systematic and rigorous path to the state-to-state cross sections and the various rate phenomena.[9] However, progress along this totally *ab initio* route is much too slow for the practitioner in atomic and molecular physics, fluid dynamics, or chemical kinetics. Progress in these fields is to a large extent dependent upon experiment, from which regularities and correlations can be inferred. As the theory becomes better implemented, these *a posteriori* empiricisms will gradually be replaced by *a priori* predictions.

With the development of modern electronic and vacuum instrumentation, a wealth of new and detailed experimental data on atomic and molecular collision cross sections has been forthcoming, covering the gamut from elastic to inelastic to reactive scattering.[10,11] The venerable molecular beam technique,[12] stemming from L. Dunoyer in 1911 and first seriously applied to molecular scattering by O. Stern in the twenties, has, since the fifties, been transformed into a versatile, general methodology for the study of intermolecular collisions.[10,13] Measurements of the angular and energy distributions of the scattering of atoms, molecules, and ions are performed as a function of collision energy (routinely), and even of initial quantum state (for suitable systems). The 1977 symposium volume, *State-to-State Chemistry*,[14] emphasizes the chemical implications of collision cross sections, as does the Eyring compendium, *Physical Chemistry–An Advanced Treatise*.[15]

The field of intermolecular collisions is one characterized by a strong admixture of theory with experiment. This interdependence, this close interaction, can be recognized as a recurring theme running through most of the literature on molecular dynamics.[10,11,16,17] It will be seen throughout this book as well.

A number of excellent texts on formal collision theory are available, the ones most relevant to the subject of this book being *Quantum Mechanics of Molecular Rate Processes*, by Levine,[6] and *Molecular Collision Theory*, by Child.[18] The present volume, however, has a different intent than that of a textbook. First, it is somewhat restricted in scope to a subclass of intermolecular interactions (namely, atom–molecule collisions) for which the theory is at present tractable. Second, and most important, it is directed specifically toward the needs of the experimentalist in the field of molecular scattering. In what follows we shall try to delineate the experimentalist's "need to know."

2. The Experimentalist's "Need to Know"

In the field of atomic and molecular physics the successful experimentalist is required to know more than a modicum of theory. Even casual inspection of the literature immediately reveals the strong interaction, even dependence, of experiment upon theory, from concept through execution to analysis.[19-24]

Experimentalists in the area of molecular collisions are required to have expertise in many advanced laboratory techniques, since the research usually requires fairly elaborate, often costly physical equipment and complex electronic instrumentation for data acquisition. Yet all this technical enterprise is for naught unless the experiment can provide interpretable numbers which have at least some connected theoretical significance. There are several levels at which theory plays a role in such research.

First, one requires some assurance that the *magnitude* of the phenomenon is such that it could be observed with an existing apparatus or a more sensitive one under design. With every advance in the state of the art in the laboratory, there is an accompanying widening of the scope of possible experiments in the field. Nevertheless, at each stage of advancement one always requires a "ball-park estimate" (i.e., of the expected "signal level" of the effect to be studied), based on theory, before any experiment can even be seriously contemplated. The signal-to-noise (S/N) ratio is the iron hand which rules in the laboratory. The problem of noise and its reduction is the job of the experimenter. But for the estimation of the signal, the experimentalist in atom–molecule collisions needs to know the approximate cross sections or rates for the process in question. This is the first level at which the theory is called upon, i.e., in experiment design.

The next level is at the point of data transformation. Here one requires algorithms for the conversion of these laboratory "signals" into more meaningful. "scientific" quantities such as differential or integral cross sections,[13,22,24] state-to-state rate constants,[14] etc. Often the transformation procedure is nontrivial, for example in the conversion of scattered intensity data from the laboratory to the center-of-mass system or the deconvolution of observed excitation functions from residual relative velocity distributions in the molecular beam scattering field. In other types of experiments, especially those involving bulk systems (i.e., laser excitation and collisional relaxation), problems arise in the solution of the coupled rate equations to yield the desired, individual, state-to-state rate coefficients even when real-time observations, say, of internal state populations, are available. This is the second level at which theory is called upon, i.e., in data analysis, *per se*.

Then, after the scientific results are in hand, the experimentalist has a need to know how to estimate the parameters of some simple theoretical model which can account, at least in a phenomenological way, for the measured collision cross sections or rates. Theory must provide some kind of practical algorithm that enables the experimentalist to "reproduce" his data (or at least the trends or systematics). The penultimate goal is a graphical or tabular presentation in which the experimental data serve as the arbiter of various theoretical–computational approximation methods.

Finally, of course, one faces the ultimate goal, that of an *ab initio* theoretical calculation, in effect a prediction of this same body of data, which lies within a narrow band of experimental uncertainty, and which would thereby obviate the need for further experiments of that type.

In the field of atom–atom collisions at low energies (at least for ground-state, spherically symmetric atoms) the theory of elastic scattering is secure.[(2,18,22)] This implies that the experimentalist can infer a great deal about the interatomic potential[(8,25)] from measured atom–atom scattering cross sections.[(26)] This is especially so if the angular and energy resolution of the experiments is sufficient to observe semiclassical and quantum effects.[(27)] A very significant recent advance has been the development of a theoretically sound, yet practical inversion scheme to deduce the potential directly from the elastic scattering cross section.[(28)]

For the more general case of the interaction of open-shell atoms, in which several interatomic potentials play a role in determining the scattering, interpretation of the observations is more complicated. The experimentalist has a need to know optimum theoretical procedures for data interpretation. And for collisions at higher energy the concomitant inelastic processes must be taken into account, so a more complicated theoretical–computational program involving the interaction of excited-state and ground-state potentials is required. There is a need for an effective procedure for inversion of such experimental data.

Of course, the ideal way to approach the scattering problem is to begin with *ab initio* potential energy surfaces accurately computed from first principles. Chapter 2, by Schaefer, reviews the present position in this field. Except for the simplest few-electron systems such computations are extremely expensive and thus rare; one must therefore resort to semiempirical methods following the examples of London, Eyring, and Polanyi[(3)] from four decades ago. However, there has been much progress since the introduction of the "LEPS" potential,[(8,29-32)] as semiempirical methods have been improved. The present state of the art is described in detail by Kuntz in Chapter 3, with special attention to the "diatomics-in-molecules" approach.[(32)]

In spite of this ever-increasing ability to estimate potential energy surfaces one must generally resort to experiment to determine the essential features of an intermolecular potential. (Even for diatomic molecules, "spectroscopic accuracy" is rarely obtained from *ab initio* computations.)

Returning now to the simplest case in the subject of molecular collisions, in Chapter 4 Pauly summarizes the state of the art in the interpretation of experiments on the *elastic* scattering of spherically symmetrical atoms and molecules.[28,33-35] Specific recommendations on optimal theoretical–computational procedures are presented to the experimentalist.

In the more general case of elastic scattering by *anisotropic* potentials the theory is somewhat more complicated, essentially because of the problem of angular momentum coupling. The experimentalist in this field usually employs state selection of the molecules such that certain preferred alignments of the molecular axis with respect to the relative velocity of the colliding pair can be achieved.[36] The goal is to deduce from the small difference in cross sections for scattering from different orientations the noncentral part of the interaction potential.

However, the exact quantal treatment of the scattering problem requires the solution of close-coupled Schrödinger equations,[37] even at collision energies below the threshold for rotational or vibrational excitation of the molecule. In the absence of an inversion procedure, the experimentalist must postulate the anisotropic potential and test its accuracy by computation of the observed orientation effects in the scattering. Clearly there is a need for convenient and reliable approximation methods to obviate the repetitive application of the tedious close-coupled computations. In Chapter 5, Stolte and Reuss review the theoretical procedures, appraise the various approximations, and suggest procedures for data analysis which the experimentalist in the field needs to know.

For atom–molecule collisions at thermal and epithermal energies, *inelastic* processes are always concomitant with elastic scattering. As the collision energy is increased, first pure rotational (R) excitation,[38] then vibrational (V) + rotational,[39] and ultimately electronic (E) + vibrational + rotational excitation[40] and chemically (C) reactive scattering[41-44] take place. The theoretical framework to deal accurately with these respective processes becomes successively more complex: $R < R + V < R + V + E < R + V + E + C$. Thus approximation methods become increasingly more important in the same sequence. In order for the experimentalist[34,45] to interpret results in these areas there is a considerable "need to know" of theory.

A brief introduction to *inelastic* scattering theory[46] is presented in Chapters 6 and 7. In Chapter 6, Light reviews the essence of the essentially exact (quantal) treatment of inelastic scattering theory. Brief descriptions of the more promising of the various approximation methods for inelastic atom–molecule scattering are presented in Chapter 7 by Kouri.

Chapters 8–10 deal entirely with *rotationally* inelastic scattering,[38,46] starting with a review of the quantum theory approach by Secrest (Chapter 8), then a detailed presentation of quantal approximation methods by Kouri (Chapter 9), followed by an explicit description of quasiclassical trajectory[47] methods by Pattengill (Chapter 10). The experimentalist should find that most of the theory needed to account for the data on collision-induced rotational transitions is now well in hand.

Chapters 11 and 12 are directed toward the experimentalist's "need to know" of the subject of *vibrationally* inelastic scattering of molecules by atoms.[46,48] Chapter 11, by Secrest, reviews the quantal treatment of vibrotational transitions of a diatomic molecule induced by collisions with an atom. In Chapter 12 Gentry presents a detailed description of the more useful of the currently available classical and semiclassical approximation methods to deal with vibrational excitation cross sections and vibrational energy transfer rates.[40] The common feature is that at least one degree of freedom in the collision system is treated classically. The problem of the concomitant rotational transitions is also included. The relative merits of the quasiclassical trajectory method and other semiclassical approximations[49] are discussed. Here the experimentalist has a need to know how to choose among the various approximations. The goal is to extract the maximum information (on the potential) from the inelastic scattering cross sections (or energy transfer rates).

Up to this point the experiments considered could be understood on the basis of a single, *adiabatic* potential surface. Chapter 13, by Child, deals with *nonadiabatic* transitions[50] in atom–atom and atom–diatomic collisions. This becomes a coupled-states problem and there is a question regarding the choice of diabatic vs. adiabatic formulation to execute the computations. Considerable judgment is required in order to optimize the theoretical–computational approach once it becomes necessary to go beyond the Landau–Zener approximation.[16,50] The experimentalist dealing with electronic transitions[51-58] and curve-crossing systems needs to know more than a little scattering theory.

Chapters 14–16 deal with the theoretical approach to chemically *reactive* collisions,[59] focusing particularly upon simple atom–molecule exchange reactions. The general quantal theory[60] is outlined briefly by Light in Chapter 14. Following this, Wyatt reviews the current status of various approximate quantal methods in Chapter 15. Then, in Chapter 16, Truhlar and Muckerman present a detailed description of various feasible classical trajectory[47,61-63] and semiclassical approximation methods[49] intended to deal with the reactive and nonreactive scattering of atom–molecule systems.[61-67] The hope is to help the experimentalist, who needs to know how to go from an assumed (empirical, semiempirical, or *ab initio*) potential hypersurface to a calculated set of state-to-state reaction cross sections, suitably averaged to allow comparison with experimental results.[68-70] The choice of approximation procedures depends to a great extent upon the degree of detail of the experimental data, and here again there is a strong interplay of theory with experiment.

Experimentalists in the field of reactive molecular scattering often find it useful to classify the various systems according to their "microscopic reaction mechanism,"[1] i.e., reactions characterized by a so-called *direct-*

mode interaction ("fast") or a *complex-mode* interaction involving a relatively long-lived ($\tau \gtrsim 1$ psec) intermediate.[1] Schematically, one can write, for an atom–diatomic system:

$$\text{Direct-mode:} \qquad A + BC \xrightarrow{\tau \lesssim 1\ \text{psec}} AB + C \quad (\text{and/or } AC + B)$$

$$\text{Complex-mode:} \qquad A + BC \rightleftarrows \overset{\tau \gtrsim 1\ \text{psec}}{[ABC]} \rightarrow AB + C \quad (\text{and/or } AC + B)$$

The angular distribution of the products from a direct-mode reaction is asymmetric (anisotropic), whereas that of the products from the decay of a "persistent complex" is symmetric (forward–backward) in the center-of-mass system.

Chapters 17 and 18 are concerned primarily with *direct-mode* reactions. Chapter 17 consists of a detailed description of the recently developed accurate quantal procedures for the calculation of reaction cross sections, by Wyatt. Chapter 18, by Truhlar and Dixon, presents a thorough discussion of classical theories of direct-mode reactions.

Chapter 19, by Light, deals with *complex-mode* reactions, reviewing the status of various statistical theories[71] of formation and decay of long-(and short-)lived complexes.[72-76] The manifestation of the complex-mode mechanism is found in the angular and recoil energy distribution of the reaction products and their state of polarization. The experimentalist has a need to know how to analyze data on these attributes and their energy dependence, both translational and internal.

At higher energies (e.g., above a few eV) a new reaction path always becomes energetically accessible, namely, collision-induced *dissociation* (CID)[77] of the diatomic partner:

$$A + BC \rightarrow A + B + C$$

The threshold energy for this process is the bond dissociation energy of the diatomic. Chapter 20 by Diestler outlines the present, rather primitive state of the quantal treatment of collision-induced dissociation, while Chapter 21, by Kuntz, describes various practical approximation methods to estimate CID cross sections, both via classical trajectory procedures such as those of Chapter 16, and by a number of dynamical models for the CID process. Once again the experimentalist needs guidance for both planning and interpretation of experiments in which CID is concomitant with reaction.

One of the recently developed tools intended to aid in the analysis of detailed experimental (or computational) data on atom–molecule collision cross sections and rates is the *information-theoretic* methodology of Levine, Bernstein, and co-workers.[78-80] In its most primitive application the method reduces to "surprisal analysis," providing a convenient procedure for compaction of dynamical data. Here one examines the dependence of the surprisal, a logarithmic measure of the deviation from prior expectation (on statistical grounds), upon the key observables (such as the internal or translational energy, scattering angle, etc.) and finds relatively simple behavior. In many cases a very few "surprisal parameters" suffice to reproduce a large body of experimental (or computer-generated) dynamical data.

Information theory can establish the framework on which to "hang" detailed experimental data. Surprisal *analysis* allows interpolation and even extrapolation in favorable cases. Surprisal *synthesis* goes further, toward prediction of dynamical results from only a limited information base (e.g., from the low-order moments of a product state energy distribution to the most probable distribution itself). In Chapter 22, Levine and Kinsey provide a "handbook" of information-theoretic procedures, codifying the basic material helpful to an experimentalist in the field of atom–molecule collisions.

In conclusion, it should be clear from this section that the modern experimentalist in molecular collisions has a considerable, even formidable, "need-to-know" of theoretical methods, ranging over a wide area from quantum chemistry and potential energy surfaces to scattering theory, statistical dynamics, and information theory. The remaining sections will present *examples* of various classes of experiments that have characterized this field in the past and a few of those that are now at the limit of present experimental methodology.

3. *Overview of Experiments in Atom–Molecule Collisions*

In what follows several of the "standard" classes of experiments in atom–molecule collisions will be outlined. However, specific illustrations will be deferred for presentation in later sections.

Bulk gas phase measurements involve "translationally averaged" collisions, from which one deduces rates of the various inelastic and reactive processes at a given temperature, i.e., $k_{i\to f}(T)$. (Here k is the specific rate coefficient for the process of conversion from an initial state i to a final state f at the temperature T.) Because of the loss of information due to the thermal averaging, these rate data (from so-called "macro" or bulk experiments) are of less intrinsic interest from a theoretical viewpoint than experiments carried out under single-collision or beam conditions (i.e., "micro-level" experiments) which yield cross sections at a given energy, i.e., $\sigma_{i\to f}(E)$. (Here σ is the cross section for the $i \to f$ transition at a total energy E.)

It is usually considered most appropriate to compare theory with experiment at the level of the cross section, rather than the rate. Thus in most of this book the emphasis is upon the calculation of state-to-state cross sections, with less attention devoted to the problems engendered by integration over translational and internal energies yielding rate coefficients.[69] The kinds of experiments to be outlined below are all of the single-collision type, involving atoms and molecules; they are ordinarily categorized as molecular beam scattering.[13,41,81]

From a theoretical point of view, results of such experiments can be ordered in a hierarchy of interpretive complexity, from elastic to inelastic to reactive scattering. However, it turns out that the difficulty of the experiments is essentially unrelated to that of the theory. In fact, it happens that reactive scattering cross sections are often more readily measured than elastic, whereas the greatest experimental difficulty is encountered in the detections and measurement of inelastic molecular collisions. In what follows, however, the canonical hierarchy will be used to order the material.

3.1. *Elastic Scattering*[13,33-35]

The most primitive measurements are of the beam-gas attenuation type, yielding values of the total scattering cross section, which usually happens to be dominated by the elastic scattering component. More refined experiments are carried out in the crossed-beam configuration, with one or both beams being velocity-selected (or of narrowed speed distribution, via the supersonic nozzle technique[82,83]). Here one measures the dependence of the integral elastic cross section upon the relative velocity, or relative translational energy. This leads to information on the long-range attractive force between the molecules as well as on the characteristics of the potential well. By measuring the *angular* distribution of the elastic scattering, further, more detailed information on the interaction potential can be extracted.[1,2,10,11,24,26,28,33-35,41,81,84-86] The data analysis procedures are discussed in great detail by Pauly in Chapter 4 of this volume.

In certain specialized cases it has been possible to prepare state-selected and polarized beams of diatomics, e.g., alkali halide molecules or H_2, and study the influence of the direction of polarization with respect to the relative velocity vector in crossed-beam experiments. Basically one measures the ratio $\sigma_{\parallel}/\sigma_{\perp}$, where the subscripts denote the relative polarization. The technique was introduced by Bennewitz *et al.*[87] and further exploited by Reuss and co-workers, as described in Chapter 5, to yield quantitative information on the *noncentral* part of the (anisotropic) atom–molecule potential. The experiments require inhomogeneous electric (or magnetic) fields for the process of rotational (and projection) quantum number selection[36,41] as well as velocity selection, and thus represent quite an experimental challenge.

Figure 1 shows a schematic view of an apparatus developed by Reuss and co-workers[88] to carry out such studies. The theoretical analysis of typical experimental results is presented in Chapter 5.

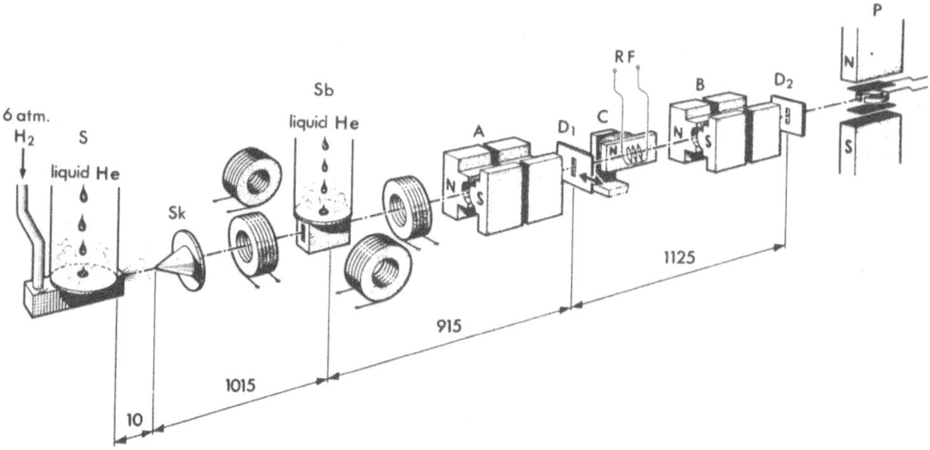

Figure 1. Schematic diagram (view) of the molecular beam apparatus[88] of Reuss and coworkers for the measurement of total collision cross sections for oriented H_2 beam molecules scattered by various gases. The H_2 beam is formed by emission from a cooled nozzle source S through a skimmer Sk; it passes through a cooled scattering box Sb, which is surrounded by magnetic field coils to establish the orientation axis. The collimated beam is state-selected by Rabi deflection fields A and B, between which is the RF transition field (C). The intensity of the transmitted beam is measured by a Penning detector P. Distances are in millimeters.

3.2. *Inelastic Scattering*[14,41,89,90]

By observing the translational energy change in a collision, inelastic processes may be detected. To measure specific inelastic (ultimately, state-to-state) cross sections by the energy change method one requires, at the least, a narrow distribution of relative velocities and high-resolution velocity analysis or time-of-flight detection, over a range of scattering angles. Figure 2 is a schematic diagram of an apparatus developed by Toennies and coworkers[89] which has produced some of the most significant measurements of differential rotational and vibrotational excitation cross sections for low-energy collisions of atomic ions with diatomic molecules. Other methods have been utilized, but space does not permit their elaboration here. The methods and results have been thoroughly discussed in the literature.[89,90] A bibliography on energy transfer collisions of atoms and molecules is available[91] and several reviews of the theory have appeared.[38,41,48,89,90-92] Chapters 6–12 should suffice to provide the experimentalist with state-of-the-art theory.

3.3. *Electronic Excitation and Curve Crossing*[10,16-18,50-58]

Even for simple atom–molecule systems there are a large variety of collision experiments dealing with electronically nonadiabatic transitions (most of which involve the construct of potential curve crossing). These include energy transfer (E ↔ E, E↔ V, E ↔ T, etc.) as well as electron transfer (ionization, etc.). The experiments are too numerous even to outline, but examples can readily be found of each of the processes shown in Table 1. Because of the diversity of these collision processes and the associated experimental arrangements for their study it is not possible to point to a single prototypical apparatus. However, Figure 3 shows a schematic view of an instrument that has been extensively used at the FOM-Instituut in Amsterdam by J. Los and co-workers to study the collisional ionization process. It has provided measurements of the angular and recoil energy distribution of CI cross sections over a wide range of relative translational energy. With some modification it has been used to measure reactive (atom transfer) cross sections, branching ratios for the formation of the various negative ions, and other attributes.

A great many other beam meachines have been used for other studies of collision-induced electronic transitions, but space does not permit listing them here. The reader is referred to the recent experimental review literature(24,54,56) for further details.

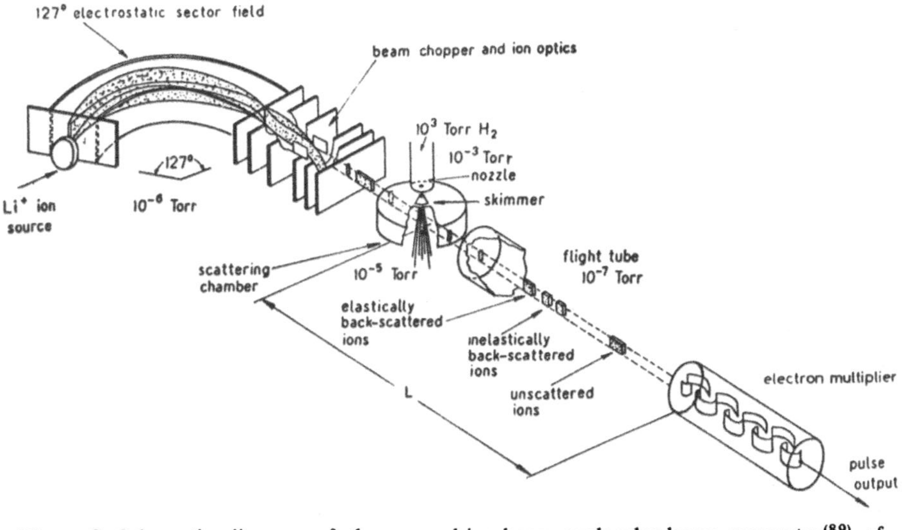

Figure 2. Schematic diagram of the crossed-ion-beam–molecular-beam apparatus(89) of Toennies and co-workers for the measurement of inelastic energy transfer collisions via the TOF technique. The Li^+ ion beam energy is set by the electrostatic sector field; the beam is electrically chopped (pulsed). The ion beam intersects a crossed beam of H_2 from the nozzle–skimmer system. The flight path of the pulses of Li^+ ions is variable ($0.50 \leq L \leq 1.7$ m).

Table 1. Some Electronically Nonadiabatic Collision Processes

Process		Designation
Excitation and Energy Transfer[a]		
A + BC	→ $A^* + BC$	T → E
	→ $A + BC^*$	T → E
$A^* + BC$	→ $A + BC^*$	E → E′
	→ $A + BC^{\dagger}$	E → V′
	→ $A + BC$	E → T
$A + BC^*$	→ $A^* + BC$	E → E′
	→ $A + BC^{\dagger}$	E → V
	→ $A + BC$	E → T
Ionization, Electron, and Atom Transfer[b]		
A + BC, $A^* + BC$, $A + BC^*$	→ $A^+ + BC^-$	CI
	→ $A^+ + BC + e^-$	PI
	→ $A + B^+ + C^-$	IPF
	→ $AB^+ + C^-$	AI
	→ $AB^+ + C + e^-$	API
	(and others with appropriate sign permutations)	

[a] Here the asterisk denotes electronic excitation, the dagger denotes vibrational (and rotational) excitation, and the prime denotes transfer to partner.

[b] Here CI is the collisional ionization, PI the Penning ionization, IPF the ion pair formation, AI the associative ionization, and API the associative Penning ionization process.

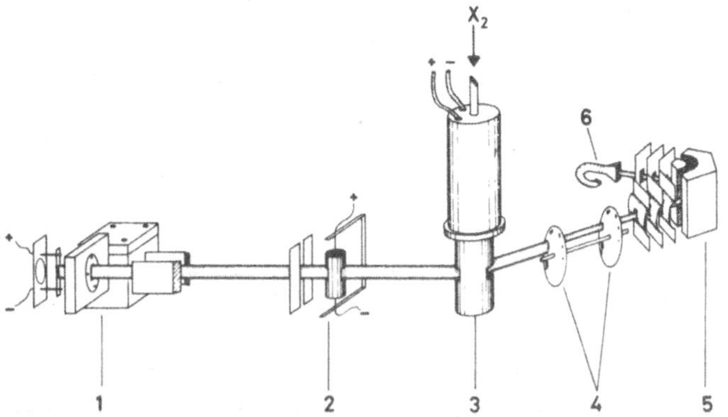

Figure 3. Schematic diagram (view) of the differential scattering apparatus[138] of Los and co-workers for the measurement of ion pair formation from alkali–halogen collisions. (1) Source of fast alkali atom beam (via charge exchange from ions); (2) alkali beam monitor; (3) scattering region ($P_{X_2} \simeq 3 \times 10^{-4}$ Torr); (4) diaphragm collimators; (5) energy analyzer; (6) ion counter.

3.4. Reactive Scattering (13,24,41,43,44,59,66,93-95)

The simplest experiment in reactive scattering consists of the observation of product molecules resulting from the controlled collision of reagent molecules. This is usually accomplished by means of an apparatus that makes use of a crossed-beam configuration, with a selective detector (e.g., a mass spectrometer) capable of measuring the angular distribution of the number density or flux of the product species. More advanced experiments involve measurement of the velocity and/or internal state distribution of the products, and make use of more-or-less monoenergetic reagent beams, sometimes state-selected. The goal of such experiments is to determine the

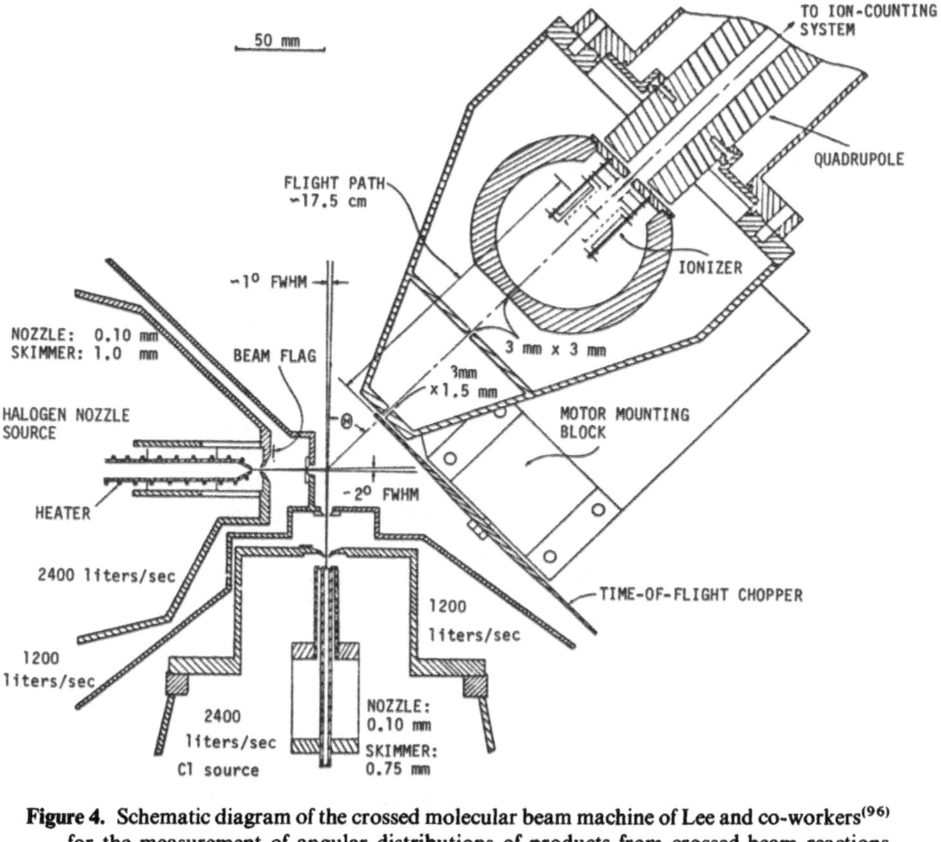

Figure 4. Schematic diagram of the crossed molecular beam machine of Lee and co-workers[96] for the measurement of angular distributions of products from crossed-beam reactions of Cl with halogen molecules. The apparatus employs two separately pumped nozzle beam systems, a TOF chopper, a rotatable ultrahigh vacuum detector chamber with electron bombardment quadrupole mass filter analyzer, and an ion counter.

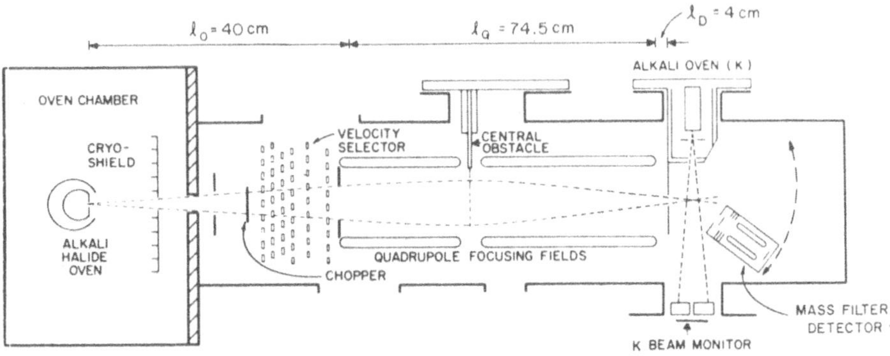

Figure 5. Schematic diagram of the crossed molecular beam of Bernstein and co-workers[99] with state selection and beam focusing of the alkali halide, a crossed alkali beam, and a rotatable quadrupole mass filter detector.

detailed differential state-to-state reaction cross section as a function of energy.[1,14]

Figure 4 is a diagram of a modern molecular beam machine for the study of reactive atom–molecule scattering,[96] as developed by Lee, Herschbach, and co-workers.[97] This apparatus produces laboratory product flux (velocity–angle) contour maps from which detailed differential cross sections in the center-of-mass system[98] can be obtained as a function of the relative translational energy of the reagents.

Figure 5 is a schematic diagram of a crossed-beam apparatus developed by Bernstein and co-workers, utilized for reactive scattering measurements with state-selected (polar) reagent molecules.[99] It employs mechanical velocity selection[100] and electric quadrupole focusing and state selection.[86,101]

Examples of experimental results obtained in several reactive scattering laboratories are presented below.

4. *Experimental Examples*

In this section a few examples of the results of experiments on atom–molecule collisions are presented. Included are some from quite early work in the field, to compare with recent results. This helps to highlight the impact of technical development upon the quality of experiments, and thus indirectly to theoretical understanding, which is based to a greater or lesser extent upon precise, detailed experimental results.

4.1. Elastic Scattering (26,28,33,64)

One of the earliest experiments showing the angular distribution of the scattering of atoms by molecules was that of Knauer[102] in 1933. Figure 6 displays Knauer's results for the H_2–Hg system. Evident is the strong "forward" scattering peak (i.e., small deflections) due to collisions at large impact parameter, dominated by the long-range attractive forces.[103,104] Because of the thermal beam velocity distributions no rainbow or interference structure was observable. Simple classical calculations[105,106] of the differential cross section reproduce the main observed feature, i.e., the rapid drop-off of the angular distribution.

With the advent of beam velocity selection it became possible to resolve some of the interesting theoretically predicted[105,107,108] "semiclassical structure" in the differential elastic scattering cross section. Figure 7 shows one of the first observations[109] of the now well-known "rainbow effect" in heavy-particle scattering.

For molecular systems of small reduced mass at low relative velocities, where the de Broglie wavelength λ ($\equiv h/\mu v$) is comparable to the collision radii, one expects to observe interference effects in the angular distribution of the scattering.[1,2,10,27] The first evidence of such quantum interferences was noted in the scattering of a velocity-selected Li atom beam by a crossed Hg beam.[110] Results are shown in Figure 8. Only an undulatory pattern is observed due to the broadening effects of the residual distribution of relative velocity and angular resolution. The "contraction" of the interference pattern with increasing velocity (i.e., wave number) is as expected.

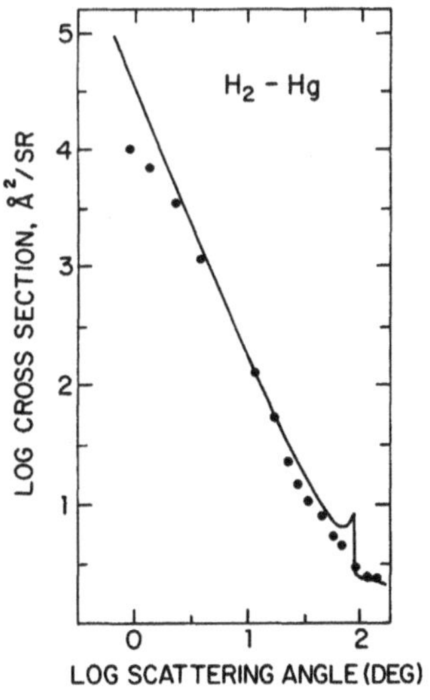

Figure 6. One of the earliest experimental observations (1933) of an atom–molecule scattering angular distribution, by Knauer.[102] The solid curve is from a classical calculation based upon an assumed Lennard-Jones (12,6) potential function.[106]

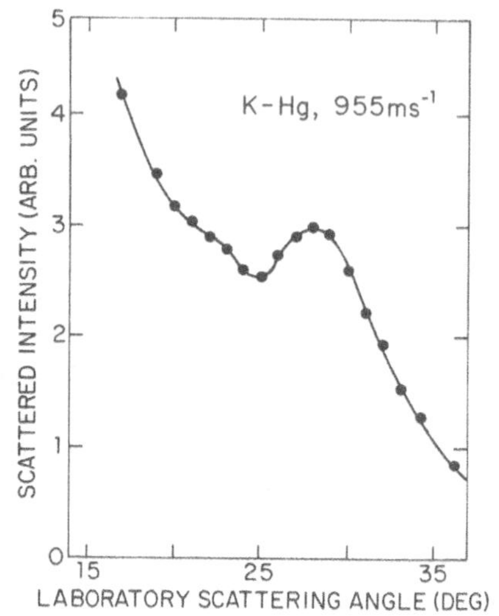

Figure 7. One of the earliest experimental observations (1962) of the rainbow structure in angular distribution of scattering (for an atom–atom system).[109] A velocity-selected K beam was crossed with a Hg beam and the intensity of the scattered K measured as a function of angle. The primary rainbow is clearly seen and there is a hint of an unresolved supernumerary rainbow at a somewhat smaller angle.

Figure 9 shows the results of a much improved experiment[111] by Pauly and co-workers in Göttingen, in which all anticipated structural features of the angular distribution have been resolved. Similar results were obtained at several energies. Through the use of the Buck–Pauly inversion procedure[28,112-114] an accurate, unique potential function $V(R)$ was thereby determined for the ground electronic state of the LiHg molecule,

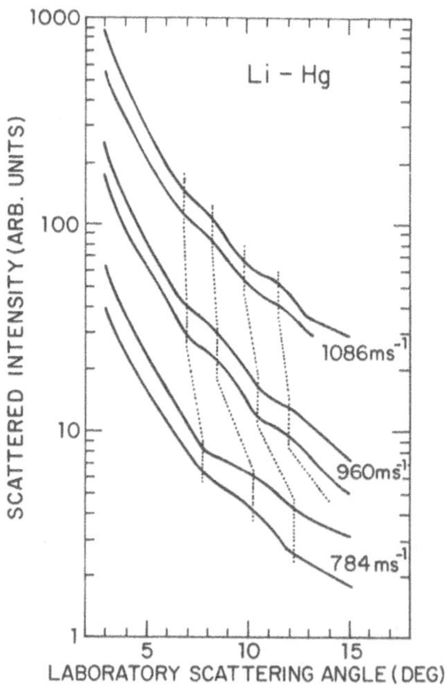

Figure 8. First observation (1960) of quantum interferences (diffraction undulations) in the angular distribution of scattering (for an atom–atom system).[110] Velocity-selected beams of Li were crossed with a Hg beam and the scattered intensity of the Li measured vs. angle (as in Figure 7). Duplicate curves are shown at the three indicated values of the average relative speed. The dotted lines denote the estimated positions of the extrema.

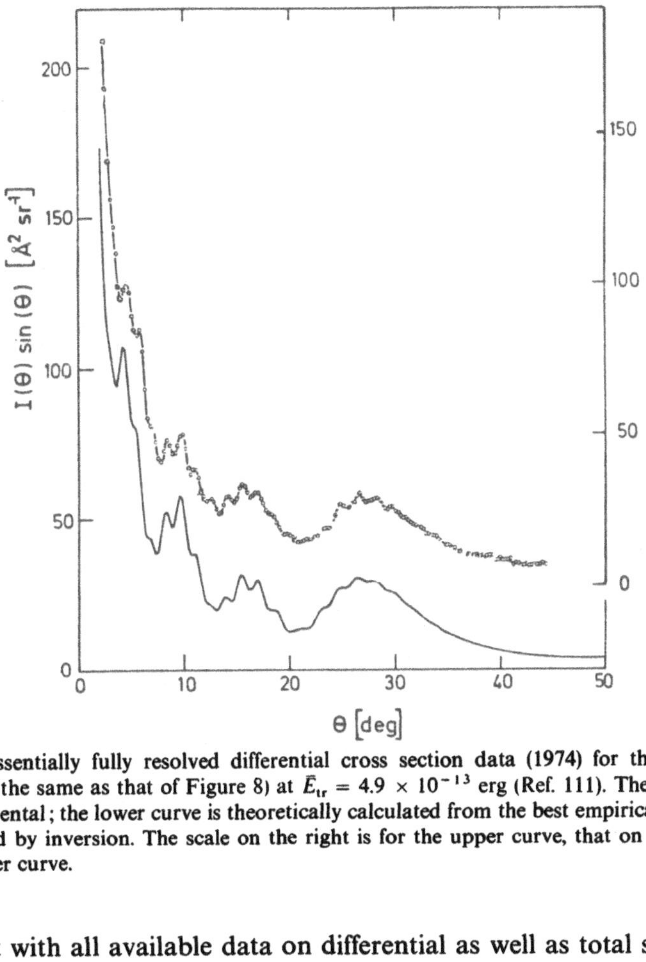

Figure 9. Essentially fully resolved differential cross section data (1974) for the Li + Hg system (the same as that of Figure 8) at $\bar{E}_{tr} = 4.9 \times 10^{-13}$ erg (Ref. 111). The points are experimental; the lower curve is theoretically calculated from the best empirical potential obtained by inversion. The scale on the right is for the upper curve, that on the left for the lower curve.

consistent with all available data on differential as well as total scattering cross sections over a wide energy range.

The technique of measurement of total elastic scattering cross sections (via beam attenuation)(1,12,13,21,33,34,41,84,104) has been advancing steadily in the wake of technical improvements over the years. One of the significant early findings, by Pauly,(114) was that the velocity dependence of the total cross section Q accorded well with the simple Massey–Mohr(103) equation, which, for the case of a long-range potential dominated by an inverse sth power ($C_s R^{-s}$) term can be written

$$Q \propto (C_6/\hbar v)^{-2/(s-1)}$$

Figure 10 shows a log–log plot of $Q(v)$ as observed for the K–N_2 system.(114) The line through the data points is of the theoretical $-2/5$ slope appropriate for the $s=6$ case.

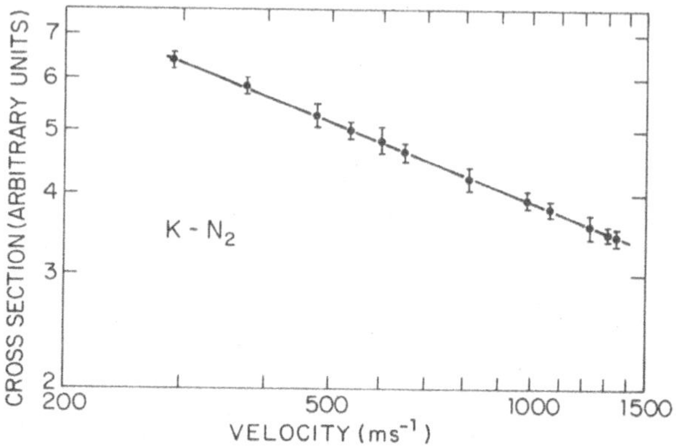

Figure 10. Earliest observation (1960) of the velocity dependence of the total cross section (here for atom–molecule scattering) for the K–N_2 system.[114] The points are experimental; the line is of the theoretical Massey–Mohr[103] slope of $-2/5$ appropriate for a C_6R^{-6} long-range (attractive) potential.

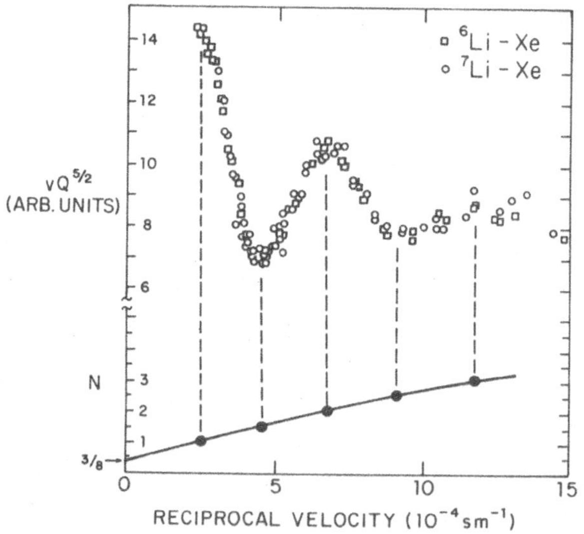

Figure 11. One of the earliest experimental observations (1963) of the glory extrema in the total cross section (for an atom–atom system).[116] The ordinate is $vQ^{5/2}$ (which, in the absence of glories, would be velocity-independent), the abscissa is the reciprocal of the relative velocity. The lower curve shows the usual plot[27,117] to index (N) the glories.

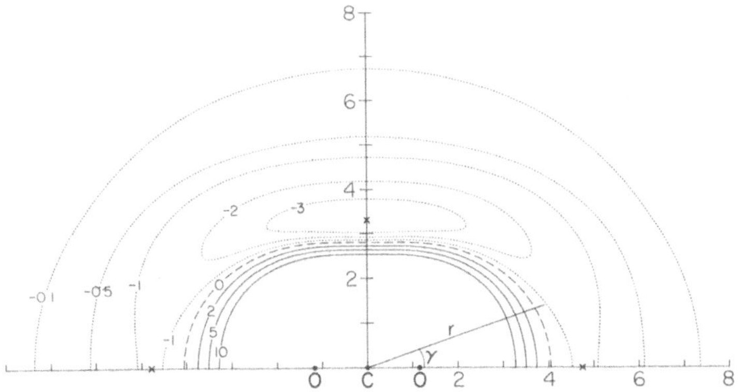

Figure 12. A contour plot of an empirically determined (1978) anisotropic potential energy surface for the He–CO_2 system.[119] Solid contours: positive energies; dotted contours: negative; dashed curve: $V(R, \gamma) = 0$. Energies in meV, distances in angströms.

A quantal treatment[115] of the elastic scattering of heavy-particle beams, taking into account the short-range repulsion as well as the long-range attraction between the collision partners, pointed to the existence of an oscillatory deviation from the monotonic Massey–Mohr velocity dependence. This so-called "glory effect"[107] was soon thereafter observed, for a number of atom–atom and atom–molecule systems, by Rothe and co-workers.[116-118] Figure 11 shows glory extrema plots for the isotopic ^{6}Li–Xe, ^{7}Li–Xe systems.[118] From such data, estimates of the potential well characteristics can be obtained, via now well-known methods (see Chapters 4 and 5).

By combining data on well-resolved differential cross sections and absolute total cross sections over a wide range of relative velocity it is now possible to obtain accurate potential energy curves, at least for spherically symmetrical systems. For many atom–molecule systems, however, the interaction potentials $V(R, \gamma)$ are sufficiently anisotropic as to require further experiments (via polarized beams or rotational inelasticity measurements) to evaluate the angle dependence $V_R(\gamma)$ and even the orientation-averaged radial dependence $\bar{V}(R)$.

Figure 12 shows such an anisotropic potential, for the He–CO_2 system, deduced from scattering studies by Kuppermann and co-workers.[119] Chapter 5 reviews methods and recent experimental results in this field, i.e., scattering by anisotropic potentials.

4.2. Rotationally Inelastic Scattering[34,41,89,90]

The earliest direct observations of rotational transitions induced by collision were those of Toennies,[120] using rotationally state-selected and

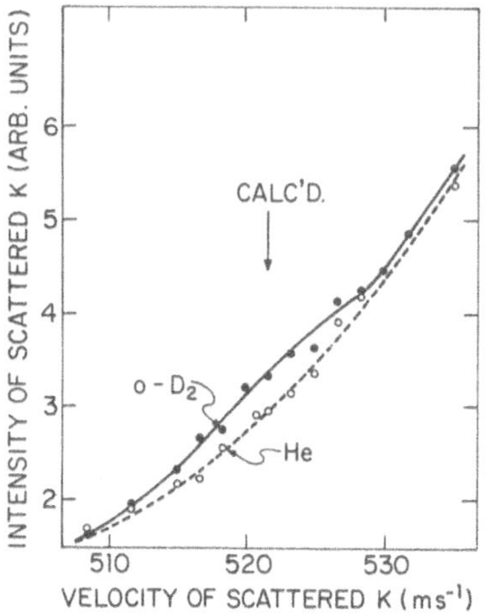

Figure 13. First observation of inelastic energy transfer in a crossed-beam experiment (in 1964) by the velocity change method.[121] Plotted is the velocity distribution of K scattered by o-D_2 and He (the "control") at a laboratory angle $\Theta = 3.5°$. Velocity-selected K beam, $v = 724$ m sec^{-1}; He, D_2 beams thermal, 190 K. The arrow indicates the predicted position of the inelastic peak, here observed only as an unresolved hump.

focused beams of TlF scattered by various atoms (and simple molecules) and refocused (state-analyzed). Such data provided approximate values of integral state-to-state cross sections, and showed the great range of magnitudes of these cross sections, ranging from a few to hundreds of Å^2 (in some cases comparable with the elastic scattering cross sections).

The first application of the velocity change technique was that of

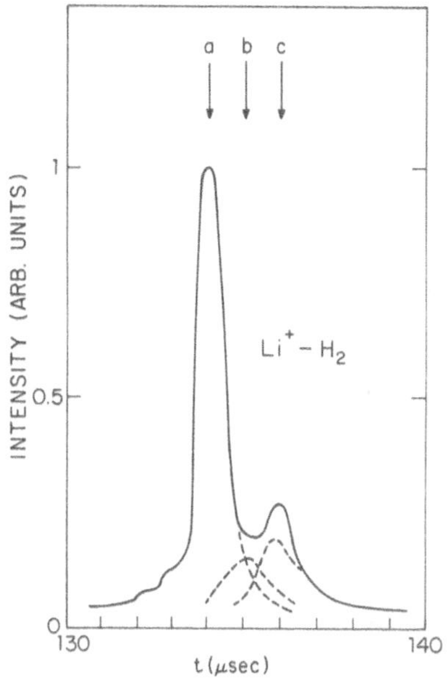

Figure 14. First *resolved* collision-induced rotational transition observed[122] (1973) in the TOF spectrum of Li^+ scattered by H_2 for $\Theta = 5°$ and a path length of 1.15 m. Arrows denote predicted arrival times for Li^+ due to (a) elastic scattering ($j = 0 \rightarrow 0$, $j = 1 \rightarrow 1$); (b) inelastic scattering ($j = 0 \rightarrow 2$); (c) inelastic scattering ($j = 1 \rightarrow 3$).

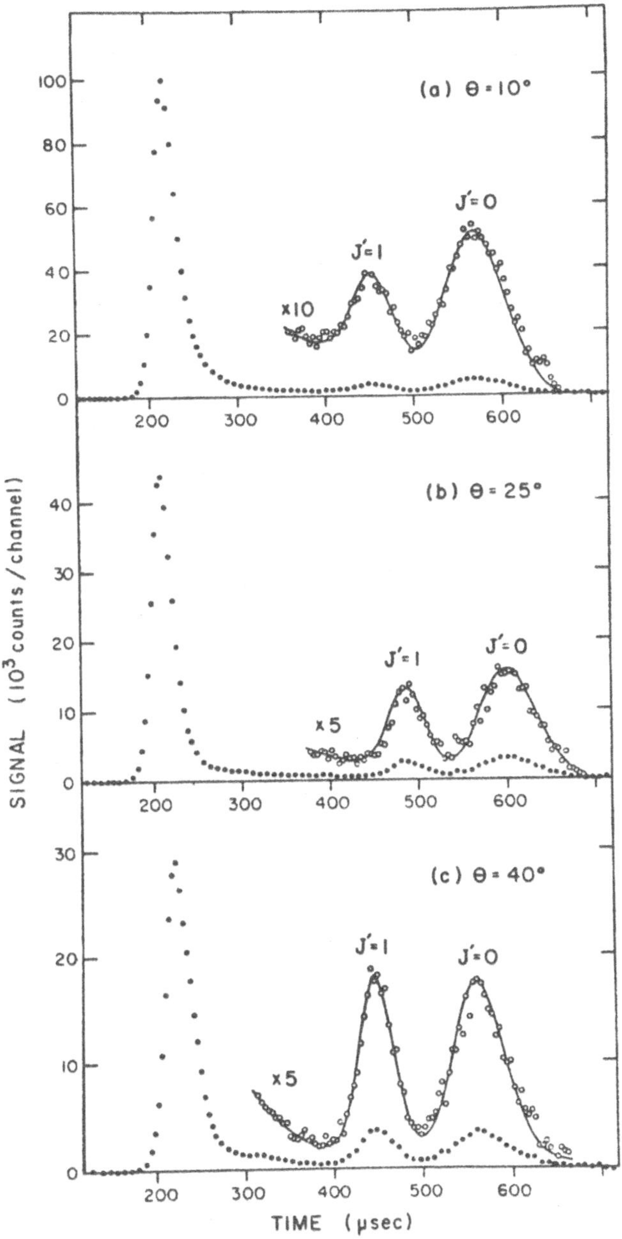

Figure 15. Recent (1977) fully resolved TOF spectra of HD scattered from He at various specified laboratory angles,[123] at a path length of 0.48 m. The solid curves are theoretical peak shapes for the HD transitions $j = 0 \rightarrow j'$ (indicated).

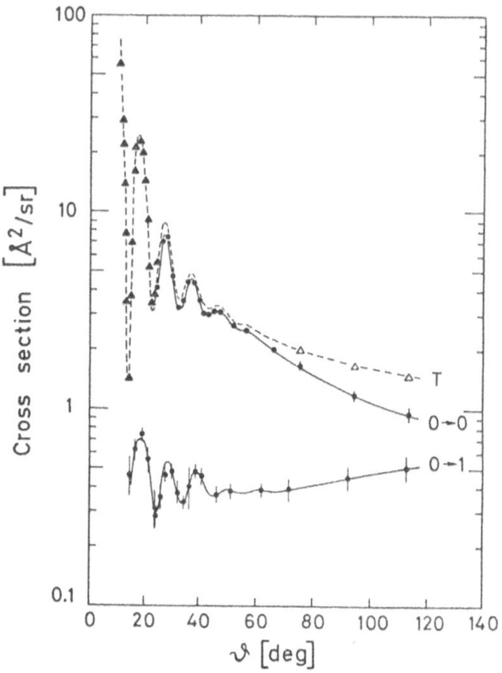

Figure 16. Resolved differential cross sections (in the c.m. system) for the transitions $j = 0 \rightarrow j' = 0$ and 1 for HD scattered from D_2[124] at $\bar{E}_{tr} = 45.4$ meV (1978). Solid curves: best experimental representation; solid triangles: measured total differential cross sections; open triangles: total of separate differential cross sections ($0 \rightarrow 0$ plus $0 \rightarrow 1$); dashed curve: calculated total differential cross section based on an *isotropic* potential.

Blythe *et al.*,[121] who used a mechanical velocity selector and analyzer system to observe a rotational transition in D_2 induced by collision with K. Figure 13 shows an experimental velocity spectrum with a hump indicative of the rotational deexcitation of D_2 ($j = 2 \rightarrow 0$), at the specified angle of scattering.

Technological advances led to apparatus capable of ever greater resolution (e.g., that of Toennies *et al.*[89] shown in Figure 2). Figure 14 shows a time-of-flight spectrum for the H_2–Li^+ system, in which, for the first time, the $\Delta j = 2$ transition was essentially fully resolved, at several scattering angles.[122] Subsequent experiments by the Toennies group have provided a substantial body of differential state-to-state cross sections in a form directly suited for comparison with theory.

The ultimate in present-day instrumentation is the pulsed molecular beam apparatus of Gentry and Giese,[123] which has produced the best-resolved data in the field to date, as shown in Figure 15. Time-of-flight spectra of HD scattered by He show fully resolved single-quantum transitions. From the fitted peak shapes it was possible to determine the ratio of differential cross sections for the transition $j = 0 \rightarrow 1$ vs. $0 \rightarrow 0$, at several scattering angles. Figure 16 shows fully resolved diffraction oscillations in rotationally inelastic differential cross sections for the HD–D_2 system, as observed by Buck *et al.*[124] For such systems, whose potential energy surfaces are fairly accurately known semiempirically, this new kind of detailed data serves as an excellent target for theoretical computation, used in an iterative mode to establish the potentials experimentally.

4.3. Vibrationally Inelastic Scattering[89,90,125]

The first direct observation of vibrational energy transfer in neutral beam collisions was that of Dittner and Datz[126] for the systems H_2, $D_2 + M$ ($M \equiv K, Na$). Similar work, but for alkali *ions* rather than atoms had been reported earlier.[126] The first observations of angular distributions for resolved vibrational excitations were reported by Gentry and co-workers in 1971[127] for $H^+ + H_2(v = 0) \rightarrow H^+ + H_2(v' = 1,2,3)$. More extensive studies using Li^+ ion beams were carried out by Toennies and co-workers,[89,90,128] providing resolved (state-to-state) vibrational excitation cross sections vs. scattering angle over a wide energy range. Figure 17 shows representative data at two collision energies, with measurements on H_2 excitations up to $\Delta v = 4$ presented.

Similar results are not yet available for neutral–neutral molecular systems, although they will surely be forthcoming in the near future. Such experiments will have transformed the field from one denoted "vibrational relaxation,"[129] referring thereby to bulk systems, to one which can truly

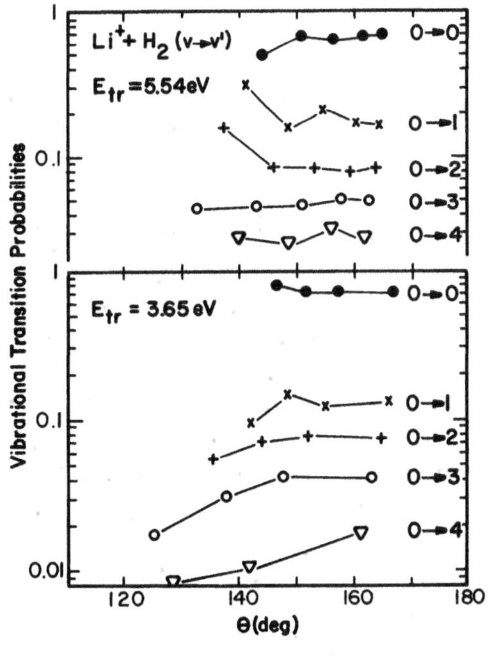

Figure 17. Early observations of vibrational transition probabilities (1973) as a function of c.m. scattering angle θ, for H_2 excited by collision with Li^+ at the two specified collision energies.[122] Results scaled such that the ordinate is the ratio of the differential cross section for the given transition to the sum for all observed transitions.

be designated "vibrationally inelastic scattering," appropriate for micro-level collisional data.

4.4. *Electronic Excitation and Charge Transfer*[10,13,19,20,45,51,54-58,130-134]

The field of electronically nonadiabatic transitions is so broad that it is not possible to do justice to the range of experiments with only a few examples. Those to be presented were chosen because they lie on the interesting border between physical and chemical transformations (from charge transfer to mass transfer). The very interesting results on collisional excitation and electronic energy transfer are, perforce, omitted in favor of studies of ionizing collisions.

The low-energy collisional ionization process was first reported in 1969 by Los, DeVries, and co-workers[135] and by Helbing and Rothe.[136]

Figure 18 shows experimental results[137] on the angular distribution of the ions formed in the alkali–halogen atom "reaction"

$$Na + I \rightarrow Na^+ + I^-$$

at two energies, as specified. The main structural features are well accounted for[137] by an extended Landau–Zener theory.

Figure 19 presents similar results[138] for an atom–*molecule* collision, namely,

$$K + Br_2 \rightarrow K^+ + Br_2^-$$

Here a classical trajectory simulation involving "potential surface-hopping" has captured most of the important features of the observed ion product angular distribution.[138]

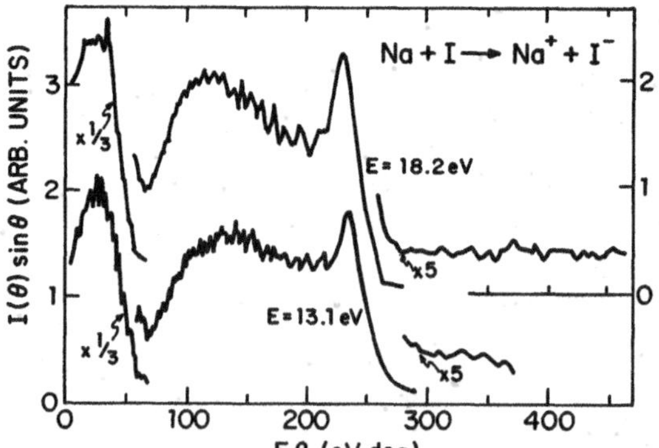

Figure 18. Differential cross sections (c.m. system)[137] observed (1973) for ion pair formation from Na + I collisions at the two specified translational energies E plotted vs. the variable $E\theta$ (often designated τ). The structure is essentially that expected on the basis of an extended Landau–Zener theory.

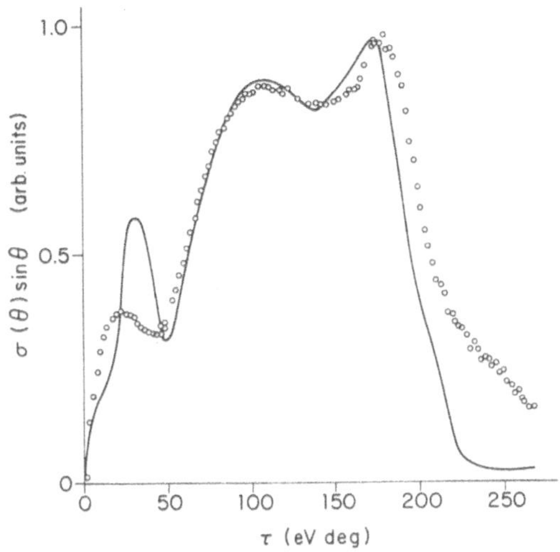

Figure 19. Results analogous to those of Figure 18, observed (1977) for the atom–molecule collision $K + Br_2$ at $E = 15$ eV.[138] Solid lines calculated using adiabatic potential surfaces. The abcissca is $\tau \equiv E\theta$.

4.5. *Reactive Atom–Molecule Scattering*[13,14,24,41-44,59,66,93-95,139-144]

The first crossed-beam experiment in which reactive scattering was definitely observed was that of Taylor and Datz[145] in 1955, who studied the alkali-atom–hydrogen-halide reaction

$$K + HBr \rightarrow KBr + H$$

The laboratory angular distribution of the KBr product showed a peak at an angle intermediate between the crossed beams of K and HBr. However, because of the large mass of the KBr relative to that of the recoiling H atom the angular distribution of the KBr is very nearly that of the centroid (as discussed in a subsequent paper[146]). Therefore little information on the center-of-mass (c.m.) angular distribution of products could be gleaned from the data.

It was not until 1961 that an experiment revealed significant information on the *shape* of the differential reaction cross section in the c.m. system. Herschbach *et al.*[147] observed the angular distribution of the reactively scattered KI from the alkali–alkyl-halide reactions

$$K + RI \rightarrow KI + R \qquad (R \equiv CH_3, C_2H_5)$$

The experiments were carried out using crossed thermal beams of K and RI. From the results it was deduced that the c.m. differential cross section for KI formation was "backward-peaked" relative to the incident K velocity vector in the c.m. system, i.e., that after the K atom strikes the RI molecule the KI product "rebounds" backward.

Figure 20 shows the first set of "raw" experimental data[147] on the

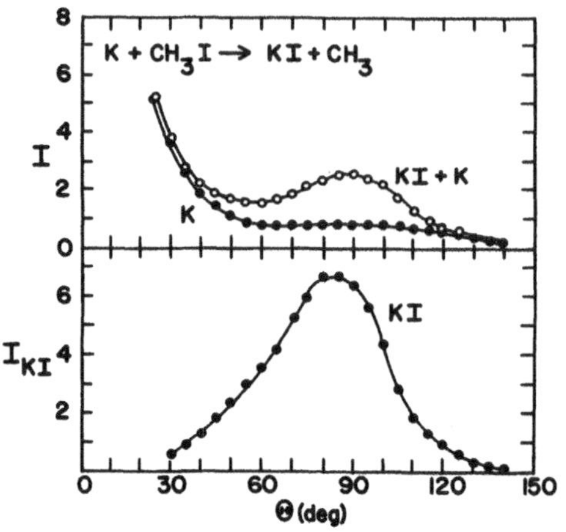

Figure 20. The earliest observation (1961) of the angular distribution of reactive scattering from crossed beams (K + CH_3I) leading to information on the *shape* of the c.m. distribution of products.[147] Top: laboratory angular distribution of nonreactively scattered K and the sum of KI + K scattering. Bottom: difference of these, yielding angular distribution of KI flux in the lab system. The peak corresponds to backscattering of the KI with respect to the incident K direction in the c.m. system.

laboratory *angular distribution* of the KI from the K + CH_3I reaction. The peak near 85° (with respect to the incident K beam) corresponds to backscattering ($\theta = 180°$) in the c.m. system. Unfortunately there is little that can be learned from such data about the product *recoil velocity distribution*, i.e., the energy disposal in the reaction (which is exoergic by ~ 1 eV).

The direct measurement of product *velocity* distributions was first reported in 1965 by Grosser *et al.*[148] in an experiment involving a velocity-selected K beam (crossing a thermal HBr beam) and a product velocity analyzer-flux detector. Figure 21 shows the velocity analysis (at a fixed angle of scattering), resolving the slow-moving KBr product from the fast, elastically scattered K atoms. Once the technique of recoil analysis was developed it was possible to determine the partitioning of the exoergicity between products' relative translational motion and internal excitation.

Figure 22 is a flux (velocity–angle) *contour map* (a detailed differential

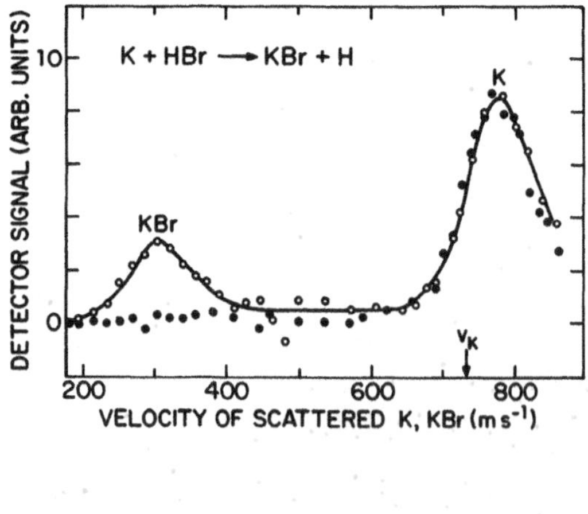

Figure 21. First observation (1965) of velocity analysis of reactive scattering from crossed beams (K + HBr).[148] Velocity-selected K beam ($v_K = 732$ m sec^{-1}, indicated by arrow); thermal HBr (260 K); $\Theta = 25°$. Open circles denote results with detector sensitized to both K and KBr; solid circles, only K detected. Difference signal near 300 m sec^{-1} is KBr flux.

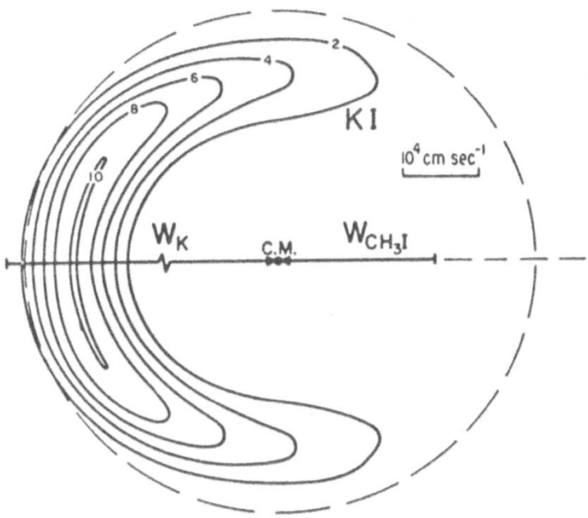

Figure 22. C.m. contour map of flux (velocity–angle) distribution (1973) for KI from the reaction K + CH_3I (Ref. 149) at $\bar{E}_{tr}$ = 0.12 eV, showing the backscattering and high recoil of the KI product.

reaction cross section in the c.m. system) for the K + CH_3I reaction, as deduced from such product velocity analysis experiments.[(149)] Detailed reviews and theoretical analyses of experimental product recoil distributions for this and related reactions have been published.[(150,151)]

These alkali–alkyl-halide reactions are examples of the "*direct-mode*" of reaction, where the interaction time for the combined system is less than a characteristic molecular rotation time ($\sim$ 1 psec), consistent with strongly anisotropic c.m. product distributions. As mentioned in Section 2, the other extreme is the "*complex-mode*" of reaction, discovered experimentally in 1967 by Herschbach and co-workers.[(152)] Here one expects the c.m. angular distributions to be symmetrical in the forward–backward hemispheres[(152,153)] and the recoil energy distributions to have characteristic shapes governed by statistical–theoretical considerations.[(154,155)]

Figure 23 presents the c.m. angular and recoil velocity distributions of the OBr product from the reaction

$$O + Br_2 \rightarrow OBr + Br$$

as deduced experimentally by Parrish and Herschbach.[(156)] The results show that the reaction proceeds via a long-lived intermediate complex. They are in excellent quantitative accord with theoretical calculations based on the assumptions of a statistical (transition-state) theory.[(154)]

The exchange reaction of H + H_2 is probably the single most thoroughly studied elementary reaction in the theoretical chemical reaction dynamics literature, as reviewed by Truhlar and Wyatt.[(157)] Unfortunately it is a

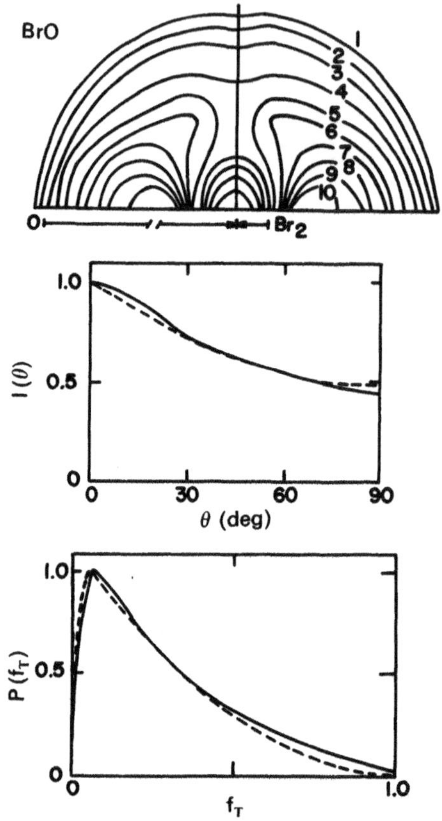

Figure 23. Detailed results (1973) on the reactive scattering of O + Br_2 (Ref. 156) at $\bar{E}_{tr}$ = 3.0 kcal mol^{-1}. Top: c.m. contour map of BrO flux (velocity–angle) distribution. Middle: Angular distribution at fixed BrO recoil velocity. Zero degrees refers to the direction of the incident O atom in the c.m. system. Bottom: Distribution of product relative translational energy E' expressed as the fraction $f_T \equiv E'/E_{tot}$ where $E_{tot} \simeq 13$ kcal mol^{-1}. Solid curves: experimental. Dashed curves: calculated from RRKM–Herschbach statistical complex model.[154]

difficult reaction to study in detail by crossed-beam methods. Here, in a sense, experiment is lagging rather than leading theory. Nevertheless, some reliable experimental data are available, as illustrated by the angular distribution for HD formation, i.e., hydrogen atom abstraction from H_2 by a deuterium atom, as measured by Geddes *et al.*[158]:

$$D + H_2 \rightarrow HD + H$$

Figure 24 shows the HD c.m. angular distribution. It is backward-peaked (in the c.m. system) with respect to the incident D atom. Although there is a considerable range of uncertainty in the width of the backward cone it is evidently very narrow. *Ab initio* theoretical computations by Schatz and Kuppermann,[159] Kuppermann and Schatz,[160] and Elkowitz and Wyatt[161] have yielded essential agreement with the experimental observations.

Most of the experiments reviewed thus far have been carried out in the thermal or epithermal energy range with essentially ground-state reagents. There is, of course, a great deal of interest in the influence of both

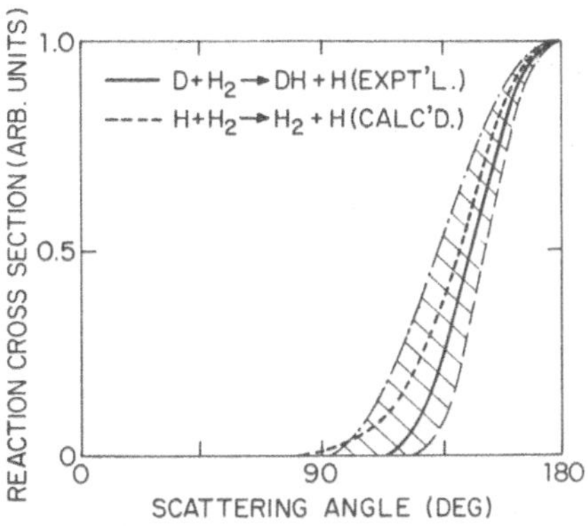

Figure 24. C.m. angular distribution (1972) of reactive scattering of D + H_2,[158] showing backscattering of HD product with respect to the incident D atom direction in the c.m. system. Solid curve: experimental,[158] with cross-hatched zone of uncertainty. Short-dashed curve: from *ab initio* quantal scattering computation for the H + H_2 reaction.[159,160]

relative translational energy and internal energy of the reagent molecules upon the reaction probability.[67,78,79] With the advent of lasers, a whole new area of "state-to-state" chemistry[14] is now becoming accessible to experiment, opening up the possibility of a much tighter coupling between theory and experiment. However, it is not possible here to elaborate on these developments.

Nevertheless, one example will be presented on the translational energy dependence of an important class of reactions, namely that of alkali atoms with halogen molecules, as exemplified by the well-studied K + Br_2 system.[133-136,138,143] Even at very low energies, elastic and inelastic collisions are accompanied by reaction to form the neutral products:

$$K + Br_2 \rightarrow KBr + Br$$

At energies above the thermodynamic threshold for ion pair formation the cross section for the formation of neutrals declines rapidly, and the ionization reaction dominates:

$$K + Br_2 \rightarrow K^+ + Br_2^-$$

(followed by K + $Br_2 \rightarrow K^+ + Br^- +$ Br).

Figure 25 shows the total (nonelastic) scattering cross section vs. $\bar{E}_{tr}$ (the average relative translational energy), from below 0.1 eV to 10^4 eV, as determined from a variety of experiments, most of which were carried out at the FOM laboratory in Amsterdam.[134] Clearly the neutral-product "channel" is of importance only over a limited energy range, below ~ 10 eV. This finding is expected[1] to be quite general; as the energy increases and more competing "channels" become energetically accessible (including dissociative processes) the branching fraction for atom transfer reactions becomes negligible.

Extensive theoretical-computational studies have been carried out by the FOM group and others in an attempt to systematize the vast panorama

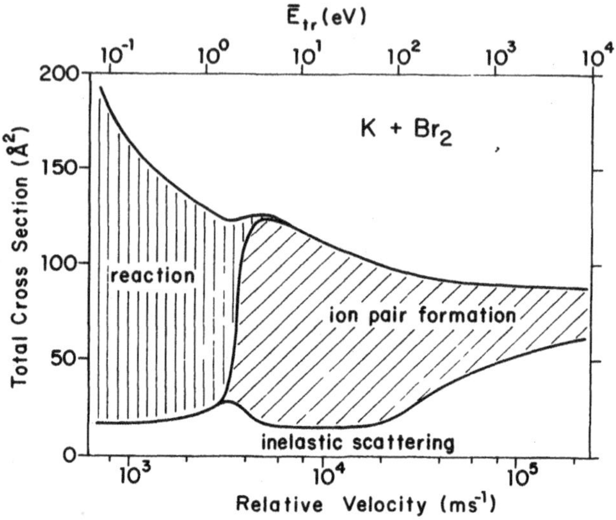

Figure 25. Overall summary plot (1978) of experimentally deduced[134] total (nonelastic) scattering cross section for the K + Br_2 system from $E_{tr} < 10^{-1}$ eV to 10^4 eV, and the various indicated contributions to this total cross section.

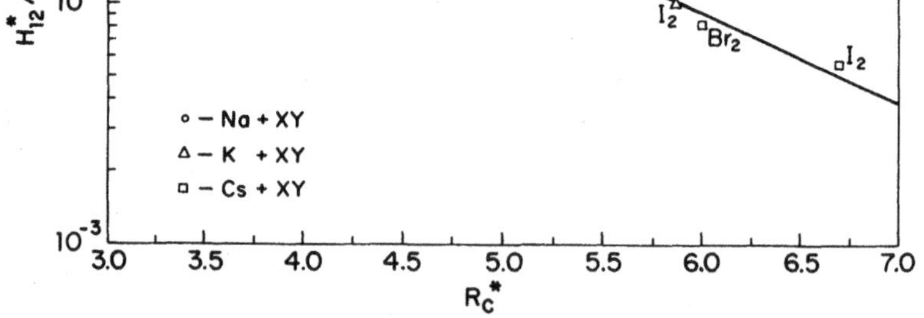

Figure 26. Empirical[134] "reduced" Landau–Zener coupling matrix element H_{12}^* as a function of reduced crossing distance $R_c^* \equiv R_c(2I)^{1/2}$, where R_c is the crossing point and I the alkali ionization potential, and $H_{12}^* \equiv 1.73R_c^* \exp(-0.875R_c^*)$. Here H_{12} is the usual coupling matrix element evaluated at R_c, expressed in terms of H_{12}^* by $H_{12} \equiv H_{12}^*(I \cdot \mathrm{EA})^{1/2}$, where EA is the vertical electron affinity of the halogen. This "universal" function (1978) makes possible calculation of ion pair formation cross sections over a wide energy range for all the systems indicated.

of data on the M + XY reactions (M ≡ alkali; X, Y ≡ halogen atoms).[134] A unified picture based on an extended harpoon model involving a set of potential curve-crossing parameters accounts for most of the observations. A "universal" Landau–Zener matrix element correlation was deduced[138] which makes it possible to relate results for different M, X, Y systems. Figure 26 is a plot of a "reduced" matrix element H_{12}^* as a function of a "reduced" crossing distance R_c^*, for M ≡ Na, K, and Cs and X, Y ≡ I, Cl, Br.

4.6. *Collision-Induced Dissociation*[77,126,127,162-167]

The earliest direct studies of collision-induced dissociation (CID) were those of Tully *et al.*[162] on alkali halide molecules in collision with rare gases and by Wexler and co-workers[163,164] on the analogous process with thallium halides, in which ion pair formation was observed immediately above threshold. From such experiments, and from those on other endoergic processes, one obtains, of course, an upper bound on the endoergicity as well as the post-threshold energy dependence of the cross section[165,166]:

$$\sigma_{\text{CID}} = A_i (E - E_0)^n / E_{\text{tr}}$$

Here the coefficient A_i is a weak function of the internal energy of the diatomic, $E - E_0$ the energy excess over threshold, and E_{tr} the relative translational energy. The power $n \leq 2.5$ is dependent upon the system, as is the coefficient A_i, and both are of theoretical interest.[165-167]

The preceding material is by no means a representative sample of the current state of experimental work in the field of atom–molecule collisions. It is nothing more than a collection of examples showing the development of certain of the experimental techniques which have led to results that can serve as a testing ground for theory. In the following section the question will be raised as to the information content of various levels of experimental data on atom–molecule collision cross sections.

5. *Information Content of Atom–Molecule Collision Cross Sections*[78-80,168-170]

As a "guide for the experimentalist" in atom–molecule collisions this book must address itself, in part, to the question of the information content of the hard-won experimental data. The experimenter contemplating the construction of ever-more sophisticated apparatus may ask if the additional detail about to be uncovered from the new experiments has enough theoretical "value" to justify the effort. What new scientifically significant information can be *expected* from the results of the proposed experiments (assuming the most favorable scenario in which the apparatus performs according to design specifications)? What are the very meanings of the terms "information content," "value," "significance"?

First of all, as stated earlier, if accurate *ab initio* atom–molecule potential surfaces were available and if accurate quantal scattering calculations could be carried out economically and conveniently, the need for most of the collision experiments on simple atom–molecule systems would be obviated by the theory. This is essentially the goal of most of the authors of the present book, i.e., to put the experimenter "out of business," as it were. (Of course, the experimentalist would quickly respond by attacking ever-more complicated polyatomic systems, thereby always keeping a respectable distance ahead of the theorist!) However, in the interim period before the advent of this "*ab initio* age," the experimentalist seeks to make a contribution toward the advancement of knowledge. One such contribution is the confirmation of validity of various theoretical approximation schemes, including estimates of potential energy surfaces and semiclassical (or classical) approximation methods to deal with the scattering. Another is to discover new observable subtleties which may have escaped the attention of the theorist and which provide new information on the interparticle interactions.

There has been some progress made on the subject of the information content of collision cross section data. Formal information theory has been applied to this question,[168,169] but with only limited success. However, it is now possible to determine the information content associated with a given angular distribution[168] or a detailed differential cross section[169] and to ascertain the incremental gain in information upon improvement in the angular and/or energy resolution of a given experiment. However, the formal information content or entropy deficiency (see Chapter 22) of the given experimental distribution is not directly correlated with the value or significance of the experiment, perceived subjectively.

An example is provided by consideration of Figure 27, which compares the angular distribution of the reaction cross section for two related direct-mode reactions:

$$K + I_2 \rightarrow KI + I$$
$$K + CH_3I \rightarrow KI + CH_3$$

The first reaction is of the "stripping" type,[43] characterized by strong "forward-peaked" scattering and large reaction cross section.[171] The second is of the "rebound" type,[43] primarily backscattering in nature and of only modest reaction cross section. Although the differential reaction cross sections might appear to be similar in information content, it turns out that the formal measure of the information of the angular distribution (the so-called[79,80] entropy deficiency ΔS) is appreciably greater for the second reaction than for the first.

An even more interesting illustration with a somewhat nonintuitive result is provided by consideration of the simplest case of an elastic scattering angular distribution. In Reference 168 it was shown that the strong

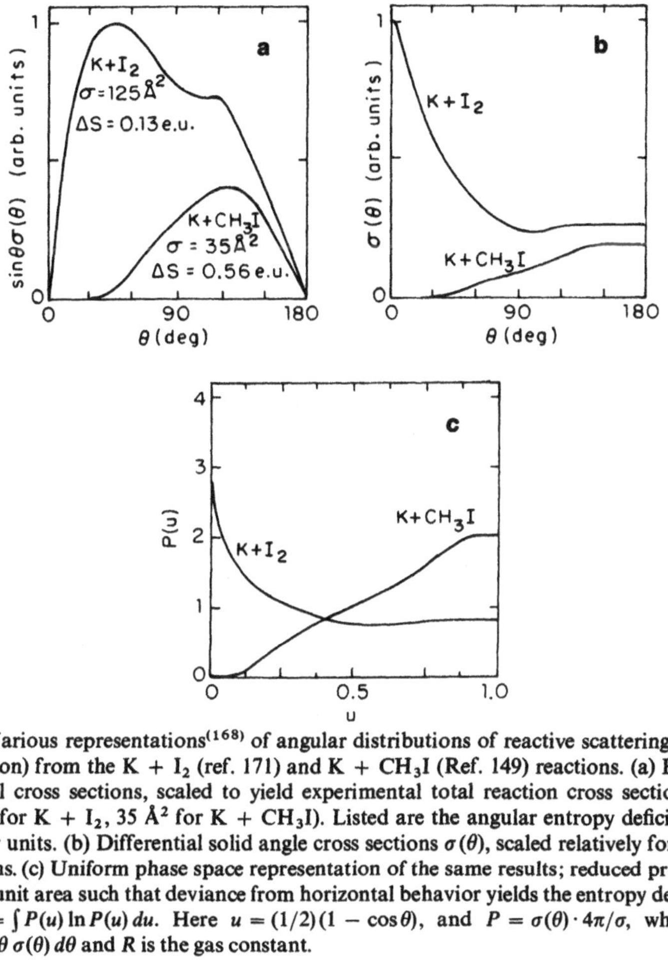

Figure 27. Various representations[168] of angular distributions of reactive scattering (i.e., KI formation) from the K + I_2 (ref. 171) and K + CH_3I (Ref. 149) reactions. (a) Polar differential cross sections, scaled to yield experimental total reaction cross sections ($\sigma \cong$ 125 Å^2 for K + I_2, 35 Å^2 for K + CH_3I). Listed are the angular entropy deficiencies in entropy units. (b) Differential solid angle cross sections $\sigma(\theta)$, scaled relatively for the two reactions. (c) Uniform phase space representation of the same results; reduced probability plot of unit area such that deviance from horizontal behavior yields the entropy deficiency: $\Delta S/R = \int P(u) \ln P(u)\, du$. Here $u = (1/2)(1 - \cos\theta)$, and $P = \sigma(\theta)\cdot 4\pi/\sigma$, where $\sigma = \int 2\pi \sin\theta\, \sigma(\theta)\, d\theta$ and R is the gas constant.

forward-scattered peak ("shadow scattering"[2,27]) contributes more to the formal information content than the supernumerary rainbow structure, which in turn contributes more than the quantal diffraction oscillations in the angular distribution. Now the experimenter already knowns that the low-angle scattering yields only the long-range force constant (the C_6 for neutral–neutral systems), whereas the rainbow data provide knowledge of the potential well (mainly the well depth ε, but also an estimate of the size parameter r_{min}). The periodicity of the rapid oscillation is governed by the reduced wave number[27] $A \equiv kr_0$ [where r_0 is the zero of the potential and $k = 2\pi/\lambda = (2\mu E_{tr})^{1/2}/\hbar$ the wave number]:

$$\Delta\theta = \pi/A\beta_0$$

where β_0 is the reduced impact parameter β at the zero of the deflection function [i.e., $\chi(\beta_0) = 0$ (here $\beta \equiv b/r_0$ as usual)], which is close to unity at high energies. Thus this periodicity essentially yields the characteristic size parameter of the potential. In practice, very accurate values of r_0 (and quite accurate r_{min}) are obtained from such high-resolution data. Yet the formal information content of the oscillatory structure is small,[168] presumably because of the redundancy associated with diffraction oscillations.

From the viewpoint of the experimenter it is important to develop better means of determining the "value" or "significance" of new, detailed observations, not merely the formal information content. Future theoretical research in this area and in the information-theoretic approach to collision theory (cf. Chapter 22) would be of great interest to the experimentalist.

6. *Future Theoretical Demands of the Experimentalist*

The impact of theory upon experiment in the field of atom–molecule collisions is great.[139] The goal of most current experimentation in this area is to make detailed, quantitative observations which not only test the qualitative predictions of the theoretical framework but also serve to evaluate the parameters in the theory which are least well known. Ultimately one would like to be able to abstract from the theory the essence in the form of appropriate dynamical models, with which to account for and "understand" the main experimental features.

As the field of molecular collisions progresses, and ever-more detailed measurements on elementary processes become possible, the experimentalist demands more of the theory. It is clear that most of the results of experiments on simple atom–molecule systems could by now be "predicted" more or less quantitatively if the relevant interaction potentials were known *ab initio*, since the "roadmap" from the potential to the cross sections is well marked. Yet practical, accurate methods of calculating such potentials are still wanting, and in their absence much too much effort is devoted to the rather fruitless task of deducing empirical potentials by "inversion" of scattering and spectroscopic data (and then showing them to be consistent with the same data base via scattering calculations!). The potential energy surface is the *sine qua non*, underlying (and pervading) the entire field. Thus the first demand of the experimentalist is addressed to the quantum chemist: provide the *algorithms*, if not the potential surfaces themselves, to deal with collisional aspects of *any* triatomic system:

$$\mathrm{A + BC} \rightleftharpoons \mathrm{A}^{\mathrm{B}}\mathrm{C} \begin{array}{l} \nearrow\!\!\!\swarrow\ \mathrm{AB + C} \\ \searrow\!\!\!\nwarrow\ \mathrm{AC + B} \end{array}$$

The second demand is for improved (efficient) algorithms to carry out the scattering calculations themselves, *given* the potentials. The computational procedures should be "stratified" so that the detail provided is not much more than that required to compare with the available experimental data. A global solution to the scattering problem is both costly and unnecessary for many applications. The selection among various approximation methods should be made according to systematic procedures built in to the computing program after due consideration of various inequalities (i.e., $\Delta E \ll E$), or the magnitude of certain key dimensionless variables. It is time for a workmanlike approach to the problem of the molecular scattering computation.[(172)]

The third demand is for extension of theoretical methods to deal effectively with polyatomic systems. At some point one must come to grips with the large and ever-growing body of data on energy transfer[(129,173-175)] as well as chemically reactive collisions[(44,66,93-95,149)] of polyatomics. Ultimately, practical theoretical–computational means of dealing with all aspects of the scattering of polyatomic molecules must be developed.

The fourth demand is for innovative theoretical input to the planning of experiments. What new *kind* of observations on what *kind* of systems would be most informative, most provocative? Are there overlooked *areas* of experimentation, even on simple atom–molecule systems? (The answer is, of course, yes!)[(176,177)]

Finally, there is the purely pragmatic side. There are many potential (if not yet fully realized) practical implications[(178)] and applications of research in molecular collisions.[(179)] Theory provides the basis for the design of not only new *experiments* but new *devices* as well. As an example, the technological development of the chemical laser[(180)] has been very much influenced by theoretical considerations, involving state-to-state reaction dynamics and relaxation processes.[(14,181)]

Recently there has been considerable interest in the interaction of strong laser fields with atoms and molecules in collision, a subject which is strongly interfaced with basic scattering theory.[(182-184)] The possibility of altering an activation energy barrier by laser irradiation of the colliding reagents[(183)] is an interesting one, with possible practical applications.

In conclusion, it is clear from the preceding discussion that there is a strong interdependency of theory and experiment in the field of molecular collisions. The subsequent chapters are intended to make the theory more accessible to the experimenter. The aim is to help bridge the small remaining gap between researchers dealing with complementary aspects of the same discipline, atom–molecule collisions.

ACKNOWLEDGMENT

The author's research program in the field of molecular collisions has received its principal support from the National Science Foundation, whose assistance is gratefully acknowledged.

References

1. R.D. Levine and R.B. Bernstein, *Molecular Reaction Dynamics*, Clarendon Press, Oxford (1974).
2. S. Chapman and T.G. Cowling, *The Mathematical Theory of Non-Uniform Gases*, Cambridge University Press, Cambridge (1939).
3. S. Glasstone, K.J. Laidler, and H. Eyring, *Theory of Rate Processes*, McGraw-Hill, New York (1941).
4. N.F. Mott and H.S.W. Massey, *The Theory of Atomic Collisions*, Oxford University Press, Oxford (1933, 1949, 1965).
5. J.O. Hirschfelder, C.F. Curtiss, and R.B. Bird, *Molecular Theory of Gases and Liquids*, J. Wiley and Sons, New York (1954); second printing (1964).
6. R.D. Levine, *Quantum Mechanics of Molecular Rate Processes*, Clarendon Press, Oxford (1969).
7. K. Smith, *The Calculation of Atomic Collision Process*, Wiley-Interscience, New York (1972).
8. W.A. Lester, Jr., editor, *Potential Energy Surfaces in Chemistry*, IBM, San Jose, Calif. (1970).
9. W.H. Miller, editor, *Dynamics of Molecular Collisions, Parts A and B*, Plenum Press, New York (1976).
10. H.S.W. Massey, *Electronic and Ionic Impact Phenomena*, Vol. 3, *Slow Collision of Heavy Particles*, Clarendon Press, Oxford (1971).
11. M.R.C. McDowell, editor, *Atomic Collision Processes*, North-Holland Publ. Co., Amsterdam (1964).
12. N.F. Ramsey, *Molecular Beams*, Clarendon Press, Oxford (1956).
13. M.A.D. Fluendy and K.P. Lawley, *Chemical Applications of Molecular Beam Scattering*, Chapman and Hall, London (1973).
14. P.R. Brooks and E.F. Hayes, editors, *State-to-State Chemistry*, ACS Symposium Series No. 56, American Chemical Society, Washington, D.C. (1977).
15. H. Eyring, D. Henderson, and W. Jost, editors, *Physical Chemistry—An Advanced Treatise*, Vols. VIA, VIB, Academic Press, New York (1974).
16. E.E. Nikitin, *Theory of Elementary Atomic and Molecular Processes in Gases*, Clarendon Press, Oxford (1974).
17. H. Hartmann, editor, *Chemische Elementarprozesse*, Springer-Verlag, Berlin (1968).
18. M.S. Child, *Molecular Collision Theory*, Academic Press, New York (1974).
19. E.W. McDaniel, *Collision Phenomena in Ionized Gases*, J. Wiley and Sons, New York (1964).
20. J.B. Hasted, *Physics of Atomic Collisions*, Elsevier, New York (1964); second edition (1972).
21. J. Zorn, editor, *Experiments with Molecular Beams, Selected Reprints*, Vol. 1, American Institute of Physics, New York (1965).
22. J. Ross, editor, *Molecular Beams*, Interscience-Wiley, New York, *Adv. Chem. Phys.* **10** (1966).

23. C. Schlier, editor, *Molecular Beams and Reaction Kinetics*, Academic Press, New York (1970).
24. K.P. Lawley, editor, *Molecular Scattering: Physical and Chemical Applications*, J. Wiley, New York, *Adv. Chem. Phys.* **30** (1975).
25. J.O. Hirschfelder, editor, *Intermolecular Forces*, Wiley-Interscience, New York, *Adv. Chem. Phys.* **12** (1967).
26. R.B. Bernstein and J.T. Muckerman, Determination of intermolecular forces via low-energy molecular beam scattering, Chap. 8 of Ref. 25, *Adv. Chem. Phys.* **12**, 389–486 (1967).
27. R.B. Bernstein, Quantum effects in elastic molecular scattering, Chap. 3 of Ref. 22, *Adv. Chem. Phys.* **10**, 75–134 (1966).
28. U. Buck, Elastic scattering, in Ref. 24, *Adv. Chem. Phys.* **30**, 313–388 (1975).
29. M. Karplus, Potential energy surfaces, in Ref. 23, pp. 320–348 (1970); G.G. Balint-Kurti, Potential energy surfaces for chemical reaction, in Ref. 24, *Adv. Chem. Phys.* **30**, 137–183 (1975); D.G. Truhlar and R.E. Wyatt, $H + H_2$: Potential energy surfaces and elastic and inelastic scattering, *Adv. Chem. Phys.* **36**, 141–204 (1977).
30. H. Eyring and S.H. Lin, Potential energy surfaces, in *Physical Chemistry—An Advanced Treatise*, Vol. VIA, *Kinetics of Gas Reactions*, W. Jost, editor, Academic Press, New York (1974), pp. 121–186; P.J. Kuntz, Features of potential energy surfaces and their effect on collisions, in Ref. 9, Part B, Chap. 2, pp. 53–120 (1976).
31. *Faraday Discuss. Chem. Soc.* No. 62, Potential energy surfaces (1977).
32. J.C. Tully, Diatomics in molecules, in *Semiempirical Methods of Electronic Structure Calculation, Part A*, G.A. Segal, editor, pp. 173–200, Plenum, New York (1977).
33. H. Pauly and J.P. Toennies, The study of intermolecular potentials with molecular beams at thermal energies, in *Advances in Atomic and Molecular Physics*, Vol. 1, D.R. Bates and I. Estermann, editors, Academic Press, New York (1965), pp. 201–344.
34. H. Pauly and J.P. Toennies, Beam experiments at thermal energies, in *Methods of Experimental Physics*, B. Bederson and W.L. Fite, Jr., editors, Vol. 7A, *Atomic and Electronic Physics: Atomic Interactions*, Chap. 3.1, pp. 227–341, Academic Press, New York (1968).
35. H. Pauly, Collision processes, theory of elastic scattering, in *Physical Chemistry—An Advanced Treatise*, W. Jost, editor, Vol. VIB, Chap. 8, pp. 553–628, Academic Press, New York (1975).
36. J. Reuss, Scattering from oriented molecules, in Ref. 24, *Adv. Chem. Phys.* **30**, 389–415 (1975).
37. W.A. Lester, Jr., The N coupled-channel problem, in Ref. 9, Part A, Chap. 1, pp. 1–32 (1976).
38. G.G. Balint-Kurti, The theory of rotationally inelastic molecular collisions, in *Int. Rev. of Science, MTP, Physical Chemistry, Series II*, A.D. Buckingham and C.A. Coulson, editors, Butterworths, Boston (1975), pp. 283–326.
39. K. Takayanagi, The production of rotational and vibrational transitions in encounters between molecules, in *Advances in Atomic and Molecular Physics*, D.R. Bates and I. Estermann, editors, Vol. 1, Academic Press, New York (1965), pp. 149–194.
40. E.E. Nikitin, Theory of energy transfer in molecular collisions, in *Physical Chemistry–An Advanced Treatise*, Vol. VIA, *Kinetics of Gas Reactions*, W. Jost, editor, Academic Press, New York (1974), pp. 187–226.
41. J.P. Toennies, Molecular beam scattering experiments on elastic, inelastic and reactive collisions, in Ref. 15, Vol. VIA, Chap. 5, pp. 228–381 (1974).
42. J. Ross and E.F. Greene, The study of chemical reactions in crossed molecular beams, in *Energy Transfer in Gases*, Solvay Institute, 12th Chemistry Conference, Interscience, New York (1962), pp. 363–396.
43. D.R. Herschbach, Reactive scattering in molecular beams, in Ref. 22, *Adv Chem. Phys.* **10**, 319–393 (1965).

44. J.P. Toennies, Molecular beam studies of chemical reactions, in Ref. 17, pp. 157–218 (1968).
45. R.F. Stebbings, Some new experimental methods in collision physics, in *Advances in Atomic and Molecular Physics,* D.R. Bates and I. Estermann, editors, Vol. 4, Academic Press, New York (1968), pp. 299–330.
46. R.J. Cross, Jr., Vibrationally and rotationally inelastic scattering, in Ref. 23, pp. 86–107 (1970).
47. R.N. Porter and L.M. Raff, Classical trajectory methods in molecular collisions, in Ref. 9, Part B, Chap. 1, pp. 1–52 (1976).
48. H.K. Shin, Vibrational energy transfer, in Ref. 9, Part A, Chap. 4, pp. 131–210 (1976).
49. M.S. Child, Semiclassical methods in molecular collision theory, in Ref. 9, Part B, Chap. 4, pp. 171–215 (1976).
50. J.C. Tully, Nonadiabatic processes in molecular collisions, in Ref. 9, Part B, Chap. 5, pp. 217–267 (1976).
51. F.J. de Heer, Experimental studies of excitation in collisions between atomic and ionic systems, in *Advances in Atomic and Molecular Physics,* D.R. Bates and I. Estermann, editors, Vol. 2, Academic Press, New York (1966), pp. 328–384.
52. R.S. Berry, Chemi-ionization, in Ref. 23, pp. 193–228 (1970).
53. R.S. Berry, Theory of electronic excitation, in Ref. 23, pp. 229–248 (1970).
54. V. Kempter, Electronic excitation in collisions between neutrals, in Ref. 24, *Adv. Chem. Phys.* **30,** 417–462 (1975).
55. R.A. Mapleton, *Theory of Charge Exchange,* Wiley-Interscience, New York (1972).
56. A.P.M. Baede, Charge transfer between neutrals at hyperthermal energies, in Ref. 24, *Adv. Chem. Phys.* **30,** 463–535 (1975).
57. E.E. Muschlitz, Jr., Collisions of electronically excited atoms and molecules, in Ref. 22, *Adv. Chem. Phys.* **10,** 171–193 (1966).
58. R.F. Stebbings, Charge transfer, in Ref. 22, *Adv. Chem. Phys.* **10,** 195–246 (1966).
59. R. Grice, Reactive scattering, in Ref. 24, *Adv. Chem. Phys.* **30,** 247–312 (1975).
60. D.A. Micha, Quantum theory of reactive molecular collisions, in Ref. 24, *Adv. Chem. Phys.* **30,** 7–75 (1975).
61. D.L. Bunker, Trajectory studies, in Ref. 23, pp. 355–371 (1970).
62. M. Karplus, Special results of trajectory studies, in Ref. 23, pp. 372–391 (1970).
63. R.N. Porter, Molecular trajectory calculations, *Ann. Rev. Phys. Chem.* **25,** 317–355 (1974).
64. J. Ross and E.F. Greene, Elastic scattering of reactive systems, in Ref. 23, pp. 86–107 (1970); E.F. Greene, A.L. Moursund, and J. Ross, Elastic scattering in chemically reactive systems, in Ref. 22, *Adv. Chem. Phys.* **10,** 135–169 (1966).
65. D.R. Herschbach, Molecular beam studies of internal excitation of reaction products, *Appl. Opt. Suppl.* **2,** 128–144 (1965).
66. D.R. Herschbach, Reactive scattering, *Faraday Discuss. Chem. Soc.* **55,** 233–251 (1973).
67. J.C. Polanyi, Molecular beam scattering, *Faraday Discuss. Chem. Soc.* **55,** 389–409 (1973); Some concepts in reaction dynamics, *Acc. Chem. Res.* **5,** 161–168 (1972).
68. K.E. Shuler, Reaction cross sections, rate coefficients and nonequilibrium kinetics, in Ref. 17, pp. 1–22 (1968).
69. J. Ross, J.C. Light, and K.E. Shuler, Rate coefficients, reaction cross sections and microscopic reversibility, in *Kinetic Processes in Gases and Plasmas,* A.R. Hochstim, editor, Academic Press, New York (1969), pp. 281–320.
70. J.C. Polanyi and J.L. Schreiber, The dynamics of bimolecular reactions, in Ref. 15, Vol. VIA, Chap. 6, pp. 383–449 (1976).
71. P. Pechukas, Statistical approximations in collision theory, in Ref. 9, Part B, Chap. 6, pp. 269–322 (1976).
72. R.A. Marcus, Unimolecular reaction rate theory, in Ref. 17, pp. 109–116 (1968).
73. J. Troe, Unimolecular reactions: Experiments and theories, in Ref. 9, Vol. VIB, pp. 835–929 (1975).

74. W.L. Hase, Dynamics of unimolecular reactions, in Ref. 9, Part B, Chap. 3, pp. 121–169 (1976).
75. P.J. Robinson and K.A. Holbrook, *Unimolecular Reactions*, Wiley-Interscience, New York (1972).
76. W. Forst, *Theory of Unimolecular Reactions*, Academic Press, New York (1973).
77. G.W. McClure and J.M. Peek, *Dissociation in Heavy Particle Collisions*, Wiley-Interscience, New York (1972).
78. R.D. Levine and R.B. Bernstein, Energy disposal and energy consumption in elementary chemical reactions: The information-theoretical approach, *Acc. Chem. Res.* **7**, 393–400 (1974).
79. R.B. Bernstein and R.D. Levine, Role of energy in reactive molecular scattering: An information-theoretic approach, in *Advances in Atomic and Molecular Physics*, D.R. Bates and I. Estermann, editors, Vol. 11, Academic Press, New York (1975), pp. 215–297.
80. R.D. Levine and R.B. Bernstein, Thermodynamic approach to collision processes, in Ref. 9, Part B, Chap. 7, pp. 323–364 (1976).
81. R.B. Bernstein, Molecular beam scattering at thermal energies, *Science* **144**, 141–150 (1964).
82. J.B. Anderson, R.P. Andres, and J.B. Fenn, High intensity and high energy molecular beams, *Adv. At. Mol. Phys.* **1**, 345–389 (1965).
83. J.B. Anderson, R.P. Andres, and J.B. Fenn, Supersonic nozzle beams, in Ref. 22, *Adv. Chem. Phys.* **10**, 275–317 (1966).
84. R.B. Bernstein, Elastic scattering of atoms and molecules in the thermal range, in Ref. 11, pp. 895–913 (1964).
85. R.B. Bernstein, Molecular beam scattering as a route to the intermolecular potential, in Molecular Forces, *Pontif. Accad. Sci. Scr. Varia* **31**, 25–71 (1967).
86. C. Schlier, Intermolecular forces, *Ann. Rev. Phys. Chem.* **20**, 191–218 (1969).
87. H.G. Bennewitz, K.H. Kramer, W. Paul, and J.P. Toennies, Messung der Anisotropie des van der Waals-Potentials durch Streuung von Molekülen in definiertem Quantenzustand, *Z. Phys.* **177**, 84–110 (1964).
88. L. Zandee and J. Reuss, The anisotropy of the intermolecular potential for H_2–inert gas systems, determined from total collision cross section measurements with state selected molecules; The experimental set-up and results, *Chem. Phys.* **26**, 327–343 (1977).
89. J.P. Toennies, Rotationally and vibrationally inelastic scattering of molecules, *Chem. Soc. Rev.* **3**, 407–411 (1974).
90. J.P. Toennies, The calculation and measurement of cross sections for rotational and vibrational excitation, *Ann. Rev. Phys. Chem.* **27**, 225–260 (1976).
91. J.R. Rumble, Jr. and J.W. Gallagher, *Bibliography of Energy Transfer Collisions of Atoms and Molecules*, publication of the Low Energy Atomic Collision Cross Section Information Center, Joint Institute for Laboratory Astrophysics, Univ. of Colorado, Boulder, Colorado (1977–1978).
92. D. Secrest, Theory of rotational and vibrational energy transfer in molecules, *Ann. Rev. Phys. Chem.* **24**, 379–406 (1974).
93. J.L. Kinsey, Molecular beam reactions, in MTP International Review of Science, Phys. Chem. Ser. 1, J.C. Polanyi, editor, Vol 9, pp. 173–212, *Reaction Kinetics*, Butterworths, London (1972).
94. Y.T. Lee, Reactive scattering of atoms and molecules: Crossed molecular beam experiments, in *Physics of Electronic and Atomic Collisions*, T.R. Govers and F.J. de Heer, editors, North-Holland Publ. Co., Amsterdam (1972), pp. 357–372.
95. J.M. Farrar and Y.T. Lee, Chemical dynamics, *Ann. Rev. Phys. Chem.* **25**, 357–385 (1974).
96. J.J. Valentini, Y.T. Lee, and D.J. Auerbach, A crossed molecular beam study of the translational energy dependence of $Cl + Br_2 \rightarrow BrCl + Br$ reaction, *J. Chem. Phys.* **67**, 4866–4872 (1977).

97. Y.T. Lee, J.D. McDonald, P.R. LeBreton, and D.R. Herschbach, Molecular beam reactive scattering apparatus with electron bombardment detector, *Rev. Sci. Instrum.* **40,** 1402–1408 (1969).
98. T.T. Warnock and R.B. Bernstein, Transformation relationships from center-of-mass cross section and excitation functions to observable angular and velocity distribution of scattered flux, *J. Chem. Phys.* **49,** 1878–1886 (1968).
99. S.Stolte, A.E. Proctor, W.M. Pope, and R.B. Bernstein, Rotational energy effect upon the branching fraction for reactive delay of the CsF + K collision complex, *J. Chem. Phys.* **66,** 3468–3481 (1977).
100. H.U. Hostettler and R.B. Bernstein, Improved slotted disk type velocity selector for molecular beams, *Rev. Sci. Instrum.* **31,** 872–877 (1960).
101. E.E. Bromberg, A.E. Proctor, and R.B. Bernstein, Pure-state molecular beams: Production of rotationally, vibrationally and translationally selected CsF beams, *J. Chem. Phys.* **63,** 3287–3294 (1975).
102. F. Knauer, Uber die Streuung von Molekularstrahlen in Gasen I, *Z. Phys.* **80,** 80–99 (1933); II, *ibid.* **90,** 559–566 (1934).
103. H.S.W. Massey and C.B.O. Mohr, Free paths and transport phenomena in gases and the quantum theory of collisions II. The determination of the laws of force between atoms and molecules, *Proc. R. Soc. (London)* **A144,** 188–205 (1934).
104. H. Pauly, Streuversuche an Molekularstrahlen and zwischenmolekulare Kräfte, *Fortschr. Phys.* **9,** 613–687 (1961).
105. E.A. Mason, Scattering of low velocity molecular beams in gases, *J. Chem. Phys.* **26,** 667–677 (1957).
106. H.U. Hostettler and R.B. Bernstein, Comparison of exp-6, L.-J.(12,6) and Sutherland potential functions applied to the calculation of differential scattering cross sections, *J. Chem. Phys.* **31,** 1422–1423 (1959).
107. K.W. Ford and J.A. Wheeler, Semiclassical description of scattering and application of semiclassical scattering analysis, *Ann. Phys. (N.Y.)* **7,** 259–322 (1959).
108. E.A. Mason and L. Monchick, Supernumerary rainbows in molecular scattering, *J. Chem. Phys.* **41,** 2221–2222 (1964).
109. F.A. Morse and R.B. Bernstein, Velocity dependence of the differential cross sections for the scattering of atomic beams of K and Cs by Hg, *J. Chem. Phys.* **37,** 2019 (1962).
110. H.U. Hostettler and R.B. Bernstein, Observation of quantum effects in the scattering of a monoenergetic Li beam by a crossed Hg beam, *Phys. Rev. Lett.* **5,** 318–320 (1960).
111. U. Buck, H.O. Hoppe, F. Huisken, and H. Pauly, Intermolecular potentials by the inversion of molecular beam scattering data IV. Differential cross sections and potential for LiHg, *J. Chem. Phys.* **60,** 4925–4929 (1974).
112. U. Buck and H. Pauly, Determination of intermolecular potentials by inversion of molecular beam scattering data, *J. Chem. Phys.* **51,** 1662–1664 (1969).
113. U. Buck, Determination of intermolecular potentials by the inversion of molecular beam scattering data I. The inversion procedure, *J. Chem. Phys.* **54,** 1923–1928 (1971).
114. H. Pauly, Streuversuche an Molekularstrahlen bei verschiedenen Strahlgeschwindigkeiten, *Z. Naturforsch.* **15a,** 277–278 (1960).
115. R.B. Bernstein, Quantum calculations of the velocity dependence of the differential and total cross sections for elastic scattering of molecular beams, *J. Chem. Phys.* **34,** 361–365 (1961).
116. E.W. Rothe, P.K. Rol, S.M. Trujillo, and R.H. Neynaber, Velocity dependence of the total cross section for the scattering of Li and K by Xe, *Phys. Rev.* **128,** 659–662 (1962).
117. E.W. Rothe, P.K. Rol, and R.B. Bernstein, Interaction potentials from the velocity dependence of total atom–atom scattering cross sections, *Phys. Rev.* **130,** 2333–2338 (1963).
118. P.K. Rol and E.W. Rothe, Velocity dependence of the total cross section for atom–atom scattering at thermal energies, *Phys. Rev. Lett.* **9,** 494–495 (1962).

119. M.H. Keil, G.A. Parker, and A. Kuppermann, An anisotropic intermolecular potential for He–CO_2, *Chem. Phys. Lett.* **59,** 443–448 (1978).
120. J.P. Toennies, Molekularstrahlmessungen von Stossquerschnitten für Übergänge zwischen definierten Rotationszugständen, *Z. Phys.* **182,** 257–277 (1965).
121. A.R. Blythe, A.E. Grosser, and R.B. Bernstein, Observation of rotational de-excitation of D_2 using crossed molecular beams, *J. Chem. Phys.* **41,** 1917–1921 (1964).
122. H.E. Van den Bergh, M. Faubel, and J.P. Toennies, Measurement of differential cross sections for individual rotational quantum transitions in the scattering of Li^+ by H_2 at $E_{c.m.} = 0.6$ eV, *Faraday Discuss. Chem. Soc.* **55,** 203–210 (1973).
123. W.R. Gentry and C.F. Giese, Resolved single-quantum rotational excitation in HD + He collisions: First results from a unique pulsed molecular beam apparatus, *J. Chem. Phys.* **67,** 5389–5391 (1977).
124. U. Buck, F. Huisken, and J. Schleusener, Diffraction oscillations in rotationally inelastic differential cross sections: HD + D_2, *J. Chem. Phys.* **68,** 5654–5655 (1978).
125. H. Pauly, Inelastic scattering: Molecular beam studies of inelastic collision processes, *Faraday Discuss. Chem. Soc.* **55,** 191–202 (1973).
126. P.F. Dittner and S. Datz, Collision-induced vibrational excitation and dissociation of hydrogen with alkali atoms and ions from 2 to 50 eV, *J. Chem. Phys.* **54,** 4228–4233 (1971); Vibrational excitation and fragmentation of molecular hydrogen from slow K^+ collisions, *J. Chem. Phys.* **49,** 1969–1970 (1968).
127. H. Udseth, C.F. Giese, and W.R. Gentry, Differential cross sections for excitation of resolved vibrational states in $H^+ + H_2(v = 0)$ nonreactive collisions, *J. Chem. Phys.* **54,** 3642–3643 (1971); Transition probabilities and differential cross sections for vibrational excitation in collisions of H^+ with H_2, HD, and D_2, *Phys. Rev. A* **8,** 2483–2493 (1973).
128. W.D. Held, J. Schöttler, and J.P. Toennies, Experimental cross sections for individual vibrational quantum transitions in low energy central collisions of Li^+ with H_2, *Chem. Phys. Lett.* **6,** 304–306 (1970).
129. R.G. Gordon, W. Klemperer, and J.I. Steinfeld, Vibrational and rotational relaxation, *Ann. Rev. Phys. Chem.* **19,** 215–250 (1968).
130. S. Wexler, Associative and non-associative ion pair formation by fast atoms, *Ber. Bunsenges. Phys. Chem.* **77,** 606–627 (1973).
131. E.E. Muschlitz, Jr., Associative ionization reactions of electronically excited particles, *Ber. Bunsenges. Phys. Chem.* **77,** 628–632 (1973).
132. A. Niehaus, Penning ionization, *Ber. Bunsenges. Phys. Chem.* **77,** 632–640 (1973).
133. A.P.M. Baede, Charge transfer between neutrals at hyperthermal energies, *Adv. Chem. Phys.* **30,** 463–535 (1975).
134. J. Los and A.W. Kleyn, Ion pair formation, in *Alkali Halide Vapors,* P. Davidovits and D. McFadden, editors, Academic Press, New York (1979), to be published.
135. A.P.M. Baede, A.M.C. Moutinho, A.E. DeVries, and J. Los, Total cross sections for charge transfer between alkali atoms and halogen molecules, *Chem. Phys. Lett.* **3,** 530–531 (1969).
136. R.K.B. Helbing and E.W. Rothe, Crossed-beam study of ionization of Cs by Br_2, *J. Chem. Phys.* **51,** 1607–1609 (1969).
137. G.A. Delvigne and J. Los, Rainbow, Stueckelberg oscillations and rotational coupling on the differential cross section of Na + I → $Na^+ + I^-$, *Physica* **67,** 166–196 (1973).
138. J.A. Aten and J. Los, The influence of the rate of bond stretching on the differential cross sections for ion-pair formation in K + XY collisions, *Chem. Phys.* **25,** 47–54 (1977).
139. R.B. Bernstein, Molecular reaction dynamics: Impact of theory on experiment, *Israel J. Chem.* **14,** 79–88 (1975).
140. D.R. Herschbach, Molecular dynamics of chemical reactions, *Pure Appl. Chem.* **47,** 61–73 (1976).

141. R.B. Bernstein, Molecular reaction dynamics by the crossed molecular beam technique: A few recent examples, *Pure Appl. Chem.* **50,** 781–783 (1978).
142. R.B. Bernstein, Reaction dynamics by molecular beams, in *Advances in Laser Chemistry,* A.H. Zewail, editor, Springer, Berlin (1978), pp. 384–396.
143. R.R. Herm, Reactive single collision studies with metal atoms or metal halides, in *Alkali Halide Vapors,* P. Davidovits and D. McFadden, editors, Academic Press, New York (1979), to be published.
144. J.C. Polanyi, Molecular beam scattering, *Faraday Discuss. Chem. Soc.* **55,** 389–409 (1973).
145. E.H. Taylor and S. Datz, Study of chemical reaction mechanisms with molecular beams. The reaction of K with HBr, *J. Chem. Phys.* **23,** 1711–1718 (1955).
146. S. Datz, D.R. Herschbach, and E.H. Taylor, Collision mechanics in crossed Maxwellian beams, *J. Chem. Phys.* **35,** 1549–1559 (1969).
147. D.R. Herschbach, G.H. Kwei, and J.A. Norris, Reactive scattering in crossed molecular beams. K atoms with CH_3I and C_2H_5I, *J. Chem. Phys.* **34,** 1842–1843 (1961).
148. A.E. Grosser, A.R. Blythe, and R.B. Bernstein, Internal energy of reaction products by velocity analysis. I. Scattered KBr from the crossed molecular beam reaction K + HBr, *J. Chem. Phys.* **42,** 1268–1273 (1965).
149. A.M. Rulis and R.B. Bernstein, Molecular beam study of the K + CH_3I reaction: Energy dependence of the detailed differential reactive cross section, *J. Chem. Phys.* **57,** 5497–5515 (1973); R.B. Bernstein and A.M. Rulis, Translational energy dependence of product energy and angular distribution for the K + CH_3I reaction, *Faraday Discuss. Chem. Soc.* **55,** 293–298 (1973).
150. E. Pollak and R.D. Levine, Surprisal analysis of products' translational energy distributions in molecular collisions, *Chem. Phys.* **21,** 61–80 (1977).
151. R.B. Bernstein and B.E. Wilcomb, On the products' translational distribution for the reactions of CH_3I and CF_3I with K and Ba, *J. Chem. Phys.* **67,** 5809–5817 (1977).
152. W.B. Miller, S.A. Safron, and D.R. Herschbach, Exchange reactions of alkali atoms with alkali halides: A collision complex mechanism, *Faraday Discuss. Chem. Soc.* **44,** 108–122 (1967).
153. R. Wolfgang, Energy and chemical reactions II. Intermediate complexes *vs.* direct reactions, *Acct. Chem. Res.* **3,** 48–54 (1970).
154. S.A. Safron, N.D. Weinstein, D.R. Herschbach, and J.C. Tully, Transition state theory for collision complexes: Product translational energy distributions, *Chem. Phys. Lett.* **12,** 564–568 (1972).
155. R.A. Marcus, On the theory of energy distributions of products of molecular beam reactions involving transient complexes, *J. Chem. Phys.* **62,** 1372–1384 (1975).
156. D.D. Parrish and D.R. Herschbach, Molecular beam chemistry: Persistent collision complexes in reactions of oxygen atoms with bromine molecules, *J. Am. Chem. Soc.* **95,** 6133–6134 (1973).
157. D.G. Truhlar and R.E. Wyatt, History of H_3 kinetics, *Ann. Rev. Phys. Chem.* **27,** 1–43 (1976).
158. J. Geddes, H.F. Krause, and W.L. Fite, Atom–molecule reaction D + $H_2 \rightarrow$ HD + H studied by molecular beams, *J. Chem. Phys.* **56,** 3298–3307 (1972).
159. G.C. Schatz and A. Kuppermann, Quantum mechanical reactive scattering for three-dimensional atom plus diatom systems II. Accurate cross sections for H + H_2, *J. Chem. Phys.* **65,** 4668–4682 (1976).
160. A. Kuppermann and G.C. Schatz, Quantum mechanical reactive scattering: An accurate three-dimensional calculation, *J. Chem. Phys.* **62,** 2502–2504 (1975).
161. A. Elkowitz and R.E. Wyatt, Quantum mechanical cross sections for the three-dimensional hydrogen exchange reaction, *J. Chem. Phys.* **62,** 2504–2506 (1975).
162. F.P. Tully, Y.T. Lee, and R.S. Berry, Crossed molecular beam study of collision-induced dissociation of alkali halides, *Chem. Phys. Lett.* **9,** 80–84 (1971).

163. E.K. Parks and S. Wexler, Collisional dissociative ionization of TlBr by accelerated Xe and Kr atoms, *Chem. Phys. Lett.* **10,** 245–249 (1971).
164. E.K. Parks, J.G. Kuhry, and S. Wexler, Collision-induced ion-pair formation of the thallium halides: TlF and Tl_2F_2, *J. Chem. Phys.* **67,** 3014–3028 (1977).
165. R.D. Levine and R.B. Bernstein, Collision-induced dissociation: A simplistic optical model analysis, *Chem. Phys. Lett.* **11,** 552–556 (1971).
166. C. Rebick and R.D. Levine, Collision-induced dissociation: A statistical theory, *J. Chem. Phys.* **58,** 3942–3952 (1973).
167. A.F. Wagner and E.K. Parks, A classical statistical theory for chemical reactions, *J. Chem. Phys.* **65,** 4343–4361 (1976).
168. M.B. Faist, R.D. Levine, and R.B. Bernstein, Angular entropy: The information content of molecular scattering angular distributions, *J. Chem. Phys.* **66,** 511–523 (1977).
169. R.B. Bernstein and M.B. Faist, Information content of detailed differential reaction cross sections (molecular beam scattering velocity-angle maps) vs. resolved product state distributions, *J. Chem. Phys.* **65,** 5436–5444 (1976).
170. R.D. Levine, Information theory approach to molecular reaction dynamics, *Ann. Rev. Phys. Chem.* **29,** 59–92 (1978).
171. K.T. Gillen, A.M. Rulis, and R.B. Bernstein, Molecular beam study of the $K + I_2$ reaction: Differential cross section and energy dependence, *J. Chem. Phys.* **54,** 2831–2851 (1971).
172. R.G. Gordon, Rational selection of methods for molecular scattering calculations, *Faraday Discuss. Chem. Soc.* **55,** 22–29 (1973).
173. R.D. Levine and J. Jortner, editors, *Molecular Energy Transfer,* Halsted Press, J. Wiley and Sons, New York (1976).
174. S. Lemont and G.W. Flynn, Vibrational state analysis of electronic-to-vibrational energy transfer processes, *Ann. Rev. Phys. Chem.* **28,** 261–282 (1977).
175. B.L. Earl, L.A. Gamss, and A.M. Ronn, Laser-induced vibrational energy transfer kinetics: Methyl and methyl-d_3 halides, *Acc. Chem. Res.* **11,** 183–190 (1978).
176. R.N. Zare and P.J. Dagdigian, Tunable laser fluorescence method for product state analysis, *Science* **185,** 739–747 (1974).
177. P.R. Brooks, Reactions of oriented molecules, *Science* **193,** 11–16 (1976).
178. S.H. Bauer, How energy accumulation and disposal affect the rates of reactions, *Chem. Rev.* **78,** 147–184 (1978).
179. R.B. Bernstein, Potential practical applications of basic research in molecular dynamics, *Israel J. Chem.* **9,** 615–636 (1971).
180. R.W. Gross and J.F. Bott, editors, *Handbook of Chemical Lasers,* John Wiley, New York (1976).
181. R.B. Bernstein, State-to-state cross sections and rate constants for reactions of neutral molecules, in Ref. 14, pp. 3–21 (1977).
182. A.M.F. Lau and C.K. Rhodes, Control of collisionless and collisional processes by nonresonant laser fields, *Phys. Rev. A* **15,** 1570–1587 (1977).
183. T.F. George, I.H. Zimmermann, J.M. Yuan, J.R. Laing, and P.L. DeVries, A new concept in laser-assisted chemistry: The electron-field representation, *Acc. Chem. Res.* **10,** 449–455 (1977).
184. T.F. George, editor, *Theoretical Aspects of Laser Radiation and Its Interaction with Atomic and Molecular Systems,* National Science Foundation Workshop Report, University of Rochester, June, 1977.

2

Interaction Potentials I: Atom–Molecule Potentials

HENRY F. SCHAEFER III

1. Current State of Ab Initio Electronic Structure Theory

The concept of the potential energy surface (or hypersurface) is without doubt one of the most fundamental in all of science.(1) The minima on a potential energy surface correspond to the equilibrium geometries of the various isomers. Saddle points or "transition states" play an important role in the determination of the rates of the various chemical reactions which may occur on a surface or hypersurface.

With this background in mind it is clear that the most important practical use of electronic structure is its ability, at least in principle, to predict the shapes of potential energy surfaces. Until quite recently, however, the promise of *ab initio* potential energy surfaces vastly exceeded the delivery of surfaces suitable for dynamical studies. And, to be quite candid, this is still the case today. The reader may recall that the first genuinely useful *ab initio* potential surface for H_3, a very simple system indeed, did not appear(2) until 1968. Furthermore, even today, ten years later, rather exhaustive configuration interaction (CI) studies(3) on $F + H_2$, another simple system, continue to result in values of the barrier height ~ 1 kcal greater than experiment.

Why are potential energy surfaces so difficult to describe theoretically? First, because a relatively large number of points, i.e., separate quantum-mechanical calculations, must be treated. For a triatomic A + BC system, a minimum of 50 geometries is necessary for a description of reactants,

HENRY F. SCHAEFER III • Department of Chemistry, University of California, Berkeley, California 94720.

saddle point region, and products. For simple systems with more degrees of freedom (e.g.,[4] $CH_2 + H_2$), the number of required geometries can be much greater.

In striking contrast to this are many other problems in chemistry for which only a *single* geometry need be considered. One such example is the vertical electronic spectra of molecules; systems as large as benzene,[5] thioacetone,[6] and glyoxal[7] have recently been treated rather carefully, i.e., with reasonable basis sets and electron correlation considered explicitly. A problem for which even rather crude *ab initio* methods can be useful is the prediction of the ionization potentials or photoelectron spectra of molecules, the ethyl chlorophyllide *a* system (340 electrons!) being an impressive recent example.[8] So it is clear that there has been tremendous progress in electronic structure theory over the past decade,[9] but this progress is typically more apparent for chemical problems which do not require an entire potential surface.

Secondly, some of the A + BC systems of most interest to dynamicists and experimentalists are notoriously difficult to precisely describe by *ab initio* methods. For example, most simple repulsive potential surfaces ($H + H_2$, $F + H_2$, etc.) are very sensitive to the effects of electron correlation.[10,11] In striking contrast, there are seemingly more complicated systems, such as the methyl isocyanide isomerization

$$CH_3NC \rightarrow CH_3CN \tag{1}$$

which are quite reasonably described within the Hartree–Fock approximation.[12] Electron correlation[13] is also relatively unimportant for such diverse questions as the determination of hydrogen bond energies and the equilibrium structures of molecules. As examples we note a recent study[14] of the molecular complex tetrathiofulvalene–tetracyanoquinodimethane (208 electrons) and a full geometry determination[15] for but-1-yn-3-ene (fourteen independent structural parameters).

2. *Philosophy: Judicious Synthesis of Theory and Experiment*

The critical problem, then, is that contemporary atom–molecule collision theory often requires,[16] at a very high level of reliability, the entire potential energy surface—not just one, or even a few points. We have been convinced for about five years that *ab initio* electronic structure should not even attempt (except for the very simplest systems) to predict the entire potential surface. Instead we have advocated a judicious synthesis of theory and experiment to arrive at a complete, working potential surface.

The motivation for this philosophy is the observation that certain parts of potential surfaces are known experimentally to a very high accuracy, while other parts are not known at all. As an example, consider the

$$H + ClF \rightarrow \begin{cases} HCl + F & (2a) \\ HF + Cl & (2b) \end{cases}$$

system. The spectroscopic constants of the $^{35}Cl^{19}F$ molecule are known quite accurately[17] ($\omega_e = 786.34\ cm^{-1}$, $\omega_e x_e = 6.23\ cm^{-1}$, $r_e = 1.6281$ Å) except for the dissociation energy, which is either 60.3 or 59.0 kcal.[18] Since the best one could hope for from a state-of-the-art *ab initio* calculation on ClF would be $10\ cm^{-1}$ accuracy for ω_e, $0.1\ cm^{-1}$ for $\omega_e x_e$, 0.002 Å for r_e, and 1 kcal for the dissociation energy, it is fruitless to expend much theoretical effort on the reactants end of the surfaces. Similarly, the spectroscopic constants of the product molecules HCl and HF are known more accurately than can be predicted theoretically. Our philosophy, then, is to choose some suitable analytical form, perhaps the London–Eyring–Polanyi–Sato (LEPS) form[19] (although this might not be sufficiently flexible) and use the majority of its adjustable parameters to precisely reproduce the experimental spectroscopic constants of the three diatomics ClF, HCl, and HF.

At this point the reader may quite properly ask "If theory can't even reproduce the experimental diatomic potential curves, why bother to continue?" The answer is that although the diatomic potential curves are often known very accurately from spectroscopy, little else is usually known about reactive A + BC potential surfaces. If we are fortunate, one additional piece of information is available, the activation energy. For the H + ClF system there will be a barrier for the collinear HClF approach and another barrier for the collinear HFCl approach. These barriers and the accompanying saddle point geometries are truly critical to the dynamics of the H + ClF reaction and are accessible to theory. The expected state-of-the-art accuracy of ±1–2 kcal in barrier heights and ±0.02 Å in saddle point geometrical parameters is not breathtaking, but it can be extremely useful. An additional critical feature of the H + ClF surface which is amenable to theory is the manner in which the two collinear barriers connect as one goes to nonlinear geometries from each direction.

The above "compromise" procedure is of obvious real value, since it can yield features of the potential surface not amenable to direct experimental inquiry. Perhaps equally important, it drastically reduces the number of surface points to be investigated by the theoretician. In the H + ClF example, ~15 points are needed for a characterization (including force constants) of each collinear saddle point, and an additional 20 points should suffice for a qualitative description of the anisotropy of the surface.

3. Brief Survey of Methods

Before moving on to a survey of *ab initio* potential surfaces studied to date, it is of critical importance that we have a realistic picture of the reliability that may be expected from the different levels of theory. Assuming that a particular calculation is technically correct (e.g., all the two-electron integrals have been computed accurately), two factors will determine its predictive reliability: first, the basis set of one-electron functions, and secondly the degree to which electron correlation has been incorporated in the wave function.

3.1. Basis Sets

To solve the Hartree–Fock equations exactly one must either expand the orbitals ϕ in a complete set of analytic basis functions or obtain strictly numerical (i.e., tabulated) orbitals. The former approach is impossible from a practical point of view for systems with more than two electrons, and the latter has been accomplished only for atoms, and very recently[20] for a few diatomic molecules. Therefore the *exact* solution of the Hartree–Fock equations is abandoned for polyatomic molecules. Instead one adopts an incomplete (but reasonable) set of analytic basis functions and solves for the best variational (i.e., lowest energy) single-configuration wave function. Following the Chicago group we refer to such a wave function as being of self-consistent-field (SCF) quality. As one goes to very large basis sets, then, it is reasonable to refer to the resulting SCF wave function as near Hartree–Fock.[21]

For large chemical systems, at present only *minimum basis sets* (MBS) can be used in *ab initio* theoretical studies. By the term "large," we include

Table 1. Minimum Basis Set of Slater Functions for the Carbon Atom. The Orbital Exponents Zeta are Optimum for the ^3P Ground State of the Carbon Atom

Label	Analytical form	Exponent ζ
$1s$	$(\zeta^3/\pi)^{1/2}\exp(-\zeta r)$	5.673
$2s$	$(\zeta^5/3\pi)^{1/2} r\exp(-\zeta r)$	1.608
$2p_x$	$(\zeta^5/\pi)^{1/2} x\exp(-\zeta r)$	1.568
$2p_y$	$(\zeta^5/\pi)^{1/2} y\exp(-\zeta r)$	1.568
$2p_z$	$(\zeta^5/\pi)^{1/2} z\exp(-\zeta r)$	1.568

Table 2. Double Zeta Contracted Gaussian Basis Set (9*s*5*p*/4*s*2*p*) for the Carbon Atom ^3P Ground State[a]

Function type	Orbital exponents α	Coefficients[b]
s	4232.61	0.002029
s	634.882	0.015535
s	146.097	0.075411
s	42.4974	0.257121
s	14.1892	0.596555
s	1.9666	0.242517
s	5.1477	1.000000
s	0.4962	1.000000
s	0.1533	1.000000
p	18.1557	0.018534
p	3.9864	0.115442
p	1.1429	0.386206
p	0.3594	0.640089
p	0.1146	1.000000

[a] Primitive Gaussian *s* functions are of the form $(2\alpha/\pi)^{3/4} \times \exp(-\alpha r^2)$.
[b] The contracted basis functions are separated by lines.

molecular systems having more than 25 first-row atoms. A minimum basis set includes one function for each orbital occupied in the ground state of each atom included in the molecule. For the first row atoms B, C, N, O, and F, this means that a minimum basis includes 1*s*, 2*s*, $2p_x$, $2p_y$, and $2p_z$ functions.

Traditionally, minimum basis sets have been composed of Slater functions, such as those seen in Table 1 for the carbon atom. However, experience has shown that the evaluation of the molecular integrals arising from the Slater functions employed is extremely time consuming.[22] Therefore each Slater function in an MBS is typically replaced by a linear combination of three or four Gaussian functions.[23] The resulting chemical predictions obtained with such STO-3G or STO-4G basis sets are usually indistinguishable from the corresponding Slater function results.

Minimum basis sets are inadequate for many types of chemical predictions. Therefore a basis twice as large, and appropriately designated double zeta (DZ), is often used in theoretical studies. Here, however, it is not as fruitful to expand each Slater function as a linear combination of Gaussians. Instead one uses Gaussian functions $x^l y^m z^n \exp(-\alpha r^2)$ directly in atomic SCF calculations and then *contracts*[24] them according to the atomic results. Perhaps the most widely used contracted Gaussian DZ basis sets are those of Dunning,[25] and his carbon atom basis is given in Table 2. This basis has 9*s* and 5*p* original (or primitive) Gaussian functions and is contracted to 4s and 2p. Thus the basis may be designated C(9s5p/4s2p).

Just as the double zeta basis logically follows the minimum set, the logical extension of the DZ set involves the addition of polarization functions.[26] Polarization functions are of higher orbital angular momentum than the functions occupied in the atomic SCF wave function. That is, for carbon *d*, *f*, *g*,... functions will be polarization functions. Fortunately, *d* functions are far more important than *f*, *f* functions are far more important than *g*, and so on. For most chemical applications a double zeta plus polarization (DZ + P) basis including a single set of five *d* functions ($d_{x^2-y^2}, d_{z^2}, d_{xy}, d_{xz}, d_{yz}$) will be adequate for first-row atoms.

3.2. *The Problem of Electron Correlation*

3.2.1. *The Concept*

In the Hartree–Fock approximation one solves equations for the behavior of each electron in the averaged field of the remaining $(n-1)$ electrons. Unfortunately, electrons respond to each other in an instantaneous manner via Coulomb's law. That is, the motions of the $[n(n-1)]/2$ pairs of electrons are correlated, and it is precisely this *electron correlation* which is neglected in the Hartree–Fock model. The *correlation energy* is conveniently defined[13] as

$$E_c = E_{\text{exact}} - E_{\text{HF}}$$

the difference between the (nonrelativistic) exact energy and the Hartree–Fock limit energy. For this reason, quantum-mechanical calculations are sometimes judged by the percentage of the correlation energy obtained. For example, Meyer's pioneering theoretical study[27] of the water molecule yielded ~84% of the correlation energy.

3.2.2. *Configuration Interaction (CI)*

The most obvious way to go beyond the Hartree–Fock approximation is to include more than a single configuration in a variational manner.[28] For a molecule even as small as water, the number of possible configurations becomes very large with a reasonable (say, DZ + P) basis set. Therefore one immediately looks for ways of eliminating unimportant configurations. A logical and widely accepted method of classifying configurations rests on the number of orbitals by which a configuration differs from the Hartree–Fock reference configuration. A *single excitation* ϕ_{Si} differs by one orbital from the reference configurations, a double excitation ϕ_{Dj} by two, and so on. Thus an arbitrary CI wave function is of the form

$$\psi_{\text{CI}} = c_0\psi_{\text{HF}} + \sum_i c_{Si}\phi_{Si} + \sum_j c_{Dj}\phi_{Dj} + \sum_k C_{Tk}\phi_{Tk} + \sum_l c_{Ql}\phi_{Ql} + \cdots \quad (4)$$

where c_{Tk} is the coefficient of the *k*th triple excitation.

The CI coefficients are determined to yield the lowest possible total energy by solving the secular equation or eigenvalue problem

$$\mathbf{HC} = \mathbf{EC} \tag{5}$$

The Hamiltonian matrix $\mathbf{H}$ in turn is composed of elements

$$H_{ij} = \int \phi_i H \phi_j d\tau \tag{6}$$

and these Hamiltonian matrix elements H_{ij} are linear combinations of one- and two-electron integrals.[28]

The most obvious reason for a classification of configurations according to (4) is that Hamiltonian matrix elements

$$\int \psi_{\mathrm{HF}} H \phi_{Tk} d\tau = \int \psi_{\mathrm{HF}} H \phi_{Ql} d\tau = 0 \tag{7}$$

between the SCF reference configuration and all triple, quadruple, quintuple, ... excitations are identically zero. Thus if we take the SCF wave function as a zeroth-order wave function in perturbation theory, then all triple and higher excitations make no contribution to the wave function to first order. For this reason, CI including all single and double ($S + D$) excitations is the goal of a number of the most powerful theoretical methods[29-33] for the treatment of electron correlation. For the water molecule with a DZ + P basis and the $1a_1$ (or oxygen $1s$) orbital constrained to be always doubly occupied, there are 1026 single and double excitations. Calculations of this type are now relatively routine, requiring only ~ 10 min on the newer minicomputers.[34-36]

For small molecules single and double (S + D) excitations give rise to almost all of the correlation energy. For the water molecule, for example, S + D CI accounts for ~ 95–96% of the correlation energy.[28,37] However, as molecular size increases, it is now widely accepted[29,38] that higher excitations become more important. For example, Sasaki[39] has recently estimated that for the pyrrole molecule C_4H_5N, S + D CI will account for only 80% of the correlation energy. It should be noted that many of the important higher excitations involve the core (carbon and nitrogen $1s$ in the case of pyrrole) orbitals, and should be relatively unimportant for chemical predictions. Thus it is likely that S + D CI will account for perhaps 90% of the critical *valence* correlation energy of molecules the size of pyrrole.

Perhaps the most successful variational method for treating triple and quadruple excitations includes single and double excitations relative to several reference configurations.[40,41] The motivation for this approach is the observation[42] that an important quadruple excitation invariably differs by two orbitals from an important double excitation. For small molecules this multiple reference procedure (MR S + D CI) is capable of extremely reliable predictions.[43]

Among nonvariational procedures for treating higher excitations, the coupled electron pair approximation (CEPA)[38] has proven to be quite useful.[29,31] A simpler, but still useful estimate of the "unlinked cluster" effects is Davidson's approximation[44]

$$\Delta E_Q \sim (1 - c_0^2)\,\Delta E_D \tag{8}$$

which gives the correlation energy due to quadruple excitations in terms of c_0, the coefficient of the SCF configuration in the S + D CI, and ΔE_D, the correlation energy due to double excitations. The form of Davidson's correction is rather satisfying, since it shows in a transparent manner that quadruple excitations become more important as (a) the SCF configuration becomes a smaller fraction of the S + D CI wave function, and as (b) the correlation energy due to double excitations increases.

Before leaving this section, a simpler but very widely used type of CI wave function should be mentioned. Based in part on the perturbation theory ideas of Silverstone and Sinanoglu,[45] the *first-order wave function* was introduced by Schaefer and Harris[46] in 1968. Such a CI wave function was designed to include "all orbital occupancies in which at most one electron is assigned to an orbital beyond the valence shell." Since in a typical molecule there are several valence orbitals not occupied in the Hartree–Fock approximation, the first-order wave function does allow for higher than double excitations, although the total number of configurations is usually much less than the S + D CI. A modified form (designated POL–CI) of the first-order wave function has been used very successfully by Goddard and co-workers[47] in conjunction with their generalized valence bond (GVB) orbitals.[48] The essentials of this method for the selection of configurations have also been partially utilized by Wahl and co-workers in their important MCSCF[49] and CI[50] studies. For a recent assessment of the first-order method, the reader is referred to Liu's recent research[51] on the CO molecule.

4. Examples

The specific systems discussed here are by no means the only ones studied to date theoretically. However, those chosen are intended to include some of the most reliable quantum-mechanical studies and some of the systems of greatest interest to experimentalists and theoretical dynamicists.

4.1. *Nonreactive*

4.1.1. Li^+-H_2

Among nonreactive systems, one for which the potential energy surface has been carefully studied, is the Li^+-H_2 system. Li^++H_2 is also of special interest because it has been the subject of a number of experimental studies[52] of rotational and vibrational excitation. The first *ab initio* potential surface reported for $Li^+ + H_2$ was that of Lester,[53,54] whose work was done within the single-configuration Hartree–Fock approximation. Lester chose a flexible basis set, labeled Li(9*s*3*p*/5*s*3*p*), H(6*s*3*p*/6*s*3*p*), and of better than double zeta plus polarization quality. Regions of the surface pertinent to both rotational and vibrational excitation were considered.

Perhaps the most interesting feature of the surface is the presence of an attractive well, illustrated in Figure 1. This attraction is most pronounced for broadside (or C_{2v}) approaches, and the figure shows that for H_2 fixed near its equilibrium geometry the well is ~0.24 eV deep. As the H_2 internuclear separation is increased the well becomes deeper, being ~0.35 eV in magnitude for $r(H-H) = 2.0$ bohrs. The angular dependence of the Lester surface is especially interesting since the seemingly similar Li^+-N_2 surface has exactly the opposite anisotropy. That is, Staemmler[55] has shown that the attractive well is strongest for collinear Li^+-N_2 and weakest for the C_{2v} approach. As Staemmler has noted, these differences may be qualitatively explained in terms of electrostatic considerations. That is, the molecular quadrupole moment of N_2 is negative, while that of H_2 is positive and smaller in magnitude.

The other important theoretical study of the Li^+-H_2 surface is that of Kutzelnigg, Staemmler, and Hoheisel.[56] They used a slightly smaller basis set than Lester, but included electron correlation via the independent electron pair approximation (IEPA). Unfortunately, the predictions of this study are probably not as rock-solid as one thought, since IEPA has been shown in recent years to have some serious weaknesses. Fortunately, for the Li^+-H_2 system the SCF and IEPA surfaces are qualitatively very similar, although for H_2 fixed at its equilibrium geometry the Kutzelnigg well depth is slightly smaller, 0.22 eV.

Both the Lester and Kutzelnigg potential surfaces have been fit to analytical forms suitable for dynamical studies. Given the latter surfaces, Schaefer and Lester[57] have carried out exhaustive quantum-mechanical studies of inelastic scattering of H_2 by Li^+. As expected, these close-coupling studies show that most dynamical features are independent of which of the two surfaces was used. However there are quantitative differences for the small multiquantum rotational and vibrational-rotation transitions between the Lester and Kutzelnigg surfaces. Finally it should be noted

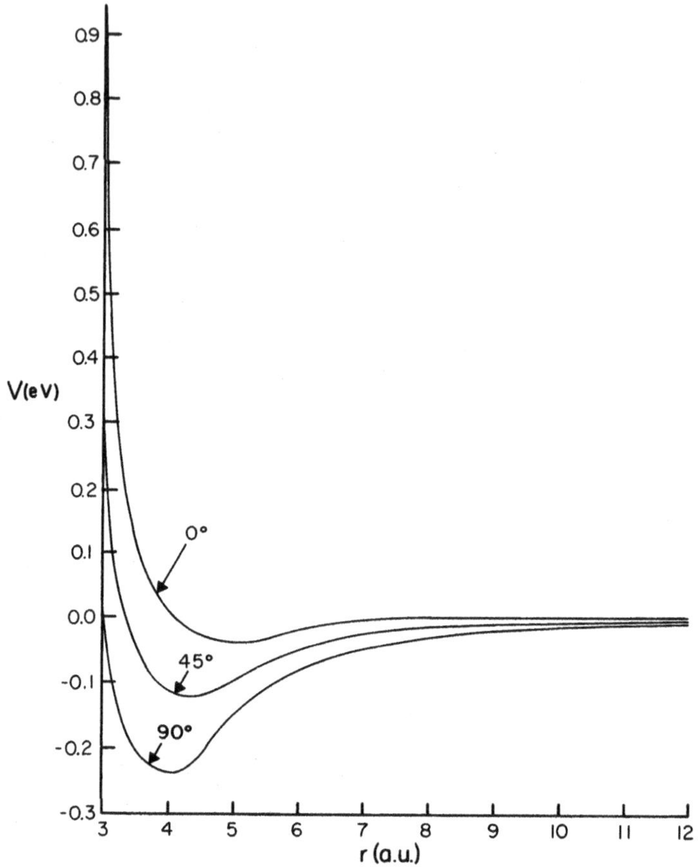

Figure 1. Li^+-H_2 interaction potential[53] for fixed H–H internuclear separation of 1.4 bohrs. The collinear approach is labeled 0° and the C_{2v} approach 90°.

that Barg, Kendall, and Toennies[58] have made extensive studies of the Li^+-H_2 system using classical dynamics. In all of the above dynamical studies the agreement with experimental cross sections is quite good and it seems fair to say that we now have a good fundamental understanding of rotational and vibrational excitation of H_2 by Li^+.

4.1.2. He–H_2CO

Among several other nonreactive systems which have been the subject of theoretical study, He–H_2CO stands out. The motivation for the studies of Garrison and co-workers was the observation of anomalous absorption in interstellar formaldehyde. The observed absorption is anomalous because it implies an excitation temperature for two rotational states lower than both the background radiation temperature (~ 2.7 K) and the expected kinetic temperature (10–20 K). In 1969, to explain the interstellar observations, Townes and Cheung[59] proposed a pumping model involving rotational excitation of H_2CO followed by radiative decay. To test the Townes model Garrison determined the pertinent part of the He–H_2CO potential surface, carried out close coupling calculations of the rotational excitation cross sections, and solved the equations of statistical equilibrium for the lowest energy levels.

In their first paper, Garrison, Lester, and Schaefer[60] investigated the He–H_2CO potential surface within the Hartree–Fock approximation. The *smallest* basis set used was of rather high quality: He(6*s*2*p*/3*s*2*p*), H(5*s*1*p*/3*s*1*p*), C(9*s*5*p*1*d*/5*s*3*p*1*d*), O(9*s*5*p*1*d*/5*s*3*p*1*d*), and a considerably larger set (including two sets of *d* functions on C and O) was tested extensively. The *ab initio* surface for He–H_2CO distances > 11 bohrs was in good agreement with perturbation theory, another indication of basis set adequancy at the onset of the nonoverlap region.

Unlike Li^+–H_2, however, the He–H_2CO system is not described in a qualitatively reasonable manner with the Hartree–Fock approximation. This is of course because the dispersion energy, which is a correlation effect, is of vastly greater relative importance for neutral systems than ionic ones. For this reason the above authors and Siegbahn[61] carried out extensive CI calculations for the He–H_2CO systems. In fact for the low symmetry (C_s) points considered, no less than 37,779 configurations were included variationally, one of the largest calculations ever attempted.

As expected the CI potential surface for He approaching a rigid formaldehyde is rather different from the SCF surface. An interesting comparison is illustrated in Figure 2, which shows how much deeper is the attractive well in the CI surface. To be more specific, the maximum well depth is attained for in-plane approaches of He and lies in the range 35–40 K at center-of-mass separation of 7.5 bohrs.

The dynamical studies alluded to above were completed with Miller and Green[62] and very clearly established that the 6- and 2-cm doublets of H_2CO are cooled by a collisional pump. Thus their rather rigorous theoretical study confirmed the conclusion of several previous crude-model calculations.[63] However, the detailed molecular mechanism determined by Garrison *et al.*[62] was definitely unexpected. They found that transitions from the $j = 1$ to the $j = 3$ doublets are the prime determiners of the cooling.

4.2. Reactive

4.2.1. $H + H_2$

The H_3 system has played a tremendously important role in the development of the theory of modern chemical kinetics.[1] A discussion of H_3

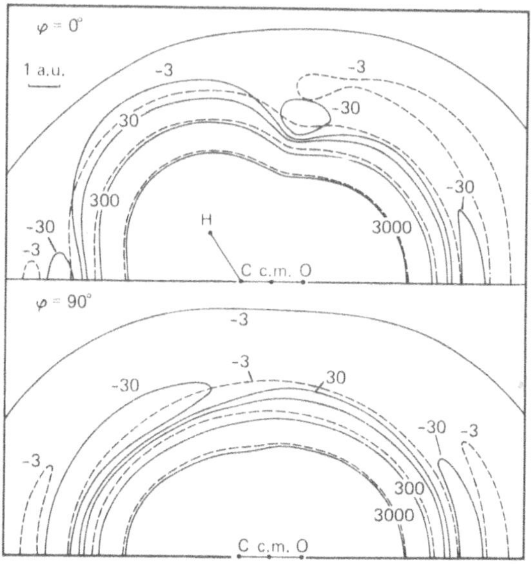

Figure 2. Contour plots of the interaction potential for He lying in the plane of H_2CO ($\phi = 0°$) and He approaching in the bisector plane ($\phi = 90°$). Solid lines refer to the CI surface,[61] dashed lines to the SCF surface.[60] Energies are in degrees Kelvin and c.m. denotes the center of mass of the rigid formaldehyde molecule.

is particularly pertinent here since a recently reported[64] potential surface for H_3 appears to be of nearly quantitative accuracy. However, we must also note that one of the very first (22 years ago!) genuinely *ab initio* calculations on a polyatomic system was that of Boys, Cook, Reeves, and Shavitt[65] for H_3. Concentrating on the saddle point region they predicted a barrier of 15.7 kcal and transition state geometry r(H–H) = 1.78 bohrs. The latter geometry prediction is remarkably similar to the most accurate theoretical result to date, and the barrier is respectable although ~ 60% too high. One of the most interesting results of the Boys study from a historical viewpoint was its finding that the H_3 surface does not have a "dimple" or "lake" in the saddle point region. The latter feature had been predicted by the early pioneers[66] of theoretical kinetics but has since been shown to have been a "mirage," in the words of Professor Eyring.

The second landmark *ab initio* paper in the H_3 saga[67] was that of Shavitt, Stevens, Minn and Karplus (SSMK),[2] published in 1968. Using a double zeta plus polarization basis set, they carried out full CI calculations (with optimized orbital exponents) for this three-electron model system. For general geometries (C_s point group) this means that a total of 680 configurations were included. A collinear barrier height of 11.0 kcal was obtained and at the time thought to be considerably larger than the much less than the S + D CI. A modified form (designated POL–CI) of 120° and 150° H–H–H angles and constrained barriers of 18.2 and 12.6 kcal were found, verifying the linear transition state predicted by the London–Eyring–Polanyi model.[1,66]

The most accurate quantum-mechanical study of any polyatomic potential surface to date begins with Liu's collinear surface.[68] A very large basis set of Slater functions was used, ($4s3p2d$) centered on each H atom. Using approximate natural orbitals, a set of 672 configurations (out of a total of 14,949 in the full CI) was selected for the most accurate studies. In this way a barrier height of 9.8 kcal was predicted, with saddle point geometry r(H–H) = 1.757 bohrs. It is interesting to note that Liu's saddle point total energy is only 10.3 kcal above the *exact* energy of separated $H + H_2$, setting an upper bound to the barrier. In addition Liu estimated that the true barrier is no less than 9.5 kcal, i.e., notably higher than the 9.1 kcal assumed in the widely used Porter–Karplus (PK) semiempirical surface.[69] Another comparison of the three surfaces is given in Figure 3, where it is seen that the Liu surface is energetically intermediate between those of SSMK[2] and PK.[69]

The noncollinear part of the H_3 surface has been very carefully examined by Siegbahn and Liu.[64] They used a H($9s3p1d/4s3p1d$) basis of contracted Gaussian functions. For the linear symmetric ($D_{\infty h}$ point group) H_3 it was possible to carry out full CI calculations and the predicted barrier (9.80 kcal) and saddle point geometry (1.757 bohrs) were identical to Liu's Slater basis predictions.[68] For the actually noncollinear points, 14 of the original 57 molecular orbitals were deleted with the help of a transformation to approximate natural orbitals. Full CI with this truncated basis yielded a barrier only 0.10 kcal higher and was considered adequate for the nonlinear surface. Siegbahn and Liu's 156 points on the nonlinear

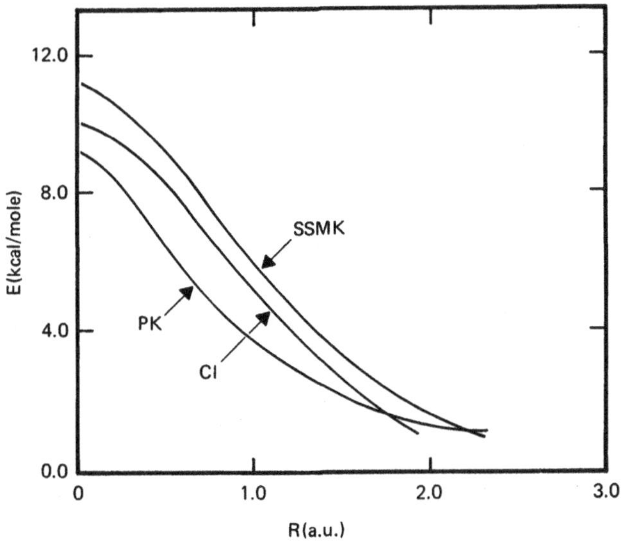

Figure 3. Energy profiles along the reaction paths for linear H_3. SSMK refers to the potential surface of Reference 2, PK to Reference 69, and CI to Liu's very accurate *ab initio* surface.

surface were believed to lie within 0.1 kcal of the exact clamped-nuclei limit, when the energy of the saddle point is taken to be zero.

A very recent and important result is the remarkable analytical fit of the Siegbahn–Liu *ab initio* surface obtained by Truhlar and Horowitz.[70] Using only 15 linear and 8 nonlinear parameters, Truhlar was able to fit all 267 *ab initio* points with a root-mean-square error of 0.17 kcal and a maximum absolute deviation of 0.55 kcal. Perhaps most impressive, this semiempirical valence bond potential surface precisely reproduces the barrier height (semiempirical 9.802 kcal) and position (semiempirical 1.7570 bohrs). In addition the force constants at the transition state agree with the *ab initio* values to better than 1%. The Truhlar–Horowitz work clearly implies that in the future we can be confident that even the most accurate *ab initio* A + BC surfaces will be amenable to fitting with reasonably simple analytical forms. Finally it is encouraging to note that the spherical average of the Siegbahn–Liu–Truhlar–Horowitz interaction potential is in good agreement with the recent experimental estimate of Toennies and co-workers.[71]

4.2.2. Fluorine–Hydrogen Systems

During the past decade, there has been a genuine explosion of interest in the reactions of atomic and molecular fluorine and hydrogen. As is usually the case, experimental studies have led the way, due in considerable measure to the impact of the HF laser. In addition to traditional kinetics,[72] the most successful experimental techniques have been infrared chemiluminescence,[73] chemical lasers,[74] crossed molecular beams,[75] and laser-induced fluorescence.[76] This experimental research has yielded a great deal of important information concerning these elementary reactions. In addition, this work has stimulated a keen interest in the detailed understanding on the molecular level of how these simple reactions occur.

Given the above background, it is hardly surprising that there has been considerable theoretical activity directed toward fluorine–hydrogen systems. The amount of detailed experimental information now available for fluorine–hydrogen reactions is sufficient to challenge the most sophisticated theorists of both the electronic structure and dynamic schools. Since dynamical considerations are taken up in different chapters of the present volume, we will be concerned here with the potential-surface half of the theoretical problem. Clearly, however, we must keep in mind the essential complementarity of these two pieces of the puzzle.

4.2.2.1. $F + H_2 \rightarrow FH + H$. Three progressively more reliable potential surfaces have been reported for this key reaction, which has become the focus of a considerable amount of scientific research. In the first,[77] a double zeta (DZ) basis set was adopted. Using the DZ basis and both SCF and configuration interaction (CI) methods, the results in Table 3

Table 3. Some Features of *Ab Initio* and Semiempirical Potential Energy Surfaces for $F + H_2 \rightarrow FH + H$

Year	Surface	Saddle point geometry (Å)		Barrier height (kcal/mole)	Exothermicity (kcal/mole)
		r(F–H)	r(H–H)		
1971	Double zeta (DZ) Self-consistent field (SCF)[a]	1.06	0.81	34.3	−0.6
1971	DZ configuration interaction (CI)[a]	1.37	0.81	5.7	20.4
1972	DZ + P SCF[b]	1.18	0.836	29.3	13.2
1972	DZ + P CI[b]	1.54	0.767	1.7	34.4
1972	Muckerman V[c]	1.54	0.76	1.1	31.8
1974	Polanyi–Schreiber SE-I[d]	1.43	0.777	2.2	31.0
1976	Large basis[e] Extended CI	1.48	0.778	3.3	31.3
	Experiment	—	—	1.6[f]	31.5 ± 0.5

[a] Reference 77.
[b] Reference 10.
[c] Reference 79.
[d] Reference 80.
[e] Reference 3.
[f] Experimental activation energy; see Reference 72.

were obtained. It seems reasonable to conclude that the SCF results obtained with this basis are in poor agreement with experiment, while the CI results are much improved but even yet in only fair agreement. This conclusion was initially a surprise to us, since such techniques often yield rather reliable results, for example, in the prediction of equilibrium molecular structures.(78)

The most obvious (in light of previous theoretical research(13)) extension of the DZ basis is the addition of polarization functions, i.e., a set of *d* functions on fluorine and sets of *p* functions on each H atom. The basis thus obtained is labeled DZ + P and the ensuing results are summarized in Table 3.(10) There we see once again that the SCF approximation yields a barrier height much larger than experiment. And in fact even if one goes to a complete set of one-electron functions, the Hartree–Fock limit barrier height will be ~ 25 kcal larger than the true barrier. The discovery of this inherent inability of the SCF approximation to even qualitatively describe repulsive potential surfaces must be viewed as one of the most important developments of our research to date.

Fortunately, the judicious use of CI yields a chemically reasonable DZ + P surface for $F + H_2$ (Figure 4). In particular, the predicted classical barrier height of 1.66 kcal is in fortuitously close agreement with the experimental activation energy. Further, the exothermicity of the BOPS(10)

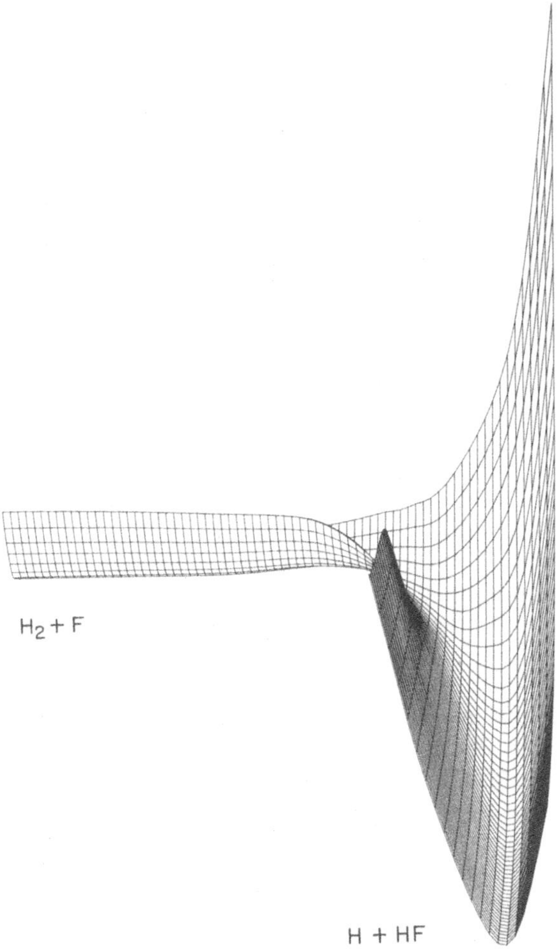

Figure 4. Perspective plots of the FH_2 potential energy surface[10] from the entry and exit channels.

(Bender–O'Neil–Pearson–Schaefer) surface is ~ 3 kcal greater than the experimental value. Thus, the primary significance of the BOPS CI surface was that it appeared to be the first qualitatively correct *ab initio* surface for a chemical reaction more complicated than the prototype[2] $H + H_2$ system.

The earliest indication of the qualitative correctness of the BOPS surface came from the classical trajectory studies of Muckerman.[79] Muckerman varied several features of the London–Eyring–Polanyi–Sato (LEPS) surface for $F + H_2$ to get the best agreement between predicted and experimental[73,74] FH vibrational energy distributions. Although this work was done completely independent of the BOPS *ab initio* study, Muckerman's "best" semiempirical surface (his surface V, summarized in Table 3) has a saddle point position essentially indistinguishable from the BOPS surface. Since the saddle point position is probably the most critical surface feature not directly accessible to experimental determination, this concurrence was especially significant.

Perhaps the next important development was the fitting of the BOPS surface to an LEPS form by Polanyi and Schreiber.[80] It should be noted that this fitting was carried out against the present author's advice, our feeling being that the BOPS surface was not sufficiently accurate to be used without adjustment in dynamical studies. Nevertheless, Polanyi and Schreiber fit the 232 collinear points to the LEPS form with an ensuing rms deviation of less than 1 kcal per point. This exercise in itself is rather striking confirmation of the LEPS form, which has only *seven* adjustable parameters. The most obvious weakness of the BOPS–FIT surface was a spurious 0.4 kcal attractive well in the entrance valley. In addition, of course, the BOPS–FIT surface incorporated the known 3 kcal BOPS error in the predicted exothermicity.

Collinear classical trajectory studies were then performed using the BOPS–FIT surface. By comparing with the experimental vibrational distribution, Polanyi and Schreiber (PS) concluded that although the BOPS surface was qualitatively reasonable, it does have a rather serious failing. That is, PS concluded that the BOPS surface drops too rapidly from the "shoulder" into the exit valley. It was noted, of course, that the known error of 3 kcal in the exothermicity is a major factor in this rapid drop. Our current feeling is that while this criticism of the BOPS surface may well prove to be at least partially valid, the use of classical (rather than quantum-mechanical) dynamics and one (rather than three) dimension detracts from the strength of the PS conclusion.

Polanyi and Schreiber also suggested a "best" adjusted LEPS surface, their SE–I, and this surface is also summarized in Table 3. There it is seen that the SE–I surface has its saddle point position somewhat later than either BOPS or Muckerman V. Nevertheless, it seems clear that the primary features of all three surfaces are rather similar. Both Muckerman[79] and Polanyi and Schreiber[80] have argued that the agreement between their best surfaces and the *ab initio* BOPS surface supports their use of the generalized LEPS form. We concur.

In an attempt to resolve some of the controversy arising from the BOPS surface, we decided in 1974 to attempt to determine an *ab initio* surface of sufficient reliability to be used directly, or with modest amounts of scaling, in dynamical studies. Recent developments[81] in the fully quantum-mechanical treatment of A + BC reactions were another motivation for this research. However, C. F. Bender discovered very quickly that apparent improvements in the quality of wave functions used led to a deterioriation of the predicted FH_2 potential energy surface.

After an interval of contemplation, this research was continued in collaboration with Steven Ungemach and Bowen Liu (IBM, San Jose). Their conclusions have been presented[3] at the latest General Discussion of the Faraday Division of the Chemical Society (September 1976). Using a basis set more than twice as large as DZ + P and very large CI, USL (Ungemach, Schaefer, and Liu) predict a barrier of 3.3 kcal, with a suggested error of no more than one kcal. Thus, it now appears that the true barrier height is notably greater than the experimental activation energy. This idea is qualitatively supported by USL's additional prediction that there may be as much as 1.5 kcal *less* zero point vibrational energy at the saddle point than for separated $F + H_2$. Finally, it is worth noting that the USL saddle point geometry is about halfway between the Muckerman-V–BOPS results and the PS–SE-I prediction.

It should be clear from the above that the $F + H_2$ potentlal energy surface is likely to remain the source of much controversy for years to come. More generally, it seems apparent that the closer one looks at a particular problem, as both theoretical and experimental techniques become more sophisticated, the more fine structure comes to light.

4.2.2.2. H + FH → HF + H. As mentioned above, efforts to develop the HF chemical laser as a practical device have contributed greatly to the interest in elementary fluorine–hydrogen systems. In this context, essentially every process leading to vibrational relaxation of HF must be precisely characterized. One of the simplest such processes,

$$H + FH(v') \rightarrow HF(v) + H \tag{9}$$

has been the subject of several studies, both experimental[76] and theoretical. Clearly, the barrier height or activation energy for this simple exchange reaction will play a critical role in the dynamics.

Since semiempirical potential surfaces have been so successful in describing the $F + H_2$ dynamics, it is not completely unreasonable to assume that the same LEPS surfaces might be suitable for H + FH. After all, the latter reaction merely corresponds to a different channel of the FH_2 surface. And, in fact, this is precisely the procedure adopted by Thompson[82] and by Wilkins[83] in their classical trajectory studies of H + HF.

Table 4 summarizes the predictions of a number of semiempirical FH_2 surfaces for the H + FH barrier. The final entries indicate that rather reliable calculations,[84] using a better than double zeta plus polarization basis set, predict a barrier of 49 kcal. Bender, Garrison, and Schaefer[84] (BGS) concluded in their paper that the true collinear barrier is no less than 40 kcal. Thus, it is seen that the two "best" semiempirical $F + H_2$ surfaces, Muckerman V and Polanyi–Schreiber SE–I, fail miserably for the collinear

Table 4. Barrier Height and Saddle Point Geometry for H + FH → HF + H. The Saddle Point is Assumed to Occur for a Linear Symmetric H—F—H Geometry

Type of potential energy surface	Authors	r(H–F) (Å)	Barrier (kcal/mole)
Bond-energy bond-order (BEBO)	Johnston[a]	1.10	6.8
London–Eyring–Polanyi–Sato (LEPS)	Muckerman I[b]	1.04	1.0
LEPS	Jaffe and Anderson[c]	1.04	−5.2
LEPS	Muckerman II[b]	1.04	1.0
	III	1.05	1.7
	IV	1.05	2.3
LEPS	Wilkins[d]	1.04	1.4
LEPS	Thompson[e]	1.12	28.6
Semiempirical valence bond	Blais and Truhlar[f]	1.10	14.0
Diatomics-in-molecules	Tully I[g]	1.05	14.4
	II	1.09	13.1
LEPS	Muckerman V[b]	1.04	1.2
LEPS	Polanyi and Schreiber[h]	1.05	3.5
A priori methods			
Self-consistent field	Bender, Garrison, and Schaefer[i]	1.12	67.8
Configuration interaction		1.14	49.0
Coupled electron pair approximation	Botschwina and Meyer[j]	1.14	44.9
Configuration interaction	Wadt and Winter[k]	1.14	47.6

[a] H. S. Johnston, *Gas Phase Reaction Rate Theory*, Ronald Press, New York (1966).
[b] Reference 79.
[c] R. L. Jaffe and J. B. Anderson, Classical trajectory analysis of the reaction F + H_2 → HF + H, *J. Chem. Phys.* **54**, 2224–2236 (1971).
[d] Reference 83.
[e] Reference 82.
[f] N. C. Blais and D. G. Truhlar, Monte Carlo trajectories: Dynamics of the reaction F + D_2 on a semi-empirical valence bond potential energy surface, *J. Chem. Phys.* **58**, 1090–1108 (1973).
[g] J. C. Tully, Diatomics-in-molecules potential energy surfaces. I. First-row triatomic hydrides, *J. Chem. Phys.* **58**, 1396–1410 (1973).
[h] Reference 80.
[i] Reference 84.
[j] Reference 85.
[k] Reference 86.

H + FH channel. This is perhaps the strongest evidence to date for the importance of *ab initio* information in potential energy surface calibration.

Two very recent papers support the qualitative conclusions of BGS. Using a very large basis, Meyer[(85)] has predicted a collinear barrier of 45 kcal/mole for H + FH → HF + H from the coupled electron pair approximation. Attempts to estimate the true barrier via error analysis lead to a value of ~40 kcal. Secondly, Wadt and Winter[(86)] have found that the use of diffuse basis functions yields a surface that is quite "flat" with respect to the HFH bond angle. In fact, the true saddle point may occur for a bond angle less than 180° and yield a barrier slightly lower than 40 kcal.

The primary physical (or chemical, according to one's preference) conclusion is that at energies below 35 kcal the atom exchange mechanism cannot be a significant contributor to the vibrational relaxation of HF by hydrogen atoms. This conclusion seems to be given support by the recent experiments of Heidner.[76] Specifically, he finds that the cross section for vibrational relaxation increases enormously between $v = 2$ and $v = 3$. It may be more than coincidental that the $v = 3$ state of HF has enough internal energy to surmount a barrier of 35 kcal/mole.

4.2.2.3. $H + F_2$, $F + HF$. Polanyi and co-workers[87] have made truly impressive infrared chemiluminescence studies of this highly exothermic reaction. They find that the ratios of population of the HF product vibrational states ($v = 1$ up to $v = 9$) are 12:13:25:35:78:100:40:26: < 16.

Electronic structure theory has been applied[88,89] in a manner analogous to the first four columns of Table 3, and these results are summarized in Table 5. A significant difference between these $H + F_2$ results and the earlier $F + H_2$ studies[10,77] is that the DZ CI barrier is lower than that obtained with the more reliable DZ + P CI. Since (for reasons discussed above) it is difficult to determine the true barrier, it must be concluded that both methods yield barriers roughly compatible with the experimental activation energy. However, the superiority of the DZ + P basis does become apparent with an inspection of the calculated exothermicities.

Both the $F + H_2$ and $H + F_2$ surfaces were found to be in harmony with the LEPS model in their angular characteristics. That is, the true

Table 5. Some Features of *Ab Initio* Potential Surfaces[88,89] for $H + F_2 \rightarrow HF + F$

Surface	Saddle point geometry (Å)		Barrier height (kcal)	Exothermicity (kcal)
	r(H–F)	r(F–F)		
DZ SCF	1.56	1.49	12.2	132.4
DZ CI	2.05	1.57	1.0	88.3
DZ + P SCF	1.61	1.41	13.9	130.1
DZ + P CI	1.68	1.50	4.1	99.0
Experiment	—	—	2.4 ± 0.2[a]	102.5 ± 2.8

[a] Experimental activation energy. R. G. Albright, A. F. Dodonov, G. K. Lavrovskaya, I. I. Morosov, and V. L. Tal'roze, Mass-spectrometric determination of rate constants for H-atom reactions with Cl_2 and F_2, *J. Chem. Phys.* **50**, 3632–3633 (1969).

Table 6. Summary of Theoretical Potential Energy Surfaces [90] for the $F + HF \rightarrow FH + F$ Exchange Reaction

Surface	Saddle point geometry r(HF) (Å)	Barrier height (kcal/mole)
DZ SCF	1.087	53.8
DZ CI	1.126	21.8
DZ + P SCF	1.083	53.7
DZ + P CI	1.099	23.9

saddle points do occur for collinear geometrical arrangements. It is worth noting[77,88] that the two surfaces become repulsive at about the same rate as the angle of approach is bent away from 180°. For the $F + H_2$ system, e.g., it is found[77] that constraining the F–H–H angle to be 90° yields a barrier 11 kcal higher than that obtained for θ(F–H–H) = 180°.

Another part of the HF_2 surface that has been carefully explored is the collinear F + HF exchange reaction. This is another process of vital importance for an adequate understanding of the HF chemical laser. The results of O'Neill, Schaefer, and Bender[90] (OSB) are summarized in Table 6. There it is seen that the most reliable predicted barrier height is 24 kcal. On this basis, the true barrier was estimated to be ≥ 18 kcal. This result is much greater than the 6 kcal predicted[91] by the BEBO method. However, good agreement is found with Thompson's LEPS surface[92] calibrated for use on the $H + F_2$ reaction dynamics. The primary conclusion one can draw from these results is that the atom exchange mechanism is unlikely to contribute to the vibrational relaxation of the $v = 1$ state of HF. However, $v = 2$ relaxation could be greatly enhanced by atom exchange.

4.2.3. $N^+ + H_2$

Simple ion–molecule reactions have provided some of the most fascinating examples to date of the interplay between different potential surfaces of a single chemical system.[93] Especially noteworthy in this regard are the molecular beam studies of Mahan and co-workers,[94-96] who have carefully investigated, among other systems, the $C^+ + H_2$, $N^+ + H_2$, and $O^+ + H_2$ reactions. These reactions are particularly appealing as prototypes, since they are sufficiently simple to be studied by both electronic structure theory and classical or semiclassical dynamics. In addition, the use of qualitative electronic correlation diagrams[93,97] has proven

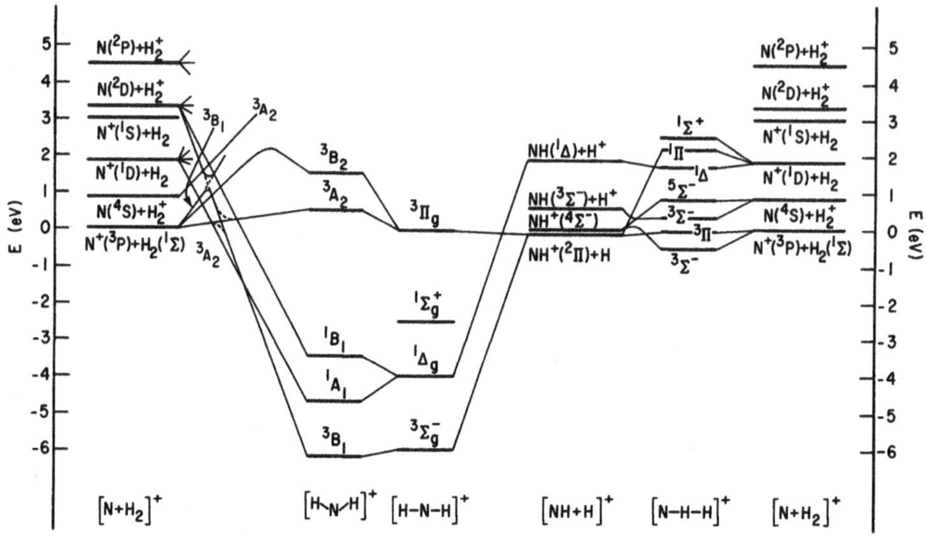

Figure 5. Correlation diagram of Fair and Mahan[95] for the $N^+ + H_2$ system.

very helpful in understanding these simple reactions, and alternatively the experiments may serve as testing grounds for simple molecular orbital theory.

A reasonable starting point for our discussion is the $N^+ + H_2$ electronic state correlation diagram of Fair and Mahan.[95] This diagram is reproduced with their permission in Figure 5. As discussed by Fair and Mahan (and elsewhere[98,99] in regard to the $C^+ + H_2$ reaction) the key feature in the interpretation of low-energy (say less than ~ 3 eV) molecular beam results is the intersection of two low-lying potential energy surfaces. For the $N^+ + H_2$ case (in C_{2v} symmetry) these are the $^3B_1$ and $^3A_2$ surfaces. The $^3B_1$ state is known to be the ground state of NH_2^+, the nitrenium ion, while the $^3A_2$ state is less understood. However, on the basis of orbital symmetry considerations[94] and earlier theoretical work[98] on $C^+ + H_2$, the $^3B_1$ surface is expected to be quite repulsive as the N^+ initially approaches H_2. The deep well of the $^3B_1$ surface is "protected" from N^+–H_2 collisions on the same surface by means of this large barrier. However, the $^3A_2$ surface should be either much less repulsive or attractive as N^+ approaches H_2. And since the two surfaces are both of $^2A''$ symmetry as soon as the N^+ ion moves off the H_2 perpendicular bisector, the C_{2v} crossing of surfaces becomes an avoided intersection. If there are points along this crossing of $^3B_1$ and $^3A_2$ surfaces which lie at energies near or below the $N^+ + H_2$ asymptote, then there exists a barrier-free pathway

$$^3A_2 \rightarrow {}^3A'' \rightarrow {}^3B_1 \qquad (10)$$

for the formation of ground state NH_2^+ from separated N^+ plus H_2. Such a pathway for the analogous situation with respect to $C^+ + H_2$ was demonstrated unequivocally in the important theoretical work of Pearson and Roueff.[99]

In their communication Pearson and Roueff brought to light a critical ingredient in the proper theoretical treatment of this problem. That is, polarization functions (*d* functions on carbon and *p* functions on the hydrogen atoms in their case) critically affect the energy at which the seam or line of intersection occurs. Their finding is pertinent to the present discussion since in 1975 Gittins and Hirst[100] reported single-configuration self-consistent-field (SCF) results for $N^+ + H_2$ using a basis set that is quite well chosen and flexible but lacks polarization functions. Gittins and Hirst concluded that access to the deep $^3B_1$ potential well may be possible with only a small barrier, on the order of 4 kcal/mole. By comparison of the effects of polarization functions in the $C^+ + H_2$ system,[98,99] it would appear likely that this barrier should disappear completely.

This hypothesis has been tested in the more recent work of Bender, Meadows, and Schaefer[101] on the N^+–H_2 system. Using a double zeta plus polarization basis, CI was carried out including all single and double excitations. In addition, a few calculations with a much larger basis set were performed to demonstrate the qualitative reliability of the DZ + P set.

Bender's most important conclusion, illustrated in Figure 6, concerns the intersection of the $^3A_2$ and $^3B_1$ potential surfaces. There we see that the $^3A_2$ surface is attractive all the way into the region of the $^3B_1$ well. It is noteworthy that the line of intersection for the N^+–H_2 system passes through much lower relative energies (-50 kcal as opposed to -15) than the corresponding line of intersection for the C^+–H_2 system.[99] A naive interpretation of this comparison would suggest that at low energies one should observe *more* complex formation for the N^+ than the C^+ reaction. At this point, however, we believe that detailed dynamical studies, along the lines described in later chapters of this volume, are called for.

4.2.4. $H + Li_2, F + Li_2$

Lee, Gordon, and Herschbach[102] have reported crossed molecular beam studies of the reactions of H and D atoms with K_2, Rb_2, and Cs_2. From their results a number of important qualitative conclusions were drawn. For example, only a small fraction of the available energy appeared in product translation. Thus it seems likely that either the alkali hydride product MH is vibrationally and/or rotationally excited or that the alkali atom M is electronically excited. In addition, Lee *et al.*[102] concluded

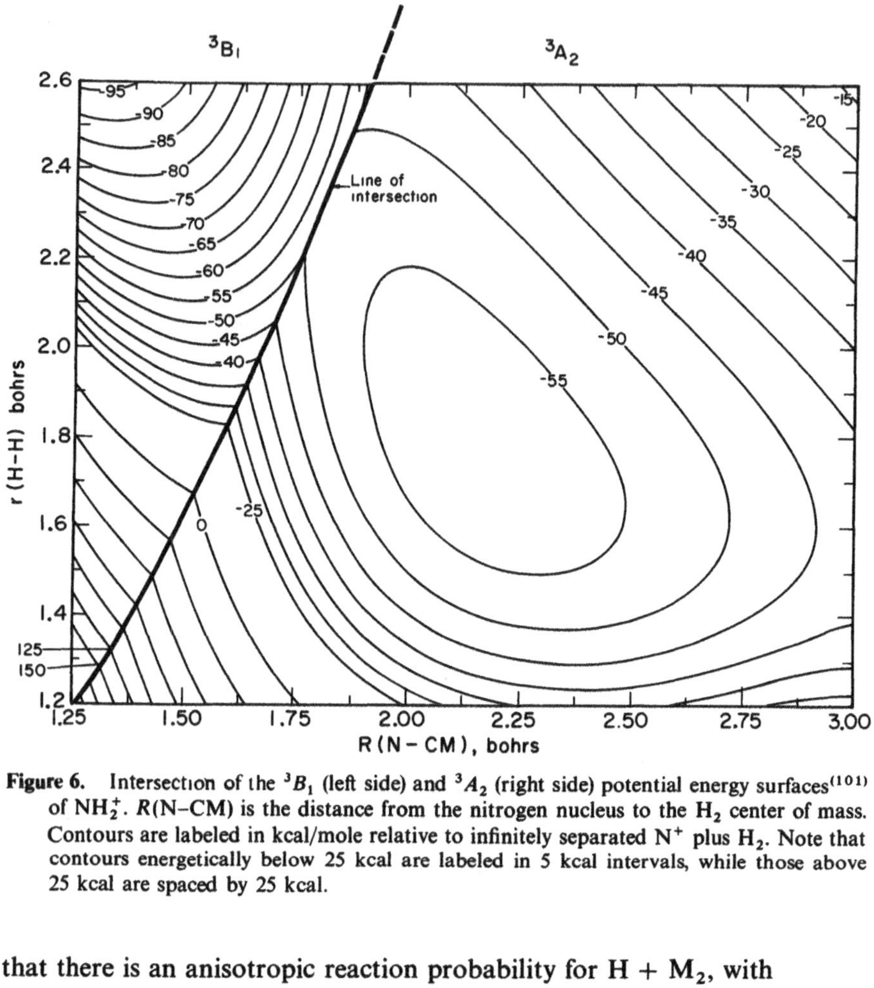

Figure 6. Intersection of the $^3B_1$ (left side) and $^3A_2$ (right side) potential energy surfaces[101] of NH_2^+. R(N–CM) is the distance from the nitrogen nucleus to the H_2 center of mass. Contours are labeled in kcal/mole relative to infinitely separated N^+ plus H_2. Note that contours energetically below 25 kcal are labeled in 5 kcal intervals, while those above 25 kcal are spaced by 25 kcal.

that there is an anisotropic reaction probability for H + M_2, with

$$\begin{array}{c} \mathrm{H} \\ \mathrm{M}\text{——}\mathrm{M} \end{array}$$

configurations more likely to lead to reaction.

The hydrogen atom plus alkali dimer reactions are of interest to theoreticians first because of the opportunity for fruitful interaction with experiment and second because these systems are among the simplest for which the dynamics frequently must be described in terms of more than a single potential energy surface.[103] For the simplest such reaction at thermal energies only a single channel is energetically open:

$$\mathrm{H} + \mathrm{Li}_2 \rightarrow \mathrm{LiH}(^1\Sigma) + \mathrm{Li}(^2S), \qquad \Delta H = -30\,\text{kcal/mole} \qquad (11)$$

However, under the conditions utilized by Lee *et al.*, namely, ~10 kcal/mole of H or D atom translational energy and 6 ± 3 kcal/mole of alkali dimer vibrational excitation, a second pathway is possible:

$$H + Li_2 \rightarrow LiH(^1\Sigma) + Li(^2P), \qquad \Delta H = +\,12\,\text{kcal/mole} \tag{12}$$

From either an experimental or theoretical viewpoint, then, it will be of great interest to determine the relative importance of these two competing reaction pathways. Herschbach's work, though not definitive, suggested the former path (11) to be the dominant one.

The potential beauty of a theoretical treatment of this reaction lies in the opportunity to study the product energy distribution as a function of initial conditions. While the molecular beam experimentalist will do splendidly to study this reaction under one particular set of circumstances, we are free to study it under whichever circumstances appear most interesting. For example, the ratio of LiH + Li to LiH + Li* can be studied as a function of H atom translational energy or as a function of Li_2 vibrational energy. Even if the LiH + Li* pathway is shown to be unimportant, reaction pathways. Herschbach's work, though not definitive, suggested the former path (11) to be the dominant one.

The potential surfaces required for a thorough study are those arising from LiH + Li(^2S) and LiH + Li(^2P). For general geometry (C_s symmetry), three of these surfaces will be of $^2A'$ irreducible representation and one of $^2A''$. The first and most exhaustive study to date of the HLi_2 ground state potential energy surface was that of Siegbahn and Schaefer.[105] They adopted a contracted basis set of size Li(8*s*3*p*/4*s*3*p*), H(5*s*1*p*/3*s*1*p*), i.e., of slightly better than DZ + P quality. A necessary feature of this HLi_2 basis is the presence of a relatively diffuse *s* function on hydrogen to describe the H^- character in LiH. For this system of three valence electrons, it was possible to carry out *full* valence CI studies, which included 5175 configurations for general geometries (point group C_s). Thus, the only significant deterrent to the accuracy of such a theoretical treatment will be the incompleteness of the basis set of molecular orbitals. It should be clear, of course, that full CI treatments of this type are not feasible for systems with more than a few electrons to be correlated.

The most critical feature of the lowest potential surface of H + Li_2 is the fact that HLi_2 is a conspicuously bound molecule. At its equilibrium geometry r(H–Li) = 1.72 Å, θ(LiHLi) = 95°, HLi_2 is bound by 22 kcal relative to the separated products LiH + Li. In this regard it is noteworthy that unlike the H_3 and FH_2 surfaces discussed above, electron correlation is *not* critical for many of the features of the HLi_2 surface. This seems to be a fairly general attribute of attractive potential surfaces, and provides justification for Pearson's SCF study[106] of the related F + Li_2 surface. Actually one might suspect that HLi_2 and FLi_2 could be quite different,

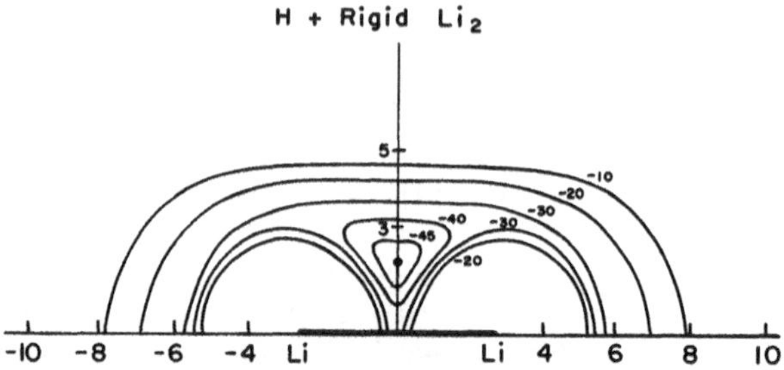

Figure 7. Contour map of one part [r(Li–Li) fixed at 5.27 bohrs] of the potential surface[105] for HLi_2. Distances from the Li_2 bond midpoint are given in bohrs. The contours are labeled in kcal/mole relative to separated $H + Li_2$.

since fluorine is so much more electronegative than hydrogen. However, the two ground state surfaces are rather similar, with the deeper (~ 34 kcal[106]) FLi_2 potential well consistent with its more ionic character. Given in Figure 7 is a contour map of the potential surface for the approach of H atom to the Li_2 molecule.

Many of the excited state HLi_2 potential energy surfaces have been characterized theoretically by England, Sabelli, and Wahl.[107] They used a slightly larger basis set than Siegbahn, in conjunction with MCSCF and modest CI methods. Wahl and co-workers find the first excited state of HLi_2 to be of $^2B_2$ symmetry, and this state is bound by ~ 20 kcal relative to its lowest dissociation limit, Li(^2P) + LiH. These authors also discuss the geometrical positions of "electron jumps," which take place at the crossing point of diabatic ionic and covalent surfaces. These electron jump positions are quite important to the dynamics,[103] and the England predictions should be rather reliable, since the semiempirical method of Kahn, Hay, and Shavitt[108] was employed.

4.2.5. H + ClH, H + BrH

It seems fitting to conclude the present review with a discussion of a problem for which theory and experiment are in apparent conflict. The word *apparent* should be stressed of course, since it is possible that the experimental data may in themselves be technically correct, but their interpretation faulty. The conflict in question began with the publication in 1975 of the BGS prediction[84] that the barrier for H + FH hydrogen exchange barrier is ≥ 40 kcal.* This result indirectly suggested a sizeable barrier for the analogous chlorine reaction

$$H + ClH \rightarrow HCl + H \qquad (13)$$

As we shall see, this rough and ready prediction has since been sup-

* See Note Added in Proof on p. 78.

ported by the reliable theoretical studies of both Botschwina and Meyer[85] and Dunning.[109]

Before moving on to the details of the *ab initio* studies, however, we should note briefly the pertinent experiments. First the mass spectroscopic analysis[110] of deuterium-chloride–hydrogen mixtures photolyzed at 1849 Å can be interpreted to imply that the activation energy for the exchange reaction (13) is 2.2 kcal less than that for the abstraction reaction

$$H + DCl \rightarrow HD + Cl \tag{14}$$

However kinetic studies[111] imply that the latter barrier should be only 3.5 kcal, and hence that the exchange barrier is only 1.3 ± 0.4 kcal. This conclusion appears to be strengthened by a second set of photochemical experiments,[112] which yield an exchange barrier of 1.0 ± 1.3 kcal, again very small. More recently, McDonald and Herschbach[113] have reported crossed molecular beam studies of reaction (13). Although their results certainly do not require a barrier as low as 1 kcal/mole, the fact that the reaction occurs at a mean collision energy of ~9 kcal has been taken by some to require that the barrier for exchange be ≤9 kcal.

In the theoretical study of Botschwina and Meyer,[85] a large and flexibly contracted Cl(11*s*7*p*2*d*/8*s*6*p*2*d*), H(5*s*2*p*/4*s*2*p*) basis set was chosen. CI including all important (following the PNO–CI format) single and double excitations was corrected for higher excitations using CEPA. The predicted collinear barrier is 22.1 kcal, a result which in our opinion should be no more than 6 kcal above the exact collinear barrier. However, using a semiempirical procedure for estimating errors, Botschwina and Meyer suggest that the true barrier is 10–15 kcal. Thus although the *ab initio* prediction of 22.1 kcal is unambiguous, we feel that Botschwina and Meyer's error estimates are too large.

The latter view is shared by Dunning,[109] who studied the problem using a nearly comparable basis set. Accounting for some of the effects of higher excitations in a variational manner, Dunning predicts an *ab initio* barrier of 25.3 kcal and estimates the true barrier to be in the range 14–22 kcal. We concur with Dunning's conclusions and further suggest ~ 18 kcal as the best estimate of the true barrier. In our opinion this prediction, based on theory, is sufficiently solid that experiments[110,113] suggesting notably smaller barriers (i.e., ≤ 9 kcal) should be carefully reexamined.

Finally we note that the analogous bromine exchange reaction has also been studied by Botschwina and Meyer.[114] They predict SCF and CEPA barriers of 36.7 and 17.2 kcal, and again suggest a large error for their correlated result. Although the precise values of the classical barriers for these H + XH exchange reactions must be considered in doubt at present [we suggest 40(HFH), 18(HClH), and 12(HBrH) kcal] it is clear that the barriers decrease steadily as the halogen atom becomes heavier.

5. Concluding Remarks

Before closing, it is well to reiterate that the examples noted above sample only a fraction of the reliable *ab initio* potential energy surfaces now available for atom–molecule collisions. One such omitted surface is that of Lester and Krauss[115] for Li + HF, which was studied by SCF theory. However, each theoretical study has its strengths and weaknesses and a tabulation of surfaces without adequate discussion might be counterproductive. It may be helpful to the reader for me to identify a few of the centers from which reliable potential energy surfaces (which incorporate the effects of electron correlation) are currently appearing. These include (in alphabetical order): Bochum (W. Kutzelnigg), Bonn (R. J. Buenker and S. D. Peyerimhoff), Columbus (I. Shavitt), Kaiserslautern (W. Meyer), Karlsruhe (R. Ahlrichs), Livermore (C. F. Bender and N. W. Winter), Los Alamos (T. H. Dunning and P. J. Hay), Lund (B. Roos and P. Siegbahn), Moffett Field (S. R. Langhoff), Pasadena (W. A. Goddard), Pittsburgh (J. A. Pople), San Jose (B. Liu, A. D. McLean, and M. Yoshimine), Seattle (E. R. Davidson), and Urbana (C. E. Dykstra).

ACKNOWLEDGMENT

In the course of writing this review, the author was supported by the U. S. National Science Foundation, Grant No. CHE-7622621.

References

1. S. Glasstone, K.J. Laidler, and H. Eyring, *The Theory of Rate Processes*, McGraw-Hill Book Company, New York (1941).
2. I. Shavitt, R.M. Stevens, F.L. Minn, and M. Karplus, Potential-energy surface for H_3, *J. Chem. Phys.* **48**, 2700–2713 (1968).
3. S.R. Ungemach, H.F. Schaefer, and B. Liu, Concerning *ab initio* potential energy surfaces for F + H_2, *Faraday Discuss. Chem. Soc.* **62**, 330–332 (1977).
4. C.W. Bauschlicher, K. Haber, H.F. Schaefer, and C.F. Bender, Concerted non-least-motion pathway for the singlet methylene insertion reaction $CH_2(^1A_1) + H_2 \rightarrow CH_4$, *J. Am. Chem. Soc.* **99**, 3610–3614 (1977).
5. P.J. Hay and I. Shavitt, *Ab initio* configuration interaction studies of the π-electron states of benzene, *J. Chem. Phys.* **60**, 2865–2877 (1974).
6. P.J. Bruna, R.J. Buenker, and S.D. Peyerimhoff, Theoretical prediction of the electronic spectrum of thioacetone and comparison with related systems, *Chem. Phys.* **22**, 375–382 (1977).
7. C.E. Dykstra, R.R. Lucchese, and H.F. Schaefer, Electron correlation effects on the excitation energies of the lowest triplet states of glyoxal, *J. Chem. Phys.* **67**, 2422–2426 (1977).
8. D. Spangler, R. McKinney, R.E. Christoffersen, G.M. Maggiori, and L.L. Shipman, *Ab initio* calculations on large molecules using molecular fragments. Preliminary in-

vestigation of ethyl chlorophyllide *a* and related molecules, *Chem. Phys. Lett.* **36**, 427–431 (1975).
9. H.F. Schaefer, Molecular electronic stucture theory: 1972–1975, *Ann. Rev. Phys. Chem.* **27**, 261–290 (1976).
10. C.F. Bender, S.V. O'Neil, P.K. Pearson, and H.F. Schaefer, Potential energy surface including electron correlation for $F + H_2 \rightarrow FH + H$: refined linear surface, *Science* **176**, 1412–1414 (1972).
11. B. Liu, *Ab initio* potential energy surface for linear H_3, *J. Chem. Phys.* **58**, 1925–1937 (1973).
12. D. H. Liskow, C. F. Bender, and H. F. Schaefer, Some features of the $CH_3NC \rightarrow CH_3CN$ potential surface, *J. Chem. Phys.* **57**, 4509–4511 (1972); P. Pulay and H.F. Schaefer, unpublished (1976).
13. H.F. Schaefer, *The Electronic Structure of Atoms and Molecules: A Survey of Rigorous Quantum Mechanical Results*, Addison-Wesley Publishing Company, Reading, Mass. (1972).
14. F. Cavallone and E. Clementi, Electronic structure of the TTF-TCNQ complex, *J. Chem. Phys.* **63**, 4304–4307 (1975).
15. W.J. Hehre and J.A. Pople, Molecular orbital theory of the electronic structure of organic compounds XXVI. Geometries, energies, and polarities of C_4 hydrocarbons, *J. Am. Chem. Soc.* **97**, 6941–6955 (1975).
16. W.H. Miller, *Modern Theoretical Chemistry*, Vols. 1 and 2, Plenum Press, New York (1976).
17. B. Rosen, *Spectroscopic Data Relative to Diatomic Molecules*, Pergamon Press, Oxford (1970).
18. A.G. Gaydon, *Dissociation Energies and Spectra of Diatomic Molecules*, Chapman and Hall Ltd., London (1968).
19. J.C. Polanyi and J.L. Schreiber, The dynamics of bimolecular reactions, in *Physical Chemistry: An Advanced Treatise*, H. Eyring, W. Jost, and D. Henderson, editors, Vol. 6A, Academic Press, New York (1974), Chap. 6, pp. 383–487.
20. E.A. McCullough, The partial-wave self-consistent-field method for diatomic molecules: Computational formalism and results for small molecules, *J. Chem. Phys.* **62**, 3991–3999 (1975).
21. A.C. Wahl, Analytical self-consistent-field wave functions and computed properties for homonuclear diatomic molecules, *J. Chem. Phys.* **41**, 2600–2611 (1964).
22. R.M. Pitzer and W.N. Lipscomb, Calculation of the barrier to internal rotation in ethane, *J. Chem. Phys.* **39**, 1995–2004 (1963).
23. W.J. Hehre, R.F. Stewart, and J.A. Pople, Self-consistent-field molecular-orbital methods I. Use of Gaussian expansion of Slater-type atomic orbitals, *J. Chem. Phys.* **51**, 2657–2664 (1969).
24. J.L. Whitten, Gaussian lobe function expansions of Hartree–Fock solutions for the first-row atoms and ethylene, *J. Chem. Phys.* **44**, 359–364 (1966).
25. T.H. Dunning, Gaussian basis functions for use in molecular calculations I. Contraction of (*9s5p*) atomic basis sets for first-row atoms, *J. Chem. Phys.* **53**, 2823–2833 (1970).
26. A.D. McLean and M. Yoshimine, Ground states of linear molecules: dissociation energies and dipole moments in the Hartree–Fock approximation, *Int. J. Quantum Chem.* **1S**, 313–326 (1967).
27. W. Meyer, Ionization energies of water from PNO–CI calculations, *Int. J. Quantum* Chem. **5S**, 341–348 (1971).
28. I. Shavitt, in *Modern Theoretical Chemistry*, Vol. 4, H.F. Schaefer, editor, Plenum Press, New York (1977), pp. 189–276.
29. W. Meyer, PNO–CI studies of electron correlation effects I. Configuration expansion by means of nonorthogonal orbitals, and application to the ground state and ionized states of methane, *J. Chem. Phys.* **58**, 1017–1035 (1973).

30. B. Roos, A new method for large-scale CI calculations, *Chem. Phys. Lett.* **15**, 153–159 (1972).
31. R. Ahlrichs, H. Lischka, V. Staemmler, and W. Kutzelnigg, PNO–CI (pair natural orbital configuration interaction) and CEPA–PNO (coupled electron pair approximation with pair natural orbitals) calculations of molecular systems I. Outline of the method for closed-shell states, *J. Chem. Phys.* **62**, 1225–1234 (1975).
32. C.E. Dykstra, H.F. Schaefer, and W. Meyer, A theory of self-consistent electron pairs. Computational methods and preliminary applications, *J. Chem. Phys.* **65**, 2740–2750 (1976).
33. J.A. Pople, R. Seeger, and R. Krishnan, Variational configuration interaction methods and comparison with perturbation theory, *Int. J. Quantum. Chem.* **11S**, 149–163 (1977).
34. H.F. Schaefer and W.H. Miller, Large scale scientific computation via minicomputer, *Comput. Chem.* **1**, 85–90 (1976).
35. R.R. Lucchese, B.R. Brooks, J.H. Meadows, W.C. Swope, and H.F. Schaefer, BERKELEY: an "open ended" configuration interaction (CI) program designed for minicomputers, *J. Comput. Phys.* **26**, 243–251 (1978).
36. A.L. Robinson, Computational chemistry: getting more from a minicomputer, *Science* **193**, 470–472 (1976); W.G. Richards, Minicomputers for quantum chemists, *Nature* **266** (5597), 18 (1977).
37. R.J. Bartlett and I. Shavitt, *Chem. Phys. Lett.* **50**, 190–198 (1978).
38. J. Cizek, On the use of the cluster expansion and the technique of diagrams in calculations of correlation effects in atoms and molecules, *Adv. Chem. Phys.* **14**, 35–89 (1969); J. Paldus, J. Cizek, and I. Shavitt, Correlation problems in atomic and molecular systems IV. Extended coupled-pair many-electron theory and its application to the BH_3 molecule, *Phys. Rev.* **A5**, 50–67 (1972).
39. F. Sasaki, Effectiveness of configuration interaction calculations for large molecules, Contributions from the Research Group on Atoms and Molecules, No. 13, K. Hijikata and E. Ishiguro, editors, Dept. of Physics, Ochanomizu University, Tokyo 112, Japan (1977), pp. 39–43.
40. P.S. Bagus, B. Liu, A.D. McLean, and M. Yoshimine, in *Wave Mechanics: The First Fifty Years*, W.C. Price, S.S. Chissick, T. Ravensdale, editors, Butterworth Publishers, London (1973), pp. 99–118.
41. R.J. Buenker and S.D. Peyerimhoff, Energy extrapolation in CI calculations, *Theor. Chim. Acta* **39**, 217–228 (1975).
42. O. Sinanoglu, Many-electron theory of atoms and molecules, *Proc. Natl. Acad. Sci. USA* **47**, 1217–1226 (1961).
43. G.C. Lie, J. Hinze, and B. Liu, Valence excited states of CH, *J. Chem. Phys.* **59**, 1872–1898 (1973).
44. S.R. Langhoff and E.R. Davidson, Configuration interaction calculations on the nitrogen molecule, *Int. J. Quantum Chem.* **8**, 61–72 (1974).
45. H.J. Silverstone and O. Sinanoglu, Many-electron theory of nonclosed-shell atoms and molecules II. Variational theory, *J. Chem. Phys.* **44**, 3608–3617 (1966).
46. H.F. Schaefer and F.E. Harris, Metastability of the ^1D state of the nitrogen negative ion, *Phys. Rev. Lett.* **21**, 1561–1563 (1968).
47. P.J. Hay, T.H. Dunning, and W.A. Goddard, Configuration interaction studies of O_3 and $O_3{}^+$. Ground and excited states, *J. Chem. Phys.* **62**, 3912–3924 (1975).
48. W.J. Hunt, P.J. Hay, and W.A. Goddard, Self-consistent-field procedures for generalized valence bond wavefunctions. Applications to H_3, BH, H_2O, C_2H_6, and O_2, *J. Chem. Phys.* **57**, 738–748 (1972).
49. W.J. Stevens, G. Das, A. C. Wahl, M. Krauss, and D. Neumann, Study of the ground state potential curve and dipole moment of OH by the method of optimized valence configurations, *J. Chem. Phys.* **61**, 3686–3699 (1974).
50. G.D. Gillespie, A.U. Khan, A.C. Wahl, R.P. Hosteny, and M. Krauss, The electronic

structure of nitrogen dioxide I. Multiconfiguration self-consistent-field calculation of the low-lying electronic states, *J. Chem. Phys.* **63**, 3425–3444 (1975).
51. K. Kirby-Docken and B. Liu, Theoretical study of molecular dipole moment functions I. The X $^1\Sigma^+$ state of CO, *J. Chem. Phys.* **66**, 4309–4316 (1977).
52. J.P. Toennies, Rotationally and vibrationally inelastic scattering of molecules, *Chem. Soc. Rev.* **3**, 407–441 (1974).
53. W.A. Lester, Interaction potential between Li^+ and H_2 I. Region appropriate for rotational excitation, *J. Chem. Phys.* **53**, 1511–1515 (1970).
54. W.A. Lester, Interaction potential between Li^+ and H_2 II. Region appropriate to vibrational excitation, *J. Chem. Phys.* **54**, 3171–3179 (1971).
55. V. Staemmler, *Ab initio* calculation of the potential energy surface of the system N_2Li^+, *Chem. Phys.* **7**, 17–29 (1975).
56. W. Kutzelnigg, V. Staemmler, and C. Hoheisel, Computer potential hypersurface (including electron correlation) of the system Li^+/H_2, *Chem. Phys.* **1**, 27–44 (1973).
57. J. Schaefer and W.A. Lester, Theoretical study of inelastic scattering of H_2 by Li^+ on SCF and CI potential energy surfaces, *J. Chem. Phys.* **62**, 1913–1924 (1975).
58. G.D. Barg, G.M. Kendall, and J.P. Toennies, Quasi-classical calculations of elastic and rotationally and vibrationally inelastic differential cross sections for $Li^+ + H_2$, *Chem. Phys.* **16**, 243–268 (1976).
59. C.H. Townes and A.C. Cheung, A pumping mechanism for anomalous microwave absorption in formaldehyde in interstellar space, *Astophys. J. Lett.* **157**, L103–L108 (1969).
60. B.J. Garrison, W.A. Lester, and H.F. Schaefer, A Hartree–Fock interaction potential between a rigid asymmetric top and a spherical atom: (H_2CO,He), *J. Chem. Phys.* **63**, 1449–1454 (1975).
61. B.J. Garrison, W.A. Lester, P. Siegbahn, and H.F. Schaefer, Effect of electron correlation on the H_2CO–He interaction potential, *J. Chem. Phys.* **63**, 4167–4170 (1975).
62. B.J. Garrison, W.A. Lester, W.H. Miller, and S. Green, Cooling of the 6-centimeter and 2-centimeter doublets of interstellar H_2CO by collision: an accurate quantum mechanical calculation, *Astrophys. J. Lett.* **200**, L175–L177 (1975).
63. N.J. Evans, B. Zuckerman, G. Morris, and T. Sato, Interstellar H_2CO I. Absorption studies, dark clouds, and the cosmic background radiation, *Astrophys. J.* **196**, 433–456 (1975).
64. P. Siegbahn and B. Liu, An accurate three-dimensional potential energy surface for H_3, *J. Chem. Phys.* **68**, 2457–2465 (1978).
65. S.F. Boys, G.B. Cook, C.M. Reeves, and I. Shavitt, Automatic fundamental calculations of molecular structure, *Nature* **178**, 1207–1209 (1956).
66. J. Hirschfelder, H. Eyring, and B. Topley, Reactions involving hydrogen molecules and atoms, *J. Chem. Phys.* **4**, 170–187 (1936).
67. D.G. Truhlar and R.E. Wyatt, History of H_3 kinetics, *Ann. Rev. Phys. Chem.* **27**, 1–43 (1976).
68. B. Liu, *Ab initio* potential energy surface for linear H_3, *J. Chem. Phys.* **58**, 1925–1937 (1973).
69. R.N. Porter and M. Karplus, Potential energy surface for H_3, *J. Chem. Phys.* **40**, 1105–1115 (1964).
70. D.G. Truhlar and C.J. Horowitz, Functional representation of Liu and Siegbahn's accurate *ab initio* potential energy calculations for $H + H_2$, *J. Chem. Phys.* **68**, 2466–2476 (1978).
71. R. Gengenbach, Ch. Hahn, and J.P. Toennies, Molecular beam measurements of the $D + H_2$ potential and recalibration of the reactive cross section, *J. Chem. Phys.* **62**, 3620–3630 (1975).
72. R. Foon and M. Kaufman, Kinetics of gaseous fluorine reactions, *Prog. React. Kinet.* **8**, 81–160 (1975).
73. D.J. Douglas and J.C. Polanyi, Effect of changing reagent energy on reaction dynamics

VII. Dependence of product energy distribution on reagent rotational excitation in $F + H_2(J) \rightarrow HF + H$, *Chem. Phys.* **16**, 1–8 (1976), and references therein.
74. R.D. Coombe and G.C. Pimentel, Effects of rotation on the vibrational energy distributions in the reaction $F + H_2$, *J. Chem. Phys.* **59**, 1535–1536 (1973).
75. T.P. Schafer, P.E. Siska, J.M. Parson, F.P. Tully, Y.C. Wong, and Y.T. Lee, Crossed molecular beam study of $F + D_2$, *J. Chem. Phys.* **53**, 3385–3387 (1970).
76. R.F. Heidner, and J.F. Bott, Vibrational deactivation of $HF(v = 1)$ and $DF(v = 1)$ by H and D atoms, *J. Chem. Phys.* **63**, 1810–1817 (1975).
77. C.F. Bender, P.K. Pearson, S.V. O'Neil, and H.F. Schaefer, Potential energy surfaces including electron correlation for the chemical reaction $F + H_2 \rightarrow FH + H$. I. Preliminary surface, *J. Chem. Phys.* **56**, 4626–4631 (1972).
78. W.A. Lathan, L.A. Curtiss, W.J. Hehre, J.B. Lisle, and J.A. Pople, Molecular orbital structures for small organic molecules and cations, *Prog. Phys. Org. Chem.* **11**, 175–261 (1974).
79. J.T. Muckerman, Chemical dynamics of the reaction of fluorine atoms with hydrogen molecules II. Dependence on the potential energy surface, *J. Chem. Phys.* **56**, 2997–3006 (1972).
80. J.C. Polanyi and J.L. Schreiber, Distribution of reaction products (theory). Investigation of an *ab initio* energy surface for $F + H_2 \rightarrow HF + H$, *Chem. Phys. Lett.* **29**, 319–322 (1974).
81. A.L. Robinson, Chemical dynamics: accurate quantum calculations at last, *Science* **191**, 275–276 (1976).
82. D.L. Thompson, Monte Carlo classical trajectory calculation of the rates of H- and D-atom vibrational relaxation of HF and DF, *J. Chem. Phys.* **57**, 4170–4173 (1972).
83. R.L. Wilkins, Monte Carlo calculations of reaction rates and energy distribution among reaction products. II. $H + HF(v) \rightarrow H_2(v') + F$ and $H + HF(v) \rightarrow HF(v') + H$, *J. Chem. Phys.* **58**, 3038–3046 (1973).
84. C.F. Bender, B.J. Garrison, and H.F. Schaefer, A critical test of semiempirical FH_2 potential energy surfaces: the barrier height for $H + FH \rightarrow HF + H$, *J. Chem. Phys.* **62**, 1188–1190 (1975).
85. P. Botschwina and W. Meyer, PNO–CEPA calculation of collinear potential energy barriers for thermoneutral exchange reactions, *Chem. Phys.* **20**, 43–52 (1977).
86. W.R. Wadt and N.W. Winter, Accurate characterization of the transition state geometry for the $HF + H' \rightarrow H + H'F$ reaction, *J. Chem. Phys.* **67**, 3068–3073 (1977).
87. J.C. Polanyi and J.J. Sloan, Energy distribution among reaction products VII. $H + F_2$, *J. Chem. Phys.* **57**, 4988–4998 (1972).
88. S.V. O'Neil, P.K. Pearson, H.F. Schaefer, and C.F. Bender, On the $H + F_2 \rightarrow HF + F$ reaction. An *ab initio* potential energy surface, *J. Chem. Phys.* **58**, 1126–1131 (1973).
89. C.F. Bender, C.W. Bauschlicher, and H.F. Schaefer, Saddle point geometry and barrier height for $H + F_2 \rightarrow HF + F$, *J. Chem. Phys.* **60**, 3707–3708 (1974).
90. S.V. O'Neil, H.F. Schaefer, and C.F. Bender, Barrier height for the exchange reaction $F + HF \rightarrow FH + F$, *Proc. Natl. Acad. Sci. USA* **71**, 104–106 (1974).
91. D.G. Truhlar, P.C. Olson, and C.A. Parr, Computed bond energies and vibrational frequencies for ClHCl, BrHBr, and IHI, including isotope effects and anharmonicity, *J. Chem. Phys.* **57**, 4479–4483 (1972).
92. D.L. Thompson, Monte Carlo classical trajectory calculation of the rates of F-atom vibrational relaxation of HF and DF, *J. Chem. Phys.* **57**, 4164–4169 (1974).
93. B.H. Mahan, Electronic structure and chemical dynamics, *Acct. Chem. Res.* **8**, 55–61 (1975).
94. B.H. Mahan and T.M. Sloane, Dynamics of the $C^+–H_2$ reaction, *J. Chem. Phys.* **59**, 5661–5675 (1973).
95. J.A. Fair and B.H. Mahan, Dynamics of the reaction of N^+ with H_2 II. Reactive scattering at relative energies below 3 eV, *J. Chem. Phys.* **62**, 515–519 (1975).

96. K.T. Gillen, B.H. Mahan, and J.S. Winn, Dynamics of the O^+–H_2 reaction I. Reactive scattering of $O^+ (^4S_{3/2})$ at relative energies below 15 eV, *J. Chem. Phys.* **58**, 5373–5384 (1973).
97. R.B. Woodward and R. Hoffmann, *The Conservation of Orbital Symmetry*, Verlag Chemie, Weinham/Bergstr., West Germany (1970).
98. D.H. Liskow, C.F. Bender, and H.F. Schaefer, Potential energy surfaces related to the ion molecule reaction $C^+ + H_2$, *J. Chem. Phys.* **61**, 2507–2513 (1974).
99. P.K. Pearson and E. Roueff, A low-energy passage of $C^+ + H_2 \rightarrow CH_2{}^+$, *J. Chem. Phys.* **64**, 1240–1241 (1976).
100. M.A. Gittins and D.M. Hirst, Mn *ab initio* potential energy surface for the reaction $N^+ + H_2 \rightarrow NH^+ + H$, *Chem. Phys. Lett.* **35**, 534–536 (1975).
101. C.F. Bender, J.H. Meadows, and H.F. Schaefer, Potential energy surfaces for ion–molecule reactions. Intersection of the $^3A_2$ and $^3B_1$ surfaces of $NH_2{}^+$, *Faraday Discuss. Chem. Soc.* **62**, 59–66 (1977).
102. Y.T. Lee, R.J. Gordon, and D.R. Herschbach, Molecular beam kinetics: Reactions of H and D atoms with diatomic alkali molecules, *J. Chem. Phys.* **54**, 2410–2423 (1971).
103. R.K. Preston and J.C. Tully, Effects of surface crossing in chemical reactions: The H_3^+ system, *J. Chem. Phys.* **54**, 4297–4303 (1971); Trajectory surface hopping approach to nonadiabatic molecular collisions: The reaction of H^+ with D_2, *J. Chem. Phys.* **55**, 562–572 (1971).
104. R.D. Levine and R.B. Bernstein, Energy disposal and energy requirements for elementary chemical reactions, *Discuss. Faraday Soc.* **55**, 100–112 (1973).
105. P. Siegbahn and H.F. Schaefer, Potential energy surfaces for $H + Li_2 \rightarrow LiH + Li$. Ground state surface from large scale configuration interaction, *J. Chem. Phys.* **62**, 3488–3495 (1975).
106. P.K. Pearson, W.J. Hunt, C.F. Bender, and H.F. Schaefer, Simplest halogen atom plus alkali dimer potential surface: $F + Li_2 \rightarrow LiF + Li$, *J. Chem. Phys.* **58**, 5358–5363 (1973).
107. W.B. England, N.H. Sabelli, and A.C. Wahl, A theoretical study of Li_2H. I. Basis set and computational survey of excited states and possible reaction paths, *J. Chem. Phys.* **63**, 4596–4605 (1975).
108. L.R. Kahn, P.J. Hay, and I. Shavitt, Theoretical study of curve crossing: *ab initio* calculations on the four lowest $^1\Sigma^+$ states of LiF, *J. Chem. Phys.* **61**, 3530–3546 (1974).
109. T.H. Dunning, The barriers for abstraction and exchange in H + HCl, *J. Chem. Phys.* **66**, 2752–2753 (1977).
110. G.O. Wood, Isotope exchange vs. abstraction for H + DCl, *J. Chem. Phys.* **56**, 1723–1727 (1972).
111. A.A. Westenberg and N. de Haas, Atom–molecule kinetics using ESR detection IV. Results for $Cl + H_2 = HCl + H$ in both directions, *J. Chem. Phys.* **48**, 4405–4415 (1968).
112. A.E. de Vries and F.S. Klein, Ultraviolet-induced isotope exchange in gaseous mixtures of HCl and D_2 and of DCl and H_2, *J. Chem. Phys.* **41**, 3428–3435 (1969).
113. J.D. McDonald and D.R. Herschbach, Molecular beam kinetics: Exchange reactions of deuterium atoms with hydrogen halides, *J. Chem. Phys.* **62**, 4740–4744 (1975).
114. P. Botschwina and W. Meyer, A PNO– EPA calculation of the barrier height for the collinear atom exchange reaction H′ + BrH → H′Br + H, *J. Chem. Phys.* **67**, 2390–2391 (1977).
115. W.A. Lester and M. Krauss, Interaction potential between Li and HF, *J. Chem. Phys.* **52**, 4775–4781 (1970).

Note Added in Proof

A critical experimental study [F. E. Bartoszek, D. M. Manos, and J. C. Polanyi, Effect of changing reagent energy X. Vibrational threshold energies for the alternative reaction paths $HF(v) + D \rightarrow F + HD$ and $\rightarrow H + DF$. *J. Chem. Phys.* **69**, 933–935 (1978)] has been reported since this manuscript was prepared. Using the technique of chemiluminescence depletion with mass spectrometry, Polanyi and co-workers have determined the activation energy for $H + FH \rightarrow HF + H$ exchange to be ~ 45 kcal. This provides a direct confirmation of the BGS prediction[84] and indirectly supports the theoretical results discussed in Section 4.2.5.

3

Interaction Potentials II: Semiempirical Atom–Molecule Potentials for Collision Theory

P. J. KUNTZ

1. Introduction

1.1. Potential Surfaces for Collision Theory

This chapter treats a particular type of semiempirical method suitable for the calculation of potential energy surfaces (PES) which are needed in atom–molecule collision processes. Such surfaces need to be accurate and readily calculable over a wide range of geometrical configurations from the composite molecule to the separated fragments. Hence, semiempirical methods designed primarily to handle equilibrium geometries are not considered here.

Semiempirical methods often have little predictive value, so wherever an accurate *ab initio* PES is at hand for a particular system, it should be used. In fact it is probably better to use an accurate functional fit to *ab initio* points than to use semiempirical methods at all, provided that this representation of the potential has the requisites listed below (Section 1.2). In practice, this option arises in only a few cases; most of the time another representation for the PES must be sought, and it is here that semiempirical methods are useful.

Semiempirical techniques can be implemented in two ways, depending on how the PES is to be represented in the dynamical calculation. The

P. J. KUNTZ • Hahn–Meitner Institut für Kernforschung, Glienicker Str. 100, 1000 Berlin 39, West Germany.

first way arises when the PES is represented by an arbitrary function so that computation of the PES at a particular geometry is very fast. One starts by calibrating the parameters in the semiempirical method so that it approximately agrees with experimental and *ab initio* information at a limited number of molecular configurations. Having done this, one proceeds to calculate the PES semiempirically at a large number of points in configuration space, including those where no *ab initio* information is available. This set of PES points is then fitted to some arbitrary function to give a representation of the potential surface which is suitable for dynamical calculations.[1] The major drawback here is that a considerable effort is required in fitting the calculated points to the arbitrary function; this is not a trivial problem but will not be discussed further here.

The second implementation of semiempirical methods is to use them directly in the dynamical calculations. They still need to be "calibrated" with *ab initio* points or experimental data (i.e., data about the PES, not kinetics data), but there is no need to fit them with arbitrary functions. The price paid for this is the increased time in evaluating the potential at a given configuration. In many systems, however, this is not a serious limitation.[2] In this article I shall emphasize a method which can be used in this fashion, since the overwhelming proportion of dynamical studies on realistic potential surfaces have adopted this course.

1.2. Requisites for the Potential Energy Surface and Its Representation

1.2.1. Physical Requirements

Any representation of a PES, be it a calibrated semiempirical Ansatz or a fitted function, must be physically reasonable. The checklist which follows is meant to provide the minimum requisites of a reasonable PES.

(a) The PES must have the same symmetry properties as the molecule to which it corresponds, e.g., a PES for the triatomic AB_2 should be invariant to rotation about the C_{2v} symmetry axis.

(b) The PES should provide a good description of the molecule in those geometries where experimental information (or an *ab initio* calculation) is available. For an atom–molecule collision system this includes the reactant and product channels (e.g., A + BC, AB + C, AC + B, A + B + C) and the strong-coupling region where A, B, C are very close to one another (and especially configurations where the molecule is bound with respect to all asymptotic channels).

(c) The PES must behave reasonably in the region between the strong-coupling region and the asymptotic regions, as this region is of prime importance for collision processes even though experimental information about it may be lacking.

The above requirements seem obvious enough, but it is remarkable how many calculations have been done with potentials which are inadequate with respect to these points. In fact, point (c) enormously restricts the possible semiempirical methods which can be employed for collision processes.

1.2.2 Computational Requirements

Whether a classical or quantal treatment of the collision process is used, it is normally necessary to evaluate the potential function a large number of times. A classical trajectory calculation also demands a knowledge of the derivatives at a large number of molecular configurations. Obviously, one must be able to calculate the PES very quickly; this is especially true if derivatives must be calculated numerically.

In a collision process involving several potential energy surfaces, the nonadiabatic coupling[3] must also be computed fairly often. It is then a decided advantage to have a method which delivers this along with the potential function and its derivatives.

1.3. Selection of Methods

In a recent review of potential surfaces for chemical reactions, G. Balint-Kurti divides non-*ab-initio* methods among three groups: (i) semitheoretical, (ii) semiempirical, and (iii) empirical.[4] He regards *ab initio* methods to be those which are as rigorous as possible. Semitheoretical methods are essentially *ab initio* calculations where specific, reasonable approximations are made. Semiempirical methods are also based on the *ab initio* approach but they contain gross and often difficult-to-assess approximations. At the bottom of the heap come the empirical approaches which are based on physical models (e.g., electrostatics) or are thinly disguised fitting procedures.

The semitheoretical group contains Hückel theory and those methods which are based on Hartree–Fock SCF theory: MINDO, CNDO, INDO, etc. These methods are described in a book by Murrell and Harget.[5] They utilize parametrization of the integrals which occur in Hartree–Fock SCF theory in order to simplify the task of computing the electronic structure of molecules. Hückel theory[6] and INDO[7] have been used in collision theory but the methods in this group are not generally recommended for two main reasons[8]: (i) Since they are based on the Hartree–Fock approximation, they are often incapable of describing properly the separating fragments [requirements (b) and (c) of Section 1.2 above]. (ii) The parameters in the methods are chosen to give the best results near the most stable molecular configurations; these regions need not be the most important for collision processes. A third objection to these methods is that they may be too slow to be used directly in a dynamical calculation.

The purely empirical methods can be very useful in collision theory, but they are not very general. An example is the Rittner potential for alkali halide systems.[9] These methods usually apply only to simple potential surfaces and do not treat the nonadiabatic coupling between them. Nevertheless, two-surface processes are often treated empirically by parametrizing the elements of a 2×2 matrix.[10,11] We shall consider this briefly in Section 3.

The semiempirical category includes atoms-in-molecules,[4] diatomics-in-molecules (DIM),[12-21] certain valence-bond (VB) methods[21-30] and pseudopotential methods.[31,32] Of these the DIM and VB appear most frequently in collision theory work. The DIM method and simplifications which can be derived from it (e.g., the London equation for H_3-like systems[33]) are considered the method of choice by many. Indeed, in a recent review of DIM[16] Tully exuberates, "In five years DIM will be a widely used, quantitatively accurate, and predictive tool of the quantum chemist." Whether such ebullient optimism is justified remains to be seen but one must admit that the method is singularly suited to molecular collision theory and beguilingly practical. It satisfies the physical and computational requirements listed above and it often yields derivatives and nonadiabatic coupling directly. It has been applied to single-[2] and multiple-[34,35] surface processes and to cases where spin–orbit coupling is important.[14] For this reason, the bulk of this chapter will illustrate the implementation of DIM. There will be no attempt at a critical evaluation of the method, since in my opinion there is very little work published on which a true critique of the method could be based. It appears to work very well for many systems (e.g., H_3^+)[34-36] and to have difficulty with others (e.g., H_4).[18] In cases where the method is not too good it is unfortunately not clear whether this is due to the DIM method itself or to the way in which it was implemented. Very often a very small basis set has been used in cases where a larger basis might be necessary. Further studies are necessary before a good assessment of this method can be made but it is at the moment unrivaled for collision processes, particularly those between diatomic molecules.

2. *The Method of Diatomics-in-Molecules (DIM)*

2.1. *Introduction*

The DIM method was invented by F. O. Ellison,[12] who made use of the fact that a polyatomic Hamiltonian operator which consists only of *one-* and *two-body* operators can be expressed *exactly* in terms of the Hamiltonians of diatomic and atomic subsystems. Let A, B, C, ... be N atoms which make up the N-atom molecule ABC.... The atoms are assigned electrons $n_{A\alpha}, n_{B\beta}, n_{C\gamma}, \ldots$ such that the sum of the electrons equals the total number of electrons in the polyatomic molecule. In general, there are many possible assignments (e.g., for H_2F there is HHF, H^+H^-F, HH^+F^-, etc.), and each one defines a decomposition of the Hamiltonian operator:

$$H_{\alpha\beta\gamma\cdots}^{(\mathrm{ABC}\cdots)} = \sum_{S=1}^{N-1} \sum_{L=S+1}^{N} H_{\sigma\lambda}^{(SL)} - (N-2) \sum_{S=1}^{N} H_{\sigma}^{(S)} \tag{1}$$

$H_{\sigma\lambda}^{(SL)}$ is the Hamiltonian operator for diatomic molecule SL having $(n_{S\sigma} + n_{L\lambda})$ electrons; $H_{\sigma}^{(S)}$ is an atomic Hamiltonian.

Equation (1) is exact and provides the foundation for the DIM approach, which is to set up a Hamiltonian matrix for the polyatomic molecule, **H**, whose matrix elements are expressed *entirely* in terms of quantities pertaining to the diatomic and atomic subsystems. The diatomic and atomic fragment quantities may be obtained from *ab initio* calculations or from experiment (e.g., spectroscopy). The object of the method is to find the eigenvalues of the matrix **H**, which are obtained by solving the secular equation

$$\mathbf{HC} = \mathbf{SCE} \tag{2}$$

where **E** is diagonal, **C** is the eigenvector matrix, and **S** is the overlap matrix of the polyatomic basis functions.

The DIM procedure is outlined in Section 2.2 and in Section 2.3 an example is worked out in detail for a fairly complex system. Section 2.4 illustrates the application to simpler systems for which the DIM **H** matrix can be constructed by angular momentum coupling techniques alone, and Section 2.5 mentions the treatment of nonadiabatic coupling and spin–orbit coupling. For discussions about the theory of DIM and about its general value and use, the reader is referred to the original papers[8,12-15] and to a recent review by Tully.[16]

2.2. General Formulation

2.2.1. Defining the Scope of the Problem

The outline given here follows fairly closely the paper of Steiner, Certain, and Kuntz,[13] who present the theory in a way which is easily applied to complex systems. It is assumed that one is prepared to do a valence-bond calculation on the atomic and diatomic systems into which the polyatomic molecule can be partitioned. Even though such calculations are not necessary for many simple systems, the following procedure furnishes a systematic construction of the DIM **H** matrix in a form convenient for programming on a computer.

The polyatomic molecule is considered to be made up of N atoms A, B, C, ... with nuclear charges $Z_A, Z_B, Z_C, \ldots$. The first step is to decide on how many different partitionings of the electrons among the N atoms there are going to be. Each partitioning is denoted by specifying the atomic species A^{+q_α}, B^{+q_β}, ... with the symbols $\alpha, \beta, \gamma, \ldots$, where $q_\alpha = Z_A - n_{A\alpha}$ is the total charge on atom A and $n_{A\alpha}$ is the number of electrons assigned to A in the particular partitioning. For example, some suitable partitionings for molecule ABCD would be ABCD, A^+BC^-D, $A^+B^+C^-D^-$, etc.; obviously the total number of electrons for each partitioning must be the same:

$$n_{A\alpha} + n_{B\beta} + \cdots + n_{D\delta} = n \tag{3}$$

The larger the number of partitionings, the more complicated is the DIM calculation because (i) the **H** matrix is larger, and (ii) more atomic and diatomic information is required as input.

The minimum number of partitionings is determined by the physical system under consideration. A DIM treatment of the process $K + Br_2 \rightarrow K^+ + Br_2^-$ would require the partitionings KBrBr, K^+Br^-Br, K^+BrBr^-. The process $F + H_2 \rightarrow HF + H$ would require only FHH; however, a better description of the PE surface might result from considering F^-H^+H and F^-HH^+ in addition.

2.2.2. The DIM Basis Set

The collision process defines a minimum size of basis, which must be able to describe all of the states of the system which are accessible in the process. If the DIM procedure is used in an *ab initio* fashion (i.e., all atomic and diatomic properties are calculated), then an increase in the basis will tend to produce better results. Semiempirical implementation of the method, on the other hand, may not respond so favorably to an increase of basis.

The polyatomic basis Φ_I consists of antisymmetrized products of atomic basis functions:

$$\Phi_I = \mathscr{A}_n \phi_I = \mathscr{A}_n A_{\alpha i} B_{\beta j} C_{\gamma k} \cdots \tag{4}$$

Here $\mathscr{A}_n$ is the n-electron antisymmetrizer and $A_{\alpha i}$, $B_{\beta j}$, $C_{\gamma k}$, ... are atomic basis functions for the species $\alpha, \beta, \gamma, \ldots$. The basic set for species α consists of $M_{A\alpha}$ functions indexed by $i = 1, 2, 3, \ldots, M_{A\alpha}$. Species $\beta, \gamma, \ldots$ have their basis functions indexed by $j, k, \ldots$.

The atomic basis functions may be taken as orbital products, but this is not necessary. They should be defined relative to a common fixed coordinate system. It is necessary to know how these basis functions behave upon rotation through angles (θ, ϕ). Let $A'_{\alpha i}$, $B'_{\beta j}$, ... be the functions into which the $A_{\alpha i}$, $B_{\beta j}$, ... are transformed upon rotation:

$$\begin{aligned} A'_{\alpha i} &= \sum_s A_{\alpha s} R_\alpha^{si}(\theta, \phi) \\ B'_{\beta j} &= \sum_t B_{\beta t} R_\beta^{tj}(\theta, \phi) \end{aligned} \tag{5}$$

or $\mathbf{A}'_\alpha = \mathbf{A}_\alpha \mathbf{R}_\alpha$, $\mathbf{B}'_\beta = \mathbf{B}_\beta \mathbf{R}_\beta$, where $\mathbf{A}_\alpha$ and $\mathbf{B}_\beta$ are row vectors.

It is assumed that the rotation matrices $\mathbf{R}(\theta, \phi)$ are known. If the atomic basis functions are chosen to be eigenfunctions of $\underset{\wedge}{\mathbf{L}}^2$ and $\underset{\wedge}{\mathbf{L}}_z$ for the atom, then the functions $\mathbf{R}(\theta, \phi)$ are built up from the standard rotation matrices[37]

$$\underset{\wedge}{\mathbf{R}}(\theta_1, \theta_2, \theta_3) \,|\, JM\rangle = \sum_{M'} |\, JM'\rangle \, R^{(J)}_{M'M}(\theta_1, \theta_2, \theta_3) \tag{6}$$

$$R^{(J)}_{MM'}(\theta_1, \theta_2, \theta_3) = \exp(-i\theta_1 M)\, r^{(J)}_{MM'}(\theta_2) \exp(-i\theta_3 M') \tag{7}$$

For example, if one of the atomic functions corresponded to a P state ($L = 1$), the associated matrix $\mathbf{r}(\theta)$ would be

$$\mathbf{r}^{(1)}(\theta) = \begin{array}{c|ccc} M \backslash M' & 1 & 0 & -1 \\ \hline 1 & (1+\cos\theta)/2 & -(\sin\theta)/2^{1/2} & (1-\cos\theta)/2 \\ 0 & (\sin\theta)/2^{1/2} & \cos\theta & -(\sin\theta)/2^{1/2} \\ -1 & (1-\cos\theta)/2 & (\sin\theta)/2^{1/2} & (1+\cos\theta)/2 \end{array} \tag{8}$$

Note that the functions need not be eigenvectors of $\underset{\wedge}{\mathbf{L}}^2$ and $\underset{\wedge}{\mathbf{L}}_z$. For a triatomic system it may be more convenient to choose the atomic functions to have the symmetry of the C_s point group. Using the P functions once again as an example, functions P_x, P_y, and P_z could be formed from P_{-1}, P_0, P_{+1}:

$$(P_x, P_y, P_z) = (P_1, P_0, P_{-1})\, 2^{-1/2} \begin{bmatrix} -1 & -1 & 0 \\ 0 & 0 & 2^{1/2} \\ 1 & -1 & 0 \end{bmatrix} \tag{9}$$

i.e., $(P_x, P_y, P_z) = (P_1, P_0, P_{-1})\,\mathbf{W}$.

The new functions then transform among themselves under rotation as

$$\underset{\wedge}{\mathbf{R}}(\theta_1, \theta_2, \theta_3)(P_x, P_y, P_z) = (P_x, P_y, P_z)\, \mathbf{W}^{-1}\mathbf{R}^{(J)}\mathbf{W} \tag{10}$$

For example, when $\theta_1 = \theta_3 = 0$, then

$$\mathbf{W}^{-1}\mathbf{R}^{(J)}\mathbf{W} = \mathbf{W}^{-1}\mathbf{r}^{(J)}(\theta)\,\mathbf{W} = \begin{bmatrix} \cos\theta & 0 & \sin\theta \\ 0 & 1 & 0 \\ -\sin\theta & 0 & \cos\theta \end{bmatrix} \tag{11}$$

The atomic secular equation for species α is (dropping the index A for the moment)

$$\mathbf{H}_\alpha \mathbf{c}_\alpha = \mathbf{S}_\alpha \mathbf{c}_\alpha \mathbf{E}_\alpha \tag{12}$$

where

$$\mathbf{H}_\alpha = \langle \mathscr{A}_\alpha \mathbf{A}_\alpha \,|\, H_\alpha \,|\, \mathbf{A}_\alpha \rangle \tag{13}$$

$$\mathbf{S}_\alpha = \langle \mathscr{A}_\alpha \mathbf{A}_\alpha \,|\, \mathbf{A}_\alpha \rangle \tag{14}$$

and $\mathbf{E}_\alpha$ is diagonal. The $M_\alpha \times M_\alpha$ matrix $\mathbf{c}_\alpha$ diagonalizes both $\mathbf{H}_\alpha$ and $\mathbf{S}_\alpha$:

$$\mathbf{c}_\alpha^\dagger \mathbf{S}_\alpha \mathbf{c}_\alpha = \mathbf{1} \qquad \text{and} \qquad \mathbf{c}_\alpha^\dagger \mathbf{H}_\alpha \mathbf{c}_\alpha = \mathbf{E}_\alpha \tag{15}$$

The basic approximation in the DIM procedure is to express $\underset{\wedge}{H}_\alpha(\mathscr{A}_\alpha \mathbf{A}_\alpha)$ in terms of a linear transformation of the $\mathscr{A}_\alpha \mathbf{A}_\alpha$:

$$\underset{\wedge}{H}_\alpha \mathscr{A}_\alpha \mathbf{A}_\alpha \rightarrow \mathscr{A}_\alpha \mathbf{A}_\alpha \mathbf{b}_\alpha \tag{16}$$

where

$$\mathbf{b}_\alpha = \mathbf{S}_\alpha^{-1}\mathbf{H}_\alpha = \mathbf{c}_\alpha \mathbf{E}_\alpha \mathbf{c}_\alpha^{-1} = \mathbf{c}_\alpha \mathbf{E}_\alpha \mathbf{c}_\alpha^\dagger \mathbf{S}_\alpha \tag{17}$$

The matrices $\mathbf{b}_\alpha^{(A)}$, $\mathbf{b}_\beta^{(B)}$, $\mathbf{b}_\gamma^{(C)}$, ... constitute the atomic fragment information which is required by DIM through the atomic terms in the partitioning of equation (1). These fragment matrices may be computed entirely by *ab initio* methods, or they can be constructed from experimental $\mathbf{E}_\alpha$, $\mathbf{E}_\beta$, $\mathbf{E}_\gamma$, ... matrices and calculated $\mathbf{c}_\alpha$, $\mathbf{c}_\beta$, $\mathbf{c}_\gamma$, ... matrices.

In order to obtain the diatomic input required through the use of equation (1), a diatomic basis is required for each pair of atomic species. Taking the pair AB as an example, consider the basis describing the diatomic species $(\alpha\beta)$. In general, several pairs of atomic species α and β are compatible with the diatomic $(\alpha\beta)$, e.g., for $(\alpha\beta) = \text{HF}$, one could have the pairs $(\alpha = \text{H},\ \beta = \text{F})$ and $(\alpha = \text{H}^+,\ \beta = \text{F}^-)$. Therefore, the diatomic basis functions $\mathbf{AB}_{(\alpha\beta)l}$ (indexed by l) are constructed from *direct products* of the bases $\mathbf{A}_\alpha$ and $\mathbf{B}_\beta$ for all pairs α and β which are consistent with the species $(\alpha\beta)$ (i.e., $n_{\mathrm{A}\alpha} + n_{\mathrm{B}\beta} = n_{(\mathrm{AB})(\alpha\beta)}$); the size of this basis is $\sum_{\alpha,\beta\rightarrow(\alpha\beta)} M_{\mathrm{A}\alpha} \times M_{\mathrm{B}\beta}$. The diatomic secular equation must be solved in this basis:

$$\mathbf{H}_{\alpha\beta}\mathbf{c}_{\alpha\beta} = \mathbf{S}_{\alpha\beta}\mathbf{c}_{\alpha\beta}\mathbf{E}_{\alpha\beta} \tag{18}$$

From $\mathbf{c}_{\alpha\beta}$ and $\mathbf{E}_{\alpha\beta}$ the matrix $\tilde{\mathbf{b}}_{\alpha\beta}$ is defined

$$\tilde{\mathbf{b}}_{\alpha\beta} = \mathbf{S}_{\alpha\beta}^{-1}\mathbf{H}_{\alpha\beta} = \mathbf{c}_{\alpha\beta}\mathbf{E}_{\alpha\beta}\mathbf{c}_{\alpha\beta}^{-1} \tag{19}$$

The diatomic fragment information is contained in the matrices $\tilde{\mathbf{b}}_{\alpha\beta}^{(AB)}$, $\tilde{\mathbf{b}}_{\beta\gamma}^{(BC)}$, $\tilde{\mathbf{b}}_{\alpha\gamma}^{(AC)}$, ... which are constructed either semiempirically or entirely nonempirically, analogously to the atomic fragment matrices.

In setting up the DIM $\mathbf{H}$ matrix it is necessary to know the way in which the functions $\mathbf{AB}_{(\alpha\beta)l}$, $\mathbf{BC}_{(\beta\gamma)m}$, $\mathbf{AC}_{(\alpha\gamma)n}$, ... transform upon rotation of the atomic functions through the angles (θ, ϕ). The reason for this is that the computation of the matrices $\tilde{\mathbf{b}}_{(\alpha\beta)}^{(AC)}$, etc., is done only *once* in a coordinate frame suitable for the diatomic molecule. In the DIM calculation, however, the orientation of the diatomic fragments with respect to the common coordinate axes of the basis functions depends upon the particular geometric configuration of the polyatomic molecule. Therefore, before the replacement of $\underset{\wedge}{H}_{\alpha\beta}\mathscr{A}_{\alpha\beta}\mathbf{AB}_{(\alpha\beta)}$ by a diatomic fragment can be made, the functions $\mathbf{AB}_{(\alpha\beta)}$ must be rotated so that the basis functions $\mathbf{A}_\alpha$ and $\mathbf{B}_\beta$ have the same orientation with respect to the diatomic molecule as they do in the calculation of the matrix $\tilde{\mathbf{b}}_{\alpha\beta}$. The rotation of the atomic basis is given by equations (5). The diatomic rotation matrix is, therefore, the direct product of these, $\mathbf{R}_\alpha \otimes \mathbf{R}_\beta = \mathbf{R}_{\alpha\beta}$,

$$R_{\alpha\beta}^{st,ij} = R_\alpha^{si} R_\beta^{tj} \tag{20}$$

$$\mathrm{AB}'_{(\alpha\beta)l} = \mathrm{A}'_{\alpha i}\mathrm{B}'_{\beta j} = \sum_{s,t} \mathrm{A}_{\alpha s}\mathrm{B}_{\beta t} R_{\alpha\beta}^{st,ij} \tag{21}$$

The correct diatomic replacement is therefore

$$\begin{aligned} \underset{\wedge}{\mathbf{H}}_{\alpha\beta}\mathscr{A}_{\alpha\beta}\mathbf{AB}_{(\alpha\beta)} &= \underset{\wedge}{\mathbf{H}}_{\alpha\beta}\mathscr{A}_{\alpha\beta}\mathbf{AB}'_{(\alpha\beta)}\mathbf{R}_{\alpha\beta}^{-1} \\ &\to \mathscr{A}_{\alpha\beta}\mathbf{AB}'_{(\alpha\beta)}\tilde{\mathbf{b}}_{\alpha\beta}\mathbf{R}_{\alpha\beta}^{-1} \\ &= \mathscr{A}_{\alpha\beta}\mathbf{AB}_{(\alpha\beta)}\mathbf{R}_{\alpha\beta}\tilde{\mathbf{b}}_{\alpha\beta}\mathbf{R}_{\alpha\beta}^{-1} \\ &= \mathscr{A}_{\alpha\beta}\mathbf{AB}_{\alpha\beta}\mathbf{b}_{\alpha\beta} \end{aligned} \tag{22}$$

i.e.,

$$\mathbf{b}_{\alpha\beta} = \mathbf{R}_{\alpha\beta}\tilde{\mathbf{b}}_{\alpha\beta}\mathbf{R}_{\alpha\beta}^{-1} \tag{23}$$

$\mathbf{b}_{\alpha\beta}$ will depend on the angles of rotation.

The polyatomic basis functions are given by equation (4). The index I is a compound label which specifies (i) the ordinal number of the polyatomic basis function in the set; (ii) the atomic species $\alpha, \beta, \gamma, \ldots$ and the diatomic species $\alpha\beta, \beta\gamma, \alpha\gamma, \ldots$; (iii) the label of the atomic basis functions $i, j, k, \ldots$; and (iv) the label of the diatomic basis functions l for diatomic AB, m for diatomic BC, n for diatomic AC, etc. Note that the order of atomic functions in equation (4) is taken as the standard order for the basis set.

2.2.3. The DIM Hamiltonian Matrix

After choosing the atomic species and listing the atomic, diatomic, and polyatomic functions, it is necessary to use the fragment matrices to build up the polyatomic DIM Hamiltonian matrix $H_{I'I} = \langle \mathscr{A}_n \phi_{I'} | \underset{\wedge}{\mathbf{H}} | \phi_I \rangle$. Each basis function I defines the DIM decomposition (1) of H. Therefore

$$\begin{aligned} H_{I'I} = \langle \mathscr{A}_n \phi_{I'} | &\underset{\wedge}{\mathbf{H}}_{\alpha\beta}\mathscr{A}_{\alpha\beta} + \underset{\wedge}{\mathbf{H}}_{\beta\gamma}\mathscr{A}_{\beta\gamma} + \underset{\wedge}{\mathbf{H}}_{\alpha\gamma}\mathscr{A}_{\alpha\gamma} + \cdots \\ &- \underset{\wedge}{\mathbf{H}}_{\alpha}\mathscr{A}_{\alpha} - \underset{\wedge}{\mathbf{H}}_{\beta}\mathscr{A}_{\beta} - \underset{\wedge}{\mathbf{H}}_{\gamma}\mathscr{A}_{\gamma} - \cdots | \phi_I \rangle \end{aligned}$$

i.e.,

$$\mathbf{H} = \mathbf{SB} \tag{24}$$

where

$$\mathbf{S} = \langle \mathscr{A}_n \phi | \phi \rangle \tag{25}$$

and

$$\mathbf{B} = \mathbf{B}_{\mathrm{AB}} + \mathbf{B}_{\mathrm{BC}} + \mathbf{B}_{\mathrm{AC}} + \cdots - \mathbf{B}_{\mathrm{A}} - \mathbf{B}_{\mathrm{B}} - \mathbf{B}_{\mathrm{C}} - \cdots \tag{26}$$

The matrices $\mathbf{B}$ are defined in terms of the atomic and diatomic fragment matrices. The $(I'I)$ element of matrix $\mathbf{B}_A$ is given by

$$B_A^{(I'I)} = b_\alpha^{(i'i)} \delta_{\alpha'\alpha} \delta_{\beta'_{j'}, \beta_j} \delta_{\gamma_{k'}, \gamma_k} \cdots \tag{27}$$

In this expression, $\delta_{\alpha\alpha'} = 0$ if $\alpha' \neq \alpha$ and is otherwise 1; $\delta_{\beta'_{j'}, \beta_j} = 0$ if $\beta' \neq \beta$ *or* if the atomic function label j' for species β' does not equal the label j associated with the species β. The symbols are written in this manner merely to emphasize that j is associated with β, k with γ, etc. Once the composite labels I have been listed in a table, the matrix $\mathbf{B}_A$ can be constructed very quickly merely by comparing all pairs of rows in the table; this is particularly convenient for automatic programming. The atomic matrices $\mathbf{B}_B$, $\mathbf{B}_C$, ... are constructed similarly.

The diatomic matrices are built up in a similar fashion. Here, however, a phase factor of ± 1 must be included in the expression for the matrix elements. It arises only when (i) the atomic species making up the particular diatomic species in $\phi_{I'}$ differ from those in ϕ_I, and (ii) if the atoms making up the diatomic under consideration are not immediate neighbors in the standard order defining ϕ_I and $\phi_{I'}$. Consider the matrix $\mathbf{B}_{AT'}$ where the atomic species for T is labeled by τ, the atomic function index by r, and the diatomic function index by u. The charge transfer from atom A to T in going from $\phi_{I'}$ to ϕ_I is

$$\Delta q = |n_{A\alpha'} - n_{A\alpha}| \tag{28}$$

The number of electrons between function A_α (or $A_{\alpha'}$) and T_τ (or $T_{\tau'}$) in the function ϕ_I (or $\phi_{I'}$) in standard order is simply

$$\Delta e = n_{B\beta} + n_{C\gamma} + \cdots + n_{S\sigma} \tag{29}$$

where the sum extends over all atoms which lie between A and T in the standard order. This is true because we need only consider the case where $\beta' = \beta, \gamma' = \gamma, \ldots, \sigma' = \sigma$. The phase factor is then

$$P = (-1)^{\Delta e \Delta q} \tag{30}$$

The matrix $\mathbf{B}_{AT}$ can now be expressed in terms of $\mathbf{b}_{\alpha\tau}$:

$$\mathbf{B}_{AT}^{(I'I)} = b_{\alpha\tau}^{(u'u)} P \prod_{S \neq A,T} \delta_{\sigma'_{r'}, \sigma_r} \tag{31}$$

where the running atom-index S for the product is over all atoms except A and T, and where r is the function index of species σ on atom S. Equation (31) shows that matrix elements will exist only between those functions $\phi_{I'}$ and ϕ_I for which all atomic species and their function labels (other than A and T) are identical. For example, the matrix $\mathbf{B}_{AC}$ is

$$\mathbf{B}_{AC}^{(I'I)} = b_{\alpha\gamma}^{(n'n)} (-1)^{n_{B\beta} |n_{A\alpha} - n_{A\alpha'}|} \delta_{\beta_j \beta'_j} \cdots \tag{32}$$

The DIM **H** matrix as constructed here is often quite large. Considerable reduction in size can be achieved by defining a new basis set whose functions are eigenvectors of total spin and also perhaps of some symmetry operators:

$$\Gamma = \phi \mathbf{T} \tag{33}$$

The new matrices are

$$\mathbf{H}' = \mathbf{S}'\mathbf{B}' = \mathbf{T}^{\dagger}\mathbf{H}\mathbf{T} \tag{34}$$

$$\mathbf{S}' = \mathbf{T}^{\dagger}\mathbf{S}\mathbf{T} \tag{35}$$

$$\mathbf{B}' = \mathbf{T}^{-1}\mathbf{B}\mathbf{T} \tag{36}$$

2.2.4. The DIM Eigenvalues

In general, the matrix $\mathbf{B} = \mathbf{S}^{-1}\mathbf{H}$ will not be Hermitian, but it has the same eigenvalues as $\mathbf{H}$. The best procedure is to diagonalize $\mathbf{B}$ itself[38] so that the overlap matrix $\mathbf{S}$ does not need to be calculated. This procedure, suggested by Tully,[15,16] could lead to complex eigenvalues,[38] but work done so far has indicated this does not often occur.

The conventional treatment of overlap[12] is to form a symmetric $\mathbf{H}$ matrix:

$$\mathbf{H} \leftarrow \tfrac{1}{2}(\mathbf{B}^{\dagger}\mathbf{S}^{\dagger} + \mathbf{S}\mathbf{B}) \tag{37}$$

This procedure is known to lead to serious difficulties[13] and is not recommended. Tully[15] has also tried using a symmetrized B matrix:

$$\mathbf{B}_{\text{sym}} = \tfrac{1}{2}(\mathbf{B} + \mathbf{B}^{\dagger}) \tag{38}$$

This avoids the serious difficulties of equation (37) but does not give such good results; however, if DIM potentials were used to fit *ab initio* points, equation (38) would be a viable alternative to diagonalizing an unsymmetrical $\mathbf{B}$ and is much superior to equation (37).

2.3. A Specific Example: FH_2

The exposition of Section 2.2 is difficult to assimilate because of its generality and the proliferation of labels. It is much easier to demonstrate the method by reference to a specific problem. I have chosen the system FH_2 because it is not too complicated but it includes most of the difficulties which one would encounter in a typical DIM calculation. Ionic structure H^+F^- and covalent structure HF are included to demonstrate the effect of having two different electron partitionings among atoms corresponding to a single diatomic species. The procedure is presented as a sequence of steps which one would follow in applying the method to any molecule.

2.3.1. Define the Coordinate System

Axes xyz form a right-handed coordinate system with atom F at the origin; one H atom, designated hereafter as a, lies on the z axis, and the other H atom, designated by b, lies in the first or second quadrant of the xz plane. The atoms H_a, H_b, F are also denoted by A, B, C, the internuclear distances by R_{AB}, R_{BC}, R_{AC}, respectively, and the angle ABC by θ.

2.3.2. Define the Atomic Basis Functions and Fragment Matrices

A short-form notation is used throughout. For the atom F this is

$$|x| = |1s^2\, 2s^2\, 2px\, 2p_y^2\, 2p_z^2| \qquad |z| = |1s^2\, 2s^2\, 2p_x^2\, 2p_y^2\, 2p_z|$$
$$|y| = 1s^2\, 2s^2\, 2p_x^2\, 2p_y\, 2p_z^2| \qquad |F^-| = |1s^2\, 2s^2\, 2p_x^2\, 2p_y^2\, 2p_z^2|$$

For H_a, $|a| = |1s|$; for H_b, $|b| = |1s|$. A bar over the symbols x, y, z, a, or b indicates a value of the spin quantum number $m_s = -1/2$; no bar indicates $m_s = +1/2$. The number of electrons for the various species are $n_{AH} = n_{BH} = 1$, $n_{AH^+} = n_{BH^+} = 0$, $n_{CF} = 9$, $n_{CF^-} = 10$. The atomic basis functions on centers A, B, C and their labels are given in Table 1.

The rotation matrices [equation (5)] for all species are unity except for $\gamma = F$, where the six functions transform as the components of the ^2P rstate of F. The matrix $\mathbf{R}_F(\theta)$ is constructed from rotation matrix $\mathbf{r}^{(1)}(\theta)$, transformed as in equation (11) to give (using the labeling from Table 1)

$$\mathbf{R}_F(\theta) = \begin{bmatrix} c & 0 & -s & 0 & 0 & 0 \\ 0 & c & 0 & -s & 0 & 0 \\ s & 0 & c & 0 & 0 & 0 \\ 0 & s & 0 & c & 0 & 0 \\ 0 & 0 & 0 & 0 & 1 & 0 \\ 0 & 0 & 0 & 0 & 0 & 1 \end{bmatrix} \qquad (39)$$

where $c = \cos\theta$, $s = \sin\theta$.

Table 1. Atomic Basis Functions

α	i	$A_{\alpha i}$	β	j	$B_{\beta j}$	γ	k	$C_{\gamma k}$
H	1	$\|a\|$	H	1	$\|b\|$	F	1	$\|z\|$
H	2	$\|\bar{a}\|$	H	2	$\|\bar{b}\|$	F	2	$\|\bar{z}\|$
H^+	1		H^+	1		F	3	$\|x\|$
						F	4	$\|\bar{x}\|$
						F	5	$\|y\|$
						F	6	$\|\bar{y}\|$
						F^-	1	$\|F^-\|$

Table 2. Diatomic Fragments for AB ($g = 2^{-1/2}$)

l	$\alpha\beta$	α	β	i	j	$AB_{(\alpha\beta)l}$	$\mathbf{c}_{(\alpha\beta)}$				$\mathbf{b}_{\alpha\beta}$			
1	H_2	H	H	1	1	$\lvert ab\rvert$	1				T			
2	H_2	H	H	2	1	$\lvert \bar{a}b\rvert$		g	g			$(S+T)/2$	$(T-S)/2$	
3	H_2	H	H	1	2	$\lvert a\bar{b}\rvert$		$-g$	g			$(T-S)/2$	$(S+T)/2$	
4	H_2	H	H	2	2	$\lvert \bar{a}\bar{b}\rvert$				1				T
						$\mathrm{diag}(\mathbf{E}_{H_2}) =$	T	S	T	T				
1	H_2^+	H	H^+	1	1	$\lvert a\rvert$	g	g			$(G+U)/2$	$(G-U)/2$		
2	H_2^+	H^+	H	1	1	$\lvert b\rvert$	g	$-g$			$(G-U)/2$	$(G+U)/2$		
3	H_2^+	H	H^+	2	1	$\lvert \bar{a}\rvert$			g	g			$(G+U)/2$	$(G-U)/2$
4	H_2^+	H^+	H	1	2	$\lvert \bar{b}\rvert$			g	$-g$			$(G-U)/2$	$(G+U)/2$
						$\mathrm{diag}(\mathbf{E}_{H_2^+}) =$	G	U	G	U				

The atomic fragment matrices are all diagonal and have the following diagonal elements: $\mathbf{b}_{H^+}^{(A)} = E_{H^+}$, $\mathbf{b}_{H}^{(A)} = (E_H, E_H)$, $\mathbf{b}_{F}^{(C)} = E_F(1, 1, 1, 1, 1, 1)$, and $\mathbf{b}_{F}^{(C)} = E_{F^-}$, with $\mathbf{b}_{\beta}^{(B)} = \mathbf{b}_{\alpha}^{(A)}$. Note that the atomic fragment matrices are only diagonal because of the way we chose the basis set. Nondiagonal matrices could arise in more complicated systems (e.g., O atoms).

2.3.3. Define the Diatomic Basic and Fragment Matrices

The diatomic species which arise are HF, HF^-, H_2, and H_2^+. The basis functions for molecule AB, labeled by l, are listed in Table 2. The eigenvectors are written as column vectors over their respective eigenvalues: T is the energy of the H_2 $^3\Sigma_u^+$ state, S is the energy of the H_2 $X\ ^1\Sigma_g^+$ state, G is the energy of the H_2^+ $X\ ^2\Sigma_g^+$ state, and U is the energy of the H_2^+ $^2\Sigma_u^+$ state.

The corresponding information for molecule BC is included in Table 3. The eigenvalues for HF are $E_1(^3\Sigma^+)$, $E_2(X\ ^1\Sigma^+)$, $E_5(^3\Pi)$, $E_6(^1\Pi)$, and E_7 (a "mixed" $^1\Sigma$ state); E_1, E_2, E_5, and E_6 could be replaced by experimental values, but E_7 would best be calculated. E_- is the $X\ ^2\Sigma^+$ state of HF^-. The symbols used in the matrix $\bar{\mathbf{b}}_{HF}$ are defined as

$$Q_2 = \tfrac{1}{2}[(1 - \lambda_1\lambda_2)E_1 + (E_2 - \lambda_1\lambda_2 E_7)]/(1 - \lambda_1\lambda_2) \tag{40}$$

$$J_2 = \tfrac{1}{2}[(1 - \lambda_1\lambda_2)E_1 - (E_2 - \lambda_1\lambda_2 E_7)]/(1 - \lambda_1\lambda_2) \tag{41}$$

$$Q_5 = \tfrac{1}{2}(E_5 + E_6) \tag{42}$$

$$J_5 = \tfrac{1}{2}(E_5 - E_6) \tag{43}$$

$$J_1 = g\lambda_1(E_2 - E_7)/(1 - \lambda_1\lambda_2) \tag{44}$$

$$J_3 = g\lambda_2(E_2 - E_7)/(1 - \lambda_1\lambda_2) \tag{45}$$

$$Q_1 = (E_7 - \lambda_1\lambda_2 E_2)/(1 - \lambda_1\lambda_2) \tag{46}$$

Table 3. Diatomic Fragments for BC ($g = 2^{-1/2}$)

m	$\beta\gamma$	β	γ	j	k	$(\mathrm{BC})_m$	$\mathbf{c}_{(\beta\gamma)}$													$\mathbf{b}_{\beta\gamma}$												
1	HF	H	F	1	1	$\lvert bz\rvert$	1													E_1												
2	HF	H	F	1	2	$\lvert b\bar{z}\rvert$		g	g										$g\lambda_2$		Q_2	J_2										$-J_3$
3	HF	H	F	2	1	$\lvert \bar{b}z\rvert$		$-g$	g										$-g\lambda_2$		J_2	Q_2										J_3
4	HF	H	F	2	2	$\lvert \bar{b}\bar{z}\rvert$				1													E_1									
5	HF	H	F	1	3	$\lvert bx\rvert$					1													E_5								
6	HF	H	F	1	4	$\lvert b\bar{x}\rvert$						g	g												Q_5	J_5						
7	HF	H	F	2	3	$\lvert \bar{b}x\rvert$						$-g$	g												J_5	Q_5						
8	HF	H	F	2	4	$\lvert \bar{b}\bar{x}\rvert$								1													E_5					
9	HF	H	F	1	5	$\lvert by\rvert$									1													E_5				
10	HF	H	F	1	6	$\lvert b\bar{y}\rvert$										g	g												Q_5	J_5		
11	HF	H	F	2	5	$\lvert \bar{b}y\rvert$										$-g$	g												J_5	Q_5		
12	HF	H	F	2	6	$\lvert \bar{b}\bar{y}\rvert$												1													E_5	
13	HF	H^+	F^-	1	1	$\lvert F^-\rvert$		λ_1											1		J_1	$-J_1$										Q_1
						$\mathrm{diag}(\mathbf{E}_{\mathrm{HF}}) =$	E_1	E_2	E_1	E_1	E_5	E_6	E_5	E_5	E_5	E_6	E_5	E_5	E_7													
1	HF^-	H	F^-	1	1	$\lvert bF^-\rvert$	1	0												E_-												
2	HF^-	H	F^-	2	1	$\lvert \bar{b}F^-\rvert$	0	1													E_-											
						$\mathrm{diag}(\mathbf{E}_{\mathrm{HF^-}}) =$	E_-	E_-																								

where $g = 2^{-1/2}$. The matrix $\tilde{\mathbf{b}}_{\alpha\gamma}$ is exactly the same as $\tilde{\mathbf{b}}_{\beta\gamma}$; the basis functions in Table 3 need only have a replacing b in order to represent the functions $AC_{(\alpha\gamma)n}$. Of course, the matrix $\tilde{\mathbf{b}}_{\beta\gamma}$ is evaluated at a distance $R_{HF} = R_{BC}$, whereas $\tilde{\mathbf{b}}_{\alpha\gamma}$ is evaluated at $R_{HF} = R_{AC}$. In general, all elements of the matrices $\tilde{\mathbf{b}}_{\alpha\gamma}$, $\tilde{\mathbf{b}}_{\beta\gamma}$, and $\tilde{\mathbf{b}}_{\alpha\beta}$ depend on internuclear distance. In the present example, the diatomic eigenvalues S, G, U, E_i and the numbers λ_1 and λ_2 are all functions of internuclear distance.

The matrices $\tilde{\mathbf{b}}$ in Tables 2 and 3 are constructed from $\mathbf{c}$ and $\mathbf{E}$ via equation (19). The matrices $\mathbf{c}$ would normally come from diagonalizing the diatomic H matrix in the basis shown and they would satisfy equation (18). The matrix $\mathbf{c}_{HF}$ in Table 3 is written in an unnormalized form [it does not satisfy (18)] for ease of display. This causes no difficulty since the matrix $\mathbf{b}_{HF}$, by virtue of the definition in equation (19), is invariant to the normalization of $\mathbf{c}_{HF}$. Notice that many of the elements of $\mathbf{c}$ arise from symmetry or through angular momentum coupling[39] (the numbers g and 1); other elements (the numbers λ_1 and λ_2) arise because more than one valence-bond structure has been used to describe a particular diatomic state (in this case the $^1\Sigma^+$ states of HF). If this configuration interaction were absent, the matrices $\mathbf{c}$ could be written down without doing a valence-bond calculation on the diatomic systems. The procedure for doing this is described in Section 2.4.

2.3.4. Compute the Rotated Fragment Matrices

The atomic fragment matrices for this problem are diagonal and invariant to rotation; the diatomic matrices are not. The matrix $\tilde{\mathbf{b}}_{\alpha\gamma}$ may be used directly, since atoms A and C lie on the z axis, which is the orientation used in constructing $\tilde{\mathbf{b}}_{\alpha\gamma}$, i.e.,

$$\mathbf{b}_{\alpha\gamma} = \tilde{\mathbf{b}}_{\alpha\gamma} \tag{47}$$

The matrix $\mathbf{b}_{\beta\gamma}$, however, must be obtained from $\tilde{\mathbf{b}}_{\beta\gamma}$ via equation (23). The matrix $\mathbf{b}_{HF^-} = \tilde{\mathbf{b}}_{HF^-}$; the rotation matrix $\mathbf{R}_{HF}$ is obtained from $\mathbf{R}_F$ [equation (39)] using Table 3 and equation (20):

$$R_{HF}^{(m,m')} = R_F^{(k,k')}\delta_{jj'} \tag{48}$$

For example, to find element $R_{HF}^{(2,6)}$, look in Table 3 at the rows $m = 2$ and $m' = 6$ from which one sees $k = 2$, $j = 1$, $k' = 4$, $j' = 1$; $\delta_{jj'} = 1$ so that the required element is $R_F^{(2,4)} = -\sin\theta$. The matrix $\mathbf{R}_{HF}$ is as follows: the diagonal elements (1, 1) to (8, 8) equal $\cos\theta$. The diagonal elements (9, 9) to (13, 13) equal 1; the off-diagonal elements (5, 1), (6, 2), (7, 3), and (8, 4) equal $\sin\theta$, and the elements (1, 5), (2, 6), (3, 7), and (4, 8) equal $-\sin\theta$; the other elements are zero. The matrix $\mathbf{b}_{HF}^{(BC)}$ is given in Table 4.

Table 4. Nonzero Elements of $\mathbf{b}_{\mathrm{HF}}^{(\mathrm{BC})}$ ($s = \sin\theta$, $c = \cos\theta$)

Elements	Symbol	Value	Elements	Symbol	Value
(1, 1), (4, 4)	b^{11}	$E_1c^2 + E_5s^2$	(6, 6), (7, 7)	b^{66}	$Q_2s^2 + Q_5c^2$
(1, 5), (5, 1), (4, 8), (8, 4)	b^{15}	$(E_1 - E_5)sc$	(6, 7), (7, 6)	b^{67}	$J_2s^2 + J_5c^2$
(2, 2), (3, 3)	b^{22}	$Q_2c^2 + Q_5s^2$	(6, 13)	$-J_3s$	$-J_3s$
(2, 3), (3, 2)	b^{23}	$J_2c^2 + J_5s^2$	(7, 13)	J_3s	J_3s
(2, 6), (6, 2), (3, 7), (7, 3)	b^{26}	$(Q_2 - Q_5)sc$	(9, 9), (12, 12)	E_5	E_5
(2, 7), (7, 2), (3, 6), (6, 3)	b^{27}	$(J_2 - J_5)sc$	(10, 10), (11, 11)	Q_5	Q_5
(2, 13)	$-J_3c$	$-J_3c$	(10, 11), (11, 10)	J_5	J_5
(3, 13)	J_3c	J_3c	(13, 2)	J_1c	J_1c
(5, 5), (8, 8)	b^{55}	$E_1s^2 + E_5c^2$	(13, 3)	$-J_1c$	$-J_1c$
			(13, 6)	J_1s	J_1s
			(13, 7)	$-J_1s$	$-J_1s$
			(13, 13)	Q_1	Q_1

2.3.5. *Construct the Triatomic Basis*

This basis is the triple direct product of the atomic basis functions for A, B, and C, and consists of $M_{\mathrm{H}}^{(\mathrm{A})} \times M_{\mathrm{H}}^{(\mathrm{B})} \times M_{\mathrm{F}}^{(\mathrm{C})} + M_{\mathrm{H}}^{(\mathrm{A})} \times M_{\mathrm{H}^-}^{(\mathrm{B})} \times M_{\mathrm{F}^-}^{(\mathrm{C})} + M_{\mathrm{H}^+}^{(\mathrm{A})} \times M_{\mathrm{H}}^{(\mathrm{B})} \times M_{\mathrm{F}^-}^{(\mathrm{C})} = 2 \times 2 \times 6 + 2 \times 1 \times 1 + 1 \times 2 \times 1 = 28$ functions. We could use all of these functions, but a considerable effort can be saved by recognizing that the DIM **H** matrix in this basis has many degenerate eigenvalues, since all of the spin components of S^2 differing only in M_s have the same energy (i.e., we are neglecting spin–orbit coupling). The triatomic H_2F states corresponding to the basis are the ${}^4A'$, ${}^4A''$, ${}^2A'$, and ${}^2A''$ states. The possible M_s values are $\frac{3}{2}, \frac{1}{2}, -\frac{1}{2}, -\frac{3}{2}$. We are therefore

Table 5. Triatomic Basis and Genealogy

Triatomic functions				(AB) C				(AC) B				(BC) A			
I	Φ_I	M_s	σ_h	$\alpha\beta$	γ	l	k	$\alpha\gamma$	β	n	j	$\beta\gamma$	α	m	i
1	$\|ab\bar{z}\|$	1/2	A'	H_2	F	1	2	HF	H	2	1	HF	H	2	1
2	$\|a\bar{b}z\|$	1/2	A'	H_2	F	3	1	HF	H	1	2	HF	H	3	1
3	$\|\bar{a}bz\|$	1/2	A'	H_2	F	2	1	HF	H	3	1	HF	H	1	2
4	$\|ab\bar{x}\|$	1/2	A'	H_2	F	1	4	HF	H	6	1	HF	H	6	1
5	$\|a\bar{b}x\|$	1/2	A'	H_2	F	3	3	HF	H	5	2	HF	H	7	1
6	$\|\bar{a}bx\|$	1/2	A'	H_2	F	2	3	HF	H	7	1	HF	H	5	2
7	$\|aF^-\|$	1/2	A'	H_2^+	F^-	1	1	HF^-	H^+	1	1	HF	H	13	1
8	$\|bF^-\|$	1/2	A'	H_2^+	F^-	2	1	HF	H	13	1	HF^-	H^+	1	1
9	$\|ab\bar{y}\|$	1/2	A''	H_2	F	1	6	HF	H	10	1	HF	H	10	1
10	$\|a\bar{b}y\|$	1/2	A''	H_2	F	3	5	HF	H	9	2	HF	H	11	1
11	$\|\bar{a}by\|$	1/2	A''	H_2	F	2	5	HF	H	11	1	HF	H	9	2
12	$\|abz\|$	3/2	A'	H_2	F	1	1	HF	H	1	1	HF	H	1	1
13	$\|abx\|$	3/2	A'	H_2	F	1	3	HF	H	5	1	HF	H	5	1
14	$\|aby\|$	3/2	A''	H_2	F	1	5	HF	H	9	1	HF	H	9	1

free to consider only those functions with $M_s > 0$ and can thereby take 14 of the 28 functions. These functions together with their diatomic and atomic genealogy and their symmetry and M_s values are listed in Table 5.

2.3.6. *Construct the Atomic Matrices* **B**

The matrix $\mathbf{B}_A$ is constructed from equation (27) and $\mathbf{B}_B$ and $\mathbf{B}_C$ from similar equations with the help of Table 5. Consider, for example, element $B_C^{(2,12)}$: $I' = 2, I = 12$. From rows 2 and 12 of Table 5, $\alpha' = \alpha = H = \beta' = \beta$, $\gamma' = \gamma = F$, $i' = i = 1$, $j' = 2$, $k' = k = 12$; hence, because $j' \neq j$, $B_C^{(2,12)} = 0$ even though $b_H^{(12,12)} \neq 0$. A cursory inspection of Table 5 shows that $\mathbf{B}_A$, $\mathbf{B}_B$, and $\mathbf{B}_C$ are diagonal. $\mathbf{B}_B$ has diagonal elements E_H except for $B_B^{(7,7)} = E_{H^+}$; $\mathbf{B}_A$ is the same as $\mathbf{B}_B$ but with elements (7, 7) and (8, 8) interchanged. $\mathbf{B}_C$ has E_F for all diagonal elements except $B_C^{(7,7)}$ and $B_C^{(8,8)}$ which are both equal to E_{F^-}.

2.3.7. *Construct the Diatomic Matrices* **B**

These are built-up from equation (31). For example, to find $B_{BC}^{(7,2)}$ look at rows 7 and 2 of Table 5 under the columns headed by (BC)A. $I' = 7$, $I = 2, i' = i = 1\,(\delta_{ii'} = 1), m' = 13, m = 2$. Under columns γ and β it can be seen that there is a charge transfer $\Delta q = 1$ in going from function $I = 2$ to $I = 7$. The number of electrons between functions A and C in the standard order (ABC) is $n_{B\beta} = n_{BH} = 1$; i.e., $\Delta e = 1$ so that the phase factor from equation (30) is $P = -1$. Therefore

$$B_{BC}^{(7,2)} = -b_{\beta\gamma}^{(m'm)} = -b_{HF}^{(13,2)} = -J_1c \tag{49}$$

where $b_{HF}^{(13,2)}$ was obtained from Table 4.

The matrix $\mathbf{B}_{AB}$ is block diagonal consisting of six 1×1 blocks of value T for elements (1, 1), (4, 4), (9, 9), (12, 12), (13, 13), and (14, 14); three 2×2 blocks the same as the 2–3 block of $\mathbf{b}_{H_2}$ in Table 2 for the blocks 2–3, 5–6, and 10–11; and one 2×2 block the same as the 1–2 block of $\mathbf{b}_{H_2^+}$ in Table 2 for the block 7–8. All other elements are zero. The matrices $\mathbf{B}_{AC}$ and $\mathbf{B}_{BC}$ are given in Tables 6 and 7.

2.3.8. *Find the DIM Eigenvalues*

The DIM eigenvalues are found from the **B** matrix via equation (26), which in the FH_2 example reduces to

$$\mathbf{B} = \mathbf{B}_{AB} + \mathbf{B}_{BC} + \mathbf{B}_{AC} - \mathbf{B}_A - \mathbf{B}_B - \mathbf{B}_C \tag{50}$$

In calculating the energies needed to define the elements of **B** it is convenient to set the energy of a particular atomic species equal to zero for each atom occurring in the polyatomic system. Here, $E_H = E_F = 0$ so that E_{H^+} and

Table 6. DIM Matrix $\mathbf{B}_{BC}$ (Symbols as in Table 4)

	1	2	3	4	5	6	7	8	9	10	11	12	13	14
1	b^{22}	b^{23}		b^{26}	b^{27}		$-J_3c$							
2	b^{23}	b^{22}		b^{27}	b^{26}		J_3c							
3			b^{11}			b^{15}								
4	b^{26}	b^{27}		b^{66}	b^{67}		$-J_3s$							
5	b^{27}	b^{26}		b^{67}	b^{66}		J_3s							
6			b^{15}			b^{55}								
7	J_1c	$-J_1c$		J_1s	$-J_1s$		Q_1							
8								E_-						
9									Q_5	J_5				
10									J_5	Q_5				
11											E_5			
12												b^{11}	b^{15}	
13												b^{15}	b^{55}	
14														E_5

Table 7. DIM Matrix $\mathbf{B}_{AC}$ [Symbols as in Equations (40)–(46)]

	1	2	3	4	5	6	7	8	9	10	11	12	13	14
1	Q_2		J_2					J_3						
2		E_1												
3	J_2		Q_2					$-J_3$						
4				Q_5		J_5								
5					E_5									
6				J_5		Q_5								
7							E_-							
8	$-J_1$		J_1					Q_1						
9									Q_5		J_5			
10										E_5				
11									J_5		Q_5			
12												E_1		
13													E_5	
14														E_5

E_{F^-} are the ionization potential of H and the negative of the electron affinity of F, respectively. The diatomic energies are found from the sum of the atomic energies to which they dissociate plus the atom–atom interaction, V. For example, for H_2^+ in the $^2\Sigma_u^+$ state,

$$U = E_H + E_{H^+} + V_u(R_{AB}) \tag{51}$$

Because of the elements involving J_1 and J_3, the matrix **B** is non-Hermitian, so a suitable method must be used in diagonalizing it. The book by Wilkinson[38] has information on methods to do this.

If the eigenvalues of **B** are required very often, as is usual in collision problems, then it is convenient to reduce the 14×14 matrix to block diagonal form. Inspection of the columns headed M_s and σ_h in Table 5 shows that the **B** matrix ought to block off into $^4A'$, $^4A''$, $^2A'$, and $^2A''$ parts. This can be done by transforming the triatomic basis via equation (33) to yield eigenfunctions of the total spin operator. A coupling scheme is used which couples the spin of A to that of B and then the result to C:

$$\begin{aligned} |(s_A, s_B)\, S_{AB}, s_C; S, M\rangle = \sum_{m_A, m_B, m_C} & |s_A m_A s_B m_B s_C m_C\rangle \\ & \times \langle s_A m_A s_B m_B \,|\, s_A s_B S_{AB} M_{AB}\rangle \\ & \times \langle S_{AB} M_{AB} s_C m_C \,|\, S_{AB} s_C S M\rangle \end{aligned} \tag{52}$$

The kets on the right-hand-side are the triatomic functions in Table 5 for which $s_A = s_B = s_C = 1/2$. The transformation **T** which results is shown in Table 8, where $a = 3^{-1/2}$ and $g = 2^{-1/2}$, the matrix elements having been obtained from tables of vector-coupling coefficients.[38,40] The result of applying this transformation to **B** via equation (36) is a block-diagonal matrix whose structure is in Figure 1. The elements of the various blocks are listed below where the following conventions are used to save space: (i) $E_H = E_F = 0$; (ii) a symbol with a bar over it is evaluated at internuclear distance R_{AC}, those without the bar at R_{BC} if they refer to HF or HF^-, or at R_{AB} if they refer to H_2 or H_2^+; (iii) the definition for Q_3 is

$$Q_3 = (E_2 - \lambda_1\lambda_2 E_7)/(1 - \lambda_1\lambda_2) \tag{53}$$

$^4A''(1 \times 1)$:

$$T + E_5 + \bar{E}_5$$

$^4A'(2 \times 2)$:

$$(1, 1) = T + E_1 c^2 + E_5 s^2 + \bar{E}_1$$

$$(1, 2) = (2, 1) = (E_1 - E_5)\, sc$$

$$(2, 2) = T + E_1 s^2 + E_5 c^2 + \bar{E}_5$$

Table 8. Triatomic Basis Transformation T

	1	2	3	4	5	6	7	8	9	10	11	12	13	14
1	a			a/g										
2	a		g	$-ag$										
3	a		$-g$	$-ag$										
4		a				a/g								
5		a			g	$-ag$								
6		a			$-g$	$-ag$								
7							1							
8								1						
9											a		a/g	
10											a	g	$-ag$	
11											a	$-g$	$-ag$	
12									1					
13										1				
14														1

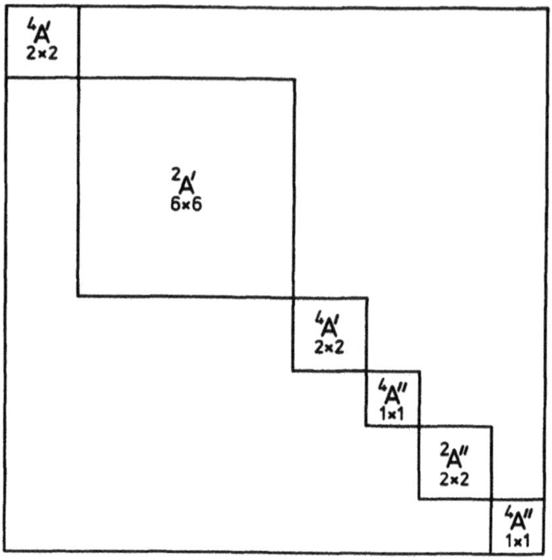

Figure 1. Schematic diagram of the block-diagonal **B** matrix for DIM calculation of FH_2 eigenvalues. Each block is labeled by the spin multiplicity, by the irreducible representation of the C_s group to which it corresponds, and by its size.

$^2A'(2 \times 2)$:

$$(1,1) = S + (3E_5 + E_6 + 3\bar{E}_5 + \bar{E}_6)/4$$
$$(1,2) = (2,1) = (E_5 - E_6 + \bar{E}_6 - \bar{E}_5)/(4a)$$
$$(2,2) = T + (E_5 + 3E_6 + \bar{E}_5 + 3\bar{E}_6)/4$$

$^2A'(6 \times 6)$:

$$(1,1) = S + [c^2(3E_1 + Q_3) + s^2(3E_5 + E_6) + 3\bar{E}_1 + \bar{Q}_3]/4$$
$$(1,2) = (2,1) = [c^2(E_1 - Q_3) + s^2(E_5 - E_6) - \bar{E}_1 + \bar{Q}_3]/(4a)$$
$$(1,3) = (3,1) = sc[3(E_1 - E_5) + Q_3 - E_6]/4$$
$$(1,4) = (4,1) = sc(E_1 - Q_3 - E_5 + E_6)/(4a)$$
$$(1,5) = gcJ_3$$
$$(1,6) = g\bar{J}_3$$
$$(2,2) = T + [c^2(E_1 + 3Q_3) + s^2(E_5 + 3E_6) + \bar{E}_1 + 3\bar{Q}_3]/4$$
$$(2,3) = (1,4) = (3,2)$$
$$(2,4) = (4,2) = sc[(E_1 - E_5) + 3(Q_3 - E_6)]/4$$
$$(2,5) = -gcJ_3/a$$
$$(2,6) = g\bar{J}_3/a$$

$$(3,3) = S + [s^2(3E_1 + Q_3) + c^2(3E_5 + E_6) + 3\bar{E}_5 + \bar{E}_6]/4$$
$$(3,4) = (4,3) = [s^2(E_1 - Q_3) + c^2(E_5 - E_6) - (\bar{E}_6 - \bar{E}_5)]/(4a)$$
$$(3,5) = gsJ_3$$
$$(3,6) = (6,3) = (4,6) = (6,4) = 0$$
$$(4,4) = T + [s^2(E_1 + 3Q_3) + c^2(E_5 + 3E_6) + \bar{E}_5 + 3\bar{E}_6]/4$$
$$(4,5) = -gsJ_3/a$$
$$(5,1) = -gcJ_1$$
$$(5,2) = gcJ_1/a$$
$$(5,3) = -gsJ_1$$
$$(5,4) = gsJ_1/a$$
$$(5,5) = (G + U)/2 + Q_1 + \bar{E}_- - E_{H^+} - E_{F^-}$$
$$(5,6) = (6,5) = (G - U)/2$$
$$(6,1) = -g\bar{J}_1$$
$$(6,2) = -g\bar{J}_1/a$$
$$(6,6) = (G + U)/2 + E_- + \bar{Q}_1 - E_{H^+} - E_{F^-}$$

The DIM eigenvalues for the $^4A'$ and $^2A''$ states of H_2F can be written down immediately:

$$E_\pm(^4A') = T + (E_1 + E_5 + \bar{E}_1 + \bar{E}_5)/2 \pm [(E_1 - E_5)^2 + (\bar{E}_1 - \bar{E}_5)^2 + 2(E_1 - E_5)(\bar{E}_1 - \bar{E}_5)\cos 2\theta]^{1/2}/2 \tag{54}$$

$$E_\pm(^2A'') = (S + T + E_5 + E_6 + \bar{E}_5 + \bar{E}_6)/2 \pm [(S - T)^2 + (E_5 - E_6)^2 + (\bar{E}_5 - \bar{E}_6)^2 + (T - S)(E_5 - E_6) - (T - S)(\bar{E}_5 - \bar{E}_6) - (E_5 - E_6)(\bar{E}_5 - \bar{E}_6)]^{1/2}/2 \tag{55}$$

Equation (55) is just the London equation,[33] which also results whenever three ^2S states are coupled to give a doublet state.

2.4. Simple Systems: An Alternative Formulation

Many PE (potential energy) surfaces used in collision theory are the result of diagonalizing a 2 × 2 matrix, as for example equations (54) and (55). Another example is obtained from the FH_2 system if the configuration interaction parameters λ_1 and λ_2 are set to zero, blocking off the $^2A'$ part of the **B** matrix into a 4 × 4 covalent part and a 2 × 2 ionic part. The ionic part arises whenever a singlet atom (here F^-) is coupled to a doublet atom (here H) and its singlet ion (H^+). The eigenvalues of the $F^-H_2^+$ system are[41]

$$E_\pm = [(G + U) + (E_7 + E_-) + (\bar{E}_7 + \bar{E}_-) - 2(E_H - E_{H^+} - E_{F^-})]/2$$
$$\pm[(G - U)^2 + (E_7 - E_-)^2 + (\bar{E}_7 - \bar{E}_-)^2 - 2(E_7 - E_-)(\bar{E}_7 - \bar{E}_-)]^{1/2}/2 \tag{56}$$

In general, if there are no configuration interaction parameters, then the elements of the **c** matrices can usually be obtained by symmetry or from spin-coupling alone. Examples of such systems are H_3, H_3^+, $Ne_2H_2^+$, HeH_3^+, and FH_2 with no ionic structures. The spin-recoupling techniques[17] that one can use to obtain the DIM **B** matrix in these cases are considered here. These provide an *alternative* method for simpler systems.

Each assignment of atomic species to A, B, C, ... defines a separate subset of basis functions. This subset corresponds to the spin coupling of a standard pair group

$$G_0 = \{(A, B)(C, D) \cdots (X, Y)Z\} \tag{57}$$

in which the spins of the designated pairs are first coupled to each other before further coupling of all spins together. This means that the basis functions may be considered (in the DIM approximation) as eigenfunctions of the spins $s_{AB}, s_{CD}, \ldots, s_{XY}, s_Z$. Other groups G_i can be defined such that in the sequence $G_0, G_1, G_2, \ldots$, no pair appearing in G_j will have appeared in any G_i for $i < j$. The DIM **B** matrix may then be written as

$$\mathbf{B} = \sum_{ij} \mathbf{B}_{ij}^{(0)} + \sum_{ij} \mathbf{B}_{ij}^{(1)} + \cdots \tag{58}$$

where the first sum is over all pairs in the pair-group G_0, the second sum over all pairs in the group G_1, etc. Each matrix $\mathbf{B}_{ij}^{(k)}$ then has the form

$$\mathbf{B}_{ij}^{(k)} = \mathbf{R}_i\mathbf{R}_j\mathbf{U}_k\mathbf{H}_{ij}\mathbf{U}_k^{-1}\mathbf{R}_j^{-1}\mathbf{R}_i^{-1} \tag{59}$$

where U_k is a matrix of spin-recoupling coefficients[39] which relates the spin-coupling G_k to G_0:

$$|\mathbf{G}_k\rangle = |\mathbf{G}_0\rangle\,\mathbf{U}_k \tag{60}$$

The matrix $\mathbf{H}_{ij}$ is the diatomic **H** matrix and in general will have the form $\mathbf{C}_{ij}\mathbf{E}_{ij}\mathbf{C}_{ij}^{-1}$, where $\mathbf{E}_{ij}$ is diagonal and $\mathbf{C}_{ij}$ mixes the functions corresponding to the different species assignments.

As an example, consider the system H_3^+. The three species assignments are HHH^+, HH^+H, H^+HH. Take the pair group G_0 as $[(A, B)C]$, G_1 as $[A, (B, C)]$, and G_2 as $[B, (A, C)]$. The angular momentum states corresponding to these are $|(s_As_B)S_{AB}, s_C; S\rangle$, $|s_A, (s_Bs_C)S_{BC}; S\rangle$, and $|s_B, (s_As_C)S_{AC}; S\rangle$.

Indexing the vectors by S_{AB}, S_{BC}, S_{AC}, respectively, the transformations of equation (60) are

$$|G_1, S_{BC}\rangle = \sum_{S_{AB}} |G_0, S_{AB}\rangle\, U_1(S_{AB}; S_{BC}) \tag{61}$$

$$|G_2, S_{AC}\rangle = \sum_{S_{AB}} |G_0, S_{AB}\rangle\, U_2(S_{AB}; S_{AC}) \tag{62}$$

i.e.,

$$\begin{aligned} U_1(S_{AB}; S_{BC}) &= \langle (s_A s_B) S_{AB}, s_C; S \,|\, s_A, (s_B s_C) S_{BC}; S\rangle \\ &= (-1)^{s_A + s_B + s_C - S}[(2S_{AB}+1)(2S_{BC}+1)]^{1/2} \\ &\quad \times \begin{Bmatrix} s_A & s_B & S_{AB} \\ s_C & S & S_{BC} \end{Bmatrix} \end{aligned} \tag{63}$$

$$\begin{aligned} U_2(S_{AB}; S_{AC}) &= \langle (s_A s_B) S_{AB}, s_C; S | s_B, (s_A s_C) S_{AC}; S\rangle \\ &= (-1)^{s_A + s_B - S_{AB}} \langle (s_B s_A) S_{AB}, s_C; S | s_B, (s_A s_C) S_{AC}; S\rangle \\ &= (-1)^{2s_A + 2s_B + s_C - S - S_{AB}} \\ &\quad \times [(2S_{AB}+1)(2S_{AC}+1)]^{1/2} \begin{Bmatrix} s_B & s_A & S_{AB} \\ s_C & S & S_{AC} \end{Bmatrix} \\ &= (-1)^{s_A + s_B - S_{AB}}\, U_1(S_{AB}; S_{AC}) \end{aligned} \tag{64}$$

The quantities in braces are the 6-j coefficients.[37,39]

For H_3^+ the spin recoupling is trivial, each species assignment giving the 1×1 unit matrix; no rotation of the basis is required either. The only thing that couples the functions are the matrices $\mathbf{C}_{AB}$, $\mathbf{C}_{BC}$, and $\mathbf{C}_{AC}$, which have the form

$$\mathbf{C}_{AB} = \begin{bmatrix} 1 & & \\ & g & g \\ & g & -g \end{bmatrix}, \quad \mathbf{C}_{BC} = \begin{bmatrix} g & g & \\ g & -g & \\ & & 1 \end{bmatrix}, \quad \mathbf{C}_{AC} = \begin{bmatrix} g & & g \\ & 1 & \\ g & & -g \end{bmatrix}$$

The energy matrices are diagonal: $\mathbf{E}_{AB} = (E, G, U)$, $\mathbf{E}_{BC} = (G, U, E)$, $\mathbf{E}_{AC} = (G, E, U)$, $\mathbf{E}_A = (0, 0, E_{H^+})$, $\mathbf{E}_B = (0, E_{H^+}, 0)$, $\mathbf{E}_C = (E_{H^+}, 0, 0)$, taking $E_H = 0$. Letting $Q = 0.5(G + U)$, $J = 0.5(G - U)$, the $\mathbf{B}$ matrix is

$$\begin{aligned} \mathbf{B}_{11} &= E_{AB} + Q_{BC} + Q_{AC} - E_{H^+} & \mathbf{B}_{22} &= Q_{AB} + Q_{BC} + E_{AC} \\ \mathbf{B}_{12} &= \mathbf{B}_{21} = J_{BC} & \mathbf{B}_{23} &= \mathbf{B}_{32} = J_{AB} \\ \mathbf{B}_{13} &= \mathbf{B}_{31} = J_{AC} & \mathbf{B}_{33} &= Q_{AB} + E_{BC} + Q_{AC} \end{aligned}$$

Here E is the $H_2(X\,^1\Sigma_g^+)$ energy or the $H_2(^3\Sigma_u^+)$ energy depending on whether the $^1A'$ or $^3A'$ state of H_3^+ is desired.

As a final simple system, consider coupling two doublet atoms A and B to the triplet atom C (the same spin scheme found in H_2O) to form the triplet state of ABC. Here the matrices **C** are unity and no rotation is required. The basis functions ϕ_1 and ϕ_2 are taken to correspond to the triplet and singlet states of AB and we need the transformation to the quartet and doublet states of AC and BC. The spin-recoupling matrices are then

$$\mathbf{U}_1 = \begin{bmatrix} a & a/g \\ a/g & -a \end{bmatrix} \qquad \mathbf{U}_2 = \begin{bmatrix} a & +a/g \\ -a/g & +a \end{bmatrix} \tag{65}$$

Denoting the quartet and doublet interaction energies by V_Q and V_D, and the singlet and triplet interactions by V_S and V_T, the diatomic energy matrices are $\mathbf{E}_{AB} = (V_Q, V_D)$, $\mathbf{E}_{BC} = (V_T, V_S)$, and $\mathbf{E}_{AC} = (\bar{V}_Q, \bar{V}_D)$. Formation of the **B** matrix by equation (59) and diagonalization then produces the eigenvalues for the polyatomic triplet state:

$$E_{\pm} = Q_{AB} + Q_{BC} + Q_{AC} \pm [J_{AB}^2 + J_{BC}^2 + J_{AC}^2 - \tfrac{2}{3}(J_{BC} + J_{AC})J_{AB} - \tfrac{14}{9}J_{BC}J_{AC}]^{1/2} \tag{66}$$

where the Q's and J's are $Q_{AB} = 0.5(V_S + V_T)$, $J_{AB} = 0.5(V_T - V_S)$, $Q_{BC} = 0.5(V_Q + V_D)$, $J_{BC} = 0.5(V_Q - V_D)$, $Q_{AC} = 0.5(\bar{V}_Q + \bar{V}_D)$, $J_{AC} = 0.5(\bar{V}_Q - \bar{V}_D)$.

Equations for handling 4- and 5-atom systems by this method are given by Eaker and Parr,[17] and a general treatment of spin recoupling applied to DIM is given by Pickup.[19]

2.5. Coupling

2.5.1. Spin–Orbit Coupling

The treatment of spin–orbit coupling can be included in the general formulation above through the addition of the spin–orbit terms in the atomic and diatomic Hamiltonians used in computing the fragment matrices. These terms involve only one- and two-body operators so that the DIM partitioning of the Hamiltonian [equation (1)] applies to them as well. The only difference to the formulation given above is that the atomic fragment matrices would no longer be diagonal in the basis chosen.

In some simple systems it is possible with angular momentum coupling alone to construct unambiguously an atomic basis set whose functions are eigenvectors of $\underset{\sim}{S}^2$, $\underset{\sim}{L}^2$, $\underset{\sim}{J}^2$, and $\underset{\sim}{J}_z$. An example is the FH_2 system without the ionic interactions. The atomic fragment matrices will be diagonal in this new basis, and the effect of spin–orbit coupling can be introduced by using atom energies which have the spin–orbit splittings present. The two-body terms in the spin–orbit Hamiltonian can then be neglected, as suggested by Tully,[14] and the diatomic energies adjusted to go to separated-atom limits which include the spin–orbit coupling. This provides an empirical way of introducing spin–orbit coupling into the polyatomic system. This procedure would not be so easy to implement if the atomic fragment matrices could not be diagonalized by angular momentum coupling alone. A calculation on FH_2 including the required basis set transformation is described in a paper by Tully,[14] so will not be reproduced here.

2.5.2. Nonadiabatic Coupling

In many collision processes it is necessary to include coupling between two adiabatic states which is brought about through the motion of the atomic nuclei. The dominant coupling[3,14] is brought about by the matrix $\langle \boldsymbol{\Psi} \mid \nabla \boldsymbol{\Psi} \rangle$. ∇ is the gradient operator with respect to the nuclear coordinates, and $\boldsymbol{\Psi}$ are the DIM polyatomic eigenvectors (row vectors) given in terms of the matrix $\mathbf{C}$ of equation (2):

$$\boldsymbol{\Psi} = \phi \mathbf{C} \tag{67}$$

In the DIM approach one further assumes that

$$\nabla \boldsymbol{\Psi} = \phi \nabla \mathbf{C} \tag{68}$$

$$\begin{aligned} \langle \boldsymbol{\Psi} \mid \nabla \boldsymbol{\Psi} \rangle &= \mathbf{C}^{\dagger} \mathbf{S} \nabla \mathbf{C} \\ &= \mathbf{C}^{-1} \nabla \mathbf{C} \end{aligned} \tag{69}$$

Denoting the ith column of $\mathbf{C}$ by $\mathbf{C}_i$, the coupling between states i and j is

$$\langle \boldsymbol{\Psi}_i \mid \nabla \boldsymbol{\Psi}_j \rangle = (\mathbf{C}^{-1})_i^{\dagger} \nabla \mathbf{C}_j \tag{70}$$

To evaluate this, start with the expressions

$$\mathbf{B} \mathbf{C}_j = \mathbf{S}^{-1} \mathbf{H} \mathbf{C}_j = E_j \mathbf{C}_j \tag{71}$$

$$(\mathbf{C}^{-1})_i^{\dagger} \mathbf{B} = E_i (\mathbf{C}^{-1})_i^{\dagger} \tag{72}$$

$$(\mathbf{C}^{-1})_i^{\dagger} \mathbf{B} \mathbf{C}_j = E_i \delta_{ij} \tag{73}$$

Differentiation of (73) followed by application of (71) and (72) and the condition $(\mathbf{C}^{-1})_i \mathbf{C}_j = \delta_{ij}$ yields

$$E_i = (\mathbf{C}^{-1})_i^{\dagger} \nabla \mathbf{B} \mathbf{C}_i \tag{74}$$

$$(\mathbf{C}^{-1})_i^{\dagger} \nabla \mathbf{C}_j = (\mathbf{C}^{-1})_i^{\dagger} \nabla \mathbf{B} \mathbf{C}_j / (E_j - E_i) \tag{75}$$

Hence, both the derivatives of the eigenvalues and the nonadiabatic coupling can be obtained directly from the derivatives of $\mathbf{B}$ and the eigenvectors. This is particularly advantageous for classical trajectory surface-hopping calculations.[3,34,35]

3. Methods Related to DIM

In closing I should like to comment on a few semiempirical methods which have enjoyed considerable application to collision problems. The most important of these are based on valence-bond theories and are implemented in the same spirit that DIM is applied: polyatomic potential surfaces are obtained from diatomic and atomic input energies.

3.1. The LEPS Method[4]

This method is based on the London expression[33] [equation (55)] and requires as input two potential curves (usually one bound and one repulsive) for each diatomic pair. In the conventional method for ABC, the three attractive curves ($i = 1, 2, 3$) are represented by Morse functions specified by three parameters, D_i, β_i, R_i^0:

$$V_i(R) = D_i x_i(x_i - 2) \tag{76}$$

$$x_i = \exp -\beta_i(R - R_i^0) \tag{77}$$

The repulsive curves are represented by functions similar to (76) but with the minus sign changed to a plus sign and with a value D_i' substituted for D_i. The parameters D_i, β_i, and R_i^0 ($i = 1, 2, 3$) *must* be chosen to describe the diatomic molecules AB, BC, and AC correctly. The parameters D_i' are then obtained empirically, often being used to fit the triatomic potential. The parameter D_i' is related to the Sato parameter s_i, which appears in the older literature by

$$D_i' = D_i[(1 - s_i)/(1 + s_i)]/2 \tag{78}$$

The LEPS equation is still useful for trajectory calculations and can be used for 4-atom reactions as well.

3.2. Method of Blais and Truhlar[22]

These authors used valence-bond arguments to develop a DIM-like theory for FH_2 which explicitly took into account the directional character of the *p* orbitals on F. They obtained a 2×2 secular equation involving the singlet and triplet Σ and Π states of HF and an angle specifying the orientation of the *p* orbitals. To apply the method they minimized the total energy of the FH_2 molecule with respect to the orbital orientation.

The energy expression they derived is easily shown to result from neglect of off-diagonal elements of the DIM **B** matrix for this problem.[21] In general, these elements will not be small. Also, although their secular equation is very simple, it is necessary to minimize the energy with respect to the orbital angle at every molecular configuration, which makes the method more cumbersome to use than the original 4×4 DIM secular equations.

3.3. *Valence-Bond Methods*

3.3.1. *Porter–Karplus Surface for H_3*[28]

This method starts with valence-bond theory and also eliminates integrals through their relation to the energies of H_2 in the $X\ {}^1\Sigma_g^+$ and ${}^3\Sigma_u^+$ states. However, it introduces orbital overlap and parametrizations of some integrals. It is more flexible than the simplest DIM treatment of H_3 and may offer useful representations of M_3 systems, but it has so far not been generalized. Pederson and Porter[29] give an extended version of this method which takes explicit account of three-center integrals. Derivatives of the potential in this extension must be computed numerically, which might limit somewhat the applicability of the method to classical trajectories.

3.3.2. *Valence-Bond Methods with Transferable Parameters*

A step in the direction of obtaining a set of consistent semiempirical parameters for a chemical system was taken by Raff *et al.*[23] in a calculation of potential surfaces for H_2I_2. They use the DIM expression for H_4 in terms of singlet and triplet energies for each pair of atoms. The singlet curves for all diatomics are represented by Morse functions with appropriate spectroscopic parameters. The triplet curves for HI and I_2 are expressed in terms of overlap integrals, an inverse R term and atomic ionization potentials, electron affinities, and a core-repulsion integral for each atom. These parameters are kept fixed for calculations of three-atom surfaces H_2I, HI_2. This approach has been used in dynamical calculations on several systems.[24-26]

Having transferable parameters in semiempirical theories for collision processes is a useful goal. Tully has outlined a program to achieve this with DIM by computing standard atomic and diatomic fragment matrices to be used in all calculation.[16] No proper assessment of semiempirical methods for collision processes can be done until this program has been carried out.

3.4. *Simple Approach to Nonadiabatic Coupling*

In systems where nonadiabatic effects are essential to the collision process, a 2×2 secular equation is set up with parametrized matrix elements.[10,11] The eigenvalues are written as

$$E_{\pm} = 0.5(H_{11} + H_{22}) \pm 0.5[(H_{11} - H_{22})^2 + 4H_{12}^2]^{1/2} \qquad (79)$$

The diagonal elements H_{11} and H_{22} are diabatic potential energy surfaces and H_{12} is an interaction element. An example is the collision of alkali atoms with halogens to form ions $K + Br_2 \rightarrow KBr + Br$ or $\rightarrow K^+ + Br_2^-$. These processes start off on a covalent surface H_{11} and finish on an ionic surface H_{22}. Equation (79) is very useful as a first approach since it behaves in the correct manner in the region of avoided crossings, an advantage when doing surface-hopping calculations. It is much simpler to implement than a DIM calculation but is considerably more arbitrary.

ACKNOWLEDGMENT

I should like to thank Dr. John Kendrick, Mrs. G. Snoei, and Mrs. H. Gadewoltz for their assistance in preparing the manuscript and Mr. Paul P. Paech for his astute comments and inspirational advice.

References

1. L.M. Raff, Theoretical investigations of the reaction dynamics of polyatomic systems: $T + CH_4$, *J. Chem. Phys.* **60**, 2220–2243 (1974).
2. W.N. Whitton and P.J. Kuntz, Trajectory calculations of the effectiveness of reagent vibration in the $H_2^+ + He \rightarrow HeH^+ + H$ or $He + H + H^+$ reactions, *J. Chem. Phys.* **64**, 3624–3633 (1976).
3. J.C. Tully, Nonadiabatic processes in molecular physics, in *Dynamics of Molecular Collisions, Part A* (Volume 2 of *Modern Theoretical Chemistry*), W.H. Miller, editor, Plenum Press, New York (1976).
4. G.G. Balint-Kurti, Potential energy surfaces for chemical reactions, *Adv. Chem. Phys.* **30**, 137–183 (1975).
5. J.N. Murrell and A.J. Harget, *Semi-empirical Self-Consistent Field MO Theory of Molecules*, Wiley-Interscience, London (1972).
6. D.J. Malcolme-Lawes, Hydrogen isotopic exchange reactions at high energies, *J. Chem. Soc. Faraday Trans. 2*, **71**, 1183–1199 (1975).
7. M. MacGregor and R.S. Berry, Formation of HCO^+ by associative ionization of $CH + O$, *J. Phys.* **B6**, 181–196 (1973).
8. J.C. Tully, Diatomics-in-molecules potential surfaces I, *J. Chem. Phys.* **58**, 1396–1410 (1973).
9. E.S. Rittner, Binding energy and dipole moment of alkali halide molecules, *J. Chem. Phys.* **19**, 1030–1035 (1951).
10. C. Evers, Trajectory surface hopping study of $M + I_2$ collisions (M = Na, K, Cs), *Chem. Phys.* **21**, 355–371 (1977).
11. B. Garetz, M. Rubinson, and J.I. Steinfeld, Classical trajectory surface hopping applied to CID, *Chem. Phys. Lett.* **28**, 120–124 (1974).
12. F.O. Ellison, A method of diatomics-in-molecules, *J. Am. Chem. Soc.* **85**, 3540–3544 (1963).
13. E. Steiner, P.R. Certain, and P.J. Kuntz, Extended diatomics-in-molecules calculations, *J. Chem. Phys.* **59**, 47–55 (1973).

14. J.C. Tully, DIM potential-energy surfaces II. Non-adiabatic and spin-orbit interactions, *J. Chem. Phys.* **59**, 5122–5134 (1973).
15. J.C. Tully and C.M. Truesdale, DIM potential surfaces III. Non-Hermitian formulation, *J. Chem. Phys.* **65**, 1002–1007 (1976).
16. J.C. Tully, Diatomics in molecules, in *Semiempirical Methods of Electronic Structure Calculation, Part B* (Volume 7 of *Modern Theoretical Chemistry*), G.A. Segal, editor, Plenum Press, New York (1977).
17. C.W. Eaker and C.A. Parr, Optimization of diatomic state mixing in diatomics-in-molecules theory: The CH_n potential surfaces, *J. Chem. Phys.* **64**, 1322–1332 (1976).
18. C.W. Eaker and C.A. Parr, Optimized diatomics-in-molecules potentials for H_3 and H_4, *J. Chem. Phys.* **65**, 5155–5160 (1976).
19. B.T. Pickup, The symmetric group and the method of diatomics-in-molecules: An application to small lithium clusters, *Proc. R. Soc. (London) A* **333**, 69–87 (1973).
20. R. Polak, Diatomics-in-molecules study of the H_4^+ ion, *Chem. Phys.* **16**, 353–359 (1976).
21. D.L. Miller and R.E. Wyatt, Comparison of DIM and simple VB PE surfaces for FH_2, *Chem. Phys. Lett.* **38**, 410–416 (1976).
22. N.C. Blais and D.G. Truhlar, Dynamics of F + D_2 on a semiempirical VB potential surface, *J. Chem. Phys.* **58**, 1090–1108 (1973).
23. L.M. Raff, L. Stivers, R.N. Porter, D.L. Thompson, and L.B. Sims, Semi-empirical VB calculation of the H_2I_2 potential, *J. Chem. Phys.* **52**, 3449–3457 (1970).
24. R.N. Porter, L.B. Sims, D.L. Thompson, and L.M. Raff, Classical dynamical investigations of reaction mechanism in 3-body hydrogen halogen systems, *J. Chem. Phys.* **58**, 2855–2869 (1973).
25. N.C. Blais and D.G. Truhlar, Monte Carlo trajectories: The reaction H + $Br_2 \rightarrow$ HBr + Br, *J. Chem. Phys.* **61**, 4186–4203 (1974).
26. J.M. White, Trajectory study of reactions in HBr–Br_2 systems, *J. Chem. Phys.* **58**, 4482–4495 (1973).
27. J.C. Whitehead and R. Grice, London–Eyring–Polanyi potential surfaces for alkali trimers, *Mol. Phys.* **26**, 267–280 (1973); Errata *Mol. Phys.* **28**, 324 (1975).
28. R.N. Porter and M. Karplus, Potential energy surface for H_3, *J. Chem. Phys.* **40**, 1105–1115 (1964).
29. L. Pederson and R.N. Porter, Modified semi-empirical approach to the H_3 potential-energy surface, *J. Chem. Phys.* **47**, 4751–4757 (1967).
30. A.C. Yates and W.A. Lester, A new PE surface for H_3 and its implications for chemical reaction, *Chem. Phys. Lett.* **24**, 305–309 (1974).
31. A.C. Roach and M.S. Child, Potential energy surfaces for K + NaCl $\rightarrow$ KCl + Na, *Mol. Phys.* **14**, 1–15 (1968).
32. W.S. Struve, Semiempirical pseudopotential surfaces for chemical reactions, *Mol. Phys.* **25**, 777–803 (1973).
33. F. London, Quantenmechanische Deutung des Vorgangs der Aktivierung, *Z. Elektrochem.* **35**, 552–555 (1929).
34. R.K. Preston and R.J. Cross, Jr., Charge exchange and chemical reaction: D_2^+ + H, *J. Chem. Phys.* **59**, 3616–3622 (1973).
35. J.R. Krenos, R.K. Preston, R. Wolfgang, and J.C. Tully, Molecular beam and trajectory study of H^+ + H_2, *J. Chem. Phys.* **60**, 1634–1659 (1974).
36. C.W. Bauschlicher, S.V. O'Neil, R.K. Preston, H.F. Schaefer III, and C.F. Bender, Avoided intersection of potential surfaces: The H^+ + H_2, H_2^+ + H system, *J. Chem. Phys.* **59**, 1286–1292 (1973).
37. A. Messiah, *Quantum Mechanics*, J. Wiley and Sons, New York (1962).
38. J.H. Wilkinson, *The Algebraic Eigenvalue Problem*, Oxford University Press, London, (1965).
39. A.R. Edmonds, *Angular Momentum in Quantum Mechanics*, Princeton University Press, Princeton, N.J., (1960).

40. M. Abramowitz and I.A. Stegun, *Handbook of Mathematical Functions,* Dover, New York (1965).
41. P.J. Kuntz, Use of DIM in fitting *ab initio* potential surfaces: The system HeH_2^+, *Chem. Phys. Lett.* **16,** 581–583 (1972).

4

Elastic Scattering Cross Sections I: Spherical Potentials

H. PAULY

1. Introduction

It was clearly realized more than fifty years ago that the experimental investigation of atomic collision processes with molecular beams provides an ideal method for the determination of intermolecular potentials. Excellent pioneering experiments were performed by Stern, Rabi, Knauer, Frazer, and others,[1-3] and the first quantum-mechanical calculations of cross sections were carried out by Massey and Mohr.[4] In spite of this successful beginning, a decisive breakthrough was not achieved.

The main reason for this was that the experimental techniques were not adequate to solve the complicated problems. On the other hand, a quantitative calculation of cross sections was also not possible because of the large numerical effort involved, although the basis of the quantum-mechanical treatment was already well developed. Even simple classical calculations such as the determination of the angle of deflection as a function of the impact parameter for simple realistic potentials, which nowadays may be performed with programmable pocket calculators, could not be performed because of the large numerical effort. A correct interpretation of the experimental results, therefore, was not possible. Moreover, the lack of information about the general behavior of the cross sections made it impossible to estimate the experimental conditions required for resolving and measuring specific phenomena of the scattering or to direct the research to special cross section features.

H. PAULY • Max-Planck-Institut für Strömungsforschung, 3400 Göttingen, Postfach 867 West Germany.

Only twenty years ago, an essential improvement of the experimental techniques and the development of computers led to a resumption of scattering experiments, and a large number of experimental and theoretical investigations have been performed since then. For reviews of this work the reader is referred to the literature.[5-16] The most extensive investigations have been performed on elastic scattering from spherically symmetric potentials, since this is the simplest collision process. The experimental techniques are highly developed and exceed in many cases the requirements necessary for the investigation of elastic collisions since they are also used to study inelastic and reactive collisions. The same holds for the numerical treatment of the elastic cross section data, for which a number of methods have been developed; these methods are more or less universally applicable and complete each other.

It is the aim of this chapter to give a description of these methods within the frame of a discussion of the theoretical treatment of elastic atomic collision processes.

As a prerequisite for this, Section 2 gives a survey of the various potential models that have been applied to the interpretation of cross section data, after a brief introduction of the concept of an intermolecular potential. Because of the large number of multiparameter potential models and splined functionalities which have been adopted in recent years, we restrict ourselves to the most frequently used models and typical examples of groups of potential models.

Section 3 gives the definition of the quantities that can be measured in an elastic scattering experiment. These quantities refer always to an ideal experiment, which has an infinitely high angular and energy resolution. A real experiment shows more or less deviations from this, and the corresponding influence on the observable quantities is briefly discussed.

Section 4 contains the results of the classical scattering theory. Although the classical scattering theory is not sufficient to describe the collision process and may even fail in certain regions, it provides not only a good survey of the general behavior of the measurable quantities, but also permits the derivation of a number of formulas which are an important basis for the semiclassical description of the scattering. Furthermore, many useful and simple relations can be derived from classical mechanics which can be used for a first evaluation of experimental data and a rough estimate of potential parameters. Such estimates of potential parameters are very important for a number of evaluation procedures, where they may be used as initial values of iteration methods.

There follows in Section 5 a description of the quantum-mechanical treatment of the scattering. The resulting expressions for the cross sections do not allow a simple intuitive interpretation due to the large number of contributing partial waves. They are discussed by means of some numerical results. The problems connected with the numerical solution of the Schrödinger equation are also discussed briefly.

Section 6 gives a survey of the semiclassical treatment of the elastic atomic collision process. With the aid of the semiclassical theory the differences between the classical and quantal treatments can be explained and more insight into the physics of the collision process is obtained. Furthermore, a number of refined formulas for the simple evaluation of cross section data and for a determination of initial values for the potential parameters may be gained.

Finally, in Section 7, the most important methods for the determination of potentials from highly resolved, accurate cross sections are discussed.

2. *Intermolecular Potential*

2.1. *The Concept of an Intermolecular Potential*

The theoretical treatment of a collision between two atoms or molecules requires the solution of the Schrödinger equation for the combined electronic and nuclear motions. In the limit as the two particles come together slowly, the electronic and nuclear motions can be separated (Born–Oppenheimer approximation). This separation yields for each specific electronic state a wave equation describing the nuclear motion and which contains the electronic energy V as a function of the internuclear distance r. This potential $V(r)$ is obtained by solving the Schrödinger equation for the electronic motion at all internuclear distances r. The development of numerical methods to calculate such solutions and thus determine "*ab initio*" potentials is a domain of quantum chemistry.[17] The treatment of the nuclear motion, on the other hand, is the subject of scattering theory and will be discussed—as far as the potential is spherically symmetric and the collision is elastic—in this chapter.

2.2. *General Behavior of the Intermolecular Potential*

A typical intermolecular potential V as a function of the internuclear distance is shown in Figure 1. It is nonmonotonic, attractive at large and repulsive at small distances, and spherically symmetric if we consider two atoms in S ground states. For chemically inert partners the depth of the well ε may vary by two or three orders of magnitude, 9.24×10^{-4} eV for the He–He interaction being a representative value for the lower limit, 0.74 eV for the $^1\Sigma^+$ potential of Na_2 for the upper limit of ε. The equilibrium distance r_m varies by hardly more than a factor of 2, and lies typically between 3 and 6 Å. The attractive forces at long distances can be attributed to the coupling between the electron motion in the two atoms produced by the long-range correlation of the motion of the electrons in the two atoms (interaction of instantaneous multipole moments). They are called dispersion forces because the perturbation terms are expressed in terms of the same

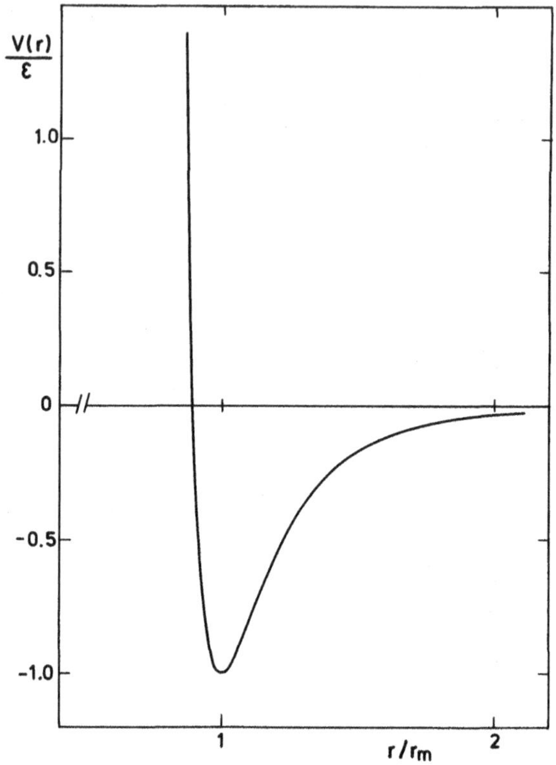

Figure 1. General behavior of the intermolecular potential.

oscillator strengths as appear in the equations for the dispersion of light. The strong repulsion at short distances can be attributed to a repulsion of the overlapping electron clouds, due to the Pauli principle, and to the electrostatic repulsion of the nuclei.

Despite our qualitative understanding of the intermolecular potential, an exact solution of the Schrödinger equation for the electronic motion and thus an accurate quantum-mechanical calculation of potential curves is still not possible for most systems.[17] Reliable calculations have only been carried out for a few simple systems, for instance H–H,[18] H–He, Li–He,[19] He–He,[20,21] and Li–Li.[22]

In the asymptotic regions of the potential, the situation is much better. At large internuclear distances r where electron overlap can be neglected, perturbation theory yields a series expansion in r^{-2n}, with $n \geq 3$:

$$V_a(r) = -C_6 r^{-6} - C_8 r^{-8} - C_{10} r^{-10} - \cdots \tag{1}$$

The expansion coefficients can be expressed in terms of physical properties of the separate atoms,[23] and reliable methods for their calculation are available.[24,25] The only disadvantage of this expansion is its divergence,[26-29] which has recently been overcome by refined theories.[30-34]

On the other extreme of small separations the potential is purely repulsive and approximate methods are well developed.[35,36] Best results are obtained by the Hartree–Fock self-consistent-field theory,[37-40] especially if intra-atomic correlation effects are taken into account. In many cases the numerical results can be well fitted by a simple exponential term

$$V_r(r) = A\exp(-ar) \tag{2}$$

or, with higher accuracy, by

$$V_r(r) = A\exp(-ar + br^2) \tag{3}$$

At intermediate distances in the vicinity of the potential minimum both theories break down and the theoretical treatment becomes difficult. Estimates of the potential well depth and the equilibrium distance r_m, obtained by simply joining the above-mentioned theoretical long- [equation (1)] and short-range [equation (2)] calculations together in the form

$$V(r) = A\exp(-ar) - C_6 r^{-6} - C_8 r^{-8} - C_{10} r^{-10} \tag{4}$$

can be subject to large errors.

Hepburn *et al.*[41] and Ahlrichs *et al.*,[39] therefore, have suggested the introduction of a semiempirical correction term to equation (4) to improve its predictive value:

$$\begin{aligned} V(r) &= A\exp(-ar) - \exp\left\{-\left(\gamma\frac{r}{r_m} - 1\right)^2\right\}\sum_{s=3}^{5} C_{2s}r^{-2s} \qquad \text{for } r < \gamma r_m \\ V(r) &= A\exp(-ar) - \sum_{s=3}^{5} C_{2s}r^{-2s} \qquad \text{for } r \geq \gamma r_m \end{aligned} \tag{5}$$

with $\gamma = 1.28$. Since all constants (A, a, C_6, C_8, C_{10}) are determined by theory, r_m is not a parameter. It is determined by an iterative procedure. The shape of the correction factor and its parameters were obtained from the difference between the simple potential given by equation (4) and the exact quantum-mechanical results for the accurately known ${}^3\Sigma_u^+$ H_2 potential. The method has been found to give accurate predictions for inert-gas systems with a small number of electrons, where the atomic correlation energy is almost independent from the interatomic distance. The correction term, however, has no real physical basis and thus is not universally applicable. Another method for estimating the potential in the vicinity of the minimum has recently been proposed by Tang and Toennies.[42] They derive two correction terms to the simple approximation [equation (4)] which take account of the correct behavior of the dispersion terms in the overlap region. One term corrects for the error made in not truncating the asymptotic expansion at the smallest term. The second term is derived from the semiclassical Drude model and corrects for the influence of electron overlap in the dispersion potential. With these corrections it is possible to obtain good agreement with experimentally determined values of potential well depths and equilibrium distances, as has been shown for a number of rare-gas–rare-gas atom and alkali–rare-gas atom combinations.

Even more accuracy may be achieved by a potential model derived by Feltgen *et al.*[43] They use equation (3) for the repulsive part of the potential with the constants A, a, and b being fitted to *ab initio* calculations. The dispersion potential includes terms up to r^{-12} and up to third order, and is modified by a parameterized correction which results from a convergent expansion[30-34] of the dispersion forces.

As we will discuss in Section 6 the most common method for determining intermolecular potentials from molecular beam scattering data is based on a comparison of experimental data with results of quantum-mechanical calculations for an assumed potential model. The adjustable parameters of the model are varied until a best fit to the experimental data is achieved. A selection of potential models, therefore, is discussed in the following section.

2.3. Potential Models Used in the Evaluation of Scattering Cross Sections

In this section we will discuss a number of potential models commonly used in the interpretation of molecular beam scattering data. We write the potential in the "corresponding states" form, with

$$V(r) = \varepsilon U(x), \qquad x = r/r_m \tag{6}$$

where r_m is the equilibrium distance and ε is the potential well depth. The position of the zero of the potential x_0 depends on r_m and the shape of the potential $U(x)$.

2.3.1. Basic Potential Models

2.3.1.1. The Lennard-Jones (n, 6) Potential:

$$U(x) = \frac{6}{n-6}\left(x^{-n} - \frac{n}{6}x^{-6}\right) \tag{7}$$

This simple model has been most commonly applied in the past for the interpretation of cross section data as well as for the evaluation of virial and transport data. Although the accuracy which has been achieved by molecular beam techniques has clearly demonstrated the inappropriateness of this simple model, it is still useful for a first estimate of potential parameters from cross sections and for an interpretation of the phenomena that occur in elastic scattering. The inverse sixth-power attraction represents the main contribution of the dispersion interaction; the repulsive part of the potential is approximated by an inverse power law. The simple mathematical form allows us to calculate a number of quantities necessary for the interpretation of scattering data in closed form, at least within certain approximations (see below). In many cases the exponent n which is the only shape parameter of this model, is chosen to be 12, primarily for further mathematical convenience. The main disadvantage of this potential as well as of all other simple two- or three-parameter models is that it is too inflexible to reproduce accurate scattering results.[44-46] Furthermore there is a strong coupling between the attractive and the repulsive part of the potential, so that a change in n also changes the attractive region of the potential.

Further properties are as follows :

1. In the limit of large reduced energies ($K \to \infty$) it is possible to calculate the following quantities in closed form:
   a. the classical deflection function $\Theta(\beta)$;
   b. the phase shifts $\delta(\beta)$;
   c. in the case $n = 12$ the classical differential scattering cross section in the small angle and rainbow region.
2. In the limit $K \to 0$ it is possible to derive analytic expressions for
   a. the s-wave phase shift within the WKB approximation, and thus the number of bound vibrational states, and
   b. in the special case $n = 10$ this may also be done quantum-mechanically[47] exact.

2.3.1.2. The Buckingham–Slater Potential:

$$U(x) = \frac{6}{\gamma - 6}\left\{\exp[\gamma(1 - x)] - \frac{\gamma}{6}x^{-6}\right\} \tag{8}$$

This potential is very similar to the L.-J. $(n, 6)$ potential, the inverse power repulsion being replaced by an exponential repulsion. Although this should make it superior to the L.-J. $(n, 6)$ potential due to the more realistic repulsion, there is, in general, no essential improvement. At very short distances this potential behaves unrealistically in that it goes to $-\infty$ at the origin. This is, however, unimportant for the interpretation of scattering data in the thermal energy region. As in the case of the L.-J. potential, the short- and long-range regions of the potential are not independent.

Further properties are as follows:

In the limit of large reduced energies it is possible to calculate
a. the classical deflection function $\Theta(\beta)$,
b. the phase shifts $\delta(\beta)$.

2.3.1.3. The Morse Potential:

$$U(x) = \exp[2\gamma(1 - x)] - 2\exp[\gamma(1 - x)] \tag{9}$$

This model has often been used in the interpretation of spectroscopic data, since the energies of the vibrational levels with zero angular momentum can be calculated in closed form. Because of its unrealistic long-range behavior, it has not been used in the interpretation of scattering data in this simple form. It is, as is described below, however, often used in more complicated potentials which consist of different functional forms for different regions.

Further properties are as follows:

1. For large K analytic expressions can be derived for
   a. the classical deflection function $\Theta(\beta)$,
   b. the phase shifts $\delta(\beta)$.
2. For $K \to 0$ analytic expressions can be derived for
   a. the s-wave phase shift (quantum-mechanically exact and in the WKB approximation),
   b. the energies of the bound states with zero angular momentum (quantum-mechanically exact and in the WKB approximation).

2.3.2. Modifications of the Basic Potentials and Piecewise Analytic Potentials

2.3.2.1. Modifications of the Lennard-Jones Potential:

(i) The Kihara modification[48-51]:

$$U(x) = \frac{6}{n - 6}\left[\frac{(1 - \alpha)^n}{(x - \alpha)^n} - \frac{n}{6}\frac{(1 - \alpha)^6}{(x - \alpha)^6}\right] \tag{10}$$

This modification of the Lennard-Jones potential has two shape parameters: the hard-core radius parameter α, and the repulsive term exponent n. By setting $\alpha = 0$ the L.-J. $(n, 6)$ potential is recovered. For $\alpha \neq 0$ the asymptotic behavior includes x^{-7} and higher-order terms:

$$U_a(x) \to -\frac{n}{n-6}(1-\alpha)^6\left(x^{-6} + 6\alpha x^{-7} + 21\alpha^2 x^{-8} + 56\alpha^3 x^{-9} + \cdots\right) \tag{11}$$

There exists a high correlation between α and n, so that practically only one of them is truly independent. Furthermore, the long-range part of the potential is strongly coupled with the short-range part.

(ii) The Maitland–Smith modification[52]:

$$U(x) = \frac{6}{n-6}\left[x^{-n} - \frac{n}{6}x^{-6}\right] \tag{12}$$

with

$$n = n(x) = 13 + \gamma(x - 1)$$

In this modification, the repulsive term exponent n of the L.-J. $(n, 6)$ potential is regarded as a function of the internuclear separation. Again, the asymptotic behavior includes x^{-7} and higher-order terms:

$$U_a(x) = -x^{-6} - \frac{6}{\gamma}x^{-7} - \frac{6(\gamma-7)}{\gamma^2}x^{-8} - \frac{6(\gamma-7)^2}{\gamma^3}x^{-9} - \cdots \tag{13}$$

On the other hand, the function is easy to manipulate mathematically. It has been applied to inert-gas data and the obtained agreement with the experimental values is as good as that obtained with more complicated functions. If $n(x)$ becomes exactly 6, a singularity can appear. This presents in general no problem since it occurs at very small separations $x = 1 - 7/\gamma$. For the heavier inert gases, the values of γ were found to lie between 8 and 10. For Ne_2, however, a value of $\gamma = 2$ was obtained.

(iii) The Klein–Hanley modification[53]:

$$U(x) = \frac{8 - 2c_6}{n-8}x^{-n} - c_6 x^{-6} - \frac{6c_6 + n - nc_6}{n-8}x^{-8} \tag{14}$$

This modification of the L.-J. $(n, 6)$ potential includes the second dispersion term, but the coefficient of this term is fixed by the other parameters, so that the shape parameters are n and c_6. There is again a very high parameter correlation between n and c_6, indicating that the short- and long-range regions are not independent.

(iv) The Düren–Schlier modification[45,46,54]:

$$U(x) = f(x) - [f(x) + 1]\Gamma_1 \exp\left\{-\frac{(x - x_1)^2}{\gamma_1^2}\right\} \tag{15}$$

where

$$f(x) = x^{-12} - 2x^{-6}$$

is the simple Lennard-Jones (12, 6) potential. This model makes it possible to modify the Lennard-Jones (12,6) potential at any arbitrary region without disturbing its asymptotic behavior. Γ_1 is the amplitude of the Gauss modification, x_1 its location, and γ_1 the width of the modification. Usually $x_1 = 1$ is used in order to widen the potential well. It is also possible to modify the potential in more than one region:

$$U(x) = f(x) - [f(x) + 1] \sum \Gamma_i \exp\left\{-\frac{(x - x_i)^2}{\gamma_1^2}\right\} \tag{16}$$

In this way very flexible potential models are obtained. The number of adjustable parameters increases with each modification by 3.

2.3.2.2. Modifications of the Slater–Buckingham Potential:

(i) The Buckingham–Corner potential[55-57]:

$$U(x) = \left(\frac{8 - 2c_6}{\gamma - 8}\right) \exp\{\gamma(1 - x)\} - c_6 x^{-6} - \frac{6c_6 + \gamma - \gamma c_6}{\gamma - 8} x^{-8} \tag{17}$$

This modification is equivalent to the Klein–Hanley modification of the L.-J. (n, 6) potential. It includes the second dispersion term, but its coefficient is fixed by the two parameters γ and c_6.

More precisely, the above form is usually used for $x > 1$, i.e., beyond the minimum. For $x < 1$, the attractive x^{-6} and x^{-8} terms are multiplied by an exponential factor $\exp[-4(x - 1)^3]$, which rapidly removes the x^{-6} and x^{-8} terms but maintains continuity of U and several derivatives at the minimum. In this way, the unphysical singularity at small distances of the Buckingham–Slater model is removed. The remarks about the coupling between the short- and long-range parts made in the discussion of the Klein–Hanley modification hold for the Buckingham–Corner potential as well.

(ii) The modification of Heller[58,59]:

$$U(x) = \frac{10 - 4c_6 - 2c_8}{\gamma - 10} \exp\{\gamma(1 - x)\} - c_6 x^{-6} - c_8 x^{-8}$$
$$\frac{6c_6 + \gamma - c_6 + 8c_8 - \gamma c_8}{\gamma - 10} x^{-10} \tag{18}$$

As early as 1941 Heller suggested an extension of the Buckingham potential to include the r^{-6}, r^{-8}, and r^{-10} dispersion terms. This modification has three adjustable parameters, i.e., γ, c_6, and c_8. The coefficient of the third dispersion term is fixed by the other parameters. All shape parameters are strongly correlated and the short- and long-range region are not independent.

2.3.2.3. Piecewise Analytical Potentials. Another approach to obtain highly flexible potential models for fitting accurate scattering data is the use of different functional forms for different regions of the potential, which are joined smoothly. This procedure has been used most extensively for rare-gas atom scattering data.[15,60-63] Due to the many possibilities to join two or more potentials together, a large variety of compound potentials has been used. Therefore, we restrict ourselves to a few examples.

(i) Two-piece Lennard-Jones potential[44]:

$$U(x) = \frac{m}{n_1 - m}\left(x^{-n_1} - \frac{n_1}{m} x^{-m}\right) \qquad \text{for } x < 1$$
$$U(x) = \frac{6}{n_2 - 6}\left(x^{-n_2} - \frac{n_2}{6} x^{-6}\right) \qquad \text{for } x \geq 1 \tag{19}$$

Here the attractive and repulsive regions are decoupled. The attractive part of the potential can be changed without changing the repulsive part and vice versa. Available shape parameters are n_1, m, and n_2; n_1 and m, however, are correlated in the vicinity of the minimum, so that only one of these parameters is available for the determination of the shape of the repulsive wall near the minimum.

(ii) The Morse–Morse–van-der-Waals (MMV) potential[64]:

$$U(x) = \exp\{2\gamma(1 - x)\} - 2\exp\{\gamma(1 - x)\}$$
$$\gamma = \begin{cases} \gamma_1 & \text{for } 0 < x \leq 1 \\ \gamma_2 & \text{for } 1 < x < x_c \end{cases} \tag{20}$$
$$U(x) = -c_6 x^{-6} - c_8 x^{-8} - c_{10} x^{-10} \qquad \text{for } x \geq x_c$$

The value of x_c is determined by the separation at which the second Morse curve joins reasonably smoothly onto the theoretical curve for the long-range van der Waals term. Again the attractive and repulsive parts of the potential are decoupled. Available shape parameters are γ_1, γ_2, c_6, c_8, and c_{10}.

(iii) The Morse–spline–van-der-Waals (MSV) potential[65]:

$$\begin{aligned} U(x) &= \exp[2\gamma(1-x)] - 2\exp[\gamma(1-x)] && \text{for } x \le x_1 \\ U(x) &= s_1 + (x-x_1)[s_2 + (x-x_2)\{s_3 + (x-x_1)s_4\}] && \text{for } x_1 < x < x_2 \\ U(x) &= -c_6x^{-6} - c_8x^{-8} - c_{10}x^{-10} && \text{for } x \ge x_2 \end{aligned} \tag{21}$$

The s_i ($i = 1, 2, 3, 4$) are spline coefficients, which are fixed by smoothness conditions at the spline points x_1 and x_2. The first spline point x_1 is taken at the inflection point of the Morse curve, $x_1 = 1 + (\ln 2)/\gamma$. The second spline point x_2 may be treated as an adjustable parameter but is usually held fixed. Remaining shape parameters then are γ, governing the shape of the repulsive and well regions, and the dispersion series coefficients c_6, c_8, and c_{10}.

(iv) The exponential–spline–Morse–spline–van-der-Waals (ESMSV) potential[62]: The MSV potential described above may be modified by using an exponential law at short distances, in order to get a more realistic description of the potential in this region. The result is

$$\begin{aligned} U(x) &= A\exp\{\gamma_1(1-x)\} && \text{for } x \le x_1 \\ U(x) &= \exp[a_1 + (x-x_1)\{a_2 + (x-x_2) \\ &\quad \times (a_3 + (x-x_1)a_4)\}] && \text{for } x_1 < x < x_2 \\ U(x) &= \exp\{2\gamma_2(1-x)\} - 2\exp\{\gamma_2(1-x)\} && \text{for } x_2 \le x \le x_3 \\ U(x) &= s_1 + (x-x_3)[s_2 + (x-x_4) \\ &\quad \times \{s_3 + (x-x_3)s_4\}] && \text{for } x_3 < x < x_4 \\ U(x) &= -c_6x^{-6} - c_8x^{-8} - c_{10}x^{-10} && \text{for } x \ge x_4 \end{aligned} \tag{22}$$

The different regions of the potential are largely decoupled and can be varied independently. Adjustable parameters are A, γ_1, γ_2, c_6, c_8, and c_{10}.

(v) The Barker–Pompe potential[66]:

$$U(x) = \exp\{\gamma(1-x)\}\sum_{i=0}^{N} A_i(x-1)^i - c_6x^{-6} - c_8x^{-8} - c_{10}x^{-10}$$

with

$$\begin{aligned} A_0 &= c_6 + c_8 + c_{10} - 1 \\ A_1 &= A_0 - 6c_6x^{-6} - 8c_8 - 10c_{10} \end{aligned} \tag{23}$$

Available parameters are γ, the A_i coefficients ($i = 2, 3, \ldots, N$), and the dispersion constants c_{2i+4} ($i = 1, 2, 3$). To prevent a spurious maximum in the potential at small distances, the dispersion term is actually used in the form

$$U_a(x) = -\sum_{s=3}^{5} c_{2s}(x^{2s} + \delta)^{-1}$$

where δ has always been given the value 0.01. Thus the potential has $N + 3$ adjustable shape parameters. $N = 5$ gives sufficient flexibility and usually the values of c_6, c_8, and c_{10} are chosen to agree with the best quantum-mechanical estimates, so that 5 shape parameters remain.

2.3.3. The Simons–Parr–Finlan (SPF) Modified Dunham Expansion[67,68]

All the available multiparameter and piecewise analytical potentials discussed above can be represented in a compact way by the Dunham[69] expansion as modified by Simons *et al.*[70] This was shown by Bickes and Bernstein.[67] The expansion is written in the form

$$U(x) = -1 + b_0(1 - x^{-1})^2\left[1 + \sum_{n=1}^{N} b_n(1 - x^{-1})^n\right] \qquad \text{for } x < x_f \tag{24}$$

$$U(x) = -c_6x^{-6} - c_8x^{-8} - c_{10}x^{-10} \qquad \text{for } x \geq x_f$$

Smoothness conditions at $x = x_f$ are used to fix the highest-order coefficients b_N and b_{N-1}. The radius of convergence of the series is $0.5 \leq x < \infty$. This restriction, however, is only important if high-energy molecular beam data are considered. The joining point x_f is taken such that $U(x_f) \geq -0.05$ and can be treated in the same fashion as, for instance, in the case of the MSV potential. As we will discuss in Section 7, Bickes and Bernstein[68] have shown how the modified Dunham expansion (with ar^{-6} long-range tail) can be incorporated into a systematic, iterative procedure for the analysis of differential elastic-scattering experiments. This and the high flexibility of the potential make-it superior to the multiparameter potentials discussed before.

3. *Definitions of the Quantities That Can Be Measured in Elastic-Scattering Experiments. Influence of Experimental Conditions*

If we consider a parallel beam of atoms of homogeneous velocity passing through a hypothetical stationary gas target of n particles per cm^3, the amount of beam current strength dI lost in traversing a distance dx from a point P will be given by

$$dI = nI\sigma\, dx \tag{25}$$

where I is the current strength in P and σ the total scattering cross section. Hence, on integration, we find

$$I = I_0 \exp(-n\sigma x) \tag{26}$$

where I_0 is the unattenuated current strength. Actually, the target particles are not at rest before the collision, and equation (26) has to be replaced by

$$I = I_0 \exp(-n\sigma_{\text{eff}}\, x) \tag{27}$$

The effective cross section σ_{eff} is obtained by averaging the cross section $\sigma(g)$ (g is the relative velocity) over the velocity distribution $f(\mathbf{v}_2)$ of the target particles[71]:

$$\sigma_{\text{eff}}(v_1) = \int f(\mathbf{v}_2)(g/v_1)\,\sigma(g)\, d\mathbf{v}_2 \tag{28}$$

v_1 is the velocity of the projectile particles. If the projectile beam has a finite velocity spread, described by the distribution function $f(v_1)$, the effective total cross section must also be averaged over $f(v_1)$:

$$\bar{\sigma}_{\text{eff}} = \iint f(\mathbf{v}_2)(g/v_1)\,\sigma(g) f(v_1)\, d\mathbf{v}_2\, dv_1 \tag{29}$$

If the velocity dependence of σ can be expressed by a power law $[\sigma(g) \sim g^{-2/(s-1)}]$ (see Section 6), equation (28) may be written in the form

$$\sigma_{\text{eff}}(v_1) = \sigma(v_1)\, F(x) \tag{30}$$

where $x = v_1/c_2$, and c_2 is the most probable velocity of the target particles. Correction functions have been calculated for different experimental situations.[71]

The total scattering cross section includes all collisions without a specification of the distribution in angle of the scattered atoms. If $p(\theta, \phi)\, d\Omega$ is the probability that, in an elastic collision, the atom is scattered into the solid angle $d\Omega = \sin\theta\, d\theta\, d\phi$, then $p(\theta, \phi)\,\sigma\, d\Omega$ is called the differential cross section. It is usually written

$$d\sigma = p(\theta, \phi)\,\sigma\, d\Omega = I(\theta, \phi)\, d\Omega \tag{31}$$

and is the number of particles scattered into the solid angle $d\Omega$ per unit time divided by the incident flux density. For spherically symmetric potentials $I(\theta, \phi)$ becomes independent of the azimuthal angle ϕ.

A comparison with theoretical differential cross sections requires a transformation of the values measured in the laboratory system (θ, ϕ) into the center-of-mass system (ϑ, φ). This transformation is particularly simple when we consider a crossed-beam arrangement with uniform velocities in both beams. This case has been treated in the literature.[72-77] For a table of formulas for the transformation see Pauly and Toennies.[9,11]

The comparison between empirical and theoretical cross section data, however, is considerably complicated due to the finite velocity spread in the beams, due to the finite dimensions of the target volume and due to the finite detector area.

Averaging over the velocity distributions in both beams (6 integrals), the spatial distribution of the target (3 integrals) and the detector area (2 integrals) leads to an 11-dimensional integral, which usually cannot be evaluated exactly, since it requires too much computer time. The approximations depend on the experimental setup used in the measurements. If, for instance, the dimensions of the scattering volume and the detector area can be neglected, only the velocity averaging is necessary.[78-80] If, on the other hand, the velocity spread (in magnitude and direction) can be neglected, only the averaging over the finite angular resolution may be necessary.[81]

The finite width of the projectile beam and of the detector have also an influence on total cross section measurements. According to the definition [equation (31)] of the total scattering cross section

$$\sigma = \int I(\theta, \phi)\, d\Omega \tag{32}$$

all scattered particles make a contribution. Because of the finite dimensions of the beam and the detector, this is not entirely the case in practice. A fraction of beam particles, although scattered, strike the detector. As a result, the measured effective scattering cross section is too small by an amount given by

$$\Delta\sigma = \int_{\Delta\Omega} I(\theta, \phi)\, d\Omega \tag{33}$$

where the integral extends over the solid angle $\Delta\Omega$, which characterizes the resolving power of the apparatus. $\Delta\sigma$ depends both on the geometry of the apparatus and on the interaction potential [which determines $I(\theta, \phi)$].[82,83]

4. *Classical Scattering Theory*

4.1. *Basic Formulas*

Although the exact treatment of atomic collision has to follow along quantum-mechanical methods, classical mechanics often yields in many respects a good approximation. It is especially useful, as already mentioned, for estimating the general behavior or the order of magnitude of the cross sections in a given experiment. Moreover, the knowledge of the classical treatment is the necessary basis for an understanding of the semiclassical theory, which plays an important role in the interpretation of atomic collision processes.

The two-particle collision problem is formally reduced to a one-particle problem by a transformation to the center-of-mass system (cf. Section 3). In the center-of-mass system we have to do with the scattering of a particle of the reduced mass $\mu = m_1 m_2/(m_1 + m_2)$ which is moving with the relative momentum $p = \mu g$ (where g is the relative velocity) with respect to an infinitely heavy stationary particle. From the conservation of energy and angular momentum we obtain the angle of deflection Θ as a function of the reduced impact parameter $\beta = b/r_m$:

$$\Theta = \pi - 2\beta \int_{x_0}^{\infty} x^{-2}\left[1 - \frac{U(x)}{K} - \frac{\beta^2}{x^2}\right]^{-1/2} dx \tag{34}$$

x_0 is the reduced distance of closest approach, which is determined by the largest positive root of the equation

$$1 - [U(x_0)/K] - \beta^2/x_0^2 = 0 \tag{35}$$

For a fixed reduced energy K, equation (34) relates an angle of deflection to every reduced impact parameter β. This function $\Theta(\beta)$ is called the classical deflection function and can adopt—contrary to the center-of-mass system angle of deflection ϑ—positive and negative sign. All incident particles with impact parameters between b and $b + db$ will be scattered into an angular range between ϑ and $\vartheta + d\vartheta$ (because of the spherical symmetry of the potential we have no dependence on the azimuthal angle φ).

If the incident flux density is j_0, the number of particles per unit time incident upon the ring of width db at a distance b from the scattering center is $2\pi j_0 b\, db$. This is equal to the number of particles scattered into the solid angle element $d\omega$.

Thus we obtain

$$\begin{aligned} d\sigma &= 2\pi I(\vartheta) \sin\vartheta\, d\vartheta \\ &= 2\pi b |db/d\vartheta| d\vartheta \end{aligned} \tag{36}$$

or

$$I(\vartheta) = (\sin\vartheta)^{-1}\, b(\vartheta)\, |db/d\vartheta|$$

Figure 2 shows classical trajectories as a function of the reduced impact parameter β for a realistic potential; the lower half of the figure shows the corresponding deflection function $\Theta(\beta)$. Its main features are as follows:

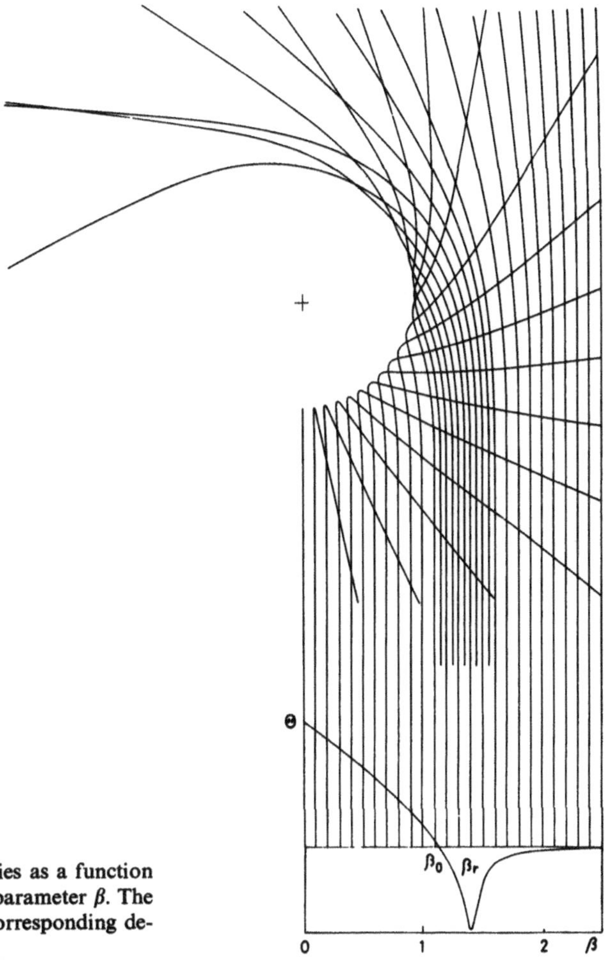

Figure 2. Classical trajectories as a function of the reduced impact parameter β. The lower half shows the corresponding deflection function $\Theta(\beta)$.

(a) A positive and a negative branch as the impact parameter varies, due to net repulsion and net attraction, respectively, in the collision (see the trajectories in Figure 2);

(b) A minimum at $\beta = \beta_r$ due to the beginning influence of the repulsive potential core on the particle trajectory (called rainbow angle in analogy to optics);

(c) A zero at $\beta = \beta_0$ corresponding to a cancellation of attractive and repulsive forces in the collision.

As a consequence of this type of deflection function, β is, in a certain angular range which depends on the reduced energy K, a three-valued function of

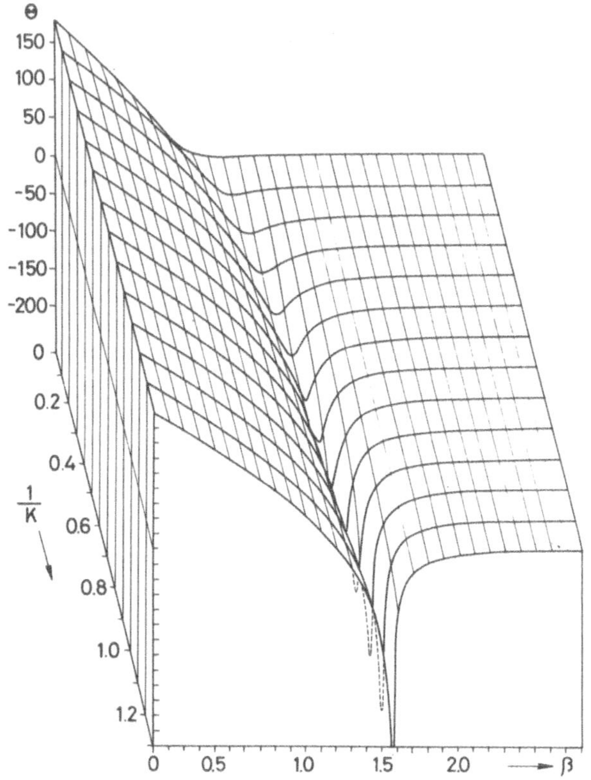

Figure 3. The classical deflection function $\Theta(\beta, K)$ as a function of β and $1/K$ [L.-J. (12, 6) potential].

ϑ. The differential cross section, therefore, is the sum of three contributions:

$$I(\vartheta) = (\sin\vartheta)^{-1} \sum_{i=1}^{3} b_i \left|\frac{db_i}{d\vartheta}\right| = r_m^2 (\sin\vartheta)^{-1} \sum_{i=1}^{3} \beta_i \left|\frac{d\beta_i}{d\vartheta}\right| \tag{37}$$

Figure 3 shows a three-dimensional representation of the classical deflection function $\Theta(\beta, 1/K)$. As can be expected from simple considerations, the attractive part of the potential loses more and more influence with increasing reduced energy K. On the other hand, the rainbow angle increases with decreasing K. For reduced energies smaller than a given K_{crit} (the exact value depends on the shape of the potential), the rainbow angle goes to infinity (orbiting collisions, see Section 4.2.5).

4.2. *Differential Cross Section*

4.2.1. *Small-Angle Scattering*

For small angles the main contribution to the differential scattering cross section arises from the asymptotic branch of the deflection function at large impact parameters. Compared with the contributions at larger angles, both β and $|d\beta/d\vartheta|$ become large. Considering only these impact parameters and the attractive part of the potential in the form $U_a(x) = -c_s x^{-s}$, the deflection angle ϑ may be obtained approximately from an

Table 1. The Functions $f(n)$, $g_0(n)$, $g_1(n)$, and $g_2(n)$

n	$f(n)$	$g_0(n)$	$g_1(n)$	$g_2(n)$
2	1.570796			
3	1			
4	0.785398			
5	0.666667			
6	0.589049			
7	0.533333	2.371088	0.071455	−0.059484
8	0.490874	2.327184	0.120494	−0.036913
9	0.457143	2.292587	0.155798	−0.017037
10	0.429515	2.264616	0.182178	0
11	0.406349	2.241533	0.202485	0.014507
12	0.386563	2.222158	0.218505	0.026877
13	0.369408	2.205663	0.231402	0.037473
14	0.354350	2.191449	0.241965	0.046606
15	0.340992	2.179075	0.250746	0.054529
16	0.329039	2.168204	0.258140	0.061447
17	0.318260	2.158578	0.264436	0.067526
18	0.308474	2.149995	0.269350	0.072898
19	0.299538	2.142293	0.274547	0.077673
20	0.291337	2.135345	0.278655	0.081939
21	0.283773	2.129043	0.282272	0.085769
22	0.276770	2.123302	0.285478	0.089224
23	0.270260	2.118051	0.288337	0.092352
24	0.264189	2.113228	0.290898	0.095196
25	0.258510	2.108785	0.293205	0.097791
26	0.253181	2.104677	0.295292	0.100167
27	0.248169	2.100868	0.297187	0.102350
28	0.243444	2.097326	0.298916	0.104361
29	0.238978	2.094025	0.300497	0.106219
30	0.234749	2.090940	0.301949	0.107940
Limiting values for large n ($n \to \infty$)				
n	$\dfrac{4n-1}{4(n-1)}\left(\dfrac{\pi}{2n}\right)^{1/2}$	2	$\dfrac{1}{3}$	$\dfrac{3}{20}$

expansion[84] of equation (34):

$$\Theta(\beta) = -(c_s/K)(s-1) f(s)\,\beta^{-s} \tag{38}$$

with

$$f(s) = (\pi^{1/2}/2)\,\Gamma\left(\frac{s-1}{2}\right) \Big/ \Gamma\left(\frac{s}{2}\right)$$

(see Table 1).

Taking higher-order terms into account[85-88] the following series is obtained:

$$\Theta(\beta) = -\sum_{v=0}^{\infty} \left(\frac{c_s}{K}\right)^{v+1} \beta^{-s(v+1)} f_v(s) \tag{39}$$

with

$$f_v(s) = \pi^{1/2}\Gamma\left(\frac{(v+1)s+1}{2}\right) \Big/ \Gamma(v+2)\,\Gamma\left((v+1)\frac{s}{2} - v\right)$$

The equations (38) and (39) are also valid for repulsive potentials. In this case the opposite sign has to be used for the potential constant c_s.

With equation (38) the differential scattering cross section at small angles becomes

$$I(\vartheta) = r_m^2 [(s-1)/s]\,[f(s)\,c_s/K]^{2/s}\,\vartheta^{-2(s+1)/s} \tag{40}$$

The classical differential scattering cross section diverges as ϑ goes to zero.

4.2.2. *Glory Scattering*

Superimposed on the singularity at $\vartheta = 0$ resulting from the long-range attractive forces according to equation (40) there is a second singularity coming from collisions with a reduced impact parameter β_0. This singularity as well as all others that are the result of $\sin\vartheta = 0$ in equation (36) are designated "glory effect" in analogy to optics.[89] Additional glory singularities may occur, if the deflection function passes through $\pm\pi$, $\pm 2\pi$, etc., at nonzero impact parameter (for example, in the case of orbiting). The condition $\Theta = -2n\pi$ $(n = 0, 1, 2, \ldots)$ yields glory singularities in the forward direction; $\Theta = -(2n+1)\pi$ $(n = 0, 1, 2, \ldots)$ yields singularities in the backward direction. As an example we consider here the forward glory at $\vartheta = 0$ for $\beta = \beta_0$. In the vicinity of β_0 the deflection function can be approximated by

$$\Theta = \gamma_0(\beta - \beta_0) \tag{41}$$

(see Figure 3).

The glory contribution to the differential cross section in the forward direction is the sum of two equal contributions for $\Theta > 0$ and $\Theta < 0$. Thus we get

$$I_g(\vartheta) = 2\beta_0^2 r_m^2/\gamma_0 \vartheta \tag{42}$$

Values of β_0 and γ_0 as a function of the reduced energy K are tabulated for a Kihara potential by Düren and Pauly.[50]

4.2.3. *Rainbow Scattering*

Another singularity in the classical differential scattering cross section is known as the rainbow effect because of its analogy to the formation of rainbows by light scattering. It arises when the slope $d\Theta/d\beta$ of the deflection function vanishes. Near the rainbow angle ϑ_r the deflection function may be expanded in the form

$$\Theta(\beta) = \vartheta_r + \tilde{q}(\beta - \beta_r)^2 \tag{43}$$

with

$$\tilde{q} = \tfrac{1}{2}[d^2\Theta/d\beta^2]_{\vartheta=\vartheta_r} \tag{44}$$

From this expansion the rainbow part of the differential cross section can be easily derived. On the bright side of the rainbow ($\vartheta < \vartheta_r$) the rainbow cross section will be

$$I_r(\vartheta) = r_m^2(2\sin\vartheta_r)^{-1}\,|q(\vartheta - \vartheta_r)|^{-(1/2)} \tag{45}$$

whereas for the dark side ($\vartheta > \vartheta_r$) the rainbow cross section will be zero.

Numerical values of the quantities needed to describe the classical rainbow scattering, ϑ_r, β_r, and $\tilde{q}$, are tabulated by Schlier[49] and Mason *et al.*[90]

4.2.4. *Large-Angle Scattering*

The positive branch of the deflection function which describes the scattering at large angles (small impact parameters) is mainly determined by the repulsive part of the potential. If this repulsive part of the potential is approximated by an inverse power law $U_r(x) = c_n x^{-n}$, the deflection function may be expanded[48,91,92] in the form

$$\Theta = \pi - \sum_{v=0}^{\infty} g_v(n)\left(\frac{c_n}{K}\right)^{-(2v+1)/n} \beta^{(2v+1)} \tag{46}$$

where

$$g_v(n) = (-1)^v \frac{2\pi^{1/2}}{n}\Gamma\left(\frac{2v+1}{n}\right) \Big/ \Gamma(v+1)\,\Gamma\left(\frac{2v+1}{n} - v + \frac{1}{2}\right)$$

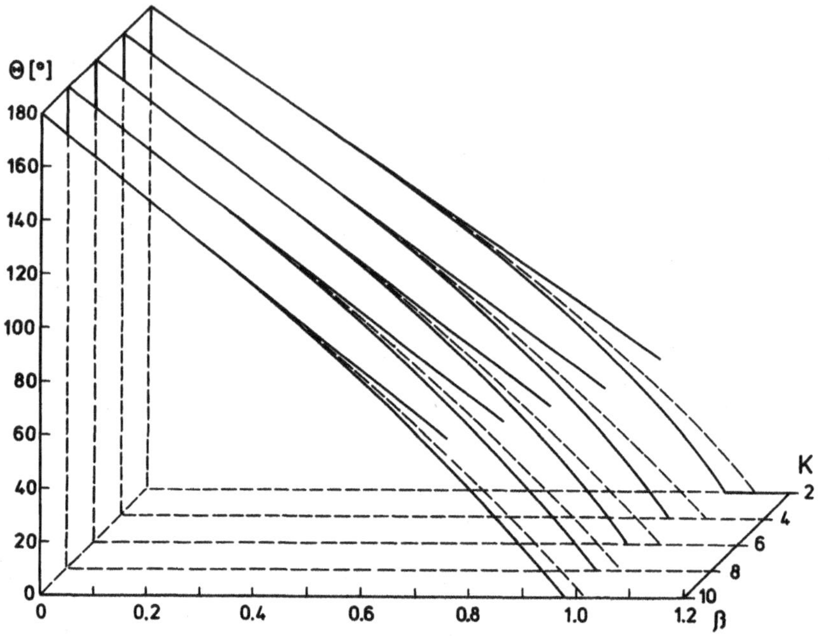

Figure 4. Repulsive branch of the deflection function $\Theta(\beta)$ [calculated for a L.-J. (12, 6) potential] for various reduced energies K (full lines). The dashed lines are calculated using two terms ($v = 0$ and $v = 1$) of the expansion given by equation (46). The straight lines correspond to the linear approximation [equation (47)]. The deviation between the full-line curves and the dashed curves is mainly due to the neglect of the attractive potential part in equation (46).

(see Table 1). Using only the first term of this expansion yields

$$\Theta = \pi - g_0(n)\,(K/c_n)^{(1/n)}\,\beta \tag{47}$$

From this we obtain for the differential scattering cross section at large angles:

$$I(\vartheta) \approx r_m^2\,[g_0(n)]^{-2}\,[c_n/K]^{(2/n)} \tag{48}$$

The differential cross section at large angles is nearly angle independent and it depends only weakly on the reduced energy K. If the next term of the series (46) is taken into account, a good qualitative approximation of the whole repulsive branch of the deflection function is obtained (see Figure 4).

Leibfried and Plesser[92] have shown that equation (46) diverges for reduced impact parameters less than $\beta_c = n^{1/2}(c_n/2K)^{1/n}\,|n-2|^{(1/n)-(1/2)}$, whereas the small-angle expansion equation (39) diverges for reduced impact parameters greater than β_c. An exact series solution of the classical deflection function for a repulsive inverse power potential, which converges at all impact parameters, has been derived by Gislason.[93]

4.2.5. Orbiting Collisions

With decreasing reduced energy K the rainbow angle ϑ_r tends to infinity. The relative position vector makes many revolutions during the collision and orbiting is said to occur. To examine this effect further it is convenient to introduce a reduced classical relative angular momentum l^* into equation (34):

$$l^* = \mu g b (2\mu \varepsilon r_m^2)^{-1/2} = K^{1/2}\beta \tag{49}$$

where μ is the reduced mass and g the relative velocity. The classical deflection function equation (34) may be written

$$\Theta(l^*) = \pi - 2l^* \int_{x_0}^{\infty} x^{-2}[K - U_{\text{eff}}(x, l^*)]^{-1/2}\, dx \tag{50}$$

with

$$U_{\text{eff}}(x, l^*) = U(x) + l^{*2}/x^2$$

Orbiting occurs at that value of l^* for which the height of the centrifugal barrier [which will arise for potentials with an attractive part stronger than r^{-2} (see Figure 5)] equals the incident reduced energy K. Thus the

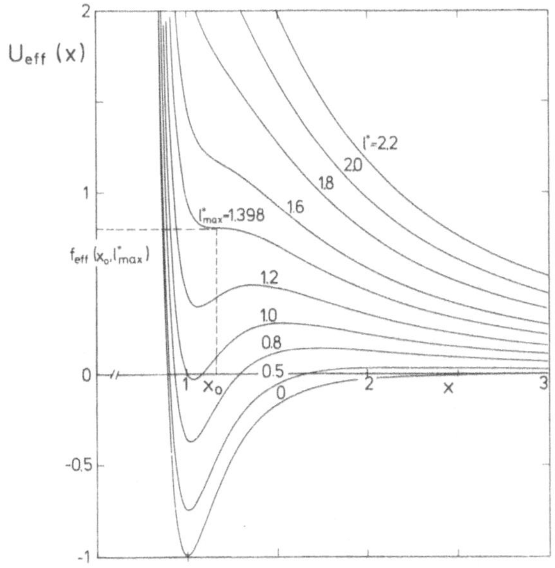

Figure 5. Reduced effective potential $U_{\text{eff}}(x, l^*)$ as a function of the reduced distance x for various reduced angular momenta $l^* = K^{1/2}\beta$, calculated for a L.-J. (12, 6) potential. For $l^* > 1.398$ the effective potential is monotonic. For $l^* = 1.398$ the point of inflection is characterized by $K = 0.8$. For $K < 0.8$ orbiting occurs. (Adapted from Reference 138.)

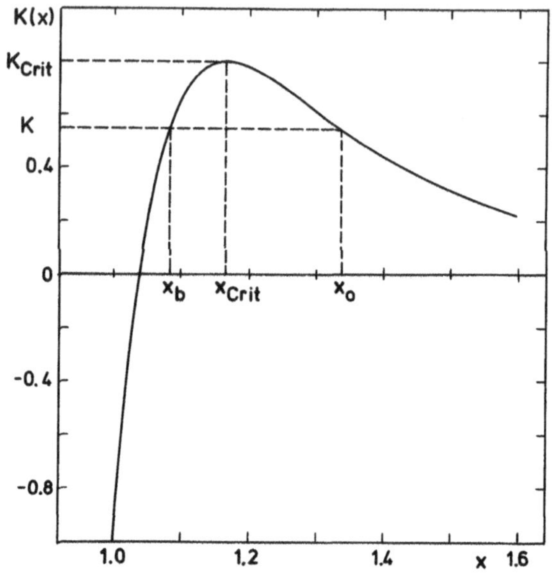

Figure 6. Functional relation between the reduced orbiting radius and the reduced energy [L.-J. (12, 6) potential].

following conditions have to be satisfied:

$$K = U(x) + l^{*2}/x^2 \tag{51}$$

$$U'_{\text{eff}}(x) = 0 \tag{52}$$

where the prime denotes differentiation with respect to x. These can be combined to give

$$K(x) = U(x) + (x/2)\,U'(x) \tag{53}$$

With this expression it is possible to find the orbiting radius as a function of energy. The typical behavior of $K(x)$ is shown in Figure 6. At positive energies we can see that there are two possible orbiting radii, x_b and x_0, which correspond to the stable periodic orbit at the bottom of $U_{\text{eff}}(x)$ and the unstable periodic orbit around the top of the maximum of the centrifugal barrier. There is also a critical energy, K_{crit} (which we call the orbiting threshold), with the corresponding radius x_{crit} and angular momentum l^*_{crit} above which there can be no orbiting.

In the case of the L.-J. $(n, 6)$ potential we can easily derive from equations (51)–(53), the critical quantities

$$K_{\text{crit}}(n) = 2\left(\frac{4}{n-2}\right)^{6/(n-6)}$$

$$x_{\text{crit}}(n) = \left(\frac{n-2}{4}\right)^{1/(n-6)} \tag{54}$$

$$l^*_{\text{crit}}(n) = \left(\frac{4}{n-2}\right)^{2/(n-6)} \left\{\frac{3[(n-2)(n-4)-8]}{(n-6)(n-2)}\right\}^{1/2}$$

Ford and Wheeler[89] have shown that the deflection function $\Theta(\beta)$ can be expanded in the region of the orbiting singularity with the result

$$\Theta(l^*) = \begin{cases} C + \alpha \ln\left(\dfrac{l^*}{l_0^*} - 1\right), & l^* > l_0^* \\ D + 2\alpha \ln\left(1 - \dfrac{l^*}{l_0^*}\right), & l^* < l_0^* \end{cases} \tag{55}$$

where C and D are constants depending on the potential within the barrier. α is given by

$$\alpha = [2l_0^*/x_0^4 U''_{\text{eff}}]^{1/2} \tag{56}$$

This parameter is roughly the square root of the radius of the curvature of the centrifugal barrier and gives a measure of the strength of the orbiting singularity (small α means a narrow spike and large α a broad spike in the deflection function).

The classical differential cross section arising from orbiting will be the sum of two parts from $l^* < l_0^*$ and $l^* > l_0^*$. Each part will be a superposition of an infinite number of terms arising from successive intervals $\Theta = 0$ to $-\pi$, $-\pi$ to -2π, etc. Each term in the series will be much smaller than the preceding term. The major contribution will be

$$I(\vartheta) = \beta_0^2 (\alpha \sin\vartheta)^{-1} [\exp\{-(\vartheta + C)/\alpha\} + \tfrac{1}{2}\exp\{-(\vartheta + D)/2\alpha\}] \tag{57}$$

showing an exponential decrease with increasing angle. Successive terms may be obtained from equation (57) by replacing ϑ by $2\pi - \vartheta$, $2\pi + \vartheta$, $4\pi - \vartheta$, $4\pi + \vartheta$, etc.

A simple orbiting model has frequently been used to calculate capture cross sections in ion–molecule reactions.[94] In this model it is assumed that all collisions with impact parameter $b \leq b_{\text{orb}}$ lead to a reaction, whereas all collisions with impact parameters $b > b_{\text{orb}}$ lead to elastic scattering. Furthermore it is assumed, that the height and the position of the centrifugal barrier are determined by the attractive part of the potential alone. If this is represented by an inverse power law $U(x) = -c_s x^{-s}$, the effective potential becomes

$$U_{\text{eff}}(x) = -c_s x^{-s} + l^{*2} x^{-2} \tag{58}$$

From equation (52) we find the position of the barrier maximum x_{orb}

$$x_{orb} = \left(\frac{sc_s}{2l^{*2}} \right)^{1/(s-2)} \tag{59}$$

Using equation (51) we obtain the reduced angular momentum l^*_{orb} for which orbiting occurs:

$$l^*_{orb} = \left[\frac{s}{s-2} K \right]^{(s-2)/2s} \left(\frac{c_s}{2} \right)^{1/s} \tag{60}$$

Together with equation (49) this yields for the reduced orbiting impact parameter β_{orb}:

$$\beta_{orb} = \left(\frac{s}{s-2} \right)^{(s-2)/2s} \left(\frac{sc_s}{2K} \right)^{1/s} \tag{61}$$

In the case of the ion–molecule interaction we have $s = 4$, and the total reaction cross section in this model becomes

$$\sigma_r = \pi r_m^2 \beta_{orb}^2 = \pi r_m^2 \left(\frac{4c_s}{K} \right)^{1/2} \tag{62}$$

4.2.6. *Summary of the Classical Results for the Differential Scattering Cross Section and Limits of Validity*

Figure 7 shows the characteristic behavior of the classically calculated differential scattering cross section, weighted with sin ϑ, as a function of the angle with the reduced energy K as parameter [the calculations have been performed for a L.-J. (12, 6) potential]. This figure confirms the results of the various approximations discussed before: The monotonic decrease at small angles proportional to $\vartheta^{-7/3}$, which is followed by the rainbow singularity. The angular position of this singularity depends only on the reduced energy K, roughly proportional to $1/K$. On the dark side of the rainbow the cross section shows a sharp drop to a nearly angle- and energy-independent value. Tables of classically calculated differential scattering cross sections for a Buckingham potential are given by Mason.[95]

As we will discuss in Sections 5 and 6, the classical description is modified by quantum mechanics. The most essential modifications are as follows:

1. All classical singularities discussed above (singularity in forward direction, glory and rainbow singularities) are replaced by finite maxima of the cross section [if the potential $V(r)$ vanishes faster than with r^{-2} asymptotically]. The width of these finite maxima depends on the de Broglie wavelength $\lambda = h/\mu g$. In the case of forward scattering, for example, it is described by a limiting angle $\vartheta_{lim} = \lambda/2\pi a$, where a is the collision radius

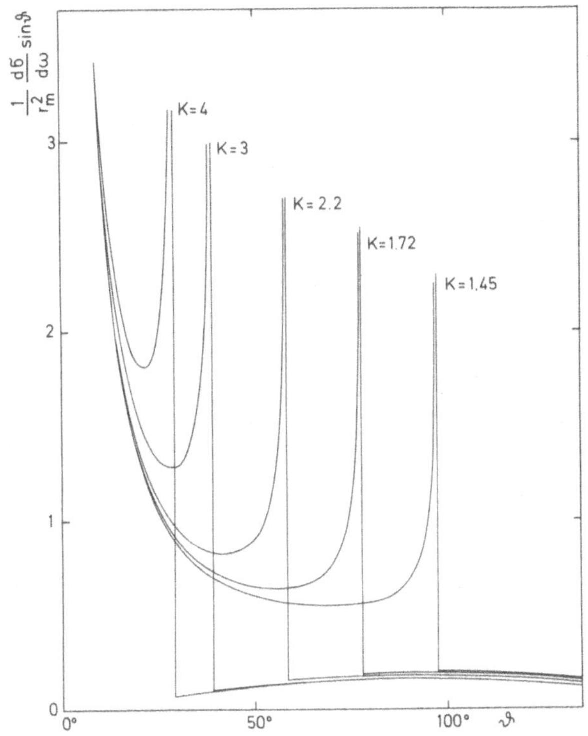

Figure 7. Classical differential cross section (weighted with $\sin\vartheta$) as a function of the angle of deflection ϑ in the center-of-mass system for different reduced energies [L.-J. (12, 6) potential].

(largest impact parameter which makes an observable contribution to the scattering). In other words, classical mechanics breaks down for scattering angles smaller than ϑ_{lim}. This has also been experimentally confirmed by high-resolution differential cross section measurements.[96,97]

2. For angles smaller than the rainbow angle, where three impact parameters lead to the same angle of deflection, a marked interference structure is obtained. If, however, the cross section is averaged over the interference oscillations, the classical result is generally observed.

Thus the classical results are especially well suited for comparison with low-resolution differential cross section measurements, and the following information is obtainable:

1. The differential scattering cross section shows a $\vartheta^{-\alpha}$ dependence [equation (40)] with $\alpha = 2(s + 1)/s$; α can be directly determined from small-angle scattering data and thus it is possible to find the exponent s governing the distance dependence of the attractive part of the poten-

tial.[77,96,97] However, since

$$\frac{\Delta\alpha}{\alpha} = (s+1)^{-1}\frac{\Delta s}{s} \tag{63}$$

this method is not very accurate. To obtain s within given error limits, the accuracy in α has to be roughly an order of magnitude better.

2. As has been shown above, the rainbow angle ϑ_r depends strongly on the reduced energy K, the dependence on the potential shape being rather weak. The observation of ϑ_r yields, therefore, for a given potential the well depth ε. This value is, within limits of 10–15%, independent of the potential shape.

3. As follows from the reduced notation, the parameter r_m enters into the differential scattering cross section only as a factor. This means that r_m can only be inferred from an absolute measurement of the cross section.

4. Although the rainbow effect is smeared out if beams with Maxwellian velocity distribution are used to obtained low-resolution cross section data, there remains a small effect which allows a rough estimate of the potential well depth ε.[98]

4.3. Total Elastic Cross Sections

Although the total elastic cross section for a potential which is of unbounded range does not exist, we may define an "incomplete" total scattering cross section

$$S(\alpha, g) = 2\pi \int_{\alpha}^{\pi} I(\vartheta, g)\sin\vartheta\, d\vartheta = 2\pi r_m^2 \int_0^{\beta(\alpha)} \beta\, d\beta \tag{64}$$

(with $\alpha > \vartheta_{\text{lim}}$), that accounts for all scattering with angles larger than a certain angle α. If β_1, β_2, and β_3 are the three impact parameters which belong to α, then

$$S(\alpha)/\pi r_m^2 = \beta_3^2 + \beta_1^2 - \beta_2^2 \tag{65}$$

Figure 8 shows the incomplete total cross section $S(\alpha)$ as a function of the reduced energy K for various angles of resolution α. From this figure it is obvious that we can distinguish two energy regions, defined by $\alpha < \vartheta_r(K)$ and $\alpha > \vartheta_r(K)$.

For $\alpha < \vartheta_r$ we obtain from equation (65)

$$S(\alpha)/\pi r_m^2 = \beta_3^2 + (\beta_1 - \beta_2)(\beta_1 + \beta_2) \tag{66}$$

which reduces to

$$S(\alpha)/\pi r_m^2 = \beta_3^2 + 2\beta_0\Delta\beta = S_a + \Delta S \tag{67}$$

if α is small. β_0 is the reduced glory impact parameter.

The incomplete cross section consists of a contribution S_a, which is

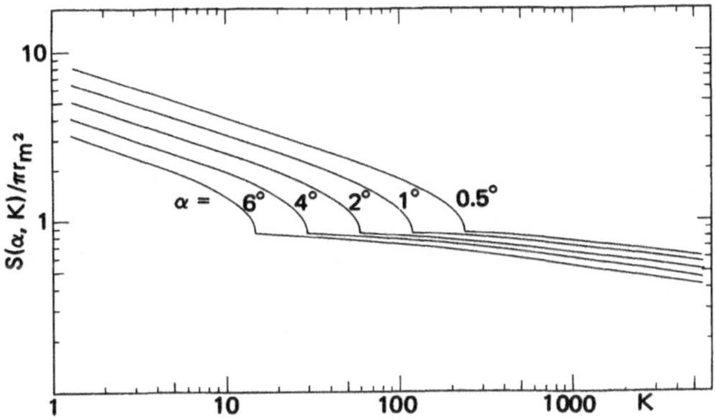

Figure 8. Incomplete total scattering cross section $S(\alpha, K)/\pi r_m^2$ as a function of the reduced energy K for various angles of resolution [L.-J. (12, 6) potential].

almost entirely determined by the attractive part of the potential and a glory contribution ΔS, which tends to zero for $\alpha \to 0$. Consequently, for small angles α the classical incomplete total cross section has the limit

$$S(\alpha) = \pi r_m^2 \beta_3^2 \tag{68}$$

The integration of equation (40) yields

$$S(\alpha) = \pi r_m^2 [(s-1) f(s)(c_s/K)]^{2/s} \alpha^{-2/s} \tag{69}$$

Once more the divergence of the total scattering cross section for $\alpha \to 0$ becomes apparent.

For $\alpha > \vartheta_r(K)$ only the scattering due to the repulsive branch of the deflection function is observed. Equation (65) has to be replaced by

$$S(\alpha) = \pi r_m^2 \beta_1^2 \tag{70}$$

where β_1 may also be calculated from equation (40), taking only into account the repulsive part of the potential ($U_r = -c_n x^{-n}$). The result is

$$S(\alpha) = \pi r_m^2 [(n-1) f(n)(c_n/K)]^{2/n} \alpha^{-2/n} \tag{71}$$

An approximate treatment of the transition region between equations (69) and (71) is also possible with the aid of the first-order momentum approximation. This is discussed in Section 4.5.

Equation (71) has been frequently used in the interpretation of incomplete total scattering cross section measurements in the energy range from several hundred to several thousand eV.[99] From relative measurements of the energy dependence of the incomplete cross section it is possible to determine the exponent n of the distance dependence of the repulsive potential (or the exponent γ if an exponential law is used). For an evaluation of the constant $C_n = \varepsilon r_m^n c_n$ an absolute measurement of the cross section is required.

4.4. Identical Particles

If the two colliding particles are indistinguishable, then the flux in a given direction, which consists both of projectile particles and of recoil particles associated with projectile particles flying off in the opposite direction (in the center-of-mass system), is measured as a whole. Consequently, the center-of-mass scattering cross section is the sum of the cross sections for the two opposite directions,

$$I_i(\vartheta) = I(\vartheta) + I(\pi - \vartheta) \tag{72}$$

Thus the differential cross section for identical particles is symmetric about $\pi/2$.

4.5. First-Order Momentum Approximation and Results for the Basic Potentials

If the reduced energy K is large, the momentum transferred during the collision is small. Under these circumstances the interacting potential can be treated as a perturbation. In zero order the incoming particle moves with its initial velocity along a straight line (z direction) defined by the impact parameter, and the angle of deflection ϑ is simply given by the ratio p_y/p_z, if $p_z = \mu g$ is the momentum of the relative motion and p_y the momentum transferred during the collision perpendicular to p_z. Since

$$p_y = \int_{-\infty}^{\infty} F \cos\varphi \, dt \tag{73}$$

where $F \cos\varphi$ is the component of the force in y direction, we obtain with $dt = dz/g$ and $\cos\varphi = b/r$

$$\Theta = -2\frac{b}{E}\int_b^{\infty} \frac{dV}{dr}(r^2 - b^2)^{-1/2}\, dr \tag{74}$$

or, in reduced notation,

$$\Theta = -(2\beta/K)\int_{\beta}^{\infty} \frac{dU}{dx}(x^2 - \beta^2)^{-1/2}\, dx \tag{75}$$

It can be shown that this approximation is equivalent to the small-angle approximation discussed in Section 4.2.1. Although its range of validity is limited to large values of K and large impact parameters (small angles), so that the straight-line approximation is expected to hold, equation (75) describes at least qualitatively the whole attractive branch of the deflection function over a wide range of reduced energies. In the case $K = 2$, for instance, where the exact rainbow angle for a L.-J. (12, 6) potential is 65.3°, the momentum approximation makes an error of 10%, and this error decreases rapidly with increasing reduced energy. This approximation, therefore, is often very useful to get a first estimate of potential parameters from experimental data, especially since the integration of equation (75) can be performed analytically for the basic potentials. The essential results are summarized in the following.

For the Lennard-Jones $(n, 6)$ potential we obtain from equation (75)

$$\Theta = \frac{3\pi n}{16(n-6)K}\left\{\frac{32(n-1)f(n)}{\pi n}\beta^{-n} - 5\beta^{-6}\right\} \tag{76}$$

where $f(n)$ is defined by equation (38) (see Table 1). From this equation the characteristic quantities which describe the rainbow scattering can be easily determined (see Table 2).

In the special case $n = 12$ it is also possible to derive a closed expression for the differential scattering cross section as far as it is determined by the attractive branch of the deflection function. The result is

$$\begin{aligned} I(\vartheta)\sin\vartheta = {} & \frac{r_m^2}{24[100 - (154/\pi)K\vartheta]^{1/2}}\left(\frac{3\pi}{4}\right)^{1/3}\vartheta^{-4/3}K^{-1/3} \\ & \times \{(10 + [100 - (154/\pi)K\vartheta]^{1/2})^{4/3} \\ & + (10 - [100 - (154/\pi)K\vartheta]^{1/2})^{4/3}\} \end{aligned} \tag{77}$$

This formula shows again the monotonic decrease of the differential scattering cross section at small angles $I(\vartheta) \sim \vartheta^{-7/3}$ followed by the rainbow singularity at $\vartheta = \vartheta_r = 50\pi/77K$, where equation (77) reduces to equation (45) if the values of Table 2 for the rainbow parameters are used.

With the aid of equation (76) it is also possible to calculate the incom-

Table 2. Characteristic Quantities of the Rainbow Scattering (First-Order Momentum Approximation)

Potential	β_r	$\lvert\vartheta_r\rvert$	$\bar{q}$
L.-J. $(n, 6)$	$\left[\frac{16(n-1)}{15\pi}f(n)\right]^{1/(n-6)}$	$\frac{(n-1)f(n)}{K}\beta_r^{-n}$	$3n\lvert\vartheta_r\rvert\beta_r^{-2}$
L.-J. (12, 6)	1.0631	$\frac{2.0400}{K}$	$\frac{64.9782}{K}$
Buckingham $(\gamma, 6)$	$1 + \frac{2}{2(\gamma-6)+3}\ln\frac{8}{15}\left(\frac{2\gamma}{\pi}\right)^{1/2}$	$\frac{e^{\gamma(1-\beta_r)}}{K}\left(\frac{\pi\beta_r\gamma}{2}\right)^{1/2}\frac{(\gamma\beta_r-6)}{(\gamma-6)}$	$3\gamma\lvert\vartheta_r\rvert\frac{(\gamma\beta_r-7)}{\beta_r(\gamma\beta_r-6)}$
Morse	$1 + \frac{1}{2\gamma}\ln 2$	$\frac{(\pi\beta_r\gamma)^{1/2}}{2K}$	$\gamma^2\lvert\vartheta_r\rvert$

plete total scattering cross section (see Section 4.3) in the transition region. For a L.-J. (12, 6) potential the result is, for $\alpha < \vartheta_r$,

$$\frac{S(\alpha)}{\pi r_m^2} = \left(\frac{231}{160}\right)^{1/3} \left\{ \left[1 - \left(1 - \frac{\alpha}{\vartheta_r}\right)^{1/2}\right]^{-1/3} + \left[1 + \left(1 + \frac{\alpha}{\vartheta_r}\right)^{1/2}\right]^{-1/3} - \left[1 + \left(1 - \frac{\alpha}{\vartheta_r}\right)^{1/2}\right]^{-1/3} \right\} \tag{78}$$

and for $\alpha > \vartheta_r$

$$\frac{S(\alpha)}{\pi r_m^2} = \left(\frac{231}{160}\right)^{1/3} \left[1 + \left(1 + \frac{\alpha}{\vartheta_r}\right)^{1/2}\right]^{-1/3}$$

ϑ_r can be calculated from Table 2.

For the Buckingham (γ, 6) potential we obtain from equation (75)

$$\Theta(\beta) = K^{-1}\{[6\gamma/(\gamma - 6)]\, e^{\gamma}\beta K_0(\gamma\beta) - [15\pi\gamma/16(\gamma - 6)]\, \beta^{-6}\} \tag{79}$$

where K_0 is the modified Bessel function of zeroth order. In the range of validity of the momentum approximation we have $\gamma\beta \gg 1$, so that $K_0(\gamma\beta)$ may be replaced by its asymptotic value

$$K_0(\gamma\beta) \sim \left(\frac{\pi}{2\gamma\beta}\right)^{1/2} e^{-\gamma\beta} \tag{80}$$

Thus we finally obtain

$$\Theta(\beta) = K^{-1}\left[\frac{6}{\gamma - 6}\left(\frac{\pi\gamma\beta}{2}\right)^{1/2} e^{\gamma(1-\beta)} - \frac{15\pi}{16}\left(\frac{\gamma}{\gamma - 6}\right)\beta^{-6}\right] \tag{81}$$

From this expression the approximate values of the rainbow parameters shown in Table 2 are obtained.

Similarly we obtain for the Morse potential

$$\Theta(\beta) = 2\gamma\beta K^{-1}[\exp(2\gamma)\, K_0(2\gamma\beta) - \exp(\gamma)\, K_0(\gamma\beta)] \tag{82}$$

or, making use of equation (80),

$$\Theta(\beta) = [(2\pi\gamma\beta)^{1/2}/K]\,[2^{-1/2}\exp(2\gamma(1 - \beta)) - \exp(\gamma(1 - \beta))] \tag{83}$$

5. *Quantal Treatment*

5.1. *Introduction*

In quantum mechanics it is not possible to describe the trajectory of a single particle, since the uncertainty principle does not allow an exact specification of the position $\mathbf{r}(t)$ and the momentum $\mathbf{p}(t)$ at a given time t.

Instead of having the initial conditions $\mathbf{r}(t_1)$ and $\mathbf{p}(t_1)$, we have a wave packet described by a wave function $\psi_i(\mathbf{r}, t_1)$, which yields the probability with which the particle may be found at a certain position $\mathbf{r}(t_1)$ at the initial time t_1. This wave packet must be constructed in such a way that it represents the initial expectation values

$$\langle \mathbf{r} \rangle_i = \langle \psi_i | \mathbf{r} | \psi_i \rangle$$

and

$$\langle \mathbf{p} \rangle_i = \langle \psi_i | \mathbf{p} | \psi_i \rangle$$

The equation of motion is the time-dependent Schrödinger equation

$$H|\psi(\mathbf{r}, t)\rangle = i\hbar \frac{\partial}{\partial t} |\psi(\mathbf{r}, t)\rangle \tag{84}$$

with

$$H = \frac{\hbar^2}{2\mu} \nabla^2 + V(r)$$

In a field-free region the most general solution of this equation can be constructed from plane waves

$$\psi(\mathbf{r}, t) = \int d\mathbf{p}' a_{\mathbf{p}'} \exp\left\{ \frac{i}{\hbar} [(\mathbf{p} \cdot \mathbf{r}) - E't] \right\} \tag{85}$$

with

$$E' = \frac{p'^2}{2\mu}$$

From equation (85) any initial wave packet may be constructed by choosing appropriate amplitudes $a_{\mathbf{p}'}$. Then its evolution with time due to the dispersion of the de Broglie waves in the scattering field can be calculated by solving the time-dependent Schrödinger equation (84). The result is the final state $\psi_f(\mathbf{r}, t_f)$ at the time t_f after the collision.

5.2. Stationary Scattering Theory and Partial-Wave Analysis

The foregoing procedure is considerably simplified if we discuss the scattering of a monoenergetic beam of particles. In this case the initial wave packet degenerates to a simple plane wave, since the requirement of being monoenergetic is incompatible with the localization in space which is required to study the motion of the wave packet in time. The monoenergetic limit of the wave packet is obtained from equation (85) if the amplitude $a_{\mathbf{p}'}$ is replaced by a δ function:

$$a_{\mathbf{p}'} = \delta(\mathbf{p}' - \mathbf{p})$$

This yields for the wave function in the field-free space

$$\psi(\mathbf{r}, t) = \exp\left\{\frac{i}{\hbar}[(\mathbf{p}\cdot\mathbf{r}) - Et]\right\} \tag{86}$$

In the presence of a scattering field the plane wave is no longer a solution. If the scattering field is time independent, however, a separation of time and space is possible:

$$\psi(\mathbf{r}, t) = \psi_k(\mathbf{r}) \exp\left\{-\frac{i}{\hbar}Et\right\} \tag{87}$$

Substituting equation (87) into equation (84) yields a time-independent Schrödinger equation

$$\left[\nabla^2 - \frac{2\mu}{\hbar^2}V(r) + k^2\right]\psi_k(\mathbf{r}) = 0 \tag{88}$$

with

$$\mathbf{p} = \mathbf{k}\hbar, \qquad \mathbf{k}^2 = \frac{2\mu}{\hbar^2}E$$

This is the starting point of the so-called stationary scattering theory. It can be visualized with the following picture: The incoming plane wave represents a steady beam of particles. The beam is scattered and at each angle a time-independent scattered beam intensity is observed.

From this picture it is easy to obtain the asymptotic behavior of the wave function $\psi_k(r)$. Asymptotically ψ_k must consist of an incoming plane wave and an outgoing spherical wave centered about the origin of the scattering field. The amplitude of the scattered wave must depend on two angles ϑ and φ of a polar coordinate system. Thus we obtain

$$\psi_k(\mathbf{r}) \xrightarrow[r\to\infty]{} e^{i(\mathbf{k}\cdot\mathbf{r})} + \frac{e^{ikr}}{r}f(\vartheta, \varphi) = \psi_0 + \psi_s \tag{89}$$

From this asymptotic behavior of the wave function the cross sections can be obtained by considering the particle flux, as in classical mechanics. The particle flux density is

$$j = \frac{\hbar}{2i\mu}(\psi_k^* \nabla\psi_k - \psi_k \nabla\psi_k^*) \tag{90}$$

The incident plane wave ψ_0 yields a steady particle flux density

$$\mathbf{j}_0 = \frac{\hbar\mathbf{k}}{\mu} \tag{91}$$

whereas the scattered wave ψ_s yields a flux density

$$\mathbf{j}_s = \frac{\hbar\mathbf{k}}{\mu r^2}|f(\vartheta, \varphi)|^2 \tag{92}$$

The flux density due to the cross term of incident and scattered waves can be shown to vanish everywhere except in the forward direction.

Following the definition of the differential scattering cross section

$$d\sigma = \frac{\text{particle flux scattered into solid angle } d\omega}{\text{incident particle flux density}} \tag{93}$$

we obtain

$$\frac{d\sigma}{d\omega} = I(\vartheta, \varphi) = |f(\vartheta, \varphi)|^2 \tag{94}$$

and

$$\sigma = \int |f(\vartheta, \varphi)|^2 \, d\omega \tag{95}$$

For central scattering fields, the scattering amplitude becomes independent of the azimuthal angle φ.

The solution of the time-independent Schrödinger equation (88) is expanded in terms of Legendre polynomials:

$$\psi_k(r) = \sum_{l=0}^{\infty} (2l + 1)\, i^l \left(\frac{\alpha_l}{kr} \right) \chi_l(r)\, P_l(\cos\vartheta) \tag{96}$$

The radial wave functions $\chi_l(r)$ obey the differential equation

$$\frac{d^2\chi_l}{dr^2} + \left[k^2 - \frac{2\mu}{\hbar^2} V(r) - \frac{l(l+1)}{r^2} \right] \chi_l = 0 \tag{97}$$

or, in reduced notation,

$$\frac{d^2\chi_l}{dx^2} + \left[A^2 - BU(x) - \frac{l(l+1)}{x^2} \right] \chi_l = 0 \tag{98}$$

where

$$A = kr_m = \frac{\mu g r_m}{\hbar}$$

is the reduced wave number and

$$B = 2\mu\varepsilon r_m^2/\hbar^2$$

is a quantum parameter, which often is referred to as "potential well capacity," since the number of bound vibrational states with zero angular momentum is proportional to $B^{1/2}$.

The asymptotic form of the radial wave function follows from equation (96):

$$\chi_l(r) \to \sin(kr - l\pi/2 + \delta_l) \tag{99}$$

Since in the field-free case the asymptotic behavior of the solution of equation (97) is

$$\chi_l^{(f)} \rightarrow \sin(kr - l\pi/2) \tag{100}$$

the phase shifts δ_l describe the influence of the potential.

The coefficients of the expansion (96) are determined in such a way, that the asymptotic behavior (89) is obtained.*

This yields for the expansion coefficients $\alpha_l = \exp(i\delta_l)$, and for the scattering amplitude

$$f(\vartheta) = -\frac{i}{2k}\sum_{l=0}^{\infty}(2l+1)\left[S_l(k) - 1\right]P_l(\cos\vartheta) \tag{101}$$

where the S matrix can be expressed by the phase shifts in the form

$$S_l(k) = e^{2i\delta_l(k)} \tag{102}$$

Making use of the orthogonality property of the Legendre polynomials, we obtain for the total scattering cross section

$$\sigma = \frac{4\pi}{k^2}\sum_{l=0}^{\infty}(2l+1)\sin^2\delta_l \tag{103}$$

A comparison of this formula with equation (101) shows the validity of the optical theorem

$$\sigma = (4\pi/k)\,\mathrm{Im}(f(0)), \tag{104}$$

which is valid under more general conditions and which was first derived for the scattering of light by spherically symmetric obstacles.[100]

Equations (102) and (103) can be used to calculate the cross sections for collisions between unlike atoms in which one is in a $^1S_0$ state and the other may be either in a $^1S_0$ or $^2S_{1/2}$ state, yielding the single molecular state $^1\Sigma^+$ or $^2\Sigma^+$, respectively, so that only one potential is involved in the scattering process. If the colliding atoms are in other electronic states, more than one molecular state exists and the corresponding potentials will all contribute to the scattering if the beams are not selected with respect to their states. The cross sections become sums of contributions from each state, properly weighted according to the multiplicity of the molecular state, if inelastic effects (transitions between the potentials) can be neglected.

The numerical calculation of cross sections according to equations (102) and (103) is straightforward. At specified values of the reduced wave number A and the quantum parameter B one has to solve the radial Schrödinger equation for each angular momentum l with the boundary condition $\chi_l(0) = 0$. The radial wave function is then compared with a spherical Bessel function of the same order (which is the solution of the field-free equation) and the asymptotic phaseshift δ_l obtained. Finally the summations

* For details see standard textbooks on quantum mechanics.

of equations (101) and (103) are carried out and the cross sections determined. The necessary computer time for such calculations depends on the number of angular momenta l involved. This number is approximately given by $l_{max} = ka$, where a describes the size of the scattering field. For collision pairs with light reduced mass, such as He–He for instance, l_{max} is of the order of magnitude of 10–200 in the energy range accessible with thermal beams. With increasing reduced mass, however, l_{max} increases and typical values lie between several hundred and several thousand. This makes the numerical integration of the radial wave equations very time consuming (for an outline of the computational procedure see, for example, References 101–105). Fortunately the accuracy of the available approximate methods increases also with increasing mass. For heavier systems, therefore, the phase shifts are usually calculated using the JWKB approximation (see Section 6) or, if higher accuracy is required, with more refined approximations.[105,106] In the case of orbiting, however, the numerical solution of the radial wave equations remains, in general, necessary. For large angular momenta, where the phase shifts are small, they may be calculated in closed form using the high-energy or Born approximation (see Section 6).

Instead of using the numerical integration of the radial Schrödinger equation for an exact evaluation of the phase shifts, it is also possible to integrate a nonlinear, first-order differential equation derived for the phase shifts by Calogero[107,108]:

$$\frac{d\gamma_l}{dx} = \frac{A}{K} U_{\text{eff}}(x) \sin^2(Ax + \gamma_l) \tag{105}$$

The asymptotic solution of this equation yields the phase shift δ_l:

$$\delta_l = \lim_{x \to \infty} \gamma(x) + l\pi/2 \tag{106}$$

This method has the advantage that $\gamma_l(x)$—apart from a few oscillations at small distances—approaches its asymptotic value monotonically. The numerical integration, therefore, can be performed with a larger step width than in the case of the Schrödinger equation.[109]

5.3. *Examples of Numerical Results*

5.3.1. *Differential Cross Sections*

As a result of exact numerical calculations, Figure 9 shows the phase shifts δ_l as a function of l for different values of the reduced wave number A and for a fixed value of the quantum parameter $B = 729$. The calculations have been performed for a Lennard-Jones $(\infty, 4)$ potential.[110] The discontinuities (each of magnitude π) in the curves belonging to small A values

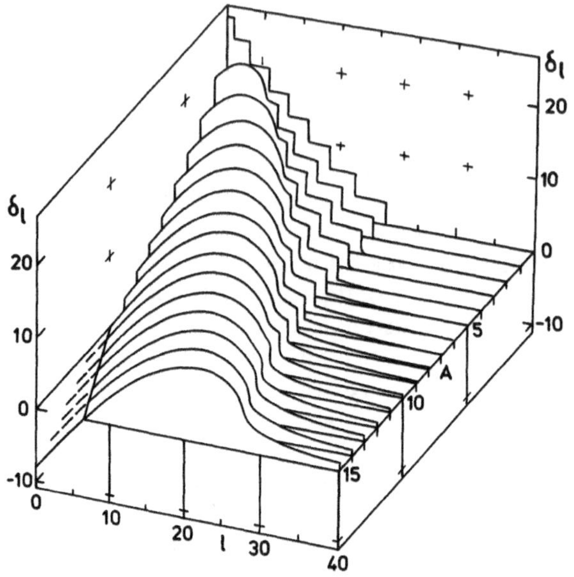

Figure 9. Dependence of the phase shift δ_l upon the angular momentum quantum number l and the reduced wave number A, calculated for a L.-J. (∞, 4) (Sutherland) potential for a fixed value of the quantum parameter $B = 729$. (Adapted from Reference 110.)

correspond to resonance levels within the barrier (see Section 5.4). This corresponds to the classical orbiting.

Figure 10 shows a plot of the differential scattering cross section $I(\vartheta)\sin\vartheta$ as a function of the angle of deflection calculated for a Lennard-Jones (12, 6) potential (with a reduced energy $K = 4$; the reduced wave numbers are $A = 400, 300$, and 200).

Similar to the classical result, which is also shown in Figure 10, two angular regions can be distinguished. At low angles ($\vartheta < \vartheta_r$), where in the classical case three impact parameters lead to the same angle of deflection, we observe a marked interference structure, whereas at angles larger than the classical rainbow angle the interferences die out and we obtain a monotonic behavior of the cross section, similar to the classical result. The increase of the angular separation of the interference oscillations with decreasing reduced wave number is demonstrated from the three cross section curves (a), (b), and (c).

The oscillations with large angular separation are called the rainbow effect in analogy to optics. The first maximum near the classical rainbow is the primary rainbow; the other maxima are called supernumerary rainbows.

Oscillations with small angular separation are observed superimposed on the rainbow structure. They are discussed, together with the rainbow oscillations, in Section 6 in terms of interfering partial waves.

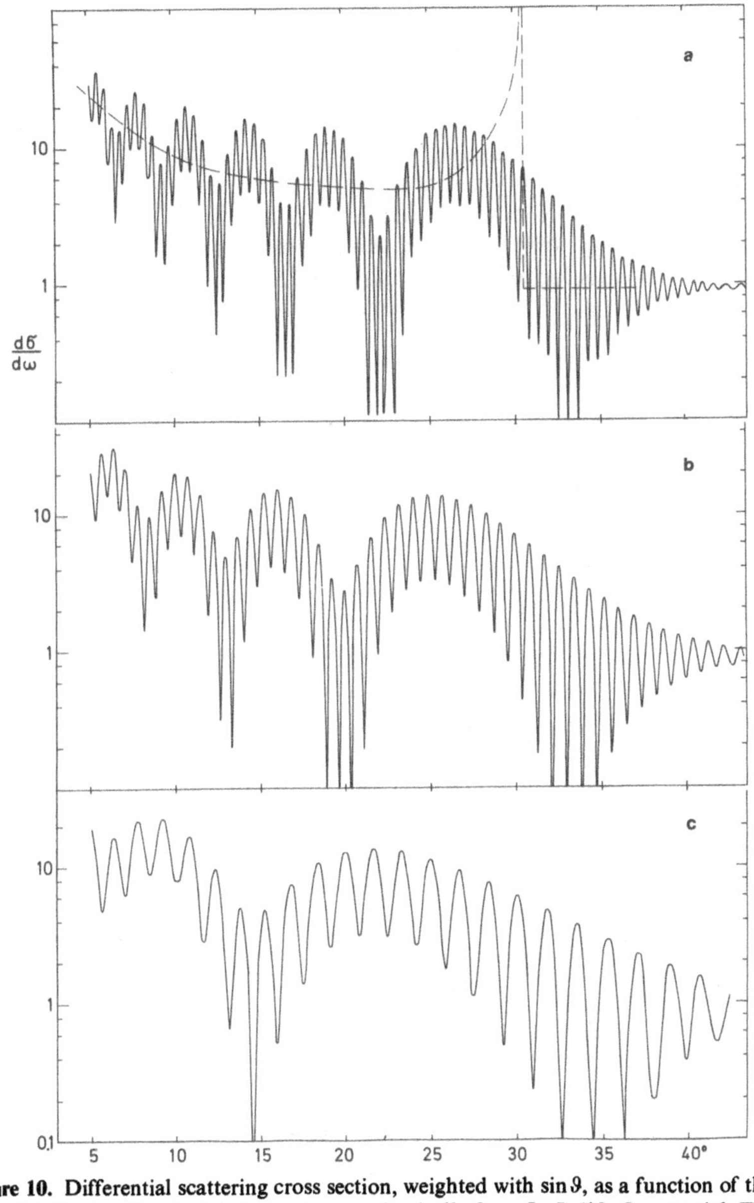

Figure 10. Differential scattering cross section, weighted with $\sin\vartheta$, as a function of the angle of deflection ϑ, calculated quantum mechanically for a L.-J. (12, 6) potential. The three curves (a), (b), and (c) refer to the same reduced energy $K = 4$ but different values of the reduced wave number A: (a) $A = 400$; (b) $A = 300$; (c) $A = 200$. The corresponding classical differential cross section is also shown [dashed curve in part (a)]. (From Reference 136.)

If we average over all oscillations, the classical cross section is a good approximation to the scattering, except in the direct vicinity of the classical rainbow singularity (see Figure 10). This is the reason transport properties usually can be treated by classical mechanics. An exception are collision partners with a very small attractive well (small value of the quantum parameter B) at low velocities (small values of the reduced wave number A), for instance, He–He at low temperatures. In these cases the angular separation of the interference extrema become large enough so that quantum effects become observable in spite of the averaging processes which have to be made in the calculation of the transport cross sections.

5.3.2. Total Scattering Cross Section

General Behavior. The velocity dependence of the total scattering cross section is shown in Figure 11. Here the total cross section is plotted against a reduced velocity g/g_c, with $g_c = 2\varepsilon r_m/\hbar$. Two velocity regions may be distinguished. At low velocities $g < g_c$ an oscillatory structure is observed, since both attractive and repulsive forces contribute to the scattering, giving rise to interference oscillations (referred to as glory undulations in analogy to optics), which are closely related to the rainbow structure in the differential scattering cross section.[44,111] These undulations are shown in Figure 12 in more detail for two special examples.[112]

The total number m of glory maxima is equal to the number n_0 of bound vibrational states with zero angular momentum of the collision pair, as has been shown by Bernstein.[113,114] It is therefore possible to calculate m in closed form for a Morse and a Lennard-Jones potential. The results

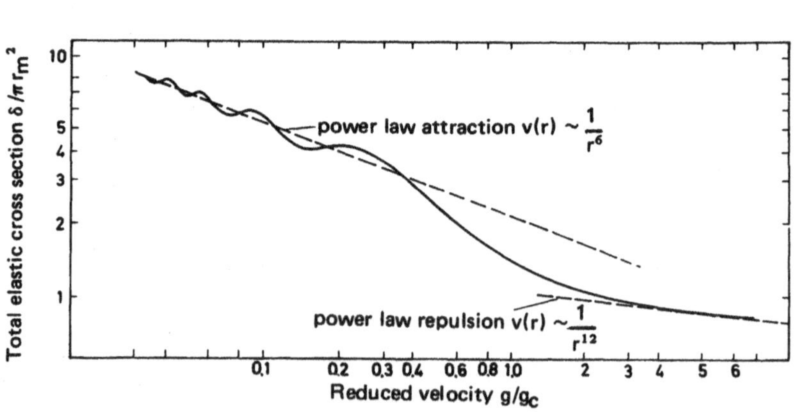

Figure 11. Total elastic scattering cross section as a function of the reduced relative velocity $g\hbar/2\varepsilon r_m$ [L.-J. (12, 6) potential]. The dashed lines are the semiclassical approximations discussed in Section 6. The undulations for $g < g_c$ are displayed in more detail Figure 12. (From Reference 136.)

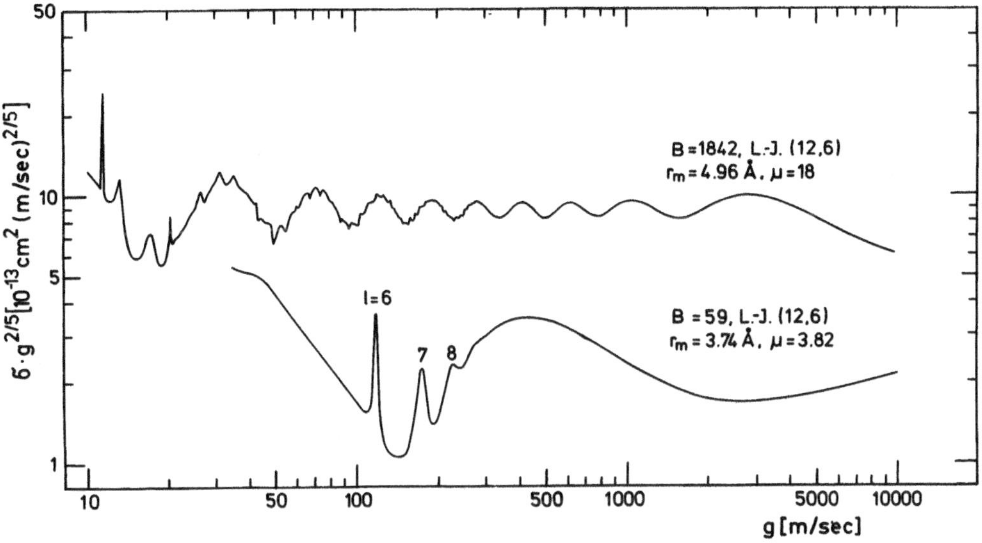

Figure 12. Two examples of the total elastic scattering cross section σ (weighted with $g^{2/5}$) as a function of the relative velocity. The factor $g^{2/5}$ removes the monotonic decrease of the cross section. A L.-J. (12, 6) potential has been used with $B = 1842$ (upper curve) and $B = 59$ (lower curve). There are 9 bound vibrational states with zero angular momentum in the first case, compared to only one in the second case. At low velocities orbiting resonances are superimposed on the glory structure. (From Reference 112.)

can be represented by the relation

$$m = n_0 = aB^{1/2} - b \tag{107}$$

from which n_0 is obtained by rounding the values calculated by the above equation off to the next integer. For values of a and b see Table 3.

Table 3 includes also values for the L.-J. (12, 6), Buckingham (12, 6), and Buckingham (15, 6) potential, obtained from numerical solutions of the wave equation.(115)

Due to the Levinson theorem,(116) the s wave phase shift in the limit $K \to 0$ is

$$\lim_{K\to 0} \delta_s = n_0 \pi \tag{108}$$

The values of Table 3, therefore, are also useful to estimate the behavior of the s phase shift at extremely low energies.

For velocities $g > g_c$, the total scattering cross section becomes a monotonic function of the velocity, since the collision is governed mainly by the repulsive part of the potential. A more detailed discussion of the cross section and an explanation of the different interference effects can be given with the aid of the semiclassical approximation (see Section 6).

Table 3. Number n_0 of Bound Vibrational States of the Collision Pair (Total Number of Glory Maxima m in the Total Scattering Cross Section)

	Quantum result		WKB approximation	
Potential	*a*	*b*	*a*	*b*
Morse	$1/\gamma$	1/2	$1/\gamma$	1/2
L.-J. (10, 6)[a]	$5/8 \cdot 3^{1/2}$	5/8	$5/8 \cdot 3^{1/2}$	1/2
L.-J. (n, 6)	—	—	$\frac{[n(n-6)]^{1/2}}{2\pi(n-2)}\left(\frac{2n}{3}\right)^{[2/n-6]}\frac{\Gamma^2(1+2/(n-6))}{\Gamma(1+4/(n-6))}$	1/2
L.-J. (12, 6)	0.2387	0.6302	0.23858	0.5
Buckingham (12, 6)	0.2443	0.6203	—	—
Buckingham (15, 6)	0.2262	0.6276	—	—

[a] See Mahan and Lapp.[47]

5.4. *Resonance Scattering*

At very low velocities (energies near and below the orbiting threshold) the total scattering cross section may show sharp spikes superimposed on the glory oscillations, due to quasibound states of the effective potential (see Figure 12). To explain these resonances, we first consider an interaction potential with a repulsive barrier which extends to infinity, as is shown in Figure 13a. Supposing that the potential barrier is sufficiently high, there may be a number of bound states of given angular momentum lying above the zero-energy asymptote.

If this barrier is replaced by a finite one, as in Figure 13b, the bound states no longer exist as stationary states but decay in time, due to tunneling through the barrier. They are called quasibound, resonating, or decaying states and their energy eigenvalues E_r are more or less shifted and broadened depending on the barrier thickness, as compared with those of the bound states. Since in the stationary scattering theory the time dependence of the total wave function is

$$\psi(\mathbf{r}, t) = \psi_k(\mathbf{r}) \exp\{-(i/\hbar) Et\} \tag{109}$$

the probability $\psi^*\psi$ of finding a particle at a certain position is independent of time, as long as only real values of the energy are considered. Thus decaying states require a full time-dependent description or, in the stationary theory, complex energies. Assuming that the spacings of the decaying levels are large compared with the width of the levels, we put at the resonance

$$E = E_r - (i/2)\,\Gamma \tag{110}$$

with E_r and Γ real and $\Gamma > 0$. Now the probability of finding a particle at a certain position becomes time dependent:

$$\psi^*\psi = \psi_k^*\psi_k \exp\{-(\Gamma/\hbar)\,t\} \tag{111}$$

The probability of decay per unit time is $\omega = \Gamma/\hbar$. The width of the level, according to the uncertainty principle, is Γ. We now investigate the consequences of complex energies for the radial wave function. The asymptotic behavior of χ_l [see equation (99)] may be written

$$\chi_l(r) \xrightarrow[r\to\infty]{} A_l(E)\,e^{-ikr} + A_l^*(E)\,e^{ikr} \tag{112}$$

where $A_l(E)$ are complex functions of the complex energy

$$A_l(E) = e^{-i(\delta_l - l\pi/2)} \tag{113}$$

such that $A_l[E_r - (1/2)\,i\Gamma] = 0$ and only the outgoing wave e^{ikr} remains, describing the particles leaving the potential well through the barrier which are responsible for the decay. In the vicinity of $E_r - (i/2)\,\Gamma$ we expand A_l, obtaining

$$A_l(E) = A_l\left(E_r - \frac{i}{2}\Gamma\right) + a_l\left[E - \left(E_r - \frac{i}{2}\Gamma\right)\right] + \cdots \tag{114}$$

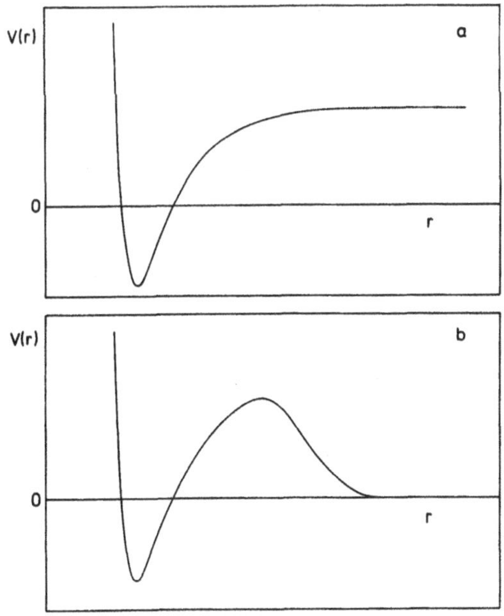

Figure 13. (a) Potential $V(r)$ with a repulsive barrier that extends to infinity. (b) Potential $V(r)$ with a repulsive barrier of finite extent. (From Reference 136.)

with

$$a_l = \left(\frac{dA_l}{dE}\right)_{E=E_r-(i/2)\Gamma}$$

Using equations (113) and (114), we obtain for the S matrix

$$S_l(E) = e^{il\pi}(a_l^*/a_l)\left[1 - i\Gamma\Big/\left(E - E_r + \frac{i}{2}\Gamma\right)\right] \tag{115}$$

If we write the phase shift δ_l as the sum

$$\delta_l = \tau_l + \eta_l \tag{116}$$

with

$$e^{2i\tau_l} = \exp(il\pi)\,(a_l^*/a_l) \tag{117}$$

the scattering matrix becomes

$$S_l(E) = e^{2i\tau_l}\left\{1 - i\Gamma\Big/\left(E - E_r + \frac{i}{2}\Gamma\right)\right\} \tag{118}$$

τ_l is a potential or background scattering phase shift which varies slowly with energy, and η_l is small except in regions near the resonance energy. In the neighborhood of the resonance levels we obtain from equation (118)

$$\delta_l = \tau_l + \arctan[\Gamma/2(E_r - E)] \tag{119}$$

If the energy E of the incident particles increases from below E_r to above E_r, the phase shift δ_l changes by π. On putting equation (119) into equation (101) (if E_r is the only resonance level for l), we have

$$f(\vartheta) = (2ik)^{-1}\sum_{l=0}^{\infty}(2l+1)\,[e^{2i\tau_l} - 1]\,P_l(\cos\vartheta)$$
$$-(2k)^{-1}(2l+1)\,e^{2i\tau_l}\frac{\Gamma}{[E - E_r + (i/2)\Gamma]}P_l(\cos\vartheta) \tag{120}$$

The first sum is called the potential scattering amplitude, while the second term is called the resonance scattering amplitude. The differential scattering cross section will contain contributions from each, and will also contain an interference term. The resonance cross section $I_l(\vartheta)$ becomes

$$I_l(\vartheta) = \frac{(2l+1)^2}{4k^2}\frac{\Gamma^2}{(E - E_r)^2 + \Gamma^2/4}[P_l(\cos\vartheta)]^2 \tag{121}$$

The partial total cross section

$$\sigma_l = (4\pi/k^2)(2l+1)\sin^2(\tau_l + \eta_l) \tag{122}$$

becomes

$$\sigma_l = \frac{4\pi}{k^2}(2l+1)\left[(\sin^2\tau_l)\frac{(E-E_r)^2}{(E-E_r)^2+\Gamma^2/4} + \cos^2\tau_l\frac{\Gamma^2/4}{(E-E_r)^2+\Gamma^2/4} + \sin^2\tau_l\frac{\Gamma(E-E_r)}{(E-E_r)^2+\Gamma^2/4}\right] \tag{123}$$

which reduces to the well-known Breit–Wigner[117] formula if the potential phase shift τ_l is small.

The shape of a resonance in the total cross section depends, according to equation (123), on the value of the background phase shift τ_l. Figure 14 shows typical resonance shapes for constant background phase shifts $\tau_l = 0$, $\pi/4$, $\pi/2$, and $3\pi/4$.

The occurrence of resonance scattering is the usual case in atom–atom scattering, since the repulsive barrier necessary for the containment of decaying states is provided by the centrifugal term. This is illustrated by Figure 15, which shows the interatomic potential (for the H–Kr interaction) as a function of the internuclear distance together with the effective potentials for angular momenta $l = 4$, $l = 5$, and $i = 6$.[118] The quasibound states are the continuation of the bound rotational states to positive energies; the energy levels are displayed in the right half of Figure 15.

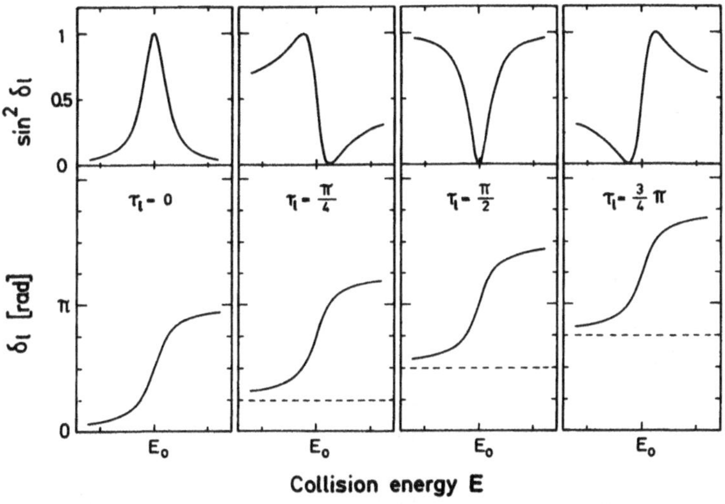

Figure 14. Partial total cross section $\sigma_l \sim \sin^2\delta_l$ as a function of the collision energy E calculated for different background phase shifts $\tau_l = 0$, $\pi/4$, $\pi/2$, and $3\pi/4$ (upper half) and energy dependence of the phase shift δ_l in the vicinity of a resonance (lower half). (From Reference 118.)

Figure 15 also shows the total scattering cross section as a function of the collision energy and it is clearly seen that the total cross section shows a resonance if the collision energy coincides with the energy of a quasibound state.

Figure 16 illustrates the behavior of the radial wave function $\chi_l(r)/r$ in the vicinity of a resonance.[118] In the resonance case the radial wave function obtains large values within the region of the potential well, indicating clearly the high probability $\chi_l^* \chi_l$ of finding the colliding particles close together, corresponding to the existence of the decaying state.

Results of a numerical study of the phase shift behavior due to barrier penetration and resonance effects are discussed by Bernstein *et al.*[119] for a Lennard-Jones (12, 6) potential. The connection between the inverse process, predissociation, and the pattern of resonances also has been investigated.[120,121]

To observe resonances in atom–atom or atom–molecule collisions, several conditions have to be fulfilled:

(i) The reduced collision energy must be below an upper limit, which is approximately given by the orbiting threshold.

(ii) The number of partial waves which contribute to the cross section must be small, so that the resonance cross section makes a considerable contribution to the total cross section. This requires a small wave number,

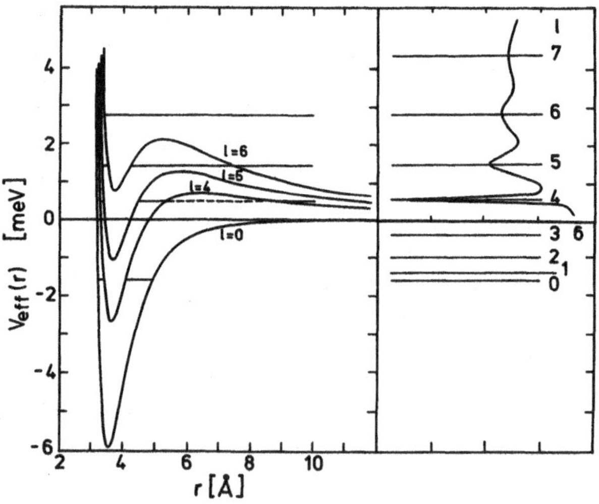

Figure 15. Left side: L.-J. (12, 6) potential (parameters correspond to H–Kr) and effective potentials for angular momentum quantum numbers $l = 4, 5$, and 6. Only one vibrational state ($v = 0$) exists. The horizontal lines correspond to the energies of the quasibound states with the quantum numbers $v = 0, l = 4, 5$, and 6. These quasibound states can be observed as resonances in the energy dependence of the total scattering cross section (right side of the figure). (Adapted from Reference 118.)

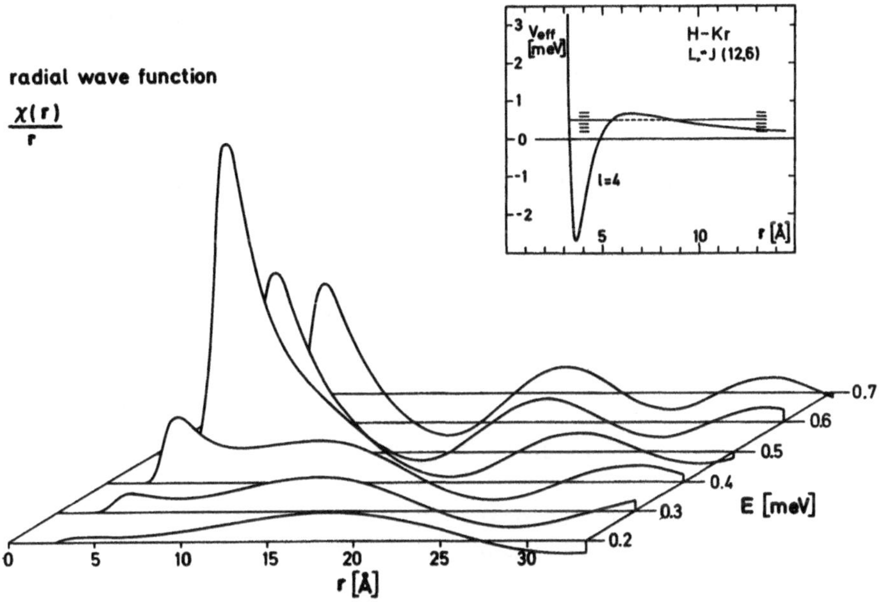

Figure 16. The radial wave function $\chi_l(r)/r$ as a function of the distance r for several energies in the vicinity of the resonance $v = 0, l = 4$ (see Figure 15) at E_r. For $E = E_r$ the large amplitude of χ_l in the well region shows the existence of a decaying state. The energies are indicated in the effective potential shown in the upper half of the figure. (Adapted from Reference 118.)

which in turn requires a small relative velocity and a low reduced mass for the collision pair.

(iii) The spread in the relative velocity must be as small as possible. At low velocities, this spread usually comes from the target beam and can be reduced by choosing an approximate intersection angle between the two molecular beams.[11] For further experimental details see References 122–126.

5.5. *Identical Particles*

In the case of identical particles, the total wave function has to be symmetric (Bose statistics) or antisymmetric (Fermi statistics) with respect to the coordinates. Thus the asymptotic form of the wave function can be written

$$\psi \xrightarrow[r\to\infty]{} e^{ikz} \pm e^{-ikz} + \frac{e^{ikr}}{r}[f(\vartheta) \pm f(\pi - \vartheta)] \tag{124}$$

The positive sign (symmetric wave function) is valid for Bose statistics, the negative sign (antisymmetric wave function) for Fermi statistics. The differential scattering cross section becomes (if we normalize, as before, to one incoming particle)

$$I(\vartheta) = \tfrac{1}{2}|f(\vartheta) \pm f(\pi - \vartheta)|^2 \tag{125}$$

and contains an additional interference term if we compare with the classical result [equation (72)]. Since $P_l(-\cos\vartheta) = (-1)^l P_l(\cos\vartheta)$, the partial-wave expansion of the cross sections becomes

$$I(\vartheta) = (4k)^{-2}\left|\sum_{l=0}^{\infty} \omega_l(2l+1)\left[e^{2i\delta_l} - 1\right] P_l(\cos\vartheta)\right|^2 \tag{126}$$

and

$$\sigma = (4\pi/k^2) \sum_{l=0}^{\infty} \omega_l(2l+1)\sin^2\delta_l \tag{127}$$

with the following relations for the weight factor ω_l.

(i) For bosons (symmetric wave function):

$$\omega_l = \begin{cases} 2 & \text{for } l \text{ even} \\ 0 & \text{for } l \text{ odd} \end{cases}$$

(ii) For fermions (antisymmetric wave function):

$$\omega_l = \begin{cases} 0 & \text{for } l \text{ even} \\ 2 & \text{for } l \text{ odd} \end{cases}$$

In the following the cross sections for the two cases are designated as $I_s(\vartheta)$, σ_s (bosons) and $I_a(\vartheta)$, σ_a (fermions), respectively.

Equations (126) and (127) can be used for a calculation of the cross sections only if the two colliding particles are in the same spin state, so that the total spin of the two-particle system has a definite value. Consequently, these cross sections can only be measured in a molecular beam experiment with two state-selected beams (if the spin of the particles is unequal to zero). Without spin selection, the cross sections become a sum of a symmetric and an antisymmetric contribution, weighted with the statistical weights

Table 4. Statistical Weights g_s and g_a for the Scattering of Identical Particles without Spin Selection

Spin s of the individual particles	g_s	g_a
Half-integer	$s/(2s+1)$	$(s+1)/(2s+1)$
Integer	$(s+1)/(2s+1)$	$s/(2s+1)$

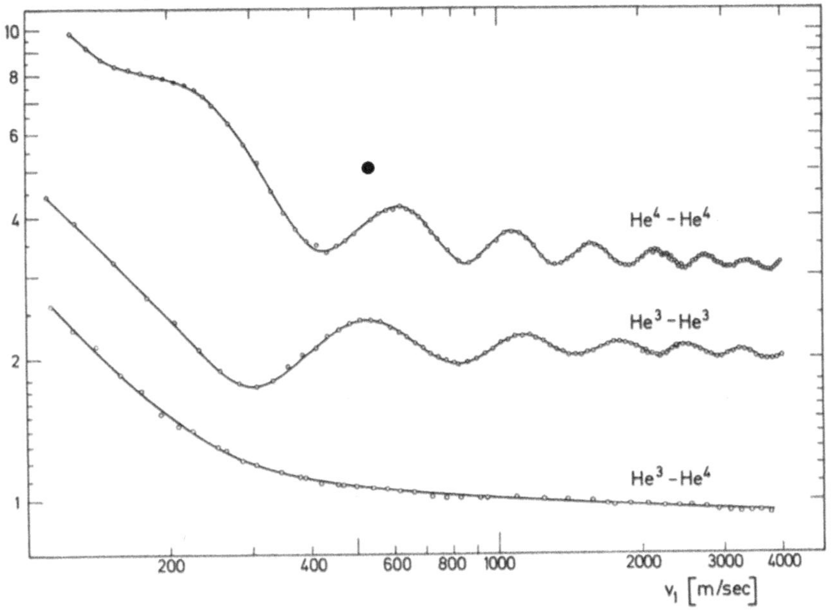

Figure 17. Measured velocity dependence of the total scattering cross section for He–He together with quantum-mechanical calculations for a refined model potential (full lines). (Adapted from Reference 43.)

g_i shown in Table 4:

$$I(\vartheta) = g_s I_s(\vartheta) + g_a I_a(\vartheta) \tag{128}$$

and

$$\sigma = g_s \sigma_s + g_a \sigma_a$$

For an example of the interference effects due to identical particles, Figure 17 shows the measured velocity dependence of the ^4He–^4He and He^3–He^3 total scattering cross section,(127,128) together with calculated curves.(129) This system is best suited to demonstrate these oscillations, since no glory undulations exist due to the low B value of this system. In the velocity dependence of the He^3–He^4 total scattering cross section no oscillations are observed.

6. *Semiclassical Approximation*

6.1. *General Assumptions and Introductory Remarks*

The semiclassical analysis of the scattering not only illustrates the relations between quantum theory and classical mechanics, but also is of high practical value in the analysis of atomic scattering processes. In particular, it provides a way to gain qualitative understanding of the results of the detailed quantum-mechanical calculations discussed in Section 5, and to survey quickly the probable results of a variation of potential parameters or of the collision energy. Since the pioneering work of Ford and Wheeler[89] theories of increasing sophistication have been developed[130-134] and the results are discussed in detail by Berry and Mount.[135] Surveys on semiclassical scattering theory may also be found in several review articles or books.[9-13,16,84,136] We first describe the general assumptions and mathematical approximations introduced into the exact quantum-mechanical expressions for the cross sections and then we will discuss the cross section features, similar as in the classical case.

The partial-wave sum for the scattering amplitude [equation (101)] can be transformed into a rapidly converging sum over integrals, using a modification of the Poisson sum rule.[130] This yields

$$f(\vartheta) = -\frac{i}{k}\sum_{m=-\infty}^{\infty} e^{im\pi}\int_0^{\infty} \lambda\left[e^{2i\delta_{\lambda-1/2}} - 1\right] e^{2im\pi\lambda} P_{\lambda-1/2}(\cos\vartheta)\, d\lambda \quad (129)$$

where $\lambda = l + 1/2$ is now a continuous variable.*

Into equation (129) we introduce the following approximations[89]:

(a) Equation (129) is a sum over integrals, the integrands of which oscillate rapidly over large regions of the variable λ. The main contributions arise from stationary points on the positive real axis, whenever these exist, and therefore the integrals may be evaluated by the method of stationary phase.

(b) The Legendre polynomials are replaced by their asymptotic expressions:

For $\sin\vartheta \lesssim 1/\lambda$,

$$P_{\lambda-1/2}(\cos\vartheta) \cong J_0(\lambda\vartheta) \quad (130)$$

and for $\sin\vartheta \gtrsim 1/\lambda$

$$P_{\lambda-1/2}(\cos\vartheta) \cong \left[2/(\pi\lambda\sin\vartheta)\right]^{1/2} \sin(\lambda\vartheta + \pi/4) \quad (131)$$

(c) The phase shift $\delta_{\lambda-1/2}$ is replaced by its JWKB approximate value[137-140]

$$\delta_{\lambda-1/2} = \int_{x_0}^{\infty}\left(A^2 - BU(x) - \frac{\lambda^2}{x^2}\right)^{1/2} dx - \int_{\lambda/A}^{\infty}\left(A^2 - \frac{\lambda^2}{x^2}\right)^{1/2} dx \quad (132)$$

where x_0 is the largest zero of the radicand (classical turning point). This approximation requires for its validity that the variation of the potential

* The connection between λ and the reduced impact parameter β is $\lambda = \beta A$.

$\Delta V = k^{-1} dV/dr$ with the local wavelength

$$k^{-1} = \frac{\hbar}{[2\mu(E - V)]^{1/2}}$$

has to be a small fraction of the available kinetic energy:

$$\frac{1}{k(E - V)} \left| \frac{dV}{dr} \right| \ll 1 \tag{133}$$

or, in reduced notation,

$$B^{-1/2} [K - U(x)]^{-3/2} \left| \frac{dU}{dx} \right| \ll 1 \tag{134}$$

This relation (134) shows the importance of the quantum parameter B for estimating the "classical" or "quantum" behavior of a collision pair.

The JWKB phase shift is simply related to the classical deflection function Θ:

$$\frac{d\delta(\lambda)}{d\lambda} = \tfrac{1}{2}\Theta(\lambda) \tag{135}$$

This relation is known as the semiclassical equivalence relationship.

6.2. *Special Features of the Differential Cross Section*

6.2.1. *Interference Effects*

We consider the differential cross section for angles ϑ with $0 < \vartheta < \pi$. Introducing equation (131) into equation (129) and making use of the identity

$$\sum_{l=0}^{\infty} (2l + 1) P_l(\cos\vartheta) = 0 \qquad \text{for } \vartheta \neq 0, \pi \tag{136}$$

we obtain

$$f(\vartheta) = -k^{-1}(2\pi \sin\vartheta)^{-1/2} \sum_{m=-\infty}^{\infty} e^{-im\pi} \int_0^{\infty} \lambda^{1/2} [e^{i\varphi_+^{(m)}(\lambda)} - e^{i\varphi_-^{(m)}(\lambda)}] \, d\lambda \tag{137}$$

where

$$\varphi_{\pm}^{(m)}(\lambda) = 2\delta(\lambda) + 2m\pi\lambda \pm \lambda\vartheta \pm \pi/4$$

The stationary points are determined by

$$\frac{d\varphi_{\pm}^{(m)}(\lambda)}{d\lambda} = 0 \tag{138}$$

or, using the semiclassical equivalence relationship equation (135),

$$\Theta(\lambda) = \pm\vartheta - 2m\pi \tag{139}$$

For realistic interatomic potentials, the classical deflection angle is always smaller than π (except for zero impact parameter). Thus the integrals with $m > 0$ have no stationary points and vanish in the stationary phase approximation. In order to simplify the following discussion, we first assume that Θ is never less than $-\pi$. Then all integrals with $m < 0$ have also no stationary points and we obtain for the semiclassical scattering amplitude

$$f(\vartheta) = -k^{-1}(2\pi \sin\vartheta)^{-1/2} \int_0^\infty \lambda^{1/2}\left[e^{i\varphi_+(\lambda)} - e^{i\varphi_-(\lambda)}\right] d\lambda \tag{140}$$

with

$$\varphi_\pm(\lambda) = \varphi_\pm^{(0)}(\lambda) = 2\delta(\lambda) \pm \lambda\vartheta \pm \pi/4$$

Figure 18 shows as an example the classical deflection function $\Theta(\beta)$, the phase shift curve $\delta(\beta)$ versus the reduced impact parameter $\beta = \lambda/A$ for a fixed reduced energy, and the two functions φ_+ and φ_- for a given angle which is indicated in the classical deflection function. If ϑ is smaller than the rainbow angle ϑ_r, φ_+ has two points of stationary phase, whereas φ_- has one point of stationary phase. The three angular momenta which belong to the stationary points correspond to the three classical impact parameters leading into the same angle of deflection ϑ. If $\vartheta > \vartheta_r$, then φ_+ no longer has stationary points.

Classically, the differential cross section is the sum over these three contributions. In the semiclassical theory, however, the scattering amplitude is a sum over three contributions,

$$f(\vartheta) = \sum_{i=1}^{3} f_i(\vartheta) \tag{141}$$

each of which belongs to a certain point of stationary phase. In the differential cross section, therefore, we get interference terms in addition to the classical values.

If the points of stationary phase are well separated from each other, we can expand $\varphi(\lambda)$ in the vicinity of a stationary point λ_i to give

$$\varphi(\lambda) = \varphi(\lambda_i) + \tfrac{1}{2}\varphi''(\lambda_i)(\lambda - \lambda_i)^2 \tag{142}$$

Introducing equation (142) into (140) and performing the integration the contribution of the stationary point λ_i to the scattering amplitude $f_i(\vartheta)$ becomes

$$f_i(\vartheta) = k^{-1}\lambda_i^{1/2}\left[|\varphi''(\lambda_i)| \sin\vartheta\right]^{-1/2} e^{i[\varphi(\lambda_i) - \alpha]} \tag{143}$$

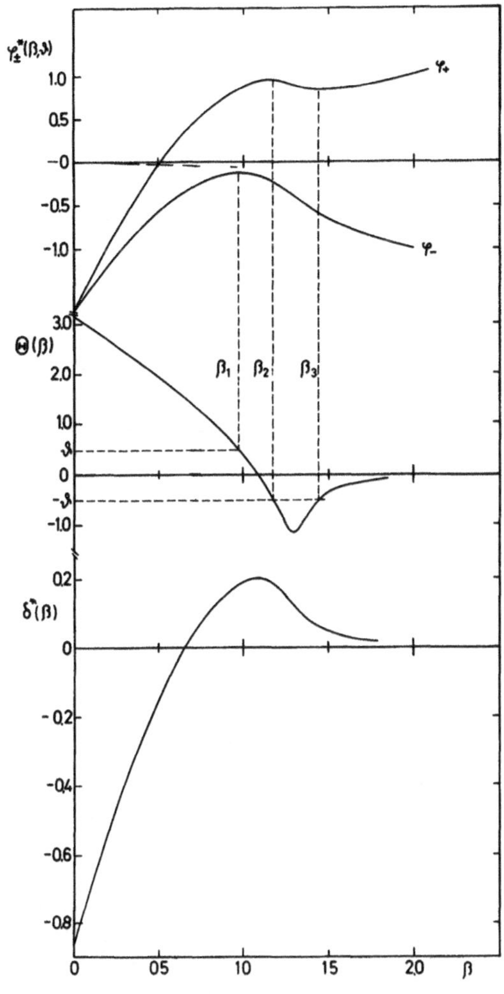

Figure 18. Typical behavior of the classical deflection function $\Theta(\beta)$ (in the center of the figure), the reduced phase shift function $\delta^*(\beta) = \delta(\beta)/A$ (lower half), and the reduced functions $\varphi^*_\pm(\beta, \vartheta) = \varphi_\pm(\beta, \vartheta)/A$ (upper half) with $\vartheta = 28.6°$, as is indicated by the dashed lines in the deflection function. The curves are calculated for a L.-J. (12, 6) potential with $K = 2$. (Adapted from Reference 136.)

where the phase α depends on the sign of $\varphi''(\lambda_i)$:

$$\alpha = \begin{cases} \pi/4 & \text{for} \quad \varphi''(\lambda_i) < 0 \\ 3\pi/4 & \text{for} \quad \varphi''(\lambda_i) > 0 \end{cases}$$

Since

$$\varphi''(\lambda_i) = \frac{1}{A}\left(\frac{d\Theta}{d\beta}\right)_{\beta=\beta_i}$$

equation (143) may be written

$$f_i(\vartheta) = r_m(\sin\vartheta)^{-1/2}\left(\beta_i\left|\frac{d\beta_i}{d\vartheta}\right|\right)^{1/2} e^{i[\varphi(\beta_i)-\alpha]} \tag{144}$$

The differential cross section according to equation (141) becomes

$$I(\vartheta) = (r_m^2/\sin\vartheta)\,[I_{cl}(\vartheta) + 2f_2 f_3 + 2f_1 f_2 + 2f_1 f_3] \tag{145}$$

where $I_{cl}(\vartheta)$ is the classical differential cross section. The interference terms are

$$\begin{aligned} f_2 f_3 &= (I_2 I_3)^{1/2}\cos[2\delta(\lambda_2) - 2\delta(\lambda_3) - (\lambda_2 - \lambda_3)\,\vartheta - \pi/2] \\ f_1 f_2 &= (I_1 I_2)^{1/2}\cos[2\delta(\lambda_1) - 2\delta(\lambda_2) - (\lambda_1 + \lambda_2)\,\vartheta] \\ f_1 f_3 &= (I_1 I_3)^{1/2}\cos[2\delta(\lambda_1) - 2\delta(\lambda_3) - (\lambda_1 + \lambda_3)\,\vartheta + \pi/2] \end{aligned} \tag{146}$$

$I_k(\vartheta)$ is the contribution of β_k to the classical differential cross section. The most rapidly changing part in these terms is $(\lambda_2 - \lambda_3)\,\vartheta$, $(\lambda_1 + \lambda_2)\,\vartheta$, or $(\lambda_1 + \lambda_3)\,\vartheta$, respectively. Thus, to a first approximation, the angular separation of the interference oscillations is

$$\Delta\vartheta_{2;3} = 2\pi|\lambda_2 - \lambda_3|^{-1} = 2\pi(A|\beta_2 - \beta_3|)^{-1} \tag{147}$$

and

$$\Delta\vartheta_{1;2,3} = 2\pi|\lambda_1 + \lambda_{2,3}|^{-1} = 2\pi(A|\beta_1 + \beta_{2,3}|)^{-1}$$

These are the two types of oscillations in the calculated cross section curve, shown in Figure 10.

The interference of f_2 and f_3 yields oscillations with large angular separations (supernumerary rainbows), whereas the interference of f_1 with f_2 and of f_1 with f_3 yield the superimposed rapid oscillations.

If the angle ϑ exceeds the classical rainbow angle, only one contribution to the scattering amplitude remains (in φ_-) and the interferences die out behind the primary rainbow.

If the angle of deflection becomes less than $-\pi$ (in the case of orbiting) additional stationary points in the integrals with $m > 0$ occur. These correspond to the additional impact parameters, which classically lead to the same angle of deflection in this case. Thus further contributions to the scattering amplitude and differential cross section arise.

6.2.2. *Rainbow Scattering*

As the angle of deflection approaches the rainbow angle ϑ_r the two points λ_2 and λ_3 of stationary phase in $\varphi_+(\lambda)$ come closer together and finally meet at $\lambda_2 = \lambda_3 = \lambda_r$. The two contributions to the scattering amplitude f_2 and f_3, therefore, cannot be calculated independently. φ_+ has a reversing point at λ_r and must be expanded in the following way:

$$\varphi(\lambda) = \varphi(\lambda_r) + \varphi'(\lambda_r)(\lambda - \lambda_r) + \frac{\varphi'''}{3!}(\lambda - \lambda_r)^3 \tag{148}$$

Using equations (140), (135), and the expansion of the classical deflection function in the vicinity of the rainbow angle, equation (43), we obtain

$$\varphi(\lambda) = \Delta + (\vartheta + \vartheta_r)(\lambda - \lambda_r) + \frac{\tilde{q}}{3A^2}(\lambda - \lambda_r)^3 \tag{149}$$

with

$$\Delta = 2\delta(\lambda_r) + \lambda_r\vartheta + \pi/4$$

Substitution into equation (140) yields for the rainbow amplitude $f_r(\vartheta)$

$$f_r(\vartheta) = -\frac{r_m\lambda_r^{1/2}e^{i\Delta}}{A(2\pi\sin\vartheta)^{1/2}} \times \int_{-\infty}^{\infty} \exp\left\{i\left[(\vartheta_r + \vartheta)(\lambda - \lambda_r) + \frac{\tilde{q}}{3A^2}(\lambda - \lambda_r)^3\right]\right\} d\lambda \tag{150}$$

This integral may be expressed by the Airy integral $\mathrm{Ai}(x)$ with the final result

$$f_r(\vartheta) = r_m\left(\frac{2\pi\beta_r}{A\sin\vartheta}\right)^{1/2}\left(\frac{\tilde{q}}{A^2}\right)^{1/3} e^{i\Delta}\mathrm{Ai}(x) \tag{151}$$

where

$$x = \left(\frac{\tilde{q}}{A^2}\right)^{-1/3}(\vartheta - \vartheta_r) \tag{152}$$

If we neglect the interference of the rainbow amplitude $f_r(\vartheta)$ with the amplitude $f_1(\vartheta)$ of the repulsive branch of the deflection function, the differential cross section near the rainbow angle has the form

$$I_r(\vartheta) = \frac{2\pi\beta_r r_m^2 A^{1/3}}{\tilde{q}^{2/3}\sin\vartheta}\mathrm{Ai}^2(x) \tag{153}$$

Since for $x \to -\infty$

$$\mathrm{Ai}(x) = \pi^{-1/2}(-x)^{-1/4}\cos\left[\tfrac{2}{3}(-x)^{3/2} - \pi/4\right] \tag{154}$$

the rainbow interference structure reduces to the general case equation (146) obtained when the two points of stationary phase are well separated.

An extension of the foregoing treatment of the rainbow scattering which includes a cubic term in the Taylor-series expansion of the classical deflection function about its minimum has been given by Greene *et al.*[141]

A more general treatment of the rainbow scattering which yields a uniform expression for the rainbow cross section $|f_2(\vartheta) + f_3(\vartheta)|^2$ valid for all angles has been derived by Berry.[130] He obtains for the rainbow cross section

$$I_r(\vartheta) = P^2\mathrm{Ai}^2(-|z|) + Q^2\mathrm{Ai}'^2(-|z|) \tag{155}$$

with

$$P = \pi^{1/2}|z|^{1/4}\{[I_2(\vartheta)]^{1/2} + [I_3(\vartheta)]^{1/2}\}$$
$$Q = \pi^{1/2}|z|^{-1/4}\{[I_3(\vartheta)]^{1/2} - [I_2(\vartheta)]^{1/2}\}$$
$$z = \{\tfrac{3}{4}[2\delta(\beta_2) - 2\delta(\beta_3) + A\vartheta(\beta_2 - \beta_3)]\}^{2/3}$$

$I_i(\vartheta)$ $(i = 1, 2)$ is the classical differential cross section. If the contribution of the repulsive branch of the deflection function (β_1) is included, the complete differential cross section in the uniform approximation becomes

$$I(\vartheta) = I_1 + I_r + 2I_1^{1/2}[Q\mathrm{Ai}'(-|z|)\cos\alpha - P\mathrm{Ai}(-|z|)\sin\alpha]$$

with

$$\alpha = 2\delta(\beta_1) - \delta(\beta_2) - \delta(\beta_3) - A\vartheta[\beta_1 + \tfrac{1}{2}(\beta_2 + \beta_3)] - \pi/4 \quad (156)$$

A comparison of this uniform approximation with exact calculations of $I(\vartheta)$ shows that equation (156) yields highly accurate results if the reduced wave numbers and angles are not too small ($A > 50, \vartheta > 2\pi/A$).[142]

With the aid of equation (155) it is possible to construct the attractive branch of the deflection function from experimental observations of the rainbow scattering (see Section 7).

6.2.3. Orbiting Collisions

For energies below the orbiting threshold we have three classical turning points giving rise to the possibility of tunneling resonances. The JWKB phase shift, therefore, contains an additional contribution from the region between the two inner turning points due to the penetration of the centrifugal barrier. Berry and Mount[135] (see also References 143–145) have derived this contribution and obtain for the phase shift

$$\delta(\lambda) = \bar{\delta}(\lambda) - \frac{\eta}{2} + \arctan\left[\tan\left(\frac{S}{\hbar} - \frac{\eta}{2}\right)\frac{(1 + e^{-2\Gamma/\hbar})^{1/2} - 1}{(1 + e^{-2\Gamma/\hbar})^{1/2} + 1}\right] \quad (157)$$

where S is the classical action between the two inner turning points r_2 and r_3:

$$S = \hbar A \int_{x_3}^{x_2}\left(1 - \frac{U(x)}{K} - \frac{\beta^2}{x^2}\right)^{1/2} dx \quad (158)$$

Γ is the modulus of the action in the classically forbidden region between r_2 and r_1:

$$\Gamma = -i\hbar \int_{x_2}^{x_1}\left(1 - \frac{U(x)}{K} - \frac{\beta^2}{x^2}\right)^{1/2} dx \quad (159)$$

η is a small correction factor[132,146] and $\bar{\delta}(\lambda)$ is the one-turning-point phase shift according to equation (132).

For energies near the barrier top (e.g., $\Gamma/\hbar \ll 1$), equation (157) reduces to

$$\delta(\lambda) = \bar{\delta}(\lambda) - \frac{\eta}{2} \quad \arctan\left\{\tan\left(\frac{S}{\hbar}\right)[3 + 2\cdot 2^{1/2}]^{-1}\right\} \tag{160}$$

whereas for energies far from the barrier top (i.e., $\Gamma/\hbar \gg 1$) we obtain

$$\delta(\lambda) = \bar{\delta}(\lambda) + \arctan\left[\tan\left(\frac{S}{\hbar}\right)\frac{e^{-2\Gamma/\hbar}}{4}\right] \tag{161}$$

Berry[131] has shown that the phase shift equation (157) is insensitive to the potential for $r < r_1$ except very near to resonance energies where S satisfies a Bohr quantization condition $S = [n + \frac{1}{2}]\,\pi\hbar$. Every time S satisfies this condition the phase shift displays a resonant jump of π with a width proportional to $e^{-\Gamma/\hbar}$ similar to that found in the exact quantum-mechanical treatment (see Section 5.4 and Figure 9).

In the semiclassical limit the jumps in the phase shift are sharp and thus have negligible effect on the scattering since this involves $\exp(2i\delta_l)$. Between the resonances the phase shift is then essentially given by the one-turning-point phase shift modulo π. Near the barrier top $\bar{\delta}_l$ will jump catastrophically as the turning point jumps from r_1 to r_3. This jump, however, will be essentially a multiple (the number of bound states supported in that effective potential) of π and again has negligible effect on $\exp(2i\delta_l)$.

In the semiclassical description, therefore, all effects associated with orbiting arise solely from the fact that an infinite number of classical paths contribute to the scattering.

In particular, if we return to the general expression for the scattering amplitude equation (137), both $\varphi_{+}^{(m)}$ and $\varphi_{-}^{(m)}$ have two stationary points λ_i for each value of m, corresponding to the various classical paths leading to the same angle of deflection (see Section 4.2.5). These paths can be divided into four groups according to whether λ_i exceeds λ_0 or not and whether the path emerges after a "repulsive" $[\Theta(\lambda_i) = \vartheta - 2m\pi]$ or "attractive" $[\Theta(\lambda_i) = -\vartheta - 2m\pi]$ deviation.

The scattering amplitude, therefore, consists of four infinite series which are convergent because the amplitudes $|d\Theta/d\lambda|^{-1}$ diminish exponentially as m increases and λ_i approaches the singularity at λ_0. The net result on the differential cross section depends on the shape of the deflection function in the vicinity of λ_0. If $\Theta(\lambda)$ exhibits a narrow spike, the effect on $I(\vartheta)$ may not be noticed, whereas the effect of a broad spike, for which $|d\Theta/d\lambda|^{-1}$ hardly changes on successive circuits, is to produce a set of angular resonances.[89]

It should be noted that these results apply to the semiclassical limit.

Here the tunneling resonances (see Section 5.4) do not make a significant contribution. For light, quantal systems, however, they do play an important role. We come back to this point when we discuss the total scattering cross section.

6.2.4. Large-Angle Scattering

For angles larger than the rainbow angle ϑ_r, rapid interference oscillations are still observed, although only one branch of the deflection function contributes to the scattering. These oscillations are due to the exponentially decreasing tail of the rainbow amplitude $f_r(\vartheta)$. For $\vartheta > \vartheta_r$ we have

$$f(\vartheta) = f_1(\vartheta) + f_r(\vartheta) \tag{162}$$

where $f_1(\vartheta)$ is the contribution corresponding to the repulsive branch of the deflection function.

Substituting the asymptotic expression of the Airy integral for $x \to \infty$

$$\mathrm{Ai}(x) \approx (4\pi)^{-1/2} x^{-1/4} \exp(-\tfrac{2}{3}x^{3/2}) \tag{163}$$

into equation (151), we obtain

$$f_r(\vartheta) = r_m \left(\frac{\beta_r}{2A \sin \vartheta}\right)^{1/2} \left(\frac{\tilde{q}}{A^2}\right)^{5/12} (\vartheta - \vartheta_r)^{-1/4} \exp(i\Delta) \times \exp\left[-\frac{2}{3}\left(\frac{\tilde{q}}{A^2}\right)^{-1/2} (\vartheta - \vartheta_r)^{3/2}\right] \tag{164}$$

where Δ is defined in equation (149).

The differential cross section consists of three contributions: The classical large-angle cross section $|f_1(\vartheta)|^2$, the exponentially decreasing tail of the rainbow scattering $|f_r(\vartheta)|^2$, and the interference term

$$f_1 f_r = \frac{r_m^2}{\sin \vartheta}\left(\frac{\beta_r \beta_1}{2A}\left|\frac{d\beta_1}{d\vartheta}\right|\right)^{1/2} \left(\frac{\tilde{q}}{A^2}\right)^{5/2} (\vartheta - \vartheta_r)^{-1/4} \times \exp\left[-\frac{2}{3}\left(\frac{\tilde{q}}{A^2}\right)^{-1/2} (\vartheta - \vartheta_r)^{3/2}\right] \cos \phi \tag{165}$$

with

$$\phi = 2[\delta_r - \delta(\beta_1)] + A\vartheta(\beta_r + \beta_1) + \pi/4 \tag{165a}$$

The amplitude of the interference term also decreases exponentially.

For angles large enough so that equation (47) can be used to calculate the repulsive branch of the deflection function $\Theta(\beta_1)$ we obtain, using the semiclassical equivalence relationship, for the phase shift

$$\delta(\beta) = \delta_s + A(\pi/2)\beta_1 - \gamma_s A(\beta_1/2)^2 \tag{166}$$

with

$$\gamma_s = g_0(n)\left(\frac{K}{c_n}\right)^{1/n}$$

[see equation (47)]; δ_s is the s-wave phase shift.

From equation (166) together with equations (47) and (165a), the angular separation of the interference oscillations can be estimated to be

$$\Delta\vartheta = 3\pi/2A\beta_r \tag{167}$$

For angles very close to π, $f_1(\vartheta)$ cannot be calculated from equation (144), since the approximation equation (131) for the Legendre polynomials breaks down and has to be replaced by equation (130). We obtain for this case $[(\pi - \vartheta)A \ll 1]$

$$f_1(\pi - \alpha) = k^{-1}\int_0^\infty \lambda e^{2i\delta(\lambda) - i\pi\lambda} J_0(\lambda\alpha)\, d\lambda \tag{168}$$

with $\alpha = \pi - \vartheta$.

Using $\delta(\lambda) = \delta_s + \pi/2\lambda - (\gamma_s/4A)\lambda^2$ [compare with equation (166)] the integration can be easily performed to yield

$$\frac{f_r(\pi - \alpha)}{r_m} = |\gamma_s|^{-1}\, e^{-\alpha^2 A/2|\gamma_s|} e^{2i\delta_s} \tag{169}$$

and

$$|f_1(\pi - \alpha)|^2 = r_m^2 \gamma_s^{-2} e^{-\alpha^2 A/|\gamma_s|} \approx r_m^2 \gamma_s^{-2} \tag{170}$$

which is again identical to the classical result [equation (48)].

6.2.5. *Glory Scattering*

6.2.5.1. Forward Glories. Near the forward direction the Legendre polynomials in the series of integrals equation (129) can be replaced by equation (130). This yields for the scattering amplitude

$$f(\vartheta) = -\frac{i}{k}\sum_{m=-\infty}^{\infty} e^{-im\pi}\int_0^\infty \lambda(e^{2i\delta(\lambda)} - 1)\, e^{2im\pi\lambda} J_0(\lambda\vartheta)\, d\lambda \tag{171}$$

The regions of stationary phase are characterized by

$$\frac{d\delta(\lambda)}{d\lambda} = -m\pi \quad \text{or} \quad \Theta(\lambda) = -2m\pi \tag{172}$$

In the vicinity of $\Theta = -2m\pi$ we expand the phase shift in the form

$$\delta = \delta_m - m\pi(\lambda - \lambda_m) - \frac{\gamma_m}{4A}(\lambda - \lambda_m)^2 \tag{173}$$

Substituting equation (173) into (171) and evaluating the integral finally yields

$$f_g(\vartheta) = -\frac{i}{k}\sum_{m=0}^{\infty} e^{i\chi_m}\lambda_m\left(\frac{2\pi A}{|\gamma_m|}\right)^{1/2} J_0(\lambda_m\vartheta) \tag{174}$$

with

$$\chi_m = 2\delta_m + m\pi(\lambda_m - 1) - \pi/4 \tag{174a}$$

If the rainbow angle ϑ_r is smaller than 2π only the forward glory according to $m = 0$ remains. The contribution of this to the differential cross section is

$$I_g(\vartheta) = \frac{2\pi r_m^2}{A|\gamma_0|}\lambda_0^2 J_0^2(\lambda_0\vartheta) = \frac{2\pi r_m^2 A\beta_0^2}{|\gamma_0|} J_0^2(A\beta_0\vartheta) \tag{175}$$

When the intensity is averaged over several oscillations of the Bessel function by way of the formula $\langle J_0^2(x)\rangle = 1/\pi x$, then equation (175) reduces to the classical formula, equation (42).

If the deflection function passes through multiples of -2π (orbiting or near orbiting), additional forward glories arise in complete analogy to the classical scattering. For each angle $\Theta = -2m\pi$ there are two points of stationary phase λ_{m1} and λ_{m2}, and the result for the scattering amplitude is

$$f_g(\vartheta) = -\frac{i}{k}\sum_{v=1}^{2}\sum_{m=1}^{m} e^{i\chi_v}\lambda_{mv}\left(\frac{2\pi A}{|\gamma_{mv}|}\right)^{1/2} J_0(\lambda_{mv}\vartheta) \tag{176}$$

with

$$\chi_v = 2\delta_{mv} + m\pi(\lambda_{mv} - 1) - \pi/4$$

6.2.5.2. Backward Glories. To consider angles in the vicinity of $\vartheta = \pi$ we put $\vartheta = \pi - \alpha$ (with α small) and obtain

$$f(\pi - \alpha) = -\frac{i}{k}\sum_{m=0}^{\infty} e^{i\chi_m}\int_0^{\infty} \lambda e^{i[2\delta(\lambda)+(2m-1)\pi\lambda]} J_0(\lambda\alpha)\,d\lambda \tag{177}$$

with

$$\chi_m = (2m - 1)\pi/2$$

The points of stationary phase are given by

$$\Theta(\lambda) = -(2m - 1)\pi \tag{178}$$

In the vicinity of these points the phase shift can be written as

$$\delta(\lambda) = \delta_m - (2m - 1)\pi(\lambda - \lambda_m) - (\gamma_m/4A)(\lambda - \lambda_m)^2 \tag{179}$$

The contribution of $m = 0$ is the normal backward scattering amplitude $f_1(\vartheta)$ near $\vartheta = \pi$ due to head-on collisions, which has been discussed in section 6.2.4 [see equation (169)].

For $m > 0$ the relevant collisions have deflection angles of $-\pi$, -3π, etc., which occur in the orbiting case. Similarly as for the forward scattering amplitude the glory contribution to the backward amplitude becomes

$$f_g(\pi - \alpha) = -\frac{i}{k}\sum_{\nu=1}^{2}\sum_{m=1}^{\infty} e^{i\chi_m}\lambda_{m\nu}\left(\frac{2\pi A}{|\gamma_{m\nu}|}\right)^{1/2} J_0(\lambda_{m\nu}\alpha) \tag{180}$$

with

$$\chi_m = 2\delta_{m\nu} + \pi(2m-1)(\lambda_{m\nu} - 1/2) - \pi/4 \tag{180a}$$

Similar to the rainbow scattering, the above approximations are valid only very near the critical angles, but fail to merge smoothly with the semiclassical formulas discussed in Section 6.2.1 away from these angles. A uniform approximation for the glory scattering, which has been derived by Berry,[133] removes this disadvantage.

6.2.6. *Small-Angle Scattering (Forward Diffraction Peak)*

Equation (171), which describes the forward scattering amplitude, has, for $m = 0$, a second region where the phase $\delta(\lambda)$ is slowly varying with λ. This is due to trajectories of large angular momentum which pass through the long-range tail of the potential and emerge near the forward direction.

To evaluate this contribution, we only need to consider the asymptotic behavior of the phase shifts, which is determined by the long-range attractive part of the potential. For an inverse power potential $U(x) = -c_s x^{-s}$, equation (38) together with equation (135) yield for the phase shift

$$\delta(\lambda) = a\lambda^{-(s-1)} \tag{181}$$

with

$$a = \frac{f(s)}{2}\left(\frac{c_s}{K}\right)k^s$$

Herewith the scattering amplitude becomes

$$f(\vartheta) = -\frac{i}{k}\int_0^\infty \lambda[\exp(2ia\lambda^{-(s-1)}) - 1]\, J_0(\lambda\vartheta)\, d\lambda \tag{182}$$

Unfortunately, this integral cannot be evaluated in closed form in terms of known functions of analysis. An approximate evaluation[9] yields

$$f(\vartheta) = Ar_m\left[c_s f(s)\frac{A}{K}\right]^{2/(s-1)}$$
$$\times\left[\frac{i}{x}J_1(x) + \frac{i}{4(s-2)}G_{2s-4}(x) + \frac{1}{(s-3)}G_{s-3}(x)\right] \tag{183}$$

with

$$x = \left(\frac{c_s f(s) k^s}{K} \right)^{1/(s-1)} \cdot \vartheta$$

and

$$G_n(x) = n x^n \int_0^\infty J_0(t)\, t^{-n-1}\, dt.$$

For very small angles only the first terms in the expansion of $J_1(x)$ and $G_n(x)$ are required. The differential cross section then can be approximated by

$$I(\vartheta) = A^2 r_m^2 \left[c_s f(s) \frac{A}{K} \right]^{4/(s-1)} g_1(s) \exp\left\{ -A^2 \left[c_s f(s) \frac{A}{K} \right]^{2/(s-1)} g_2(s)\, \vartheta^2 \right\} \tag{184}$$

with $g_1(6) = 0.4275$ and $g_2(6) = 0.6091$.[97,147]

In contrast to the classical result $I(\vartheta)$ remains finite for $\vartheta = 0$. The complete formal solution to the problem of the shape of the diffraction peak is the uniform approximation of Berry.[133] It is not suited, however, to practical applications, since the functions involved cannot be expressed in terms of known functions.

6.3. Special Features of the Total Elastic Scattering Cross Section

From the forward scattering amplitude equation (174) (for $m = 0$) and equation (183) we can easily obtain an expression for the total scattering cross section using the optical theorem equation (104) with the result

$$\sigma = \sigma_{\mathrm{MM}} + \Delta\sigma$$

with

$$\sigma_{\mathrm{MM}} = \frac{2s-3}{s-2} \left[c_s f(s) \frac{g_c}{g} \right]^{2/(s-1)}$$

$$\Delta\sigma = 4\beta_0 \left(\frac{2\pi}{A|\gamma_0|} \right)^{1/2} \cos\left(2\delta_0 - \frac{\pi}{4} \right) \tag{185}$$

and

$$\frac{g_c}{g} = \frac{A}{K} = \frac{2\varepsilon r_m}{\hbar g}$$

The total scattering cross section contains a contribution σ_{MM} (first derived by Massey and Mohr[4]) due to the attractive long-range part of the potential. The presence of a short-range repulsive potential introduces an additional contribution $\Delta\sigma$ due to the glory effect.[148] σ_{MM} is a monotonic function of the velocity, whereas the glory contribution oscillates with the velocity. The result, equation (185), can also be derived directly from equation (103) for the total scattering cross section, using the same approximation as in the case of the scattering amplitude. Since it is possible to integrate

$$\mathrm{Im}(f(0)) = \frac{1}{k}\int_0^\infty \lambda \sin^2(a\lambda^{-(s-1)})\,d\lambda \tag{186}$$

[a is defined in equation (181)] in closed form [this is not possible for $f(\vartheta)$ in the case of $\vartheta \neq 0$], a more refined expression σ_{LL} instead of σ_{MM} is obtained.[137,149] The ratio σ_{LL}/σ_{MM} depends only on s;

$$\frac{\sigma_{LL}}{\sigma_{MM}} = \frac{2(s-1)}{2s-3}\,\frac{\Gamma\left(\dfrac{s}{s-1}\right)\Gamma\left(\dfrac{2s-3}{s-1}\right)}{\Gamma\left(\dfrac{s+1}{s-1}\right)} \tag{187}$$

For $s = 6$ we have $\sigma_{LL}/\sigma_{MM} = 1.0709$.

Exact quantum-mechanical calculations of total cross sections for a monotonic potential show that the constant obtained with the Landau–Lifshitz approximation is correct within 1.5%.[150]

The glory contribution to the total cross section is due to the fact that the deflection function passes through zero at the glory impact parameter b_0. Thus, according to equation (173) (for $m = 0$), the phase shift curve $\delta(\beta)$ exhibits a broad maximum around β_0 providing a significant fraction of nonrandom phases at intermediate angular momenta. The maximum phase δ_0 increases with decreasing wave number k (see, for example, Figure 9), and if the attractive well is sufficiently deep, δ_0 can pass successively through multiples of $\pi/2$, giving rise alternately to positive and negative incremental contributions $\Delta\sigma$ to the Massey–Mohr cross section σ_{MM}. From the observed velocities g_N of the extrema the maximum phase δ_0 can be directly determined as a function of the relative velocity g. The condition for an extremum follows from equation (185):

$$\delta_0(k_N) = \pi(N - \tfrac{3}{8}) \tag{188}$$

where $N = 1, 2, 3, \ldots$ are indices for maxima, while $N = 1.5, 2.5, 3.5, \ldots$ refer to minima in $\sigma(g)$. The first maximum $N = 1$ is the maximum at the highest velocity.

For any assumed potential the velocity dependence of the maximum phase shift may be evaluated numerically, and thus the locations of the extrema in $\sigma(g)$ determined. For large reduced energies δ_0 may be expanded in the following way[151,152]:

$$\delta_0(K) = \frac{g_c}{g}\left(a_1 - \frac{A_1}{K} + \frac{A_2}{K^2} + \cdots\right) \tag{189}$$

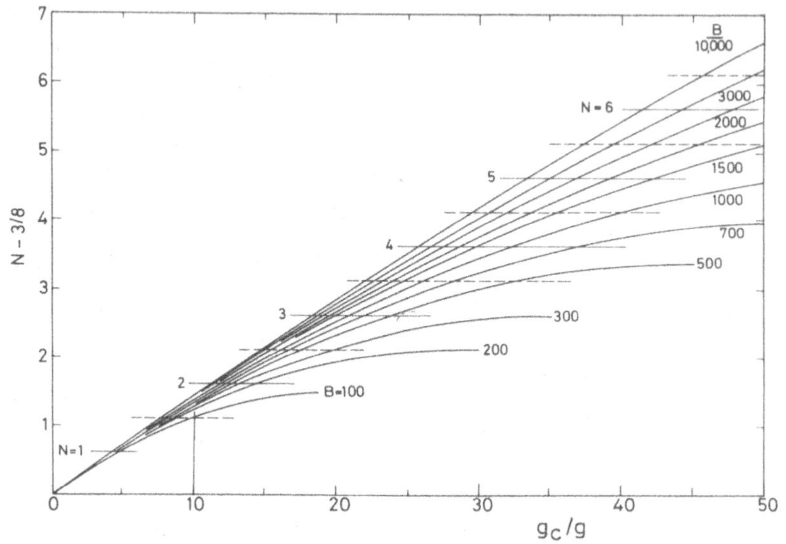

Figure 19. $N - 3/8$ vs. g_c/g for various values of B [L.-J. (12, 6) potential]. The horizontal lines corresponding to different extrema intersect the curves at the appropriate values of g_c/g_N. (Adapted from Reference 114.)

A physical interpretation of the expansion coefficients has been given by Greene and Mason.[153] The coefficient a_1 measures the area of the intermolecular potential well along the glory trajectory, whereas A_1 and A_2 give measures of the square and cube of the well, respectively. Numerical values of the expansion coefficients are tabulated by Bernstein and O'Brien,[151] Greene and Mason,[153] and Bernstein and LaBudde.[154]

If the reciprocal velocities g_N of the extrema are plotted against $N - 3/8$ a straight line is obtained in the first approximation. Figure 19 shows calculated values of g_c/g plotted against $N - 3/8$ for various values of the potential well capacity B. With increasing N, deviations from the straight line occur due to the higher-order terms in equation (189). As B is increased, the deviations occur at higher N values. The limiting value of δ_0 for $K \to 0$ is given by the theorem of Levinson $\delta_0 = n_0\pi$, where n_0 is the number of bound states with zero angular momentum in the potential well (see Table 3).

To analyze experimental data, the product $g^{2/(s-1)}\sigma(g)$ (usually with $s = 6$) is plotted as a function of $1/g$, yielding an undulatory curve, which is symmetric about a horizontal mean line and in which the extrema are nearly uniformly spaced. From this plot the extremum velocities can be obtained and plotted versus $N - 3/8$. The characteristic velocity g_c can be found from the initial slope of the resulting line, and thus the product εr_m can be determined.

More refined methods for the evaluation of potential parameters from total cross section measurements using the above expansion are discussed by Bernstein *et al.*[152,154]

From equation (185) (with $s = 6$) it follows that the relative amplitudes of the oscillations in the velocity dependence of the total scattering cross section are given by

$$\Delta\sigma/\sigma_{MM} = G(K)\, B^{-9/10} \tag{190}$$

where

$$G(K) = 5.5069\beta_0(K)\,|\gamma_0|^{-1/2}\,K^{-1/20}$$

Numerical calculations[50,51] show that β_0 is only weakly dependent on K. The same result holds approximately for γ_0 for reduced energies in the range $1 < K < 10$. The function $G(K)$, therefore, is nearly independent of K in this region. This follows also from Figure 20, which shows $G(K)$ for a Lennard-Jones (12, 6) potential. In this region of reduced energies the relative amplitude of the glory extrema depends on the potential parameters ε and r_m and the reduced mass μ in the following way:

$$\Delta\sigma/\sigma_{MM} \sim \mu^{-1/2}\varepsilon^{-2/5}r_m^{-9/5} \tag{191}$$

Equation (191) shows that the oscillations are expected to be largest for light atoms. With increasing K, however, γ_0 and consequently $G(K)$ depend

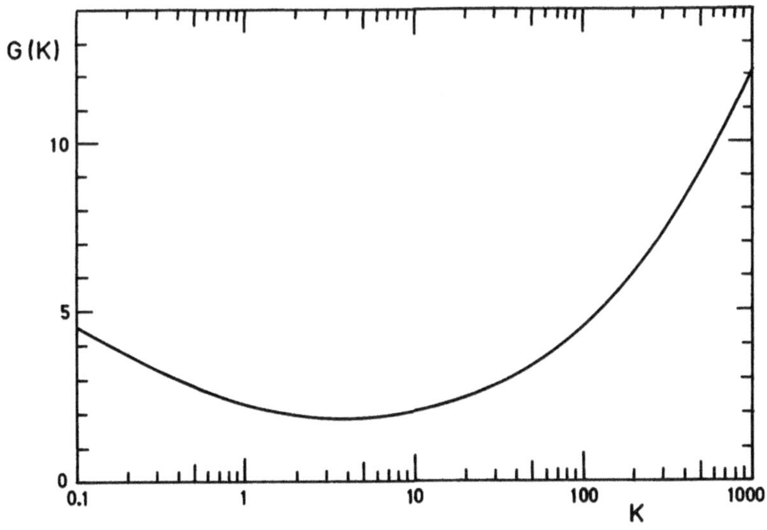

Figure 20. The function $G(K)$ which determines the relative glory amplitudes [L.-J. (12, 6) potential]. (Adapted from Reference 112.)

more strongly on K. In the high-energy limit both β_0 and γ_0 can be expanded in inverse powers of the reduced energy to yield

$$\beta_0 = B_0 + \frac{B_1}{K} + \frac{B_2}{K^2} + \cdots, \qquad \frac{\gamma_0}{2} = \frac{C_0}{K} + \frac{C_1}{K^2} + \frac{C_2}{K^3} + \cdots \tag{192}$$

From these expansions a similar expansion for the relative amplitude is obtained:

$$\begin{aligned} \Delta\sigma &= g^{1/2}(R_0 + R_1/K + \cdots) \\ R_0 &= (2\pi r_m)^{3/2}\,(\hbar/\varepsilon)^{1/2}\,(B_0/C_0^{1/2}) \end{aligned} \tag{193}$$

In this limit the relative amplitude becomes independent of the reduced particle mass. The coefficients R_j can be interpreted as a measure of the force in the vicinity of the distance of closest approach.[155]

The essential assumption made in the calculation of the forward scattering amplitude and of the total scattering cross section [equation (185)] is the following: The maximum phase shift δ_0 must be sufficiently large, so that the region of stationary phase in the vicinity of β_0 is well separated from the asymptotic tail of the phase shift curve and thus can be treated separately. An estimate of the lower limit on the relative velocity for which the treatment is valid yields the condition $\delta_0 > 1/2$. For a L.-J. $(n, 6)$ potential, using the high-energy approximation discussed in Section 6.4, this condition becomes

$$\frac{g}{g_c} < \frac{3\pi}{16}\frac{n}{n-1}\left(\frac{32f(n)(n-1)}{5\pi n}\right)^{-5/(n-6)} \tag{194}$$

For larger velocities, an approximate method for the calculation of the total cross section has been derived by Bernstein[156] for L.-J. $(n, 6)$ and a Buckingham $(\gamma, 6)$ potential. For even higher velocities, where the repulsive part of the potential completely determines the scattering, a treatment similar to that which leads to equation (185) is possible. For an inverse power repulsive potential $U_r = c_n x^{-n}$ the result is

$$\sigma_{\mathrm{LL}} = p(n)\left[c_n f(n)\frac{g_c}{g}\right]^{2/(n-1)} \tag{195}$$

with

$$p(n) = \frac{2(n-1)}{n-2}\,\frac{\Gamma\left(\dfrac{n}{n-1}\right)\Gamma\left(\dfrac{2n-3}{n-1}\right)}{\Gamma\left(\dfrac{n+1}{n-1}\right)}$$

6.4. *Identical Particles*

As an example for the semiclassical treatment of the scattering of identical particles we consider the velocity dependence of the total scattering cross section in the case of spinless bosons. Due to the symmetrization of the wave function the scattering amplitude becomes $f(\vartheta) + f(\pi - \vartheta)$. Two contributions account for $f(\vartheta)$: The forward diffraction amplitude, equation (183), and the forward glory amplitude, equation (174). $f(\pi - \vartheta)$ is given by equation (169). Applying the optical theorem, the forward diffraction amplitude yields σ_{LL}, the forward glory amplitude yields the glory contribution $\Delta\sigma$. Thus we obtain for the total cross section

$$\sigma = \sigma_{\mathrm{LL}} + \Delta\sigma - \frac{4\pi r_m}{A|\gamma_s|}\cos\left(2\delta_s - \frac{\pi}{2}\right) \tag{196}$$

Superimposed on the usual glory undulations of the total scattering cross section are oscillations which are due to the symmetrization of the wave function. The amplitudes of these oscillations decrease with $1/A$. The position of the extrema is given by the velocity dependence of the s-phase δ_s. These oscillations can best be seen in the He–He scattering, since this system does not show the normal glory structure, due to the small potential well capacity B (see Figure 17).[127,128,157]

6.5. *High-Energy Approximation*

6.5.1. *Brief Outline of the Method*

To estimate the various effects discussed above it is often useful to have a simple approximation for the phase shift function $\delta(\beta)$. This is provided by an approximation which is completely analogous to the momentum approximation discussed in Section 4.5 for the classical deflection function. For large values of the reduced energy K and reduced impact parameters β which are not too small we can expand equation (132) with the result

$$\delta(\beta) \approx \frac{A}{2K}\int_\beta^\infty \frac{xU(x)}{(x^2 - \beta^2)^{1/2}}\,dx = \frac{A}{2K}\int_\beta^\infty U(z)\,dz \tag{197}$$

where z is the direction of the incoming particle.

Other derivations of equation (197) are also possible.[149,158] Similar to the equivalent expression for the classical deflection function equation (75), equation (197) describes, at least qualitatively, the positive branch of the phase shift function over a wide range of energies. We compile, therefore, in the following section a number of results for the basic potentials.

6.5.2. *Results for the Basic-Potential Models*

6.5.2.1. Lennard-Jones (*n*, 6) *Potential.* The solution of equation (197) yields for a L.-J. (*n*, 6) potential

$$\delta = \frac{3\pi}{32}\frac{n}{n-6}\frac{A}{K}\beta^{-5}\left(1 - \frac{32f(n)}{\pi n}\beta^{-n+6}\right) \tag{198}$$

$f(n)$ is defined by equation (38). From equation (198) we obtain the glory parameters β_0 and γ_0, and δ_0:

$$\beta_0 = \left(\frac{32f(n)(n-1)}{5\pi n}\right)^{1/(n-6)}$$

$$\gamma_0 = -\frac{30\pi n}{32K}\left(\frac{32f(n)(n-1)}{5\pi n}\right)^{-7/(n-6)} \tag{199}$$

$$\delta_0 = \frac{A}{K}\frac{3\pi n}{32(n-1)}\left(\frac{32f(n)(n-1)}{5\pi n}\right)^{-5/(n-6)}$$

6.5.2.2. Buckingham Potential. The phase shift function $\delta(\beta)$ becomes

$$\delta = -\frac{A}{K}\left[\frac{3}{(\gamma-6)}e^{\gamma}K_1(\beta\gamma) - \frac{3\pi}{32}\left(\frac{\gamma}{\gamma-6}\right)\beta^{-5}\right] \tag{200}$$

where $K_1(x)$ is the modified Bessel function of first order. Since $\gamma\beta \gg 1$ in the range of validity of the high-energy approximation we can use the asymptotic expansion for $K_1(x)$,

$$K_1(x) \rightarrow \left(\frac{\pi}{2x}\right)^{1/2} e^{-x} \tag{201}$$

so that we finally obtain

$$\delta(\beta) = -\frac{A}{K(\gamma-6)}\left[\frac{1}{2}\left(\frac{\pi\beta}{\gamma}\right)^{1/2} e^{\gamma(1-\beta)} - \frac{3\pi\gamma}{32}\beta^{-5}\right] \tag{202}$$

6.5.2.3. Morse Potential. In a similar way we obtain for the Morse potential

$$\delta(\beta) = -\frac{A}{K}e^{\gamma}\beta\left[\frac{e^{\gamma}}{2}K_1(2\gamma\beta) - K_1(\gamma\beta)\right] \tag{203}$$

or using the asymptotic expression for the Bessel function equation (201)

$$\delta(\beta) = -\frac{A}{K}\left(\frac{\pi\beta}{2\gamma}\right)^{1/2}\left(\frac{e^{2\gamma(1-\beta)}}{2\cdot 2^{1/2}} - e^{\gamma(1-\beta)}\right) \tag{204}$$

From equation (204) it is easy to calculate the glory data β_0 and δ_0:

$$\beta_0 = 1 - \frac{\ln 2^{1/2}}{\gamma}$$
$$\delta_0 = \frac{A}{K}\left[\frac{\pi}{4\gamma}\left(1 + \frac{\ln 2^{1/2}}{\gamma}\right)\right]^{1/2} \tag{205}$$

7. *Methods for the Evaluation of Potentials from Experimental Scattering Data*

7.1. *General Survey*

In this section we shall try to give a summary of the recent work on methods for the evaluation of interatomic potentials from experimental elastic scattering data.

In the early stage of beam research on intermolecular potentials, the evaluation of the potential was achieved using the various classical and semiclassical approximations described in Sections 4 and 6, each of which allows the description of a certain scattering phenomenon such as "rainbow scattering" or "glory scattering," etc., in a certain angle or velocity range. In this way two- or three-parameter potential models were determined. These methods are still adequate if only data with low precision or low resolution are available.

If precise, highly resolved cross section data are available, however, these procedures are no longer sufficient and may even lead to contradictory results due to the inflexibility of the simple models.[44]

Figure 21 illustrates the different methods that have been applied in order to determine intermolecular potentials from scattering data. The most direct way from the experimental data to the potential is provided by the inversion procedures developed several years ago using semiclassical methods.[159] They yield the potential—designated "experimental potential" in Figure 21—independent from any assumptions or models. Inversion procedures have been successfully applied to a number of collision pairs where precise differential cross section data, especially in the rainbow region, were available. In the special case of the He–He interaction, it was also possible to invert total scattering cross section data, again with the help of semiclassical relationships.[127,128]

Recent work concentrates on exact quantum-mechanical inversion procedures.[160] Considerable progress has already been achieved, and first results are already available.[161]

The other well-known method for the determination of intermolecular potentials from measured cross sections is the trial and error method, in Figure 21 characterized by the following loop: A model potential with

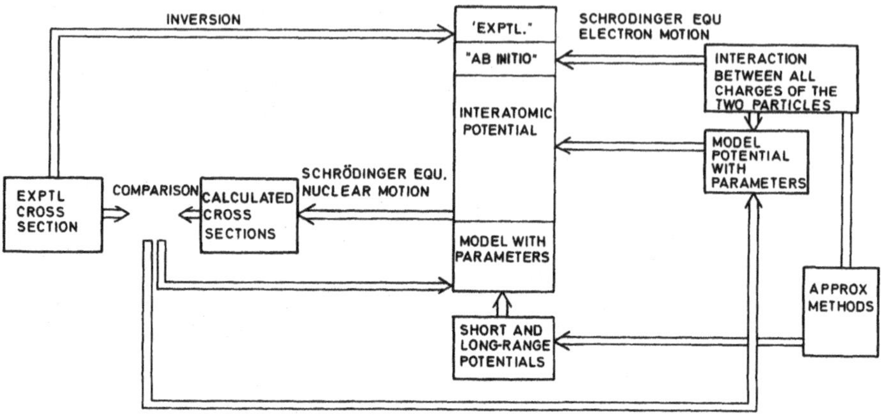

Figure 21. Schematic diagram to illustrate the various methods which have been applied in order to determine intermolecular potentials from scattering data. A detailed discussion is given in the text.

adjustable parameters is assumed and the cross sections are calculated by solving the Schrödinger equation for the nuclear motion by an exact or approximate method. The calculated cross sections are then compared with the experimental ones, the parameters changed, and the corresponding loop passed through until best agreement between experimental and calculated data is achieved. As has been discussed in Section 2.3 a large number of empirical potential models have been proposed, including multiparameter and splined functionalities which contain up to 12 adjustable parameters.

Bickes and Bernstein[67] showed that the Simons–Parr–Finlan modified Dunham expansion (see Section 2.3.3) used in conjunction with a regression program which takes advantage of the hierarchy of the expansion coefficients can systematically analyze differential scattering cross sections to obtain an optimized potential yielding a best fit to the experimental data. The program passes through the above-mentioned loop repeatedly, using as the initial values for the expansion coefficients of each pass the converged values from the previous pass.

Ab initio calculations of interatomic potentials, on the other hand, have to start with the known Coulomb interactions between all charges in the two-particle system and to solve the Schrödinger equation for the electron motion. Despite the recent progress in quantum-chemical calculation procedures for this many-particle problem, *ab initio* calculations of the entire potential curve are still restricted to a small number of light systems, where the number of electrons involved is small[17] (see Section 2).

The situation is better if we consider the asymptotic regions of the potential. For many collision pairs, the long-range expansion coefficients C_6, C_8, and C_{10} of the dispersion forces are available, and also often the short-range part of the potential is known by self-consistent-field calculations, so that the number of free parameters in the potential can be reduced.

The methods for determination of potentials from scattering data outlined so far have one common disadvantage. They yield a potential which is only valid for the special system under investigation. The parameters of the model potential which determine its shape have in general no relation to the physical properties of the two atoms involved. Therefore it is not possible to transfer the potential, its shape, or some of the parameters to other collision pairs consisting of atoms with similar chemical properties.

Many recent investigations, therefore, are concerned with the question of how the information on intermolecular potentials gathered so far both from experiment and theory can be used to estimate the potentials of collision pairs for which no experimental information is available. Some of these attempts have been discussed in Section 2.

Another possibility to determine potentials with more general validity is the "pseudopotential" method,[(162)] which is also indicated schematically in Figure 21. Instead of making a parameterized model for the atom–atom potential and determining the parameters by comparison with the experimental data, it is also possible to make a parameterized model for the interaction between all charges involved. With this model the Schrödinger equation of the electron motion has to be solved in order to obtain the atom–atom potential. The parameters of the pseudopotential can be determined by passing through the following (see Figure 21): select a pseudopotential → solution of the Schrödinger equation for the electron motion → solution of the Schrödinger equation for the nuclear motion → cross sections → comparison with the experimental data → modification of the parameters of the pseudopotential. This has several advantages: (i) The parameters in the pseudopotential can be physically understood, (ii) they may be valid for a whole class of chemically similar atoms, and (iii) finally the method also yields—depending on the basic set used—the potentials between electronically excited atoms. This is a very important fact, with regard to both scattering experiments at higher collision energies and experiments with laser-excited atoms.

In the following sections we will discuss the above-mentioned methods in some detail.

7.2. *Semiclassical Inversion Procedures*

We restrict the following discussion to the semiclassical inversion procedures which have been adopted to molecular beam measurements, since exact quantum-mechanical procedures are only now beginning to be applied to experimental scattering data.[(160,163,164)]

Since the phase shifts are a function of the angular momentum and the energy, the inverse problem provides solutions if the phase shift $\delta_l(E)$ for a fixed angular momentum l is known for all energies or if for a fixed energy the phase shifts are known for all values of l.

The second case is more important in atomic collision processes. As we have discussed in Sections 5 and 6, many angular momenta contribute to the scattering so that it is usually not possible to obtain the energy dependence of a single phase shift of given angular momentum from the experimental observations. An exception, however, has been discussed in Section 6.4. Here it is possible, due to a collective behavior of the phase shifts in the vicinity of a stationary point, to measure directly the energy dependence of the s-wave phaseshift. We therefore discuss the semiclassical inversion procedure for this case first.

7.2.1. Determination of the Repulsive Part of the Potential from the s-Phase as a Function of the Energy

The determination of the energy dependence of the s-phase shift from the measured total cross section as a function of velocity for identical particles makes an inversion procedure for the evaluation of the repulsive part of the interaction potential possible. In a continuum version of the Rydberg–Klein–Rees method[165,166] for obtaining $V(r)$ from diatomic rotation–vibration spectra it is possible to determine $r(V)$ for $V > 0$ from the empirically determined function $\delta_s(E)$.[167] The expression for the JWKB phase shift can be inverted to yield the classical turning point

$$r_0(E) = (2k)^{-1} + r_1(E) + r_2(E)$$

$$r_1(E) = -\frac{2}{\pi}\left(\frac{\hbar}{2\mu}\right)^{1/2} \pi \int_{-1/2}^{n(0)} [E - E(n)]^{-1/2}\, dn \tag{206}$$

$$r_2(E) = -\frac{2}{\pi}\left(\frac{\hbar}{2\mu}\right)^{1/2} \int_{\delta_s(0)}^{\delta_s(E)} [E - E(\delta_s)]^{-1/2}\, d\delta_s$$

The potential energy at $r = r_0$ is then

$$V(r_0) = E - \hbar^2/8\mu r_0^2 \tag{207}$$

The first integral $r_1(E)$ is a contribution from the bound states, where $n(E)$ is given by

$$[n(E) + \tfrac{1}{2}]\,\pi = \left(\frac{2\mu}{\hbar^2}\right)^{1/2} \int_{r_\alpha}^{r_\beta} \left[E - V(r) - \frac{\hbar^2}{8\mu r^2}\right]^{1/2} dr \tag{208}$$

r_α and r_β are the classical turning points. For purely repulsive potentials we have $r_1(E) = 0$. For a Morse potential $r_1(E)$ can be calculated in closed form, yielding

$$r_1(E) = -(r_m/2\gamma)\arcsin[1 + E/\varepsilon]^{-1/2} \tag{209}$$

For other potentials a numerical calculation is required. The contribution $r_1(E)$ shows that the potential well must be known in order to determine the repulsive part of the potential. For $E \gg \varepsilon$, however, $r_1(E)$ becomes a small correction to $r_2(E)$, so that an approximate knowledge of the potential well is sufficient to determine $V(r)$ in the region $V > 0$ with adequate accuracy.[127,128] The function $\delta_s(E)$ can be determined according to equation (196) from the observed oscillatory structure in the energy dependence of the total cross section.

7.2.2. *Determination of the Potential from the Phase Shift Function or the Deflection Function at a Fixed Energy*

The determination of the potential from differential scattering cross section measurements can be performed in two steps:

(i) The determination of the phase shift function $\delta(\beta)$ or the classical deflection function $\Theta(\beta)$ from the experimental data. Due to the semiclassical equivalence relationship both quantities provide the same information. We therefore restrict the following considerations to the classical deflection function $\Theta(\beta)$.

(i) The determination of the potential from $\delta(\beta)$ or $\Theta(\beta)$. The solution of the second step is straightforward.[168-170] Returning to the expression equation (34) for the classical deflection function and substituting

$$s(x) = x[1 - U(x)/K]^{1/2} \tag{210}$$

we obtain

$$\Theta(\beta) = \pi - 2\beta \int_{x_0}^{\infty} x^{-1}[s^2(x) - \beta^2]^{-1/2}\, dx \tag{211}$$

It can be shown that $ds(x)/dx > 0$ if the reduced energy K lies above the orbiting threshold. With this restriction, therefore, $x(s)$ exists. With

$$I(s) = \log\left[\frac{x(s)}{s}\right] \tag{212}$$

equation (211) reduces to

$$\Theta(\beta) = -2\beta \int_{\beta}^{\infty} (s^2 - \beta^2)^{-1/2} I'(s)\, ds \tag{213}$$

This is an Abel integral equation, which can be inverted to yield

$$I(s) = \pi^{-1} \int_{s}^{\infty} \Theta(\beta)(\beta^2 - s^2)^{-1/2}\, d\beta \tag{214}$$

From equations (210) and (212) we obtain finally

$$U(x) = K[1 - \exp\{-2I(s)\}] \tag{215}$$

with

$$x(s) = s \exp[I(s)]$$

The first step, the determination of the classical deflection function from the measured differential scattering cross section $I(\vartheta)$, is complicated by the fact that both attractive and repulsive forces contribute to the scattering. For angles smaller than the rainbow angle we have no one-to-one connection between angle and impact parameter. From equation (37) we may obtain

$$\begin{aligned} S(\vartheta) &= 2\pi \int_{\vartheta}^{\pi} I(\vartheta) \sin \vartheta \, d\vartheta \\ &= \pi r_m^2 \{\beta_1^2(\vartheta) + h(\vartheta - \vartheta_r)[\beta_3^2(\vartheta) - \beta_2^2(\vartheta)]\} \end{aligned} \tag{216}$$

where $h(\vartheta - \vartheta_r)$ is a step function:

$$h(\vartheta - \vartheta_r) = 0 \qquad \text{for } \vartheta \leq \vartheta_r$$

$$h(\vartheta - \vartheta_r) = 0 \qquad \text{for } \vartheta > \vartheta_r$$

Only the construction of the repulsive branch $\beta_1(\vartheta)$ of the deflection function, which corresponds to angles larger than ϑ_r, is simple. From equation (216) we find

$$\beta_1(\vartheta) = [S(\vartheta)/\pi r_m^2]^{1/2} \tag{217}$$

where $\beta_i(\vartheta)$ $(i = 1, 2, 3)$ are the inverse functions of the three branches of $\Theta(\beta)$.

For impact parameters corresponding to angles smaller than ϑ_r an exact determination of the deflection function or the phase shift function from the differential scattering cross section is impossible, since both attractive and repulsive forces contribute to the scattering and, therefore, no one-to-one connection between angle and impact parameter exists.

All practical methods that have been proposed use a parametrization of either the phase shift function or the S matrix or the deflection function. The unknown parameters are determined by a minimization procedure on measured and calculated data. We will discuss several examples in the following section.

7.2.3. *Determination of the Phase Shift Function $\delta(\beta)$ or the Classical Deflection Function $\Theta(\beta)$ from an Analysis of Differential Cross Section Data*

Brackett *et al.*,[171] for instance, start with a parameterization of the phase shift function, which is represented by a sum of exponential functions.

Vollmer[172] describes $\delta(\beta)$ by a superposition of functions for which the integral equation (214) can be solved analytically. Klingbeil[173] proposed a method in which all phase shifts are treated as free parameters, except those for large angular momenta, where the approximation equation (181) holds. Although this method seems to be not suited for atomic collision processes due to the large number of phase shifts contributing to the scattering, Klingbeil was able to show that the minimalization procedure is rapidly converging if good estimates (within 10%) of the starting parameters are available (e.g., 20 iterations for 126 phase shifts).

A different procedure has been developed by Remler.[174] He starts with a decomposition of the scattering amplitude into a repulsive and an attractive part,

$$\begin{aligned} f(\vartheta) &= f_a(\vartheta) + f_r(\vartheta) \\ &= (ik)^{-1} \sum_{l=0}^{\infty} (2l+1)\left[S_r S_a - 1\right] P_l(\cos\vartheta) \end{aligned} \tag{218}$$

with

$$f_a(\vartheta) = (2ik)^{-1} \sum_{l=0}^{\infty} (2l+1)\left[S_a - 1\right] P_l(\cos\vartheta) \tag{219}$$

and

$$f_r(\vartheta) = (2ik)^{-1} \sum_{l=0}^{\infty} (2l+1)\left[S_a(S_r - 1)\right] P_l(\cos\vartheta) \tag{220}$$

where $S = \exp[2i(\delta_a + \delta_r)] = S_a S_r$ is the S matrix.

In the next step the S matrix for the attractive part is parameterized by

$$S_a = \prod_{\nu=1}^{N} \left(\frac{\lambda^2 - \lambda_\nu^{*2}}{\lambda^2 - \lambda_\nu^2} \right) \tag{221}$$

where λ is the complex angular momentum with $\mathrm{Re}(\lambda) = l + 1/2$ and λ_ν is the position of the νth pole in the complex angular momentum plane. The scattering amplitude can then be calculated using the Regge–Watson–Sommerfeld transformation[175-176] with the result

$$f_a(\vartheta) = (2ik)^{-1} \sum_{\nu=1}^{N} \frac{\pi}{\cos \pi\lambda_\nu} P_{\lambda-1/2}(-\cos\vartheta)(\lambda_\nu^2 - \lambda_\nu^{*2}) \prod_{\substack{i=1 \\ \nu \neq i}}^{N} \left(\frac{\lambda_\nu^2 - \lambda_i^{*2}}{\lambda_\nu^2 - \lambda_i^2} \right) \tag{222}$$

Now the following procedure is adopted. Starting with N poles placed on a small circle with center at λ_ν in the complex λ plane, the number of these poles N and the real and imaginary part of the central pole λ_ν are derived from semiclassical quantities. The rainbow angle ϑ_r is given by $2N/\mathrm{Im}(\lambda_\nu)$, the corresponding angular momentum l_ν by $\mathrm{Re}(\lambda_\nu)$, and the width of the deflection function at $\vartheta = \vartheta_r/2$ approximately by $2\mathrm{Im}(\lambda_\nu)$.

With these values the series equation (222) and the differential cross section are calculated, the results compared with the experimental data, and the input information varied until best agreement between experimental and calculated values is achieved. The number of poles required for a good fit has been found to lie between 5 and 16.[177] The method is not applicable to the repulsive part f_r of the scattering amplitude due to the large number of poles required here. A disadvantage of this method is the wrong asymptotic behavior of the phase shift which is a result of the choice of the parameterization.

Another method which has found wide application in the analysis of atom–atom scattering data has been developed by Buck.[178-184]

The starting point is the uniform approximation equation (155) for the rainbow cross section. Because of the behavior of Ai^2 and of the smaller amplitude of the second term in this equation, the rainbow maxima z_N (N integer) are given just by the zeros of Ai′, whereas the minima z_N (N half-integer) are given by the zeros of Ai. These are tabulated in the literature.[185] The numbering starts with the primary rainbow; z_N is an implicit function of ϑ_N, the observed extremum angles:

$$z_N = \{\tfrac{3}{4}[2\delta(b_2(\vartheta_N)) - 2\delta(b_3(\vartheta_N)) + k\vartheta_N(b_2(\vartheta_N) - b_3(\vartheta_N))]\}^{2/3} \tag{223}$$

Now the deflection function is constructed from simple mathematically invertible functions:

(i) In the region of the minimum:

$$\Theta(b) = -\vartheta_r + q(b_{2,3} - b_r)^2 \tag{224}$$

$$z_N = k^{2/3}q^{-1/3}(\vartheta_r - \vartheta_N) \tag{225}$$

(ii) In the region next to the minimum:

$$\Theta(b) = -\gamma_0(b_2 - b_0), \qquad \gamma_0 > 0$$

$$\Theta(b) = -c_1 b_3^{-c_2} \tag{226}$$

$$z_N = \left(\frac{3}{4}\right)^{2/3}\left[2\delta_0 + kb_0(\vartheta_N) + \frac{k}{2\gamma_0}\vartheta_N^2 - \frac{kc_1^\kappa}{1-\kappa}\vartheta^{(1-\kappa)}\right]^{2/3} \tag{227}$$

$$\kappa = c_2^{-1}$$

δ_0 is the maximum phase shift.

(iii) In the asymptotic region: The formulas (ii) can be used with $c_1 = 15\pi C_6/16E$ and $c_3 = 6$ in equation (226), where C_6 is the van der Waals constant.

For each observed rainbow extremum ϑ_N we obtain an equation

$z_N = z_N(b(\vartheta_N))$. This system of equations contains the unknown coefficients ϑ_r, q, b_r, δ_0, b_0, γ_0, c_1, c_2, and C_6. The continuity conditions at the ends of the region of validity of the individual functions reduce the number of unknowns by two. The maximum phase shift δ_0 can be determined from the extrapolation of the z_N vs. ϑ_N curves for $\vartheta = 0$, or more precisely from the oscillatory behavior of the velocity dependence of the total cross section. The van der Waals constant C_6 can be determined from the absolute value of the total cross section or from *ab initio* calculations (see Section 2). Thus five constants remain to be determined from the foregoing system of equations, which can be solved by a least-squares fit procedure, provided that the number of observed rainbow extrema exceeds the number of unknown quantities.

Two other procedures similar to the one described above have been proposed by Miller[186] and Boyle.[187] Miller introduces monotonic parameterized expansions $p(x)$ and their inverse functions to describe the deflection function, whereas Boyle first calculates the deflection function from an assumed potential. The $\Theta(b)$ so obtained is then adjusted in an iterative procedure so that the observed features of the measured differential cross section are well described.

7.2.4. *The Inverse Problem in the High-Energy Approximation*

As has been shown in Section 4.5, the classical deflection function may be expanded in terms of powers of the reciprocal energy [equation (75)]

$$K\Theta(\beta) = -2\beta \int_0^\infty (x^2 - \beta^2)^{-1/2} \frac{dU}{dx} dx + O(K^{-1}) \tag{228}$$

Neglecting the higher-order terms, equation (228) is an Abel integral equation similar to equation (214) which can be directly inverted to yield the potential

$$U(x) = \frac{K}{\pi} \int_x^\infty (\beta^2 - x^2)^{-1/2} \Theta(\beta)\, d\beta \tag{229}$$

For high energies, the influence of the attractive well is very small, so that $\Theta(\beta)$ can be directly determined by integration of the differential cross section [equations (216) and (217)]. The high-energy approximation has another advantage, first discussed by Smith *et al.*[88] If we introduce a new variable $\tau = E\vartheta$ into equation (228), this equation can be written

$$\tau(K\beta) = \tau_0(\beta) + K^{-1}\tau_1(\beta) + \cdots \tag{230}$$

Such an expansion is also possible for the impact parameter

$$\beta(\tau, K) = \beta_0(\tau) + K^{-1}\beta_1(\tau) + \cdots \tag{231}$$

and the reduced cross section

$$\rho(\tau, K) = \vartheta \sin \vartheta I(\vartheta, K) = \rho_0(\tau) + K^{-1}\rho_1(\tau) + \cdots \tag{232}$$

Equation (228) now takes the form

$$U(x) = \pi^{-1} \int_x^{\infty} (\beta^2 - x^2)^{-1/2}\, \tau_0(\beta)\, d\beta \tag{233}$$

$\beta_0(\tau)$ is obtained from equations (216) and (217), yielding

$$\beta_0^2(\tau) = 2\pi \int_\tau^{\infty} \rho_0(\tau')(\tau')^{-1}\, d\tau' \tag{234}$$

$\tau_0(\beta)$ is the inverse function of $\beta_0(\tau)$.

In this way it is possible to reduce differential cross section data which have been taken at various energies to one single curve, a fact which is especially important if the data taken at one energy are incomplete or affected by larger errors.

7.3. *The Trial and Error Method and Regression Procedures*

Most of the interatomic potentials that have been determined so far from scattering data have been found by the trial and error method. A functional form for the potential is assumed (see Section 2.3) and its parameters are optimized to fit the experimental results. Although this method has several disadvantages, it is often the only possibility to interpret scattering data. Among these disadvantages are the lack of uniqueness of the resulting potential[188] and the restriction of the potential obtained from the experiment to the assumed form. Furthermore, there is no point-to-point correspondence between the derived potential and the measurements, making it difficult to determine the sensitivity of the potential to the data. Finally, the numerical effort increases rapidly with increasing number of parameters.

In order to obtain an objectification of the fit procedure, a χ^2-minimum method is usually applied which yields also information on correlated errors. For a detailed description of such a procedure see, for example, Düren *et al.*[45,46]

Bickes and Bernstein[68] have described a systematic least-squares nonlinear regression analysis making use of the Simons–Parr–Finlan modified Dunham expansion (see Section 2.3.3). A linearization model[189] is used that provides linearization by a Taylor expansion of the calculated differential cross section $I(\theta, \gamma_i)$ at the laboratory angle θ about the first estimates of the regression parameters $\gamma_{0,i}$. Thus

$$\hat{I}(\theta, \gamma) = I(\theta, \gamma_0) + \sum_{i=1}^{p} \left(\frac{\partial \hat{I}(\theta, \gamma)}{\partial \gamma_i} \right)_{\gamma = \gamma_0} (\gamma_i - \gamma_{0,i}) \tag{235}$$

where p is the number of free regression parameters. This equation is solved yielding revised estimates of the free parameters γ_i. The procedure is then iterated until the solution converges, i.e., until

$$\sum [I(\theta) - a\hat{I}(\theta, \gamma)]^2 < \eta$$

where η is a given convergence criterion and $I(\theta)$ is the measured intensity. The parameter a is included for comparison of the laboratory intensities, where η is a given convergence criterion and $I(\theta)$ is the measured intentity. sities.

The optimal procedure for the use of the regression method is one in which the regression is carried out in the following stepwise manner: for a given set of starting parameters ε, r_m, x_f, b_i ($i = 0, 1, 2, \ldots, N$) (see Section 2.3.3), all parameters are fixed except ε and r_m, which are free for the first pass. After convergence, these new ε and r_m values are used as the starting point for the second pass, in which b_0 and b_1 as well as ε and r_m are free parameters. In the third pass, then, ε, r_m, b_0, b_1, b_2, and b_3 are treated as free parameters, using as the initial estimates the converged values from the previous pass, etc. In this way unique potentials which produce excellent fits to both computer-generated differential cross sections and experimental angular distributions are obtained. In general, three passes have been shown to yield results of sufficient accuracy, each pass itself consisting of three or four iterations.

One prerequisite is necessary. The initial values of r_m and ε cannot be too different from the "correct" ones. Differences of 20% and 50%, respectively, in these parameters are sufficient to cause the iterative scheme to converge to a grossly incorrect potential. In this case, the difficulty is indicated by the very poor fit obtained. With the use of the semiclassical relationships discussed in Section 6, however, initial fitting with any one of simpler functional forms may be used to obtain values of r_m and ε within the convergence region of the procedure.

7.4. The Use of Pseudopotentials

In the pseudopotential method[162] the interaction between all electrons and nuclei of the two particles is described by a model which describes the interaction as well as possible, and which takes account of the special electronic configuration of the collision pair. This pseudopotential which may contain several adjustable parameters, is introduced into the equation of the electronic motion, which then is solved numerically to yield the interatomic potential. With this potential the scattering cross sections are calculated and compared with the experimental results. The parameters of the pseudopotential are then changed and the cycle repeated until sufficient agreement between measured and calculated values is obtained.

The advantages have already been pointed out in the introductory survey (Section 7.1). The pseudopotential is usually valid for a whole class of chemically similar atoms, the number of parameters, which have a physical meaning, is smaller than needed in the usual trial and error method, and finally the potentials between electronically excited atoms may also be obtained. The method has been applied to the interaction of alkali atoms with rare-gas atoms[36,190,191] and with mercury atoms.[192-194] The alkali atom is especially suited for this method, since it may be described by the valence electron and the alkali core. Thus the pseudopotential takes account of the interaction between the valence electron and the multipoles which are induced in the closed-shell atom by the charges of the electron and the alkali core, and of the repulsion between the valence electron and the electrons of the closed-shell atom, and finally of the repulsion between the alkali core and the closed-shell atom. Düren[194] has recently calculated the potentials for the interaction of ground-state alkali atoms with ground-state mercury atoms and for excited alkali atoms with ground-state mercury atoms using this method. The comparison between the calculated ground-state alkali–mercury potentials and the potentials which have been determined from direct inversion of scattering data[179-181] shows excellent agreement, not only for the potential well depths ε and the equilibrium distances r_m, but also for the shape, as can be seen from Figure 22, which shows a comparison of the potential shape for CsHg. Whereas five to seven

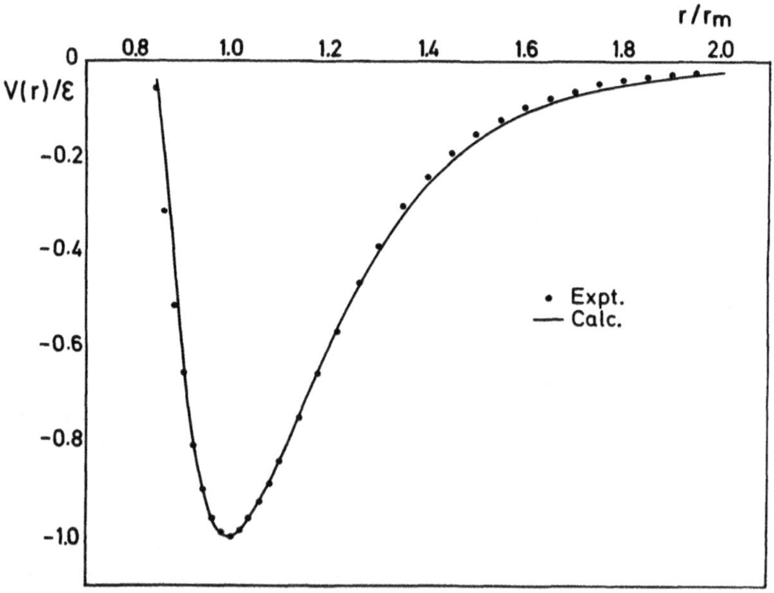

Figure 22. Comparison of the shape of the Cs–Hg potential, calculated by the pseudopotential method (full line) with the shape of the potential determined from differential cross section measurements by inversion (points).

parameters were necessary to fit a model potential to the data obtained by inversion, only two parameters are contained in the pseudopotential, and the same choice of these two parameters describes the ground-state potentials for Na–Hg, K–Hg, and Cs–Hg equally well. This shows that this method is very powerful if a suitable pseudopotential can be found.

References

1. (a) K.F. Smith, *Molecular Beams*, Methuen, London (1955); (b) N.F. Ramsey, *Molecular Beams*, Clarendon Press, Oxford (1956).
2. H. Pauly, Streuversuche an Molekularstrahlen und zwischenmolekulare Kräfte, *Fortschr. Phys.* **9**, 613–687 (1961).
3. I. Estermann, *Recent Research in Molecular Beams*, Academic Press, New York (1959).
4. H.S.W. Massey and C.B.O. Mohr, Free paths and transport phenomena in gases and the quantum theory of collisions I. The rigid sphere model, *Proc. R. Soc. London* **A141**, 434–453 (1933); Free paths and transport phenomena in gases and the quantum theory of collisions II. The determination of the laws of force between atoms and molecules, *Proc. R. Soc. London.* **A144**, 188–205 (1934).
5. W. Fite and S. Datz, Chemical research with molecular beams, *Ann. Rev. Phys. Chem.* **14**, 61–88 (1963).
6. V.B. Leonas, The present state and some new results of the molecular-beam method, *Sov. Phys.-Usp. [English Translation]* **7**, 121–144 (1964).
7. R.B. Bernstein, in *Atomic Collision Processes*, M.R.C. McDowel, editor, North-Holland, Amsterdam (1964), pp. 895–913.
8. R.B. Bernstein, Molecular beam scattering at thermal energies, *Science* **144**, 141–150 (1964).
9. H. Pauly and J.P. Toennies, The study of intermolecular potentials with molecular beams at thermal energies, *Adv. At. Mol. Phys.* **1**, 195–344 (1965).
10. R.B. Bernstein, in *Molecular Beams*, J. Ross, editor, Wiley-Interscience, New York (1966), pp. 75–134.
11. H. Pauly and J.P. Toennies, Beam experiments at thermal energies, *Methods of Exp. Phys.* **7A**, 227–341 (1968).
12. R.B. Bernstein and J.T. Muckerman, in *Intermolecular Forces*, J.O. Hirschfelder, editor, Interscience Publishers, New York (1967), pp. 389–486.
13. D. Beck, Elastic scattering of nonreactive systems, Proceedings of the International School of Physics, Enrico Fermi Course XLIV, *Molecular Beams and Reaction Kinetics*, Ch. Schlier, editor, Academic Press, New York (1970), pp. 15–49.
14. Ch. Schlier, Intermolecular forces, *Ann. Rev. Phys. Chem.* **20**, 191–218 (1969).
15. U. Buck, Elastic scattering, *Adv. Chem. Phys.* **30**, 313–388 (1975).
16. K.P. Lawley and M.A. Fluendy, *Chemical Applications of Molecular Beam Scattering*, Chapmann and Hall, London (1973).
17. H.F. Schaefer III, Atom–molecule potentials, Chap. 2 of Reference 16.
18. W. Kołos and L. Wolniewicz, Variational calculation of the long-range interaction between two ground-state hydrogen atoms, *Chem. Phys. Lett.* **24**, 457–460 (1964).
19. G. Das and A.C. Wahl, Theoretical study of the van-der-Waals-forces in alkali–noble-gas systems, *Phys. Rev. A* **4**, 825–833 (1971).
20. P. Bertoncini and A.C. Wahl, *Ab initio* calculation of the helium–helium $^1\Sigma^+$ potential at intermediate and large separations, *Phys. Rev. Lett.* **25**, 991–994 (1970); *J. Chem. Phys.* **58**, 1259–1261 (1973).
21. D.R. McLaughlin and H.F. Schaefer III, Interatomic correlation energy and the

van-der-Waals attraction between two helium atoms, *Chem. Phys. Lett.* **12,** 244–248 (1971).
22. M.L. Olson and D.D. Konowalow, Accurate potential energy curves for the $^3\Sigma_u^+$ and $b\ ^3\Sigma_g^+$ states of Li_2, *Chem. Phys.* **21,** 393–399 (1977).
23. A. Dalgarno and W.D. Davison, The calculation of van der Waals interactions, *Adv. At. Mol. Phys.* **2,** 1–32 (1966).
24. P.W. Langhoff and M. Karplus, *The Padé Approximant and Theoretical Physics*, Academic Press, New York (1970).
25. A. Dalgarno, *Advances in Chemical Physics*, J.O. Hirschfelder, editor, Vol. 12, Interscience Publishers, New York (1967), pp. 143–166.
26. F.C. Brooks, Convergence of intermolecular force series, *Phys. Rev.* **86,** 92–97 (1952).
27. A. Dalgarno and J.T. Lewis, The representation of long range forces by series expansions. I: The divergence of the series. II: The complete perturbation calculation of long range forces, *Proc. Phys. Soc.* **A69,** 57–64 (1956).
28. R.H. Young, Divergence of the R^{-1} expansion for the second order H–H interaction, *Int. J. Quantum Chem.* **9,** 47–50 (1975).
29. R. Ahlrichs, Convergence properties of the intermolecular force series ($1/R$-expansion), *Theor. Chim. Acta.* **41,** 7–15 (1976).
30. H. Kreek and W.J. Meath, Charge-overlap effects. Dispersion and induction forces, *J. Chem. Phys.* **50,** 2289–2302 (1969).
31. D.D. Richardson, Dispersion contribution to two-atom interaction energy: Multipole interactions, *J. Phys.* **A8,** 1828–1841 (1975).
32. N. Jacobi and Gy. Csanak, Dispersion forces at arbitrary distances, *Chem. Phys. Lett.* **30,** 367–372 (1975).
33. A. Koide, A new expansion for dispersion forces and its application *J. Phys. B* **9,** 3173–3183 (1976).
34. M. Battezatti and V. Magnasco, Nonexpanded multipoles for dispersion energy, *J. Chem. Phys.* **67,** 2924–2925 (1977).
35. A.A. Abrahamson, Born–Mayer-type interatomic potential for neutral ground-state atoms with $Z = 2$ to $Z = 105$, *Phys. Rev.* **178,** 76–79 (1969).
36. W.E. Baylis, Semiempirical pseudopotential calculation of alkali–noble-gas interatomic potentials, *J. Chem. Phys.* **51,** 2665–2679 (1969).
37. H. Margenau and N.R. Kestner, *Theory of Intermolecular Forces*, Pergamon Press, Oxford (1969).
38. P.R. Certain and L.W. Bruch, *MTP International Review of Science, Physical Chemistry Series One*, Vol. 1, Butterworth Publishers, London (1972), pp. 113–165.
39. R. Ahlrichs, R. Penco, and G. Scoles, Intermolecular forces in simple systems, *Chem. Phys.* **19,** 119–130 (1977).
40. W.J. Stevens, A.C. Wahl, M.A. Gardner, and A.M. Karo, *Ab initio* calculation of the neon $^1\Sigma_g^+$ potential at intermediate separations, *J. Chem. Phys.* **60,** 2195–2196 (1974).
41. J. Hepburn, G. Scoles, and R. Penco, A simple but reliable method for the prediction of intermolecular potentials, *Chem. Phys. Lett.* **36,** 451–456 (1975).
42. K.T. Tang and J.P. Toennies, A simple theoretical model for the van-der-Waals potential at intermediate distances I. Spherically symmetric potentials, *J. Chem. Phys.* **66,** 1496–1506 (1977).
43. R. Feltgen, W. Meyer, and E.-A. Reinsch, Berechnung des He_2 $X\ ^1\Sigma_g^+$-Potentials aus experimentellen und ab-initio-Daten über eine geeignete Potentialparametrisierung, *Verh. Dtsch. Phys. Ges.* **2/1978,** p. 419.
44. U. Buck und H. Pauly, Interferenzen bei atomaren Stossprozessen und ihre Interpretation durch ein modifiziertes Lennard-Jones-Potential, *Z. Phys.* **208,** 390–417 (1968).
45. R. Düren, G.P. Raabe, and Ch. Schlier, Genaue Potentialbestimmung aus Streumessungen: Alkali-Edelgas-Systeme, *Z. Phys.* **214,** 410–421 (1968).

46. R. Düren, G.P. Raabe, and Ch. Schlier, Genaue Potentialbestimmung aus Streumessungen: Alkali-Edelgas-Systeme, Sitzungsberichte der Heidelberger Akademie der Wissenschaften, *Math. Nat. Klasse*, **3.** Abhandlung (1968).
47. G.D. Mahan and M. Lapp, Bound states of alkali and noble gas atoms, *Phys. Rev.* **179,** 19–27 (1969).
48. T. Kihara, Virial coefficients and models of molecules in gases, *Rev. Mod. Phys.* **25,** 831–843 (1953).
49. Ch. Schlier, Bemerkungen zum "Regenbogeneffekt" bei der Streuung von Atomen, *Z. Phys.* **173,** 352–358 (1963).
50. R. Düren and H. Pauly, Numerische Berechnungen zum Resonanzverhalten der Streuquerschnitte bei atomaren Stössen I, *Z. Phys.* **175,** 227–234 (1963).
51. R. Düren and H. Pauly, Numerische Berechnungen zum Resonanzverhalten der Streuquerschnitte bei atomaren Stössen II. *Z. Phys.* **177,** 146–150 (1964).
52. G.C. Maitland and E.B. Smith, A simplified representation of intermolecular potential energy, *Chem. Phys. Lett.* **22,** 443–446 (1973).
53. M. Klein and H.J.M. Hanley, *m*-6-8 potential function, *J. Chem. Phys.* **53,** 4722–4723 (1970).
54. R. Düren and W. Gröger, A determination of the Na–Ar-groundstate potential from differential cross section measurements, *Chem. Phys. Lett.* **56,** 67–70 (1978).
55. R.A. Buckingham and J. Corner, Tables of second virial and low-pressure Joule–Thomson coefficients for intermolecular potentials with exponential repulsion, *Proc. R. Soc.* **A189,** 118–129 (1948).
56. O. Halpern and R.A. Buckingham, Symmetry effects in gas kinetics I. The helium isotopes, *Phys. Rev.* **98,** 1626–1631 (1955).
57. J.O. Hirschfelder, C.F. Curtiss, and R.B. Bird, *Molecular Theory of Gases and Liquids*, John Wiley and Sons, New York (1954).
58. R. Heller, Theory of some van-der-Waals molecules, *J. Chem. Phys.* **9,** 154–163 (1947).
59. J.P. Toennies, On the validity of a modified Buckingham potential for the rare gas dimers at intermediate distances, *Chem. Phys. Lett.* **20,** 238–241 (1973).
60. A.L.J. Burgmans, J.M. Farrar, and Y.T. Lee, Attractive well of He–He from ^{3}He–^{4}Hr differential elastic scattering measurements, *J. Chem. Phys.* **64,** 1345–1350 (1976).
61. J.M. Parson, P.E. Siska, and Y.T. Lee, Intemolecular potentials from crossed-beam differential elastic scattering measurements IV. Ar + Ar, *J. Chem. Phys.* **56,** 1511–1516 (1972).
62. P.E. Siska, J.M. Parson, T.P. Schafer, and Y.T. Lee, Intermolecular potentials from crossed beam differential elastic scattering measurements III. He + He and Ne + Ne, *J. Chem. Phys.* **55,** 5762–5770 (1971).
63. J.P. Toennies, in *Physical Chemistry, An Advanced Treatise*, H. Eyring, D. Henderson, and W. Jost, editors, Vol. 6A, Academic Press, New York (1974), pp. 227–381.
64. G.C. Maitland, The determination of the intermolecular potential energy function of neon from spectroscopic, equilibrium and transport data, *Mol. Phys.* **26,** 513–528 (1973).
65. J.M. Farrar, T.P. Schafer, and Y.T. Lee, in *AIP Conf. Proc.* **11,** 279–295 (1975) [*Transport phenomena*, H.C. Wolfe, editor, American Institute of Physics, New York (1973)].
66. J.A. Barker and A. Pompe, Atomic interactions in argon, *Aust. J. Chem.* **21,** 1683–1694 (1968).
67. R.W. Bickes, Jr., and R.B. Bernstein, The Simons–Parr–Finlan modified Dunham expansion: A generalized potential model for the analysis of differential elastic molecular beam scattering cross sections, *Chem. Phys. Lett.* **26,** 457–462 (1974).
68. R.W. Bickes, Jr., and R.B. Bernstein, The SPF–Dunham expansion for the potential well: A regression model for systematic analysis of differential elastic beam scattering cross sections, *J. Chem. Phys.* **66,** 2408–2420 (1977).
69. J.L. Dunham, The energy levels of a rotating vibrator, *Phys. Rev.* **41,** 721–731 (1932).

70. G. Simons, R.G. Parr, and J.M. Finlan, New alternative to the Dunham potential for diatomic molecules, *J. Chem. Phys.* **59**, 3229–3234 (1973).
71. K. Berkling, R. Helbing, K. Kramer, H. Pauly, Ch. Schlier, and P. Toschek, Effektive Stossquerschnitte bei Streuversuchen, *Z. Phys.* **166**, 406–428 (1962).
72. R. Helbing, Über den Einfluss der Bewegung der Streuteilchen auf die Messung differentieller Streuquerschnitte, Report from the Institute of Applied Physics, University of Bonn (1963).
73. R. Helbing, Transformations and averaging processes for differential cross-section measurements at thermal energies, *J. Chem. Phys.* **48**, 472–477 (1968).
74. D.R. Herschbach, Reactive collisions in crossed molecular beams, *Discuss. Faraday Soc.* **33**, 149–161 (1962).
75. A. Russek, Effect of target gas temperature on the scattering cross section, *Phys. Rev.* **120**, 1536–1542 (1960).
76. S. Datz, D.R. Herschbach, and E.T. Taylor, Collision mechanics in crossed Maxwellian molecular beams, *J. Chem. Phys.* **35**, 1549–1559 (1961).
77. F.A. Morse and R.B. Bernstein, Velocity dependence of the differential cross sections for the scattering of atomic beams of K and Cs by Hg, *J. Chem. Phys.* **37**, 2019–2027 (1962).
78. T.T. Warnock and R.B. Bernstein, Transformation relationship from center-of-mass cross section and excitation functions to observable angular and velocity distributions of scattered flux, *J. Chem. Phys.* **49**, 1878–1886 (1968).
79. R. Düren, Some aspects of the averaging effects in measured differential cross sections, Report 102/1975, Max-Planck-Institut für Strömungsforschung, Göttingen.
80. R. Düren, W.-D. Held, and D. Secrest, *VI International Symposium on Molecular Beams*, Book of Abstracts, A.E. de Vries, editor, FOM-Institute for Atomic and Molecular Physics, Noordwijkerhout, Amsterdam (1977), pp. 107–110.
81. F. von Busch, Correction of differential cross sections for finite angular resolution, *J. Phys. B* **8**, 1440–1452 (1975).
82. F. von Busch, Der Einfluss des Winkelauflösungsvermögens auf die Messung totaler atomarer Streuquerschnitte, *Z. Physik* **193**, 412–425 (1966).
83. R. Helbing, Messungen und quantenmechanische Berechnungen der Energieabhängigkeit des totalen Streuquerschnittes für die Streuung von H_2- und D_2-Molekularstrahlen an Edelgasen, Thesis, University of Bonn (1966).
84. H.S.W. Massey, E.H.S. Burhop, and H.B. Gilbody, *Electronic and Ionic Impact Phenomena*, Vol. III, Clarendon Press, Oxford (1971).
85. H.M. Mott-Smith, Collision trajectories for inverse power intermolecular potentials, *Phys. Fluids* **3**, 721–724 (1960).
86. C. Lehmann und G. Leibfried, Higher order momentum approximations in classical collision theory, *Z. Phys.* **172**, 465–487 (1963).
87. T. Kihara, Theoretisches zur Streuung von Molekularstrahlen in Gasen, *Proc. Phys. Math. Soc. Japan* **25**, 73–86 (1943).
88. F.T. Smith, R.P. Marchi, and K.G. Dedrik, Impact expansions in classical and semiclassical scattering, *Phys. Rev.* **150**, 79–150 (1966).
89. K.W. Ford and J.A. Wheeler, Semiclassical description of scattering, *Ann. Phys. N.Y.* **7**, 259–286 (1959); Application of semiclassical scattering analysis, *Ann. Phys. N.Y.* **7**, 287–322 (1959).
90. E.A. Mason, R.J. Munn, and J.F. Smith, Molecular rainbows for the 12-6 potential in the Airy approximation, *J. Chem. Phys.* **44**, 1967–1968 (1966).
91. G. Leibfried, *Bestrahlungseffekte in Festkörpern*, Teubner, Stuttgart (1965).
92. G. Leibfried and Th. Plesser, The convergence behavior of expansions in the classical collision theory, *Z. Phys.* **187**, 411–420 (1965).
93. E.A. Gislason, Exact series solution of classical deflection angle for repulsive inverse power potential, *J. Chem. Phys.* **56**, 2493–2497 (1972).
94. P. Langevin, Une formule fondamentale de théorie cinétique, *Ann. Chimie Phys.* **5**, 245–288 (1905).

95. E.A. Mason, Scattering of low velocity molecular beams in gases, *J. Chem. Phys.* **26**, 667–677 (1957).
96. H. Pauly, Die Kleinwinkelstreuung bei Stössen zwischen neutralen Atomen, *Z. Phys.* **157**, 54–64 (1959).
97. R. Helbing and H. Pauly, Differentielle Streuquerschnitte bei der Wechselwirkung neutraler Moleküle. Messungen der Kleinwinkelstreuung, *Z. Phys.* **179**, 16–37 (1964).
98. E. Gersing, E. Hundhausen, and H. Pauly, Differentielle Streuquerschnitte bei der Wechselwirkung neutraler Moleküle. Messungen der Grosswinkelstreuung, *Z. Phys.* **171**, 349–370 (1963).
99. I. Amdur and J.E. Jordan, in *Molecular Beams, Advances in Chemical Physics*, J. Ross, editor, Vol. 10, Interscience, New York (1966), pp. 29–73.
100. H.C. van de Hulst, On the attenuation of plane waves by obstacles of arbitrary size and form, *Physica (Utrecht)* **XV**, 740–746 (1949).
101. R.B. Bernstein, Quantum mechanical (phase shift) analysis of differential elastic scattering of molecular beams, *J. Chem. Phys.* **33**, 795–804 (1960).
102. W.A. Lester, Jr., in *Methods in Computational Physics*, B. Adler, S. Fernbach, and M. Rotenberg, editors, Vol. 10, Academic Press, New York (1971), pp. 211–241.
103. R.G. Gordon, in *Methods in Computational Physics*, B. Adler, S. Fernbach, and M. Rotenberg, editors, Vol. 11, Academic Press, New York (1971), pp. 81–109.
104. W.I. Newman and W.R. Thorson, Rapid numerical solution of the one-dimensional Schrödinger-equation, *Can. J. Phys.* **50**, 2997–3022 (1972).
105. K.H. Katerbau, Untersuchung verschiedener Verfahren zur Phasenberechnung für die Molekularstrahlstreuung, Diplom-thesis, University of Freiburg (1965).
106. M. Rosen and D.R. Yennie, A modified WKB approximation for phase shifts, *J. Math. Phys.* **5**, 1505–1515 (1964).
107. F. Calogero, *Variable Phase Approach to Potential Scattering*, Academic Press, New York (1967).
108. F. Calogero, *Scattering Theory*, Gordon and Breach, New York (1969).
109. B. Lantzsch, Messung der Geschwindigkeitsabhängigkeit integraler Querschnitte für die Streuung eines thermischen H-Strahls an Edelgasen und leichten Molekülen und Bestimmung der Wechselwirkungspotentiale, Report 120/1974, Max-Planck-Institut für Strömungsforschung, Göttingen.
110. S. Tani and M. Inokuti, Standing-wave modification of Regge trajectories and its application to atom–atom and ion–atom scattering at thermal energies, *J. Chem. Phys.* **61**, 4422–4432 (1974).
111. U. Buck, K.A. Köhler, and H. Pauly, Messungen der Glorienstreuung des totalen Streuquerschnitts von Na-Hg, *Z. Phys.* **244**, 180–189 (1971).
112. R. Feltgen, Weitere Messungen der Geschwindigkeitsabhängigkeit des totalen Streuquerschnittes and Systemen mit kleiner reduzierter Masse, Thesis, Bonn University, 1970.
113. R.B. Bernstein, Extrema in velocity dependence of total elastic cross sections for atomic beam scattering: Relation to di-atom bound states, *J. Chem. Phys.* **37**, 1880–1881 (1962).
114. R.B. Bernstein, Semiclassical analysis of the extrema in the velocity dependence of total elastic scattering cross sections: Relation to the bound states, *J. Chem. Phys.* **38**, 2599–2609 (1963).
115. H. Harrison and R.B. Bernstein, Iterative method for solution of the one-dimensional wave equation: Eigenvalues and eigenfunctions for L-J (12, 6) and exponential (α, 6) interatomic potentials, *J. Chem. Phys.* **38**, 2135–2143 (1963).
116. N. Levinson, On the uniqueness of the potential in a Schrödinger equation for a given asymptotic phase, *Kgl. Danske Videnskabernes Selskab Matematisk-Fysiske Meddelelser* **25**, 1–29 (1949).
117. G. Breit and E.P. Wigner, Capture of slow neutrons, *Phys. Rev.* **49**, 519–531 (1936).
118. G. Wolf, Untersuchung langlebiger Stosskomplexe bei der Streuung von atomarem

und molekularem Wasserstoff an Edelgasatomen und CF_4, Max-Planck-Institut für Strömungsforschung, Report 20/1976 (see also J.P. Toennies, W. Welz, and G. Wolf to be published).
119. R.B. Bernstein, C.F. Curtiss, S. Iman-Rahajoe, and H.T. Wood, Numerical evaluation of barrier penetration and resonance effects on phase shifts, *J. Chem. Phys.* **44**, 4072–4081 (1966).
120. R.B. Bernstein, Long-range interatomic forces from predissociation data and resonances in atomic scattering, *Phys. Rev. Lett.* **16**, 385–388 (1966).
121. Th. Waech and R.B. Bernstein, Calculated spectrum of quasibound states for $H_2(^1\Sigma_g^+)$ and resonances in H + H scattering, *J. Chem. Phys.* **46**, 4905–4911 (1967).
122. A. Schütte, D. Bassi, F. Tomasini, and G. Scoles, Orbiting resonances in the scattering of H atoms by mercury at thermal energies, *Phys. Rev. Lett.* **29**, 979–982 (1972).
123. A. Schütte, D. Bassi, F. Tomasini, and G. Scoles, Energy dependence of high resolution integral-collision-cross-section measurements for elastic scattering of H and D atoms by Hg at thermal energies, *J. Chem. Phys.* **62**, 600–605 (1975).
124. J.P. Toennies, W. Welz, and G. Wolf, Observation of orbiting resonance in the integral cross section of H–Xe, *J. Chem. Phys.* **61**, 2461–2462 (1974).
125. J.P. Toennies, W. Welz, and G. Wolf, Observation of orbiting resonances in H_2–rare-gas scattering, *J. Chem. Phys.* **64**, 5305–5307 (1976).
126. K.A. Köhler, R. Feltgen, and H. Pauly, Measurements of the resonance scattering of the systems H–Hg and H_2–Hg and determination of the isotopic mercury hydride potential, *Phys. Rev. A* **15**, 1407–1413 (1977).
127. R. Feltgen, K.A. Köhler, H. Pauly, F. Torello, and H. Vehmeyer, High resolution measurements of the He-He total scattering cross section for reduced collision energies between 0.2 and 200, *Proceedings of the 4th International Symposium on Molecular Beams,* Cannes, 1973, pp. 304–320.
128. R. Feltgen, H. Pauly, F. Torello, and H. Vehmeyer, Determination of the He^4–He^4 repulsive potential up to 0.14 eV by inversion of high-resolution total-cross-section measurements, *Phys. Rev. Lett.* **30**, 820–823 (1973).
129. R. Feltgen, private communication (1978).
130. M.V. Berry, Uniform approximation for potential scattering involving a rainbow, *Proc. Phys. Soc.* **89**, 479–490 (1966).
131. M.V. Berry, Semi-classical scattering phase shifts in the presence of metastable states, *Proc. Phys. Soc.* **88**, 285–292 (1966).
132. W.H. Miller, Uniform semiclassical approximations for elastic scattering and eigenvalue problems, *J. Chem. Phys.* **48**, 464–467 (1968).
133. M.V. Berry, Uniform approximations for glory scattering and diffraction peaks, *J. Phys.* **B2**, 381–392 (1969).
134. K.E. Mount, Potential scattering cross sections in the semiclassical limit, *J. Phys.* **B6**, 1397–1411 (1973).
135. M.V. Berry and K.E. Mount, Semiclassical approximations in wave mechanics, *Rep. Prog. Phys.* **35**, 315–397 (1972).
136. H. Pauly, in *Physical Chemistry,* H. Eyring, D. Henderson, and W. Jost, editors, Vol. 6B, Academic Press, New York (1975), pp. 553–628.
137. L.D. Landau and E.M. Lifschitz, *Quantum Mechanics,* Pergamon Press, London (1959).
138. N.F. Mott and H.S.W. Massey, *Theory of Atomic Collision,* Clarendon Press, Oxford (1949).
139. N. Fröman and P. Fröman, *JWKB Approximation,* North-Holland Publ. Co., Amsterdam (1965).
140. T.Y. Wu and T. Ohmura, *Quantum Theory of Scattering,* Prentice-Hall, Englewood Cliffs, N.J. (1962).
141. E.F. Greene, G.P. Reck, and J.L.J. Rosenfeld, Analysis of elastic scattering data in the semiclassical approximation, *J. Chem. Phys.* **46**, 3693–3697 (1967).

142. J.M. Mullen and B.S. Thomas, On an evaluation of the accuracy of the uniform semiclassical approximation for differential elastic scattering cross sections, *J. Chem. Phys.* **58**, 5216–5221 (1973).
143. C.F. Curtiss and R.S. Powers, Jr., An expansion of binary collision phase shifts in powers of *h*, *J. Chem. Phys.* **40**, 2145–2150 (1964).
144. S. Choi and J. Ross, On the semiclassical expansion of scattering phase shifts, *J. Chem. Phys.* **40**, 2151–2153 (1964).
145. C.F. Curtiss, Effect of three turning points on semiclassical scattering, *J. Chem. Phys.* **42**, 2267–2269 (1965).
146. J.N.L. Connor, On the semi-classical description of molecular orbiting collisions, *Mol. Phys.* **15**, 621–631 (1968); On the phase shift for orbiting collisions, *Mol. Phys.* **16**, 525–526 (1969).
147. E.A. Mason, J.T. Vanderslice, and C.J.G. Raw, Quantum effects in small-angle molecular-beam scattering *J. Chem. Phys.* **40**, 2153 (1964).
148. R.B. Bernstein, Quantum calculations of the velocity dependence of the differential and total cross sections for elastic scattering of molecular beams, *J. Chem. Phys.* **34**, 361–365 (1961).
149. L.I. Schiff, Approximation method for high-energy potential scattering, *Phys. Rev.* **103**, 443–453 (1956).
150. R.B. Bernstein and K.H. Kramer, Comparison and appraisal of approximation formulas for total elastic molecular scattering cross sections, *J. Chem. Phys.* **38**, 2507–2511 (1963).
151. R.B. Bernstein and T.J.P. O'Brien, Extrema-effect in total elastic molecular beam scattering cross sections for characterization of the potential well, *Discuss. Faraday Soc.* **40**, 35–44 (1965).
152. R.B. Bernstein and T.J.P. O'Brien, Potential well characteristics from the energy dependence of the glory extrema in total elastic scattering cross sections, *J. Chem. Phys.* **46**, 1208–1209 (1967).
153. E.F. Greene and E.A. Mason, Physical interpretation of glory undulations in scattering cross sections, *J. Chem. Phys.* **57**, 2065–2069 (1972).
154. R.B. Bernstein and R.A. LaBudde, On the analysis of glory scattering data for the extraction of information on the interatomic potential well, *J. Chem. Phys.* **58**, 1109–1117 (1973).
155. E.F. Greene and E.A. Mason, Interpretation of glory undulations in scattering cross sections: Amplitudes, *J. Chem. Phys.* **59**, 2651–2655 (1973).
156. R.B. Bernstein, Total elastic cross sections for L-J (n, 6) and exp-(α, 6) potentials, *J. Chem. Phys.* **38**, 515–522 (1963).
157. M.G. Dondi, G. Scoles, F. Torello, and H. Pauly, Energy dependence of the elastic total collision cross section of identical molecules: ^{4}He, *J. Chem. Phys.* **51**, 392–397 (1969).
158. R.J. Glauber, *High Energy Collision Theory*, Lectures in Theoretical Physics, Vol. 1, Interscience Publishers, New York (1959), pp. 315–414.
159. U. Buck, The inversion of molecular scattering data, *Rev. Mod. Phys.* **46**, 369–389 (1974).
160. K. Chadan and P.C. Sabatier, *Inverse Problems in Quantum Scattering Theory*, Springer, New York (1977).
161. M. Shapiro, R.B. Gerber, U. Buck, and J. Schleusener, Deconvolution of differential cross sections obtained from molecular beam experiments, *J. Chem. Phys.* **67**, 3570–3576 (1977); U. Buck, private communication (1978).
162. J.N. Bardsley, Pseudopotentials in atomic and molecular physics, *Case Studies in Atomic Physics* **4**, 299–368 (1974).
163. R.B. Gerber and M. Shapiro, A numerical method for the determination of atom–atom scattering amplitudes from the measured differential cross sections, *Chem. Phys.* **13**, 227–233 (1976).

164. M. Shapiro and R.B. Gerber, Extraction of interaction potentials from the elastic scattering amplitude: an accurate quantum mechanical procedure, *Chem. Phys.* **13**, 235 (1976).
165. O. Klein, Zur Berechnung von Potentialkurven für zweiatomige Moleküle mit Hilfe von Spektraltermen, *Z. Phys.* **76**, 226–235 (1932).
166. E.A. Mason and L. Monchick, *Adv. Chem. Phys.* **12**, 351–354 (1967).
167. L.H. Miller, *Sixth International Conference on the Physics of Electronic and Atomic Collisions,* I. Amdur, editor, The MIT Press, Cambridge, Massachusetts (1969), pp. 947–950.
168. J.B. Keller, I. Kay, and J. Shmoys, Determination of the potential from scattering data, *Phys. Rev.* **102**, 557–559 (1956).
169. O.B. Firsov, Determination of forces acting between atoms with use of the differential cross section of elastic scattering, *Zh. Eksp. Teor. Fiz.* **24**, 279–283 (1953).
170. G.H. Lane and E. Everhart, Ion-atom potential energy functions obtained from keV scattering data *Phys. Rev.* **120**, 2064–2069 (1960).
171. J.W. Brackett, C.R. Mueller, and W.A. Sanders, Direct determination of scattering phase shifts from differential cross sections, *J. Chem. Phys.* **39**, 2564–2580 (1963).
172. G. Vollmer, Inverse problem in atom–atom scattering in WKB approach, *Z. Phys.* **226**, 423–434 (1969).
173. R. Klingbeil, Determination of interatomic potentials by the inversion of elastic differential cross section data I. An inversion procedure, *J. Chem. Phys.* **56**, 132–136 (1972).
174. E.A. Remler, Complex-angular-momentum analysis of atom–atom scattering experiments, *Phys. Rev. A* **3**, 1949–1954 (1971).
175. A. Sommerfeld, *Vorlesungen über theoretische Physik,* Vol. VI, Dieterich'sche Verlagsbuchhandlung, Leipzig (1947).
176. V. De Alfaro and T. Regge, *Potential Scattering,* Wiley, New York (1965).
177. W.G. Rich, S.M. Bobbio, R.L. Champion, and L. Doverspike, Inversion problem for ion–atom differential elastic scattering, *Phys. Rev. A* **4**, 2253–2260 (1971).
178. U. Buck and H. Pauly, Determination of intermolecular potentials by inversion of molecular beam scattering data, *J. Chem. Phys.* **51**, 1662–1664 (1969).
179. U. Buck, Determination of intermolecular potentials by the inversion of molecular beam scattering data I. The inversion procedure, *J. Chem. Phys.* **54**, 1923–1928 (1971).
180. U. Buck and H. Pauly, Determination of intermolecular potentials by the inversion of molecular beam scattering data II. High resolution measurements of differential scattering cross sections and the inversion of the data for Na–Hg, *J. Chem. Phys.* **54**, 1929–1936 (1971).
181. U. Buck, M. Kick, and H. Pauly, Determination of intermolecular potentials by the inversion of molecular beam scattering data III. High resolution measurements and potentials for K–Hg and Cs–Hg, *J. Chem. Phys.* **56**, 3391–3397 (1972).
182. U. Buck, H.O. Hoppe, F. Huisken, and H. Pauly, Intermolecular potentials by the inversion of molecular beam scattering data. IV. Differential cross sections and potential for LiHg, *J. Chem. Phys.* **60**, 4925–4929 (1974).
183. U. Buck, F. Huisken H. Pauly, and J. Schleusener, Intermolecular potentials by the inversion of differential cross sections V. ArKr, *J. Chem. Phys.* **68**, 3334 (1978).
184. H. Pauly, in *Atomic Physics 2 (Proceedings of the II International Conference on Atomic Physics),* P.G.H. Sandars, editor, Plenum Press, New York (1971), pp. 155–169.
185. M. Abramowitz and I.A. Stegun, *Handbook of Mathematical Tables,* Dover, New York (1967).
186. W.H. Miller, WKB solution of inversion problems for potential scattering, *J. Chem. Phys.* **51**, 3631–3638 (1969).
187. J.F. Boyle, Semiclassical inversion of rainbow scattering data, *Mol. Phys.* **22**, 993–1011 (1971).

188. T.J.P. O'Brien and R.B. Bernstein, Investigation of the Hylleraas method for determining the potential-energy function from the phase shift, *J. Chem. Phys.* **51**, 5112–5117 (1969).
189. N.R. Draper and H. Smith, *Applied Regression Analysis*, Wiley, New York (1966).
190. W.E. Baylis, JILA Report No. 100, 1969.
191. J. Pascale and J. Vandeplanque, Excited molecular terms of the alkali–rare-gas atoms pairs, *J. Chem. Phys.* **60**, 2278–2289 (1974).
192. R. Düren. A pseudopotential calculation of the Na–Hg system including the lowest *P*-state of Na, *Chem. Phys. Lett.* **39**, 481–484 (1976).
193. R. Düren. The interpretation of experimental scattering cross sections with pseudopotential calculations for the alkali-mercury interaction, *J. Phys. B* **10**, 3467–3481 (1977).
194. R. Düren. Adiabatic potentials and diabatic matrix elements for the alkali-mercury systems Na, K, Rb and Cs, Report 12/1977, Max-Planck-Institut für Strömungsforschung, Göttingen.

5

Elastic Scattering Cross Sections II: Noncentral Potentials

S. STOLTE AND J. REUSS

1. Introduction

The title of this chapter seems to be self-contradictory, since noncentral potentials are usually and rightly considered as the source of inelastic effects. However, the angular-dependent part of the intermolecular potential (AIP) causes, too, an orientational dependence of all collision processes including the elastic scattering. It is this aspect which has been exploited in recent years to obtain valuable information on the AIP. Prerequisite for experiments in this direction is state selection, by which certain preferential orientations of the interacting molecules are produced.

Measurements of the total integral cross section σ_{tot} with state-selected molecules have been undertaken for TlF,[1] CsF,[2] H_2,[3] NO,[4,5] and Na_2.[6] Recently, the H_2–He and H_2–Ne results were discussed in terms of a new semiempirical approach to construct the intermolecular potential (IP) from its long- and short-range behavior.[7] Besides probing the AIP at large distances the anisotropy of σ_{tot} turned out to be a very sensitive indicator of the precise form of the AIP around the minimum of the potential.

Differential cross sections for state-selected LiF have been measured by Tsou *et al.*[8] On the other hand, the influence of noncentral potentials on the rainbow structure of numerous systems like CH_3I–Na have been investigated, without employing state selection.[9] Experiments with tetrahedral molecules are discussed by Buck *et al.*[10]

The aim of this chapter is to discuss the theoretical instruments with

S. STOLTE AND J. REUSS • Fysisch Laboratorium, Katholieke Universiteit, 6525 ED Nijmegen, The Netherlands.

which the above phenomena can be handled. Inelastic processes are treated in other chapters of this book and in recent review articles by Toennies.[11] In Section 2, noncentral potentials will be presented in a general form suited for the subsequent discussion of calculations. In Section 3, some general formula and close-coupling calculations will be shortly touched upon. Section 4 is dedicated to the distorted-wave approximation (dwa) for noncentral potentials. Section 5 summarizes different sudden-approximation approaches, which are employed in differential and total cross section calculations in Section 6. It turns out that only the space-fixed approach gives reliable results of orientation-dependent effects; however, its selection rule causes doubts as to whether this approach is apt to describe higher-order effects. The situation is reviewed in Section 7. A glossary of abbreviations is given at the end.

2. *Angular-Dependent Potentials*

2.1. *The General Form*

Before the problem of two colliding molecules can be handled in calculating the relevant cross sections, the intermolecular potential (IP) must be known. We first investigate, therefore, the general form of the terms which play a role in the scattering process.

Let us consider two arbitrarily shaped molecules ($i = 1$ and 2). The rotation functions[12] $D^{n_i}_{\nu_i\kappa_i}(\alpha_i, \beta_i, \gamma_i)$ form a complete basis set to describe any angular dependence of molecule (i) in terms of the Euler angles (α_i, β_i, γ_i) taken with respect to a space-fixed frame. For fixed spatial orientation and distance of the molecules (1) and (2), the AIP remains a function of the direction $\hat{\mathbf{R}}$ of the intermolecular position vector pointing from molecule (1) to molecule (2); this angular dependence can generally be expressed with the help of spherical harmonic functions $C^n_\nu(\beta, \alpha)$ (β and α stand for the polar angles of $\hat{\mathbf{R}}$). Consequently the IP can be expressed as a sum over terms, where a function depending solely on the distance R is multiplied by $C^n_\nu(\beta, \alpha) D^{n_1}_{\nu_1\kappa_1}(\alpha_1, \beta_1, \gamma_1) D^{n_2}_{\nu_2\kappa_2}(\alpha_2, \beta_2, \gamma_2)$. The IP is a scalar under spatial rotations of the whole system, yielding

$$V(1, 2, \mathbf{R}) = \sum_{\{v\}} V^{n,n_1,n_2,\kappa_1,\kappa_2}(R) \sum_{\nu_1,\nu_2,\nu} D^{n_1}_{\nu_1\kappa_1}(\alpha_1, \beta_1, \gamma_1) D^{n_2}_{\nu_2\kappa_2}(\alpha_2, \beta_2, \gamma_2) \times C^n_\nu(\beta, \alpha) \begin{pmatrix} n_1 & n_2 & n \\ \nu_1 & \nu_2 & \nu \end{pmatrix} (-1)^{\nu_1+\nu_2} \tag{1}$$

where $\{v\}$ stands for $\{n_1, n_2, n, \kappa_1, \kappa_2\}$.

The isotropic part $\bar{V}$ of the potential (IIP) is obtained from the right-hand side of equation (1) by taking the spherical average over all angles.

If molecules (1) and (2) possess cylindrical symmetry, equation (1) reduces to[5]

$$V(1,2,\mathbf{R}) = \sum_{\{n_1,n_2,n\}} V^{n_1,n_2,n}(R) \sum_{\nu_1,\nu_2,\nu} C^{n_1}_{\nu_1}(\beta_1,\alpha_1)\, C^{n_2}_{\nu_2}(\beta_2,\alpha_2) \times C^{n}_{\nu}(\beta,\alpha) \begin{pmatrix} n_1 & n_2 & n \\ \nu_1 & \nu_2 & \nu \end{pmatrix} \tag{2}$$

In the case of a cylindrically symmetric molecule and an atom, equation (1) yields

$$V(1,2,\mathbf{R}) = \sum_n \frac{V^n(R)}{(2n+1)^{1/2}} (-1)^n \sum_\nu C^n_\nu(\beta,\alpha)\, C^{n*}_\nu(\beta_1,\alpha_1) \tag{3}$$

For the purpose of the discussion below we define the general terms

$$\mathrm{GT} = D^{n_1}_{\nu_1\kappa_1}(\text{Euler angles for 1})\, D^{n_2}_{\nu_2\kappa_2}(\text{Euler angles for 2})\, C^n_\nu(\hat{\mathbf{R}}) \times \begin{pmatrix} n_1 & n_2 & n \\ \nu_1 & \nu_2 & \nu \end{pmatrix} (-1)^{\nu_1+\nu_2} f(R) \tag{4}$$

2.2. The Long-Range Terms

The interaction between two molecules can be considered in the two extreme situations where overlap effects dominate (the short-range IP) and where the two molecules slightly perturb each other by electrostatic multipole, induction, and dispersion forces (the long-range IP). Extensive reviews on the long-range IP are given by Buckingham[13] and Margenau and Kestner.[14] The long-range intermolecular potential is written as $V_{\text{long range}} = V_{\text{elec}} + V_{\text{ind}} + V_{\text{disp}}$, i.e., all terms are simply additive. If both molecules are at least cylindrically symmetric (symmetry axis $\hat{\mathbf{r}}_i$) one obtains for the electrostatic terms,[5] with $\cos\vartheta_i = \hat{\mathbf{r}}_i \cdot \hat{\mathbf{R}}$ and φ_i for the azimuth angle around $\hat{\mathbf{R}}$,

$$\begin{aligned} V_{D_1D_2} &= \frac{\mu_1\mu_2}{R^4}[-2\cos\vartheta_1\cos\vartheta_2 + \sin\vartheta_1\sin\vartheta_2\cos(\varphi_1-\varphi_2)] \\ &= -\frac{\mu_1\mu_2}{R^3}(30)^{1/2} \sum_{\nu_1,\nu_2,\nu} C^1_{\nu_1}(\hat{\mathbf{r}}_1)\, C^1_{\nu_2}(\hat{\mathbf{r}}_2)\, C^2_\nu(\hat{\mathbf{R}}) \begin{pmatrix} 1 & 1 & 2 \\ \nu_1 & \nu_2 & \nu \end{pmatrix} \end{aligned} \tag{5}$$

$$\begin{aligned} V_{D_iQ_j} &= \frac{3/2\mu_i Q_j}{R^4}(-1)^j[\cos\vartheta_i(3\cos^2\vartheta_j - 1) \\ &\qquad - 2\sin\vartheta_i\sin\vartheta_j\cos\vartheta_j\cos(\varphi_1-\varphi_2)] \\ &= \frac{3/2\mu_i Q_j}{R^4}(-1)^j\, 2\left(\frac{35}{3}\right)^{1/2} \sum_{\nu_1,\nu_2,\nu} C^1_{\nu_1}(\hat{\mathbf{r}}_i)\, C^2_{\nu_2}(\hat{\mathbf{r}}_j)\, C^3_\nu(\hat{\mathbf{R}}) \begin{pmatrix} 1 & 2 & 3 \\ \nu_1 & \nu_2 & \nu \end{pmatrix} \end{aligned} \tag{6}$$

for $i = 1, j = 2$, and $i = 2, j = 1$,

$$V_{Q_1Q_2} = \frac{3/4Q_1Q_2}{R^5}[1 - 5\cos^2\vartheta_1 - 5\cos^2\vartheta_2 + 17\cos^2\vartheta_1\cos^2\vartheta_2 + 2\sin^2\vartheta_1\sin^2\vartheta_2\cos^2(\varphi_1 - \varphi_2) - 16\sin\vartheta_1\cos\vartheta_1\sin\vartheta_2\cos\vartheta_2\cos(\varphi_1 - \varphi_2)]$$

$$= \frac{3/4Q_1Q_2}{R^5}4(70)^{1/2}\sum_{\nu_1,\nu_2,\nu} C^2_{\nu_1}(\hat{\mathbf{r}}_1)\,C^2_{\nu_2}(\hat{\mathbf{r}}_2)\,C^4_\nu(\hat{\mathbf{R}})\begin{pmatrix}2 & 2 & 4\\ \nu_1 & \nu_2 & \nu\end{pmatrix} \tag{7}$$

$$V_{D_iO_j} = \frac{1/2\mu_iO_j}{R^5}[\cos\vartheta_i\cos\vartheta_j(12 - 20\cos^2\vartheta_j) + \sin\vartheta_i\sin\vartheta_j(15\cos^2\vartheta_j - 3)\cos(\varphi_1 - \varphi_2)]$$

$$= \frac{-1/2\mu_iO_j}{R^5}12(7)^{1/2}\sum_{\nu_1,\nu_2,\nu} C^1_{\nu_1}(\hat{\mathbf{r}}_i)\,C^3_{\nu_2}(\hat{\mathbf{r}}_j)\,C^4_\nu(\hat{\mathbf{R}})\begin{pmatrix}1 & 3 & 4\\ \nu_1 & \nu_2 & \nu\end{pmatrix} \tag{8}$$

for $i = 1, j = 2$, and $i = 2, j = 1$. The multipole moments μ_i, Q_i, and O_i can be found in reference 15.

The contribution of the induction potential V_{ind} to $V_{\text{long range}}$ consists of two terms depending on the polarizabilities α_i:

$$V_{\text{ind}} = V_{1-2} + V_{2-1} \tag{9}$$

with[5]

$$V_{i-j} = \frac{-1/2\mu_i^2\alpha_j}{R^6}(3\cos^2\vartheta_i + 1) + \frac{6\mu_iQ_i\alpha_j}{R^7}(-1)^j\cos^3\vartheta_i + \cdots$$

$$= \frac{-1/2\mu_i^2\alpha_j}{R^6}2\left[1 + \sum_\nu C^2_\nu(\hat{\mathbf{r}}_i)\,C^{2*}_\nu(\hat{\mathbf{R}})\right]$$

$$+ \frac{6\mu_iQ_j\alpha_j}{R^7}(-1)^j\left[\frac{2}{5}\sum_\nu C^3_\nu(\hat{\mathbf{r}}_i)\,C^{3*}_\nu(\hat{\mathbf{R}}) + \frac{3}{5}\sum_\nu C^1_\nu(\hat{\mathbf{r}}_i)\,C^{1*}_\nu(\hat{\mathbf{R}})\right] + \cdots \tag{10}$$

Finally the R^{-6}-dependent part of the dispersion potential as originally derived by London[16] yields

$$V_{\text{disp}} = \frac{-1}{6R^6}\{(A - B_1 - B_2 + C)[\sin\vartheta_1\sin\vartheta_2\cos(\varphi_2 - \varphi_1) - 2\cos\vartheta_1\cos\vartheta_2]^2 + 3(B_1 - C)\cos^2\vartheta_1 + 3(B_2 - C)\cos^2\vartheta_2 + (B_1 + B_2 + 4C)\} \tag{11}$$

For the determination of the constants A, B_i, and C in equation (11) one may resort to the (Drude) oscillator model or to the (Unsöld) two-level approximation, for which one obtains, respectively;

$$
\begin{aligned}
A &= \frac{3}{2}\frac{e\hbar}{m_e^{1/2}}\frac{\alpha_{\parallel 1}\alpha_{\parallel 2}}{(\alpha_{\parallel 1}/N_1)^{1/2} + (\alpha_{\parallel 2}/N_2)^{1/2}} \\
A &= \frac{3}{2}\frac{U_1 U_2}{U_1 + U_2}\alpha_{\parallel 1}\alpha_{\parallel 2} \\
B_1 &= \frac{3}{2}\frac{e\hbar}{m_e^{1/2}}\frac{\alpha_{\parallel 1}\alpha_{\perp 2}}{(\alpha_{\parallel 1}/N_1)^{1/2} + (\alpha_{\perp 2}/N_2)^{1/2}} \\
B_1 &= \frac{3}{2}\frac{U_1 U_2}{U_1 + U_2}\alpha_{\parallel 1}\alpha_{\perp 2} \\
B_2 &= \frac{3}{2}\frac{e\hbar}{m_e^{1/2}}\frac{\alpha_{\perp 1}\alpha_{\parallel 2}}{(\alpha_{\perp 1}/N_1)^{1/2} + (\alpha_{\parallel 2}/N_2)^{1/2}} \\
B_2 &= \frac{3}{2}\frac{U_1 U_2}{U_1 + U_2}\alpha_{\perp 1}\alpha_{\parallel 2} \\
C &= \frac{3}{2}\frac{e\hbar}{m_e^{1/2}}\frac{\alpha_{\perp 1}\alpha_{\perp 2}}{(\alpha_{\perp 1}/N_1)^{1/2} + (\alpha_{\perp 2}/N_2)^{1/2}} \\
C &= \frac{3}{2}\frac{U_1 U_2}{U_1 + U_2}\alpha_{\perp 1}\alpha_{\perp 2}
\end{aligned}
\tag{12}
$$

In equation (12) α_i stands for the averaged polarizability (values can be found in Reference 17), $\Delta\alpha_i = \alpha_{\parallel i} - \alpha_{\perp i}$ for the difference of the polarizability along and perpendicular to the molecular symmetry axis, U_i for the average excitation energy (usually taken equal to the ionization energy), and N_i for the number of electrons in the valence or outer subshell of the molecule.

In spherical tensor notation one obtains[5]

$$
\begin{aligned}
V_{\text{disp}} = \frac{-C_{6,\text{disp}}}{R^6}\Bigg\{ & 1 + q_{2,6}\sum_{\nu} C_\nu^2(\hat{\mathbf{r}}_1)\, C_\nu^{2*}(\hat{\mathbf{R}}) + p_{2,6}\sum_{\nu} C_\nu^2(\hat{\mathbf{r}}_2)\, C_\nu^{2*}(\hat{\mathbf{R}}) \\
& + r_{2,6}\Bigg[\frac{1}{5}\sum_{\nu} C_\nu^2(\hat{\mathbf{r}}_1)\, C_\nu^{2*}(\hat{\mathbf{r}}_2) + \frac{1}{7}(70)^{1/2}\sum_{\nu_1,\nu_2,\nu} C_{\nu_1}^2(\mathbf{r}_1)\, C_{\nu_2}^2(\mathbf{r}_2)\, C_\nu^2(\hat{\mathbf{R}}) \\
& \times \begin{pmatrix} 2 & 2 & 2 \\ \nu_1 & \nu_2 & \nu \end{pmatrix} + \frac{108}{70}(70)^{1/2}\sum_{\nu_1,\nu_2,\nu} C_{\nu_1}^2(\hat{\mathbf{r}}_1)\, C_{\nu_2}^2(\hat{\mathbf{r}}_2)\, C_\nu^4(\hat{\mathbf{R}}) \\
& \times \begin{pmatrix} 2 & 2 & 4 \\ \nu_1 & \nu_2 & \nu \end{pmatrix}\Bigg]\Bigg\}
\end{aligned}
\tag{13}
$$

with

$$\begin{aligned} C_{6,\mathrm{disp}} &= \tfrac{1}{6}(A + B_1 + B_2 + 3C) \\ q_{2,6} &= \tfrac{1}{6}(A + B_1 - B_2 - C)/C_{6,\mathrm{disp}} \\ p_{2,6} &= \tfrac{1}{6}(A - B_1 + B_2 - C)/C_{6,\mathrm{disp}} \\ r_{2,6} &= \tfrac{1}{6}(A - B_1 - B_2 + C)/C_{6,\mathrm{disp}} \end{aligned} \tag{14}$$

The two-level approximation yields

$$\begin{aligned} &C_{6,\mathrm{disp}} = \frac{3}{2}\,\frac{U_1 U_2}{U_1 + U_2}\,\alpha_1\alpha_2 \\ &q_{2,6} = \frac{\Delta\alpha_1}{3\alpha_1} \qquad p_{2,6} = \frac{\Delta\alpha_2}{3\alpha_2} \qquad r_{2,6} = \frac{\Delta\alpha_1}{3\alpha_1}\,\frac{\Delta\alpha_2}{3\alpha_2} \end{aligned} \tag{15}$$

The $C_{6,\mathrm{disp}}$ constant derived using the oscillator model [equations (12) and (14)] coincides (within a few percent) with the Slater–Kirkwood value[16]:

$$C_{6,\mathrm{disp}} = \frac{(3/2)\, e\hbar\, \alpha_1\alpha_2/m_e^{1/2}}{(\alpha_1/N_1)^{1/2} + (\alpha_2/N_2)^{1/2}} \tag{16}$$

Normally equation (16) yields results closest to the actual value of $C_{6,\mathrm{disp}}$; the $C_{6,\mathrm{disp}}$ result of equation (15) yields too small a value.[18] However, the $q_{2,6}$, $p_{2,6}$, and $r_{2,6}$ values are smaller in the oscillator model than in the two-level approximation. Moreover, $q_{2,6}$ is a function of molecule (2) in the oscillator model; its value can vary by 20%.

The AIP acquires an R^{-7} dispersion term as lowest-order contribution, when one of the molecules possesses tetrahedral symmetry [e.g., molecule (2)]; this extra term depends on the dipole–quadrupole polarizability A_2 and is given by[13]

$$V_{\mathrm{disp}}^{(1)} = \frac{-C_{6,\mathrm{disp}}}{R^7}\, 8\, \frac{A_2}{\alpha_2} \cos\theta_x \cos\theta_y \cos\theta_z \tag{17}$$

The orientation of the tetrahedron is associated with a cube such that the tetrahedrally disposed atoms are at the corners (1, 1, 1), (1, −1, −1), (−1, 1, −1), and (−1, −1, 1). The Cartesian axes are then the x, y, z axes of the cube, and θ_x (θ_y, θ_z) of equation (17) is the angle between the x axis (y, z axis) of the cube and the intermolecular position vector $\hat{\mathbf{R}}$. The transformation of equation (17) to the form of equation (1) yields[10]

$$\begin{aligned} V_{\mathrm{disp}}^{(1)} = \frac{-C_6}{R^7}\,\frac{A_2}{\alpha_2}\, 8 \left(\frac{7}{30}\right)^{1/2} (+i) \sum_{\nu_2,\nu} [D_{\nu_2 2}^3(\alpha_2,\beta_2,\gamma_2) - D_{\nu_2 -2}^3(\alpha_2,\beta_2,\gamma_2)] \\ \times\, C_\nu^3(\beta,\alpha) \begin{pmatrix} 0 & 3 & 3 \\ 0 & \nu_2 & \nu \end{pmatrix} (-1)^{\nu_2} \end{aligned} \tag{18}$$

If molecule (1), for instance, which we now assume to be a diatom, possesses an open electron shell, an extra azimuthal angle dependence appears in the AIP.[19-21] As far as the dispersion interaction is concerned this extra term has been estimated for NO by Nielson *et al.*[22]; it is caused by the orientation of the unpaired electron lobe with respect to the plane which goes through the center of molecule (2) and the axis of the open-shell molecule (1). Denoting this azimuthal angle by δ, one obtains for a Π orbital[20,22]:

$$V_{\text{disp}}^{(\text{el})} = -\frac{D_6}{R^6}(2\cos^2\delta - 1)\, P_2^2(\hat{\mathbf{R}}\cdot\hat{\mathbf{r}}_1) \tag{19}$$

or in spherical tensor form [equation (1)]

$$V_{\text{disp}}^{(\text{el})} = -\frac{D_6}{R^6}\Big\{(30)^{1/2}\sum_{\nu,\nu_1}\begin{pmatrix} 2 & 2 & 0 \\ \nu & \nu_1 & 0 \end{pmatrix}(-1)^{\nu_1}\, C_\nu^2(\beta,\alpha)\,[D_{\nu_1 2}^2(\alpha_1,\beta_1,\gamma_1) + D_{\nu_1,-2}^2(\alpha_1,\beta_1,\gamma_1)]\Big\} \tag{20}$$

Short-range terms are difficult to discuss in a general way, as their specific form is derived often from tedious *ab initio* calculations. However, equation (4) for GT applies here also.

2.3. Eccentricity Effects

Mass eccentricity and/or isotopic substitution within one of the collision partners will not affect, of course, the form of the IP. However, for computational purposes it is convenient to calculate the IP with respect to the center of mass of molecule (1) and of molecule (2). Kreek and Le Roy[23] have derived analytical transformation formulas (see their appendix) to transform the IP in terms which are taken with respect to the center of mass of molecule (1) and of molecule (2), necessary to perform dynamical calculations.

2.4. Action Integrals

An important quantity which occurs in the semiclassical description of the scattering process to be used later in this chapter is the so-called action integral; it resembles the IP as it is felt by a molecule passing by:

$$I \equiv \int_{-\infty}^{\infty} V\, dt \tag{21}$$

I depends on the molecular orientation and the impact parameter b; the integral is taken along the classical trajectory of two collision partners. For collisions with large impact parameters straight trajectories may be assumed. The total action integral is simply the sum of all individual action integrals of all the different potential terms. Moreover, if one assumes that the molecules do not rotate during the collision and if $\bar{\mathbf{v}}$ is taken along $\hat{\mathbf{z}}$, one obtains for the potential terms of equation (21)

$$V = \frac{-C_s}{R^s}, \qquad I_s = \frac{-2f(s)\,C_s}{vb^{s-1}}$$

with

$$f(s) = \frac{\pi^{1/2}}{2}\Gamma\left(\frac{s-1}{2}\right)\Big/\Gamma\left(\frac{s}{2}\right) \quad \text{and} \quad f(6) = \frac{3}{16}\pi \tag{22}$$

$$V' = -\frac{C_t}{R^t}P_2(\hat{\mathbf{r}}\cdot\hat{\mathbf{R}})$$

$$I_t' = I_t\{-[(t-3)/(2t)]\,P_2(\cos\theta_1) + \tfrac{3}{4}[(t-1)/t]\sin^2\theta_1\cos(2\phi_1 - 2\phi)\} \tag{23}$$

and[24]

$$V' = V_{D_1D_2}, \qquad I'_{D_1D_2} = \frac{-2\mu_1\mu_2}{vb^2}\sin\theta_1\sin\theta_2\cos(\phi_1 + \phi_2 - 2\phi) \tag{24}$$

$$V' = V_{D_1Q_2}, \qquad I'_{D_1Q_2} = \frac{-2\mu_1Q_2}{vb^3}\sin\theta_1\sin^2\theta_2\cos(\phi_1 + 2\phi_2 - 3\phi) \tag{25}$$

$$V' = V_{Q_1Q_2}, \qquad I'_{Q_1Q_2} = \frac{3Q_1Q_2}{vb^4}\sin^2\theta_1\sin^2\theta_2\cos(2\phi_1 + 2\phi_2 - 4\phi) \tag{26}$$

For atom–diatom systems, action integrals for curved trajectories are discussed in Section 6.3. The azimuth angles ϕ (of the straight trajectory) and ϕ_i (of the axis of molecule i) are taken with respect to $\hat{\mathbf{z}}$, which is parallel to the direction of flight; $\cos\theta_i = \hat{\mathbf{r}}_i\cdot\hat{\mathbf{z}}$.

3. General Expressions and Close-Coupling Calculations

All cross sections can be calculated from the scattering amplitude $f(jm_j\beta \to j'm_j'\gamma;\hat{\mathbf{k}}_\gamma)$; here j and m_j are the initial rotational quantum numbers of a colliding diatomic molecule; β includes other internal quantum numbers, which characterize the initial situation, e.g., j_2 and m_2 if another diatom acts as collision partner. The dashed quantities and γ describe the final situation. For general reference see Arthurs and Dalgarno.[25]

For the differential cross section one has

$$d\sigma(jm_j\beta \to j'm_j'\gamma, \hat{\mathbf{k}}_\gamma)/d\Omega = (k_\gamma/k_\beta)\,|f(jm_j\beta \to j'm_j'\gamma, \hat{\mathbf{k}}_\gamma)|^2 \tag{27}$$

and for the total integral cross section

$$\sigma_{\text{tot}}(jm_j\beta) = (4\pi/k_\beta)\,\text{Im}\, f(jm_j\beta \to jm_j\beta, \hat{\mathbf{k}}_\beta) \tag{28}$$

The two quantities given by equations (27) and (28) are of main interest in this chapter, where we shall investigate the influence of the AIP on the differential and total cross section.

The scattering amplitude is related to the T matrix, e.g., for atom–rigid-diatom systems (initial direction $\hat{\mathbf{k}}_\beta$ not identical with quantization axis z),

$$\begin{aligned} f(jm_j\beta \to j'm_j'\gamma, \hat{\mathbf{k}}_\gamma) = {} & \frac{2\pi i}{(k_\beta k_\gamma)^{1/2}} \sum_{J,M,l,l'} i^{l-l'} \begin{pmatrix} l & j & J \\ M-m_j & m_j & -M \end{pmatrix} \\ & \times \begin{pmatrix} l' & j' & J \\ M-m_j' & m_j' & -M \end{pmatrix} (2J+1)\, Y_{l',M-m_j'}(\hat{\mathbf{k}}_\gamma) \\ & \times Y^*_{l,M-m_j}(\hat{\mathbf{k}}_\beta)\, T^J(j'l'\gamma, jl\beta) \end{aligned} \tag{29}$$

The T matrix, the elements of which appear in equation (29), is related to the S matrix and the K matrix by $S = 1 - T = (1 + iK)(1 - iK)^{-1}$.

In an uncoupled representation one has

$$\begin{aligned} f(jm_j\beta \to j'm_j'\gamma, \hat{\mathbf{k}}_\gamma) = {} & \frac{-2\pi i}{(k_\beta k_\gamma)^{1/2}} \sum_{l,m_l} \sum_{l',m_l'} (i^{l-l'})\, Y^*_{lm_l}(\hat{\mathbf{k}}_\beta)\, Y_{l'm_l'}(\hat{\mathbf{k}}_\gamma) \\ & \times [S(lm_l jm_j\beta, l'm_l' j'm_j'\gamma) - \delta_{ll'}\delta_{m_l m_l'}\delta_{jj'}\delta_{m_j m_j'}\delta_{\beta\gamma}] \end{aligned} \tag{30}$$

This uncoupled representation appears to be an easy starting point for dwa calculations (see Section 4), whereas the coupled representation is used in close-coupling calculations (cc).

If orientation-dependent effects are investigated as we do in this chapter it is helpful to know how the scattering amplitude depends on the direction of the quantization axis along which the projection quantum numbers m_j, m_j', m_l, and m_l' are chosen:

$$\begin{aligned} & f(j\tilde{m}_j\beta \to j'\tilde{m}_j'\gamma, \hat{\mathbf{k}}_\gamma) = \\ & \qquad \sum_{m_j,m_j'} D^j_{-m_j,-\tilde{m}_j}(\tilde{\alpha}, \tilde{\beta}, 0)\,[D^j_{-m_j',-\tilde{m}_j'}(\tilde{\alpha}', \tilde{\beta}', 0)]^*\, f(jm_j\beta \to j'm_j'\gamma, \hat{\mathbf{k}}_\gamma) \end{aligned} \tag{31}$$

Here, m_j and m_j' are defined along $\hat{\mathbf{z}}$, in a space-fixed coordinate system, whereas the quantum numbers $\tilde{m}_j$ and $\tilde{m}_j'$ are defined with respect to axes produced from $\hat{\mathbf{z}}$ by a rotation over the three Euler angles $(\tilde{\alpha}, \tilde{\beta}, 0)$ and $(\tilde{\alpha}', \tilde{\beta}', 0)$, respectively.[26,27]

From the optical theorem equation (28) follows immediately for the total integral cross section ($\hat{\mathbf{z}} // \hat{\mathbf{k}}_\beta$, atom–diatom case)

$$\sigma_{\text{tot}}(j\tilde{m}_j\beta; \tilde{\beta}) = \sum_{m_j} [d^j_{\tilde{m}_j m_j}(\tilde{\beta})]^2\, \sigma_{\text{tot}}(jm_j\beta) \tag{32}$$

This relation applies to the experimental situation of References 3 and 4, where the total cross section for a state-selected beam is measured with the quantization axis parallel ($\tilde{\beta} = 0$) and perpendicular ($\tilde{\beta} = \pi/2$) to the relative velocity. The measured $\sigma_{tot}(j\tilde{m}_j\beta; \pi/2)$ is related to the quantity of theoretical interest, $\sigma_{tot}(jm_j\beta)$, by equation (32).

Close-coupling (cc) calculations with special attention to orientation-dependent effects have been performed for H_2–Ar,[28] H_2–Kr,[29] H_2–H_2,[30] Ar–N_2,[26,31] and He–HCl.[32] Inelastic alignment was studied by Alexander and Dagdigian.[33]

4. The Distorted-Wave Approximation

For an atom–diatom system one derives from Arthurs and Dalgarno[25] after a sign correction for the S matrix, linearized in the AIP,

$$S^J(jl\beta, j'l'\gamma) = (k_\gamma/k_\beta)^{1/2} \exp[i(\eta_{l\beta} + \eta_{l'\gamma})] \times \left[\delta_{\gamma\beta}\delta_{jj'}\delta_{ll'} - \frac{4\mu i}{\hbar^2 k_\gamma} \int_0^\infty F_{l'\gamma}(k_\gamma R)\, F_{l\beta}(k_\beta R) \langle j'l'J\gamma | V' | jlJ\beta \rangle \, dR \right] \quad (33)$$

The functions $F_{l\beta}$ and $F_{l'\gamma}$ are nonsingular solutions of the radial Schrödinger equation, if V' is neglected so that a set of uncoupled ordinary differential equations is obtained. (Here $V' = V - \bar{V}$ and $\bar{V}$ equals the overall angle variables averaged potential V.) The boundary condition is imposed:

$$F_{l'\gamma} \xrightarrow[R \to \infty]{} \sin(k_\gamma R - l'\pi/2 + \eta_{l'\gamma})$$

With $\eta_{l'\gamma}$ the conventional phase shifts are denoted, for the uncoupled case, i.e., $V' = 0$.

The matrix element $\langle \cdots | V' | \cdots \rangle$ is still R dependent; all angle variables however, have disappeared upon integration. Explicit expressions for the matrix element can be found in Reference 34, even for diatom–diatom systems.

For dwa, it is advantageous to use the uncoupled representation for the S matrix, $S(jm_jlm_l\beta, j'm_j'l'm_l'\gamma)$, which obeys equation (33) also, if on the right-hand side the matrix element is replaced by $\langle j'm_j'l'm_l'\gamma | V' | jm_jlm_l\beta \rangle$ and if to the product of Kronecker symbols two factors are added: $\delta_{m_jm_j'}$ and $\delta_{m_lm_l'}$.

Thus one obtains for the scattering amplitude in dwa

$$f(jm_j\beta \to j'm_j'\gamma, \hat{\mathbf{k}}_\gamma) = \frac{-i}{2k_\beta} \sum_l (2l+1) [\exp(2i\eta_{l\beta}) - 1] P_l(\hat{\mathbf{k}}_\gamma \cdot \hat{\mathbf{k}}_\beta) \times \delta_{\gamma\beta}\delta_{jj'}\delta_{m_jm_j'} - \frac{8\pi\mu}{\hbar^2 k_\beta k_\gamma} \sum_{l,l',m_l,m_l'} i^{l-l'} Y^*_{lm_l}(\hat{\mathbf{k}}_\beta)\, Y_{l'm_l'}(\hat{\mathbf{k}}_\gamma) \times \exp[i(\eta_{l\beta} + \eta_{l'\gamma})] \int_0^\infty F_{l'\gamma}(k_\gamma R)\, F_{l\beta}(k_\beta R) \langle j'm_j'l'm_l'\gamma | V' | jm_jlm_l\beta \rangle \, dR \quad (34)$$

In this uncoupled representation, β and γ stand for all quantum numbers characterizing the initial and final state in addition to those explicitly displayed. For example, if two diatoms are considered as an interacting system, γ contains j_2' and m_{j_2}'. The advantage of using the uncoupled representation emerges from the straightforward determination of the matrix element $\langle \cdots | V' | \cdots \rangle$, even for the most general term [equation (4)]:

$$\begin{aligned}\langle \cdots |\mathrm{GT}| \cdots \rangle = {} & (-1)^{m_{j_1}' - \Omega_1 + m_{j_2}' - \Omega_2 - m_l'} [(2j_1 + 1)(2j_1' + 1) \\ & \times (2j_2 + 1)(2j_2' + 1)(2l + 1)(2l' + 1)]^{1/2} \\ & \times \begin{pmatrix} j_1 & j_1' & n_1 \\ m_{j_1} & -m_{j_1}' & \nu_1 \end{pmatrix} \begin{pmatrix} j_1 & j_1' & n_1 \\ \Omega_1 & -\Omega_1' & \kappa_1 \end{pmatrix} \\ & \times \begin{pmatrix} j_2 & j_2' & n_2 \\ m_{j_2} & -m_{j_2}' & \nu_2 \end{pmatrix} \begin{pmatrix} j_2 & j_2' & n_2 \\ \Omega_2 & -\Omega_2' & \kappa_2 \end{pmatrix} \begin{pmatrix} l & l' & n \\ 0 & 0 & 0 \end{pmatrix} \\ & \times \begin{pmatrix} l & l' & n \\ m_l & -m_l' & \nu \end{pmatrix} \begin{pmatrix} n_1 & n_2 & n \\ \nu_1 & \nu_2 & \nu \end{pmatrix} f(R) \end{aligned} \tag{35}$$

Here, Ω_i stands for the projection quantum number of the angular momentum defined along the molecular axis of molecule (i) described by the symmetric top rotor eigenfunction

$$(-1)^{+m_j - \Omega}[(2)^{1/2}/4\pi](2j + 1)^{1/2} D^j_{m_j\Omega}(\alpha\beta\gamma) \xrightarrow{\Omega = 0} \frac{1}{(2\pi)^{1/2}} Y_{jm_j}(\beta, \alpha)$$

Equation (35) reveals for $j_2 = j_2' = 0$, $n_2 = 0$, $n_1 = n = 2$, $j_1 = j_1'$, for example, that the scattering amplitude for transitions $m_{j_1}' \to m_{j_1}$ is equal, except for an unimportant phase factor, to that for $m_{j_1} \to m_{j_1}'$, an unrealistic feature in common with sudden approximation. Actual differences in these elastic and inelastic reorientation cross sections were calculated in cc.[29,31]

In dwa, reorientation collision cross sections ($j' = j$, $m_j' \neq m_j$) and inelastic cross sections ($j' \neq j$) can now be easily calculated with the help of equations (34) and (35); only the second term of equation (34) enters these calculations, which is linear in V'. In Figure 1, the result of such a calculation (obtained by Jacobs[35]) is displayed for a L.-J. (Lennard-Jones) potential [as given in equation (81) below] with $\varepsilon = 1.09 \times 10^{-14}$ erg, $R_m = 3.65$ Å, $q_{2,12} = 0.08$, $q_{2,6} = 0.09$; this potential approximately describes the H_2–Kr interaction.[3] For the relatively weak AIP of H_2–Kr, dwa results nearly coincide with cc calculations; interchange of m_j and m_j' does not lead to strong differences.

For the total integral cross section, the optical theorem, equation (28),

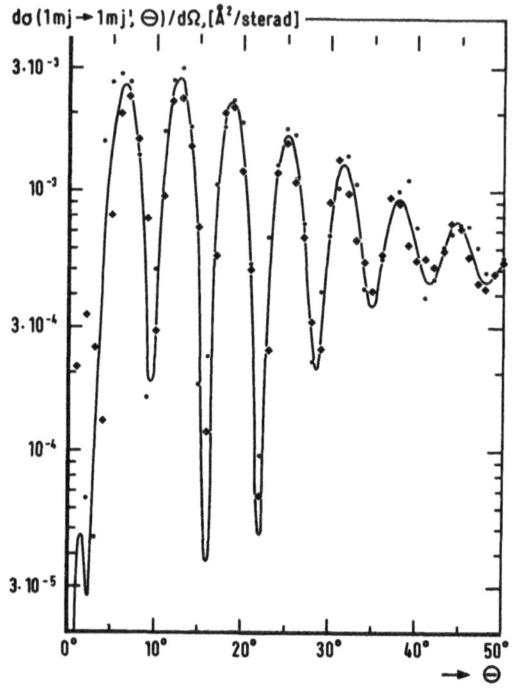

Figure 1. Differential reorientation cross section for a system resembling H_2–Kr (L.-J. potential with $\varepsilon = 1.09 \times 10^{-14}$ erg, $R_m = 3.65$ Å, $q_{2,12}=0.08$, $q_{2,6}=0.09$).[3] The drawn curve is obtained in dwa [equation (34)], for $j = j' = 1$, $\Delta m_j = 1$; points and diamonds come from a cc calculation with ($m_j = 1$, $m_j' = 0$) and ($m_j = 0$, $m_j' = 1$), respectively. All calculations were performed by M. Jacobs[35] for $v = 2600$ m/sec.

yields (with $\hat{\mathbf{k}}_\beta = \hat{\mathbf{z}}$)

$$\begin{aligned}
\sigma_{\text{tot}}(jm_j) &= \frac{4\pi}{k_\beta} \operatorname{Im} f(jm_j\beta \rightarrow jm_j\beta, \hat{\mathbf{k}}_\beta)\,(2j_2 + 1) \\
&= \frac{4\pi}{k_\beta^2} \sum_l (2l+1)\sin^2\eta_l - \frac{8\pi\mu}{\hbar^2 k_\beta^3} \sum_{l,l',m_l,m_l'} [(2l+1)(2l'+1)]^{1/2} \\
&\quad \times \sin[\eta_l + \eta_{l'} + (l-l')\pi/2] \int_0^\infty F_{l'\beta}(k_\beta R)\, F_{l\beta}(k_\beta R) \\
&\quad \times \langle jm_j l' m_l' \beta | V' | jm_j l m_l \beta \rangle \, dR\, \delta_{m_l 0} \delta_{m_l' 0}
\end{aligned} \tag{36}$$

For the matrix elements $\langle \cdots | V' | \cdots \rangle$ in equation (36) with $m_l = m_l' = 0$, the general term [equation (4)] yields, with symmetric top eigenfunctions for both molecules,

$$\begin{aligned}
\langle j_1 m_{j_1} l' 0 \beta | \mathrm{GT} | j_1 m_{j_1} l 0 \beta \rangle &= (-1)^{+m_{j_1} - \Omega_1 + m_{j_2} - \Omega_2} (2j_1 + 1)(2j_2 + 1) \\
&\quad \times [(2l+1)(2l'+1)]^{1/2} \begin{pmatrix} j_1 & j_1 & n_1 \\ m_{j_1} & -m_{j_1} & \nu_1 \end{pmatrix} \\
&\quad \times \begin{pmatrix} j_1 & j_1 & n_1 \\ \Omega_1 & -\Omega_1 & \kappa_1 \end{pmatrix} \begin{pmatrix} j_2 & j_2 & n_2 \\ m_{j_2} & -m_{j_2} & \nu_2 \end{pmatrix}
\end{aligned}$$

$$\times \begin{pmatrix} j_2 & j_2 & n_2 \\ \Omega_2 & -\Omega_2 & \kappa_2 \end{pmatrix} \begin{pmatrix} l & l' & n \\ 0 & 0 & 0 \end{pmatrix} \begin{pmatrix} l & l' & n \\ 0 & 0 & v \end{pmatrix}$$

$$\times \begin{pmatrix} n_1 & n_2 & n \\ v_1 & v_2 & v \end{pmatrix} f(R) \tag{37}$$

From equation (37) follow some "selection rules" for the integral cross section in dwa, $v_1 = v_2 = \kappa_1 = \kappa_2 = v = 0$; $l - l' =$ even/odd if n even/odd; however, $l - l' =$ odd gives no contribution to equation (36) due to the change in sign if l and l' are interchanged.[5]

The selection rules exclude contributions from the spherical top interaction, equation (18), and the $\Delta\Omega$ interaction, equation (20), to the integral cross section σ_{tot}.

The multipole terms, equations (5)–(8), give no contribution if one averages over one set of quantum numbers, say m_{j_2}; i.e., $n_2 = 0$, $n_1 = n_2$. What remains is of the form $\sum_n (2n + 1)^{-1/2} V^n(R) P_n(\hat{\mathbf{r}}_1 \cdot \hat{\mathbf{R}})$ with $n = 0, 2, 4, \ldots$; $n \le 2j_1$.

The advantage of integral cross section measurements for state-selected beams is connected to this selectiveness of the observable anisotropy effect, as has been thoroughly discussed in Reference 5.

For an AIP of the form

$$V_2(R) P_2(\hat{\mathbf{r}}_1 \cdot \hat{\mathbf{R}}) = (1/5)^{1/2} V^2(R) P_2(\hat{\mathbf{r}}_1 \cdot \hat{\mathbf{R}})$$

one obtains from equation (36)[28,36]

$$\langle \cdots | V_2 P_2(\hat{\mathbf{r}} \cdot \hat{\mathbf{R}}) | \cdots \rangle \delta_{m_l 0} \delta_{m_l' 0} = V_2(R)(2j_1 + 1)[(2l + 1)(2l' + 1)]^{1/2}$$

$$\times (-1)^{m_{j_1} - 2j_1 - \Omega_1} \begin{pmatrix} j_1 & j_1 & 2 \\ \Omega_1 & -\Omega_1 & 0 \end{pmatrix}$$

$$\times \begin{pmatrix} j_1 & j_1 & 2 \\ m_{j_1} & -m_{j_1} & 0 \end{pmatrix} \begin{pmatrix} l & l' & 2 \\ 0 & 0 & 0 \end{pmatrix}$$

$$\times \begin{pmatrix} l & l' & 2 \\ 0 & 0 & 0 \end{pmatrix} \delta_{m_l' 0} \delta_{m_l 0} \tag{38}$$

and

$$\sigma_{tot}(j_1 m_{j_1}) = \frac{4\pi}{k_\beta^2} \sum_l (2l + 1) \sin^2 \eta_l + \frac{8\pi\mu}{\hbar^2 k_\beta^3} \frac{3m_{j_1}^2 - j_1(j_1 + 1)}{(2j_1 + 3)(2j_1 - 1)}$$

$$\times \frac{j_1(j_1 + 1) - 3\Omega_1^2}{j_1(j_1 + 1)} \left\{ \sum_l \frac{l + 1}{2l + 3} \left[\frac{(2l + 1) l}{2l - 1} \sin 2\eta_l V_2^{ll} \right.\right.$$

$$\left.\left. - 3(l + 2) V_2^{ll+2} \sin(\eta_l + \eta_{l+2}) \right] \right\} \tag{39}$$

with

$$V_2^{ll'} = \int_0^\infty F_l(k_\beta R)\,F_{l'}(k_\beta R)\,V_2(R)\,dR$$

This last expression for σ_{tot} was used by Zandee *et al.*[3] and Tang and Toennies[7] to analyze the $V_2(R)$ factor for H_2–noble-gas systems. If averaged over all m_j, the factor in front of $\{\cdots\}$ vanishes.

For the coupled representation, matrix elements can be found in References 34 and 37.

5. *Sudden Approximation*

This approximation allows us to calculate the influence of the AIP in infinite order (IOS), though in an approximate way. It may be derived in two steps: first one approximates the centrifugal term in the coupled set of radial Schrödinger equations—the so-called coupled-state approximation (McGuire and Kouri[38] and Pack[39]); then one neglects excitation energies, putting $k_\beta = k_\gamma$ for all γ's.[40-42] One can also utilize the second step, retaining the full centrifugal coupling; this approximation leads to very good results but does not achieve great mathematical simplification of the problem (Khare[43]). For atom–diatom scattering the two-step approximation IOS leads to different expressions, depending on the precise way in which the centrifugal coupling is replaced; if one replaces the orbital angular momentum operator by the constant $\bar{l}(\bar{l}+1)$ with $\bar{l} = l_{\text{final}}$ [l' in equation (29)] one obtains[44-46]

$$f(jm_j \to j'm_j'; \hat{\mathbf{k}}_\gamma) = \int d\Omega_1\, Y^*_{j'm_j'}(\Omega_1)\, f(\hat{\mathbf{k}}_\gamma, \Omega_1)\, Y_{jm_j}(\Omega_1)\,(-1)^{j-j'} \tag{40}$$

As before, m_j and m_j' are defined with respect to the direction of the relative velocity, for the approaching collision partners. The spherical harmonics stand for the rotational eigenfunctions of the diatom of an atom–diatom system; they may be replaced by $(-1)^{+m_j-\Omega}(2^{1/2}/4\pi)(2j+1)^{1/2} D^j_{m_j\Omega}$ (Euler angles) for a symmetric top collision partner.

With the alternative choice $\bar{l} = l_{\text{initial}}$ [l in equation (29)], one finds[44,47]

$$f(jm_j \to j'm_j'; \hat{\mathbf{k}}_\gamma) = \sum_{v,v'} \int D^{j'*}_{-m_j',-v'}(\varphi,\theta,0)\, D^{j}_{-m_j,-v}(\varphi,\theta,0)\, Y^*_{j'v'}(\Omega_1) \times Y_{jv}(\Omega_1)\, f(\hat{\mathbf{k}}_\gamma, \Omega_1)\, d\Omega_1 \tag{41}$$

with $\hat{\mathbf{k}}_\gamma = (\theta, \varphi)$ and $\hat{\mathbf{k}}_\beta \parallel \hat{\mathbf{z}}$. Equations (40) and (41) become identical in the forward direction. For $j = j'$, transitions $m_j \to m_j'$ possess equal cross sections with those of $m_j' \to m_j$, an unrealistic feature shared with equation (34).

A priori it is unclear which choice should be preferred. In both cases one has with $k_\gamma \simeq k_\beta = k$

$$f(\hat{\mathbf{k}}_\gamma, \Omega_1) = -\frac{i}{2k}\sum_{l=0}^{\infty}(2i+1)\left[e^{2i\eta_l}-1\right]P_l(\cos\theta) \tag{42}$$

where θ is the scattering angle between $\hat{\mathbf{k}}_\gamma$ and $\hat{\mathbf{k}}_\beta$. This expression formally is equivalent to what one obtains for spherically symmetric potentials, when the phase shift η_l is independent of Ω_1. In the IOS approximation (IOS) the angle χ between the molecular axis $\hat{\mathbf{r}}_1$ and the position vector $\mathbf{R}$ between the atom and the center of the diatom remains constant, during a particular collision. For an atom(S)–diatom($^1\Sigma$) system, χ is the only angle on which the AIP depends. Consequently, η_l can be calculated just as for a spherically symmetric potential and is found parametrically depending on χ, $\eta_l = \eta_l^{\mathrm{BF}}(\chi)$. Using the JWKB approximation one finds

$$\eta_l^{\mathrm{SCBF}} = \frac{1}{\hbar}\lim_{R\to\infty}\left\{\left[\int_{r_0}^{R} dr\left[2m(E-V) - \frac{\hbar^2 l(l+1)}{r^2}\right]^{1/2}\right.\right.$$
$$\left. - \int_{b}^{R} dr\left[2mE - \frac{\hbar^2 l(l+1)}{r^2}\right]^{1/2}\right\} \tag{43}$$

where r_0 is the largest root of the first integrant and $kb = l + 1/2$.

Inserting η_l^{SCBF} into equation (42) yields $f^{\mathrm{SCBF}}(\hat{\mathbf{k}}_\gamma, \Omega_1)$. At this point it is important to realize that the assumption of fixed χ follows from the general decoupling approximations to obtain a simple solution of the Schrödinger equation. It has lost its earlier *ad hoc* character. Intuitively, one can imagine two limiting situations where a fixed χ appears sensible. First, in case of a strong AIP where the projection of j on the intermolecular position vector is a conserved quantity, in good approximation. Second, in the case where the interaction may be considered as point-contact-like so that evidently χ is conserved during the interaction. However, in both cases different χ values should be used to describe the same initial situation. For a strong AIP (IOS-SAIP), χ keeps the constant value defined at a long distance with respect to the direction of initial relative velocity, $\hat{\mathbf{k}}_\beta$. For point-contact-like interactions (IOS-PCI), the distribution of χ values should be taken as it occurs at the moment of closest approach for a noninteracting molecule; the initial distribution of χ values is rotated in such a way that its rotation during the approach of the scattering partners is undone, for the closest distance. This PCI picture must not give the impression that it would apply to (only) weak interactions, in contrast to the SAIP picture; the interaction may be strong but hard-sphere-like in duration. (Both situations are also discussed below in connection with integral cross sections for state-selected molecules.) For degeneracy-averaged total cross sections, evidently this distinction is irrelevant.

If one inserts equations (42) and (43) in equations (40) and (41) one obtains

$$f^{\text{IOS-}l'\text{-SAIP}}(jm_j \to j'm_j', \hat{\mathbf{k}}_\gamma) \quad \text{and} \quad f^{\text{IOS-}l\text{-SAIP}}(jm_j \to j'm_j', \hat{\mathbf{k}}_\gamma)$$

For the PCI case an explicit expression can be derived from Dickinson and Richards,[48] with $\hat{\mathbf{k}}_\beta$ along $\hat{\mathbf{z}}$:

$$f^{\text{IOS-PCI}}(jm_j \to j'm_j', \hat{\mathbf{k}}_\gamma) = \sum_v d^{j'}_{vm_j'}\left(\frac{\pi-\theta}{2}\right) d^{j}_{vm_j}\left(\frac{\pi-\theta}{2}\right) \times \int Y^*_{j'v}(\Omega_1)\, f^{\text{SCBF}}(\hat{\mathbf{k}}_\gamma, \Omega_1)\, Y_{jv}(\Omega_1)\, d\Omega_1 \qquad (44)$$

$f^{\text{IOS-}l'\text{-SAIP}}$ [equation (40)] does not contain reorientation cross sections, i.e., the right-hand side vanishes if $m_j \neq m_j'$; $f^{\text{IOS-}l\text{-SAIP}}(jm_j \to j'm_j, \hat{\mathbf{k}}_\gamma)$ [equation (41)] does, on the other hand, contain Δm_j transitions, which vanish in the forward and backward direction as in the exact case.

Equation (44) can now be written

$$f^{\text{IOS-PCI}}(jm_j \to j'm_j', \hat{\mathbf{k}}_\gamma) = \sum_v d^{j'}_{vm_j'}\left(\frac{\pi-\theta}{2}\right) d^{j}_{vm_j}\left(\frac{\pi-\theta}{2}\right) \times f^{\text{IOS-}l'\text{-SAIP}}(jv \to j'v, \hat{\mathbf{k}}_\gamma) \qquad (45)$$

Here, $m_j \to m_j'$ transitions are allowed, however unrealistically, in forward direction also. This means that the projection of the total angular momentum on the initial line of flight is not conserved in the PCI treatment.

An entirely different approach of calculating the scattering amplitude is based on the assumption that the direction $\hat{\mathbf{r}}_i$ of the molecular axis remains space fixed during the collision (Bernstein and Kramer,[49] Kuijpers and Reuss,[24,50] Buck and Khare[51]). In the semiclassical picture, the trajectory of the particles is assumed to be determined by the IIP. One obtains, without the restriction to the atom–diatom case,

$$\eta_l^{\text{SCSF}}(\phi) = \eta_l^{(0)} - (2\hbar)^{-1} \int_{-\infty}^{+\infty} V'(t)\, dt \qquad (46)$$

The integrand is formed by the AIP, integrated along a trajectory characterized by the impact parameter $b = [l + (1/2)]/k$ and the azimuth angle ϕ; the trajectory is determined by the IIP solely. $\eta_l^{(0)}$ stands for the phase shift resulting from the IIP solely.

Inserting this semiclassical phase shift into equation (42) yields $f^{\text{SCSF}}(\hat{\mathbf{k}}_\gamma, \Omega_1, \phi)$. Finally one has, for instance, for the atom–diatom case,

$$f^{\text{S-SF}}(jm_j \to j'm_j', \hat{\mathbf{k}}_\gamma) = \int d\Omega_1\, Y^*_{j'm_j'}(\Omega_1)\, f^{\text{SCSF}}(\hat{\mathbf{k}}_\gamma, \Omega_1, \phi)\, Y_{jm_j}(\Omega_1) \qquad (47)$$

Upon the integration $d\Omega_1$, the ϕ dependence of f^{SCSF} vanishes. Again, unrealistically, Δm_j transitions are allowed in the forward direction and reorientation cross sections do not depend on the sign of Δm_j.

Note that introducing the AIP according to equation (46) forms an approximation which is shown to be strictly valid only in first order.[48] This approach, therefore, does not earn the predicate IOS, though the AIP occurs in all orders, in equation (47). There does not exist a rigorous derivation of equation (47) other than through time-dependent perturbation theory, where one assumes right from the outset that the trajectories along which particles move in this semiclassical approach remain unchanged by the introduction of the AIP.[52]

6. *The Calculation of Cross Sections in Sudden Approximation*

6.1. *The Differential Cross Section in Sudden Approximation*

The degeneracy-averaged total differential cross section as it is measured in References 9–10, for instance, is defined as

$$\frac{d\sigma(j)}{d\Omega} = \frac{1}{2j+1} \sum_{m_j, m_j', j'} \frac{k_\gamma}{k_\beta} |f(jm_j \to j'm_j', \hat{\mathbf{k}}_\gamma)|^2 \tag{48}$$

Sudden approximation, equations (40) and (41), with $k_\gamma = k_\beta = k$, yields

$$d\sigma^{\text{IOS}}(j)/d\Omega = (4\pi)^{-1} \int d\Omega_1 |f^{\text{SCBF}}(\mathbf{k}_\gamma, \Omega_1)|^2 \tag{49}$$

The same result follows from equation (44); thus the superscripts SAIP and PCI can be dropped for what concerns the degeneracy-averaged total differential cross section.

In the space-fixed model one obtains from equation (47), similarly,

$$d\sigma^{\text{S-SF}}(j)/d\Omega = (4\pi)^{-1} \int d\Omega_1 |f^{\text{SCSF}}(\hat{\mathbf{k}}_\gamma, \Omega_1, \phi)|^2 \tag{50}$$

Again the dependence on the azimuth angle ϕ vanishes upon integration.

Equations (49) and (50) were employed by Buck and Khare[51] in a comparative study published in the aftermath of extensive experimental investigations.[9,10] The conclusion of this study is that both types of sudden approximation lead to nearly the same result, with a slight overestimation of the influence of the AIP by IOS.

Wharton and co-workers[8] have measured the total differential cross section for state-selected primary beams:

$$d\sigma(jm_j)/d\Omega = \sum_{j', m_j'} \frac{k_\gamma}{k_\beta} |f(jm_j \to j'm_j', \hat{\mathbf{k}}_\gamma)|^2 \tag{51}$$

In sudden approximation [equations (40) and (47)] the closure relation yields, with $\hat{\mathbf{k}}_\beta = \hat{\mathbf{z}}$ and $\theta = \sphericalangle\, \hat{\mathbf{k}}_\beta, \hat{\mathbf{k}}_\gamma$,

$$d\sigma^{\text{IOS-}l'\text{-SAIP/S-SF}}(jm_j)/d\Omega = \int d\Omega_1\, Y^*_{jm_j}(\Omega_1)\, Y_{jm_j}(\Omega_1)\, |f^{\text{SCBF/SF}}(\hat{\mathbf{k}}_\gamma, \Omega_1/\hat{\mathbf{k}}_\gamma, \Omega_1, \phi)|^2 \tag{52}$$

Similarly, equation (41) yields

$$d\sigma^{\text{IOS-}l\text{-SAIP}}(jm_j)/d\Omega = \sum_v \int d\Omega_1 Y^*_{jv}(\Omega_1)\, Y_{jv}(\Omega_1)\, |d^j_{m_j v}(\theta)|^2 |f^{\text{SCBF}}(\hat{\mathbf{k}}_\gamma, \Omega_1)|^2$$

From equation (44) follows

$$d\sigma^{\text{IOS-PCI}}(jm_j)/d\Omega = \sum_v \left[d^j_{vm_j}\left(\frac{\pi - \theta}{2}\right)\right]^2 \int d\Omega_1\, |Y_{jv}(\Omega_1)|^2\, |f^{\text{SCBF}}(\hat{\mathbf{k}}_\gamma, \Omega_1)|^2 \tag{53}$$

For small scattering angles, explicit expressions are derived here for the S-SF scattering amplitude and the corresponding differential cross section.

Using equations (42) and (46) with

$$\exp(2i\eta_l^{\text{SCSF}}) \simeq \exp(2i\eta_l^{(0)})\left(1 - \frac{i}{\hbar}\int_{-\infty}^{+\infty} V'\, dt\right)$$

one obtains in first order in the AIP

$$f^{\text{SCSF}}(\hat{\mathbf{k}}_\gamma, \Omega_1, \phi) = f_{\text{iso}} + f_{\text{aniso}} \tag{54}$$

with

$$f_{\text{iso}} = -\frac{i}{2k}\sum_{l=0}^{\infty}(2l+1)\left[\exp(2i\eta_l^{(0)}) - 1\right] P_l(\cos\theta)$$

$$f_{\text{aniso}} = -\frac{1}{2k}\sum_{l=0}^{\infty}(2l+1)\exp(2i\eta_l^{(0)})\left(\frac{1}{\hbar}\int_{-\infty}^{+\infty} V'\, dt\right) P_l(\cos\theta)$$

Because we will consider only small scattering angles θ, we replace $(1/\hbar)\int_{-\infty}^{+\infty} V'\, dt$ by $(1/\hbar)\, I'$, where I' corresponds to the action integrals explicitly given in Section 2, for straight-path trajectories.* The Legendre polynomials $P_l(\cos\theta)$ shall be replaced by $1 - l^2\theta^2/4$, and the isotropic phase shifts $\eta_l^{(0)}$ by $\eta_{l,\text{Born}}^{(0)}$, with $\eta_{l,\text{Born}}^{(0)} = \frac{1}{2}(l_0/l)^5$ and $l_0^5 = (3\pi/8)\,(mC_6k^4/\hbar^2)$ for an isotropic potential $V - V' = -C_6/R^6$. Summations over l are replaced by integrations.

These replacements are justified for small-angle scattering of heavy systems when only the outer part of the IP is probed. If, in addition, one

* Action integrals for curved trajectories are discussed in Section 6.3.

puts $V' = -q_{2,6}C_6P_2(\hat{\mathbf{R}}\cdot\hat{\mathbf{r}}_1)/R^6$,

$$f_{\mathrm{iso}} = (1/2k)(J_1 - \theta^2 J_3) \tag{55}$$

$$f_{\mathrm{aniso}} = (q_{2,6}/4k)(J_2 - \theta^2 J_4)\left[5/2\,\sin^2\theta_1\cos(2\phi_1 - 2\phi) - P_2(\hat{\mathbf{r}}_1\cdot\hat{\mathbf{k}}_\beta)\right] \tag{56}$$

The angles θ_1, ϕ_1 determine the space angle Ω_1 with $\hat{\mathbf{r}}_1\cdot\hat{\mathbf{k}}_\beta = \cos\theta_1$. The dimensionless integrals J_i yield[5]

$$\begin{aligned}
J_1 &= \int_0^\infty \sin\eta^{(0)}_{l,\mathrm{Born}}\exp(i\eta^{(0)}_{l,\mathrm{Born}})\,4l\,dl = l_0^2\,\Gamma\left(\frac{3}{5}\right)\exp\left(\frac{3i\pi}{10}\right)\\
J_2 &= \int_0^\infty \exp(2i\eta^{(0)}_{l,\mathrm{Born}})(l_0/l)^5\,l\,dl = \frac{1}{5}l_0^2\,\Gamma\left(\frac{3}{5}\right)\exp\left(\frac{3i\pi}{10}\right)\\
J_3 &= \int_0^\infty \sin\eta^{(0)}_{l,\mathrm{Born}}\exp(i\eta^{(0)}_{l,\mathrm{Born}})\,l^3\,dl = \frac{1}{8}l_0^4\,\Gamma\left(\frac{1}{5}\right)\exp\left(\frac{i\pi}{10}\right)\\
J_4 &= \int_0^\infty \exp(2i\eta^{(0)}_{l,\mathrm{Born}})(l_0/l)^5\,\tfrac{1}{4}l^3\,dl = \frac{1}{10}l_0^4\,\Gamma\left(\frac{1}{5}\right)\exp\left(\frac{i\pi}{10}\right)
\end{aligned} \tag{57}$$

From equation (47) follows, then in first-order sudden approximation (FOS)

$$\begin{aligned}
f^{\mathrm{FOS\text{-}SF}}(jm_j\to j'm'_j,\hat{\mathbf{k}}_\gamma) &= (1/2k)(J_1 - \theta^2 J_3)\,\delta_{jj'}\delta_{m_jm'_j}\\
&\quad + (q_{2,6}/4k)(J_2 - \theta_2 J_4)\cdot\mathrm{stf}
\end{aligned} \tag{58}$$

with the state function

$$\begin{aligned}
\mathrm{stf} &= \int \sin\theta_1\,d\theta_1\,d\phi_1\,Y^*_{j,m'_j}(\theta_1,\phi_1)\,Y_{jm_j}(\theta_1,\phi_1)\left\{\left(\frac{10\pi}{3}\right)^{1/2}\left[Y_{22}(\theta_1,\phi_1)\right.\right.\\
&\quad \left.\left.+\,Y_{2\,-2}(\theta_1,\phi_1)\right] - \left(\frac{4\pi}{5}\right)^{1/2} Y_{20}(\theta_1,\phi_1)\right\}\\
&= (-1)^{m'_j}\left[(2j+1)(2j'+1)\right]^{1/2}\begin{pmatrix}2 & j & j'\\ 0 & 0 & 0\end{pmatrix}\left[\frac{5}{6^{1/2}}\begin{pmatrix}2 & j & j'\\ 2 & m_j & -m'_j\end{pmatrix}\right.\\
&\quad \left. + \frac{5}{6^{1/2}}\begin{pmatrix}2 & j & j'\\ -2 & m_j & -m'_j\end{pmatrix} - \begin{pmatrix}2 & j & j'\\ 0 & m_j & -m'_j\end{pmatrix}\right]
\end{aligned}$$

Here, the ϕ dependence has disappeared. In contrast to the exact expression, equation (29), and to the dwa scattering amplitude, equation (34), FOS-SF gives $\Delta m_j = 2$ transitions in the forward direction, with cross sections proportional to $q^2_{2,6}$. The selection rules $\Delta m_j = 0$ and 2 are discussed in Reference 35; their approximate validity for small-angle scattering ($\theta \gtrsim l_0^{-1}$) is confirmed by close-coupling calculations.[29]

For measurements of the type performed by Tsou *et al.*[8] one obtains from equation (51) for the total differential cross section [being identical to the elastic differential cross section from equation (58) up to first order in the AIP]

$$\frac{d\sigma^{\text{FOS-SF}}(jm_j)}{d\Omega} = \frac{1}{4k^2}(J_1 - \theta^2 J_3)^2 - \frac{q_{2,6}}{8k^2}[(J_1 - \theta^2 J_3)(J_2^* - \theta^2 J_4^*) + \text{cc}]$$

$$\times (-1)^{m_j}(2j+1)\begin{pmatrix} 2 & j & j \\ 0 & 0 & 0 \end{pmatrix}\begin{pmatrix} 2 & j & j \\ 0 & m_j & -m_j \end{pmatrix} \qquad (59)$$

The same result [equation (59)] was found for the elastic differential cross sections for small scattering angles by Stolte[5] using dwa and the Landau–Lifshitz approximation (see Chapter 4).

For large angles, the S-SF approach yields[47,51]

$$f^{\text{SCSF}}(\mathbf{k}_\gamma, \Omega_1, \phi) = \sum_t (\sin\theta)^{-1/2}\left[b_t(\theta)\left|\frac{db_t(\theta)}{d\theta}\right|\right]^{1/2} \beta_t \exp\left[\frac{i}{\hbar}\int_{-\infty}^{+\infty} V'\,dt\right] \qquad (60)$$

Here the sum runs over all impact parameters b_t leading to the deflection angle θ; β_t is a complex number of unit modulus.[53] Insertion into equation (47) yields the required scattering amplitude for large scattering angles. Similar expressions have been discussed for the IOS approach.[48]

Recently, elastic reorientation effects in the differential and, principally, in the integral cross section were studied in the IOS[27] and the S-SF[35] approach. The latter was shown to yield results equivalent to what is obtained from dwa [equation (34)] for angles $\theta \gtrsim l_0^{-1}$, if the last factor of the right-hand side of equation (60) is replaced by $1 - (i/\hbar)\int_{-\infty}^{+\infty} V'\,dt$.

6.2. The Integral Cross Section in Sudden Approximation: The Nonglory Contribution

The total integral cross section can be obtained from the scattering amplitude using the optical theorem [equation (28)]

$$\sigma_{\text{tot}}(jm_j) = (4\pi/k)\,\text{Im}\, f(jm_j \to jm_j, 0)$$

The zero in the argument of the scattering amplitude indicates forward direction $\hat{\mathbf{k}}_\beta = \hat{\mathbf{k}}_\gamma$. In both IOS approximations [equations (40) and (41)] one therefore has

$$\sigma_{\text{tot}}^{\text{IOS}}(jm_j) = (4\pi/k)\,\text{Im}\left[\int Y_{jm_j}^*(\Omega_1)\, Y_{jm_j}(\Omega_1)\, f(0,\Omega_1)\, d\Omega_1\right] \qquad (61)$$

For the somewhat special case IOS-PCI, see equation (65) below.

The three approaches for the scattering amplitude $f(0,\Omega_1)$ [IOS-SAIP, equations (40) and (41); IOS-PCI, equation (44); S-SF, equation (47)] give quite different results if one selects a special (jm_j) state. This will be demonstrated first in the following by a consideration of the first-order nonglory contribution from the special IP, $V_{sp} = -(C_6/R^6)[1 + q_{2,6} P_2(\cos\theta)]$. Our conclusion shall be that S-SF appears to be the most reliable approach. Higher-order effects and results for more complicated IP's shall therefore only be discussed in S-SF.

From equation (61) follows, for $V = V_{sp}$ (with Landau–Lifshitz approximation; see Chapter 4 of this volume),

$$\sigma_{tot}^{IOS\text{-}SAIP}(jm_j) = \int Y^*_{jm_j}(\Omega_1)\, Y_{jm_j}(\Omega_1)\, d\Omega_1\, 2\pi \int b\, db\{2 - 2\cos 2\eta_l^{SCBF}\} \tag{62}$$

$$= 8.083\,\langle\{(C_6/\hbar v)[1 + q_{2,6} P_2(\cos\chi)]\}^{2/5}\rangle$$

Thus in first order,

$$\sigma_{tot}^{FOS\text{-}SAIP}(jm_j) = 8.083\,(C_6/\hbar v)^{2/5}[1 + \tfrac{2}{5} q_{2,6}\langle P_2(\cos\chi)\rangle] \tag{63}$$

The integration $\int d\Omega_1\, Y^*_{jm_j}(\Omega_1)\, Y_{jm_j}(\Omega_1)$ is replaced here by $\langle\cdots\rangle$. The last equation gives the contribution in first order of $q_{2,6}$ to the integral cross section.

Explicitly one has, for χ being fixed at its long-range value, $\cos\chi = \hat{\mathbf{r}}_1 \cdot \hat{\mathbf{k}}_\beta$, thus

$$\langle P_2(\cos\chi)\rangle = (-1)^{m_j}\begin{pmatrix} j & j & 2 \\ 0 & 0 & 0 \end{pmatrix}\begin{pmatrix} j & j & 2 \\ m_j & -m_j & 0 \end{pmatrix}(2j+1)$$

$$= -\frac{3m_j^2 - j(j+1)}{(2j+3)(2j-1)}$$

$$\simeq -\frac{1}{2} P_2\left(\frac{m_j}{j}\right), \qquad \text{for } j \gg 1 \tag{64}$$

Similarly [see equation (70) below],

$$\langle P_4(\cos\chi)\rangle = (-1)^{m_j}\begin{pmatrix} j & j & 4 \\ 0 & 0 & 0 \end{pmatrix}\begin{pmatrix} j & j & 4 \\ m_j & -m_j & 0 \end{pmatrix}(2j+1)$$

$$\simeq \frac{3}{8} P_4\left(\frac{m_j}{j}\right), \qquad \text{for } j \gg 1$$

For symmetric top rotor eigenfunctions one obtains similar expectation values.[5]

From equation (61) follows

$$\sigma_{tot}^{IOS\text{-}PCI}(jm_j) = \sum_v \int Y^*_{jv}(\Omega_1)\, Y_{jv}(\Omega_1)\, d\Omega_1 2\pi \left[d^j_{vm_j}\left(\frac{\pi}{2}\right)\right]^2$$

$$\times \int b\, db[2 - 2\cos 2\eta_l^{SCBF}] \tag{65}$$

Thus, because of

$$\sum_v \int Y^*_{jv}(\Omega_1)\, Y_{jv}(\Omega_1)\, d\Omega_1 \left[d^j_{vm_j}\left(\frac{\pi}{2}\right)\right]^2 P_{2n}(\cos\chi) = (-1)^n \frac{1}{2^{2n}}\binom{2n}{n}\langle P_{2n}(\cos\chi)\rangle$$

one has in first order, again with $\cos\chi = \hat{\mathbf{r}}_1 \cdot \hat{\mathbf{k}}_\beta$

$$\sigma^{\text{FOS-PCI}}_{\text{tot}}(jm_j) = 8.083\left(\frac{C_6}{\hbar v}\right)^{2/5}\left[1 - \frac{1}{10}q_{2,6}\langle P_2(\cos\chi)\rangle\right] \tag{66}$$

It is evident from equations (62) and (66) that the BF approaches even yield different signs for the linear $q_{2,6}$ contribution. Physically, this is satisfactory as both correspond to geometries at the moment of closest approach which are turned by $\pi/2$ with respect to each other around an axis perpendicular to the plane of scattering.

For the space-fixed case one has, from equation (61),

$$\sigma^{\text{S-SF}}_{\text{tot}}(jm_j) = \int Y^*_{jm_j}(\Omega_1)\, Y_{jm_j}(\Omega_1)\, d\Omega_1 2\pi \int b\, db[2 - 2\cos 2\eta_l^{\text{SCSF}}] \tag{67}$$

Thus in first order, using equation (46),

$$\sigma^{\text{FOS-SF}}_{\text{tot}}(jm_j) = 8.083\left(\frac{C_6}{\hbar v}\right)^{2/5} - 4\pi \int b\, db \left\langle \frac{1}{\hbar}\int_{-\infty}^{+\infty} V'\, dt \right\rangle \sin 2\eta_l^{(0)} \tag{68}$$

According to Reference 24, if straight trajectories are employed to calculate the integral $\int V'\, dt$ [see equation (23)] and if Born phase shifts $\eta^6_{l,\text{Born}} = 3\pi\mu^5 C_6 v^4/16\hbar^6 l^5$ are used, one finds

$$\sigma^{\text{FOS-SF}}_{\text{tot}}(jm_j) = 8.083\left(\frac{C_6}{\hbar v}\right)^{2/5}\left[1 - \frac{1}{10}q_{2,6}\langle P_2(\hat{\mathbf{r}}_1 \cdot \hat{\mathbf{k}}_\beta)\rangle\right] \tag{69}$$

Use is made here of the integral

$$\int_0^\infty \frac{dz \sin z}{z^{2/5}} = \Gamma\left(\frac{3}{5}\right)\sin\frac{3\pi}{10}$$

This result, equation (69), forms the most accurate description of the state dependence of the nonglory part of the integral cross section, for small $q_{2,6}$ values.[28,36] It was derived from DWA, too, which on its turn was carefully checked against CC calculations. The glory contribution is suppressed by the choice of the monotonic R dependence of the potential model. Comparison of equations (59) and (69) shows that the relative effect of the AIP on the differential cross section in the forward direction is twice that for the total collision cross section.

The body-fixed approach $\sigma^{\text{IOS-SAIP}}_{\text{tot}}$ even leads to the wrong sign of the effect, whereas $\sigma^{\text{IOS-PCI}}_{\text{tot}}$ is a factor of 2 off, with regard to the state-dependent part of σ_{tot}. Similar deviations for inelastic transition probabilities in the degeneracy-averaged case have been found by Dickinson and Richards.[54]

From here we take the justification to consider only S-SF in the following part of this section. Second-order terms are derived in Reference 24, for the same potential:

$$\sigma_{tot}^{S\text{-}SF}(jm_j) = 8.083\left(\frac{C_6}{\hbar v}\right)^{2/5}\left[1 - \frac{1}{10}q_{2,6}\langle P_2\rangle - q_{2,6}^2\left(\frac{129}{14000}\langle P_4\rangle - \frac{11}{700}\langle P_2\rangle + \frac{7}{500}\right)\right] + \cdots \quad (70)$$

To the right-hand side of equation (70) other terms must be added, if dipole–dipole, dipole–quadrupole, and quadrupole–quadrupole forces are active, representing a small perturbation of the IIP ($-C_6/R^6$). For large impact parameters b this requirement is not fulfilled, because these terms decrease with a power smaller than six (see Section 2). However, for large b, the dimensionless action integrals $(1/\hbar)\,I_{D_1D_2}$, etc., become so small that the perturbation approach of Kuijpers and Reuss[24] remains valid. Where the dimensionless AIP-action integrals become comparable to the IIP-action integral, one must, thus, test carefully whether their value is much smaller than π. In that case, the extra terms to be added to the right-hand side of equation (70) become[24], with $\theta_i = \sphericalangle\, \hat{\mathbf{r}}_i, \hat{\mathbf{k}}_\beta$,

$$\Delta\sigma_{tot}(\text{dipole 1} - \text{dipole 2}) = \frac{16}{45}\pi\mu_1^2\mu_2^2\left(\frac{8\hbar v}{3\pi C_6}\right)^{2/5}\Gamma\left(\frac{2}{5}\right)(\hbar v)^{-2}\cos\left(\frac{\pi}{5}\right)[1 - \langle P_2(\cos\theta_1)\rangle]$$

$$\Delta\sigma_{tot}(\text{dipole 1} - \text{quadrupole 2}) = \frac{16}{45}\pi\mu_1^2Q_2^2(\hbar v)^{-2}\left(\frac{8\hbar v}{3\pi C_6}\right)^{4/5}\Gamma\left(\frac{4}{5}\right)\cos\left(\frac{4\pi}{10}\right)[1 - \langle P_2(\cos\theta_1)\rangle]$$

$$\Delta\sigma_{tot}(\text{quadrupole 1} - \text{dipole 2}) = \frac{32}{63}\pi\mu_2^2\,Q_1^2(\hbar v)^{-2} \times\left(\frac{8\hbar v}{3\pi C_6}\right)^{4/5}\Gamma\left(\frac{4}{5}\right)\cos\left(\frac{4\pi}{10}\right)\left[\frac{7}{10} - \langle P_2(\cos\theta_1)\rangle + \frac{3}{10}\langle P_4(\cos\theta_1)\rangle\right]$$

$$\Delta\sigma_{tot}(\text{quadrupole 1} - \text{quadrupole 2}) = -\frac{128}{175}\pi Q_1^2Q_2^2(\hbar v)^{-2} \times\left(\frac{8\hbar v}{3\pi C_6}\right)^{4/5}\Gamma\left(\frac{6}{5}\right)\cos\left(\frac{4\pi}{10}\right)\left[\frac{7}{10} - \langle P_2(\cos\theta_1)\rangle + \frac{3}{10}\langle P_4(\cos\theta_1)\rangle\right] \quad (71)$$

In all cases, a spherically symmetric distribution of the direction $\hat{\mathbf{r}}_2$ of the molecular axis of molecule (2) was assumed.

If neither molecule is state selected, the contributions can be easily obtained from equation (71), with $\langle P_2(\cos\theta_1)\rangle = \langle P_4(\cos\theta_1)\rangle = 0$. The quadrupole–quadrupole term is negative, then, as will always be the case except for $j_1 \simeq m_1 \gg 1$ [see equation (64)]; the integral cross section thus decreases due to the presence of a quadrupole–quadrupole perturbation. However, if this quadrupole–quadrupole interaction is of overwhelming importance, the integral cross section surely must increase with increasing values of the quadrupole momenta Q_i. This shows the limitation of the perturbative approach of the multipole terms.

On the other hand, Kuijpers and Reuss[24] have shown that for pure multipole terms the S-SF approach leads to very simple analytical expressions,

$$\sigma_{\text{tot}}^{\text{S-SF}}(\text{dipole 1} - \text{dipole 2}) = 4\pi\mu_1\mu_2(\hbar v)^{-1}\langle \sin\theta_1 \sin\theta_2\rangle$$

$$\sigma_{\text{tot}}^{\text{S-SF}}(\text{dipole 1} - \text{quadrupole 2}) = 9.529(\mu_1|Q_2|/\hbar v)^{2/3}\langle(\sin\theta_1 \sin^2\theta_2)^{2/3}\rangle$$

$$\sigma_{\text{tot}}^{\text{S-SF}}(\text{quadrupole 1} - \text{quadrupole 2}) = 10.404(|Q_1Q_2|/\hbar v)^{1/2}\langle \sin\theta_1 \sin\theta_2\rangle \tag{72}$$

Here, the expectation value has to be taken with respect to the rotor eigenfunctions of molecules (1) *and* (2).

For intermediate cases, one has to solve equation (67) numerically (see Figure 2), with

$$2\eta_l^{\text{SCSF}} = 3\pi\mu^5 C_6 v^4/8\hbar^6 l^5 - (1/\hbar)[I'_6 + I'_{D_1D_2} + I'_{D_1Q_2} + I'_{D_2Q_1} + I'_{Q_1Q_2}] \tag{73}$$

The action integrals in the brackets are given in Section 2.

For atom–diatom scattering with a degeneracy-averaged cross section $\bar{\sigma}$, the most important AIP contribution comes from equation (69). In its derivation one has assumed that AIP and IIP both are proportional to R^{-6}. However, the potential is probed by the nonglory part of the integral cross section at distances about equal to

$$R_\sigma = (\bar{\sigma}/2\pi)^{1/2} \simeq (2/\pi^{1/2})(2\varepsilon R_m/\hbar v)^{1/5} R_m$$

At the velocity of the $N = 2$ glory, this distance amounts to about 1.8 R_m. It is well possible that the R^{-6} dependence is only approximately true at this distance.* Therefore the equivalent of equation (69) is given here,

* For instance, a Lennard-Jones 12-6 and a Barker–Fisher–Watts potential [J. A. Barker, R. A. Fisher, and R. D. Watts, *Mol. Phys.* **21**, 657–673 (1971)] yield an effective power of 5.9 and 3.1 at $R = 1.8R_m$, respectively.

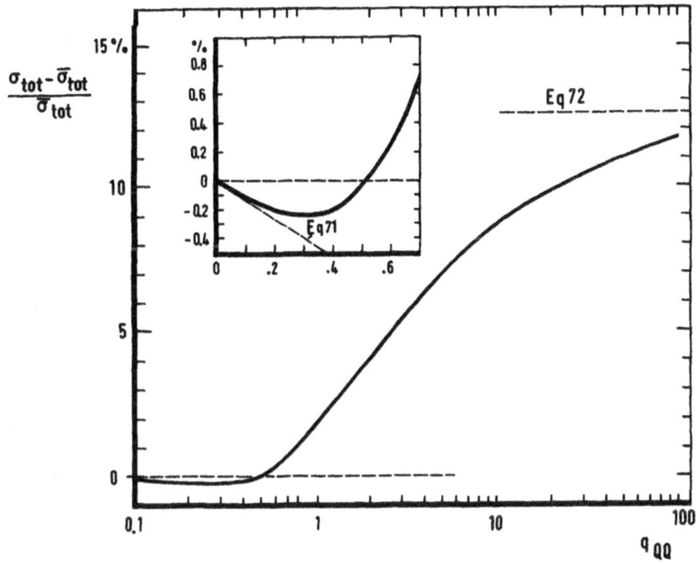

Figure 2. Quadrupole–quadrupole interaction for intermediate cases, calculated with $I' = I'_{Q_1Q_2}$ in equation (73). The S-SF value of $(\sigma_{tot} - \bar{\sigma}_{tot})/\bar{\sigma}_{tot}$ is shown as a function of the dimensionless quantity $q_{QQ} = (Q_1Q_2/\hbar v)^2/(C_6/\hbar v)^{8/5}$. The selected state of molecule (1) corresponds to NO with $j = \Omega = 3/2$ and $m_j = 1/2$.[4,5] For small (large) q_{QQ} a behavior is found as described by equation (71) [equation (72)]. For NO–CO_2 at 500 m/sec one finds $q_{QQ} = 0.27$. Note that for the insert a linear scale is used.

for a potential $V = -(C_s/R^s) - (C_t/R^t)P_2(\hat{\mathbf{r}}_1 \cdot \hat{\mathbf{R}})$,

$$\sigma_{\text{tot}}^{\text{FOS-SF}}(s,t) = 2\pi(2A_s)^{2/(s-1)}\Gamma\left(\frac{s-3}{s-1}\right)\sin\left[\frac{\pi}{2}\frac{s-3}{s-1}\right]$$

$$-\ 4\pi\frac{t-3}{2t(s-1)}2A_t(2A_s)^{-(t-3)/(s-1)}\Gamma\left(\frac{t-3}{s-1}\right)\sin\left[\frac{\pi}{2}\frac{t-3}{s-1}\right]\langle P_2(\cos\theta_1)\rangle \tag{74}$$

with $A_s = f(s)\,C_s/(\hbar v)$ and $f(s) = (\pi^{1/2}/2)\,\Gamma[(s-1)/2]/\Gamma(s/2)$ and similar expressions for A_t and $f(t)$.

This general expression shows a different velocity dependence for the zero-order term ($\sim v^{-2/(s-1)}$) and the first-order term ($\sim v^{-(s-t+2)/(s-1)}$). However, in an experiment it will be rather difficult to measure this slight difference. In Figure 3, the effect a_{ts} of an anisotropy in the IP—characterized by $q_{ts} = (C_t/R_\sigma^t)/(C_s/R_\sigma^s)$—on the total cross section is given as a function of the exponents t and s:

$$a_{ts} = [\sigma_{\text{tot}}^{\text{FOS-SF}} - \bar{\sigma}]/q_{ts}\bar{\sigma}\,\langle P_2(\cos\theta_1)\rangle$$

The effect a_{ts} varies by about a factor of 2, if s and t are changed from 4 to 8. For $s = t$, $a_{ss} = (s-3)/s(s-1)$, $s > 3$.[5]

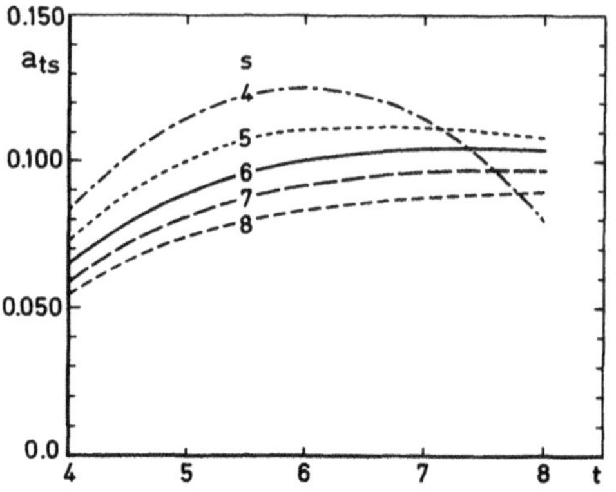

Figure 3. The effect a_{ts} of the AIP on the total collision cross section as a function of the exponents of the force laws, s and t; $a_{ts} = [\sigma_{\text{tot}}^{\text{FOS-SF}} - \bar{\sigma}]/(q_{ts}\bar{\sigma}\langle P_2(\cos\theta_1)\rangle)$ with $q_{ts} = (C_t/R_\sigma^t)/(C_s/R_\sigma^s)$, according to equation (74). The probing distance R_σ is defined as $R_\sigma = [\bar{\sigma}/2\pi]^{1/2}$.

Equation (74) permits the conclusion that for "hard" potentials, for instance $s = t \gg 1$, PCI and S-SF become identical. In the latter case one obtains

$$\sigma_{\text{tot}}^{\text{FOS-SF}} = \bar{\sigma}\left[1 - q_{ss}\frac{s-3}{s(s-1)}\langle P_2(\cos\theta_1)\rangle\right] \tag{75}$$

with

$$\bar{\sigma} = 2\pi(2A_s)^{2(s-1)}\Gamma\left(\frac{s-3}{s-1}\right)\sin\left(\frac{\pi}{2}\frac{s-3}{s-1}\right)$$

On the other hand, equation (66) can be generalized to

$$\sigma_{\text{tot}}^{\text{FOS-PCI}} = \bar{\sigma}\left[1 - q_{ss}\frac{1}{s-1}\langle P_2(\cos\theta_1)\rangle\right] \tag{76}$$

The extra factor $(s-3)/s$ for FOS-SF approximates unity for $s \gg 3$ so that equations (75) and (76) become identical.

6.3. The Total Integral Cross Section in Sudden Approximation: The Glory Contribution

Glory undulations in the integral cross section occur as a consequence of the presence of repulsion (short range) and attraction (long range) in the intermolecular interaction. The repulsion was neglected for the calculations of the last section.

For IOS-PCI one has for impact parameters in the neighborhood of the glory ray

$$\sigma_{g,\text{tot}}^{\text{IOS-PCI}}(jm_j) = 2\pi\int\sum_v Y_{jv}^*(\Omega_1)\,Y_{jv}(\Omega_1)\,d\Omega_1\,[d_{vm_j}^j(\pi/2)]^2$$
$$\times\int b\,db\{-2\cos[2\eta_m(\chi) - |\eta_m''(\chi)|\,(l - l_m(\chi)^2]\} \tag{77}$$

As before, χ denotes the BF angle, taken as its long-range position so that it coincides with the polar angle θ_1 of $\Omega_1 = (\theta_1, \phi_1)$.

The parabola approximation (Düren and Pauly[55]) has been adopted for the SCBF phase shift η_l near its maximum η_m at $l = l_m$; η''_m denotes the second derivative of η_l with respect to l, at $l = l_m$. The glory contribution $\sigma_{g,\mathrm{tot}}$ [equation (75)] must be added to equation (62) to obtain the complete integral cross section.

The integration over b yields

$$\sigma_{g,\mathrm{tot}}^{\mathrm{IOS\text{-}PCI}}(jm_j) = -2\pi \int \sum_v Y^*_{jv}(\Omega_1)\, Y_{jv}(\Omega_1)\, d\Omega_1 \left[d^j_{vm_j}\left(\frac{\pi}{2}\right)\right]^2 \times \frac{l_m(\chi)}{|\eta''_m(\chi)|^{1/2}} \frac{2\pi^{1/2}}{k^2} \cos\left[2\eta_m(\chi) - \frac{\pi}{4}\right] \tag{78}$$

This expression is now linearized with respect to the AIP; we especially replace $2\eta(\chi)$ by $2\eta_m^{(0)} - (1/\hbar)\int_{-\infty}^{+\infty} V'\,dt$ (Reference 48) and consider the second term so small that with sufficient accuracy

$$\cos\left(2\eta_m - \frac{\pi}{4}\right) = \cos\left(2\eta_m^{(0)} - \frac{\pi}{4}\right) + \frac{1}{\hbar}\int_{-\infty}^{+\infty} V'\,dt \sin\left(2\eta_m^{(0)} - \frac{\pi}{4}\right) \tag{79}$$

The same procedure was adopted by Kuijpers and Reuss[50] for the FOS-SF approach. For velocities where glories occur the argument of the cosine can easily assume values on the order of 10; consequently, a fractional change of $2\eta_m$ due to the AIP affects $\sigma_{g,\mathrm{tot}}$ much more than an equally large fractional change of the amplitude factor $l_m/|\eta''_m|^{1/2}$. Therefore we neglect the χ dependence of the amplitude factor. Values for $\eta_m^{(0)}$ and $l_m^{(0)}/|\eta''^{(0)}_m|^{1/2}$ are tabulated for L.-J. 12-6 IP's by Bernstein and LaBudde.[56] Inserting equation (79) in (78) yields for $V' = V_2(R)\, P_2(\hat{\mathbf{r}}_1 \cdot \hat{\mathbf{R}})$

$$\sigma_{g,\mathrm{tot}}^{\mathrm{FOS\text{-}PCI}}(jm_j) = -\frac{4\pi l_m^{(0)}}{k^2}\left(\frac{\pi}{|\eta''^{(0)}_m|}\right)^{1/2}\left[\cos\left(2\eta_m^{(0)} - \frac{\pi}{4}\right) - \frac{1}{2}\sin\left(2\eta_m^{(0)} - \frac{\pi}{4}\right)\left\langle \frac{1}{\hbar}\int_{-\infty}^{+\infty} V'(t)\,dt\right\rangle\right] \tag{80}$$

The IOS-SAIP approach [equations (40) and (41)] again gives qualitatively wrong results; the factor $(-1/2)$ in front of the sine term of equation (80) becomes $(+1)$ in the SAIP case, whereas the rest of equation (80) remains unchanged.

In order to achieve comparable results with those of the SF approach we specialize for an L.-J. 12-6 potential

$$V = \varepsilon\{(R_m/R)^{12}[1 + q_{2,12}\, P_2(\hat{\mathbf{R}} \cdot \hat{\mathbf{r}}_1)] - 2(R_m/R)^6[1 + q_{2,6}\, P_2(\hat{\mathbf{R}}\ \hat{\mathbf{r}}_1)]\} \tag{81}$$

The action integral can now be explicitly evaluated, for $l = l_m^{(0)}$:

$$\frac{1}{\hbar}\int_{-\infty}^{+\infty} V'(t)\,dt' = 4\varepsilon R_m \frac{1}{\hbar v} 2^{-1/6} P_2(\cos\chi)[q_{2,12}S^{(m)}(12,0,0) - q_{2,6}S^{(m)}(6,0,0)] \quad (82)$$

The dimensionless quantities $S^{(m)}(n, q, p)$ can be calculated irrespective the values of ε and R_m; they also occur in the FOS-SF approach; tables are given in Reference 57.

Thus one finds

$$\sigma_{g,\text{tot}}^{\text{FOS-PCI}}(jm_j) = -\frac{4\pi l_m^{(0)}}{k^2}\left(\frac{\pi}{|\eta_m''^{(0)}|}\right)^{1/2}\left\{A\cos\left(2\eta_m^{(0)} - \frac{\pi}{4}\right) - \frac{2(2)^{-5/6}\varepsilon R_m}{\hbar v}\sin\left(2\eta_m^{(0)} - \frac{\pi}{4}\right)\langle P_2(\cos\chi)\rangle \times[q_{2,12}S^{(m)}(12,0,0) - q_{2,6}S^{(m)}(6,0,0)]\right\} \quad (83)$$

In this approximation, $A = 1$. However, if the χ dependence of the amplitude factor is taken into account in first order of the AIP, for an L.-J. 12-6 potential one finds with the help of the series expansion of Reference 56

$$A = 1 - \frac{\langle P_2(\cos\gamma)\rangle}{8}\,\frac{(3q_{2,12} - 5q_{2,6})E - (q_{2,12} - 3q_{2,6})\cdot 3.267\,\varepsilon}{E + 3.267\,\varepsilon} \quad (84)$$

For the $N = 3$ glory, the amplitude of the extra cosine term, $A - 1$, is more than a factor 50 smaller than the amplitude of the sine factor, thus again justifying that we put $A = 1$.

Kuijpers and Reuss[50] have derived corresponding expressions for the S-SF approach:

$$\sigma_{g,\text{tot}}^{\text{S-SF}}(jm_j) = -\frac{4\pi l_m^{(0)}}{k^2}\left(\frac{\pi}{|\eta_m''^{(0)}|}\right)^{1/2}\left\langle\cos\left[2\eta_m^{(0)} - \frac{\pi}{4} - \frac{1}{\hbar}\int_{-\infty}^{+\infty} V'(t)\,dt\right]\right\rangle \quad (85)$$

For the S-SF approach one has, in the case of an atom–homonuclear-diatom system,

$$\frac{1}{\hbar}\int_{-\infty}^{+\infty} V'(t)\,dt = 4(2^{-1/6})\,\varepsilon R_m(\hbar v)^{-1}S_{\text{AIP}}^{(m)}$$

with

$$S^{(m)}_{\text{AIP}} = \sum_{i=6,12} (-1)^{a_i} \{ -\tfrac{1}{2} q_{2,i} S^{(m)}(i,0,0)$$
$$+ \tfrac{3}{2} \cos^2\theta_1 q_{2,i} [S^{(m)}(i,0,0) - S^{(m)}(i,0,2)] \tag{86}$$
$$+ \tfrac{3}{2} \sin^2\theta_1 \cos^2(\phi_1 - \phi) q_{2,i} S^{(m)}(i,0,2)\}$$

with $a_6 = 1$ and $a_{12} = 2$.*

Tables from which all $S^{(m)}$ functions can be calculated for the atom–diatom case are given in Reference 57, even for powers different from 12 and 6 in the R dependence of V'. The angles θ_1, ϕ_1 correspond to Ω_1 and are assumed to stay constant throughout a single collision.

* Equation (86) is valid for the glory ray, i.e., deflection angle $\theta = 0$. For the case of nonvanishing deflection angle, if θ is determined from a L.-J. 12-6 IIP, one has (Reference 49)

$$S_{\text{AIP}} = \sum_{i=6,12} (-1)^{a_i} q_{2,i} \left\{ \left(\frac{\pi}{5}\right)^{1/2} Y_{20}(\theta_1, \phi_1 - \phi) \left[\frac{3}{2}(1 + \cos 2\theta) S(i,2,0) \right.\right.$$
$$\left. - \frac{1}{4}(1 + 3\cos 2\theta) S(i,0,0) - \frac{3}{2}\sin 2\theta\, S(i,1,1) \right] - \left(\frac{6\pi}{5}\right)^{1/2} [Y_{21}(\theta_1, \phi_1 - \phi)$$
$$- Y_{2-1}(\theta_1, \phi_1 - \phi)] \left[\left(\frac{1}{2} - \frac{1}{2}\cos 2\theta\right) S(i,1,1) + \frac{1}{4}\sin 2\theta\, S(i,0,0) \right.$$
$$\left. - \frac{1}{2}\sin 2\theta\, S(i,2,0) \right] + \left(\frac{3\pi}{10}\right)^{1/2} [Y_{22}(\theta_1, \phi_1 - \phi) + Y_{2-2}(\theta_1, \phi_1 - \phi)]$$
$$\left. \times \left[\frac{1}{4}(3 + \cos 2\theta) S(i,0,0) - \frac{1}{2}(1 + \cos 2\theta) S(i,2,0) + \frac{1}{2}\sin 2\theta\, S(i,1,1) \right] \right\}$$

where

$$S(i,p,q) = 2\left(\frac{2^{1/6} l}{kR_m}\right)^{1-i} \int_0^{y_0} \frac{y^{i-2}\, dy}{F(y)} \cos^p[\theta'(y)] \sin^q[\theta'(y)]$$

with

$$F(y) = \{1 - y^2 - [\bar{V}(y)/E]\}^{1/2}$$

$$y = \frac{l}{kR} \quad \text{and} \quad \theta'(y) = \int_0^y \frac{dy'}{F(y')}$$

The upper limit of integration coincides with the smallest root of $F(y) = 0$. Tables of $S(i, p, q)$ and θ as functions of energy and impact parameters are available (see Reference 49).

After linearization one obtains

$$\sigma_{g,\text{tot}}^{\text{FOS-SF}}(jm_j) = -\frac{4\pi l_m^{(0)}}{k^2}\left(\frac{\pi}{|\eta_m^{(0)\prime\prime}|}\right)^{1/2}\left\{\cos\left(2\eta_m^{(0)} - \frac{\pi}{4}\right)\right.$$
$$-\frac{\varepsilon R_m 2^{-1/6}}{\hbar v}\sin\left(2\eta_m^{(0)} - \frac{\pi}{4}\right)\langle P_2(\cos\theta_1)\rangle$$
$$\times\{q_{2,12}[6S^{(m)}(12,0,2) - 4S^{(m)}(12,0,0)]$$
$$\left. - q_{2,6}[6S^{(m)}(6,0,2) - 4S^{(m)}(6,0,0)]\}\right\} \qquad (87)$$

If one replaces $S^{(m)}(n, 0, 2)$ by $S^{(m)}(n, 0, 0)$, equation (80) is obtained. For a similar observation, see Reference 54.

An extensive discussion of results obtained from equation (85) as compared to the FOS-SF approach and as compared to dwa can be found in Reference 57. A direct derivation of equation (87) starting from dwa [equation (39)] was given by Franssen and Reuss.[58] From this paper, the connection between $S^{(m)}(n, 0, 2)$ and the S matrix elements with $l' = l \pm 2$ becomes evident.

FOS-SF and dwa yield equivalent results, whereas again FOS-PCI shows rather strong deviations. If $q_{2,12}/q_{2,6} \simeq 1.5$, the $\{\cdots\}$ factor of $P_2(\cos\theta)$ in equation (87) cancels for almost the whole glory range of velocities. In equation (83), however, such a cancellation occurs for $q_{2,12}/q_{2,6} \simeq 2.2$, i.e., especially the AIP term proportional to $q_{2,6}$ is overestimated in FOS-PCI (by about a factor 1.8). For the SAIP treatment the disagreement is even worse, yielding the wrong sign for the $q_{2,6}$ and $q_{2,12}$ contributions.

Our conclusion is that the BF approach is inferior to the S-SF approach if orientation dependence of collision cross section is investigated.

7. *Conclusions*

Having confronted you, patient reader, with a large number of formulas, we feel obliged to compare all results obtained so far with the help of Table 1.

In this table we consider close-coupled calculations as being exact (indicated by X), whereas dwa and IOS calculations are compared to cc or among themselves.

The numbers on the following short comments on each approximation correspond to the table entries.

11. $\sigma(jm_j)$ in dwa, good results for not-too-large anisotropy parameters $q_{2,i}$.[3,28] There occurs a term linear in $q_{2,i}$ [equation (39)].

21. $d\sigma(jm_j)/d\Omega$ in dwa, good results for not-too-large anisotropy parameters.[35]

Table 1. The Phenomenon-Method Matrix. Each Number Entry is Commented upon in the Text in Section 7

			IOS		
			BF		SF
Cross section	CC	DWA	SAIP	PCI	
$\sigma(jm_j)$	X	11	12	13	14
$d\sigma(jm_j)/d\Omega$	X	21	22	23	24
$\sigma(j)$	X	31	32	33	34
$d\sigma(j)/d\Omega$	X	41	42	43	44
$\sigma(\Delta m_j)$	X	51	52	53	54
$d\sigma(\Delta m_j)/d\Omega$	X	61	62	63	64

31. $\sigma(j)$ in DWA, degeneracy averaged, no effect of the AIP, no glory quenching [see equation (39)].

41. $d\sigma(j)/d\Omega$ in DWA, no effect of the AIP.

51. $\sigma(\Delta m_j)$ in DWA, reorientation cross sections are realistic for not-too-large $q_{2,i}$ values. They are found, however, to be independent of the sign of Δm_j, which does not correspond with reality, for large $q_{2,i}$ values.[26,29]

61. $d\sigma(\Delta m_j)/d\Omega$ in DWA; as in the CC treatment, vanishing reorientation cross sections in the forward and backward directions; good results for small $q_{2,i}$ values, which, however, do not depend on the sign of Δm_j; this does not correspond to reality, for large $q_{2,i}$.[29]

12. $\sigma(j, m_j)$ in IOS-SAIP, even qualitatively wrong results (see Sections 6.2 and 6.3). There is evidence that SAIP works well only for small impact parameters, $b \lesssim 0.6R_m$[43]; this would be consistent with the here observed failure at glory impact parameters, $b \simeq R_m$, and further outwards.

22. $d\sigma(j, m_j)/d\Omega$ in IOS-SAIP, not to be trusted at all.

32. $\sigma(j)$ in IOS-SAIP, glory quenching calculated by Miller[59] in an early body-fixed approach. Equivalent to 33, with results different from 34 (see Reference 57, where earlier references are given).

42. $d\sigma(j)/d\Omega$ in IOS-SAIP, results identical to those of 43 and similar to those of 44 (see Reference 49).

52. $\sigma(\Delta m_j)$ in IOS-SAIP; in this approximation reorientation cross sections vanish, for $l = l_{\text{final}}$; for $l = l_{\text{initial}}$, comparison with CC calculations is needed.

62. $d\sigma(\Delta m_j)/d\Omega$ in IOS-SAIP; in this approximation reorientation cross sections vanish, for $l = l_{\text{final}}$; for $l = l_{\text{initial}}$, comparison with CC calculations is needed.

13. $\sigma(jm_j)$ in IOS-PCI, nonglory part about a factor of 2 off (see Section 6.2; glory part also badly described, see Section 6.3).

23. $d\sigma(jm_j)/d\Omega$ in IOS-PCI, not further investigated.

33. $\sigma(j)$ in IOS-PCI, equivalent to 32; contains glory quenching slightly different from 34.

43. $d\sigma(j)/d\Omega$ in IOS-PCI, equivalent to 42, results similar to 44 (see Reference 51).

53. $\sigma(\Delta m_j)$ in IOS-PCI, not yet investigated.

63. $d\sigma(\Delta m_j)/d\Omega$ in IOS-PCI; does not vanish in the forward direction, unrealistically (see Section 6.1). Thus angular momentum is not conserved.

14. $\sigma(jm_j)$ in S-SF, for small $q_{2,i}$ parameters equivalent to dwa (References 57 and 58).

24. $d\sigma(jm_j)/d\Omega$ in S-SF, for small $q_{2,i}$ parameters equivalent to dwa.

34. $\sigma(j)$ in S-SF, contains glory quenching different from 32 and 33.[57]

44. $d\sigma(j)/d\Omega$ in S-SF, nearly equivalent to 42 and 43.[51]

54. $\sigma(\Delta m_j)$ in S-SF, independent of sign of Δm_j (see, however, References 26 and 29), where for large $q_{2,i}$ values very different results were obtained for $\Delta m_j \gtrless 0$.

64. $d\sigma(\Delta m_j)/d\Omega$ in S-SF, two unrealistic features; reorientation transitions in the forward direction and independence of sign of Δm_j; for the last point see, however, References 26 and 29; the first point implies nonconservation of angular momentum. For $\theta \gtrsim l_0^{-1}$ and small $q_{2,i}$ values, equivalence with dwa has been found.[35]

It emerges from the listing that effects of the AIP on elastic and total collisions can best be treated with S-SF; however, a systematic study of whether higher-order effects are reproduced properly is still lacking; the only consolation in this respect comes from the comparative study of Buck and Khare[51] where it is shown that at least all three sudden approaches (SAIP, PCI, and SF) result in nearly the same degeneracy-averaged differential cross section. That the PCI and S-SF approaches yield reorientation transitions in the forward direction hampers light-hearted optimism. A minor warning also comes from the fact that the reorientation cross sections unrealistically depend only on the absolute value, $\Delta m_j \neq 0$; again this concerns a higher-order effect.

At a first glance the recommendation of the S-SF approximation may appear to be in striking contrast with conclusions reached by other workers in the field (see especially the contribution of Kouri to this volume) who concentrate on IOS approaches. We have shown that for the limited class of problems treated in this contribution IOS approximations give wrong answers. As for these problems the attractive part of the IP is of great influence, this failure of IOS should be seen as an extra indication that this approach is mainly valid at small impact parameters, smaller than the glory impact parameter. At large impact parameters the coupled-state assumption for the centrifugal barrier becomes highly questionable.

ACKNOWLEDGMENTS

Discussions with U. Buck, V. Khare, A. S. Dickinson, and Monique Jacobs which resulted in many improvements are gratefully acknowledged.

Glossary of Abbreviations

Abbreviations are presented in alphabetical order. The equation numbers given in brackets indicate where the abbreviation is introduced in the text or where it is important for numerical evaluation.

AIP = anistropic part of the intermolecular potential [equation (1)].
BF = body fixed, i.e., the position of the molecular symmetry axis is kept fixed with respect to the intermolecular position vector for phase shifts and action integrals [equations (43) and (82)].
CC = close coupling [equations (29) and (30)].
dwa = distorted-wave approximation [equation (33)].
GT = general term of the interaction potential [equation (4)].
FOS-PCI = first-order sudden approximation with point-contact-like interaction [equation (76)].
FOS-SAIP = first-order sudden approximation with body-fixed molecular orientation for strong anisotropic interactions [equation (63)].
FOS-SF = first-order sudden approximation with space-fixed molecular orientation [equations (58), (60), (68), (74), and (87)].
g, tot = glory part of total collision cross section [equation (77)].
I = action integral [equations (21), (26), (82), and (86)].
IIP = isotropic part of the intermolecular potential [equation (1)].
IOS = infinite-order sudden approximation [equations (44) and (61)].
IOS-l-SAIP = infinite-order sudden approximation for strong anisotropic interactions with $l = \bar{l} = l_{\text{initial}}$ [equations (41) and (52)].
IOS-l'-SAIP = infinite-order sudden approximation for strong anisotropic interactions with $l = \bar{l} = l_{\text{final}}$ [equations (40) and (52)].
IOS-PCI = infinite-order sudden approximation for point-contact-like interactions [equations (44), (45), (53), (76), and (83)].
IOS-SAIP = infinite-order sudden approximation for strong anisotropic interactions [equation (62)].
IP = intermolecular potential [equation (1)].
PCI = point-contact-like interaction, i.e., the position of the molecular symmetry axis is kept fixed with respect to the intermolecular position vector as in SAIP, but the orientation at closest approach is put equal to the real large-distance orientation [equation (44)].

SAIP = strong anisotropic interactions so that the orientation can be treated body-fixed, at all distances [equation (44)].
SCBF = semiclassical approximation with body-fixed orientation [equation (43)]
SCSF = semiclassical approximation with space-fixed molecular orientation [equation (46)].
S-SF = sudden approximation with space-fixed molecular orientation [equations (47), (50), (70)–(73), and (85)].
tot = total, i.e., integral total collision cross section [equation (28)].

References

1. H.G. Bennewitz, K.H. Kramer, W. Paul, and J.P. Toennies, Messung der Anisotropie des van der Waals-Potentials durch Streuung von Molekülen in definiertem Quantenzustand, *Z. Phys.* **177**, 84–110 (1964).
2. H.G. Bennewitz, R. Gengenbach, R. Haerten, and G. Müller, Weitere Messung zur Anisotropie der intermolekularen Wechselwirkung durch Streuung von Molekülen in definiertem Quantenzustand, I, II & III, *Z. Phys.* **226**, 139–155, 279–292 (1969); **227**, 399–411 (1969).
3. L. Zandee and J. Reuss, The anisotropy of the intermolecular potential for H_2-inert gas systems, I & II, *Chem. Phys.* **26**, 327–343, 345–365 (1977); L. Zandee, J. Verberne, and J. Reuss, The total collision cross section of oriented H_2 scattered by N_2, CO, CH_4, CCl_4, SF_6, $C(CH_3)_4$, CF_4, N_2O and CO_2; derivation of anisotropic potential parameters, *Chem. Phys.* **26**, 1–8 (1977).
4. S. Stolte, J. Reuss, and H.L. Schwartz, Orientational anisotropy in the total collision cross section of state-selected NO on CCl_4 using crossed molecular beams, *Physica (Utrecht)* **57**, 254–266 (1972); H.L. Schwartz, S. Stolte, and J. Reuss, Glory structure in the anisotropy of the total collision cross section, *Chem. Phys.* **2**, 1–5 (1973); H. Thuis, S. Stolte, and J. Reuss, Investigation of the angle dependent part of the intermolecular potential of NO-inert gas systems using crossed molecular beams, *Chem. Phys.*, to be published (1979).
5. S. Stolte, Orientational anisotropy in the total collision cross section of state selected NO molecules, Thesis, Katholieke Universiteit Nijmegen (1972).
6. A.G. Visser, Angular momentum alignment in molecular beams as studied by laser-induced fluorescence, Thesis, Rijksuniversiteit Leiden (1977).
7. K.T. Tang and J.P. Toennies, A simple theoretical model for the van der Waals potential at intermediate distances, II, *J. Chem. Phys.* **68**, 5501–5517 (1978).
8. L.Y. Tsou, D. Auerbach, and L. Wharton, Observation of high resolution differential cross sections of polarized rotationally state selected LiF on Ar, *Phys. Rev. Lett.* **38**, 20–23 (1977).
9. U. Buck, F. Gestermann, and H. Pauly, Double rainbows in atom–molecule scattering, *Chem. Phys. Lett.* **33**, 186–190 (1975); U. Buck, Untersuchungen zur Anisotropie der Wechselwirkung bei Atom-Molekül Stössen, *MPI-Bericht* **12**, Göttingen (1976); U. Buck, F. Gestermann, and H. Pauly, Messungen differentieller Streuquerschnitte bei Stössen von Na-Atomen mit mehratomigen Molekülen, *MPI-Bericht* **109**, Göttingen (1975); R. Viard, Untersuchung der elastischen Streuung von Kalium an Methyljodid, Thesis, Technische Universität Hannover (1974).
10. U. Buck, V. Khare, and M. Kick, Anisotropic potentials from rainbow scattering of sodium atoms and tetrahedral molecules, *Mol. Phys.* **35**, 65–79 (1978).
11. J.P. Toennies, The calculation and measurement of cross sections for rotational and

vibrational excitation, *Ann. Rev. Phys. Chem.* **27,** 225–260 (1976); M. Faubel and J.P. Toennies, Scattering studies of rotational and vibrational excitation of molecules, *Advances in Atomic and Molecular Physics,* Academic Press, New York (1977).
12. A.R. Edmonds, *Angular Momentum in Quantum Mechanics,* Princeton University Press, Princeton (1960); note that in Edmonds' equation 4.14 the Euler angles α, β, γ must be replaced by $-\alpha$, $-\beta$, $-\gamma$.
13. A.D. Buckingham, Permanent and induced molecular moments and long-range intermolecular forces, in *Advances in Chemical Physics,* J.O. Hirschfelder, editor, Vol. 12, Chap. 2, John Wiley & Sons, New York (1967), pp. 107–142.
14. H. Margenau and N.R. Kestner, *Theory of Intermolecular Forces,* Pergamon Press, Oxford (1969).
15. D.E. Stogryn and A.P. Stogryn, Molecular multipole moments, *Mol. Phys.* **11,** 371–393 (1966).
16. J.O. Hirschfelder, C.F. Curtiss, and R.B. Bird, *Molecular Theory of Gases and Liquids,* John Wiley & Sons, New York (1965), Chaps. 13.3 and 13.4.
17. Th.M. Miller and B. Bederson, Atomic and molecular polarizabilities—A review of recent advances, in *Advances in Atomic and Molecular Physics,* Vol. 13, Chap. 1, Academic Press, New York (1977), pp. 1–55.
18. E.A. Mason and L. Monchick, Methods for the determination of intermolecular forces, in *Advances in Chemical Physics,* J.O. Hirschfelder, editor, Vol. 12, Chap. 7, John Wiley & Sons, New York (1967), pp. 329–387.
19. K. Bergmann, H. Vilar, and W. Schlecht, Asymmetries in collision-induced rotational transitions, *Chem. Phys. Lett.* **12,** 522–525 (1972).
20. D. Poppe, Classical trajectory studies for the $Li_2(B^1\Pi_4)$–noble gas system, *Chem. Phys.* **25,** 29–45 (1977).
21. S. Green and R.N. Zare, Mechanism for collision induced transitions between Λ-doublets in $^1\Pi$-molecules. Reduction to a single scattering potential, *Chem. Phys.* **7,** 62–72 (1975).
22. G.C. Nielson, G.A. Parker, and R.T. Pack, Van der Waals interaction of Π-state linear molecules with atoms; C_6 for NO ($X\ ^2\Pi$) interactions, *J. Chem. Phys.* **64,** 2055–2061 (1976).
23. H. Kreek and R.J. Le Roy, Intermolecular potentials and isotope effects for molecular hydrogen–inert gas complexes, *J. Chem. Phys.* **63,** 338–344 (1975).
24. J.W. Kuijpers and J. Reuss, Multipole interaction in high-energy approximation calculations of the total collision cross section, *Chem. Phys.* **1,** 64–73 (1973).
25. A.M. Arthurs and A. Dalgarno, The theory of scattering by a rigid rotator, *Proc. R. Soc. London, Ser. A* **256,** 540–551 (1960)
26. M.H. Alexander, Close coupling studies of the orientation dependence of rotationally inelastic collisions, *J. Chem. Phys.* **67,** 2703–2712 (1977).
27. D.E. Fitz, A study of reorientation effects in CsF-Ar, *Chem. Phys.* **24,** 133–142 (1977).
28. J. Reuss and S. Stolte, Calculation of the total collision cross section of H_2-Ar for a particular rotational j, m_j-state of the H_2-molecules, *Physica (Utrecht)* **42,** 111–123 (1969).
29. M. Jacobs and J. Reuss, Diffraction undulations in the differential cross sections of H_2-Kr calculated for an anisotropic potential, *Chem. Phys.* **25,** 425–431 (1977).
30. J. Schäfer and W. Meyer, *MPI/Astro.* **152** (1978), München.
31. M.H. Alexander, P.J. Dagdigian, and A.E. De Pristo, Quantum interpretation of fully state selected rotationally inelastic collision experiments, *J. Chem. Phys.* **66,** 59–66 (1977); M.H. Alexander, Polarization in elastic scattering: close coupling studies on Ar-N_2, *Chem. Phys.* **27,** 229–236 (1978).
32. L. Monchick, State selected He–HCl collision cross section, *J. Chem. Phys.* **67,** 4626–4631 (1977).
33. M.H. Alexander and P.J. Dagdigian, Rotational alignment in inelastic collisions, *J. Chem. Phys.* **66,** 4126–4132 (1977).
34. S. Green, Rotational excitation in H_2–H_2 collisions, close coupling calculations, *J. Chem. Phys.* **62,** 2271–2277 (1975).

35. M. Jacobs and J. Reuss, Calculation of reorientation cross sections for atom–diatom systems, *Chem. Phys.* **33,** 443–449 (1978).
36. J. Reuss, Asymmetry-effects in molecule–molecule collisions, *Physica (Utrecht)* **31,** 597–605 (1965).
37. L. Biolsi, Glory scattering in molecular collisions: Formal expressions for the total cross section, *J. Chem. Phys.* **53,** 3900–3911 (1970).
38. P. McGuire and D.J. Kouri, Quantum mechanical close coupling approach to molecular collisions. j_z-conserving coupled states approximation, *J. Chem. Phys.* **60,** 2488–2499 (1974); P. McGuire, Elastic and inelastic angular distributions in the j_z-conserving coupled states approximation for molecular collisions, *Chem. Phys. Lett.* **23,** 575–578 (1973).
39. R.T. Pack, Space fixed vs body fixed axes in atom–diatomic molecule scattering. Sudden approximations. *J. Chem. Phys.* **60,** 633–639 (1974).
40. T.P. Tsien, G.A. Parker, and R.T. Pack, Rotationally inelastic molecular scattering. Computational tests of some simple solutions of the strong coupling problem, *J. Chem. Phys.* **59,** 5373–5381 (1973); T.P. Tsien and R.J. Parker, Rotational excitation in molecular collisions. A many-state test of the strong coupling approximation, *Chem. Phys. Lett.* **8,** 579–581 (1971); T.P. Tsien and R.T. Pack, Rotational excitation in molecular collisions: A strong coupling approximation, *Chem. Phys. Lett.* **6,** 54–56 (1970).
41. D. Secrest, Theory of angular momentum decoupling approximations for rotational transitions in scattering, *J. Chem. Phys.* **62,** 710–719 (1975).
42. L.W. Hunter, On infinite order sudden approximations for an arbitrary potential energy, *J. Chem. Phys.* **62,** 2855–2859 (1975).
43. V. Khare, private communication (1977).
44. G.A. Parker and R.T. Pack, Rotationally and vibrationally inelastic scattering in rotational IOS approximation. Ultrasimple calculation of total (differential, integral and transport) cross sections for nonspherical molecules, *J. Chem. Phys.* **68,** 1585–1601 (1978).
45. Y. Shimoni and D.J. Kouri, Quantum mechanical close coupling approach to molecular collisions: Averaged definite parity j_z approximation with Clebsch–Gordan weights, *J. Chem. Phys.* **66,** 2841–2849 (1977).
46. G.A. Parker and R.T. Pack, Identification of the partial wave parameter and simplification of the differential cross section in j_z CCS approximation in molecular scattering, *J. Chem. Phys.* **66,** 2850–2853 (1977).
47. V. Khare, On the equivalence between the space fixed and the body fixed formulations of j_z-conserving approximation, *MPI-Bericht* **112,** Göttingen (1977); *J. Chem. Phys.* **67,** 3897–3900 (1977).
48. A.S. Dickinson and D. Richards, Sudden atom–molecule collisions: CsF scattered by Ne, Ar, Kr and Xe, *J. Phys. B* **6,** 1085–1103 (1978).
49. K.H. Kramer and R.B. Bernstein, Sudden approximation applied to rotational excitation of molecules by atoms, I, *J. Chem. Phys.* **40,** 200–203 (1964); R.B. Bernstein and K.H. Kramer, Sudden approximation applied to rotational excitation of molecules by atoms. II, *J. Chem. Phys.* **44,** 4473–4485 (1966); R.W. Fenstermaker and R.B. Bernstein, Sudden approximation applied to rotational excitation of molecules by atoms, III, *J. Chem. Phys.* **47,** 4417–4426 (1967).
50. J.W. Kuijpers and J. Reuss, The total collision cross section in sudden approximation for angle dependent potentials, *Chem. Phys.* **4,** 277–283 (1974).
51. U. Buck and V. Khare, A comparison of different sudden approximations for molecular scattering, *MPI-Bericht* **118,** Göttingen (1977); *Chem. Phys.* **26,** 215–221 (1977).
52. R.D. Levine, *Quantum Mechanics of Molecular Rate Processes*, Clarendon Press, Oxford (1969).
53. M.V. Berry, Uniform approximations for glory scattering and diffraction peaks, *J. Phys. B* **2,** 381–392 (1969).
54. A.S. Dickinson and D. Richards, A semiclassical study of the body-fixed approximation

for rotational excitation in atom–molecule collsiions, *J. Phys. B: Atom. Mol. Phys.* **10**, 323–343 (1977).

55. R. Düren and H. Pauly, Numerische Berechnungen zum Resonanzverhalten der Streuquerschnitte bei atomaren Stössen. I, *Z. Phys.* **175**, 227–234 (1963).
56. R.B. Bernstein and R.A. LaBudde, On the analysis of glory scattering data for the extraction of information on the interatomic potential well, *J. Chem. Phys.* **58**, 1109–1117 (1973).
57. D. Klaassen, H. Thuis, S. Stolte, and J. Reuss, Calculation of the glory scattering with anisotropic molecule–atom interactions in sudden approximation, *Chem. Phys.* **27**, 107–116 (1978).
58. W. Franssen and J. Reuss, Calculation of anisotropic glories for atom–molecule collisions, *Physica (Utrecht)* **63**, 313–320 (1973); Erratum in *Physica (Utrecht)* **77**, 203–204 (1974).
59. W.H. Miller, Semiclassical approximation for the total cross section of atom–diatom molecule collisions; quenching of glory undulations, *J. Chem. Phys.* **50**, 3124–3126 (1969).

6

Inelastic Scattering Cross Sections I: Theory

JOHN C. LIGHT

1. Introduction

Inelastic scattering is characterized by a change in the internal state of one or both collision partners, but with no change in their chemical identities. Thus electronic, vibrational, and rotational excitation and de-excitation are inelastic events, whereas chemical reactions, charge transfer, ionization, and dissociation would be characterized as reactive events. One may also include collisions which change the orientations of the partners ($\Delta m \neq 0$), but do not change the energy state. In the following section and the next three chapters the theory of inelastic collisions of various types is given in considerable detail. In this section we shall define the relations between observables and the theoretical quantities, outline the general quantum theory of such processes and three practical (for small systems) quantum approaches, and give a brief overview of some of the alternative approximate approaches which are given in greater detail in later chapters. There are, of course, many reviews and books covering inelastic scattering; some are given in References 1–7.

2. Observables and Averaging[8–13]

Although what can be observed in a given experimental setup varies widely with the mode of preparation of the system as well as the type and

JOHN C. LIGHT • The James Franck Institute and The Department of Chemistry, The University of Chicago, Chicago, Illinois 60637.

sensitivity of detection, the information which could be obtained in the "ultimate" experiment is the set of state-to-state differential cross sections at specified total center-of-mass energy, $d\sigma(j, \Omega; i, E)/d\Omega$. In this $j(i)$ denotes the set of quantum numbers defining the final (initial) state, E is the total energy, and Ω the solid angle into which one of the products is scattered relative to the initial relative velocity vector. Although the relationships between differential cross sections and such observable quantities as rate coefficients, total cross sections, etc., have been reviewed many times before, a brief review does not seem out of place here.

The differential cross section relates the flux of particles in the specified final state at the specified solid angle to the initial flux and the initial concentration of scattering targets. Thus if $dN(j, \Omega)/d\Omega$ is the number of scattered target particles in state j observed in solid angle Ω, $d\Omega$ per unit time, the differential cross section $d\sigma(j, \Omega; i, E)/d\Omega$ is given by

$$\frac{dN(j, \Omega)}{d\Omega} = N_A(i') N_B(i'') v \Delta V \frac{d\sigma(j, \Omega; i, E)}{d\Omega} \tag{1}$$

where $N_\alpha(i')$ is the number density of particles α in the appropriate state $[i = (i'_A, i''_B)]$, ΔV is the scattering volume, and v is their relative velocity. The initial state i is the composite internal state of the two initial collision partners.

The total state-to-state cross section is merely defined by the integration over all solid angles of the differential cross section

$$\sigma(j; i, E) = \int \frac{d\sigma}{d\Omega}(j, \Omega; i, E)\, d\Omega \tag{2}$$

The rate coefficients for producing given sets of final states from given distributions of initial states are obtained by summing over final states and averaging over initial states the following state-to-state rate constant:

$$k(j, i) \equiv \int_0^\infty E\, dE\, \sigma(j; i, E)\, \rho(E) \tag{3}$$

where E is the relative energy and $\rho(E)$ is the normalized distribution of relative energies. The summation over final states and averaging over initial states is then easily carried out if the initial distribution of initial states is known.

If we denote by l_f and l_i those final and initial states to be included in these summations, by $\chi(l_i)$ the probability distribution function of the initial states, and by j' and i' the final and initial quantum numbers *not* included in the summations, we have partial or averaged state-to-state the rate constants,

$$k(j'; i') = \sum_{l_f, l_i} k(j; i)\, \chi(l_i) \tag{4}$$

Most often orientation selection of the initial and final states is not possible, and the l_f and l_i in that case refer to the quantum numbers of the internal angular momentum projections on any space-fixed axis. Other degrees of freedom may also be included in this averaging, of course, provided the initial distributions are known and the selection of final states desired is fixed. Alternatively one may average the cross sections directly using the same procedures and use the averaged cross sections to compute the averaged rate coefficients. Finally, we note that the distributions used in equations (3) and (4) must be those appropriate to the experiment, not necessarily thermal equilibrium or Boltzmann.

The means of calculating or approximating the differential cross sections, total cross sections, or averaged cross sections are the topic of this chapter and the following six chapters. In this chapter we now turn to the problem of the theory required to calculate cross sections and the general characteristics of the practical approaches to this problem.

3. *Quantum Theory of Inelastic Scattering*[1-7]

3.1. *Formal Quantum Theory*[14-19]

In a space-fixed coordinate system the formal quantum theory of inelastic scattering is quite easy to derive since the chemical entities involved in the collisions do not change, while their internal states may change. An appropriate coordinate system for inelastic scattering of two composite particles α and β thus consists of $\mathbf{R}$, a vector between the centers of mass of the particles, and $\{u_{\alpha(\beta)}\}$, the internal coordinates of particle $\alpha(\beta)$. The Hamiltonian may be partitioned

$$H = T + H_\alpha(\{u_\alpha\}) + H_\beta(\{u_\beta\}) + V(R, \{u_\alpha\}, \{u_\beta\}) \tag{5}$$

where $T = -(\hbar^2/2\mu)\nabla_R^2$ is the relative kinetic energy operator, $H_{\alpha(\beta)}$ is the Hamiltonian of $\alpha(\beta)$ as $R \to \infty$, and V is the interaction potential with $V \to 0$ as $R \to \infty$. The Schrödinger equation is then

$$(H - E)\Psi = 0 \tag{6}$$

where E is usually measured from the zero of internal energy at $R = \infty$. We shall neglect for now exchange processes that may occur between components of the α and β particles since the existence of significant exchange potentials usually implies that reactive as well as inelastic scattering can occur. This can be treated by the methods reviewed in Chapter 14 for reactive scattering. Here they cause largely unnecessary complications.

The solutions of equation (6) must satisfy appropriate scattering boundary conditions if they are to be used in the construction of the cross sections. If we are to represent specific initial and final internal states of the system we must know the separate solutions of the particle Hamiltonians, i.e., the complete orthonormal basis sets $\{\varphi_i^\alpha\}$ and $\{\varphi_i^\beta\}$:

$$H_\alpha \varphi_i^\alpha(\{u_\alpha\}) = \varepsilon_i^\alpha \varphi_i^\alpha(\{u_\alpha\}) \tag{7}$$

with a similar equation for β. The asymptotic form of the wave function must be a linear combination of the solutions of the asymptotic Hamiltonian, i.e., a linear combination of a plane-wave incoming translational state times the initial (combined) internal state plus spherical outgoing translational waves times the appropriate final internal states:

$$\Psi \xrightarrow[R\to\infty]{} e^{i\mathbf{k}_\gamma\cdot\mathbf{R}}\, \varphi_i^\alpha \varphi_j^\beta + \sum_{\gamma'} \frac{e^{ik_{\gamma'}R}}{R} f_{\gamma'\gamma}(\Theta, \Phi)\, \varphi_m^\alpha \varphi_n^\beta \tag{8}$$

where $\gamma = (i, j)$ and $\gamma' = (m, n)$, $\mathbf{k}_\gamma$ is the wave vector of the incoming state (i, j), and $f_{\gamma'\gamma}(\Theta, \Phi)$ is a function of the angular variables of $\mathbf{R}$, Θ, and Φ, and is parametrically dependent on the incoming (i, j) and outgoing (m, n) states. Equations (5)–(8), with the requirement that the solutions are also regular at the origin, i.e.,

$$\Psi \xrightarrow[R\to 0]{} c, \qquad c \text{ finite} \tag{9}$$

formally define the inelastic scattering problem completely. The differential cross sections are given explicitly in terms of equation (8) by applying the quantum flux operator evaluated over solid angle $d\Omega$:

$$d\sigma(\gamma', \Omega; \gamma, E)\, \frac{k_{\gamma'}}{k_\gamma} = |f_{\gamma'\gamma}(\Omega)|^2\, d\Omega \tag{10}$$

Equations (6)–(10), while defining the exact solution of the inelastic scattering problem, give no details as to how to solve this set of equations, the requisite properties of these solutions, the symmetry properties required of the solutions, or the limits of applicability of this particular set. In the balance of this section on the general quantum theory of inelastic scattering we shall, therefore, address these problems in more, but not great, detail. In particular we shall rewrite these equations, using the minimal symmetry and conservation properties of all systems, in the coupled differential equation form most often used in the "exact" solution. This form is also used as the basic representation of the problem for many approximate quantum approaches. We shall then define the appropriate asymptotic properties of the solutions of these equations. These include the relationships of these quantities to the cross sections, and the symmetry properties of the solutions themselves which an exact solution (and good approximate ones) should have, such as flux conservation and time reversal invariance. Finally we shall mention equivalent exact approaches in the formulation of the problem.

3.2. Angular Momentum Conservation, Parity, and Close-Coupled Equations[14,15]

Although equations (5)–(10) are elegant and simple in appearance (and could be made more so by transforming to integral equation form) there are, to say the least, significant difficulties in solving them, no matter what the appearance. This is due almost entirely to the fact that the Schrödinger equation for the simplest system capable of inelastic collisions has six independent coordinates—an internal vector for the target (composite) particle and a relative separation vector (**R**) between the projectile and target particles. Most systems are more difficult. In order to reduce the apparent dimensionality of the equations, therefore, the conservation properties of the system are utilized rigorously. Only the total angular momentum J, its z component M, and the parity are conserved in general, and the Schrödinger equation is usually decoupled to show this explicitly. [We note that equation (8) for the asymptotic wave function does not show this decoupling explicitly. Therefore the price we pay for decoupling equation (6) throughout space is that we must recouple the solutions asymptotically to obtain the angularly dependent physical scattering information required for equation (8) or (10).]

The procedures followed to transform the Schrödinger equation into a form in which the conservation laws appear explicitly as a "block diagonalization" of equation (6) are the standard (involved) transformations which perform the coupling, pairwise, of two angular momenta into their vector sum and the magnitudes of the components. These relations are given in most quantum mechanics texts,[20,21] some monographs,[1,3,7] several articles,[14,15] and a number of reviews.[22,23] What is given here is an outline of the procedure followed in deriving the equations to be solved [equivalent to equation (6)] in the inelastic scattering case.

We assume now that the eigenfunctions of H_α and H_β are complete, discrete, orthonormal basis functions with the internal angular momentum quantum numbers $j_\alpha m_\alpha$ and $j_\beta m_\beta$ specified. If we denote these basis functions (or vectors) as $|n_\alpha j_\alpha m_\alpha\rangle$ (n is for the other internal quantum numbers), and couple products of these into basis vectors for the two-particle system (α, β) as

$$|n_\alpha n_\beta j_\alpha m_\alpha j_\beta m_\beta\rangle \equiv |n_\alpha j_\alpha m_\alpha\rangle\, |m_\beta j_\beta m_\beta\rangle \tag{11}$$

then we can arrange them in order as a vector in which to evaluate $H_\alpha + H_\beta + V$ as a matrix:

$$\begin{aligned}\langle n'_\alpha n'_\beta j'_\alpha j'_\beta m'_\beta | H_\alpha + H_\beta + V |\, n_\alpha n_\beta j_\alpha m_\alpha j_\beta m_\beta\rangle &= (\mathbf{H}_\alpha + \mathbf{H}_\beta + \mathbf{V})_{\gamma'\gamma} \\ &= (\varepsilon_\gamma^\alpha + \varepsilon_\gamma^\beta)\,\delta_{\gamma\gamma'} + \mathbf{V}_{\gamma'\gamma} \end{aligned} \tag{12}$$

where γ represents the totality of the internal quantum numbers. Since the vector sum of the separate angular momenta, $\mathbf{j} = \mathbf{j}_\alpha + \mathbf{j}_\beta$, and the z components, $m = m_\alpha + m_\beta$, are "partially" conserved quantities, commuting with $H_\alpha + H_\beta$ [but not with H of equation (6)], we can transform equation (12) to a form in which the total internal angular momentum j and its z component m are diagonal asymptotically ($V \to 0$):

$$\tilde{\mathbf{W}}(R) = \mathbf{T}_j(\mathbf{H}_\alpha + \mathbf{H}_\beta + \mathbf{V})\mathbf{T}_j^\dagger \tag{13}$$

where $\mathbf{T}_j$ is a direct product of the matrix of Clebsch–Gordan coefficients $(j_\alpha j_\beta m_\alpha m_\beta | jm)$[24] and a unit operator in the other internal (vibrational) manifold of states. Thus $\tilde{\mathbf{W}}$ is the representation of $H_\alpha + H_\beta + V$ in the basis in which j, m are diagonal. Since this transformation does not affect the internal vibrational states, it is a block diagonal (in n_α, n_β) transformation. Thus the transformation (13) takes linear combinations of the basis functions of the angular momenta of the individual systems α, β and combines them into eigenfunctions of the *combined* angular momentum of the combined internal system, α and β. Note that after this operation the total angular momenta of the individual systems, j_α and j_β, are still good quantum numbers asymptotically, being mixed only by V. We note also that the transformation from one orthogonal basis to another, $\mathbf{T}$, is unitary, i.e., $\mathbf{T}\mathbf{T}^\dagger = \mathbf{I}$ where $\mathbf{I}$ is the unit matrix.

The basis in equation (11)–(13) is only for the internal states. In order to proceed we must now expand the basis to contain the orbital angular momentum functions and translational functions as well, i.e., use

$$\frac{f_\gamma(R)}{R} Y_{lm_l}(\hat{\mathbf{R}})|n_\alpha n_\beta jmj_\alpha j_\beta\rangle = \frac{f_\gamma(R)}{R}|n_\alpha n_\beta jmj_\alpha j_\beta lm_l\rangle \tag{14}$$

where γ denotes all the indices. This basis is transformed into the total angular momentum basis by coupling $jmlm_l$ to $JMjl$. In the total angular momentum basis the total Hamiltonian is, of course, block diagonal in J, M (in the absence of radiation fields).

The final space-fixed basis in the total angular momentum representation is

$$\frac{f_\gamma(R)}{R}|n_\alpha n_\beta JMjlj_\alpha j_\beta\rangle \tag{15}$$

This is obtained from the basis in equation (14) by again transforming with a matrix $\mathbf{T}^J$, which is the direct product of the matrix of Clebsch–Gordan coefficients, $(jlmm_l | JM)$, and a unit operator in j_α, j_β and the vibrational states. The vector of coefficients, $\mathbf{f}(R)$, in this basis satisfies the coupled differential equations

$$\frac{d^2}{dR^2}\mathbf{f}^J = \mathbf{W}^J\mathbf{f}^J \tag{16}$$

where $\mathbf{W}^J$ is the J, J block of the interaction matrix in this representation:

$$\mathbf{W}^J_{\gamma\gamma'} = \left[-E + \frac{l(l+1)}{R^2} + \varepsilon^{\alpha}_{\gamma} + \varepsilon^{\beta}_{\gamma} \right] \delta_{\gamma\gamma'} + \mathbf{U}^J_{\gamma\gamma'}$$

$$\mathbf{U}^J_{\gamma\gamma'} = [\mathbf{T}^J \mathbf{T}_j \mathbf{V} \mathbf{T}_j^T (\mathbf{T}^J)^T]_{\gamma\gamma'} \tag{17}$$

and $\mathbf{U}^J_{\gamma'\gamma}(R)$ is the interaction matrix between states γ', γ with the same J, M. The γ', γ now refer to the complete basis, (15). Because the Hamiltonian is independent of the absolute orientation in space, equation (16) can always be rewritten in terms of linear combinations of definite parity, reducing the coupled equations to two sets of lower dimensionality. This is discussed fully in the recent paper by Alexander and DePristo.[15] They also derive the equivalent equations in the body-fixed representation in which the "J_Z-conserving" approximation can easily be applied. The computer time involved in solving sets of equations like (16) increases as N^3, where N is the number of coupled equations. Thus the reduction in dimensionality obtained by using parity or the J, M basis is very important.

Equations of the form of (16) are, within the truncation of the basis sets, exact for inelastic scattering. Methods for their exact or approximate solution are discussed below and in detail elsewhere in this volume, as are the equivalent integral equations and methods for their solution. At this point we digress to relate the solutions at large R to the observable quantities.

3.3. *Asymptotic Forms and the* S *Matrix*

The asymptotic forms ($R \to \infty$) of the solutions to equation (16) will be dominated by the long-range behavior of the potentials in equation (17). Restricting ourselves to scattering problems in which there are no net charged particles, i.e., in the absence of Coulomb potentials, the long-range behavior is dominated by the energy and angular momentum terms, leading to Bessel function solutions. If there are N coupled equations (16), there will be N solution vectors $\mathbf{f}$ which satisfy regular boundary conditions at the origin. We must take linear combinations of these which satisfy our desired boundary conditions at $R \to \infty$, i.e., an incoming wave of unit flux and a linear combination of outgoing waves with coefficients determined by the solution of (16).

The regular solution matrix of (16) can be written (as $R \to \infty$) as

$$\mathbf{f} = -\mathbf{R}^J_4 \mathbf{f}' \tag{18}$$

where $\mathbf{f}' = (d/dR)\mathbf{f}$ and $\mathbf{R}^J_4$ is the $\mathbf{R}^J_4$ block of the $\mathbf{R}^J$ matrix.[1,25,26]* (We

* R_4 is defined in Reference 25 with the opposite sign to that used here and generally. The R-matrix propagation technique described here is taken from a paper by E. B. Stechel, R. B. Walker, and J. C. Light.[25]

shall outline below two methods by which $\mathbf{R}_4^J$ can be obtained.) Then the scattering matrix $\mathbf{S}^J$ is given by

$$\mathbf{S}^J = \mathbf{k}^{1/2}\mathbf{Q}^{-1}\mathbf{Q}^*\mathbf{k}^{-1/2} \tag{19}$$

where $\mathbf{k}$ is the diagonal wave number matrix, and $\mathbf{Q}$ is constructed from $\mathbf{R}_4^J$ and the diagonal matrix of asymptotic outgoing solutions, $\mathbf{O}$:

$$\mathbf{Q} = \mathbf{R}_4^J\mathbf{O}' - \mathbf{O} \tag{20a}$$

$$\mathbf{O}_{\gamma\gamma'} = \delta_{\gamma\gamma'} h_l^{(1)}(k_\gamma R) \tag{20b}$$

where $h_l^{(1)}$ is the Ricatti–Hankel function of the first kind. The $\mathbf{S}$ matrix defined above will be unitary if the matrix $\mathbf{R}_4^J$ is real and symmetric. This assures conservation of flux and probability as well as microscopic reversibility and is one of the most important checks on the accuracy of the calculations (see below).

The asymptotic time-independent wave function corresponding to the scattering is then given in matrix form (for the internal states) as

$$\Psi_{E,J}(\mathbf{R}) = \widetilde{\mathbf{IN}}^J - \tilde{\mathbf{O}}^J\mathbf{S}^J/R \tag{20c}$$

where now the diagonal matrices $\widetilde{\mathbf{IN}}^J$ and $\tilde{\mathbf{O}}^J$ are linear combinations of the incoming and outgoing radial function times the appropriate spherical harmonics of the polar angles of $\mathbf{R}$, $Y_l^m(\Omega)$. The linear combination is given by the transformation $\mathbf{T}$ of (15a) to the total angular momentum representation.

The observables of the scattering problem can now be constructed from the scattering matrices $\mathbf{S}^J$ for each total angular momentum J.[1] We define the (scattering) $\mathbf{T}$ matrix by

$$\mathbf{T}^J = \mathbf{I} - \mathbf{S}^J$$

and the opacity functions by

$$P_{\gamma'\gamma}^J = \frac{1}{2j+1}\sum_{l,l'} |\mathbf{T}_{\gamma',\gamma}^{J2}| \tag{21a}$$

where j and l are the initial (total) internal and orbital angular momenta, γ and γ' are the initial- and final-state labels, and l' is the final orbital angular momentum. We have assumed that the orientation of the initial internal angular momentum has not been selected, and therefore it has been averaged over. The total cross sections are obtained by further summations:

$$\sigma_{\gamma'\gamma}^J = \pi k_\gamma^{-2}(2J+1)\,P_{\gamma'\gamma}^J \tag{21b}$$

$$\sigma_{\gamma'\gamma} = \sum_{J=0}^{\infty} \sigma_{\gamma'\gamma}^J \tag{21c}$$

To obtain the angular distributions, we must reconstruct the asymptotic outgoing wave function, equation (20c), sum the contributions from all partial waves, and then take the absolute square.[1]

The above has been a brief outline of the exact quantum approach to inelastic scattering via close-coupling equations. There are, of course, severe difficulties in trying to apply them to general problems, the primary one being that there are too many quantum states involved to solve the exact equations numerically in a reasonable amount of time. It is gratifying to note, however, as detailed in other parts of this volume, that significant progress has been made in improving the algorithms used to solve such coupled equations. A number of other approaches are widely used, however, and a brief overview will be given below in Section 3.5.

3.4. Symmetry and Microscopic Reversibility

The symmetries present in the real world can be utilized in inelastic scattering calculations to reduce the complexity of the calculations and to provide constraints which any accurate calculation should satisfy.[1,11] The first of these, parity, is the result of the fact that in the absence of external fields, space is isotropic, and therefore the scattering problem can be partitioned into components symmetric and antisymmetric with respect to inversion. The utilization of this to reduce the dimensionality of the coupled equations was mentioned earlier.

The second symmetry always present in the absence of fields is time reversal invariance. Both classical and quantum equations can be reversed in time, implying a relation between "forward" and "backward" scattering events. This relation is known as microscopic reversibility and manifests itself in quantum mechanics in the unitarity of the **S** matrix. Although the unitarity of the **S** matrix can be shown from formal collision theory to follow from the Hermitian character of the Hamiltonian operator, we shall give here a brief derivation based on the finite matrix differential equation representation of the problem given above.

From equation (17) it is easily seen that for a fixed (R-independent) basis the interaction matrix $\mathbf{W}$ is real, symmetric (or at least Hermitian), and can, in some small interval i about R_i, of length h, be diagonalized by an orthogonal (or unitary) transformation:

$$\mathbf{T}_i^T \mathbf{W}_i \mathbf{T}_i = \boldsymbol{\lambda}_i^2, \qquad \mathbf{g}_i = \mathbf{T}_i^T \mathbf{f}_i \tag{22}$$

In this diagonal representation $\mathbf{g}$ the wave functions and derivatives can be related on the two sides of the interval i by diagonal matrices involving the solutions of the decoupled equations within the interval, say, trigonometric functions[25]:

$$\begin{pmatrix} \mathbf{g}_L^i \\ \mathbf{g}_R^i \end{pmatrix} = \begin{pmatrix} \mathbf{c}_i & \mathbf{s}_i \\ \mathbf{s}_i & \mathbf{c}_i \end{pmatrix} \begin{pmatrix} \mathbf{g}_L^{i'} \\ -\mathbf{g}_R^{i'} \end{pmatrix} \tag{23}$$

where

$$\mathbf{c}_{kl} = \frac{\delta_{kl}}{|\lambda_k|}\begin{cases}\cot(h|\lambda_k|), & \lambda_k^2 \le 0\\ \coth(h|\lambda_k|), & \lambda_k^2 > 0\end{cases}$$
$$\mathbf{s}_{kl} = \frac{\delta_{kl}}{|\lambda_k|}\begin{cases}1/\sin(h|\lambda_k|), & \lambda_k^2 \le 0\\ 1/\sinh(h|\lambda_k|), & \lambda_k^2 > 0\end{cases} \tag{24}$$

and $\mathbf{g}_L^i$ and $\mathbf{g}_R^i$ refer to the solution vectors on the left and right of the interval. If the solution vector in the diagonal representation in the previous $(i-1)$ interval is known in the form

$$\mathbf{g}_R^{i-1} = -\mathbf{R}_4^{i-1}\mathbf{g}_R^{i-1\prime} \tag{25}$$

then[25,27]

$$\mathbf{g}_R^i = -\mathbf{R}_4^i\mathbf{g}_R^{i\prime} \tag{26}$$

where

$$\mathbf{R}_4^i = \mathbf{c}_i - \mathbf{s}_i\mathbf{Z}_i\mathbf{s}_i \tag{27a}$$

$$\mathbf{Z}_i \equiv (\mathbf{c}_i + \mathbf{T}_i^T\mathbf{T}_{i-1}\mathbf{R}_4^{i-1}\mathbf{T}_{i-1}^T\mathbf{T}_i)^{-1} \tag{27b}$$

These relations are obtained by matching wave function and derivative vectors in the *fixed* basis, **f**, at the $i, i-1$ boundary. We shall show in a moment that the **S** matrix is unitary if $\mathbf{R}_4$ is symmetric in the asymptotic basis. It can easily be seen from the above equations that if the initial $\mathbf{R}_4$ matrix is symmetric, all subsequent ones will also be symmetric. Initially, at the origin or deep in the nonclassical region, one can take $\mathbf{R}_4^0$ to be diagonal, i.e., the logarithmic derivative matrix in a diagonal representation defining the regular solutions. Thus this type of accurate numerical solution will preserve the symmetry of the $\mathbf{R}_4$ matrix throughout the calculation to the asymptotic region where we have the final $\mathbf{R}_4$ matrix, $\mathbf{R}_4^\infty$.

That this is sufficient to guarantee the unitarity of the **S** matrix is easy to demonstrate.[1,25] Since

$$\mathbf{S} = \mathbf{k}^{1/2}\mathbf{Q}^{-1}\mathbf{Q}^*\mathbf{k}^{-1/2} \tag{28}$$

$$\mathbf{Q} = \mathbf{R}_4^\infty\mathbf{O}' - \mathbf{O} \to [i\mathbf{R}_4^\infty\mathbf{k} - \mathbf{I}]\,\mathbf{O} \qquad \text{(open channels)} \tag{29}$$

we write for open channels, realizing **O** and **k** are diagonal,

$$\begin{aligned}\mathbf{S} &= \mathbf{k}^{1/2}\mathbf{O}^{-1}(\mathbf{I} - i\mathbf{R}_4^\infty\mathbf{k})^{-1}\,\mathbf{k}^{-1/2}\mathbf{k}^{1/2}(\mathbf{I} + i\mathbf{R}_4^\infty\mathbf{k})\,\mathbf{O}^*\mathbf{k}^{-1/2}\\ &= \mathbf{O}^{-1}(\mathbf{I} - i\mathbf{K})^{-1}\,(\mathbf{I} + i\mathbf{K})\,\mathbf{O}^*\end{aligned} \tag{30}$$

where $\mathbf{K} \equiv \mathbf{k}^{1/2}\mathbf{R}_4^\infty\mathbf{k}^{1/2}$ is the **K** matrix. Therefore, since $\mathbf{O}^* = \mathbf{O}^{-1}$,

$$\mathbf{S}^\dagger\mathbf{S} = \mathbf{O}(\mathbf{I} - i\mathbf{K}^\dagger)(\mathbf{I} + i\mathbf{K}^\dagger)^{-1}\,\mathbf{O}\mathbf{O}^{-1}(\mathbf{I} - i\mathbf{K})^{-1}(\mathbf{I} + i\mathbf{K})\,\mathbf{O}^* = \mathbf{I} \tag{31}$$

and $\mathbf{S}$ is unitary. Since $\mathbf{K} = \mathbf{K}^{\dagger}$ if $\mathbf{R}_4^{\infty}$ is symmetric and real (or Hermitian), the commutation of the $(\mathbf{I} \pm i\mathbf{K})$ matrices permits the last step.

The unitarity of the S matrix guarantees flux conservation:

$$\sum_j |S_{ij}|^2 = 1 \tag{32}$$

and the fact that

$$\mathbf{S} = \mathbf{O}^{-1}(\mathbf{I} - i\mathbf{K})^{-1}(\mathbf{I} + i\mathbf{K})\mathbf{O}^* = \mathbf{O}^{-1}(\mathbf{I} + i\mathbf{K})(\mathbf{I} - i\mathbf{K})^{-1}\mathbf{O}^*$$

implies that $S_{ij} = S_{ji}\exp(i\delta_{ij})$ where δ_{ij} is a phase factor. Therefore we have the microscopic reversibility relation

$$|S_{ij}|^2 = |S_{ji}|^2 \tag{33}$$

This relation is mathematically important in that it is the result of the inherent reversibility of the dynamical equations, and physically important as it permits the immediate and exact microscopic reversibility relationship between cross sections for forward and reverse processes:

$$k_\gamma^2 \sigma_{\gamma'\gamma} = k_{\gamma'}^2 \sigma_{\gamma\gamma'} \tag{34}$$

This last relation also holds for the differential cross sections. In the presence of external fields similar relations hold if the field and orientational quantum numbers have reversed signs[11] for the forward and backward reactions.

Although some numerical methods such as the R-matrix propagation method[25] used above automatically preserve the unitarity of the $\mathbf{S}$ matrix and satisfy microscopic reversibility, such is not the case with many of the approximate methods (or even some other close-coupling techniques). In these cases the degree to which these properties are satisfied provides an important check on the reliability of the results.

3.5. *Integral Equations and Square Integrable Techniques*

There are two alternative exact approaches to the inelastic scattering problem based on the integral equations equivalent to equation (16) and the scattering boundary conditions, equation (18): The coupled integral equations approach[7,28,29] and the R-matrix approach developed originally by Wigner and Eisenbud[30] and subsequently modified.[31-35] These both have the advantage that nonlocal (or exchange) potentials can be handled with little modification of the basic approach but, except for the utilization of the R matrix in electron scattering, have not been as widely applied as the coupled differential equations approach.

If the Schrödinger equation is written as

$$(H_0 + V - E)\Psi = 0 \tag{35}$$

there are a number of corresponding integral equations depending on the

Green's function employed. In the integral equation approach one uses a zero-order Hamiltonian for which the two independent analytic solutions are known for all R and E, J (or l). These solutions are used to construct the Green's function, $(H_0 - E)^{-1}$, and the standard Lippmann–Schwinger equation

$$\Psi_{E,J}(R, \{u\}) = \Psi^0_{E,J}(R, \{u\}) - (H_0 - E)^{-1} V\Psi_{E,J} \tag{36}$$

is expanded in terms of an internal basis and is solved by noniterative means, where $\Psi^0_{E,J}$ is an appropriate solution of the unperturbed problem,

$$(H_0 - E)\Psi^0_{E,J} = 0 \tag{37}$$

The specific integral equations obtained in this way may be written in matrix form in the case where there are no nonlocal potentials, as, following Sams and Kouri,[28]

$$\mathbf{U}(R) = \mathbf{G}^1(R) + \int_0^R dR'\mathbf{G}^{21}(R|R')\mathbf{V}(R')\mathbf{U}(R') - \int_0^R dR'\mathbf{G}^{12}(R|R')\mathbf{V}(R')\mathbf{U}(R') \tag{38}$$

$$\mathbf{C} \equiv \left[\mathbf{I} - \int_0^\infty dR'\mathbf{G}^2(R')\mathbf{V}(R')\mathbf{U}(R')\right]^{-1} \times \left[\int_0^\infty dR'\mathbf{G}^2(R')\mathbf{V}(R')\mathbf{U}(R')\right] \tag{39}$$

where

$$[\mathbf{G}^1(R)]_{ij} = G^1_l(k_iR)\,\delta_{ij}$$
$$[\mathbf{G}^2(R)]_{ij} = G^2_l(k_iR)\,\delta_{ij}$$
$$[\mathbf{G}^{mn}(R|R')]_{ij} = G^m_l(k_iR_<)\,G^n_l(k_jR_>)\,\delta_{ij}, \qquad m \neq n = 1, 2$$

($R_<$, $R_>$ are the greater and lesser of R, R'), $\mathbf{V}$ is the interaction matrix in the internal basis, and G^1_l and G^2_l are the regular and irregular solutions of H_0, i.e., the lth spherical Bessel function and the corresponding Neumann function.

The $\mathbf{K}$ matrix for scattering is then given simply as $R \to \infty$ by

$$\mathbf{K} = [\mathbf{c}(R)]^{-1}[\mathbf{U}(R)(\mathbf{I} + \mathbf{C}) - \mathbf{s}(R)] \tag{40}$$

where

$$\begin{bmatrix}\mathbf{c}(R)\\ \mathbf{s}(R)\end{bmatrix}_{ij} = \begin{matrix}\cos\\ \sin\end{matrix}(k_iR - l\pi/2)\,\delta_{ij}$$

These equations may be solved noniteratively in a fashion similar to the solution of the coupled differential equations, and have been applied to a number of inelastic and reactive scattering problems.[28,29,36] As with all initial-value methods, there is often a problem maintaining the linear independence of the solution matrix $\mathbf{U}$ in the presence of closed channels. This can be resolved by several techniques.[37,38]

The square integrable (L^2) techniques take a different approach to the resolution of the Green's function of the Schrödinger equation, attempting a finite basis expansion in *all* coordinates in the scattering region, say $R \leq A$, and matching the internal wave function to the asymptotic external form at $R = A$ in order to obtain the scattering information. Extensive reviews of this method are given by Lane and Thomas,[26] and by Truhlar,[39] who has given an extensive general discussion of algebraic variational techniques. Here we shall merely outline briefly the general approach.

If we write the Schrödinger equation [equation (35)] as

$$(H_0 + V - E + L_b)\Psi_{E,J} = L_b\Psi_{E,J} \tag{41}$$

with the Bloch operator[32] L_b, defined [for each pair of internal states, $\varphi_j(\{u\})$] by

$$L_b = \frac{1}{2\mu}\delta(R - A)\left[\frac{\partial}{\partial R} - b\right]\left(\sum_j |j\rangle\langle j|\right)$$

where b is an arbitrary real number, A is the surface on which the asymptotic wave functions will be evaluated, and $|j\rangle$ is an asymptotic internal state, then the operator H_B,

$$H_B \equiv H_0 + V - E + L_b$$

is Hermitian for $R \leq A$ in any basis on $0 \leq R \leq A$. Equation (41) may be formally solved as

$$\begin{aligned}\Psi_{E,J} &= (H_0 + V - E + L_b)^{-1} L_b\Psi_{E,J} \\ &= H_B^{-1} L_b \Psi_{E,J}\end{aligned} \tag{42}$$

Now the operator H_B has a matrix representation in the direct product of a complete orthonormal basis of internal states, $|j\rangle = \varphi_j(\{u\})$ and a *discrete* complete orthonormal basis of regular (as $R \to 0$) functions of R, $f_n(R)$, on $0 \leq R \leq A$, satisfying arbitrary boundary conditions at $R = A$. Defining the (truncated) matrix representations as

$$\begin{aligned}(\mathbf{H}_B)_{nj;n'j'} &= \int_0^A dR \int d\{u\}\, f_n(R)\,\varphi_j(\{u\})\, H_B f_{n'}(R)\,\varphi_{j'}(\{u\}) \\ &= \langle nj|H_B|n'j'\rangle\end{aligned} \tag{43}$$

we can diagonalize the Hermitian matrix $\mathbf{H}_B$ by a unitary transformation $\mathbf{T}$:

$$\mathbf{T}^\dagger \mathbf{H}_B \mathbf{T} = \boldsymbol{\varepsilon} - E\mathbf{I} \tag{44}$$

where the columns of $\mathbf{T}$ are the eigenvectors and $\boldsymbol{\varepsilon} - E\mathbf{I}$ a diagonal matrix of eigenvalues.

We now use equation (42) to match the internal and external ($R \leq A$, $R \geq A$) translational wave functions for a given internal state at $R = A$. In the basis of internal $[\varphi_j(\{u\})]$ states at $R = A$, the wave function has the matrix representation

$$\boldsymbol{\Psi}_{E,J}(R = A) = \mathbf{IN}(R = A) - \mathbf{O}(R = A)\,\mathbf{S} \tag{45}$$

where $\mathbf{IN}$ and $\mathbf{O}$ are diagonal matrices of the appropriate translational functions evaluated at $R = A$. In this representation at $R = A$, equation (42) becomes

$$\mathbf{IN} - \mathbf{OS}|_{R=A} = \mathbf{R}^b[\mathbf{IN}' - \mathbf{O}'\mathbf{S} - b(\mathbf{IN} - \mathbf{OS})]|_{R=A} \tag{46}$$

where the prime denotes the derivative with respect to R evaluated at $R = A$, and the $\mathbf{R}^b$ matrix is defined in terms of the eigenvectors of $\mathbf{H}_B$:

$$[\mathbf{R}^b]_{ij} = \sum_{\lambda,n} \frac{T^*_{\lambda,ni} T_{\lambda,nj} |f_n^2(A)|}{\varepsilon_\lambda - E} \tag{47}$$

The index λ runs over all eigenvectors in the internal region, and $\mathbf{R}^b$ depends (via $\boldsymbol{\varepsilon}$ and $\mathbf{T}$) on the value of b chosen in equation (41). Since the orthogonality of the internal states, $\varphi_j(\{u\})$, is used in constructing equation (46), the parameter b may be generalized to a matrix $\mathbf{b}$ of the same dimension as the internal states, φ_j.[34] The $\mathbf{S}$ matrix is then given by

$$\mathbf{k}^{-1/2}\mathbf{S}\mathbf{k}^{1/2} = \mathbf{O}^{-1}[\mathbf{I} + \mathbf{R}^b(i\mathbf{k} + \mathbf{b})][\mathbf{I} - \mathbf{R}^b(i\mathbf{k} - \mathbf{b})]^{-1}\mathbf{O}^* \tag{48}$$

For $\mathbf{b} = 0$, we see that the R matrix defined here is equivalent formally to that used in equation (29) from the R matrix propagation method of integrating the close-coupled differential equations.[25] It is thus possible to combine close-coupling techniques and L^2 techniques for different spatial regions of the same problem.

The L^2 technique outlined above has severe problems associated with its naive use, principally very slow convergence with the size of the basis set, except at isolated energy values. A number of methods of improving the convergence rate have been proposed,[31-35] and considerable progress has been made. This technique has to date been used primarily in electron scattering where relatively few internal states are coupled and the nonlocal nature of the electronic Hamiltonian can be handled in a straightforward manner. In heavy-particle scattering the basis set size required for convergence, even with variational corrections, is large.[35] However, such methods require only a single matrix diagonalization and then little matrix manipulation at energies after the first, and thus may be well suited to intensive studies of simple systems.

4. Approximate Approaches

The basic quantities to be determined are the columns of the **S** matrix which determine the amplitudes associated with each transition in a particular collision event. The exact quantum treatments above in general give the entire **S** matrix at once—all or nothing. The approximate approaches can be aimed at reducing the difficulty in several ways: Approximating the entire set of equations (16) by a much smaller subset which contains most of the physics yet reduces the computational effort enormously [i.e., dimension-reducing approximations (DRA's)]; approximating the **S** matrix elements one at a time by solving for them from perturbation theory; and, most common, solving the dynamics by classical mechanics and identifying the classical distribution obtained with the most appropriate quantum quantities either via probabilities, cross sections, or semiclassical **S**-matrix theory. Since these topics are covered in more detail later in the book, only a brief description is given here.

4.1. Dimension-Reducing Approximations (DRA's)[15,40-45]

The DRA's have received very widespread attention in recent years, and are still under active development. The j_z-conserving or coupled-states (CS) approximation, introduced independently by McGuire and Kouri,[41] and Pack,[42] is based on the observation that in body-fixed (BF) coordinates there is kinetic energy (angular momentum) coupling but *no* potential coupling between states with different values of the projection of the internal angular momentum on the body-fixed axis. Neglect of the angular momentum coupling allows a partitioning of equation (16) into a set of equations each of greatly reduced dimensionality. This approximation is discussed in considerable detail in Chapters 7, 8, and 9.

A related DRA introduced by DePristo and Alexander,[43] is called the decoupled l-dominant approximation (DLD). In this approximation the sets of equations with a given value of $\Lambda = l + j - J$ [in the space-fixed (SF) representation] are decoupled and solved separately. In addition only values of $\Lambda \leq j$ may sometimes be retained, further reducing the number of equations to be solved. A recent comparison[15] of the exact close-coupling, CS, and DLD results for the HF–HF rotational excitation indicates that the CS results are quite accurate in this case, while the DLD may be preferred for processes dominated by large impact parameter collisions.[44]

A general "effective Hamiltonian" approach has been proposed by

Rabitz.[45] The recent review of these methods by Rabitz[40] is excellent and includes the CS and DLD approximations as well. The effective Hamiltonian approach initially proposed by Rabitz was to derive a degeneracy-averaged Hamiltonian and to omit from the coupled equations the internal orientation basis. So far it appears that the CS and DLD approximations are more work, but more accurate.

4.2. *Perturbation Theory*

The elementary perturbation theory approach (Born approximation) to determining the **S**-matrix elements suffers from a number of faults—the results are not unitary (with transition probabilities often greater than unity) and rarely accurate for the types of strongly coupled collisions we have been considering (rotational excitation, for example). However, the distorted-wave Born approximation has a better chance of success, particularly for large J where transition probabilities are small. The results can usually be improved by satisfying the unitary nature of the **S** matrix, usually by exponentiating an appropriate transition amplitude matrix obtained from perturbation theory. This approach was initially suggested by Levine[1,46] and Child,[3] but requires that an entire **S** matrix be calculated. The perturbation approach does, however, remain a viable option for certain types of processes[47] dominated by long-range collisions and rather weak transition probabilities, such as Δm transitions.

4.3. *Chemical Dynamics*[48,49]

The most widely utilized approximate approach to the dynamics of molecular collisions is to solve the classical equations of motion and to utilize the results of these calculations to determine, at least approximately, the quantities of interest. In theory this is very straightforward: one determines the Hamiltonian of the system, derives the Hamiltonian equations of motion, and integrates the set of coupled equations numerically. Since when the center-of-mass coordinates are used the number of equations is only $6N - 6$, where N is the number of particles involved (and *not* the number of quantum states), the approach is practical for systems much too large to be handled by quantum methods.

There are, however, three fundamental problems associated with this approach which limit, to some extent, its applicability. First, the coupled equations are highly nonlinear and contain multiple time scales (vibration, rotation, etc.). This necessitates rather small time steps for the numerical integration, and thus the numerical integration consumes a small but significant amount of computer time for each trajectory, say $10^{0\pm1}$ sec. Second, since each trajectory is an isolated event, a statistical ensemble of trajectories (this number of trajectories has often been on the order of

10^5 for a given system) with initial conditions chosen from the desired initial distribution must be computed. Finally, the results of the calculations must be analyzed and correlated with the physical observables of interest.

In the case where only highly averaged information (e.g., rate constants) is desired, the process works very well. Monte Carlo procedures[49] are used effectively to sample the initial distribution, and the final results of many trajectories can be obtained to yield statistically meaningful final results. In this case the only problems will lie in the use of classical dynamics itself (in which tunneling, zero-point energies, etc., play no role), and even these difficulties, if they occur, can largely be overcome by utilization of the recently developed semiclassical and imaginary time techniques[50,51] (see Chapters 10 and 16).

The more serious problems arise when more detailed (state-to-state) information is desired since in this case quantum phase interference effects may become important and the number of trajectories required to obtain statistically significant results escalates rapidly. In this case, also, the problems of proper phase averaging over the initial classical ensemble, of maintaining unitarity (or microscopic reversibility) in the approximate transition amplitudes (probabilities), and the problem of multiple electronic potential energy surfaces may become important. Semiclassical algebras, pioneered by Miller[50] and Marcus,[51] being based formally in quantum mechanics do offer a route to the resolution of these problems, but this is still a rapidly evolving field, and highly reliable techniques are not yet firmly established.[52] They nevertheless have renewed hope that an accurate alternative to the full quantum close-coupling approach may be viable.

ACKNOWLEDGMENT

This work was supported by the National Science Foundation.

References

1. R.D. Levine, *Quantum Mechanics of Molecular Rate Processes*, Oxford University Press, London, 1969.
2. B. Alder, S. Fernbach, and M. Rotenberg, editors, *Atomic and Molecular Scattering* (Volume 10 of *Methods of Computational Physics*), Academic Press, New York (1971).
3. M.S. Child, *Molecular Collision Theory*, Academic Press, New York (1974).
4. D. Secrest, Theory of rotational and vibrational energy transfer in molecules, *Ann. Rev. Phys. Chem.* **24**, 379 (1973).
5. T.F. George and J. Ross, Quantum dynamical theory of molecular collisions, *Ann. Rev. Phys. Chem.* **24**, 263 (1973).
6. W.H. Miller, editor, *Dynamics of Molecular Collision, Parts A and B* (Volumes 1 and 2 of *Modern Theoretical Chemistry*), Plenum Press, New York (1976).
7. R.G. Newton, *Scattering Theory of Waves and Particles*, McGraw-Hill, New York (1966).
8. Ch. Schlier, editor, *Molecular Beams and Reaction Kinetics*, Proceedings of the International School of Physics "Enrico Fermi," Academic Press, New York (1970).

9. R.D. Levine and R.B. Bernstein, *Molecular Reaction Dynamics*, Oxford University Press, New York (1974).
10. J. Ross, editor, *Molecular Beams*, Advances in Chemical Physics, Vol. 10, Interscience, New York (1966).
11. J. Light, J. Ross and K. Shuler, Rate coefficients, reaction cross sections, and microscopic reversibility, in *Kinetic Processes in Gases and Plasmas*, A. Hochstim, editor, Academic Press, New York (1969).
12. R.D. Levine and R.B. Bernstein, Energy disposal and energy consumption in elementary chemical reactions: The information theoretic approach, *Acc. Chem. Res.* **7**, 393 (1974).
13. M.J. Berry, Laser studies of gas phase chemical reaction dynamics, *Ann. Rev. Phys. Chem.* **26**, 259 (1975).
14. A.M. Arthurs and A. Dalgarno, The theory of scattering by a rigid rotator, *Proc. R. Soc. London* **A256**, 540 (1960).
15. M.H. Alexander and A. DePristo, Symmetry considerations in the quantum treatment of collision between two diatomic molecules, *J. Chem. Phys.* **66**, 2166 (1977).
16. D. Secrest, Amplitude densities in molecular scattering, *Methods Comput. Phys.* **10**, 243 (1971).
17. D. Secrest and W. Eastes, Calculation of rotational and vibrational transitions for the collision of an atom with a rotating vibrating diatomic oscillator, *J. Chem. Phys.* **56**, 640 (1969).
18. W.A. Lester and R.B. Bernstein, Computational procedure for the close-coupled rotational excitation problem: Scattering of diatomic molecules by atoms, *J. Chem. Phys.* **48**, 4896 (1968).
19. K. Takayanagi, Vibrational and rotational transitions in molecular collisions, *Progr. Theor. Phys. Supp.* **25**, 1 (1963).
20. A.S. Davydov, *Quantum Mechanics*, Pergamon Press, New York (1965).
21. A. Messiah, *Quantum Mechanics*, J. Wiley & Sons, New York (1962).
22. K. Takayanagi, The production of rotational and vibrational transitions in encounters between molecules, *Adv. At. Mol. Phys.* **1**, 149 (1965).
23. W.A. Lester, The *N*-coupled channel problem, in *Dynamics of Molecular Collisions, Part B* (Volume 2 of *Modern Theoretical Chemistry*), W.H. Miller, editor, Plenum Press, New York (1976), Chapter 1.
24. M.E. Rose, *Elementary Theory of Angular Momentum*, J. Wiley & Sons, New York (1957).
25. J. Light and R.B. Walker, *R*-matrix solution of coupled equations for inelastic scattering, *J. Chem. Phys.* **65**, 4272 (1976); E.B. Stechel, R.B. Walker, and J.C. Light, *J. Chem. Phys.* (1978), in press.
26. A.M. Lane and R.G. Thomas, *R*-matrix theory of nuclear reactions, *Rev. Mod. Phys.* **30**, 257 (1958).
27. D.J. Zvijac and J.C. Light, *R*-matrix theory for collinear chemical reactions, *Chem. Phys.* **12**, 237 (1976).
28. W.N. Sams and D.J. Kouri, Noniterative solutions of integral equations for scattering. II. Coupled channels, *J. Chem. Phys.* **51**, 4815 (1969).
29. R.A. White and E.F. Hayes, Quantum mechanical studies of the vibrational excitation of H_2 by Li^+, *J. Chem. Phys.* **57**, 2985 (1972).
30. E.P. Wigner and L. Eisenbud, Higher angular momenta and long range interaction in resonance reactions, *Phys. Rev.* **72**, 29 (1947).
31. P.J.A. Buttle, Solution of coupled equations by *R*-matrix techniques, *Phys. Rev.* **160**, 179 (1967).
32. C. Bloch, Une formulation unifiée de la théorie des réactions nucléaires, *Nucl. Phys.* **4**, 503 (1957).
33. P.G. Burke, A. Hibbert, and W.D. Robb, Electron scattering by complex atoms, *J. Phys. B: At. Mol. Phys.* **4**, 153 (1971).
34. E.J. Heller, *R*-matrix approach to collinear inelastic collisions: Resonances, *Chem. Phys. Lett.* **23**, 102 (1973).

35. D.J. Zvijac, E.J. Heller, and J.C. Light, Variational correction to Wigner *R*-matrix theory of scattering, *J. Phys. B: At. Molec. Phys.* **8**, 1016 (1975).
36. F.M. Chapman, Jr., and E.F. Hayes, Open and closed channel resonances in the collinear inelastic scattering of He by H_2^+, *J. Chem. Phys.* **65**, 1032 (1976); Resonances in collinear inelastic scattering of He by H_2^+ below reaction threshold, *J. Chem. Phys.* **62**, 4400 (1975).
37. See R.G. Gordon, Quantum scattering using piecewise analytic solutions, *Methods Comput. Phys.* **10**, 81 (1971).
38. R.G. Gordon, New method for constructing wavefunctions for bound states and scattering, *J. Chem. Phys.* **51**, 14 (1969).
39. D. Truhlar, Algebraic variational methods in scattering theory, in *Advances in Chemical Physics*, Vol. 25, I. Prigogine and S. Rice, editors, J. Wiley & Sons, New York (1974), pp. 211–293.
40. H. Rabitz, Effective Hamiltonians in molecular collisions, in *Dynamics of Molecular Collisions, Part A* (Volume 1 of *Modern Theoretical Chemistry*), W.H. Miller, editor, Plenum Press, New York (1976), pp. 33–80.
41. P. McGuire and D.J. Kouri, Quantum mechanical close coupling approach to molecular collisions j_z-conserving coupled states approximation, *J. Chem. Phys.* **60**, 2488 (1974).
42. R.T. Pack, Space-fixed vs body-fixed axes in atom–diatomic molecular scattering. Sudden approximations, *J. Chem. Phys.* **60**, 633 (1974).
43. A.E. DePristo and M.H. Alexander, A decoupled I-dominant approximation for ion–molecule and atom–molecule collisions, *J. Chem. Phys.* **64**, 3009 (1976).
44. A.E. DePristo and M.H. Alexander, Rotationally inelastic scattering of two HF molecules, *J. Chem. Phys.* **66**, 1334 (1977).
45. H. Rabitz, The dimensionality and choice of effective Hamiltonians for molecular collisions, *J. Chem. Phys.* **63**, 5208 (1975); Effective potentials in molecular collisions, *J. Chem. Phys.* **57**, 1718 (1972).
46. R.D. Levine, in *MTP International Review of Science, Physical Chemistry*, Series One, Vol. 1, W. Byers-Brown, editor, University Park Press, Baltimore (1972), p. 229.
47. H.K. Shin, Vibrational energy transfer, in *Dynamics of Molecular Collisions, Part A* (Volume 1 of *Modern Theoretical Chemistry*), W.H. Miller, editor, Plenum Press, New York (1976), pp. 131–210.
48. R.N. Porter and L.M. Raff, Classical trajectory methods in molecular collisions, in *Dynamics of Molecular Collisions, Part B* (Volume 2 of *Modern Theoretical Chemistry*), W.H. Miller, editor, Plenum Press, New York (1976), pp. 1–52.
49. D.L. Bunker, Classical trajectory methods, *Methods Comput. Phys.* **10**, 287 (1971).
50. W.H. Miller, Semiclassical theory of atom–diatom collisions: Path integrals and the classical *S* matrix, *J. Chem. Phys.* **53**, 1949 (1970); Classical *S* matrix for rotational excitation; quenching of quantum effects in molecular collisions, *J. Chem. Phys.* **54**, 5386 (1971).
51. R.A. Marcus, Extension of the WKB method to wavefunctions and transition probability amplitudes (*S*-matrix) for inelastic or reactive collisions, *Chem. Phys. Lett.* **7**, 525 (1970); Theory of semiclassical transition probabilities for inelastic and reactive collisions. V. Uniform approximation in multidimensional systems, *J. Chem. Phys.* **57**, 4903 (1972).
52. M.S. Child, Semiclassical methods in molecular collision theory, in *Dynamics of Molecular Collisions, Part B* (Volume 2 of *Modern Theoretical Chemistry*), W.H. Miller, editor, Plenum Press, New York (1976), pp. 176–216.

7

Inelastic Scattering Cross Sections II: Approximation Methods

DONALD JACK KOURI

1. Introduction

The theory of atom–molecule collisions has just seen a decade of dramatic advances. Probably the single factor which has stimulated the development and testing of a variety of new, accurate approximations was the development, in the 1960's, by Johnson and Secrest,(1-3) Gordon,(4) Sams and Kouri,(5-9) and Chang and Light(10) of extremely efficient numerical methods for solving large sets of coupled differential or integral equations. These methods were used to provide accurate close-coupling (CC) cross sections for some atom–diatom inelastic collision problems. These results in turn have provided reliable standards against which approximations could be tested and compared. Because the same potential surfaces could be used for both the (exact) CC and approximate calculations, differences in the results provided a quantitative measure of the accuracy of the approximation. By contrast, earlier attempts to compare approximate results *directly* with experiment had always to contend with inaccuracies in the approximation *and* in the potential surface (not to mention experimental uncertainties). The present status of available numerical methods is discussed by Secrest in Chapter 8 of this volume. It is felt that sufficient numerical tests of a number of approximations have now been carried out so that it

DONALD JACK KOURI • Departments of Chemistry and Physics, University of Houston, Houston, Texas 77004.

is possible to make recommendations on the approximations of choice for specific collision processes. In this chapter, we briefly indicate the processes which will be dealt with and give a general indication of the approximate methods which are the most promising. However, the detailed discussions of methods, validity criteria, and method selection are to be found in Chapters 8–13.

2. Rotational Excitation

The development of approximations to treat rotational degrees of freedom in atom–molecule collisions has played a central role in the theory of molecular dynamics. This is because it is the rotational degeneracy which makes a fully quantal treatment of inelastic atom–molecule collisions so difficult. Thus, in a real sense, the central question in *any* treatment of molecular dynamics is, how is the rotation of the system treated? In Chapters 8–10, the problem of pure rotational excitation is dealt with, and this discussion should be kept in mind when considering the selection of an approximation even for processes other than pure rotational transitions. The choices of methods available for rotational excitation are basically of four types. First, of course, is the full quantum-mechanical close-coupling approach, discussed in Chapter 8. Generally speaking, this approach is quite costly and should be considered only when very high accuracy is required (and when the knowledge of the intermolecular potential justifies it). Second, there are now a number of fully quantal approximations which have been developed in the last 10 years. These are considered in Chapter 9. The choice of method can be dependent on the energy of the collision, the state of the molecule involved, the range and anisotropy of the potential, and the masses of the particles. These methods also generally require the numerical solution of an effective Schrödinger equation using the methods discussed in Chapter 8, or perhaps the WKB approximation. An especially significant development has been the demonstration, by Goldflam *et al.*[11] within the infinite-order sudden approximation (IOS) and by Khare[12] within the l_z-conserving approximation, that all of the various cross sections and rates factor into sums of products of system-independent coefficients containing the internal-state information and system-dependent dynamical coefficients. The latter have been shown to be the cross sections or rates for scattering out of the $j = 0$ state of the molecule. Since the energy sudden approximation is expected to be accurate for a wide class of collision systems, this factorization has rather broad implications for theoretical calculations and for the analysis of experimental data. This is fully discussed in Chapter 9. Third are the various semiclassical approaches. Here, the choice of method can depend on how highly averaged the experimental quantity is over quantum states. These methods become most useful for high quantum states, high collision energies, and heavy-particle masses.

These are also dealt with in Chapter 9. Finally, one may elect to use a classical trajectory approach in situations where the experimental result is highly averaged. These are discussed in Chapter 10.

In selecting a method to use in analyzing an experiment, one must of course take account of the accuracy of the experimental data and of the input potential surface. If, for example, the surface is known only qualitatively and the experimental quantity has a large uncertainty, one may opt for the approximation requiring the least computational effort, provided the method is not manifestly totally inappropriate for describing the collision system. By this, we mean that the method selected should be capable of at least preserving the salient qualitative features of the scattering for the system of interest. Sufficient descriptions of the conditions under which the various approximations are valid have been given so that such method selection can be made.

3. Vibrational Excitation

As stated earlier, the development of accurate results for the rotational excitation problem has played a major role in the general development of atom–molecule collision theory. The subsequent CC treatments of simultaneous rotation–vibrational excitation, such as that for He–H_2 by Eastes and Secrest[13] and for Li^+–H_2 by Schaefer and Lester,[14] have also opened the way for the testing of various approximations and models. In addition, the ability to examine the usefulness of one-dimensional models by testing the variation of the scattering with angle of incidence has been very important. Approximation methods expected to be of particular use in rotation–vibrational excitation include the combination of the IOS for rotation and full close coupling for vibration, as discussed in Chapter 11, as well as the DECENT and INDECENT correspondence principle methods and the strong coupling correspondence principle method, both of which are discussed in Chapter 12. A variety of other approximations are also discussed and evaluated in Chapters 11 and 12. We note also that the Goldflam–Green–Kouri factorization of the IOS continues to apply to rotation–vibration excitation so that the rotational state dependence has the same form as in atom–rigid-rotor scattering. Now, however, the dynamical coefficients are the cross sections $\sigma(v_i, j = 0 \rightarrow v_f, j = L)$.[15]

4. Electronic Excitation

The approximate treatment of electronic transitions is rapidly developing and it appears that the theory is now moving beyond the collinear restriction. Recently there have become available new theoretical and CC computational studies for electronic transitions in atom–diatom collisions.[16-21] These studies are opening up this area to the adaptation of the CS and IOS approximations, and this should soon result in the ability to carry out accurate quantal calculations over a range of energies for a variety of systems. In the meantime, progress in this field has come primarily as the result of use of WKB-type approximations and classical trajectory methods combined with WKB approximations. These and other approximations are discussed in Chapter 13. It is expected that more progress in this area will be forthcoming in the near future.

ACKNOWLEDGMENTS

Supported in part by National Science Foundation Grants MPS 74-13719 and CHE 76-18970; R. A. Welch Foundation Grant E-608; and the Donors of the Petroleum Research Fund, administered by the American Chemical Society.

References

1. B.R. Johnson and D. Secrest, The solution of the nonrelativistic quantum scattering problem without exchange, *J. Math. Phys.* **7**, 2187–2195 (1966).
2. D. Secrest and B.R. Johnson, Exact quantum-mechanical calculation of a collinear collision of a particle with a harmonic oscillator, *J. Chem. Phys.* **45**, 4556–4570 (1966).
3. B.R. Johnson and D. Secrest, Quantum-mechanical calculations of the inelastic cross sections for rotational excitation of para and ortho H_2 upon collision with He, *J. Chem. Phys.* **48**, 4682–2693 (1968).
4. R.G. Gordon, New method for constructing wavefunctions for bound states and scattering, *J. Chem. Phys.* **51**, 14–25 (1969).
5. W.N. Sams and D.J. Kouri, Noniterative solutions of integral equations for scattering. I. Single channels, *J. Chem. Phys.* **51**, 4809–4814 (1969).
6. W.N. Sams and D.J. Kouri, Noniterative solutions of integral equations for scattering. II. Coupled channels, *J. Chem. Phys.* **51**, 4815–4819 (1969).
7. D. Secrest, Amplitude densities in molecular scattering, in *Methods of Computational Physics*, B. Alder, S. Fernbach, and M. Rotenberg, editors, Vol. 10, Academic Press, New York (1971), pp. 243–286.
8. R.A. White and E.F. Hayes, Quantum mechanical studies of the vibrational excitation of H_2 by Li^+, *J. Chem. Phys.* **57**, 2985–2993 (1972).
9. E.R. Smith and R.J.W. Henry, Noniterative integral-equation approach to scattering problems, *Phys. Rev. A* **7**, 1585–1590 (1973).
10. D. Chang and J.C. Light, Exponential solution of the Schrödinger equation: Potential scattering, *J. Chem. Phys.* **50**, 2517–2525 (1969).
11. R. Goldflam, S. Green, and D.J. Kouri, Infinite order sudden approximation for rotational energy transfer in gaseous mixtures, *J. Chem. Phys.* **67**, 4149–4161 (1977); R. Goldflam, D.J. Kouri, and S. Green, On the factorization and fitting of molecular scattering information, *J. Chem. Phys.* **67**, 5661–5675 (1977).
12. V. Khare, On the l_z-conserving energy sudden approximation for atom–diatom scattering, *J. Chem. Phys.* (in press).
13. W. Eastes and D. Secrest, Calculation of rotational and vibrational transitions for the collision of an atom with a rotating–vibrating diatomic oscillator, *J. Chem. Phys.* **56**, 640–649 (1972).

14. J. Schaefer and W.A. Lester, Jr., Effect of rotation on vibrational excitation of H_2 by Li^+ impact, *Chem. Phys. Lett.* **20**, 575–580 (1973).
15. S. Green, private communication.
16. S.I. Chu and A. Dalgarno, Fine structure transitions of C^+ in collisions with H_2, *J. Chem. Phys.* **62**, 4009–4015 (1975).
17. M. Baer, Adiabatic and diabatic representations for atom–diatom collisions: Treatment of the three dimensional case, *Chem. Phys.* **15**, 49–57 (1976).
18. F. Rebentrost and W.A. Lester, Jr., Nonadiabatic effects in the collision of $F(^2P)$ with $H_2(^1\Sigma_g^+)$. II. Born–Oppenheimer and angular momentum coupling in adiabatic and diabatic representations, *J. Chem. Phys.* **64**, 3879–3884 (1976).
19. F. Rebentrost and W.A. Lester, Jr., Resonant electronic-to-rotational energy transfer: Quenching of $F(^2P_{1/2})$ by H_2 ($j = 0$), *J. Chem. Phys.* **64**, 4223–4224 (1976).
20. P.L. De Vries and T.F. George, Quantum mechanical theory of a structured atom–diatom collision system: $A + BC$ ($'\Sigma$), *J. Chem. Phys.* **67**, 1293–1301 (1977).
21. D.L. Miller and R.E. Wyatt, Electronuclear basis for three-dimensional electronic nonadiabatic chemical reactions, *J. Chem. Phys.* **67**, 1302–1311 (1977).

8

Rotational Excitation I: The Quantal Treatment

DON SECREST

1. Introduction

The quantum-mechanical equations of motion for collision can be solved numerically, at least for rigid rotors, at low enough energy so that only a small number of channels are accessible. Such solutions are expensive and limited to the simplest systems. For practical solutions of collision problems one must resort to various approximations. In recent years a number of excellent approximations for rotational excitation have become available. The approximations to the fully quantal collision problem for rotational scattering are usually based on the time-independent formulation of the problem. The full quantum-mechanical equation for the collision of two rotors given in this chapter serves as a starting point for the various approximations to the quantum-mechanical treatment of rotational energy transfer. The close-coupling equations are of course the most accurate approximation to the collision problem and appear to give answers to any desired accuracy if enough terms are carried in the expansion. The close-coupling equations have the same form as that taken by almost all of the quantal approximations, and as a result the techniques used to solve the close-coupling equations are well suited to solving most systems of coupled second-order differential equations which arise in time-independent quantum scattering problems.

The methods we describe here are suitable not only to the solution of the close-coupled equations and other approximate equations arising in rotational excitation, but to those which treat vibrational excitation

DON SECREST • School of Chemical Sciences, University of Illinois, Urbana, Illinois 61801.

as well. In the study of both rotational and vibrational excitation, the presence of a long-range potential necessitates integrating the coupled system a great distance to reach the asymptotic region. A technique for overcoming this problem is discussed in Chapter 11.

There is no one method ideally suited to all scattering problems. The choice of method depends on such things as the structure of the coupling matrix, or the number of different energies at which a solution is required. Often a method is time consuming for the first energy, but much of the work expended in solving the equations for the first energy may be saved and used for other energies. Such a method would be appropriate when many energies are required, whereas some other method would be better if only a few energies are of interest. Some methods are fast for low-accuracy calculations and become time consuming at higher accuracy, while other methods which are slower at low accuracy do not have as rapid a rise in computing time as accuracy requirements increase.

This chapter is divided into two principal parts, Sections 2 and 3. In Section 2 the close-coupling equations for the collision of two molecules are given. The total angular momentum representation is discussed. The use of molecular symmetry to simplify the close-coupled equations is described and simplifications which result when a molecule is linear or one of the colliding particles is an atom are pointed out. Section 3 discusses the solution of the close-coupled equations. This section is not specific to rotational scattering, and the methods discussed here are applicable to the solution of coupled equations in general. We start with a discussion of boundary conditions which lead to scattering amplitudes in terms of the S matrix. The use of real boundary conditions in the form of the reactance matrix is discussed, and finally the solution of the coupled equations using real arithmetic is described by a representative method in common use. Methods which are applicable in a variety of different situations are discussed in Section 4.

2. *The Coupled Equations for Rotational Scattering*

The time-independent Schrödinger equation for the scattering of two rigid rotors is given by

$$(H_0 - E)\Psi = -V\psi \tag{1}$$

where

$$H = H_0 + V = H_1 + H_2 - \frac{\hbar^2}{2\mu}\nabla_R^2 + V(\mathbf{R}, \alpha_1, \beta_1, \gamma_1, \alpha_2, \beta_2, \gamma_2) \tag{2}$$

Here we have defined the Hamiltonian in the center-of-mass coordinate system with

$$\mu = \frac{m_1 m_2}{m_1 + m_2} \tag{3}$$

The H_i are the Hamiltonians for the separated systems. As a matter of convention we will take the system 1 to be the most complicated system. As we are ignoring all internal modes of the system other than rotation at present, if one of the systems is an atom, H_2 will not be present. The angular momentum eigenfunctions and eigenvalues of the two systems are assumed known, and

$$H_i \chi_{m_i \kappa_i}^{(j_i)}(\alpha_i, \beta_i, \gamma_i) = E_{j_i \kappa_i} \chi_{m_i \kappa_i}^{(j_i)}(\alpha_i, \beta_i, \gamma_i) \tag{4}$$

The energy eigenvalue depends on the angular momentum j and may or may not depend on the internal projection quantum number κ. The energy is always independent of the azimuthal quantum number m.

Equation (1) is to be solved with boundary conditions,

$$\psi \underset{R\to\infty}{\sim} e^{ik_I z} \chi_I(\Omega_1) \chi_I(\Omega_2) + \sum_f \chi_f(\Omega_1) \chi_f(\Omega_2) \frac{e^{ik_f R}}{R} f_{fI}(\theta, \phi) \tag{5}$$

Here we have used Ω_1 as argument of the rotor eigenfunction to indicate the appropriate arguments and I or f to indicate the appropriate set of initial and final quantum numbers. The wave vectors k_i have lengths defined by

$$k_i^2 = \frac{2\mu}{\hbar^2}(E - E_{j_{1_i}, \kappa_{1_i}} - E_{j_{2_i}, \kappa_{2_i}}) \tag{6}$$

The direction of the incoming wave vector gives the direction of propagation of the incoming plane wave. The differential cross section for a transition from state I to state f is given by

$$\frac{d\sigma_{fI}(\theta, \varphi)}{d\Omega} = \frac{k_f}{k_I} |f_{fI}(\theta, \varphi)|^2 \tag{7}$$

The above analysis is essentially that of Arthurs and Dalgarno.[1] Various other cross sections derivable from this one are discussed in Chapters 6 and 7. In practice the solution to equation (1) is usually found in the close-coupling formulism. This consists of expanding the wave function in some complete orthogonal set of functions of the internal coordinates, usually the separated system eigenfunctions, spherical harmonics, and unknown radial functions. Thus we write

$$\psi(\mathbf{R}, \Omega_1, \Omega_2) = \sum_i \frac{F_{iI}(R)}{R} \varphi_i(\theta, \varphi, \Omega_1, \Omega_2) \tag{8}$$

The subscript i is a compound index consisting of in general eight subscripts for two polyatomic rigid rotors to as few as four for an atom and a linear rotor. Of course if vibrational degrees of freedom are included there may be even more subscripts involved in the expansion.

The sum in equation (8) is in principle infinite, but in practice it must be truncated to a finite number of terms. One usually carries out the calculation with a comfortable number of terms and then repeats the calculation with more terms. Often as one adds more terms one finds that the result no longer changes in some figure that seems acceptable and one pronounces the result correct to that number of significant figures. A few calculations have been performed with a very large excess of terms for very simple systems and the expansion does appear to converge. In most complicated systems convergence has not been tested.

The system of equations which the radial functions F_{il} must satisfy is obtained by substituting the ψ of equation (8) into the Schrödinger equation, (1), multiplying by the complex conjugate of a φ_i and integrating over all coordinates except R. If the φ_i are eigenfunctions of the separated system Hamiltonians H_i, this leads to the close-coupled equations,

$$\left[-\frac{\partial^2}{\partial R^2}+\frac{l_i(l_i+1)}{R^2}-k_i^2\right]F_{il}(R)=-\sum_{i'}V_{ii'}(R)F_{i'l}(R) \tag{9}$$

where

$$V_{ii'}(R)=\frac{2\mu}{\hbar^2}\int\varphi_i^* V(\mathbf{R},\Omega_1,\Omega_2)\,\varphi_{i'}(\theta,\varphi,\Omega_1,\Omega_2)\,d\tau_R \tag{10}$$

Here $d\tau_R$ indicates integration over all variables except R.

The choice of the expansion set φ_i is extremely important. One could choose φ_i as simple products of the eigenfunctions of the two separated system Hamiltonians and spherical harmonics. However, since the entire Hamiltonian, equation (2), is invariant under a rotation of the coordinate system, the total angular momentum of the system is conserved. Thus if one chooses the φ_i to be eigenfunctions of the total angular momentum, the potential matrix equation (10) becomes block diagonal. This is clear since the collision cannot induce transitions among total angular momentum states. Then each block of the equation may be solved independently.

Such an expansion set may be formed easily from the individual rotation functions by coupling the angular momentum eigenfunctions using vector coupling coefficients or 3-j symbols to give,

$$\begin{aligned}\varphi^{JM}_{j_1\kappa_1 j_2\kappa_2 j_{12} l}(\theta,\varphi,\Omega_1,\Omega_2)&=\sum_{m_{12}m}\sum_{m_1m_2}\begin{pmatrix} j_1 & j_2 & j_{12}\\ m_1 & m_2 & -m_{12}\end{pmatrix}\chi^{(j_1)}_{m_1\kappa_1}(\Omega_1)\,\chi^{(j_2)}_{m_2\kappa_2}(\Omega_2)\\ &\times\begin{pmatrix} j_{12} & l & J\\ m_{12} & m & -M\end{pmatrix}Y_l^m(\theta,\varphi)\,(-1)^{-j_1+j_2-m_{12}-j_{12}+l-M}\\ &\times[(2j_{12}+1)(2J+1)]^{1/2}\end{aligned} \tag{11}$$

Here the symbol $(\because)$ is a 3-j symbol[2] and $Y_l^m(\theta, \varphi)$ are the usual spherical harmonics.[2]

If system 2 is a structureless particle, $\chi(\Omega_2)$ is a constant and $j_2 = m_2 = \kappa_2 = 0$. Thus

$$\varphi^{JM}_{j_1\kappa_1 l}(\theta, \varphi, \Omega_1) = \sum_{m_1 m} \begin{pmatrix} j_1 & l & J \\ m_1 & m & -M \end{pmatrix} \chi^{j_1}_{m_1\kappa_1}(\Omega_1) \times Y_l^m(\theta, \varphi)(-1)^{-j_1+l-M}(2J+1)^{1/2} \tag{12}$$

With φ_i chosen as in equation (11) or (12), the potential matrix in equation (9) becomes diagonal in J and independent of M. We may write the potential matrix with its indices displayed,

$$\begin{aligned} V_{ii'}(R) &\equiv V^J_{j_1\kappa_1 j_2\kappa_2 j_{12} l j_1'\kappa_1' j_2'\kappa_2' j_{12}' l'} \\ &= \sum_{n_1 n_1' n_2 n_2' n} (-1)^{j_1 - j_2 + j_{12} - l - j_1' + j_2' - j_{12}' + l'} [(2j_{12}+1)(2j'_{12}+1) \\ &\qquad \times (2l+1)(2l'+1)]^{1/2} \\ &\quad \times \begin{pmatrix} j_1 & j_2 & j_{12} \\ n_1 & n_2 & n \end{pmatrix} \begin{pmatrix} j_1' & j_2' & j_{12}' \\ n_1' & n_2' & n \end{pmatrix} \\ &\quad \times \begin{pmatrix} j_{12} & l & J \\ -n & 0 & n \end{pmatrix} \begin{pmatrix} j_{12}' & l' & J \\ -n & 0 & n \end{pmatrix} \\ &\quad \times V'_{j_1 n_1 \kappa_1 j_2 n_2 \kappa_2 j_1' n_1' \kappa_1' j_2' n_2' \kappa_2'}(R) \end{aligned} \tag{13}$$

where V' is given by

$$V'_{j_1 n_1 \kappa_1 j_2 n_2 \kappa_2 j_1' n_1' \kappa_1' j_2' n_2' \kappa_2'}(R) = \iint \chi^{(j_1)*}_{n_1\kappa_1}(\Omega_1)\chi^{(j_2)*}_{n_2\kappa_2}(\Omega_2)\, V(R, 0, 0, \Omega_1, \Omega_2) \times \chi^{(j_1')}_{n_1'\kappa_1'}(\Omega_1)\, \chi^{(j_2')}_{n_2'\kappa_2'}(\Omega_2)\, d\Omega_1\, d\Omega_2 \tag{14}$$

The integral over the rotation functions in equation (14) is done in the body-fixed coordinate system. Some of the 3-j symbols in equation (13) are a result of transforming the integral into the body-fixed frame.

Since the potential matrix is now diagonal in J, the coupled equations (9) may be solved for each value of J independently. Instead of solving the problem consisting of a large coupled system we have reduced the problem to solving a number of smaller coupled systems. Since the effort involved in solving a coupled system goes as N^3, where N is the number of coupled equations in the system, we have effected a significant saving of effort. The fact that the potential matrix and thus the radial equation, (9), are independent of M introduces an even larger saving. In the block-diagonalized potential matrix there are $2J + 1$ blocks with eigenvalue J corresponding to the $2J + 1$ values of M, $M = -J, -J + 1, \ldots, J$. These blocks are all identical and have as their solution the same radial function. Thus the equation needs to be solved for only one of these blocks. The use of eigenfunctions of the total angular momentum leads to drastic simplification of the system of equations to be solved.

The $\chi^{(j)}_{n\kappa}(\Omega)$ are in general linear combinations of rotation matrices[2] $D^{(j)}_{n\kappa}(\Omega)$. If the molecule is a linear rigid rotor, the κ quantum number will be zero and the rotation matrix degenerates into a spherical harmonic $Y^m_j(\Omega)$. For spherical top and symmetric top molecules, the simple rotation matrices are eigenfunctions of the rigid-rotor Hamiltonian. For asymmetric tops, the eigenfunctions are in general linear combinations of the rotation matrices. Even in cases where the simple rotation matrices are eigenfunctions of the molecule Hamiltonian they are not always the best functions to use for the $\chi^{(j)}_{n\kappa}(\Omega)$. If the molecule has symmetry, linear combinations of the rotation matrices which transform like the irreducible representations of the symmetry group are better suited as basis functions. For example, a molecule with tetrahedral symmetry is a spherical top, and the rotation matrices are eigenfunctions of the Hamiltonian. However, the function

$$\chi^{(3)}_{m\,-2}(\Omega) = N_{3\,-2}\left[D^{(3)}_{m\,2}(\Omega) - D^{(3)}_{m\,-2}(\Omega)\right] \tag{15}$$

transforms like the A representation of the tetrahedral group. The function

$$\chi^{(3)}_{m\,2}(\Omega) - N_{3\,-2}\left[D^{(3)}_{m\,2}(\Omega) + D^{(3)}_{m\,-2}(\Omega)\right] \tag{16}$$

transforms like one of the members to the three-dimensional T representation. Since the Hamiltonian for the entire collision problem must be invariant to the group rotations of the molecule (unless it includes nuclear spin operators which are small perturbations and are usually neglected), the interaction potential cannot couple states corresponding to different representations of the tetrahedral group. The system of coupled equations separates further into block diagonal form and each block may be independently solved. Thus the equations may be solved for all transitions involving each particular representation of the group independently. As with the total angular momentum representation the use of symmetry functions leads to simplification of the coupled system. If one of the group representations is of a higher dimension, a further simplification arises as it did with the $2J + 1$ dimensional representation of the full rotational group. For example, if the molecule has tetrahedral symmetry, the T representation is three dimensional. The potential matrix for this representation breaks up into three blocks on the diagonal, all identical. Again a solution needs to be found for only one of these blocks. It is thus of great computational value to make full use of the symmetry of the problem in solving scattering problems.

The interaction potential, whether obtained from *ab initio* calculations, semiempirically, or by pure guess, is usually expanded as a product of radial functions and angle-dependent functions. The most satisfactory expansion is in terms of rotation matrices,[3] giving

$$V(R, \theta, \varphi, \alpha_1, \beta_1, \gamma_1, \alpha_2, \beta_2, \gamma_2)$$
$$= \sum_{\substack{j_1 j_2 l \\ m_1 m_2}} V_{j_1 j_2 l}(R) \sum_{n_1 n_2 n} \begin{pmatrix} j_1 & j_2 & l \\ n_1 & n_2 & n \end{pmatrix} D^{(j_1)}_{n_1 m_1}(\alpha_1, \beta_1, \gamma_1)\, D^{(j_2)}_{n_2 m_2}(\alpha_2, \beta_2, \gamma_2)\, Y^n_l(\theta, \varphi) \tag{17}$$

If one of the molecules has symmetry, clearly the angle-dependent part of the potential will exhibit that symmetry and the linear combinations of the rotation matrices which are representations of the A representation (fully symmetric representation with character 1 for all group operations) should replace the rotation matrices in the above expression. If one of the molecules is linear, a spherical harmonic would replace the corresponding rotation matrix in equation (17). If one of the molecules had no structure (an atom or an atomic ion), the j_2 associated with it would be 0 (assuming no unpaired electron or neglect of spin interactions in the Hamiltonian) and the 3-j symbol in equation (17) would degenerate. In this case the potential is usually expanded as

$$V(R, \theta, \varphi, \alpha_1, \beta_1, \gamma_1) = \sum_{lm} V_{lm}(R) \sum_{n} D^{(l)}_{nm}(\alpha_1, \beta_1, \gamma_1)\, Y^{n*}_l(\theta, \varphi) \tag{18}$$

In equation (18) the $V(R)$ differ from those which would be obtained from equation (17) by a multiplicative constant.

The potential appearing in equation (14) is in a body-fixed coordinate system with the z axis lying along the intermolecular axis. In this coordinate system equation (17) becomes

$$V(R, 0, 0, \alpha_1', \beta_1', \gamma_1', \alpha_2', \beta_2', \gamma_2') = \sum_{m_1 m_2} V_{j_1 j_2 l}(R) \sum_{n_1 n_2} \begin{pmatrix} j_1 & j_2 & l \\ n_1 & n_2 & 0 \end{pmatrix}$$
$$\times D^{(j_1)}_{n_1 m_1}(\alpha_1', \beta_1', \gamma_1')\, D^{(j_2)}_{n_2 m_2}(\alpha_2', \beta_2', \gamma_1') \left(\frac{2l+1}{4\pi}\right)^{1/2} \tag{19}$$

Here the primed angles represent the orientation of the molecules with respect to the intermolecular axis. The angles defining the final rotation around that axis, α_1' and α_2', are not both meaningful since increasing both by the same amount corresponds to rotating the whole system around the intermolecular axis. Such a rotation has no effect on the potential. Thus the potential must depend only on the difference $\alpha_1' - \alpha_2'$. This is satisfied by equation (19). From the properties of the 3-j symbols one sees that $n_1 = -n_2$, and from the definition of the rotation matrices[2]

$$D^{(j)}_{nm}(\alpha, \beta, \gamma) = e^{in\alpha}\, d^{(j)}_{nm}(\beta)\, e^{im\gamma} \tag{20}$$

one sees that only $\alpha_1' - \alpha_2'$ contributes to the potential.

In this body-fixed coordinate system equation (18) reduces to

$$V(R, 0, 0, \alpha_1', \beta_1', \alpha_1') = \sum_{lm} V_{lm}(R)\, Y_l^m(\beta_1', \gamma_1') \tag{21}$$

In this case the dependence on the final rotation about the atom–molecule axis α_1' has no effect on the potential, again being equivalent to rotating the entire atom–molecule system around an axis. If the molecule is linear with no angular momentum around the axis, $m = 0$ and equation (21) reduces to the well-known expansion for an atom linear molecule,

$$V(R, 0, 0, \beta_1') = \sum_{l} V_l(R)\left(\frac{2l+1}{4\pi}\right)^{1/2} P_l(\cos\beta_1') \tag{22}$$

When the potential is expanded in one of these forms, equation (19), (21), or (22), the integral in equation (14) may be performed analytically. Even if the molecule has symmetry and the rotation matrices are replaced by linear combinations of rotation matrices which form representations of the symmetric representations of each molecule's symmetry group, the integrals are simple and lead to products of 3-j symbols. For two molecules, these 3-j symbols summed with those in equation (13) lead to 9-j symbols. For an atom–molecule system, the sums in equation (13) will lead to 6-j symbols.

3. *Solution of the Close-Coupling Equations*

Using the symmetry of the problem we have been able to reduce the large coupled system of equations into a number of smaller coupled systems all having the same form, equation (9), but all differing in the form of the V matrix and the values of the constants l and k_i. Though we have discussed specifically the rigid-rotor equations, many approximations such as the effective Hamiltonian methods,[4,5] of which the j_z-conserving or coupled-states method* is a special case, the vibrational approximations such as breathing sphere or the collinear approximation, and numerous others, all lead to equations of the form of equation (9). The techniques for solving these equations are all the same. Only the values of the constants and the form of the V matrix are different. The methods of approach are the same. Of course, a particular form of the V matrix may make one method better than another for that system of equations, but this is more dependent on the physical system than on the approximation which led to the coupled system. For example, there are special methods for handling long-range potentials. These may arise in rigid-rotor calculations or collinear vibrational problems or many other types of calculations. In summary, the methods

* See Chapter 9 for a discussion of the coupled-states approximation and others which lead to coupled equations of the form of equation (9).

discussed here are of a general nature and are applicable to many different situations.

Equation (9) with the transformation to the total angular momentum representation and a transformation to the symmetry functions of the molecules is a reasonably small system of coupled equations. Even when further approximations are made the equation is still of the form of equation (9). If (9) is solved with the boundary condition,

$$F_{iI}(R) \underset{R\to\infty}{\sim} \frac{1}{k_i^{1/2}}\left\{\exp\left[-i\left(k_I R - \frac{l_I\pi}{2}\right)\right]\delta_{iI} - \exp\left[i\left(k_i R - \frac{l_i\pi}{2}\right)\right]S_{iI}\right\} \tag{23}$$

the scattering amplitude $f_{fI}(\theta, \varphi)$ of equation (5) can easily be determined in terms of S_{iI} by identifying the coefficients of the eigenfunctions of the separated molecules in the expansion of the wave function and the expansion of the plane wave in the usual way.

The molecule functions χ_I in equation (5) are usually taken to be the symmetry functions of the molecule. These are the same as the $\chi_{n\kappa}^{(J)}$ which appear in equation (11) with their subscripts displayed in more detail. This is the most reasonable set of functions to use in equation (5) since the potential is incapable of causing transitions between functions of different symmetry classes. It was for this reason that these functions were selected in the first place. Identifying the χ in equation (5) with those in equation (11) one obtains for the scattering amplitude,

$$\begin{aligned} f_{f,I}(\theta,\varphi) &\equiv f_{j_1 m_1 \kappa_1 j_2 m_2 \kappa_2 j_{12} m_{12} j_{1_I}, m_{1_I}, \kappa_{1_I}, j_{2_I}, m_{2_I}, \kappa_{2_I}}(\theta,\varphi) \\ &= \sum_{\substack{l_I J l \\ j_{12_I} j_{12}}} \frac{[\pi(2l_I+1)(2j_{12_I}+1)(2j_{12}+1)]^{1/2}\, i^{l_I - l + 1}}{k_I^{1/2} k_f^{1/2}} \\ &\quad \times (2J+1)(-1)^{j_{1_I} - j_1 - j_{2_I} + j_2 + j_{12_I} - j_{12} + l - l_I + m}(\delta_{iI} - S_{iI}) \\ &\quad \times \begin{pmatrix} j_{12_I} & l_I & J \\ M & 0 & -M \end{pmatrix} \begin{pmatrix} j_{1_I} & j_{2_I} & j_{12_I} \\ m_{1_I} & m_{2_I} & -M \end{pmatrix} \\ &\quad \times \begin{pmatrix} j_{12} & l & J \\ m_{12} & m & -M \end{pmatrix} \begin{pmatrix} j_1 & j_2 & j_{12} \\ m_1 & m_2 & -m_{12} \end{pmatrix} Y_l^m(\theta,\varphi) \end{aligned} \tag{24}$$

We note that

$$m_{12} = m_1 + m_2 \tag{25}$$

and

$$M = m_{1_I} + m_{2_I} \tag{26}$$

from the properties of the 3-j symbols, and thus

$$m = m_{1_I} + m_{2_I} - m_1 - m_2 \tag{27}$$

It is then clear that the dependence of the scattering amplitude on φ goes as $e^{im\varphi}$. The absolute value of the scattering amplitude is independent of φ as it must be. In equation (24) the dependence of the S matrix on the various quantum numbers is given explicitly by

$$S_{iI} \equiv S^J_{j_1\kappa_1 j_2\kappa_2 j_{12} l, j_{1_I}\kappa_{1_I} j_{2_I}\kappa_{2_I} j_{12_I} l_I} \tag{28}$$

For atom–molecule scattering the proper formula for the scattering amplitude may be obtained from equation (24) by setting $j_2 = m_2 = j_{2_I} = m_{2_I} = 0$ and using the properties of the 3-j symbols to obtain

$$\begin{aligned} f_{f,I}(\theta,\varphi) &\equiv f_{j_1 m_1 \kappa_1, j_{1_I} m_{1_I} \kappa_{1_I}}(\theta,\varphi) \\ &= \sum_{l_I J_1} \frac{[\pi(2l_I+1)]^{1/2}}{k_I^{1/2} k_f^{1/2}} i^{l_I - l + 1}(2J+1)(-1)^{l - l_I + j_{1_I} - j_1}(\delta_{iI} - S_{iI}) \\ &\quad \times \begin{pmatrix} j_{1_I} & l_1 & J \\ m_{1_I} & 0 & M \end{pmatrix} \begin{pmatrix} j_1 & l & J \\ m_1 & m & M \end{pmatrix} Y_l^m(\theta,\varphi) \end{aligned} \tag{29}$$

The differential cross section for these m_i-dependent transitions is then given by equation (7). The so-called spin-averaged differential cross sections are obtained by summing the differential cross section in equation (7) over m_1, m_2, m_{1_I}, and m_{2_I} and dividing by $(2j_{1_I}+1)(2j_{2_I}+1)$. This is a sum over final orientations of the molecule angular momenta and an average over the initial. As shown by Arthurs and Dalgarno,[1] these sums may be performed analytically to give an apparently simpler expression in that the m_i quantum numbers no longer appear. In practice it is computationally much more efficient to compute the scattering amplitude by equation (24) or (29), as the case may be, and then summing over final states and averaging over initial states.

The scattering amplitude defined by equations (5), (24), and (29) are for molecules with their azimuthal quantum numbers representing projections along the initial approach direction in the center-of-mass coordinate system. If one is interested in state-selected differential cross sections with either the initial or final azimuthal quantum numbers state selected, or both, one in general does not prepare or measure the states with polarization in the direction of the initial approach. When state-selected cross sections are measured, a different scattering amplitude must be used in equation (7) which is related to that of equation (5) by

$$\begin{aligned} &f_{\bar f,\bar I}(\theta,\varphi,\alpha_1,\beta_1,\alpha_2,\beta_2,\alpha_{1_I},\beta_{1_I},\alpha_{2_I},\beta_{2_I}) = \\ &\sum_{\substack{m_{1_I} m_{2_I} \\ m_1\, m_2}} D^{(j_1)}_{m_1\bar m_1}(\alpha_1,\beta_1)\, D^{(j_2)}_{m_2\bar m_2}(\alpha_2,\beta_2)\, f_{fI}(\theta,\varphi)\, D^{(j_{1_I})*}_{\bar m_{1_I} m_{1_I}}(\alpha_{1_I},\beta_{1_I})\, D^{(j_{2_I})*}_{\bar m_{2_I} m_{2_I}}(\alpha_{2_I},\beta_{2_I}) \end{aligned} \tag{30}$$

Here the α_i, β_i give the orientation of the polarization z axis with respect to the center-of-mass coordinate system. The γ_i angle which gives the final orientation of the coordinate system in which the polarization was carried out has been dropped in equation (30), since it can have no effect on a measurement and clearly vanishes when the scattering amplitude of equation (30) is substituted into equation (7). If any of the four azimuthal quantum numbers are not selected but are summed or averaged over, the corresponding rotation matrix may be dropped from equation (30). It would fall out in any case when the quantum number is summed over in the cross section expression. This follows from the unitarity property of the rotation matrices,

$$\sum_n D_{mn}^{(J)}(\alpha, \beta, \gamma)\, D_{m'n}^{(J)*}(\alpha, \beta, \gamma) = \delta_{mm'} \tag{31}$$

The integral spin-averaged cross section is particularly simple:

$$\sigma_{j_1\kappa_1 j_2\kappa_2 j_{1_I}\kappa_{1_I} j_{2_I}\kappa_{2_I}} = \frac{\pi}{\kappa_I^2} \sum_{J=0}^{\infty} (2J+1) \sum_{j_{12}=|j_1-j_2|}^{j_1+j_2} \sum_{j_{12}=|j_{1_I}-j_{2_I}|} \sum_{l=|J-j_{12}|} \times \sum_{l_I=|J-j_{12_I}|} \frac{1}{(2j_{1_I}+1)(2j_{2_I}+1)} |\delta_{fI} - S_{fI}|^2 \tag{32}$$

where the meaning of the subscripts of S_{fI} is given by equation (28).

Thus the solution of equation (9) with the boundary condition of equation (23) allows us to compute the various cross sections of interest. The S matrix is complex, and a direct satisfaction of the boundary condition of equation (23) requires the use of complex arithmetic throughout the calculation. In practice one uses a real boundary condition given by

$$F_{iI}(R) \underset{R\to\infty}{\sim} \frac{1}{k_i^{1/2}}\left[\sin\left(k_I R - \frac{l_I\pi}{2}\right)\delta_{iI} + \cos\left(k_i R - \frac{l_i\pi}{2}\right) R_{iI}\right] \tag{33}$$

Here $\mathbf{R}$ is the real symmetric reactance matrix. In terms of this matrix the S matrix is given by

$$\mathbf{S} = [\mathbf{1} + i\mathbf{R}][\mathbf{1} - i\mathbf{R}]^{-1} \tag{34}$$

where $\mathbf{1}$ is a unit matrix.

Using equations (33) and (34) we may solve the coupled system of differential equations entirely in real arithmetic, and then at the end of the calculation compute the complex S matrix with a complex linear equation solver.

The essence of a practical solution of the rotational energy transfer problem is the solution of the equation (9) with the boundary condition given by equation (33). There is nothing inherent in the form of these equations to indicate that they give the solution to the rotational scattering problem. All of the rotational character of the problem is contained in the potential matrix $V_{ii'}$ of equation (9). Many other scattering problems give rise to equations of the same form as equation (9), differing from it only in the form of the potential matrix. In the following section we discuss the solution of equation (9) by various numerical techniques. While most of the methods we will discuss are of use in the solution of the rotational scattering problem, they are also applicable to many other types of scattering problem.

4. *Methods of Solution of the Coupled Scattering Equations*

Many scattering problems give rise to coupled equations generally written in the form

$$\left[-\frac{d^2}{dR^2} + \frac{l_i(l_i+1)}{R^2} - k_i^2 \right] F_{iI}(R) = -\sum_{i'} V_{ii'}(R) F_{i'I}(R) \tag{35}$$

or equivalently in matrix notation:

$$[-\mathbf{D}^2 + \mathscr{L}^2(R) - \mathbf{k}^2]\,\mathbf{F} = -\mathbf{V}\cdot\mathbf{F} \tag{36}$$

Though these equations may arise from a variety of different scattering problems, the methods of solution are to a large extent independent of the origin of the equations. There is unfortunately no one best method for solving equations of this sort. There are a number of factors that will determine which method is best suited to any particular problem. All of the methods we shall discuss will give an accurate solution to the problem, given enough computer time. Clearly, the best method is the one requiring the least computer time for a given accuracy. There are a number of factors involved in determining the speed of methods. Often methods which are fast at low precision become exceedingly slow at high precision. Some methods require a lot of computation initially but are very efficient when a large number of solutions are required for a range of energies.

I shall not attempt to discuss, or even mention, all of the methods described in the literature, but shall merely classify the various methods into broad categories and discuss at least one method in each category. The various methods in each category differ in detail, and one or another of them may be better for particular situations, but in general one method is about as good as another within each category.

There are basically two different numerical approaches in common use for solving the coupled equations. One approach is to solve the equations numerically either in their differential equation form, or the equivalent integral equation form. I shall call this approach the *approximate-solution* approach. The other approach is to approximate the potential matrix $\mathbf{V}$ in some acceptable manner and solve the coupled equations exactly. I shall refer to this as the *approximate-potential* approach.

For each of these two approaches there are two techniques for developing the solution. This leads to four categories of method which are discussed below.

The first of these techniques I shall call *solution following*. This consists of starting the solution well into the nonclassical region of the problem where the potential energy is greater than the total energy and proceeding to follow the solution step by step into the asymptotic region. This is probably the most common technique for the approximate-solution approach, and is exemplified by the DeVogelaere method developed by Lester,[6-8] the Sams and Kouri[9,10] method, the method of Choi and Tang,[11] and those of many others. This technique as applied to the approximate-potential approach is exemplified by the methods of Gordon,[12,13] Light,[14-16] and Wilson.[17]

The second technique I shall refer to as *invariant imbedding*, though all the methods I shall discuss under this heading were derived without the use of the concept of invariant imbedding. These techniques consist of solving the problem scattering from a piece of the potential. Then a connection technique is used for combining the *R* matrices for parts of the potential into *R* matrices for larger portions of the potential, until the scattering matrix for the entire potential is developed.

The approximate-solution approach with this technique was first implemented by the amplitude density[18,19] method. This method is no longer used as a numerical method except in special circumstances. The amplitude density connection formulas are still of great use, however, in connecting solutions obtained by different methods in different regions of the interaction potential. The log-derivative method of Johnson[20] is the principal method in this category still in general use.

The category defined by the approximate-potential approach and the invariant-imbedding technique has a single member, the *R*-matrix method of Light and Walker.[21] This method has been used mostly to date for reactive scattering, but is also a valuable method for inelastic scattering in certain circumstances.

The four categories are summarized in Table 1. The methods of each of these categories have properties which commend them to use in particular circumstances. The *approach* one uses is decided by the accuracy required and the number of solutions required at different energies. In general the approximate-solution approach is capable of higher accuracy at reasonable machine cost. The approximate-potential approach often gives acceptable accuracy, in many cases two or sometimes three significant figures in practical problems. The chief advantage of the approximate-potential approach is that it allows large integration step sizes. Of course high accuracy can also be obtained by taking smaller steps, but all advantage over the approximate-solution approach is lost when this is done, as the work required to improve accuracy grows much faster in the approximate-

Table 1. Methods for Solving Coupled Scattering Equations

Technique	Approach	
	Approximate solution	Approximate potential
Solution following	DeVogelaere[8] Sams and Kouri[9,10] Choi and Tang[11]	Gordon[12,13] Light[14-16] Wilson[17]
Invariant imbedding	Amplitude density[18] Log-derivative[20]	R matrix[21]

potential (AP) approach than in the approximate-solution (AS) approach. The AP approach, though it allows larger steps than the AS approach, also requires much more work per step than the AS approach for the first calculation. Thus the two approaches are of the same order of difficulty for one solution of low accuracy of a problem. The AP approach has the advantage that much of the work done for the first problem may be saved and the problem may be solved at another energy with very little effort. If a large number of solutions to the scattering problem are to be found at different energies, the AP approach is ideal. If only a few solutions are required or high accuracy is required, the AS approach can be an order of magnitude less time consuming.

The solution-following techniques are plagued by instabilities requiring measures to be taken from time to time in the progress of the solution to ensure stability. The invariant-imbedding techniques are inherently stable, though often time consuming.

4.1. The Approximate-Solution Approach in the Solution-Following Technique: The Method of Sams and Kouri

The method of Sams and Kouri[9,10] will be taken as an example of the approximate-solution approach to the solution-following technique. It differs from other methods in this category only in details and most of what we say here applies to the methods of the category in general. We have chosen to use this method as the example since it is certainly one of the best and is extremely easy to program.

By all of these methods one starts at or near the origin with a guess for the radial wave function, $F_{il}(R)$ of equation (35). The wave function is in most cases zero at the origin and usually very small in the nonclassical region (where the potential is greater than the total energy) even rather far from the origin. In practice one wants to spend as little time integrating the wave function in the nonclassical region as possible and thus starts at as large a value of R as is consistent with an acceptable answer. There appears to be no rule of thumb which determines where the integration is to start. One usually tries starting somewhere in the nonclassical region, solving the problem, and then repeating the calculation with a different starting point. Once one has found a starting point far enough into the nonclassical region the solution will be insensitive to changes in the starting point. This procedure may appear wasteful. One might feel that solving the problem several times just to find a starting place as far from the origin as possible is a waste of time. One could not go wrong starting at the origin. This is false, however! In the rotation problem we have separated the problem into many independent problems, each for a different total J. Typically one does the experimentation with $J = 0$. Once a good starting point for $J = 0$ has been found, it could serve for a number of higher J values. Thus the time saved by starting the integration as far from the origin as possible is well worth the effort. At higher J it is possible to start the integration even farther from the origin. It is sometimes profitable to determine a new starting point for, say, $J = 50$. In most problems the starting point can be determined at least approximately by solving the one-channel problem. That is to say, one truncates the V matrix to one element for purposes of performing experiments to determine the proper starting point. Once one knows where to start integration, the calculation can then be done with the full $\mathbf{V}$ matrix, perhaps checking once by moving the starting point to be sure that the location determined in the one-channel problem is good also for the many-channel problem. The one-channel problem using the ground-state matrix element is also useful for determining integration step size. It is the system ground state which has the highest kinetic energy and thus the largest wave number k_i. This in turn is the step-size-determining channel of the problem. If the step size gives good precision with the pure ground-state problem it will usually be small enough for the full problem. Thus both starting point and step size may be determined by experimentation with the one-channel problem.

One not only saves integration time by starting as far from the origin as possible, it is very often impossible to start at the origin. Usually the potential is strongly repulsive in the neighborhood of the origin. The radial wave function is a strongly decaying function as one approaches the origin. On the other hand, as one integrates away from the origin the wave functions for the various channels are strongly growing. One of these channels is more strongly growing than all of the others. The most strongly growing channel tends to dominate the solution and the solution matrix tends to become singular. This is the source of instability of the solution-following techniques and stabilization procedures must be instituted. The closer one starts to the origin the more often one must stabilize in the nonclassical region. For most potentials, there is a point in the neighborhood of the origin where the instability is so great that the currently known stabilization procedures no longer work. Thus it is impossible to start the integration at the origin for most potentials. For some particularly pathological potentials, one cannot find a point far enough from the origin that the problem can be stabilized and yet near enough to the origin to produce an acceptably accurate solution. In these cases one must use one of the invariant-imbedding techniques. This situation commonly occurs in the curve-crossing problem. It is for such problems that the log-derivative method has found great applications. These problems will be discussed in more detail later.

The method of Sams and Kouri is an integral equation method. One converts equation (35) into the equivalent integral equation,

$$F_{i_I}(R) = u_{l_I}(k_I R) - \int_0^\infty \frac{u_{l_i}(k_i R_<)\, v_{l_i}(k_i R_>)}{k_i} \sum_{i'} V_{ii'}(R)\, F_{i'I}(R) \quad (37)$$

for

$$k_i^2 = E - E_i > 0 \quad (38)$$

and

$$F_{i_I}(R) = u_{l_I}(k_i R) - \int_0^\infty \frac{u_{l_i}(k_i R_<)\, w_{l_i}^{(+)}(k_i R_>)}{|k_i|} \sum_{i'} V_{ii'}(R)\, F_{i'I}(R) \quad (39)$$

for

$$k_i^2 < 0$$

Here,

$$\begin{aligned} u_l(k_i R) &= \begin{cases} k_i R j_l(k_i R), & k_i^2 > 0 \\ |k_i|\, R j_l(k_i R)\, i^l, & k_i^2 < 0 \end{cases} \\ v_l(k_i R) &= k_i R n_l(k_i R) \\ w_l^{(+)}(k_i R) &= |k_i|\, R h_l^+(k_i R) \end{aligned} \quad (40)$$

The boundary conditions are contained in equations (37) and (39). Equation (37) is for the so-called open channels with the boundary conditions appropriate to the reactance matrix formulation, equation (33). The $j_l(R)$, $n_l(R)$, and $h_l^+(R)$ are the spherical Bessel, Neumann, and Hankel functions of Messiah.[22] The boundary conditions used in equation (39) for the closed channels are appropriate to the S-matrix formulation. In equation (39) the k_i is pure imaginary. The product $j_{l_i}(k_i R_<)\, h_{l_i}(k_i R_>)/k_i$ is real. The only reason for using reactance matrix boundary conditions in equation (37) is to achieve a real equation. Since equation (39) is real there is no need to use the reactance matrix boundary conditions, and the boundary conditions used here have superior stability properties in solving the system. The closed channels contain only a decaying exponential asymptotically, instead of a linear combination of a growing and a decaying exponential as they would with a reactance matrix boundary condition. In the final solution one is interested only in the S-matrix elements connecting the open channels. With the boundary conditions used here it is only necessary to solve equation (34) using the R-matrix elements connecting open channels. It has been shown[23] that, since the proper boundary condition was used for the closed channels, only the open-channel R matrix is needed in equation (34) to obtain the open-channel S matrix. Of course, the R matrix we obtain from the solution of equations (37) and (39) is not the standard R matrix, but is modified due to the differing boundary condition of equation (39). In addition to the added stability of this boundary condition, it has the added advantage that closed channels are not required in the solution of the complex equation (34).

Sams and Kouri[9,10] have shown how equations (37) and (39) may be converted to a Volterra integral equation which may be solved by replacing the integral by a numerical quadrature formula. This leads to the simple stepping formulas:

$$F_{iI}(R_{n+1}) = -v_{l_i}(k_i R_{n+1})\, T_{iI}(R_n) + u_{l_i}(k_i R_{n+1})\, P_{iI}(R_n) \tag{41}$$

$$G_{iI}(R_{n+1}) = \sum_{i'} V_{ii'}(R_{n+1})\, F_{i'I}(R_{n+1}) \tag{42}$$

$$T_{iI}(R_{n+1}) = T_{iI}(R_n) + u_{l_i}(k_i R_{n+1})\, G_{iI}(R_{n+1}) \frac{W_{n+1}}{k_i} \tag{43}$$

$$P_{iI}(R_{n+1}) = P_{iI}(R_n) + v_{l_i}(k_i R_{n+1})\, G_{iI}(R_{n+1}) \frac{W_{n+1}}{k_i} \tag{44}$$

In any of the solution-following techniques one could start the solution in the nonclassical region of the potential near the origin by choosing $F_{iI}(R)$ to be small there and integrate into the asymptotic region. When the solution is pursued into the asymptotic region one would find, however, that the desired boundary condition, equation (33), is not satisfied. To be able to satisfy the boundary condition it is necessary to find a complete set of solutions and, at the end of the calculation, take the proper linear combination of solutions to obtain the desired boundary conditions. Thus a complete matrix of starting conditions $F_{iI}(R)$ is chosen. A solution must be found for each value of the index I. This is easily accomplished in the Sams and Kouri method. One starts the integration well into the nonclassical region at a point we shall call R_0. At this point we take $F_{iI}(R_0)$ and $T_{iI}(R_0)$ as zero for all i and I. For a complete set of incoming waves we take $P_{iI}(R_0) = -\delta_{iI}$, a unit matrix. Then each new value of the functions is generated by successive application of equations (41)–(44). The W_n are the weights for the numerical integration formula used. The most common integration technique used is the trapezoid rule with $W_n = h/2$ for all n. Here h is the step size:

$$h = R_{n+1} - R_n \tag{45}$$

The repeated Simpson rule is also sometimes used,[23,24] though for problems with a long-range interaction potential the Simpson's rule weights sometimes lead to numerical instability. For cases where it is stable, Simpson's rule is about twice as fast as the trapezoid rule. Higher-order formulas have been used. Though they allow larger steps than the lower-order formulas they are also more work per step, and beyond Simpson's rule nothing is gained in computational speed.

As the asymptotic region is reached, the interaction potential $V_{ii'}(R)$ approaches zero. In this region T_{iI} and R_{iI} no longer change. Clearly, F_{iI} given by equation (41) does not satisfy the proper boundary condition. This can be corrected by multiplying equation (41) by the inverse of P_{iI} from the right. Then the R matrix is given by

$$\mathbf{R} = \mathbf{k}^{-1/2}\mathbf{T}\mathbf{P}^{-1}\mathbf{k}^{-1/2} \tag{46}$$

In practice there are problems with stability for most interaction potentials. The stability problem is common to the solution-following techniques whether one is using the approximate-solution approach or the approximate-potential approach. The problem is caused by the growth of the closed-channel solutions. It is important to remember that all channels are closed in the nonclassical region. It is mainly in the nonclassical region that instability develops. It is clear that in this region one of the F_{iI} will grow faster than all of the others. Since we do not know how to pick the initial wave function we choose an arbitrary linear combination of all solutions to start the problem. We sort them out at the end of the calculation, using equation (46) in the Sams and Kouri method and a similar technique in other methods. The problem comes from the fact that we are using finite arithmetic. During the solution of the problem, the F_{iI} each contains a contribution of the most strongly growing solution. Some of the weakly growing solutions may be many orders of magnitude smaller than the dominant solution and, as we are carrying only a finite number of digits in the calculation, significance is lost and the $\mathbf{P}$ matrix tends to become singular. That is to say, each column of $\mathbf{P}$ starts to look like the dominant solution. This is usually the highest energy state carried in the calculation, often a closed channel. To correct for this problem, it is only necessary to resum the solutions before they have grown too much. Various techniques for doing this have been discussed in the literature. This was an important problem when it was not well understood. Now we know that stabilization, if done properly, is not time consuming, and a rather straightforward technique is acceptable. The system of equations (41)–(44) may be stabilized by multiplying by the inverse of the $\mathbf{P}$ matrix. This is simply accomplished by computing the stabilized $\mathbf{T}$, $\tilde{\mathbf{T}}$, by

$$\tilde{\mathbf{T}} = \mathbf{T}\mathbf{P}^{-1} \tag{47}$$

and replacing $\mathbf{P}$ by the unit matrix, and $\mathbf{T}$ by $\tilde{\mathbf{T}}$. This stabilization must be done early in the nonclassical region usually after 10 or fewer integration steps, and then again many steps later. After leaving the nonclassical region, stabilization must be performed only if closed channels are retained. All channels are closed channels in the nonclassical region. In most calculations it is only necessary to stabilize four or five times during the calculation. This is not the most efficient technique of stabilizing, but since so little stabilizing needs to be done it is hardly worthwhile to use some of the more efficient methods which have been developed. The quantity $\mathbf{TP}^{-1}$ must be programmed in any case to finish the calculation as in equation (46), and this subroutine may be used for stabilizing without additional coding. In practice one does not compute $\mathbf{P}^{-1}$ and then multiply by $\mathbf{T}$ but rather solves the equation

$$\tilde{\mathbf{T}}\mathbf{P} = \mathbf{T} \tag{48}$$

for $\tilde{\mathbf{T}}$. It is useful to have an equation solver of this form, since it is desirable to solve matrix equations of the form $\mathbf{XA} = \mathbf{B}$ in scattering calculations by all of the methods we shall describe. This takes about the same time as computing $\mathbf{A}^{-1}$ and the multiplication is unnecessary. The location of the stabilization points is an experimental problem that differs for different interaction potentials. It is usually not critical exactly where the stabilization is done, but it generally must be done soon after the start of the integration. In the past people have written programs that stabilize every so many steps. When this was done it was found that it was necessary to stabilize every 10 steps or so. This was of course fallacious as only the first stabilization must be done after 10 steps. For a particular problem we have done, a stabilization was needed after 10 steps, the next 30 steps later, then 50, then after 200 steps, when no more stabilization was needed, and the asymptotic region was reached at 1500 steps. Thus only four stabilizations were needed. If we had stabilized on a regular interval we would have found it necessary to stabilize every 10 steps for 150 stabilizations and we would have found that a large percentage of the calculation time was spent stabilizing. With only four stabilizations, the time spent stabilizing is equivalent to the order of eight extra integration steps out of 1500 and is negligible.

In solving a scattering problem, one is often interested only in transitions among a certain set of states. Often transitions to states far in energy from the initial states of interest are only negligibly excited. Nonetheless it is frequently found that they must be included in the calculation to obtain accurate values for the transitions to states which are of interest. In fact, it is often necessary to carry closed channels to obtain fully converged transitions. The closed channels and the channels belonging to states which are weakly excited are necessary to describe the system properly in the perturbed region where the interaction is strong. Usually, the closed channels may be dropped after the integration has emerged from the nonclassical region of the open channels. As the interaction potential decays, more of the uninteresting open channels may be dropped. Because the time required for the calculation goes as the cube of the number of channels carried, the dropping of channels leads to considerable saving, since the full system of coupled equations needs to be treated only early in the calculation and the dimensionality of the problem may be reduced as the integration proceeds. This is particularly important when long-range potentials are involved. Once the closed channels have been dropped, stabilization is no longer necessary. The point at which each channel may be dropped is determined by trial and error.

4.2. *The Approximate-Potential Approach in the Solution-Following Technique*

The approximate potential methods are those in which the interaction potential is approximated as a piecewise simple function. The function is chosen of a form such that the Schrödinger equation may be solved analytically. In some cases the potential is represented as a step function,[16,17] or a piecewise linear[12] or piecewise quadratic function. The solution to the Schrödinger equation is found over each successive region of the potential and then combined by matching the function and the second derivative in each region to that in the previous region.

This is best illustrated by an example. Probably the simplest example is the method of Gordon.[12,13] Gordon's method is one of the most popular for solving general inelastic scattering problems, and general programs are available from QCPE.[25] Since the details of the method are readily available in the program and the literature, we shall confine ourselves to giving an outline of the method.

In implementing the approximate-potential approaches, the radial equation, (35) or (36), is written in the form

$$\mathbf{D}^2\mathbf{F} - \mathbf{W}(R)\mathbf{F} + E\mathbf{F} = 0 \tag{49}$$

where $\mathbf{W}$ includes the diagonal terms $-\mathbf{V}^2$ and $\mathscr{L}^2(R)$. Here E is the total energy of the system. A number of points R_i are selected by a technique that we shall discuss later, and the matrix $\mathbf{W}(R)$ is expanded in a power series around each of these points. In the neighborhood of the point R_i the potential matrix $\mathbf{W}$ is given by

$$\mathbf{W}(R) = \mathbf{W}(R_i) + (R - R_i)\left.\frac{d\mathbf{W}(R)}{dR}\right|_{R=R_i} + (R - R_i)^2\left.\frac{d^2\mathbf{W}(R)}{dR^2}\right|_{R=R_i} \tag{50}$$

A unitary transformation matrix $\mathbf{T}_i$ is now computed which diagonalizes $\mathbf{W}(R_i)$. When this matrix is applied to equation (49) one obtains

$$\mathbf{D}^2\mathbf{T}_i\mathbf{F} - \mathbf{T}_i\mathbf{W}(R)\,\mathbf{T}_i^\dagger\mathbf{T}_i\mathbf{F} + E\mathbf{T}_i\mathbf{F} = 0 \tag{51}$$

or

$$\mathbf{D}^2\tilde{\mathbf{F}}_i - \mathbf{U}_i(R)\,\tilde{\mathbf{F}}_i + E\tilde{\mathbf{F}}_i = 0 \tag{52}$$

where

$$\mathbf{U}_i(R) = \mathbf{T}_i\mathbf{W}(R)\mathbf{T}_i^\dagger \tag{53}$$

$\mathbf{T}_i^\dagger$ is the Hermitian conjugate of $\mathbf{T}_i$, and $\tilde{\mathbf{F}}_i$ is $\mathbf{T}_i\mathbf{F}$. Since $\mathbf{T}_i$ was chosen to degenerate, or very nearly so, the extra freedom in the choice of $\mathbf{T}_i$ may be $R = R_i$. It will in general not be diagonal for other values of R, but it should be nearly diagonal in the neighborhood of R_i.

The equations actually solved in a small region around the point R_i are

$$\mathbf{D}^2\tilde{\mathbf{F}} + [-\tilde{\mathbf{U}}(R_i) - (R - R_i)\tilde{\mathbf{U}}_i + E]\,\mathbf{F}_i = 0 \tag{54}$$

where $\tilde{\mathbf{U}}(R_i)$ and $\tilde{\mathbf{U}}_i$ are averages of the diagonal part of $\mathbf{U}(R_i)$ and $\mathbf{T}_i[d\mathbf{W}(R)/dR]|_{R=R_i}\mathbf{T}_i^\dagger$, respectively. If some eigenvalues of $\mathbf{W}(R_i)$ are degenerate, or very nearly so, the extra freedom in the choice of $\mathbf{T}_i$ may be used to force some off-diagonal values of $\mathbf{T}_i[d\mathbf{W}(R)/dR]|_{R=R_i}\mathbf{T}_i^\dagger$ to zero.

The system of equations (54) is completely decoupled as all matrices are diagonal and each equation may be solved independently. Indeed, since the potential is linear in equation (54) it may be solved analytically, the solution being a linear combination of Airy functions, or Bessel functions of the 1/3 and $-1/3$ order. This solution is correct only in the neighborhood of R_i. As one moves away from $R = R_i$ in either direction, the off-diagonal terms grow and the quadratic terms become important. An estimate of the error made in neglecting the off-diagonal terms and the quadratic terms is made using perturbation theory. In this way it is determined how far on each side of R_i the solution to equation (54) remains accurate to an acceptable tolerance. From this perturbation calculation one can determine an appropriate location for R_{i+1}. The potential must now be diagonalized at this new point. Again the solution may be found in the neighborhood of R_{i+1}, and once again the solution is a linear combination of Airy functions of the first and second kind. To get a consistent linear combination, the wave function and derivative in the neighborhood of R_{i+1} is matched to that in the neighborhood of R_i. Of course, a different matrix, $\mathbf{T}_{i+1}$, is needed to diagonalize the potential at R_{i+1} than served at R_i, so one must transform the solution in the neighborhood of R_i into the space of functions in the neighborhood of R_{i+1} before matching. This is easily accomplished by multiplying $\tilde{\mathbf{F}}_i$ by the unitary matrix,

$$\mathbf{P}_i = \mathbf{T}_{i+1}\mathbf{T}_i^\dagger \tag{55}$$

One proceeds in this way, propagating the solution from one interval to the next starting in the nonclassical region, near the origin with an arbitrary but independent set of solutions. When one arrives in the asymptotic region, the proper linear combination is formed from the complete set of solutions to match the asymptotic boundary condition, equation (33). Equation (33) is only valid when the term $l(l + 1)/R^2$ is negligible.

In the method of Sams and Kouri, the solution automatically satisfies the asymptotic boundary condition as soon as the interaction potential becomes negligible. The Gordon method is also terminated as soon as the interaction potential becomes negligible by matching the solution of that point to the asymptotic form,

$$F_i(R) \sim \frac{1}{k_{j_i}^{1/2}} \left[u_{l_i}(k_{j_i}R)\,\delta_{iI} + v_{li}(k_{ji}R)\,R_{iI} \right] \tag{56}$$

Here u_l and v_l are the Riccati–Bessel functions defined in equation (40). The asymptotic form given by equation (56) is reached as soon as the interaction potential becomes negligible, and equation (56) goes over into equation (33) as soon as $l_i(l_i + 1)/R^2$ becomes negligible with respect to $k_{j_i}^2$.

As with the method of Sams and Kouri and indeed with all solution-following techniques, the initial set of solutions, though initially linearly independent, tends to become dependent as the solution is propagated and must be stabilized by techniques similar to those used in the Sams and Kouri method. These stabilization codes are built into Gordon's program as available from QCPE.[25]

Since in each region the differential equation (54) is solved analytically, larger steps are taken than is possible using a numerical method. The largest possible step is taken in each interval, and the size of the perturbation correction is used to adjust the step size. The step is taken just so large that the corrected solution differs negligibly from the uncorrected solution. As the program is written the R and S matrices are computed both with and without the perturbation correction. The user can then easily get a feeling for the accuracy of the calculation.

In comparison with the approximate-solution methods, though larger steps are taken in Gordon's method, and other approximate-potential methods such as that of Light[16] and Wilson,[17] more work is done at each step. A matrix diagonalization must be made for each step and the solution must be transformed from the old space to the new at each step. Thus for one energy the approximate-potential methods are more time consuming than the approximate-solution methods. The approximate-potential methods become the method of choice when one wishes to solve the problem at several energies. Since the step size depends only on the accuracy that one needs to represent the potential, it is not very strongly dependent on the energy. As one goes to higher energy a slightly larger step size may be possible, but for reasonably closely spaced energies at which one wants to do the calculation one may determine the step size at the lowest energy in a series of calculations. The diagonal elements of the potential at each step are saved along with the transformation matrices $\mathbf{P}_i$ of equation (55) which are used to propagate the solution from one region to the next. Then the problem may be solved at all higher energies with no more matrix diagonalizations. The time-consuming work is performed for the first energy. The work is saved and the calculation at higher energies is then very simple. Since the perturbation correction calculation, which is time consuming, is used only to determine step size, it is not repeated at the higher energies. Thus at the higher energies only the uncorrected solutions are available. One must set the program error tolerances such that the perturbation corrections are truly negligible at the lowest energies. Since the method represents the potential by piecewise linear functions, the method is of a rather low order of accuracy. It nonetheless gives an accurate solution with an amazingly small number of steps. As one demands higher accuracy the step size becomes smaller and the accuracy does not increase rapidly as the step size is reduced. Thus if the method does not give sufficient accuracy with a modest investment of effort, one is usually advised to use an approximate-solution approach.

The approximate-potential approach works best in regions in which the potential is slowly varying. The smallest steps are used by this approach in the nonclassical region. Often one can gain speed by using an approximate-solution approach in the nonclassical region and an approximate-potential approach when the potential becomes slowly varying. Since the approximate-solution wave function is in the asymptotic space at all times, it must be transformed into the space in which the potential is diagonal to match with the first step of the approximate-potential approach. This is easily accomplished by multiplying by $\mathbf{T}_1$. The approximate-potential approach may then proceed normally into the asymptotic region.

The method of Gordon is not as easily programmed as the method of Sams and Kouri. It is advisable if one wished to use Gordon's method to obtain a copy of the program from QCPE[25] and become familiar with the program as delivered. For a straightforward application of the program, it is only necessary for the user to write a program to evaluate the potential and its first and second derivative. Sample potential programs are included with the QCPE version and the user should study this sample program and write his or her own potential programs, carefully following the programs provided. For problems with special requirements, the Gordon program is easily modified if one takes time to study the program since it is well documented in the comments, though the program write-up which accompanies the program is almost worthless.

For the pure rotation problem, the method of Light[16] is similar to the Gordon method except that the potential is approximated as a piecewise constant function with solutions given by sine and cosine rather than a piecewise linear potential with solution given by Airy functions of the first and second kind. Light used the Magnus expansion to determine step size rather than perturbation theory. The method of Light was developed for computing exchange reactions but it may also be used for inelastic scattering. Although it appears essentially the same as the method of Gordon for pure rotational transitions, there are important differences when the methods are used for vibrational scattering, and these will be discussed in Chapter 11 on vibrational scattering. Wilson has used a similar method though he determines the step size experimentally.

4.3. The Approximate-Potential Approach in the Invariant-Imbedding Technique: The R-Matrix Method

The invariant-imbedding technique was one of the earliest techniques used for solving inelastic scattering problems. Though it was developed early for elastic scattering, it had little influence on inelastic scattering, and most of the techniques for inelastic scattering were developed independent of the work on elastic scattering.[26] An exception is the log-derivative method of Johnson,[20] which was based on the invariant-imbedding techniques of Calogero.[26] This technique has the advantage of being completely stable and has led to a couple of important methods for inelastic scattering.

Since the techniques have been well discussed in the literature[18-20,23] we will give here only a heuristic account of the techniques and develop the working formulas for the methods which are in current use.

The R-matrix method is usually applied to the radial equations in the form of equation (35) or (36), though when a vibrational degree of freedom is present the R-matrix method of Light and Walker[21] takes a somewhat different form, as will be discussed in Chapter 11 on vibrational transitions. One starts by solving the problem for a part of the potential, taking the potential to be zero beyond certain limits. Then one solves the problem for another part of the potential which does not overlap with the part already solved. One then combines the two solutions to obtain the solution for the potential consisting of both parts. Proceeding in this way, concatenating parts of the potential, eventually the entire problem is solved. This can be seen clearly from a simple example. Consider just two pieces of the potential. For a portion of the radial potential, call region A the potential from the origin to the point R_a and region B from R_a to R_b. Thus if we write the Schrödinger equation for this problem,

$$\left(-\frac{\partial^2}{\partial R^2} - E_i\right)F_{iI}(R) = -\sum_j V_{ij}^{AB}(R)\,F_{jI}(R) \tag{57}$$

where

$$V_{ij}^{AB}(R) = \begin{cases} V_{ij}(R) & r < r_b \\ 0 & r > r_b \end{cases} \tag{58}$$

The potential V_{ij}^{AB} is made up of two parts V_{ij}^{A} and V_{ij}^{B} such that

$$V_{ij}^{AB} = V_{ij}^{A} + V_{ij}^{B} \tag{59}$$

and

$$V_{ij}^{A}(R) = \begin{cases} V_{ij}(R) & R < R_a \\ 0 & R > R_a \end{cases} \tag{60}$$

$$V_{ij}^{B}(R) = \begin{cases} V_{ij}(R) & R_a < R < R_b \\ 0 & R_a < R \text{ and } R > R_b \end{cases} \tag{61}$$

If we consider first the solution to the problem given by equation (57), we see that we are in the asymptotic region as soon as R is greater than R_b since the potential goes abruptly to zero at this point. For $R > R_b$, the solution is, in matrix form,

$$\mathbf{F}^{AB}(R) = k_{AB}^{-1/2}[\mathbf{u}(R) + \mathbf{v}(R)\ \mathbf{R}^{AB}] \tag{62}$$

where $\mathbf{u}(R)$ and $\mathbf{v}(R)$ are diagonal matrices consisting of the two linearly independent solutions of equation (57) in the asymptotic region. In equation (57), since we have included the $l(l+1)/R^2$ term in the potential, $\mathbf{u}(R)$ and $\mathbf{v}(R)$ can be taken to be

$$[\mathbf{u}(R)]_{ij} = \begin{cases} \delta_{ij} \sin k_i R & E_i > 0 \\ \delta_{ij} e^{k_i R} & E_i < 0 \end{cases} \tag{63}$$

$$[\mathbf{v}(R)]_{ij} = \begin{cases} \delta_{ij} \cos k_i R & E_i > 0 \\ \delta_{ij} e^{-k_i R} & E_i < 0 \end{cases} \tag{64}$$

where $k_i = E_i^{1/2}$ for $E > 0$ and $k_i = (-E_i)^{1/2}$ for $E < 0$. We note that in equation (62) both open and closed channels appear. We have chosen to represent the closed channels by decaying exponentials, and as R becomes large they will drop out of the problem naturally, as we will see later, but it is important to carry them in the intermediate stages of the problem. The choice of u and v is unimportant and we could as easily choose $\sinh(k_i R)$ and $\cosh(k_i R)$ for the closed channels, or for that matter we could choose $\mathbf{u}$ to be cosine and $\mathbf{v}$ to be sine in the open channels as long as we compute the S matrix properly from our choice. The important consideration in choosing $\mathbf{u}$ and $\mathbf{v}$ is that they be real and convenient.

Now to give an indication of the derivation of the solution to the problem we need only a physical interpretation of the problem. We shall interpret $\mathbf{R}^{AB}$ to be the magnitude of the reflected wave for a unit incoming wave. We can define an $\mathbf{R}^A$ similarly for the potential $\mathbf{V}^A$, and an $\mathbf{R}^B$ for the scattering from the potential $\mathbf{V}^B$. Thus we may consider $\mathbf{u}$ to be a unit flux moving toward the origin and $\mathbf{v}$ a unit flux moving away from the origin. This identification may be made whether we take $\mathbf{u}$ to be sines or cosines. As long as the problem is handled consistently and the S matrix is computed properly from the boundary conditions we have chosen, the rigorous derivation of the invariant-imbedding connection formulas will follow. The only requirement is that $\mathbf{u}$ and $\mathbf{v}$ be diagonal independent solutions to the problem in which the interaction potential $\mathbf{V}$ is zero.

In Figure 1 we have plotted a typical matrix element of $\mathbf{V}^A$ and $\mathbf{V}^B$ as a function of R. We are only interested in $\mathbf{R}^{AB}$, the flux reflected from the total potential. This will be composed of flux which was reflected from V^B and also flux which traversed V^B and was reflected from V^A and again traversed V^B. Thus we need more information about V^B. We need to know

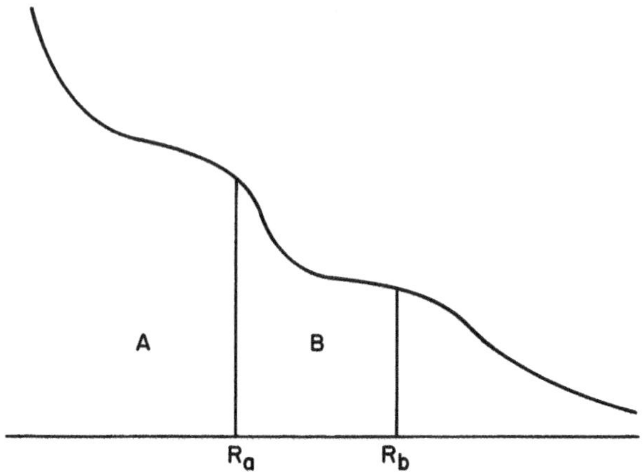

Figure 1. A typical matrix element $\mathbf{V}^A$ and $\mathbf{V}^B$ as a function of R.

the form of the wave function on the left or origin side of V^B. This is given by

$$\mathbf{F}^B(R) = \mathbf{k}^{-1/2}\mathbf{u}(R)\,\mathbf{T}^B \qquad R < R_a \tag{65}$$

Since flux from region A is incident upon region B from the left we need also a solution for the potential V^B with boundary conditions appropriate to an incident wave originating from the direction of the origin. This problem has the boundary conditions

$$\tilde{\mathbf{F}}^B(R) = \begin{cases} \mathbf{k}^{1/2}\cdot[\mathbf{v}(R) + \mathbf{u}(R)\cdot\tilde{\mathbf{R}}^B] & R < R_a \quad (66) \\ \mathbf{k}^{1/2}\cdot\mathbf{v}(R)\cdot\tilde{\mathbf{T}}^B & R > R_b \quad (67) \end{cases}$$

With these quantities defined, we see that when a unit flux is incident on region B from the right, the flux incident on A from the direction of B is

$$\mathbf{b} = \mathbf{T}^B + \tilde{\mathbf{R}}^B\cdot\mathbf{a} \tag{68}$$

Here $\mathbf{T}^B$ is the flux transmitted through V^B, the flux $\mathbf{a}$ is that arriving at B from A, $\mathbf{R}^B$ is the reflection coefficient for a unit flux from A, and $\tilde{\mathbf{R}}^B\cdot\mathbf{a}$ is the actual flux reflected back toward region A from B. The flux $\mathbf{a}$ is simply given by

$$\mathbf{a} = \mathbf{R}^A\cdot\mathbf{b} \tag{69}$$

When a unit flux is incident on V^{AB}, the flux reflected is $\mathbf{R}^{AB}$ which, in terms of the above quantities, is given by

$$\mathbf{R}^{AB} = \tilde{\mathbf{T}}^B\cdot\mathbf{a} + \mathbf{R}^B \tag{70}$$

Eliminating $\mathbf{a}$ and $\mathbf{b}$ from these matrix equations gives

$$\mathbf{R}^{AB} = \mathbf{R}^B + \tilde{\mathbf{T}}^B\cdot\,\mathbf{R}^A\cdot(\mathbf{1} - \tilde{\mathbf{R}}^B\cdot\,\mathbf{R}^A)^{-1}\cdot\mathbf{T}^B \tag{71}$$

This is of course only a heuristic derivation, and equation (71) can be rigorously derived.[23,27,28]

Using equation (71) the R matrix for the total potential may be accumulated by adding increments of the potential. This was the basis of the original amplitude density method.[18] In that method the B regions were chosen extremely small such that $\mathbf{R}^B$, $\mathbf{T}^B$, $\tilde{\mathbf{R}}^B$, and $\tilde{\mathbf{T}}^B$ could be computed in a single numerical integration step. The propagation given by equation (71) is stable and never needs to be stabilized. The method is time consuming, however, as a matrix equation must be solved at each step. The amplitude density method in its original form has fallen into disuse in recent years since faster solution-following techniques have displaced it in most cases. Nonetheless, there are situations which are so unstable that solution-following techniques will not work with any amount of stabilization. These cases may be treated by the log-derivative method,[20] which we shall discuss later.

The chief difficulties with the invariant-imbedding technique are the amount of work required at each step and the large number of steps required for the original amplitude density method or the log-derivative method. Recently the latter difficulty, the small step size, was overcome by Light and Walker[21] with their R-matrix method. They accomplished this by computing the four matrices $\mathbf{R}^B$, $\mathbf{T}^B$, $\tilde{\mathbf{R}}^B$, and $\tilde{\mathbf{T}}^B$ by the approximate potential approach. Thus they were able to take advantage of the extreme stability of the invariant-imbedding technique while retaining the large step size, or small number of steps needed in the approximate-potential approach. The work per step in this technique is comparable to that in all of the methods using the approximate-potential approach. Though the R-matrix method is new, it is being rapidly adopted by workers in the field and promises to become an important method. It has the advantage of allowing calculations at a number of energies, using much of the work done at the first energy for the higher energies as described in Gordon's method.

The method was originally derived by Light and Walker for chemical reaction calculations. For inelastic scattering, the method is a good bit simpler than that required for a reaction calculation. For pure rotational transitions, the method is extremely simple and transparent.

Suppose an R matrix $\mathbf{R}^{A_i}$ is known for a portion of the potential around the origin. (Starting the solution or finding the first $\mathbf{R}^{A_0}$ matrix will be discussed later.) The next segment of the potential $\mathbf{V}^{B_i}$ is chosen. Its size is selected by a method we will discuss later. At some point R_{i+1} near the center of region B, the potential is diagonalized by a unitary transformation T_{i+1}, just as in Gordon's method. (In the Light and Walker technique for reactive or vibrational transitions, to be discussed in Chapter 11, this step is different from that used in Gordon's method since a different expansion set is used for the potential at each step. For pure rotational transitions there is no advantage in changing the expansion set and a simple unitary transformation is used to diagonalize the potential.)

Instead of replacing the potential by a piecewise linear function as in the Gordon method, it is replaced by a piecewise constant potential. Thus after diagonalizing, the potential in region B is a constant diagonal potential. With this potential there is no scattering in region B the basis in which $\mathbf{V}^{B_{i+1}}$ is diagonal and all scattering takes place at the boundary between regions A and B.

It is convenient to choose the origin of the coordinate system at the boundary between A and B. Then inside region B as well as to the right of B, the radial wave function given by equation (62) becomes

$$\mathbf{F}^{B_i}(R) = \mathbf{k}_{B_i}^{-1/2} \cdot [\mathbf{sin}\,\mathbf{k}_{B_i}(R) + \mathbf{cos}\,\mathbf{k}_{B_i} \cdot \mathbf{R}^{B_i}] \tag{72}$$

Here $\mathbf{sin}\,\mathbf{k}_{B_i}(R)$ and $\mathbf{cos}\,\mathbf{k}_{B_i}(R)$ are just the diagonal matrices $\mathbf{u}(R)$ and $\mathbf{v}(R)$ written in a form which displays their character. For the present we will use for the closed channels hyperbolic sines and cosines, as do Light and Walker. Though the exponentials of equations (63) and (64) have useful properties in practice, the connection with the Light and Walker approach to the derivation of the R-matrix method is clearer with the use of hyperbolic functions.

When R is in the A region,

$$\mathbf{F}^{B_i}(R) = \mathbf{P}_{AB_i}\mathbf{k}_A^{-1/2} \cdot \mathbf{sin}\,\mathbf{k}_A(R) \cdot \mathbf{T}^{B_i} \tag{73}$$

Here $\mathbf{P}_{AB_i}$ is the transformation from the space of region A to the space of region B given by equation (55); k_{A_i} is the $(E_j^{A_i})^{1/2}$ matrix diagonal in region A; $\mathbf{T}^{B_i}$ and $\mathbf{R}^{B_i}$ may be found easily by matching the wave function and derivative on the boundary. With the origin of the coordinate system at the boundary, this gives

$$\mathbf{R}^{B_i} = \mathbf{0} \tag{74}$$

$$\mathbf{P}_{AB_i} \cdot \mathbf{k}_{A_i}^{1/2} \cdot \mathbf{T}^{B_i} = \mathbf{k}_{B_i}^{1/2} \tag{75}$$

Similarly, from $\tilde{\mathbf{F}}^{B_i}$ we obtain

$$\mathbf{k}_{B_i}^{1/2} \cdot \tilde{\mathbf{T}}^{B_i} = \mathbf{P}_{AB_i} \cdot \mathbf{k}_{A_i}^{1/2} \tag{76}$$

$$\tilde{\mathbf{R}}^{B_i} = \mathbf{0} \tag{77}$$

That $\mathbf{R}^{B_i}$ and $\tilde{\mathbf{R}}^{B_i}$ are zero is an artifact of our choice of origin. From equation (71) we have for $\mathbf{R}^{AB}$,

$$\mathbf{R}^{AB_i^i} = \mathbf{k}_{B_i}^{1/2} \cdot \mathbf{P}_{AB_i} \cdot \mathbf{k}_A^{-1/2} \cdot \mathbf{R}^{A_i} \cdot \mathbf{k}_{A_i}^{-1/2} \cdot \mathbf{P}_{AB_i}^{-1} \cdot \mathbf{k}_{B_i}^{1/2} \tag{78}$$

For the rotational transitions we are considering here, $\mathbf{P}_{AB_i}$ is unitary and $\mathbf{P}_{AB_i}^{-1}$ is just $\mathbf{P}_{AB_i}^{T}$. Thus it is clear that if $\mathbf{R}^{A_i}$ is symmetric, then $\mathbf{R}^{AB_i^i}$ is symmetric.

In computing $\mathbf{T}^{B_i}$ and $\tilde{\mathbf{T}}^{B_i}$ we chose the origin at the boundary between region A and region B. Therefore $\mathbf{R}^{A_i^i}$ in equation (78) must be defined with its origin on the AB boundary. If $\mathbf{R}^{AB_i^i}$ is to be used to form a new R matrix by adding still another region, the origin must be translated to the right-most boundary of region B. If we call R' the coordinate system with origin at the left-most boundary of B, R the coordinate system with the origin at the right-most boundary of B, and h the distance across region B, then we have $R' = R - h$ and

$$\begin{aligned} \mathbf{F}^{AB_i'}(R') &= \mathbf{k}_B^{-1/2} \cdot [\sin \mathbf{k}_B(R') + \cos \mathbf{k}_B(R') \cdot \mathbf{R}^{AB'}] \\ &= \mathbf{k}_{B'}^{-1/2}[\sin \mathbf{k}_B(R - h) + \cos \mathbf{k}_B(R - h) \cdot \mathbf{R}^{AB'}] \\ &= \mathbf{k}_{B_i}^{-1/2}[\sin \mathbf{k}_{B_i}(R)(\cos \mathbf{k}_{B_i}h + \sin \mathbf{k}_{B_i}h \cdot \mathbf{R}^{AB_i'}) \\ &\quad - \cos \mathbf{k}_{B_i}R(\sin \mathbf{k}_{B_i}h - \cos \mathbf{k}_{B_i}h \cdot \mathbf{R}^{AB_i'})] \end{aligned} \tag{79}$$

We define

$$\begin{aligned} \mathbf{F}^{AB_{i+1}}(R) &= \mathbf{F}^{AB_i'}(R')(\cos \mathbf{k}_{B_i}h + \sin \mathbf{k}_{B_i}h \cdot \mathbf{R}^{AB_i'})^{-1} \\ &= \mathbf{k}_{B_i}^{-1/2}[\sin \mathbf{k}_{B_i}(R) - \cos_{B_i}R(\sin \mathbf{k}_{B_i}h - \cos \mathbf{k}_{B_i}h \cdot \mathbf{R}^{AB_i'}) \\ &\quad \times (\cos \mathbf{k}_{B_i}h + \sin \mathbf{k}_{B_i}h \cdot \mathbf{R}^{AB_i'})^{-1}] \end{aligned} \tag{80}$$

or

$$\mathbf{R}^{A_{i+1}} = -(\sin \mathbf{k}_{B_i}h - \cos \mathbf{k}_{B_i}h \cdot \mathbf{R}^{AB_i'})(\cos \mathbf{k}_{B_i}h + \sin \mathbf{k}_{B_i}h \cdot \mathbf{R}^{AB_i'})^{-1} \tag{81}$$

Thus $\mathbf{R}^{A_{i+1}}$ is defined at the right-most boundary of AB and ready to be used as a new $\mathbf{R}^A$ in equation (78) for the next step. In this manner the R matrix is propagated from within the interaction potential into the asymptotic region.

The problem is started in the nonclassical region of the interaction potential. For the initial $\mathbf{R}^A$ matrix one may take either the zero matrix or a unit matrix. If the problem is started sufficiently far into the nonclassical region, it will not matter which starting value is used. When the R matrix is propagated far enough into the classical region that the potential is given by the diagonal $l(l + 1)/R^2$ term, it may be propagated for a last step with $\mathbf{u}$ and $\mathbf{v}$ given by the Riccati–Bessel functions, giving the full asymptotic R matrix. As we had noted before $\mathbf{R}^{AB'}$ is symmetric if $\mathbf{R}^A$ were symmetric. From equation (81) one can show, with a little matrix algebra, that if $\mathbf{R}^{AB'}$ is symmetric, then $\mathbf{R}^{AB}$ is symmetric. If the original choice of R in the nonclassical region is symmetric, the symmetry will be preserved and the final R matrix is symmetric as it should be.

Though equations (78) and (81) are sufficient to propagate the R matrix one step, they are not efficient equations to use numerically. They may be combined and rearranged to give

$$\begin{aligned} &[\mathbf{k}_B^{-1/2} \cdot \mathbf{R}^{AB} \cdot \mathbf{k}_B^{-1/2} + \mathbf{k}_B^{-1} \cos \mathbf{k}_B h (\sin \mathbf{k}_B h)^{-1}] \cdot [\cos \mathbf{k}_B h \sin \mathbf{k}_B h \mathbf{k}_B \\ &\quad - (\mathbf{k}_B \sin \mathbf{k}_B h \mathbf{P}_{AB})(\mathbf{k}_A^{-1/2} \mathbf{R}^A \mathbf{k}_A^{-1/2})(\mathbf{k}_B \sin \mathbf{k}_B h \mathbf{P}_{AB})^T] = \mathbf{1} \end{aligned} \tag{82}$$

Here we have used the fact that $\mathbf{P}_{AB}^{-1} = \mathbf{P}_{AB}^T$. Thus numerically we compute

$$\mathbf{X} = \mathbf{k}_B \cdot \sin \mathbf{k}_B h \cdot \mathbf{P}_{AB} \tag{83}$$

We define

$$\mathbf{Q}_i = \mathbf{k}_A^{-1/2} \cdot \mathbf{R}^A \cdot \mathbf{k}_A^{-1/2} \tag{84}$$

and propagate $\mathbf{Q}$ instead of $\mathbf{R}$ and compute $\mathbf{R}$ at the end of the calculation. Then from equation (82) we have

$$\mathbf{Y} \cdot [\mathbf{k}_B \cdot \mathbf{cos}\, \mathbf{k}_B h \cdot \mathbf{sin}\, \mathbf{k}_B h + \mathbf{X} \cdot \mathbf{Q}_i \cdot \mathbf{X}^T] = \mathbf{1} \tag{85}$$

We first form $\mathbf{X}$, which is a diagonal matrix $\mathbf{k}_B \cdot \mathbf{sin}\, \mathbf{k}_B h$ multiplied by a full matrix. Then multiply $\mathbf{Q}_i$ by $\mathbf{X}$ and $\mathbf{X}^T$, two full-matrix multiples. With the addition of the diagonal matrix $\mathbf{k}_B \cdot \mathbf{cos}\, \mathbf{k}_B h \cdot \mathbf{sin}\, \mathbf{k}_B h$ we are ready to solve equation (85) for $\mathbf{Y}$. Then $\mathbf{Q}_{i+1}$ is given by

$$\mathbf{Q}_{i+1} = \mathbf{Y} - \mathbf{k}_B^{-1} \cdot \mathbf{cos}\, \mathbf{k}h \cdot (\mathbf{sin}\, \mathbf{k}h)^{-1} \tag{86}$$

As with all methods using the approximate-potential approach, the major work is done in solving the problem for the first energy. It may then be solved at a number of higher energies easily. The matrices $\mathbf{P}_{AB_i}$, equation (55), are computed from the diagonalizing transformations and are saved for use in calculations at higher energies. The only other quantities which must be determined on the first energy calculation are, as with Gordon's method, the step sizes for each potential region. For Gordon's method, a perturbation technique was used to determine an acceptable step size. Light and Walker[21] use the Magnus method to determine an appropriate step size.[16] They have arrived at the empirical formula

$$h = \left\{ \alpha \sum_i |k_i^{2'}| \Big/ \sum_i [V_{i'i}(R_B)]^2 \right\}^{1/4} \tag{87}$$

Here $V'_{ii}(R_B)$ is the first derivative of the diagonal element of $V_{ij}(R)$ evaluated at the point R_B at which the potential is diagonalized. The value of α which gives the required accuracy for a particular problem must be determined experimentally. Light[16] gives as a rule of thumb that α values from 1 to 0.03 give an accuracy of three to four decimal places. Accuracy with α of course depends on the problem. (For problems to which I have applied this method, I could not obtain such great accuracy with these values of α.) One must experiment to determine the proper step size. It is important to determine the step size for both the present method and Gordon's method at the lowest energy of interest. Though equation (87) is only weakly dependent on energy one does obtain a larger value of h at higher energy. Using the step size determined at low energy one is certain that the accuracy will not deteriorate at higher energies. When one is making computations over a very large energy range, it may be wise to reevaluate the step size for the higher energies, as a larger step size may be permissible and the concomitant reduction in calculation time may be worth the time required to recompute the diagonalizations and connecting transformations $\mathbf{P}_{AB_i}$.

The extreme stability of the invariant-imbedding technique is required in a number of situations. There are potentials which have a structure such that the solution-following techniques become unstable and no amount of stabilizing will correct. These situations are easily handled by the amplitude density method[23] or the log-derivative method.[20] These methods, however, are slow and time consuming. The R-matrix method may be capable of handling such situations. It is not known as yet whether the R-matrix method will work in such cases as it is new and has not to my knowledge been tried. The R-matrix method has achieved a great speed advantage over the approximate-solution approaches in allowing much larger step size with only approximately twice as much effort per step. Furthermore, the extra effort is only required at the first energy. In practice it may be found that rather small steps will be required even by the R-matrix method for the potentials which lead to unstable situations. For this reason, we describe in the following section some of the potential matrix structures which lead to instability. The log-derivative method will be used as an example of the invariant-imbedding technique using the approximate-solution approach.

4.4. *The Approximate-Solution Approach in the Invariant-Imbedding Technique: The Log-Derivative Method*

The instability which arises in the solution-following techniques is most troublesome in the nonclassical region of the potential. For the rotation problem and many other problems, the nonclassical region is rather well defined, and most elements of the interaction potential matrix become nonclassical in a small range of the radial coordinate R. In most cases the wave function decays rapidly to zero just inside the nonclassical region and we may start integration close to the boundary or "turning point" as it is sometimes called. As we integrate to larger R, the wave function grows exponentially and instabilities develop. Fortunately we are usually able to stabilize and keep the instability under control until we have integrated into the classical region. It happens in some problems that one of the diagonal elements of the potential matrix will have its turning point at a much larger value of R than some other diagonal element of the potential. If we start the integration at a value of R which is small enough that all diagonal elements are in the nonclassical region, then the potential element with the large turning point will give rise to a growing exponential which will grow during a large portion of the integration. A case in which such a situation commonly occurs is the curve-crossing problem. Plotted in Figure 2 are two typical diagonal potential elements as a function of R. The horizontal line labeled E indicates the total energy for a particular calculation. The kinetic energy for the system in state i is given by the distance from V_{ii} to E. At the point where E crosses V_{ii} the kinetic energy in state i is zero. This is the classical turning point for state i. The classical turning point for state j occurs at a much smaller value of R deep into the nonclassical region of state i. Such behavior is not restricted to curve-crossing problems. In Figure 3 typical curves are plotted for an inelastic scattering problem.

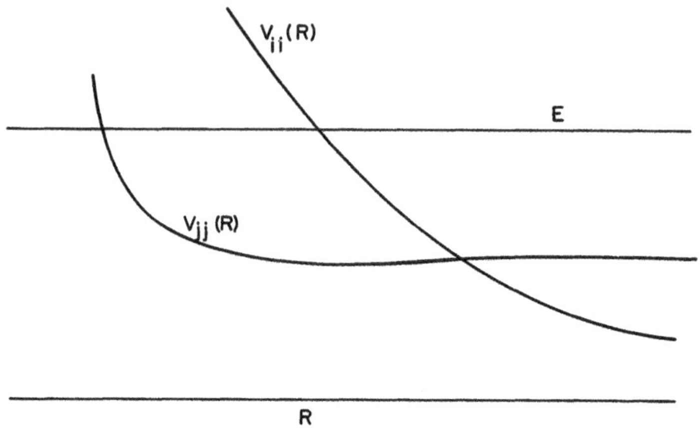

Figure 2. Two typical diagonal potential elements as a function of R. The line E is the total energy.

Here V_{ii} has three turning points, while V_{jj} has only one. During the integration of a problem in which such a configuration of potentials occurs, the kinetic energy in state i will become negative as we pass through the centrifugal hump in V_{ii}. In this region the radial wave functions will contain the growing exponential due to state i, and the solution-following techniques may exhibit extreme instability. The log-derivative method of Johnson is commonly used for the curve-crossing problem and has proven useful in the case of three turning points.[29]

The log-derivative method is an invariant-imbedding technique and may be derived in a manner similar to that used to derive the R-matrix

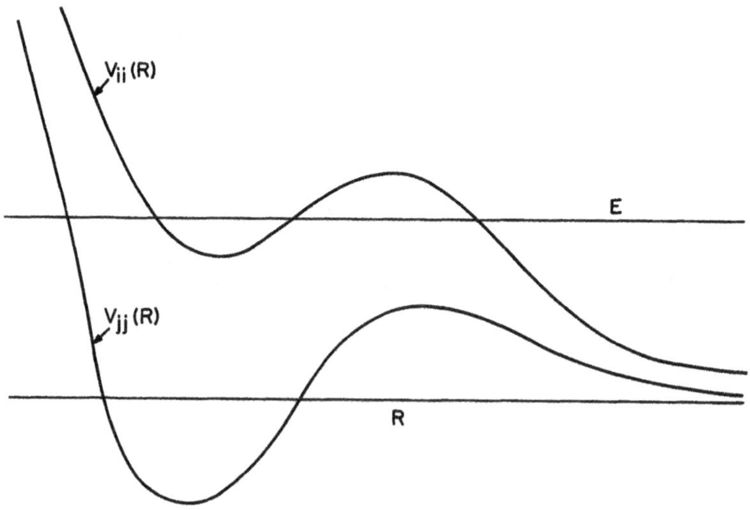

Figure 3. Two typical diagonal potential elements for an inelastic scattering problem. The line E is the total energy.

method. The method solves the equation in the form

$$\left[\frac{\partial^2}{\partial \mathbf{R}^2} + \mathbf{V}(R)\right] \cdot \mathbf{F} = 0 \tag{88}$$

Here we have included the $l(l + 1)/R^2$ term and the energy term $k_i^2 = E - E_i$, where E_i is the internal energy of the molecule [$j(j + 1)/I$ in the rotor case] in the diagonal elements of the $\mathbf{V}(R)$ matrix. Instead of propagating $\mathbf{F}$, the log-derivative method propagates the derivative of the natural logarithm of $\mathbf{F}(R)$,

$$\mathbf{Y}(R) = \frac{d}{dR} \ln \mathbf{F}(R) = \mathbf{F}'(R) \cdot \mathbf{F}^{-1}(R) \tag{89}$$

Defining $\mathbf{Z}_n = h\mathbf{Y}(R_n)$ and a step size $h = R_{n+1} - R_n$, then

$$\mathbf{Z}_n = (\mathbf{1} + \mathbf{Z}_{n-1})^{-1} \cdot \mathbf{Z}_{n-1} - \frac{h^2}{3} W_n \mathbf{U}_n \tag{90}$$

where

$$\mathbf{U}_n = \begin{cases} \mathbf{V}(R_n) & n \text{ even} \\ \left[\mathbf{1} - \frac{h^2}{6}\mathbf{V}(R_n)\right]^{-1} \mathbf{V}(R_n) & n \text{ odd} \end{cases} \tag{91}$$

and

$$W_n = \begin{cases} 1 & n = 0, N \\ 4 & n = 1, 3, 5, \ldots, N - 1 \\ 2 & n = 2, 4, 6, \ldots, N - 2 \end{cases} \tag{92}$$

The integration used here is the repeated Simpson's rule, and an even number of steps must be taken. The result is correct only at step N. Of course we do not need to know $\mathbf{Y}$ at intermediate points. When the potential has decayed to the point that the only terms which remain significant are the $l(l + 1)/R^2$ and k_i^2 terms, the $\mathbf{R}$ matrix may be determined from

$$\mathbf{R} = -[\mathbf{Y}(R_N) \cdot \mathbf{N}(R_N) - \mathbf{N}'(R_N)]^{-1} \cdot [\mathbf{Y}(R_N) \cdot \mathbf{J}(R_N) - \mathbf{J}'(R_N)] \tag{93}$$

where

$$[\mathbf{J}(R)]_{ij} = \delta_{ij} k_j^{-1/2} u_{l_j}(k_j R) \tag{94}$$

and

$$[\mathbf{N}(R)]_{ij} = \delta_{ij} k_j^{-1/2} v_{l_j}(k_j R) \tag{95}$$

Here $u_l(k_i R)$ and $v_l(k_i R)$ are Ricatti–Bessel and Riccati–Neumann functions, respectively, when $E - E_i = k_i^2 > 0$, and

$$u_l(k_i R) = k_i R^{1/2} I_{l+1/2}(k_i R) \tag{96}$$

and

$$v_l(k_i R) = k_i R^{1/2} K_{l+1/2}(k_i R) \tag{97}$$

when $E - E_i = -k_i^2 < 0$.

References

1. A.M. Arthurs and A. Dalgarno, The theory of scattering by a rigid rotator, *Proc. R. Soc. London A* **256**, 540–551 (1960).
2. A.R. Edmonds, *Angular Momentum in Quantum Mechanics,* Princeton University Press, Princeton, New Jersey (1974).
3. C.G. Gray, On the theory of multipole interactions, *Can. J. Phys.* **46**, 135–139 (1968).
4. H. Rabitz, Effective Hamiltonians in molecular collisions, in *Modern Theoretical Chemistry,* W.H. Miller, editor, Vol. 1, Plenum Press, New York (1976), pp. 33–80.
5. H. Rabitz, Effective potentials in molecular collisions, *J. Chem. Phys.* **57**, 1718–1725 (1972).
6. W.A. Lester, Jr., DeVogelaere's method for the numerical integration of second-order differential equations without explicit first derivatives: Applications to coupled equations arising from the Schrödinger equation, *J. Comp. Phys.* **3**, 322–325 (1968).
7. W.A. Lester, Jr., Calculation of cross sections for rotational excitation of diatomic molecules by heavy particle impact: Solution of the close-coupled equations, in *Methods of Computational Physics,* B. Alder, S. Fernbach, and M. Rotenberg, editors, Vol. 10, Academic Press, New York (1971), pp. 211–243.
8. W.A. Lester, Jr., The *N* coupled-channel problem, in *Modern Theoretical Chemistry,* W.H. Miller, editor, Vol. 1, Plenum Press, New York (1976), pp. 1–32.
9. W.N. Sams and D.J. Kouri, Noniterative solution of integral equations for scattering. I. Single channels, *J. Chem. Phys.* **51**, 4809–4814 (1969).
10. W.N. Sams and D.J. Kouri, Noniterative solution of integral equations for scattering. II. Coupled channels, *J. Chem. Phys.* **51**, 4815–4819 (1969).
11. B.H. Choi and K.T. Tang, Inelastic collisions between an atom and a diatomic molecule. I. Theoretical and numerical considerations for the close coupling approximation, *J. Chem. Phys.* **63**, 1775–1782 (1975).
12. R.G. Gordon, A new method for constructing wave functions for bound states and scattering, *J. Chem. Phys.* **51**, 14–25 (1969).
13. R.G. Gordon, Quantum scattering using piecewise analytic solutions, in *Methods of Computational Physics,* B. Alder, S. Fernbach, and M. Rotenberg, editors, Vol. 10, Academic Press, New York (1971), pp. 81–110.
14. C.C. Rankin and J.C. Light, Quantum solution of collinear reactive systems: $H + Cl_2 \rightarrow HCl + Cl$, *J. Chem. Phys.* **51**, 1701–1719 (1969).
15. G. Miller and J.C. Light, Quantum calculations of collinear reactive triatomic systems. II. Theory, *J. Chem. Phys.* **54**, 1635–1642 (1971).
16. J.C. Light, Quantum calculations in chemically reactive systems, in *Methods of Computational Physics,* B. Alder, S. Fernbach, and M. Rotenberg, editors, Vol. 10, Academic Press, New York (1971), pp. 111–142.
17. A.S. Cheung and D.J. Wilson, Quantum vibrational transition probabilities in atom–diatomic molecule collisions, *J. Chem. Phys.* **51**, 3448–3457 (1969).
18. B.R. Johnson and D. Secrest, The solution of the nonrelativistic quantum scattering problem without exchange, *J. Math. Phys.* **7**, 2187–2195 (1966).
19. D. Secrest and B.R. Johnson, Exact quantum-mechanical calculation of a collinear collision of a particle with a harmonic oscillator, *J. Chem. Phys.* **45**, 4556–4570 (1966).
20. B.R. Johnson, The multichannel log-derivative method for scattering calculations, *J. Comp. Phys.* **13**, 445–449 (1973).
21. J.C. Light and R.B. Walker, An *R* matrix approach to the solution of coupled equations for atom–molecule reactive scattering, *J. Chem. Phys.* **65**, 4272–4282 (1976).
22. A. Messiah, *Quantum Mechanics,* John Wiley and Sons, New York (1964).
23. D. Secrest, Amplitude densities in molecular scattering, in *Methods of Computational Physics,* B. Alder, S. Fernbach, and M. Rotenberg, editors, Vol. 10, Academic Press, New York (1971), pp. 243–286.

24. W. Eastes and D. Secrest, Calculation of rotational and vibrational transitions for the collision of an atom with a rotating vibrating diatomic oscillator, *J. Chem. Phys.* **56**, 640–649 (1972).
25. R.G. Gordon, Coupled channel scattering matrices, QCPE #187, Quantum Chemistry Program Exchange, Indiana University (1971).
26. F. Calogero, *Variable Phase Approach to Potential Scattering*, Academic Press, New York (1967).
27. V. Stocker, *A Formalism for Multiple Scattering*, Ph.D. Thesis, Urbana, Illinois (1976).
28. V. Stocker and D. Secrest, Multiple scattering from fixed scattering centers, *J. Chem. Phys.* **65**, 4857–4866 (1976).
29. R.T Pack, private communication.

9

Rotational Excitation II: Approximation Methods

DONALD JACK KOURI

1. Introduction

In this chapter, we shall describe several of the most useful quantal approximations for use in treating rotational excitation in atom–molecule collisions. In order to keep this discussion within manageable proportions, we shall *not* attempt to describe all the approximations developed in the past few years but rather focus on those which, in our admittedly prejudiced view, are the most useful, general, and computationally viable. As a result, there are a number of approximations which we will not discuss in any detail.

This discussion is organized as follows. We begin in Section 2 by discussing the highly successful CS angular momentum decoupling approximation. Then in Section 3 we discuss the IOS angular momentum decoupling approximation. Next, in Sections 4–7 we indicate some of the other alternatives when neither the CS nor IOS is appropriate. Finally, in Section 8 we deal with conditions for selecting which method will be used.

2. The CS Approximation

Tremendous progress has been made in the development and application of approximations which address the central problem of rotational excitation in atom–molecule collisions. This problem is the rapid prolifera-

---

DONALD JACK KOURI • Departments of Chemistry and Physics, University of Houston, Houston, Texas 77004.

tion of quantum channels associated with the $2j + 1$ degeneracy of the energy levels of a rotor. Because of this, the number of coupled equations one must solve increases nonlinearly with the highest rotor state j_M included in the basis set used to expand the scattering wave function. To see this, we note that for the even-parity case, the number of close-coupled equations, N_E, for an atom–heteronuclear-diatom system is given by

$$N_E = \sum_{j=0}^{j_M} (j + 1) \tag{1}$$

This yields a value $(j_M + 1)(j_M + 2)/2$, and for $j_M = 50$ one has 1326 coupled equations! For odd parity,

$$N_O = \sum_{j=1}^{j_M} j = (j_M + 1)\, j_M/2 \tag{2}$$

so that $j_M = 50$ leads to "only" 1275 coupled equations! For even the most efficient algorithms, the labor of solving coupled equations varies as the cube of the number of equations being solved, so it is clear that the computational problem rapidly becomes intractable for even modest j_M. In order to reduce this to computationally viable proportions, McGuire and Kouri[1] developed the l-labeled j_z-conserving coupled-states (CS) approximation, and, independently and simultaneously, Pack[2] developed the J-labeled centrifugal sudden (CS) approximation. Further refinements and developments of the approach and its interpretation have been reported by Tamir and Shapiro,[3] Walker and Light,[4] Secrest,[5] Shimoni and Kouri,[6-9] Parker and Pack,[10] Kouri and Shimoni,[11,12] Goldflam and Kouri,[13] Khare,[14] Kouri *et al.*,[15] and Khare *et al.*[16] Earlier studies of such approximations (but with more severe approximations in treating the dynamics of radial motion) have been made by Lawley and Ross[17] and Klar.[18] We shall not delve into the history of the development of the approach but instead employ the most current form and interpretation of the method in our discussion. Additional interpretations and forms of the approach may be found by consulting the original literature.[1-22] We note that the approach for atom–diatom collisions has been widely tested[1,3,4,6-13,23-61] and is considered the most accurate of a hierarchy of possible approximations for treating a general class of collision systems. We discuss sufficiency conditions for the method later. In the following, we will restrict our attention to atom–linear-rigid-rotor collisions but note that generalizations to more complex atom–molecule and molecule–molecule collisions have been carried out.[37,38,45,56-58]

Because of the availability of several efficient algorithms for solving both coupled differential and coupled integral equations, we shall give the CS equations in both forms. A summary of the algorithms and the question of which to use in solving a given problem is addressed in Chapter 8 of this volume.

2.1. The Basic CS Equations

We now proceed to derive the CS equations. The particular techniques we employ are new and incorporate the advances in understanding the CS method due to the works of McGuire and Kouri,[1] Secrest,[5] Kouri *et al.*,[62,63] Shimoni and Kouri,[6-9] Parker and Pack,[10] Kouri and Shimoni,[12] Goldflam and Kouri,[13] Khare,[14] Kouri *et al.*,[15] and Khare *et al.*[16] The basic Lippmann–Schwinger[64] equation describing an atom–diatom collision is given by

$$\psi_{\mathbf{k}_{j_0}}(j_0\lambda_0|\mathbf{R}\hat{\rho}) = \exp(i\mathbf{k}_{j_0}\cdot\mathbf{R})\,Y_{j_0\lambda_0}(\hat{\rho}) - \frac{2M}{\hbar^2}\sum_{\substack{j\lambda\\ l\mu}} k_j Y_{j\lambda}(\hat{\rho})\,Y_{l\mu}(\hat{\mathbf{R}})\int d\mathbf{R}'\,d\hat{\rho}'\,Y^*_{j\lambda}(\hat{\rho}')\,Y^*_{l\mu}(\hat{\mathbf{R}}') \times V(\mathbf{R}'\cdot\hat{\rho}')\,h_l^+(k_jR_>)\,j_l(k_jR_<)\,\psi_{\mathbf{k}_{j_0}}(j_0\lambda_0|\mathbf{R}'\hat{\rho}') \quad (3)$$

where j_l and h_l^+ are spherical Bessel and Hankel functions satisfying the asymptotic conditions[65]

$$\lim_{R\to\infty} h_l^+(k_jR) = \frac{\exp[i(k_jR - l\pi/2)]}{k_jR} \quad (4)$$

$$\lim_{R\to\infty} j_l(k_jR) = \frac{\sin(k_jR - l\pi/2)}{k_jR} \quad (5)$$

l and j are the relative orbital and rotor angular momentum quantum numbers, respectively, $\mu\hbar$ and $\lambda\hbar$ are the projections along an *arbitrary* z axis of the orbital and rotor angular momenta, respectively, V is the interaction between the atom and diatom producing the scattering, $\mathbf{R}$ is the vector from the center of mass of the diatom to the incoming atom, $\hat{\rho}$ is a unit vector along the diatom axis, M is the reduced mass of the incoming atom relative to the diatom center of mass, and k_j is the wave number for relative motion of the atom when the diatom is in rotor state j. It is given in terms of the diatom rotor constant B_e and the total energy E by

$$\hbar^2k_j^2/2M = E - B_e j(j+1) \quad (6)$$

In addition to describing the collision dynamics, equation (3) also contains an explicit specification of the asymptotic boundary conditions. Remembering that $R_>$ is the greater and $R_<$ is the lesser of R and R', the asymptotic form of the wave function is found from equation (3) to be

$$\lim_{R\to\infty} \psi_{\mathbf{k}_{j_0}}(j_0\lambda_0|\mathbf{R}\hat{\rho}) = \exp(i\mathbf{k}_{j_0}\cdot\mathbf{R})\,Y_{j_0\lambda_0}(\hat{\rho}) + \sum_{j\lambda}\frac{1}{R}\exp(ik_jR)\,Y_{j\lambda}(\hat{\rho})\,f(j\lambda|j_0\lambda_0|\hat{\mathbf{R}}) \quad (7)$$

where

$$f(j\lambda|j_0\lambda_0|\hat{\mathbf{R}}) = -\frac{2M}{\hbar^2}\sum_{l\mu} Y_{l\mu}(\hat{\mathbf{R}})\, i^{-l} \int d\mathbf{R}'\, d\hat{\rho}\, Y^*_{j\lambda}(\hat{\rho})\, Y^*_{l\mu}(\hat{\mathbf{R}}') \times j_l(k_j R')\, V(\mathbf{R}'\cdot\hat{\rho})\, \psi_{\mathbf{k}_{j_0}}(j_0\lambda_0|\mathbf{R}'\hat{\rho}) \tag{8}$$

We note that both the initial and final projections of the rotor angular momentum are taken along the *same* as yet unspecified axis. Indeed, it is clear from equations (3) and (7) that the primary effect of choosing a particular quantization axis is on $Y_{j_0\lambda_0}(\hat{\rho})$ and $Y_{j\lambda}(\hat{\rho})$. This is made clearer by noting that equation (3) can also be written as

$$\psi_{\mathbf{k}_{j_0}}(j_0\lambda_0|\mathbf{R}\hat{\rho}) = \exp(i\mathbf{k}_{j_0}\cdot\mathbf{R})\, Y_{j_0\lambda_0}(\hat{\rho}) - \frac{2M}{\hbar^2}\sum_{jl} k_j \frac{(2l+1)(2j+1)}{(4\pi)^2} \int d\mathbf{R}'\, d\hat{\rho}' P_j(\hat{\rho}\cdot\hat{\rho}')\, P_l(\hat{\mathbf{R}}\cdot\hat{\mathbf{R}}') \times h_l^+(k_j R_>)\, j_l(k_l R_<)\, V(\mathbf{R}'\cdot\hat{\rho}')\, \psi_{\mathbf{k}_{j_0}}(j_0\lambda_0|\mathbf{R}'\hat{\rho}') \tag{9}$$

and we see that all the vector quantities, except for $Y_{j_0\lambda_0}(\hat{\rho})$, appear as scalar products $\mathbf{k}_{j_0}\cdot\mathbf{R}$, $\hat{\mathbf{R}}\cdot\hat{\mathbf{R}}'$, and $\hat{\rho}\cdot\hat{\rho}'$. These quantities are completely independent of any rotation of the z axis. The reason $f(j\lambda|j_0\lambda_0|\hat{\mathbf{R}})$ also depends on the quantization axis chosen for the final state is due to the fact that in writing the asymptotic limit equation (7), we used the relations

$$P_l(\hat{\mathbf{R}}\cdot\hat{\mathbf{R}}') = \sum_{\mu} \frac{4\pi}{(2l+1)} Y_{l\mu}(\hat{\mathbf{R}})\, Y^*_{l\mu}(\hat{\mathbf{R}}') \tag{10}$$

and

$$P_j(\hat{\rho}\cdot\hat{\rho}') = \sum_{\lambda} \frac{4\pi}{(2j+1)} Y_{j\lambda}(\hat{\rho})\, Y^*_{j\lambda}(\hat{\rho}') \tag{11}$$

in order to obtain equation (8), and each separate piece of equations (10) and (11) is *not* invariant to rotation of the z axis. Now we expand $\exp(i\mathbf{k}_{j_0}\cdot\mathbf{R})$ in the well-known form[65]

$$\exp(i\mathbf{k}_{j_0}\cdot\mathbf{R}) = 4\pi\sum_{l\mu} i^l Y^*_{l\mu}(\hat{\mathbf{k}}_{j_0})\, Y_{l\mu}(\hat{\mathbf{R}})\, j_l(k_{j_0}R) \tag{12}$$

and similarly write the scattering wave function as

$$\psi_{\mathbf{k}_{j_0}}(j_0\lambda_0|\mathbf{R}\hat{\rho}) = 4\pi\sum_{l\mu} i^l Y^*_{l\mu}(\hat{\mathbf{k}}_{j_0})\, \psi^{l\mu}(j_0\lambda_0|\mathbf{R}\hat{\rho}) \tag{13}$$

Then substituting into equation (3), multiplying by $Y_{l_0\mu_0}(\hat{\mathbf{k}}_{j_0})$, and integrating over all incident momentum directions, yields

$$\psi^{l_0\mu_0}(j_0\lambda_0|\mathbf{R}\hat{\boldsymbol{\rho}}) = j_{l_0}(k_{j_0}R)\, Y_{l_0\mu_0}(\hat{\mathbf{R}})\, Y_{j_0\lambda_0}(\hat{\boldsymbol{\rho}}) - \frac{2M}{\hbar^2}\sum_{\substack{j\lambda\\ l\mu}} k_j Y_{j\lambda}(\hat{\boldsymbol{\rho}})\, Y_{l\mu}(\hat{\mathbf{R}}) \int d\mathbf{R}'\, d\hat{\boldsymbol{\rho}}'\, Y^*_{j\lambda}(\hat{\boldsymbol{\rho}}')\, Y^*_{l\mu}(\hat{\mathbf{R}}')\, V(\mathbf{R}'\cdot\hat{\boldsymbol{\rho}}') \times h_l^+(k_jR_>)\, j_l(k_jR_<)\, \psi^{l_0\mu_0}(j_0\lambda_0|\mathbf{R}'\hat{\boldsymbol{\rho}}') \tag{14}$$

where we have used the fact that

$$\int d\hat{\mathbf{k}}_{j_0}\, Y_{l_0\mu_0}(\hat{\mathbf{k}}_{j_0})\, Y^*_{l\mu}(\hat{\mathbf{k}}_{j_0}) = \delta_{ll_0}\,\delta_{\mu\mu_0} \tag{15}$$

We now specify our choice of quantization axis to be along the direction $\hat{\mathbf{R}}$. We note that other choices of z axis are possible.[3,10,12,15,16] The resulting *exact* equation is then

$$\psi^{l_0\mu_0}(j_0\lambda_0|\hat{\mathbf{z}}, R, \hat{\boldsymbol{\rho}}) = j_{l_0}(k_{j_0}R)\, Y_{l_0\mu_0}(\hat{\mathbf{z}})\, Y_{j_0\lambda_0}(\hat{\boldsymbol{\rho}}) - \frac{2M}{\hbar^2}\sum_{\substack{j\lambda\\ l\mu}} k_j Y_{j\lambda}(\hat{\boldsymbol{\rho}})\, Y_{l\mu}(\hat{\mathbf{z}}) \int d\mathbf{R}'\, d\hat{\boldsymbol{\rho}}'\, Y^*_{j\lambda}(\hat{\boldsymbol{\rho}}')\, Y^*_{l\mu}(\hat{\mathbf{R}}') \times h_l^+(k_jR_>)\, j_l(k_jR_<)\, V(R', \hat{\mathbf{R}}'\cdot\hat{\boldsymbol{\rho}}')\, \psi^{l_0\mu_0}(j_0\lambda_0|\mathbf{R}'\hat{\boldsymbol{\rho}}') \tag{16}$$

and we note that

$$Y_{l_0\mu_0}(\hat{\mathbf{R}}) = Y_{l_0\mu_0}(\hat{\mathbf{z}}) \tag{17}$$

$$= \left(\frac{2l_0+1}{4\pi}\right)^{1/2} \delta_{\mu_0 0} \tag{18}$$

Then equations (16)–(18) enable us to write

$$\psi^{l_0\mu_0}(j_0\lambda_0|\hat{\mathbf{z}}, R, \hat{\boldsymbol{\rho}}) = \left(\frac{2l_0+1}{4\pi}\right)^{1/2} \delta_{\mu_0 0}\psi^{l_0}(j_0\lambda_0|\hat{\mathbf{z}}, R, \hat{\boldsymbol{\rho}}) \tag{19}$$

where $\psi^{l_0}(j_0\lambda_0|\hat{\mathbf{z}}, R, \hat{\boldsymbol{\rho}})$ is given by

$$\psi^{l_0}(j_0\lambda_0|\hat{\mathbf{z}}, R, \hat{\boldsymbol{\rho}}) = Y_{j_0\lambda_0}(\hat{\boldsymbol{\rho}})\, j_{l_0}(k_{j_0}R) - \frac{2M}{\hbar^2}\sum_{\substack{j\lambda\\ l\mu}} k_j Y_{j\lambda}(\hat{\boldsymbol{\rho}})\, Y_{l\mu}(\hat{\mathbf{z}}) \int d\mathbf{R}'\, d\hat{\boldsymbol{\rho}}'\, Y^*_{j\lambda}(\hat{\boldsymbol{\rho}}')\, Y^*_{l\mu}(\hat{\mathbf{R}}') \times h_l^+(k_jR_>)\, j_l(k_jR_<)\, V(R', \hat{\mathbf{R}}', \cdot\hat{\boldsymbol{\rho}}')\, \psi^{l_0}(j_0\lambda_0|\mathbf{R}'\hat{\boldsymbol{\rho}}') \tag{20}$$

We emphasize again that equation (20) is exact and applies to a description of the collision where both λ and λ_0 are along the direction $\hat{\mathbf{R}}$. Now to introduce the CS approximation, we introduce an effective orbital angular momentum quantum number L, such that

$$h_l^+(k_jR_>)\, j_l(k_jR_<) \cong h_L^+(k_jR_>)\, j_L(k_jR_<) \tag{21}$$

This is equivalent to approximating the centrifugal potential $\mathbf{l}^2/2MR^2$ by an eigenvalue form[1] $L(L+1)\hbar^2/2MR^2$. It is also, of course, a sudden approximation[2] which essentially says that the relative kinetic energy is sufficiently large that the precise value of the centrifugal potential is unimportant. Whether this is true or not is sensitive to how the different turning points for different values of l are related.[4] If the potential is purely repulsive, then the rate of change of the turning-point position with a change in l is not large *and the approximation is expected to always be quite good.*[63] For the potentials with wells, one may encounter ranges of l's which have three turning points (all of which change relatively rapidly with l) and the approximation is not as good.[63] However, as the kinetic energy increases beyond the three-turning-point region, the approximation is expected to improve significantly.[63] We shall return to discuss the accuracy conditions in more detail later.

If we use equation (21) in equation (20), then the sums over l, μ can be done analytically using the closure relation

$$\sum_{l\mu} Y_{l\mu}(\hat{\mathbf{z}})\, Y_{l\mu}^*(\hat{\mathbf{R}}') = \delta(\hat{\mathbf{z}} - \hat{\mathbf{R}}) \tag{22}$$

which essentially results in the Green's operator being made local in the angles of the scattering vector.[66] (In fact, sudden approximations in general can be viewed as closure approximations that localize the angle dependence of the Green's function.) Combining equations (20)–(22), we can carry out the integral over the angles of $\hat{\mathbf{R}}'$, and therefore we obtain

$$\begin{aligned}\psi_L^{l_0}(j_0\lambda_0|\hat{\mathbf{z}}, R, \hat{\boldsymbol{\rho}}) = {} & Y_{j_0\lambda_0}(\hat{\boldsymbol{\rho}})\, j_{l_0}(k_{j_0}R) \\ & - \frac{2M}{\hbar^2}\sum_{j\lambda} k_j Y_{j\lambda}(\hat{\boldsymbol{\rho}}) \int d\hat{\rho}' \int_0^\infty dR' R'^2 Y_{j\lambda}^*(\hat{\boldsymbol{\rho}})\, h_L^+(k_jR_>) \\ & \times j_L(k_jR_<) V(R', \hat{\mathbf{z}}\cdot\hat{\boldsymbol{\rho}}')\, \psi_L^{l_0}(j_0\lambda_0|\hat{\mathbf{z}}, R', \hat{\boldsymbol{\rho}}')\end{aligned} \tag{23}$$

The form of equation (23) then immediately shows that the component λ_0 of rotor angular momentum will be conserved because of the fact that $\hat{\mathbf{z}}\cdot\hat{\boldsymbol{\rho}}'$ now appears in the potential in place of $\hat{\mathbf{R}}'\cdot\hat{\boldsymbol{\rho}}'$. Since $\hat{\mathbf{z}}\cdot\hat{\boldsymbol{\rho}}'$ equals $\cos\theta$, where θ is the polar angle of $\hat{\boldsymbol{\rho}}'$, the potential is independent of the azimuthal angle φ and the result is that no coupling in λ can occur.[1,2] Explicitly, we expand $\psi_L^{l_0}(j_0\lambda_0|R, \hat{\boldsymbol{\rho}})$ according to

$$\psi_L^{l_0}(j_0\lambda_0|R, \hat{\boldsymbol{\rho}}) = \sum_{j\lambda} Y_{j\lambda}(\hat{\boldsymbol{\rho}})\, \chi_{l_0L}(j\lambda|j_0\lambda_0|R) \tag{24}$$

substitute into equation (23), multiply by $Y^*_{j\lambda}(\hat{\boldsymbol{\rho}})$, and integrate over $\hat{\boldsymbol{\rho}}$ to obtain

$$\chi_{l_0L}(j\lambda|j_0\lambda_0|R) = \delta_{\lambda\lambda_0}\delta_{jj_0}j_{l_0}(k_{j_0}R) - \frac{2M}{\hbar^2}k_j\int_0^\infty dR'R'^2h_L^+(k_jR_>)\,j_L(k_jR_<)\sum_{j'\lambda'}\int d\hat{\boldsymbol{\rho}}'Y^*_{j\lambda}(\hat{\boldsymbol{\rho}}') \times V(R',\hat{\mathbf{z}}\cdot\hat{\boldsymbol{\rho}}')\,Y_{j'\lambda'}(\hat{\boldsymbol{\rho}}')\,\chi_{l_0L}(j'\lambda'|j_0\lambda_0|R') \tag{25}$$

Now as is well known,[1,2]

$$\int d\hat{\boldsymbol{\rho}}\,Y^*_{j\lambda}(\hat{\boldsymbol{\rho}})\,V(R',\hat{\mathbf{z}}\cdot\hat{\boldsymbol{\rho}})\,Y_{j'\lambda'}(\hat{\boldsymbol{\rho}}) = \delta_{\lambda\lambda'}V_\lambda(j|j'|R') \tag{26}$$

where

$$V_\lambda(j|j'|R') = \sum_t(-1)^\lambda[(2j+1)(2j'+1)]^{1/2}\begin{pmatrix} j & t & j' \\ 0 & 0 & 0\end{pmatrix}\begin{pmatrix} j & t & j' \\ -\lambda & 0 & \lambda\end{pmatrix}V_t(R') \tag{27}$$

Here $V_t(R')$ is the coefficient of $P_t(\hat{\boldsymbol{\rho}}\cdot\hat{\mathbf{R}})$ in a Legendre expansion of the potential. Then

$$\chi_{l_0L}(j\lambda|j_0\lambda_0|R) = \delta_{\lambda\lambda_0}\delta_{jj_0}j_{l_0}(k_{j_0}R) - \frac{2M}{\hbar^2}k_j\sum_{j'}\int_0^\infty dR'R'^2h_L^+(k_jR_>)\,j_L(k_jR_<)\,V_\lambda(j|j'|R') \times\chi_{l_0L}(j'\lambda|j_0\lambda_0|R') \tag{28}$$

and because there is no coupling in λ, λ_0 we can define a function $\psi^L_{l_0\lambda}(j|j_0|R)$ by

$$\chi_{l_0L}(j\lambda|j_0\lambda_0|R) = \delta_{\lambda\lambda_0}\psi^L_{l_0\lambda}(j|j_0|R) \tag{29}$$

If we substitute equation (29) into equation (28), we find that $\psi^L_{l_0\lambda}(j|j_0|R)$ satisfies

$$\psi^L_{l_0\lambda}(j|j_0|R) = \delta_{jj_0}j_{l_0}(k_{j_0}R) - \frac{2M}{\hbar^2}k_j\sum_{j'}\int_0^\infty dR'R'^2h_L^+(k_jR_>)\,j_L(k_jR_<)\,V_\lambda(j|j'|R') \times\psi^L_{l_0\lambda}(j'|j_0|R') \tag{30}$$

Now if we follow Secrest[5] and Khare[14] in picking L to equal l_0, we obtain the initial l-labeled McGuire–Kouri form of the CS equation[1]:

$$\psi_{l_0\lambda}(j\,|\,j_0|R) = \delta_{jj_0} j_{l_0}(k_{j_0}R) - \frac{2M}{\hbar^2} k_j \sum_{j'} \int_0^\infty dR' R'^2 h_{l_0}^+(k_j R_>)\, j_{l_0}(k_j R_<)\, V_\lambda(j\,|\,j'|R') \times \psi_{l_0\lambda}(j'|\,j_0|R') \tag{31}$$

This equation may be solved by the Sams–Kouri method[67,68] as refined by Secrest,[69] White and Hayes,[70,71] and Smith and Henry[72] [see also the discussion by Secrest (Chapter 8) in this volume]. If one prefers differential equation methods, the equivalent equation is

$$\left[\frac{\hbar^2}{2M}\frac{1}{R}\frac{\partial^2}{\partial R^2}R - \frac{l_0(l_0+1)\hbar^2}{2MR^2} + \frac{\hbar^2 k_j^2}{2M} - V_\lambda(j\,|\,j\,|R)\right]\psi_{l_0\lambda}(j\,|\,j_0|R) = \sum_{j'\neq j} V_\lambda(j\,|\,j'|R)\,\psi_{l_0\lambda}(j'|\,j_0|R) \tag{32}$$

Again, Secrest discusses modern efficient methods for solving such equations in this volume.

2.2. *The CS Scattering Amplitude and Boundary Conditions*

The boundary conditions to be applied in solving equation (32) can be obtained from equation (31), as we shall show momentarily. First we use the techniques of Kouri and Shimoni[12] and Kouri *et al.*[15] to derive the CS approximation to the differential scattering amplitude $f(j\lambda|\,j_0\lambda_0|\hat{\mathbf{R}})$ since it will also yield the scattered portion of $\psi_{l_0\lambda}(j\,|\,j_0|R)$. We can combine equations (13), (19), (24), and (29) to write the wave function as

$$\psi_{\mathbf{k}_{j_0}}(j_0\lambda_0|R,\hat{\rho}) = 4\pi \sum_{\substack{l_0\mu_0\\ j\lambda}} i^{l_0} Y_{l_0\mu_0}^*(\hat{\mathbf{k}}_{j_0}) \left(\frac{2l_0+1}{4\pi}\right)^{1/2} \delta_{\mu_0 0} Y_{j\lambda}(\hat{\rho}) \times \delta_{\lambda\lambda_0}\psi_{l_0\lambda}(j\,|\,j_0|R) \tag{33}$$

or upon simplifying,

$$\psi_{\mathbf{k}_{j_0}}(j_0\lambda_0|R,\hat{\rho}) = \sum_{\substack{l_0\\ j\lambda}} i^{l_0}[4\pi(2l_0+1)]^{1/2}\, Y_{l_0 0}^*(\hat{\mathbf{k}}_{j_0})\,\delta_{\lambda\lambda_0} Y_{j\lambda}(\hat{\rho})\,\psi_{l_0\lambda}(j\,|\,j_0|R) \tag{34}$$

Now the asymptotic limit of equation (34) is obtained by substituting the asymptotic limit of equation (31), which is

$$\lim_{R\to\infty} \psi_{l_0\lambda}(j\,|\,j_0|R) = \delta_{jj_0} j_{l_0}(k_{j_0}R) - \frac{2M}{\hbar^2}\frac{e^{ik_jR}}{R} i^{-l_0} \sum_{j'} \int_0^\infty dR' R'^2 j_{l_0}(k_j R') \times V_\lambda(j\,|\,j'|R')\,\psi_{l_0\lambda}(j'|\,j_0|R') \tag{35}$$

We introduce a transition amplitude $T_{l_0\lambda}(j\,|\,j_0)$ by

$$T_{l_0\lambda}(j\,|\,j_0) = \frac{4iM}{\hbar^2}(k_{j_0}k_j)^{1/2}\sum_{j'}\int_0^\infty dR'R'^2 j_{l_0}(k_jR')\,V_\lambda(j\,|\,j'|R')\,\psi_{l_0\lambda}(j'|\,j_0|R') \tag{36}$$

so that

$$\lim_{R\to\infty}\psi_{l_0\lambda}(j|j_0|R) = \delta_{jj_0}j_{l_0}(k_{j_0}R) + \frac{i}{2(k_{j_0}k_j)^{1/2}}\frac{T_{l_0\lambda}(j\,|\,j_0)\exp[i(k_jR - l_0\pi/2)]}{R} \tag{37}$$

Substituting equation (37) into equation (34) yields

$$\lim_{R\to\infty}\psi_{\mathbf{k}_{j_0}}(j_0\lambda_0|R,\hat{\boldsymbol{\rho}}) = \sum_{\substack{l_0\\ j\lambda}} i^{l_0}[4\pi(2l_0+1)]^{1/2}\,\delta_{jj_0}\delta_{\lambda\lambda_0}Y^*_{l_00}(\hat{\mathbf{k}}_{j_0})\,Y_{j\lambda}(\hat{\boldsymbol{\rho}})\,j_{l_0}(k_{j_0}R)$$
$$+\sum_{\substack{l_0\\ j\lambda}}\frac{i^{l_0}e^{ik_jR}}{2R(k_jk_{j_0})^{1/2}}\,i^{-l_0+1}T_{l_0\lambda}(j\,|\,j_0)\,[4\pi(2l_0+1)]^{1/2}$$
$$\times\; Y^*_{l_00}(\hat{\mathbf{k}}_{j_0})\,\delta_{\lambda\lambda_0}Y_{j\lambda}(\hat{\boldsymbol{\rho}}) \tag{38}$$

and we note that the first term on the right-hand side of equation (38) is just the incident plane wave times the initial state of the rotor:

$$\sum_{\substack{l_0\\ j\lambda}} i^{l_0}[4\pi(2l_0+1)]^{1/2}\,\delta_{jj_0}\delta_{\lambda\lambda_0}Y^*_{l_00}(\hat{\mathbf{k}}_{j_0})\,Y_{j\lambda}(\hat{\boldsymbol{\rho}})\,j_{l_0}(k_{j_0}R)$$
$$= \left[4\pi\sum_{l_0\mu_0} i^{l_0}Y_{l_0\mu_0}(\hat{\mathbf{z}})\,Y^*_{l_0\mu_0}(\hat{\mathbf{k}}_{j_0})\,j_{l_0}(k_{j_0}R)\right]Y_{j_0\lambda_0}(\hat{\boldsymbol{\rho}}) \tag{39}$$
$$= \exp(i\mathbf{k}_{j_0}\cdot\hat{\mathbf{z}}R)\,Y_{j_0\lambda_0}(\hat{\boldsymbol{\rho}}) \tag{40}$$

The second term on the right-hand side of equation (38) is the scattered wave and we *define* the differential scattering amplitude by requiring

$$\sum_{\substack{l_0\\ j\lambda}}\frac{ie^{ik_jR}}{2R(k_{j_0}k_j)^{1/2}}\,Y_{j\lambda}(\hat{\boldsymbol{\rho}})\,[4\pi(2l_0+1)]^{1/2}\,\delta_{\lambda\lambda_0}Y^*_{l_00}(\hat{\mathbf{k}}_{j_0})\,T_{l_0\lambda}(j\,|\,j_0)$$
$$= 4\pi\sum_{\substack{l_0\mu_0\\ j\lambda}}\frac{ie^{ik_jR}}{2R(k_{j_0}k_j)^{1/2}}\,Y_{j\lambda}(\hat{\boldsymbol{\rho}})\,\delta_{\lambda\lambda_0}Y_{l_0\mu_0}(\hat{\mathbf{z}})\,Y^*_{l_0\mu_0}(\hat{\mathbf{k}}_{j_0})\,T_{l_0\lambda}(j\,|\,j_0) \tag{41}$$
$$= \sum_{j\lambda}\frac{e^{ik_jR}}{R}\,Y_{j\lambda}(\hat{\boldsymbol{\rho}})\,f(j\lambda|\,j_0\lambda_0|\hat{\mathbf{z}}\cdot\hat{\mathbf{k}}_{j_0}) \tag{42}$$

Comparing equations (41) and (42), we see that

$$f(j\lambda|j_0\lambda_0|\hat{\mathbf{z}}\cdot\hat{\mathbf{k}}_{j_0}) = \frac{\delta_{\lambda\lambda_0}4\pi i}{2(k_{j_0}k_j)^{1/2}}\sum_{l_0\mu_0} Y_{l_0\mu_0}(\hat{\mathbf{z}})\, Y^*_{l_0\mu_0}(\hat{\mathbf{k}}_{j_0})\, T_{l_0\lambda}(j\,|\,j_0) \tag{43}$$

and using equation (10), we obtain our final result

$$f(j\lambda|j_0\lambda_0|\hat{\mathbf{z}}\cdot\hat{\mathbf{k}}_{j_0}) = \frac{\delta_{\lambda\lambda_0}i}{2(k_{j_0}k_j)^{1/2}}\sum_{l_0}(2l_0+1)\,P_{l_0}(\hat{\mathbf{z}}\cdot\hat{\mathbf{k}}_{j_0})\,T_{l_0\lambda}(j\,|\,j_0) \tag{44}$$

We note that since $\hat{\mathbf{z}} = \hat{\mathbf{R}}$, we can express this as

$$f(j\lambda|j_0\lambda_0|\hat{\mathbf{R}}\cdot\hat{\mathbf{k}}_{j_0}) = \frac{i\delta_{\lambda\lambda_0}}{2(k_{j_0}k_j)^{1/2}}\sum_{l_0}(2l_0+1)\,P_{l_0}(\hat{\mathbf{R}}\cdot\hat{\mathbf{k}}_{j_0})\,T_{l_0\lambda}(j\,|\,j_0) \tag{45}$$

which form combines the two leading interpretations of the CS approximation. It has been shown[10,12] that had we chosen the $\hat{\mathbf{z}}$ axis along $\hat{\mathbf{k}}_{j_0}$, so that λ, λ_0 are quantized along $\hat{\mathbf{k}}_{j_0}$, then equation (45) is the usual McGuire–Kouri approximation to the Arthurs–Dalgarno differential scattering amplitude.[73] In the present discussion, $\hat{\mathbf{z}} = \hat{\mathbf{R}}$ and the result is that the McGuire–Kouri[1] differential scattering amplitude is interpreted as describing transitions in magnetic states quantized along the final observation direction. Now the result in equation (37) also gives the boundary condition for solving the Schrödinger equation form of the CS, equation (32). We can express this in terms of an S matrix by writing equation (37) in the form

$$\lim_{R\to\infty}\psi_{l_0\lambda}(j\,|\,j_0|R) = \frac{1}{2ik_{j_0}}\Bigg\{-\delta_{jj_0}\exp[-i(k_{j_0}R - l_0\pi/2)] + \left(\frac{k_{j_0}}{k_j}\right)^{1/2} \times \delta_{jj_0}\exp[i(k_jR - l_0\pi/2)] - \left(\frac{k_{j_0}}{k_j}\right)^{1/2} T_{l_0\lambda}(j\,|\,j_0) \times \exp[i(k_jR - l_0\pi/2)]\Bigg\}\Bigg/R \tag{46}$$

$$= \frac{-1}{2ik_{j_0}}\Bigg\{\delta_{jj_0}\exp[-i(k_{j_0}R - l_0\pi/2)] - \left(\frac{k_{j_0}}{k_j}\right)^{1/2} \times S_{l_0\lambda}(j\,|\,j_0)\exp[i(k_jR - l_0\pi/2)]\Bigg\}\Bigg/R \tag{47}$$

so that the T and S matrices are related via

$$S_{l_0\lambda}(j\,|\,j_0) = \delta_{jj_0} - T_{l_0\lambda}(j\,|\,j_0) \tag{48}$$

One may thus use either equation (37) or (47) as a boundary condition in solving equation (32).

2.3. CS Differential and Integral Cross Sections

We are now ready to obtain the differential and integral cross sections using the result equation (45). The differential cross section for scattering from $j_0\lambda_0$ to $j\lambda$, where the axis of quantization for *both* initial and final states is the direction $\hat{\mathbf{R}}$, is given by[1,14]

$$\frac{d\sigma}{d\Omega}(j_0\lambda_0 \to j\lambda) = \frac{k_j}{k_{j_0}} |f(j\lambda|j_0\lambda_0|\hat{\mathbf{R}}\cdot\hat{\mathbf{k}}_{j_0})|^2 \tag{49}$$

$$= \frac{\delta_{\lambda\lambda_0}}{4k_{j_0}^2} \sum_{\substack{l_0 \\ l_0'}} (2l_0+1)(2l_0'+1) P_{l_0}(\hat{\mathbf{R}}\cdot\hat{\mathbf{k}}_{j_0}) P_{l_0'}(\hat{\mathbf{R}}\cdot\hat{\mathbf{k}}_{j_0}) T_{l_0\lambda}(j|j_0) \times T^*_{l_0'\lambda}(j|j_0) \tag{50}$$

The degeneracy-averaged cross section is obtained by averaging over initial projections λ_0 and summing over final projections λ, so that we have

$$\frac{d\sigma}{d\Omega}(j_0 \to j) = \frac{1}{(2j_0+1)\,4k_{j_0}^2} \sum_{\substack{l_0 l_0' \\ \lambda}} (2l_0+1)(2l_0'+1) P_{l_0}(\hat{\mathbf{R}}\cdot\hat{\mathbf{k}}_{j_0}) \times P_{l_0'}(\hat{\mathbf{R}}\cdot\hat{\mathbf{k}}_{j_0}) T_{l_0\lambda}(j|j_0) T^*_{l_0'\lambda}(j|j_0) \tag{51}$$

If we integrate equations (50) and (51) over all solid angles Ω, the integral state-to-state cross sections are

$$\sigma(j_0\lambda_0 \to j\lambda) = \frac{\pi}{k_{j_0}^2} \delta_{\lambda\lambda_0} \sum_{l_0} (2l_0+1) |T_{l_0\lambda}(j|j_0)|^2 \tag{52}$$

and

$$\sigma(j_0 \to j) = \frac{\pi}{(2j_0+1)\,k_{j_0}^2} \sum_{\substack{l_0 \\ \lambda}} (2l_0+1) |T_{l_0\lambda}(j|j_0)|^2 \tag{53}$$

Now it is generally desired to have cross sections relative to other axes of quantization rather than the final observation direction $\hat{\mathbf{R}}$. We may obtain the differential scattering amplitude quantized with respect to arbitrary axes by use of a rotation formula given by Alexander *et al.*[74] (see also References 12 and 14) as

$$f(j\nu|j_0\nu_0|\chi) = \sum_{\lambda\lambda_0} D^{j*}_{\nu\lambda}(\mathscr{R}_f)\, f(j\lambda|j_0\lambda_0|\chi)\, D^{j_0}_{\nu_0\lambda_0}(\mathscr{R}_i) \tag{54}$$

where $\chi = \hat{\mathbf{R}}\cdot\hat{\mathbf{k}}_{j_0}$, $\mathscr{R}_i$ is the rotation from the $\hat{\mathbf{R}}$ direction to the new initial quantization axis with projection ν_0, and $\mathscr{R}_f$ is the rotation from the $\hat{\mathbf{R}}$ direction to the new final quantization axis having projection ν, and the

$D^j_{\nu\lambda}$ are elements in the unitary rotation matrix[75] (i.e., the representation coefficients of the three-dimensional Lie group of rotations). As an example, the P-helicity amplitude has the initial quantization axis along the incident momentum $\hat{\mathbf{k}}_{j_0}$ and the final axis of quantization along the observation direction $\hat{\mathbf{R}}$. In this case, we take $\mathscr{R}_f$ to be the identity rotation while $\mathscr{R}_i$ is taken to be a rotation through the angle $\cos^{-1}(\hat{\mathbf{k}}_{j_0}\cdot\hat{\mathbf{R}})$. Then we obtain

$$f(j\lambda|j_0m_0|\chi) = \sum_{\lambda_0} f(j\lambda|j_0\lambda_0|\chi)\, D^{j_0}_{m_0\lambda_0}(0,\chi,0) \tag{55}$$

and using equation (45), we write this as

$$f(j\lambda|j_0m_0|\chi) = \frac{i}{2(k_{j_0}k_j)^{1/2}} \sum_{l_0} (2l_0+1)\, D^{l_0}_{00}(0,\chi,0)\, D^{j_0}_{m_0\lambda}(0,\chi,0)\, T_{l_0\lambda}(j\,|\,j_0) \tag{56}$$

Now we note that

$$D^{j_0}_{m_0\lambda}(0,\chi,0) = (-1)^{m_0-\lambda_0}\, D^{j_0}_{\lambda m_0}(0,\chi,0) \tag{57}$$

and

$$D^{l_0}_{00} D^{j_0}_{\lambda m_0} = \sum_J \langle l_0 0 j_0 \lambda | J\lambda\rangle \langle l_0 0 j_0 m_0 | J m_0\rangle\, D^J_{\lambda m_0} \tag{58}$$

Then we may write the helicity amplitude as[76]

$$f(j\lambda|j_0m_0|\chi) = \frac{i(-1)^{m_0-\lambda_0}}{2(k_{j_0}k_j)^{1/2}} \sum_{\substack{l_0\\J}} (2J+1)\frac{2l_0+1}{2J+1}\langle l_0 0 j_0\lambda|J\lambda\rangle$$

$$\times \langle l_0 0 j_0 m_0 | J m_0\rangle\, D^J_{\lambda m_0}(0,\chi,0)\, T_{l_0\lambda}(j\,|\,j_0) \tag{59}$$

and this leads us to define the helicity T matrix $T^J(j\lambda|j_0m_0)$ by[9,12,14]

$$T^J(j\lambda|j_0m_0) = \sum_{l_0} \frac{2l_0+1}{2J+1} \langle l_0 0 j_0\lambda|J\lambda\rangle \langle l_0 0 j_0 m_0|J m_0\rangle\, T_{l_0\lambda}(j\,|\,j_0) \tag{60}$$

The final result for the helicity amplitude is

$$f(j\lambda|j_0m_0|\chi) = \frac{i(-1)^{m_0-\lambda_0}}{2(k_{j_0}k_j)^{1/2}} \sum_J (2J+1)\, D^J_{\lambda m_0}(0,\chi,0)\, T^J(j\lambda|j_0m_0) \tag{61}$$

Several comments are now in order. As was first noted by Shimoni and Kouri[9,12] (in a final l-labeled CS approach), the P-helicity amplitude, equation (61), and T matrix are *not* diagonal in λ, m_0 even though the CS approximation is diagonal in λ, λ_0. The point is that even though the CS amplitude $f(j\lambda|j_0\lambda_0|\chi)$ is diagonal in λ, λ_0, *the amplitude for any other choices of quantization axis or axes will not be diagonal.* Second, had we chosen our z axis initially along $\hat{\mathbf{k}}_{j_0}$ rather than the final direction $\hat{\mathbf{R}}$, then the basic CS amplitude would be written

$$f(jm|j_0m_0|\chi) = \frac{i\delta_{mm_0}}{2(k_{j_0}k_j)^{1/2}} \sum_l (2l+1)\, D^l_{00}(0,\chi,0)\, T_{lm_0}(j|j_0) \tag{62}$$

and would now be diagonal in m, m_0. Then to obtain the helicity amplitude in this case, we would set $\mathscr{R}_i$ equal to the identity and take $\mathscr{R}_f$ equal to $(0, \chi, 0)$. Our result would then be[12]

$$f(j\lambda|j_0m_0|\chi) = \sum_m f(jm|j_0m_0|\chi)\, D^{j*}_{\lambda m}(0,\chi,0) \tag{63}$$

and using equation (45), we have [by the reality of $D^j_{\lambda m}(0,\chi,0)$]

$$f(j\lambda|j_0m_0|\chi) = \frac{i}{2(k_{j_0}k_j)^{1/2}} \sum_{l,m} (2l+1)\, D^l_{00}(0,\chi,0)\, \delta_{mm_0} T_{lm}(j|j_0)\, D^j_{\lambda m}(0,\chi,0) \tag{64}$$

$$= \frac{i}{2(k_{j_0}k_j)^{1/2}} \sum_l (2l+1)\, D^l_{00}(0,\chi,0)\, D^j_{\lambda m_0}(0,\chi,0)\, T_{lm_0}(j|j_0) \tag{65}$$

Now we again use equation (60) and write

$$f(j\lambda|j_0m_0|\chi) = \frac{i}{2(k_{j_0}k_j)^{1/2}} \sum_{l,J} (2J+1)\, D^J_{\lambda m_0}(0,\chi,0) \langle l0j\lambda|J\lambda\rangle \times \langle l0jm_0|Jm_0\rangle \frac{2l+1}{2J+1} T_{lm_0}(j|j_0) \tag{66}$$

Finally, we are led to introduce a helicity T matrix $T^J(j\lambda|j_0m_0)$ given by

$$T^J(j\lambda|j_0m_0) = \sum_l \frac{2l+1}{2J+1} \langle l0j\lambda|J\lambda\rangle \langle l0jm_0|Jm_0\rangle\, T_{lm_0}(j|j_0) \tag{67}$$

and the helicity amplitude is then given by

$$f(j\lambda|j_0m_0|\chi) = \frac{i}{2(k_{j_0}k_j)^{1/2}} \sum_J (2J+1)\, D^J_{\lambda m_0}(0,\chi,0)\, T^J(j\lambda|j_0m_0) \tag{68}$$

Comparison of equations (61) and (68) shows them to be identical *in form* except for the unimportant phase factor $(-1)^{m_0-\lambda_0}$. However, the important substantive difference lies in the two different expressions for $T^J(j\lambda|j_0m_0)$, equations (60) and (67). The former is referred to as an initial l-labeled result (CS-l_i) and was first given for the CS by Khare *et al.*,[16] while equation (67) is referred to as a final l-labeled result (CS-l_f) and was first given by Kouri and Shimoni.[9,12] Khare *et al.*[16] note that if one desires $f(jm|j_0m_0|\hat{\mathbf{R}})$, then the CS-$l_i$ should be used, and if one desires $f(j\lambda|j_0\lambda_0|\hat{\mathbf{R}})$, then the CS-$l_f$ should be used. Further, they show that for the standard *experimental* axis, both CS-l_f and CS-l_i give *identical* cross sections, and therefore the two choices are equivalent. We remark that, as noted by Kouri and Shimoni[12] and Khare,[14] the degeneracy-averaged cross sections are *independent* of the quantization axis choice. This is true both for approximate and exact cross sections. This result stems from the unitarity of the rotation matrices. Explicitly, we have from equation (54)

$$\sum_{\substack{\nu \\ \nu_0}} |f(j\nu|j_0\nu_0|\chi)|^2 = \sum_{\substack{\lambda\lambda_0 \\ \lambda'\lambda_0' \\ \nu\nu_0}} D^j_{\nu\lambda'}(\mathscr{R}_f)\, D^{j*}_{\nu\lambda}(\mathscr{R}_f)\, f(j\lambda|j_0\lambda_0|\chi) f^*(j\lambda'|j_0\lambda_0'|\chi) \times D^{j_0*}_{\nu_0\lambda'_0}(\mathscr{R}_i)\, D^{j_0}_{\nu_0\lambda_0}(\mathscr{R}_i) \tag{69}$$

and the unitarity of the $D^j_{\nu\lambda}$ implies that[75,77]

$$\sum_{\nu} D^j_{\nu\lambda'}(\mathscr{R})\, D^{j*}_{\nu\lambda}(\mathscr{R}) = \delta_{\lambda\lambda'} \tag{70}$$

We then obtain the general result

$$\sum_{\nu\nu_0} |f(j\nu|j_0\nu_0|\chi)|^2 = \sum_{\lambda\lambda_0} |f(j\lambda|j_0\lambda_0|\chi)|^2 \tag{71}$$

for *any* choice of rotations $\mathscr{R}_i$ and $\mathscr{R}_f$. This immediately shows that the degeneracy-averaged CS-l_i and CS-l_f differential and integral cross sections are identical and are given by equations (51) and (53).

2.4. CS Approximation for General Relaxation Cross Sections

In certain derivations, it is helpful to know the CS approximation to other types of T matrices besides $T^J(j\lambda|j_0m_0)$ and $T^{l_0}(j\lambda|j_0\lambda_0)$. Most useful is the CS approximation to the Arthurs–Dalgarno[73] $T^J(jl|j_0l_0)$. The exact equation is easily shown to be[62]

$$T^J(jl|j_0l_0) = \sum_{\lambda m_0} \left(\frac{2l+1}{2J+1}\right)^{1/2} \langle l0j\lambda|J\lambda\rangle\, T^J(j\lambda|j_0m_0) \left(\frac{2l_0+1}{2J+1}\right)^{1/2} \times \langle l_0 0 j_0 m_0|Jm_0\rangle \tag{72}$$

and the CS-l_i and CS-l_f results are obtained by substituting for $T^J(j\lambda|j_0m_0)$ using equations (60) and (67), so that one obtains

$$T^J(jl|j_0l_0) = \sum_{\lambda m_0} \left(\frac{2l+1}{2J+1}\right)^{1/2} \langle l0j\lambda|J\lambda\rangle \sum_L \frac{2L+1}{2J+1} \langle L0\bar{j}\lambda|J\lambda\rangle \times T_{L\bar{\lambda}}(j|j_0) \langle L0\bar{j}m_0|Jm_0\rangle \left(\frac{2l_0+1}{2J+1}\right)^{1/2} \langle l_0 0 j_0 m_0|Jm_0\rangle \tag{73}$$

Here, $\bar{j}$ equals j_0 and $\bar{\lambda} = \lambda$ for the CS-l_i, while $\bar{j}$ equals j and $\bar{\lambda} = m_0$ for the CS-l_f. In this discussion we focus primarily on the results for the CS-l_i (the CS-l_f is obtained in analogous fashion). Thus, using the identity

$$\sum_{m_0}\left(\frac{2L+1}{2J+1}\right)^{1/2}\langle L0j_0m_0|Jm_0\rangle\langle l_00j_0m_0'|Jm_0\rangle\left(\frac{2l_0+1}{2J+1}\right)^{1/2}=\delta_{Ll_0} \quad (74)$$

we obtain

$$T^J(jl|j_0l_0)=\sum_{\lambda}\left(\frac{2l+1}{2J+1}\right)^{1/2}\langle l0j\lambda|J\lambda\rangle\, T_{l_0\lambda}(j|j_0)\left(\frac{2l_0+1}{2J+1}\right)^{1/2}\langle l_00j_0\lambda|J\lambda\rangle \quad (75)$$

The CS-l_f result[12] is identical except for the appearance of $T_{l\lambda_0}(j|j_0)$ in place of $T_{l_0\lambda}(j|j_0)$. The utility of equation (75) is that many cross sections are expressed in terms of the $T^J(jl|j_0l_0)$, and therefore to obtain the corresponding CS-l_i or CS-l_f approximation, one simply substitutes with equation (75). The procedure has been used effectively by Goldflam and Kouri[13] in considering the general relaxation cross section $\sigma_n(j'_aj'_b|j_aj_b)$ of Fano.[78-81] A wide variety of relaxation-type phenomena can be expressed in terms of the $\sigma_n(j'_aj'_b|j_aj_b)$, including ir and Raman line shapes,[79,80] spin–spin and spin–rotation relaxation of NMR,[80] ordinary rotational relaxation,[78,80] and various double-resonance phenomena.[81] The exact CC expression for $\sigma_n(j'_aj'_b|j_aj_b)$ is given by

$$\begin{aligned}\sigma_n(j'_aj'_b|j_aj_b)=\frac{\pi}{k_j^2}\sum_{ll'}\sum_{JJ'}(2J+1)(2J'+1)\left[\frac{2j'_a+1}{2j_a+1}\right]^{1/2}\begin{Bmatrix} j'_a & n & j'_b\\ J' & l' & J\end{Bmatrix}\\ \times\begin{Bmatrix} j_a & n & j_b\\ J' & l & J\end{Bmatrix}[\delta_{j_aj'_a}\delta_{ll'}T^{J'*}(j'_bl'|j_bl)+\delta_{j_bj'_b}\delta_{ll'}T^J(j'_al'|j_al)\\ -T^{J'*}(j'_bl'|j_bl)\,T^J(j'_al'|j_al)]\end{aligned} \quad (76)$$

where $\hbar^2k_j^2/2M$ is the initial relative kinetic energy (the same for $j=j_a$ and j_b), $\{\because\}$ is a six-j symbol,[75] and n is the tensor rank of the operator whose correlations are under study [e.g., for dipole (ir) spectra, $n=1$ since the dipole is a first-rank tensor; for Raman spectra, $n=2$, since the polarizability is a second-rank tensor, and so on]. By specifying j'_a, j'_b, j_a, j_b, n, one selects the cross section appropriate to the process of interest.[80,81] Then by use of equation (75) in equation (76), Goldflam and Kouri[13] show that

$$\begin{aligned}\sigma_n(j'_aj'_b|j_aj_b)=\frac{\pi}{k_j^2}\left[\frac{2j'_a+1}{2j_a+1}\right]^{1/2}\sum_{\substack{l_0\\ \lambda\lambda'}}(-1)^{j_b+j'_b}(2l_0+1)\\ \times\begin{pmatrix} j_a & j_b & n\\ \lambda & -\lambda' & \lambda'-\lambda\end{pmatrix}\begin{pmatrix} j'_a & j'_b & n\\ \lambda & -\lambda' & \lambda'-\lambda\end{pmatrix}\\ \times[\delta_{j_aj'_a}T^*_{l_0\lambda'}(j'_b|j_b)+\delta_{j_bj'_b}T_{l_0\lambda}(j'_a|j_a)-T^*_{l_0\lambda'}(j'_b|j_b)\,T_{l_0\lambda}(j'_a|j_a)]\end{aligned} \quad (77)$$

In the present discussion, we have followed Khare *et al.*[16] in employing the CS-l_i; however, Goldflam and Kouri first obtained the same result with the CS-l_f version of equation (75). Thus the general CS relaxation cross section is independent of the choice of quantization axis (due to the fact that it involves a type of degeneracy average).

Before leaving the CS approximation, we note that in certain theoretical applications it is desirable to know the wave function in various representations. In this connection, we note that the same transformation between the CS T matrix and the P-helicity $T^J(j\lambda|j_0m_0)$, equations (60) and (67), also holds for the radial wave function. Thus, as shown by Kouri *et al.*[15] and Khare *et al.*,[16] respectively, $\psi^J(j\lambda|j_0m_0|R)$ is given by

$$\psi^J(j\lambda|j_0m_0|R) = \sum_l \frac{2l+1}{2J+1}\langle l0j\lambda|J\lambda\rangle\langle l0jm_0|Jm_0\rangle\,\psi_{lm_0}(j|j_0|R) \quad (78)$$

or

$$\psi^J(j\lambda|j_0m_0|R) = \sum_{l_0} \frac{2l_0+1}{2J+1}\langle l_00j_0\lambda|J\lambda\rangle\langle l_00j_0m_0|Jm_0\rangle\,\psi_{l_0\lambda}(j|j_0|R) \quad (79)$$

Of course, equation (78) results from the CS-l_f and equation (79) from the CS-l_i. To date, computational tests of these expressions have not been made, but it is felt that such expressions are of great potential significance for a CS theory of chemical reactions.

The CS theory can also be developed in a parallel fashion for molecule–molecule collisions, with a corresponding increase in the complexity of the equations to be solved. Work in this direction has been reported by Rabitz,[21] Heil and Kouri,[45] De Pristo and Alexander,[82,83] and Heil *et al.*[58] It is felt that the last of these contains the preferred approach since it constitutes an l_f-labeled CS theory, while the former are all J-labeled theories. More work on rotor–rotor collisions can be expected in the future (in particular, the l_i-labeled molecule–molecule theory). In addition, Green[57] and Garrison and Lester[56] have developed treatments of atom–top-molecule collisions, and Heil and Secrest[37,38] have analyzed atom–spherical-top collisions in the CS. Except for Green's work,[57] these treatments are based on J labeling, and thus more work on these types of collision systems should be done. However, for degeneracy-averaged integral cross sections involving low rotor states, the results obtained to date *do* indicate the CS to be quite accurate for such systems as well. We now turn to a discussion of the infinite-order sudden (IOS) approximation.

3. The IOS Approximation

3.1. Basic IOS Equations and Boundary Conditions

The infinite-order sudden approximation[84-129] can be viewed as an energy sudden approximation to the CS approximation.[2,119] In the CS, the $\mathbf{l}^2$ operator is replaced by an effective eigenvalue form $L(L+1)\hbar^2$, thereby enabling one to use the closure property of the complete set to carry out the sums over l, μ in equation (20). This effectively replaces the Green's function in the integral equation by a local operator.[66] *In the IOS, we do the same sort of approximation to the rotational energy of the diatom,* $\mathbf{j}^2/2I$; *thus we replace* $\mathbf{j}^2$ *by an effective eigenvalue form* $j(j+1)\hbar^2$. This is equivalent to replacing the wave numbers k_j everywhere in equation (23) by an effective wave number k. Then we can use the closure property of the complete set of rotor states $Y_{j\lambda}(\hat{\boldsymbol{\rho}})$,

$$\sum_{j\lambda} Y_{j\lambda}(\hat{\boldsymbol{\rho}})\, Y^*_{j\lambda}(\hat{\boldsymbol{\rho}}') = \delta(\hat{\boldsymbol{\rho}} - \hat{\boldsymbol{\rho}}') \tag{80}$$

to write equation (23) (where we have set $L = l_0$) in the form

$$\begin{aligned}\psi^{l_0}(j_0\lambda_0|\hat{\mathbf{z}}R\hat{\boldsymbol{\rho}}) = {} & Y_{j_0\lambda_0}(\hat{\boldsymbol{\rho}})\, j_{l_0}(kR) \\ & - \frac{2M}{\hbar^2} k \int_0^\infty dR' R'^2 h^+_{l_0}(kR_>)\, j_{l_0}(kR_<) \int d\hat{\boldsymbol{\rho}}' \delta(\hat{\boldsymbol{\rho}} - \hat{\boldsymbol{\rho}}') \\ & \times V(R', \hat{\mathbf{z}}\cdot\hat{\boldsymbol{\rho}}')\, \psi^{l_0}(j_0\lambda_0|\hat{\mathbf{z}}R'\hat{\boldsymbol{\rho}}')\end{aligned} \tag{81}$$

Next the integral over $d\hat{\boldsymbol{\rho}}$ may be done immediately to yield the IOS equation

$$\begin{aligned}\psi^{l_0}(j_0\lambda_0|R\hat{\mathbf{z}}\hat{\boldsymbol{\rho}}) = {} & Y_{j_0\lambda_0}(\hat{\boldsymbol{\rho}})\, j_{l_0}(kR) - \frac{2M}{\hbar^2} k \int_0^\infty dR' R'^2 h^+_{l_0}(kR_>)\, j_{l_0}(kR_<) \\ & \times V(R', \hat{\mathbf{z}}\cdot\hat{\boldsymbol{\rho}})\, \psi^{l_0}(j_0\lambda_0|R'\hat{\mathbf{z}}\hat{\boldsymbol{\rho}})\end{aligned} \tag{82}$$

with the equivalent differential equation being

$$\left[\frac{\hbar^2}{2M}\frac{1}{R}\frac{\partial^2}{\partial R^2}R - \frac{l_0(l_0+1)}{2MR^2}\hbar^2 + \frac{\hbar^2 k^2}{2M} - V(R, \hat{\mathbf{z}}\cdot\hat{\mathbf{R}})\right]\psi^{l_0}(j_0\lambda_0|R\hat{\mathbf{z}}\hat{\boldsymbol{\rho}}) = 0 \tag{83}$$

We also notice that there is now *no* integral over the relative angle $\cos^{-1}(\hat{\mathbf{z}}\cdot\hat{\boldsymbol{\rho}})$, i.e., equation (82) is local in this variable. This is analogous to the effect that there is no coupling in the projection λ in equation (28), so that the Kronecker delta $\delta_{\lambda\lambda_0}$ could be factored out. In the case of equation (82), we can now factor out the $Y_{j_0\lambda_0}(\hat{\boldsymbol{\rho}})$ in the inhomogeneity by writing

$$\psi^{l_0}(j_0\lambda_0|R\hat{\mathbf{z}}\hat{\boldsymbol{\rho}}) = Y_{j_0\lambda_0}(\hat{\boldsymbol{\rho}})\, \psi_{l_0}(R, \hat{\mathbf{z}}\cdot\hat{\boldsymbol{\rho}}) \tag{84}$$

Then substituting equation (84) into (82), we find that the function $\psi_{l_0}(R, \hat{\mathbf{z}} \cdot \hat{\boldsymbol{\rho}})$ must satisfy

$$\psi_{l_0}(R, \hat{\mathbf{z}} \cdot \hat{\boldsymbol{\rho}}) = j_{l_0}(kR) - \frac{2M}{\hbar^2} k \int_0^\infty dR' R'^2 h_{l_0}^+ (kR_>)\, j_{l_0}(kR_<) \times V(R', \hat{\mathbf{z}} \cdot \hat{\boldsymbol{\rho}})\, \psi_{l_0}(R', \hat{\mathbf{z}} \cdot \hat{\boldsymbol{\rho}}) \tag{85}$$

and in place of the differential equation (83), we have

$$\left[\frac{\hbar^2}{2M} \frac{1}{R} \frac{\partial^2}{\partial R^2} R - \frac{l_0(l_0 + 1)}{2MR^2} \hbar^2 + \frac{\hbar^2 k^2}{2M} - V(R, \hat{\mathbf{z}} \cdot \hat{\boldsymbol{\rho}}) \right] \psi_{l_0}(R, \hat{\mathbf{z}} \cdot \hat{\boldsymbol{\rho}}) = 0 \tag{86}$$

It is now clear that in the IOS, the relative angle between $\hat{\mathbf{R}}$ ($\equiv \hat{\mathbf{z}}$) and $\hat{\boldsymbol{\rho}}$ enters solely as a parameter. One carries out a scattering calculation at a fixed angle of orientation between the axis of the rotor and the vector $\mathbf{R}$ to the atom. Then the scattering information is obtained by projecting the fixed-orientation scattering amplitude with the appropriate initial and final rotor states. To see explicitly how this comes about, we note that by equations (13), (17), (19), and (84),

$$\psi_{\mathbf{k}_{j_0}}(j_0\lambda_0 | \hat{\mathbf{R}}\hat{\boldsymbol{\rho}}) = 4\pi \sum_{l_0\mu_0} i^{l_0} Y^*_{l_0\mu_0}(\hat{\mathbf{k}}_{j_0})\, Y_{l_0\mu_0}(\hat{\mathbf{z}})\, Y_{j_0\lambda_0}(\hat{\boldsymbol{\rho}})\, \psi_{l_0}(R, \hat{\mathbf{z}} \cdot \hat{\boldsymbol{\rho}}) \tag{87a}$$

$$= \sum_{l_0} (2l_0 + 1)\, i^{l_0} P_{l_0}(\hat{\mathbf{R}} \cdot \hat{\mathbf{k}}_{j_0})\, Y_{j_0\lambda_0}(\hat{\boldsymbol{\rho}})\, \psi_{l_0}(R, \hat{\mathbf{z}} \cdot \hat{\boldsymbol{\rho}}) \tag{87b}$$

To obtain the differential scattering amplitude, we must examine the asymptotic limit of $\psi_{\mathbf{k}_{j_0}}(j_0\lambda_0 | \mathbf{R}\hat{\boldsymbol{\rho}})$, which in turn requires the asymptotic limit of $\psi_{l_0}(R, \hat{\mathbf{z}} \cdot \hat{\boldsymbol{\rho}})$. This can be readily obtained from equation (85), so that

$$\lim_{R\to\infty} \psi_{l_0}(R, \hat{\mathbf{z}} \cdot \hat{\boldsymbol{\rho}}) = j_{l_0}(kR) - \frac{2M}{\hbar^2} \frac{e^{ikR}}{R} i^{-l_0} \int_0^\infty dR' R'^2 j_{l_0}(kR') \times V(R', \hat{\mathbf{z}} \cdot \hat{\boldsymbol{\rho}})\, \psi_{l_0}(R', \hat{\mathbf{z}} \cdot \hat{\boldsymbol{\rho}}) \tag{88}$$

Equation (88) suggests that we introduce the *orientation-dependent T matrix element*

$$T_{l_0}(\hat{\mathbf{z}} \cdot \hat{\boldsymbol{\rho}}) = \frac{4iM}{\hbar^2} k \int_0^\infty dR R^2 j_{l_0}(kR)\, V(R, \hat{\mathbf{z}} \cdot \hat{\boldsymbol{\rho}})\, \psi_{l_0}(R, \hat{\mathbf{z}} \cdot \hat{\boldsymbol{\rho}}) \tag{89}$$

Then the asymptotic limit of $\psi_{\mathbf{k}_{j_0}}(j_0\lambda_0 | \mathbf{R}\hat{\boldsymbol{\rho}})$ becomes

$$\lim_{R\to\infty} \psi_{\mathbf{k}_{j_0}}(j_0\lambda_0 | \mathbf{R}\hat{\boldsymbol{\rho}}) = 4\pi \sum_{l_0\mu_0} i^{l_0} Y^*_{l_0\mu_0}(\hat{\mathbf{k}}_{j_0})\, Y_{l_0\mu_0}(\hat{\mathbf{z}})\, Y_{j_0\lambda_0}(\hat{\boldsymbol{\rho}}) \times \left[j_{l_0}(kR) + \frac{i}{2kR} e^{ikR} i^{-l_0} T_{l_0}(\hat{\mathbf{z}} \cdot \hat{\boldsymbol{\rho}}) \right] \tag{90}$$

The first term in brackets yields the incoming plane wave. The scattered portion of the wave function is given by

$$\lim_{R\to\infty} \psi^{\text{scatt}}_{\mathbf{k}_{j_0}}(j_0\lambda_0|\mathbf{R}\hat{\rho}) = \frac{i}{2k}\sum_{l_0}(2l_0+1)P_{l_0}(\hat{\mathbf{z}}\cdot\hat{\mathbf{k}}_{j_0})\; Y_{j_0\lambda_0}(\hat{\rho})\, T_{l_0}(\hat{\mathbf{z}}\cdot\hat{\rho})\frac{e^{ikR}}{R} \tag{91}$$

so we conclude that we are to identify

$$\sum_{j\lambda}\frac{e^{ik_jR}}{R} Y_{j\lambda}(\hat{\rho})\, f(j\lambda|j_0\lambda_0|\hat{\mathbf{z}}\cdot\hat{\mathbf{k}}_{j_0}) = \frac{i}{2kR}e^{ikR}\sum_{l_0}(2l_0+1)P_{l_0}(\hat{\mathbf{z}}\cdot\hat{\mathbf{k}}_{j_0}) \times Y_{j_0\lambda_0}(\hat{\rho})\, T_{l_0}(\hat{\mathbf{z}}\cdot\hat{\rho}) \tag{92}$$

To do this, we can expand the angle dependence of the product $Y_{j_0\lambda_0}(\hat{\rho}) \times T_{l_0}(\hat{\mathbf{z}}\cdot\hat{\rho})$ in terms of the complete set $Y_{j\lambda}(\hat{\rho})$, according to

$$Y_{j_0\lambda_0}(\hat{\rho})\, T_{l_0}(\hat{\mathbf{z}}\cdot\hat{\rho}) = \sum_{j\lambda} Y_{j\lambda}(\hat{\rho})\, T^{l_0}_{\text{IOS}}(j\lambda|j_0\lambda_0) \tag{93}$$

where, by the orthonormality of the $Y_{j\lambda}(\hat{\rho})$, we see that

$$T^{l_0}_{\text{IOS}}(j\lambda|j_0\lambda_0) = \int d\hat{\rho}\, Y^*_{j\lambda}(\hat{\rho})\, T_{l_0}(\hat{\mathbf{z}}\cdot\hat{\rho})\, Y_{j_0\lambda_0}(\hat{\rho}) \tag{94}$$

Now it is clear that $T_{l_0}(\hat{\mathbf{z}}\cdot\hat{\rho})$ depends *only* on the azimuthal angle θ of the rotor orientation $\hat{\rho}$. As a result, the right-hand side of equation (94) vanishes unless λ equals λ_0, so that

$$T^{l_0}_{\text{IOS}}(j\lambda|j_0\lambda_0) = \delta_{\lambda\lambda_0}\int d\hat{\rho}\, Y^*_{j\lambda_0}(\hat{\rho})\, T_{l_0}(\hat{\mathbf{z}}\cdot\hat{\rho})\, Y_{j_0\lambda_0}(\hat{\rho}) \tag{95}$$

$$= \delta_{\lambda\lambda_0} T^{\text{IOS}}_{l_0\lambda}(j|j_0) \tag{96}$$

Our notation is intended to emphasize the viewpoint that the IOS is an energy sudden approximation to the CS.[2,119] Using equations (93) and (96) in equation (92), we obtain the condition

$$\sum_{j\lambda}\frac{e^{ik_jR}}{R} Y_{j\lambda}(\hat{\rho})\, f(j\lambda|j_0\lambda_0|\hat{\mathbf{z}}\cdot\hat{\mathbf{k}}_{j_0}) = \frac{i}{2kR}e^{ikR}\sum_{j\lambda} Y_{j\lambda}(\hat{\rho})\,\delta_{\lambda\lambda_0}\sum_{l_0}(2l_0+1) \times P_{l_0}(\hat{\mathbf{z}}\cdot\hat{\mathbf{k}}_{j_0})\, T^{\text{IOS}}_{l_0\lambda}(j|j_0) \tag{97}$$

and projecting with a particular $Y^*_{j\lambda}(\hat{\rho})$ and noting that $k \approx k_j$ within the sudden approximation, we obtain

$$f(j\lambda|j_0\lambda_0|\hat{\mathbf{z}}\cdot\hat{\mathbf{k}}_{j_0}) = \frac{i\delta_{\lambda\lambda_0}}{2k}\sum_{l_0}(2l_0+1)P_{l_0}(\hat{\mathbf{z}}\cdot\hat{\mathbf{k}}_{j_0})\, T^{\text{IOS}}_{l_0\lambda}(j|j_0) \tag{98}$$

as the IOS-l_i differential scattering amplitude. Just as in the case of the CS, if we recall that $\hat{\mathbf{z}} = \hat{\mathbf{R}}$, equation (98) can be written in the symmetric form

$$f(j\lambda|j_0\lambda_0|\hat{\mathbf{R}}\cdot\hat{\mathbf{k}}_{j_0}) = \frac{i\delta_{\lambda\lambda_0}}{2k}\sum_{l_0}(2l_0+1)\,P_{l_0}(\hat{\mathbf{R}}\cdot\hat{\mathbf{k}}_{j_0})\,T^{\mathrm{IOS}}_{l_0\lambda}(j|j_0) \tag{99}$$

Then if we choose $\hat{\mathbf{z}} = \hat{\mathbf{R}}$, we obtain equation (98) and the IOS differential scattering amplitude is diagonal in the projection along the final observational direction $\hat{\mathbf{R}}$.[121] If we choose $\hat{\mathbf{z}} = \hat{\mathbf{k}}_{j_0}$, then λ, λ_0 are interpreted as projections along the incident momentum (which we usually denote m, m_0, respectively), and the $f(jm|j_0m_0|\hat{\mathbf{R}}\cdot\hat{\mathbf{z}})$ becomes[119]

$$f(jm|j_0m_0|\hat{\mathbf{R}}\cdot\hat{\mathbf{z}}) = \frac{i\delta_{mm_0}}{2k}\sum_{l}(2l+1)\,P_l(\hat{\mathbf{R}}\cdot\hat{\mathbf{z}})\,T^{\mathrm{IOS}}_{lm_0}(j|j_0) \tag{100}$$

It should be obvious that except for the difference in interpretation, the *numerical* results of equations (99) and (100) will be identical when $\lambda = m$, $\lambda_0 = m_0$. Equation (99) should be used to calculate Δm transitions via equation (54), and equation (100) should be used to calculate $\Delta\lambda$ transitions via equation (54). *Again, the* IOS-l_i *and* IOS-l_f *predict identical results for the standard experimental axis of quantization. Further, for all degeneracy-averaged cross sections, equations* (99) *and* (100) *yield exactly the same results.*[12,14]

3.2. IOS Cross Sections and Factorizations

We are now prepared to obtain various cross sections in the IOS approximation. The most efficient and useful forms for these results are the factored IOS expressions obtained by Goldflam *et al.*[119,120] This results in expressions for various cross sections given in terms of sums of products of system-independent spectroscopic coefficients times a few system-dependent dynamical coefficients.[119,120] This result is based on the observation that by equations (95)–(96), $T^{\mathrm{IOS}}_{l_0\lambda}(j|j_0)$ can be expressed as

$$T^{\mathrm{IOS}}_{l_0\lambda}(j|j_0) = \int d\hat{\boldsymbol{\rho}}\,Y^*_{j\lambda_0}(\hat{\boldsymbol{\rho}})\,T_{l_0}(\theta)\,Y_{j_0\lambda_0}(\hat{\boldsymbol{\rho}}) \tag{101}$$

where $\cos\theta = \hat{\mathbf{z}}\cdot\hat{\boldsymbol{\rho}}$. Then one can expand $T_{l_0}(\theta)$ in terms of Legendre polynomials $P_L(\cos\theta)$, so that

$$T_{l_0}(\theta) = \sum_{L} P_L(\cos\theta)\,T^L_{l_0} \tag{102}$$

The significance of equation (102) is that since the IOS localizes the Green's operator in the Lippmann–Schwinger equation, the transition or T operator is also local in $\hat{\mathbf{z}}\cdot\hat{\boldsymbol{\rho}}$, and therefore in θ. As a result, it can be expanded in a Legendre expansion just like the potential, $V(R,\hat{\mathbf{R}}\cdot\hat{\boldsymbol{\rho}}) = V(R,\hat{\mathbf{z}}\cdot\hat{\boldsymbol{\rho}})$, *but the number of terms required for equation* (102) *to converge will, in general, differ from the number required to expand the potential.* This is because the IOS is not a perturbation solution to the Lippmann–Schwinger equation and therefore we are not just approximating T by V. Of course, it is clear that Born-type approximations to T will result in *similar* expansions and therefore, in similar factorizations. We feel, however, that the IOS will be more accurate in situations where the potential is strong. Given equation (102), the integral over $d\hat{\boldsymbol{\rho}}$ can be performed analytically, so that equation (101) becomes [in analogy to equation (27)][120]

$$T^{\text{IOS}}_{l_0\lambda}(j\,|\,j_0) = (-1)^{\lambda}\,[(2j+1)(2j_0+1)]^{1/2} \sum_L \begin{pmatrix} j & L & j_0 \\ 0 & 0 & 0 \end{pmatrix} \begin{pmatrix} j & L & j_0 \\ \lambda & 0 & -\lambda \end{pmatrix} T^L_{l_0} \tag{103}$$

This is a very valuable result since it represents the factored form of the IOS approximation to the CS amplitude $T_{l_0\lambda}(j\,|\,j_0)$. As a result, we may use equation (103) in *any* CS expression in order to obtain the corresponding IOS cross section. For example, the IOS approximation to the P-helicity amplitude $T^J(j\lambda|\,j_0m_0)$ is found by using equation (103) in equation (60), with the result being[120]

$$T^J_{\text{IOS}}(j\lambda|\,j_0m_0) = (-1)^{\lambda}\,[(2j+1)(2j_0+1)]^{1/2} \sum_{l_0L} \frac{2l_0+1}{2J+1} \langle l_0 0 j_0 \lambda | J\lambda\rangle \times \langle l_0 0 j_0 m_0 | J m_0\rangle \begin{pmatrix} j & L & j_0 \\ 0 & 0 & 0 \end{pmatrix} \begin{pmatrix} j & L & j_0 \\ \lambda & 0 & -\lambda \end{pmatrix} T^L_{l_0} \tag{104}$$

Similarly, Goldflam *et al.*[120] have obtained the IOS approximation to the Arthurs–Dalgarno amplitude,[73]

$$T^J_{\text{IOS}}(jl\,|\,j_0l_0) = \sum_L f_L(jl\,|\,j_0l_0|J)\,T^L_{l_0} \tag{105}$$

where the Percival–Seaton coefficient[130] $f_L(jl\,|\,j_0l_0|J)$ is given by

$$f_L(jl\,|\,j_0l_0|J) = \begin{pmatrix} j & L & j_0 \\ 0 & 0 & 0 \end{pmatrix} \frac{[(2j+1)(2j_0+1)(2l+1)(2l_0+1)]^{1/2}}{2J+1} \times \sum_{\lambda} (-1)^{\lambda} \langle l0j\lambda|J\lambda\rangle \langle l_0 0 j_0 \lambda | J\lambda\rangle \begin{pmatrix} j & L & j_0 \\ \lambda & 0 & -\lambda \end{pmatrix} \tag{106}$$

Equation (105) could also be derived directly from equation (102), if we recall that in this equation, $\cos\theta = \hat{\mathbf{R}}\cdot\hat{\boldsymbol{\rho}}$. Then by standard techniques,[130]

$$T^J_{\text{IOS}}(jl\,|\,j_0l_0) = \int d\hat{\mathbf{R}}\, d\hat{\boldsymbol{\rho}}\, \mathscr{Y}^{JM*}_{jl}(\hat{\mathbf{R}}\hat{\boldsymbol{\rho}}) \sum_L P_L(\cos\theta)\, T^L_{l_0} \mathscr{Y}^{JM}_{j_0l_0}(\hat{\mathbf{R}}\hat{\boldsymbol{\rho}}) \tag{107}$$

$$= \sum_L f_L(jl\,|\,j_0l_0|J)\,T^L_{l_0} \tag{108}$$

where the $\mathscr{Y}_{jl}^{JM}(\hat{\mathbf{R}}\hat{\rho})$ are the usual coupled representation total angular momentum functions[73]

$$\mathscr{Y}_{jl}^{JM}(\hat{\mathbf{R}}\hat{\rho}) = \sum_{\mu m} \langle l\mu jm|JM\rangle\, Y_{jm}(\hat{\rho})\, Y_{l\mu}(\hat{\mathbf{R}}) \tag{109}$$

Again, the result equation (106) can be viewed as following from the localization of the T operator in the angle between $\hat{\mathbf{R}}$ and $\hat{\rho}$. This results in $T^J(jl\,|\,j_0l_0)$ having the same *form* as the corresponding matrix element of the potential,[73]

$$V^J(jl\,|\,j_0l_0|R) = \sum_t V_t(R)\, f_t(jl\,|\,j_0l_0|J) \tag{110}$$

where $V_t(R)$ and $T_{l_0}^L$ play similar roles. It is also worth noting that equation (106) implies the usual selection rules on Δj, Δl transitions, as well as triangle inequalities on J, j, l and on J, j_0, l_0. These arise because of the Percival–Seaton coefficients. Thus the IOS ensures both conservation of parity, so that $j + l + j_0 + l_0$ is even, and that

$$|J - j| \le l \le J + j \tag{111}$$

$$|J - j_0| \le l_0 \le J + j_0 \tag{112}$$

We now wish to consider the various IOS cross sections in factored form. In this connection, we note that Goldflam *et al.*[120] have proved an optical theorem for the multipole components of the localized T operator, $T_{l_0}^L$. The result follows from the optical theorem satisfied by the CS (or IOS) T matrices:

$$2\mathrm{Re}\,T_{l_0\lambda}(j_0|j_0) = \sum_j |T_{l_0\lambda}(j\,|\,j_0)|^2 \tag{113}$$

Then it can be shown that

$$\mathrm{Re}\,T_{l_0}^0 = \sum_L \frac{1}{2L+1}|T_{l_0}^L|^2 \tag{114}$$

which result is useful in simplifying IOS expressions for certain cross sections. The reader is referred to the work of Goldflam *et al.*[120] for a detailed derivation of equation (114).

If one uses equation (103) in equation (99) and defines the system-independent spectroscopic coefficient $G_{L\lambda}(j\,|\,j_0)$ by

$$G_{L\lambda}(j\,|\,j_0) = [(2j_0+1)(2j+1)(2L+1)]^{1/2}(-1)^\lambda \times \begin{pmatrix} j & L & j_0 \\ 0 & 0 & 0 \end{pmatrix}\begin{pmatrix} j & L & j_0 \\ \lambda & 0 & -\lambda \end{pmatrix} \tag{115}$$

and the system-dependent dynamical coefficient $q_L(\hat{\mathbf{R}}\cdot\hat{\mathbf{k}}_{j_0})$ by

$$q_L(\hat{\mathbf{R}}\cdot\hat{\mathbf{k}}_{j_0}) = \frac{i}{2k(2L+1)^{1/2}}\sum_{l_0}(2l_0+1)\,P_{l_0}(\hat{\mathbf{R}}\cdot\hat{\mathbf{k}}_{j_0})\,T_{l_0}^L \tag{116}$$

then the differential scattering amplitude is expressed in factored form as

$$f(j\lambda|j_0\lambda_0|\chi) = \frac{\delta_{\lambda\lambda_0}k}{(k_{j_0}k_j)^{1/2}} \sum_L G_{L\lambda}(j\,|\,j_0)\, q_L(\chi) \tag{117}$$

where $\cos\chi$ equals $\hat{\mathbf{R}}\cdot\hat{\mathbf{k}}_{j_0}$. It is then noted that for j_0 equal to zero, one obtains

$$G_{L\lambda}(j\,|0) = \delta_{\lambda 0}\delta_{jL} \tag{118}$$

so that

$$f(j0|00|\chi) = \frac{k}{(k_0k_j)^{1/2}}\, q_j(\chi) \tag{119}$$

It immediately follows that[120]

$$f(j\lambda|j_0\lambda_0|\chi) = \delta_{\lambda\lambda_0} \sum_L \left(\frac{k_0k_L}{k_{j_0}k_j}\right)^{1/2} G_{L\lambda}(j\,|\,j_0)\, f(L0|00|\chi) \tag{120}$$

This result implies that the general definite polarization differential scattering amplitude, with both initial and final states quantized along the final observation direction $\hat{\mathbf{R}}$, is expressed in terms of the amplitudes for the transitions $j_0 = \lambda_0 = 0 \rightarrow j = L$, $\lambda = 0$, with appropriate system-independent weights. The detailed angular momentum transfer information is built into the $G_{L\lambda}(j\,|\,j_0)$, with the index L being the "interaction" multipole responsible for changing the rotor state from j_0 to j. Its importance for any given system is governed by the amplitude $f(L0|00|\chi)$. Now, for *very* weak potentials, $T_{l_0}(\hat{\mathbf{R}}\cdot\hat{\boldsymbol{\rho}})$ will be proportional to $V(\hat{\mathbf{R}}\cdot\hat{\boldsymbol{\rho}})$, and $f(L0|00|\chi)$ becomes approximately a Born-type amplitude. Of course, for stronger interactions, this is no longer true but the factorization equation (120) will still be valid so long as the closure approximations are permitted. We shall discuss this later when we consider validity criteria in more detail. We shall also see later that such factorizations depend only on the energy sudden approximation to $\mathbf{j}^2$, and not on the centrifugal sudden approximation to $\mathbf{l}^2$.

In order to find the differential scattering amplitude quantized with respect to arbitrary axes, we simply substitute equation (120) into equation (54). It is clear that the factorization remains valid, except that the spectroscopic factors become dependent now on the rotations $\mathscr{R}_i$ and $\mathscr{R}_f$. The result is

$$f(j\nu|j_0\nu_0|\chi) = \sum_L \left(\frac{k_0k_L}{k_{j_0}k_j}\right)^{1/2} \mathscr{G}_{L\nu\nu_0}(j\,|\,j_0|\mathscr{R}_i, \mathscr{R}_f)\, f(L0|00|\chi) \tag{121}$$

where

$$\mathscr{G}_{L\nu\nu_0}(j\,|\,j_0|\mathscr{R}_i, \mathscr{R}_f) = \sum_\lambda D^{j*}_{\nu\lambda}(\mathscr{R}_f)\, G_{L\lambda}(j\,|\,j_0)\, D^{j_0}_{\nu_0\lambda}(\mathscr{R}_i) \tag{122}$$

The differential cross section for the transition $j_0\lambda_0 \to j\lambda$ is given by

$$\frac{d\sigma}{d\Omega}(j_0\lambda_0 \to j\lambda) = \frac{k_j}{k_{j_0}} |f(j\lambda|j_0\lambda_0|\chi)|^2 \tag{123}$$

$$= \frac{\delta_{\lambda\lambda_0}}{k_{j_0}^2} \sum_{LL'} G_{L\lambda}(j|j_0)\, G_{L'\lambda}(j|j_0)\, f(L0|00|\chi)\, f^*(L'0|00|\chi) \times k_0(k_L k_{L'})^{1/2} \tag{124}$$

This does not appear to simplify further. The form of equation (124) is interesting in that it implies that in the amplitude for the transition $j_0\lambda_0 \to j\lambda$, there is interference between amplitudes for transitions $0, 0 \to L, 0$ and $0, 0 \to L', 0$, where $L \neq L'$. Similarly, the general $j_0 v_0 \to jv$ differential cross section can be gotten from equations (121) and (122), but is rather messy. However, computationally, these expressions still lead to significant savings of labor compared to other approaches. If one is interested in degeneracy-averaged differential cross sections, the situation is much simpler. As noted by Kouri and Shimoni[12] and Khare,[14] the various quantization schemes yield the same results so we need consider only the result of degeneracy averaging equation (124). Then summing over final projections and averaging over initial projections, it is easily shown that

$$\frac{d\sigma}{d\Omega}(j_0 \to j) = \frac{k_0^2}{k_{j_0}^2}(2j+1) \sum_L \begin{pmatrix} j_0 & L & j \\ 0 & 0 & 0 \end{pmatrix}^2 \frac{k_L}{k_0} |f(L0|00|\chi)|^2 \tag{125}$$

Finally, we note that in the IOS,

$$\frac{k_L}{k_0} |f(L0|00|\chi)|^2 = \frac{d\sigma}{d\Omega}(0 \to L) \tag{126}$$

so that

$$\frac{d\sigma}{d\Omega}(j_0 \to j) = \frac{k_0^2}{k_{j_0}^2}(2j+1) \sum_L \begin{pmatrix} j_0 & L & j \\ 0 & 0 & 0 \end{pmatrix}^2 \frac{d\sigma}{d\Omega}(0 \to L) \tag{127}$$

Thus we have succeeded in expressing the general degeneracy-averaged differential cross section in terms of a sum of simple system-independent spectroscopic factors times the degeneracy-averaged differential cross sections for scattering out of $j_0 = 0$. Following Parker and Pack[121] and Goldflam *et al.*,[120] we note that if one does not resolve the final state j, then an additional sum over final state j in equation (127) is made to yield the result

$$\sum_j \frac{d\sigma}{d\Omega}(j_0 \to j) = \frac{k_0^2}{k_{j_0}^2} \sum_L \frac{d\sigma}{d\Omega}(0 \to L) \tag{128}$$

This result implies that within the factor $k_0^2/k_{j_0}^2$, related to detailed balance, *the total differential cross section is independent of the initial state of the rotor.*

Results similar to equations (124), (127), and (128) for the integral cross sections are obtained when integrated over all solid angles. The definite polarization integral cross section $\sigma(j_0\lambda_0 \to j\lambda)$ becomes

$$\sigma(j_0\lambda_0 \to j\lambda) = \frac{\delta_{\lambda\lambda_0}\pi}{k_{j_0}^2}(2j+1)(2j_0+1)\sum_{\substack{l_0\\ LL'}}(2l_0+1)\begin{pmatrix} j & L & j_0 \\ 0 & 0 & 0 \end{pmatrix}$$
$$\times \begin{pmatrix} j & L' & j_0 \\ 0 & 0 & 0 \end{pmatrix}\begin{pmatrix} j & L & j_0 \\ \lambda & 0 & -\lambda \end{pmatrix}\begin{pmatrix} j & L' & j_0 \\ \lambda & 0 & -\lambda \end{pmatrix} T_{l_0}^{L'*}T_{l_0}^{L} \tag{129}$$

which does not appear to simplify further. Notice that it is *not* expressed as a sum of products of spectroscopic coefficients times *cross sections*, but rather the $T_{l_0}^L$ appear explicitly. Furthermore, we again see interference effects between terms arising from the amplitudes for the $0, 0 \to L, 0$ and $0, 0 \to L', 0$ transitions. The situation for the degeneracy-averaged integral cross section is much simpler, and from equation (127), we see immediately that

$$\sigma(j_0 \to j) = \frac{k_0^2}{k_{j_0}^2}(2j+1)\sum_L \begin{pmatrix} j_0 & L & j \\ 0 & 0 & 0 \end{pmatrix}^2 \sigma(0 \to L) \tag{130}$$

This result has been obtained by several authors in various ways.[25,120-123] If we sum over j, we obtain[120-122]

$$\sum_j \sigma(j_0 \to j) = \frac{k_0^2}{k_{j_0}^2}\sum_L \sigma(0 \to L) \tag{131}$$

which again implies that, apart from the detailed balance factor $(k_0^2/k_{j_0}^2)$, the total integral cross section is independent of the initial state.

In addition to factorizations of ordinary cross sections, Goldflam *et al.*[119] have also considered Fano's general relaxation cross section.[78-80] In the preceding section we gave the CS expression for $\sigma_n(j'_a j'_b | j_a j_b)$ in terms of the quantities $T_{l_0\lambda}(j\,|\,j_0)$ generated in a CS calculation. The detailed derivation is given in the work of Goldflam *et al.*[119] and the final result is

$$\sigma_n(j'_a j'_b | j_a j_b) = \frac{k_0^2}{k_j^2}\left(\frac{2j'_a+1}{2j_a+1}\right)^{1/2} \sum_{L\neq 0} \sigma(0 \to L)\;[\delta_{j_a j'_a}\delta_{j_b j'_b} - f_L(j'_a j'_b | j_a j_b | n)] \tag{132}$$

This expression gives the general relaxation cross section as a sum of the *inelastic* cross sections out of $j_0 = 0$, $\sigma(0 \to L)$, times the system-independent spectroscopic coefficients $\delta_{j_a j_a'}\delta_{j_b j_b'} - f_L(j_a' j_b' | j_a j_b | n)$. This is an interesting and important result in that it implies that within the IOS approximation for pure rotational processes, the general relaxation cross section is always real. In the case of line shapes, this means that the simplest IOS predicts zero line shifts. For most systems in the gas phase, this is probably not a bad approximation.[131,132] In addition, other treatments of line shapes have similarly resulted in expressions which involve an average of inelastic cross sections.[132] These result from the neglect of interference terms in the line shape expression,[132] and the use of the optical theorem. In the present case, Goldflam *et al.*[120] also employ the optical theorem but do *not* neglect the interference terms. We emphasize that this factorization can be employed regardless of how one has obtained the $\sigma(0 \to L)$, $L \neq 0$ (or equivalent dynamical information).

3.3. IOS Factored Rates and Transport Properties

One may also use these results to obtain thermally averaged integral cross sections and rates. The rate for scattering from j_0 to j is given by

$$R(j_0 \to j) = V\sigma(j_0 \to j) \tag{133}$$

where V is the incident relative velocity. Assuming a Boltzmann distribution for the relative velocity one obtains in the simplest approximation[120]

$$R_T(j_0 \to j) = (2j + 1)\sum_L \begin{pmatrix} j_0 & L & j \\ 0 & 0 & 0 \end{pmatrix}^2 R_T(0 \to L) \tag{134}$$

where $R_T(0 \to L)$ is the thermal rate for scattering out of $j_0 = 0$ to $j = L$. Similarly, the thermally averaged general relaxation cross sections are given, in the simplest IOS treatment, by[120]

$$\langle \sigma_n(j_a' j_b' | j_a j_b) \rangle_T = \left(\frac{2j_a' + 1}{2j_a + 1}\right)^{1/2} \sum_{L \neq 0} R_T(0 \to L) \times [\delta_{j_a j_a'}\delta_{j_b j_b'} - f_L(j_a' j_b' | j_a j_b | n)] \tag{135}$$

Parker and Pack[121] have considered transport properties within the IOS and Goldflam *et al.*[120] have given a complementary discussion of viscosity and diffusion in a manner analogous to the preceding discussion. The latter treatment shows how the information content of the viscosity and diffusion coefficient is different (and yet related) to that, say, of the integral state-to-state degeneracy-averaged cross sections or the general relaxation cross section expression equation (135). From a computational standpoint, the treatment of Parker and Pack[121] is preferred, and therefore their expression is given. Because it emphasizes the information content of the viscosity and diffusion coefficients, we also give the expression derived by Goldflam *et al.*[120] In the case of transport properties, especially simple expressions result due to the fact that these quantities involve sums over all final states. The results have similarities to equations (128) and (131) in that dependence on specific rotational states no longer enters. Following Parker and Pack,[121] we note that the transport coefficients are expressed in terms of the collision integrals defined by[133,134]

$$\Omega^{(n,s)} = \frac{1}{2}\left(\frac{k_B T}{2M\pi}\right)^{1/2} \frac{1}{(k_B T)^{s+2}} \int_0^\infty d\varepsilon \varepsilon^{s+1} e^{-\varepsilon/k_B T} Q^{(n)}(\varepsilon) \tag{136}$$

where T is the absolute temperature, k_B is Boltzmann's constant, and ε is the incident relative kinetic energy. The $Q^{(n)}(\varepsilon)$ is a generalized integral cross section given by

$$Q^{(n)}(\varepsilon) = \left[2\pi \sum_j (2j+1)\, e^{-E_j/k_B T} \times \sum_{j'} \int_{-1}^{1} d\cos\chi \frac{d\sigma}{d\Omega}(j \to j')\, \Phi_n(\varepsilon) \right] \Big/ \sum_{j''} (2j''+1)\, e^{-E_{j''}/k_B T} \tag{137}$$

where

$$E_j = j(j+1)\, B_e \tag{138}$$

$$\Phi_1(\varepsilon) = 1 - (\varepsilon'/\varepsilon)^{1/2} \cos\chi \tag{139}$$

$$\Phi_2(\varepsilon) = 1 - (\varepsilon'/\varepsilon)\cos^2\chi - \tfrac{1}{6}(1 - \varepsilon'/\varepsilon)^2 \tag{140}$$

Here ε' is the final kinetic energy, given by

$$\varepsilon' = \varepsilon + E_j - E_{j'} \tag{141}$$

and B_e is the rotor constant. The simplest IOS treatment[120,121] implies that we replace $(\varepsilon'/\varepsilon)$ by one, since it approximates E_j and $E_{j'}$ in equation (141) by $E_{\bar{j}}$. Then $\Phi_n(\varepsilon)$ becomes

$$\Phi_n(\varepsilon) = 1 - \cos^n\chi \tag{142}$$

and we may write equation (137) as

$$Q^{(n)}(\varepsilon) = \left[2\pi \sum_j (2j+1)\, e^{-E_j/kT} \times \sum_{j'} \int_{-1}^{1} d\cos\chi \frac{d\sigma}{d\Omega}(j \to j')(1 - \cos^n\chi) \right] \Big/ \sum_{j''} (2j''+1)\, e^{-E_{j''}/k_B T} \tag{143}$$

Next, we note that by equation (128) [again approximating (k_0/k_{j_0}) by one], the sum over j' leads to a result which is *independent* of j, so that we define $I(\varepsilon, \chi)$ by[121]

$$I(\varepsilon, \chi) = \sum_{j'} \frac{d\sigma}{d\Omega}(j \to j') \tag{144}$$

Then clearly, the partition function cancels in equation (143), and we obtain[121]

$$Q^{(n)}(\varepsilon) = 2\pi \int_{-1}^{1} d\cos\chi \, I(\varepsilon, \chi)(1 - \cos^n\chi) \tag{145}$$

Now by equations (89) and (99) we note that [for $k = (k_{j_0}k_j)^{1/2}$]

$$f(j\lambda | j_0\lambda_0 | \chi) = \frac{i}{2(k_{j_0}k_j)^{1/2}} \sum_{l_0} P_{l_0}(\cos\chi)(2l_0 + 1) \times \int d\hat{\rho}\, Y^*_{j\lambda}(\hat{\rho})\, T_{l_0}(\cos\theta)\, Y_{j_0\lambda_0}(\hat{\rho}) \tag{146}$$

where θ is the angle between the rotor axis and the vector $\mathbf{R}$ from the diatom center of mass to the atom. Now the degeneracy-averaged cross section is

$$\frac{d\sigma}{d\Omega}(j_0 \to j) = \frac{1}{2j_0 + 1} \frac{k_j}{k_{j_0}} \sum_{\lambda\lambda_0} |f(j\lambda | j_0\lambda_0 | \chi)|^2 \tag{147}$$

and so $I(\varepsilon, \chi)$ becomes

$$I(\varepsilon, \chi) = \frac{1}{4(2j_0 + 1)k_{j_0}^2} \sum_{\substack{j \\ \lambda\lambda_0}} \sum_{\substack{l_0 \\ l'_0}} (2l_0 + 1)(2l'_0 + 1) P_{l_0}(\cos\chi) P_{l'_0}(\cos\chi) \times \int d\hat{\rho}\, d\hat{\rho}'\, Y^*_{j\lambda}(\hat{\rho})\, Y_{j\lambda}(\hat{\rho}')\, T_{l_0}(\cos\theta)\, T^*_{l'_0}(\cos\theta')\, Y_{j_0\lambda_0}(\hat{\rho})\, Y^*_{j_0\lambda_0}(\hat{\rho}') \tag{148}$$

Then the sum over j, λ can be performed using the closure relation equation (80), the integral over $d\hat{\rho}'$ then yielding

$$I(\varepsilon, \chi) = \frac{1}{4k_{j_0}^2(2j_0 + 1)} \sum_{\substack{l_0 \\ l'_0}} (2l_0 + 1)(2l'_0 + 1) P_{l_0}(\cos\chi) P_{l'_0}(\cos\chi) \times \int d\hat{\rho} \sum_{\lambda_0} Y_{j_0\lambda_0}(\hat{\rho})\, Y^*_{j_0\lambda_0}(\hat{\rho})\, T_{l_0}(\cos\theta)\, T^*_{l'_0}(\cos\theta) \tag{149}$$

Next, we note the identity[121,135]

$$\sum_{\lambda_0} Y_{j_0\lambda_0}(\hat{\rho})\, Y^*_{j_0\lambda_0}(\hat{\rho}) = \frac{2j_0 + 1}{4\pi} \tag{150}$$

and carry out the integral over the azimuthal angle ϕ of $d\hat{\rho}$ to obtain

$$I(\varepsilon, \chi) = \frac{1}{8k_{j_0}^2} \sum_{l_0 l'_0} (2l_0 + 1)(2l'_0 + 1) P_{l_0}(\cos\chi) P_{l'_0}(\cos\chi) \times \int_{-1}^{1} d\cos\theta\, T_{l_0}(\cos\theta)\, T^*_{l'_0}(\cos\theta) \tag{151}$$

If one then defines[121]

$$f^{\varepsilon}(\theta, \chi) = \frac{i}{2k_{j_0}} \sum_{l_0} (2l_0 + 1) P_{l_0}(\cos\chi)\, T_{l_0}(\cos\theta) \tag{152}$$

then the related "differential cross section" is

$$\frac{d\sigma}{d\chi}(\theta) = |f^{\varepsilon}(\theta, \chi)|^2 \tag{153}$$

so that

$$I(\varepsilon, \chi) = \frac{1}{2}\int_{-1}^{1} d\cos\theta \frac{d\sigma}{d\chi}(\theta) \tag{154}$$

The final result, due to Parker and Pack,[121] is

$$Q^{(n)}(\varepsilon) = 2\pi \int_{-1}^{1} d\cos\chi \left[\frac{1}{2}\int_{-1}^{1} d\cos\theta \frac{d\sigma}{d\chi}(\theta)\right](1 - \cos^n\chi) \tag{155}$$

The quantity $d\sigma(\theta)/d\chi$ can be interpreted as the contribution to the differential cross section for scattering into an angle χ due to a collision in which the angle between the rotor axis and the atom remained fixed at a value θ. It is its arithmetic average that enters the generalized integral cross section $Q^{(n)}(\varepsilon)$. The reader is referred to the paper by Parker and Pack[121] for further details.

The alternate treatment of Goldflam *et al.*[120] involves again assuming that $(\varepsilon'/\varepsilon)$ and k_0/k_{j_0} are approximately one, and then using equation (128) in equation (143) to obtain

$$Q^{(n)}(\varepsilon) = \int_0^{2\pi} d\phi \int_{-1}^{1} d\cos\chi \sum_L \frac{d\sigma}{d\Omega}(0 \to L)(1 - \cos^n\chi) \tag{156}$$

We note that by equation (126), $d\sigma(0 \to L)/d\Omega$ is independent of the angle ϕ, and, defining the moments $A_n(L)$ of the cross section by

$$A_n(L) = \int_0^{2\pi} d\phi \int_{-1}^{1} d\cos\chi \frac{d\sigma}{d\Omega}(0 \to L)\, P_n(\cos\chi) \tag{157}$$

we obtain for the first two $Q^{(n)}(\varepsilon)$ the results[120]

$$Q^{(1)}(\varepsilon) = \sum_L [\sigma(0 \to L) - A_1(L)] \tag{158}$$

and

$$Q^{(2)}(\varepsilon) = \frac{2}{3}\sum_L [\sigma(0 \to L) - A_2(L)] \tag{159}$$

Here it is seen that in contrast to the state-to-state cross sections $\sigma(j_0 \to j)$ or general relaxation cross sections $\langle\sigma_n(j'_a j'_b | j_a j_b)\rangle_T$, the viscosity and diffusion generalized integral cross sections involve information *besides* the integral cross sections for scattering out of the $j = 0$ rotor state.[120] This completes our detailed discussion of the IOS approximation. We now turn to discuss the closely related l_z-conserving approximation and the decoupled l-dominant method.

4. The l_z-Conserving Energy Sudden Approximation

4.1. Basic l_z-Conserving Equations and Boundary Conditions

Chu and Dalgarno,[25] Secrest,[5] and Khare[136] have considered (both formally and computationally[25,136]) an approximation for rotational excitation which is complementary to the CS method of McGuire and Kouri[1] and Pack.[2] The approximation (in the present context) can be viewed as a closure approximation to the rotor states (i.e., an energy sudden approximation[2,66]) so that the centrifugal potential is treated in a close-coupling fashion. Thus the approximation can be used when the rotor level spacings are small compared to the relative kinetic energy[2,119] and the closed rotor states are not strongly coupled to the states involved in the transition being studied.[119] Furthermore, since the centrifugal potential operator $\mathbf{l}^2/2MR^2$ is not approximated, the approach can be used for situations where the CS or IOS are not appropriate. In particular, the method is useful when the collision energy is small compared to the potential well or there is a very long-ranged interaction which causes significant inelastic scattering.[136] The method was first suggested a number of years ago by Chase[85] for treating nuclear scattering. We consider the approximation by using equation (20) along with the closure approximation equation (84) in order to emphasize the unity of these angular momentum decoupling approximations. Alternative derivations may be found in the excellent articles by Chu and Dalgarno,[25] Secrest,[5] and Khare.[136] The result of using equation (80) in equation (14) is

$$\psi^{l_0\mu_0}(j_0\lambda_0|\mathbf{R}\hat{\rho}) = Y_{l_0\mu_0}(\hat{\mathbf{R}})\, Y_{j_0\lambda_0}(\hat{\rho})\, j_{l_0}(kR) - \frac{2M}{\hbar^2}\sum_{l\mu} kY_{l\mu}(\hat{\mathbf{R}}) \int d\mathbf{R}'\, Y^*_{l\mu}(\hat{\mathbf{R}}')$$
$$\times\, h_l^+(kR_>)\, j_l(kR_<)\, V(R', \hat{\mathbf{R}}'\cdot\hat{\rho})\, \psi^{l_0\mu_0}(j_0\lambda_0|\mathbf{R}'\hat{\rho}) \qquad (160)$$

where the complete physical wave function is given by equation (13). We note that the orientation of the rotor, $\hat{\rho}$, enters only as a parameter (i.e., the equation has been localized in $\hat{\rho}$). We can factor out the $Y_{j_0\lambda_0}(\hat{\rho})$ by writing

$$\psi^{l_0\mu_0}(j_0\lambda_0|\mathbf{R}\hat{\rho}) = Y_{j_0\lambda_0}(\hat{\rho})\, \chi^{l_0\mu_0}(\mathbf{R}\hat{\rho}) \qquad (161)$$

where $\chi^{l_0\mu_0}(\mathbf{R}\hat{\rho})$ satisfies the equation

$$\chi^{l_0\mu_0}(\mathbf{R}\hat{\rho}) = Y_{l_0\mu_0}(\hat{\mathbf{R}})\, j_{l_0}(kR) - \frac{2M}{\hbar^2}\sum_{l\mu} kY_{l\mu}(\hat{\mathbf{R}}) \int d\mathbf{R}'\, Y^*_{l\mu}(\hat{\mathbf{R}}')\, h^+_l(kR_>)$$
$$\times\, j_l(kR_<)\, V(R', \hat{\mathbf{R}}'\cdot\hat{\rho})\, \chi^{l_0\mu_0}(\mathbf{R}'\hat{\rho}) \tag{162}$$

Next, one expands $\chi^{l_0\mu_0}(\mathbf{R}\hat{\rho})$ in terms of $Y_{l\mu}(\hat{\mathbf{R}})$, so that

$$\chi^{l_0\mu_0}(\mathbf{R}\hat{\rho}) = \sum_{l\mu} Y_{l\mu}(\hat{\mathbf{R}})\, \psi^{\bar{J}}(l\mu|l_0\mu_0|R, \hat{\rho}) \tag{163}$$

and substitution of equation (163) into (162) yields

$$\psi^{\bar{J}}(l\mu|l_0\mu_0|R, \hat{\rho}) = \delta_{ll_0}\delta_{\mu\mu_0} j_{l_0}(kR) - \frac{2M}{\hbar^2} k \int d\mathbf{R}'\, Y^*_{l\mu}(\hat{\mathbf{R}}')\, h^+_l(kR_>)$$
$$\times\, j_l(kR_<)\, V(R', \hat{\mathbf{R}}'\cdot\hat{\rho}) \sum_{l'\mu'} Y_{l'\mu'}(\hat{\mathbf{R}}')\, \psi^{J}(l'\mu'|l_0\mu_0|R', \hat{\rho}) \tag{164}$$

Now in order to perform the integral over the angles $\hat{\mathbf{R}}'$ most conveniently, we use the fact that $\hat{\mathbf{R}}'\cdot\hat{\rho}$ is independent of the choice of coordinate axes, and we rotate the $Y_{l\mu}(\hat{\mathbf{R}}')$ and $Y_{l'\mu'}(\hat{\mathbf{R}}')$ to a coordinate system with the z axis along $\hat{\rho}$:

$$Y_{l\mu}(\hat{\mathbf{R}}') = \sum_{\tilde{\mu}} D_{\tilde{\mu}\mu}(\hat{\rho})\, Y_{l\tilde{\mu}}(\hat{\mathbf{R}}'') \tag{165}$$

Then equation (164) becomes

$$\psi^{\bar{J}}(l\mu|l_0\mu_0|R, \hat{\rho}) = \delta_{ll_0}\delta_{\mu\mu_0} j_{l_0}(kR) - \frac{2M}{\hbar^2} k \int d\mathbf{R}'' \sum_{\substack{\tilde{\mu}\tilde{\mu}'\\ l'\mu'}} D^{l*}_{\tilde{\mu}\mu}(\hat{\rho})\, Y^*_{l\tilde{\mu}}(\hat{\mathbf{R}}'')$$

$$\times\, h^+_l(kR_>)\, j_l(kR_<)\, V(R'', \hat{\mathbf{R}}''\cdot\hat{\mathbf{z}})\, D^{l'}_{\tilde{\mu}'\mu'}(\hat{\rho})\, Y_{l'\tilde{\mu}'}(\hat{\mathbf{R}}'')\, \psi^{\bar{J}}(l'\mu'|l_0\mu_0|R'', \hat{\rho}) \tag{166}$$

We note that[75,77]

$$\delta_{\mu\mu_0} = \sum_{\tilde{\mu}\tilde{\mu}'} D^{l_0*}_{\tilde{\mu}\mu}(\hat{\rho})\, D^{l_0}_{\tilde{\mu}'\mu_0}(\hat{\rho})\, \delta_{\tilde{\mu}\tilde{\mu}'} \tag{167}$$

and

$$\psi^{\bar{J}}(l\mu|l_0\mu_0|R, \hat{\rho}) = \sum_{\tilde{\mu}\tilde{\mu}'} D^{l*}_{\tilde{\mu}\mu}(\hat{\rho})\, \psi^{\bar{J}}(l\tilde{\mu}|l_0\tilde{\mu}'|R)\, D^{l_0}_{\tilde{\mu}'\mu_0}(\hat{\rho}) \tag{168}$$

It follows from equations (166)–(168) that

$$\psi^{\bar{J}}(l\tilde{\mu}|l_0\tilde{\mu}'|R) = \delta_{ll_0}\delta_{\tilde{\mu}\tilde{\mu}'} j_{l_0}(kR) - \frac{2M}{\hbar^2} k \sum_{l'} \int_0^\infty dR'\, R'^2 h^+_l(kR_>)\, j_l(kR_<)$$
$$\times\, V_{\tilde{\mu}}(l|l'|R')\, \psi^{\bar{J}}(l'\tilde{\mu}|l_0\tilde{\mu}'|R') \tag{169}$$

and since there is no coupling in μ, $\tilde{\mu}'$, we immediately obtain the conservation of $\tilde{\mu}$:

$$\psi^{J}(l\tilde{\mu}|l_0\tilde{\mu}'|R) = \delta_{\tilde{\mu}\tilde{\mu}'}\psi_{J\tilde{\mu}}(l\,|l_0|R) \tag{170}$$

Here $\psi_{J\tilde{\mu}}(l\,|l_0|R)$ satisfies the l_z-conserving radial equation

$$\psi_{J\tilde{\mu}}(l\,|l_0|R) = \delta_{ll_0} j_{l_0}(kR) - \frac{2M}{\hbar^2} k \sum_{l'} \int_0^\infty dR' R'^2 h_l^+(kR_>)\, j_l(kR_<) \times V_{\tilde{\mu}}(l\,|l'|R')\,\psi_{J\tilde{\mu}}(l'|l_0|R') \tag{171}$$

or in differential form,[136]

$$\left[\frac{\hbar^2}{2M}\frac{1}{R}\frac{\partial^2}{\partial R^2}R - \frac{l(l+1)\hbar}{2MR^2} + \frac{\hbar^2 k^2}{2M} - V_{\tilde{\mu}}(l|l|R)\right]\psi_{J\tilde{\mu}}(l|l_0|R) = \sum_{l' \neq l} V_{\tilde{\mu}}(l\,|l'|R)\,\psi_{J\tilde{\mu}}(l'|l_0|R) \tag{172}$$

Then combining equations (13), (161), (163), and (168), we obtain the expression for the full wave function given by[136]

$$\psi_{\mathbf{k}_{j_0}}(j_0\lambda_0|\mathbf{R}\hat{\rho}) = 4\pi \sum_{\substack{l_0\mu_0\\ l\mu\\ \tilde{\mu}}} i^{l_0} Y^*_{l_0\mu_0}(\hat{\mathbf{k}}_{j_0})\, Y_{j_0\lambda_0}(\hat{\rho})\, Y_{l\mu}(\hat{\mathbf{R}})\, D^{l*}_{\tilde{\mu}\mu}(\hat{\rho})\,\psi_{J\tilde{\mu}}(l|\,l_0|\,R)\, D^{l_0}_{\tilde{\mu}\mu_0}(\hat{\rho}) \tag{173}$$

where the axis of quantization is arbitrary. It is convenient now to specialize this result to a z axis along the incident momentum. Then we obtain, by use of equation (18) for $Y^*_{l_0\mu_0}(\hat{\mathbf{k}}_{j_0})$, the expression

$$\psi_{\mathbf{k}_{j_0}}(j_0 m_0|\mathbf{R}\hat{\rho}) = \sum_{\substack{l_0 l\\ \mu\tilde{\mu}}} i^{l_0}[4\pi(2l_0+1)]^{1/2}\, Y_{l\mu}(\hat{\mathbf{R}})\, Y_{j_0 m_0}(\hat{\rho})\, D^{l*}_{\tilde{\mu}\mu}(\hat{\rho}) \times \psi_{J\tilde{\mu}}(l\,|l_0|R)\, D^{l_0}_{\tilde{\mu}0}(\hat{\rho}) \tag{174}$$

If we examine the asymptotic limit of this expression, we require the asymptotic limit of $\psi_{J\tilde{\mu}}(l\,|l_0|R)$, which is

$$\lim_{R\to\infty} \psi_{J\tilde{\mu}}(l\,|l_0|R) = \frac{\delta_{ll_0}\sin(kR - l_0\pi/2)}{kR} + i^{-l+1}\frac{e^{ikR}}{2kR} T_{J\tilde{\mu}}(l\,|l_0) \tag{175}$$

where

$$T_{J\tilde{\mu}}(l\,|l_0) = \frac{4Mik}{\hbar^2}\sum_{l'}\int_0^\infty dR' R'^2 j_l(kR')\, V_{\tilde{\mu}}(l\,|l'|R')\,\psi_{J\tilde{\mu}}(l'|l_0|R') \tag{176}$$

The first term on the right-hand side of equation (175) gives the incident plane wave, while the second term gives the scattering, so that

$$\psi^{\text{scatt}}_{\mathbf{k}_{j_0}}(j_0 m_0|\mathbf{R}\hat{\rho}) = \sum_{\substack{ll_0\\ \mu\tilde{\mu}}} \frac{i^{l_0 - l + 1}}{2k}[4\pi(2l_0+1)]^{1/2}\, Y_{l\mu}(\hat{\mathbf{R}})\, Y_{j_0 m_0}(\hat{\rho}) \times D^{l*}_{\tilde{\mu}\mu}(\hat{\rho})\, D^{l_0}_{\tilde{\mu}0}(\hat{\rho})\, T_{J\tilde{\mu}}(l\,|l_0)(e^{ikR}/R) \tag{177}$$

We then follow Khare[136] and define the effective differential scattering amplitude $f(\hat{\mathbf{R}}, \hat{\boldsymbol{\rho}})$ by

$$f(\hat{\mathbf{R}}, \hat{\boldsymbol{\rho}}) = \frac{1}{2k} \sum_{\substack{ll_0 \\ \bar{\mu}\mu}} i^{l_0 - l + 1} [4\pi(2l_0 + 1)]^{1/2} \, T_{\bar{J}\bar{\mu}}(l\,|\,l_0) \, Y_{l\mu}(\hat{\mathbf{R}}) \, D^{l*}_{\bar{\mu}\mu}(\hat{\boldsymbol{\rho}}) \, D^{l_0}_{\bar{\mu}0}(\hat{\boldsymbol{\rho}}) \tag{178}$$

Then equation (177) for $\psi^{\text{scatt}}_{\mathbf{k}_{j_0}}(j_0 m_0|\mathbf{R}\hat{\boldsymbol{\rho}})$ can be expressed as

$$\psi^{\text{scatt}}_{\mathbf{k}_{j_0}}(j_0 m_0|\mathbf{R}\hat{\boldsymbol{\rho}}) = \frac{e^{ikR}}{R} f(\mathbf{R}, \hat{\boldsymbol{\rho}}) \, Y_{j_0 m_0}(\hat{\boldsymbol{\rho}}) \tag{179}$$

In order to obtain the differential scattering amplitude into state $Y_{jm}(\hat{\boldsymbol{\rho}})$ (remembering that k in the exponential can be expressed approximately as *any* of the k_j), we write

$$f(\hat{\mathbf{R}}, \hat{\boldsymbol{\rho}}) \, Y_{j_0 m_0}(\hat{\boldsymbol{\rho}}) = \sum_{jm} Y_{jm}(\hat{\boldsymbol{\rho}}) \left(\frac{k_j k_{j_0}}{k^2} \right)^{1/2} f(jm|\,j_0 m_0|\hat{\mathbf{R}}) \tag{180}$$

It then follows immediately that

$$f(jm|\,j_0 m_0|\hat{\mathbf{R}}) = \frac{k}{(k_{j_0} k_j)^{1/2}} \int d\hat{\boldsymbol{\rho}} \, Y^*_{jm}(\hat{\boldsymbol{\rho}}) \, f(\hat{\mathbf{R}}, \hat{\boldsymbol{\rho}}) \, Y_{j_0 m_0}(\hat{\boldsymbol{\rho}}) \tag{181}$$

which is a form of Chase's formula.[85] Also, the analogy with equations (93)–(94) of the IOS should be evident.

4.2. *Factorization of l_z-Conserving Amplitudes and Cross Sections*

It is instructive to actually use equation (178) in (181), so that the integral over $d\hat{\boldsymbol{\rho}}$ may be carried out analytically. Then we obtain

$$\begin{aligned} f(jm|\,j_0 m_0|\hat{\mathbf{R}}) = {} & \frac{1}{2(k_{j_0} k_j)^{1/2}} \sum_{\substack{ll_0 \\ J\bar{\mu}}} i^{l_0 - l + 1} [4\pi(2l_0 + 1)]^{1/2} \, Y_{lm_0 - m}(\hat{\mathbf{R}}) \, T_{\bar{J}\bar{\mu}}(l\,|\,l_0) \\ & \times \frac{[(2j+1)(2j_0+1)]^{1/2}}{2J+1} \langle l\bar{\mu} j 0|J\bar{\mu}\rangle \langle l m_0 - m j m|J m_0\rangle \\ & \times \langle l_0 \bar{\mu} j_0 0|J\bar{\mu}\rangle \langle l_0 0 j_0 m_0|J m_0\rangle \end{aligned} \tag{182}$$

and this leads to the identification of the Arthurs and Dalgarno[73] $T^J(jl\,|\,j_0 l_0)$ as

$$T^J(jl\,|\,j_0 l_0) = \sum_{\bar{\mu}} \frac{[(2j+1)(2j_0+1)]^{1/2}}{2J+1} \langle l\bar{\mu} j 0|J\bar{\mu}\rangle \langle l_0 \bar{\mu} j_0 0|J\bar{\mu}\rangle \, T_{\bar{J}\bar{\mu}}(l\,|\,l_0) \tag{183}$$

It then follows that $f(jm|j_0m_0|\hat{\mathbf{R}})$ is expressed in the usual form as[73]

$$f(jm|j_0m_0|\hat{\mathbf{R}}) = \frac{1}{2(k_{j_0}k_j)^{1/2}} \sum_{\substack{ll_0\\J}} i^{l_0-l+1}[4\pi(2l_0+1)]^{1/2} Y_{lm_0-m}(\hat{\mathbf{R}}) \times \langle lm_0 - mjm|Jm_0\rangle \langle l_0 0 j_0 m_0|Jm_0\rangle T^J(jl|j_0l_0) \tag{184}$$

This appears to be identical to the *exact* amplitude, but now $T^J(jl|j_0l_0)$ is evaluated in terms of the l_z-conserving approximate $T_{\bar{J}\bar{\mu}}(l|l_0)$ via equation (183). A simple expression for $T_{\bar{J}\bar{\mu}}(l|l_0)$ in terms of the $T^J(jl|j_0l_0)$ results in the special case where $j_0 = 0$. Then equation (183) becomes[136]

$$T^{l_0}(jl|0l_0) = \sum_{\bar{\mu}} \frac{(2j+1)^{1/2}}{2l_0+1} \langle l\bar{\mu}j0|l_0\bar{\mu}\rangle T_{\bar{J}\bar{\mu}}(l|l_0) \tag{185}$$

and this is easily inverted to yield

$$T_{\bar{J}\bar{\mu}}(l|l_0) = \sum_{j'} (2j'+1)^{1/2} \langle l\bar{\mu}j'0|l_0\bar{\mu}\rangle T^{l_0}(j'l|0l_0) \tag{186}$$

We then can express $f(\hat{\mathbf{R}}, \hat{\boldsymbol{\rho}})$ in terms of the $T^{l_0}(j'l|0l_0)$ by substituting equation (186) into equation (178):

$$f(\hat{\mathbf{R}}, \hat{\boldsymbol{\rho}}) = \frac{1}{2k} \sum_{\substack{ll_0\\ \bar{\mu}\mu\\ j'}} i^{l_0-l+1}[4\pi(2l_0+1)]^{1/2} Y_{l\mu}(\hat{\mathbf{R}}) D^{l*}_{\bar{\mu}\mu}(\hat{\boldsymbol{\rho}}) D^{l_0}_{\bar{\mu}0}(\hat{\boldsymbol{\rho}}) (2j'+1)^{1/2} \times \langle l\bar{\mu}j'0|l_0\bar{\mu}\rangle T^{l_0}(j'l|0l_0) \tag{187}$$

The advantage of this expression will be apparent momentarily. Our objective is to express $f(\hat{\mathbf{R}}, \hat{\boldsymbol{\rho}})$ in terms of the amplitude $f(jm|00|\hat{\mathbf{R}})$, so that the various cross sections can be factored in the same way as in the IOS. To do this, we now express the product $D^{l*}_{\bar{\mu}\mu}(\hat{\boldsymbol{\rho}}) D^{l_0}_{\bar{\mu}0}(\hat{\boldsymbol{\rho}})$ in terms of the $Y_{j''m''}(\hat{\boldsymbol{\rho}})$. It is easy to show that[136]

$$D^{l*}_{\bar{\mu}\mu}(\hat{\boldsymbol{\rho}}) D^{l_0}_{\bar{\mu}0}(\hat{\boldsymbol{\rho}}) = \sum_{j''m''} \frac{[4\pi(2j''+1)]^{1/2}}{2l_0+1} \delta_{\mu+m'',0} \langle l\bar{\mu}j''0|l_0\bar{\mu}\rangle \times \langle l-m''j''m''|l_0 0\rangle Y_{j''m''}(\hat{\boldsymbol{\rho}}) \tag{188}$$

and then we write equation (187) as

$$f(\hat{\mathbf{R}}, \hat{\boldsymbol{\rho}}) = \frac{1}{2k} \sum_{\substack{ll_0\\ \bar{\mu}j'\\ j''m''}} i^{l_0-l+1} \frac{4\pi}{2l_0+1} [(2j'+1)(2j''+1)(2l_0+1)]^{1/2} \times Y_{l-m''}(\hat{\mathbf{R}}) Y_{j''m''}(\hat{\boldsymbol{\rho}}) T^{l_0}(j'l|0l_0) \langle l\bar{\mu}j''0|l_0\bar{\mu}\rangle \langle l\bar{\mu}j'0|l_0\bar{\mu}\rangle \langle l-m''j''m''|l_0 0\rangle \tag{189}$$

Then we may carry out the sum over $\tilde{\mu}$ to obtain

$$f(\hat{\mathbf{R}}, \hat{\boldsymbol{\rho}}) = \frac{1}{2k} \sum_{\substack{ll_0 \\ j''m''}} i^{l_0 - l + 1} 4\pi(2l_0 + 1)^{1/2} Y_{l-m''}(\hat{\mathbf{R}}) Y_{j''m''}(\hat{\boldsymbol{\rho}}) \times \langle l - m''j''m'' | l_0 0\rangle T^{l_0}(j''l|0l_0) \tag{190}$$

Now let us reconsider equation (184) for the special case $j_0 = m_0 = 0$, whereby we find

$$f(jm|00|\hat{\mathbf{R}}) = \frac{1}{2(k_j k_0)^{1/2}} \sum_{ll_0} i^{l_0 - l + 1} [4\pi(2l_0 + 1)]^{1/2} Y_{l-m}(\hat{\mathbf{R}}) \times \langle l - mjm | l_0 0\rangle T^{l_0}(jl|0l_0) \tag{191}$$

Comparing equations (190) and (191), it is readily seen that

$$f(\hat{\mathbf{R}}, \hat{\boldsymbol{\rho}}) = \sum_{j''m''} \frac{(4\pi k_{j''} k_0)^{1/2}}{k} Y_{j''m''}(\hat{\boldsymbol{\rho}}) f(j''m''|00|\hat{\mathbf{R}}) \tag{192}$$

which result was first obtained by Khare.[136] We may now use equation (192) in equation (181) to obtain Khare's[136] l_z-conserving expression:

$$f(jm|j_0 m_0|\hat{\mathbf{R}}) = \sum_{j'm'} \left(\frac{k_{j'} k_0}{k_j k_{j_0}}\right)^{1/2} \left[\frac{(2j_0 + 1)(2j' + 1)}{2j + 1}\right]^{1/2} \times \langle j_0 m_0 j'm' | jm\rangle \langle j_0 0 j'0 | j0\rangle \times f(j'm'|00|\hat{\mathbf{R}}) \tag{193}$$

We note next that

$$\left[\frac{(2j_0 + 1)(2j' + 1)}{2j + 1}\right]^{1/2} \langle j_0 m_0 j'm' | jm\rangle \langle j_0 0 j'0 | j0\rangle = (-1)^m [(2j_0 + 1)(2j' + 1)(2j + 1)]^{1/2} \begin{pmatrix} j_0 & j' & j \\ 0 & 0 & 0 \end{pmatrix} \begin{pmatrix} j_0 & j' & j \\ m_0 & m' & -m \end{pmatrix} \tag{194}$$

so that equation (193) becomes

$$f(jm|j_0 m_0|\hat{\mathbf{R}}) = \sum_{j'm'} \left(\frac{k_{j'} k_0}{k_j k_{j_0}}\right)^{1/2} [(2j_0 + 1)(2j' + 1)(2j + 1)]^{1/2} \times (-1)^m \begin{pmatrix} j_0 & j' & j \\ 0 & 0 & 0 \end{pmatrix} \begin{pmatrix} j_0 & j' & j \\ m_0 & m' & -m \end{pmatrix} f(j'm'|00|\hat{\mathbf{R}}) \tag{195}$$

At first glance, this appears to be the same form as the IOS factored differential scattering amplitude, equations (115)–(117). However, there are important differences. Equation (117) is for the final $\hat{\mathbf{R}}$-direction quantization axis and is *diagonal* in λ, λ_0. Equation (195) is *not* diagonal in m, m_0 and the quantization axis is the incident momentum direction, $\hat{\mathbf{k}}_{j_0}$. Nevertheless, degeneracy-averaged quantities obtained via equations (117) and (195) will have the *same factored form*. The definite polarization cross section is then given by

$$\frac{d\sigma}{d\Omega}(j_0 m_0 \to jm) = \frac{k_j}{k_{j_0}} |f(jm|j_0 m_0|\hat{\mathbf{R}})|^2 \tag{196}$$

and if we use equation (195), degeneracy average, and employ the identity

$$\sum_{\substack{m \\ m_0}} (2j+1) \begin{pmatrix} j_0 & j' & j \\ m_0 & m' & -m \end{pmatrix} \begin{pmatrix} j_0 & j'' & j \\ m_0 & m'' & -m \end{pmatrix} = \delta_{j'j''} \delta_{m'm''} \tag{197}$$

we obtain

$$\frac{d\sigma}{d\Omega}(j_0 \to j) = \frac{k_0^2}{k_{j_0}^2} (2j+1) \sum_{j'} \begin{pmatrix} j_0 & j' & j \\ 0 & 0 & 0 \end{pmatrix}^2 \frac{d\sigma}{d\Omega}(0 \to j') \tag{198}$$

This l_z-conserving result due to Khare[136] is *identical* to the Goldflam *et al.*[120] IOS result, equation (127), thereby showing that the range of validity of the factorization extends beyond the IOS. We next note that summing equation (198) over j leads to the same result as equation (128), so that to within the detailed balance factor $(k_0^2/k_{j_0}^2)$, the total differential cross section is independent of initial state. Of course, these results also immediately apply to the integral cross sections, so that one also obtains equations (130)–(131) in the l_z-conserving approximation. Finally, we notice that one may also obtain a generalization of the "average differential cross section," equation (154), which arose in the IOS treatment of viscosity and diffusion. If we approximate all k_j by k and employ equations (80) and (150), it is easy to show that[136]

$$\frac{1}{2j_0+1} \sum_{\substack{jm \\ j_0 m_0}} \frac{d\sigma}{d\Omega}(j_0 m_0 \to jm) = \frac{1}{4\pi} \int |f(\hat{\mathbf{R}}, \hat{\rho})|^2 \, d\hat{\rho} \tag{199}$$

The computational details of the l_z-conserving approximation are dealt with in the excellent work by Khare.[136] Essentially, the method requires somewhat more computational effort than the CS (and considerably more than the IOS). It is also more time consuming than the DLD method of De Pristo and Alexander,[137] but should be more accurate so long as the rotational spacings are small compared to the relative kinetic energy. We now discuss briefly the DLD approach.

5. The Decoupled l-Dominant Approximation

De Pristo and Alexander[137-139] have developed a space-fixed angular momentum decoupling scheme for treating collision systems in which the centrifugal potential plays a dominant role. This is expected whenever there is a significant contribution to the inelasticity coming from the long-range part of the potential. In the long-range region, the potential varies slowly and as a result, different values of the orbital angular momentum quantum number lead to significantly different classical turning points. It was observed by Kouri and McGuire,[24] in their CS and CC studies of Li^+–H_2 collisions, that the major contribution to the scattering came from the lower values of orbital angular momentum. Essentially, one finds that for a given value of total angular momentum J, $j + l$ is approximately conserved. Therefore, in the Li^+–H_2 transition $j = 0 \rightarrow j = 2$, the most important contribution was observed to come from the $l = J \rightarrow l = J - 2$ orbital transition. Of course, this also produces the lowest centrifugal barrier and smallest classical turning point. Therefore, it permits more of the coupling potential to be sampled and dominates the inelastic scattering. De Pristo and Alexander[137] have formulated this rigorously by introducing a channel indexing in terms of the usual j quantum number and an alignment quantum number τ, where $\tau = l + j - J$. The quantum number τ measures the degree to which $\mathbf{J}$ and $\mathbf{j}$ are aligned. Since l ranges from $|J - j|$ to $J + j$, τ ranges from 0 to $2j$. The minimum value of $\tau = 0$ corresponds to $l = J - j$ (for $J > j$) and describes the situation where the rotor spins so that $\mathbf{l}$ and $\mathbf{j}$ add constructively to produce the total angular momentum $\mathbf{J}$. One can interpret this in terms of a collision dominated by nearly coplanar geometry in which the molecular and orbital motion occurs in a cooperative fashion. By considering the asymptotic behavior of the Percival–Seaton coefficients, De Pristo and Alexander[137] arrive at the coupled equations

$$\frac{\hbar^2}{2M}\left[\frac{d^2}{dR^2} + k_j^2 - \frac{(J - j + \tau)(J - j + \tau + 1)}{R^2}\right] U^{J\tau}(j|j_0|R) = \sum_{j'} V^J(j\tau|j'\tau|R)\, U^{J\tau}(j'|j_0|R) \quad (200)$$

where

$$V^J(j\tau|j'\tau|R) = \sum_t V_t(R)\, f_t(j, J - j + \tau|j', J - j' + \tau|J) \quad (201)$$

It is seen that only coupling in j occurs since the coupling in τ is neglected. Thus, in the DLD, the centrifugal potential is included exactly while the potential is approximated, in contrast to the CS, where the potential is considered exactly and the $\mathbf{l}^2$ operator is approximated. The degree of coupling is the same in the DLD as in the CS, there being one coupled equation per state j. Also, the number of τ values included ranges from zero up to j_{max}, and the number of coupled equations decreases as τ increases, just as in the CS.[1] Finally, we note that the asymptotic boundary conditions applied to the $U^{J\tau}(j\,|\,j_0|R)$ are the usual ones in the Arthurs–Dalgarno approach.[73] This method is expected to be useful in dealing with the larger angular momentum contributions to scattering in systems with large rotor constants B_e (so that the l_z-conserving or IOS are not appropriate) and long-range coupling.

6. *Exponential Distorted-Wave Approximation*

Another approximation, which can be quite useful, is based on perturbation solutions to the scattering equations. The distorted-wave method has, of course, been known for many years and we shall not give a detailed derivation here; excellent discussions can be found in the literature. (Reference 140 gives an especially good review and extensive list of references.) The most useful form of the method is probably the exponential form which eliminates the problem of unitarity breakdown. It is also felt that from a computational standpoint, the most useful form of the exponential distorted-wave (EDW) method is that in which the distorted waves are calculated within the WKB approximation. The elements of the (Arthurs–Dalgarno) S matrix are given by[140]

$$S^J(j'l'|jl) = \exp(i\eta^J_{j'l'})\,[\exp(-i\mathbf{D}^J)]_{j'l',jl}\exp(i\eta^J_{jl}) \qquad (202)$$

where the WKB phase shifts η^J_{jl} satisfy

$$\eta^J_{jl} = \lim_{R\to\infty}\left\{\int_{R_t}^{R}\left[k_j^2 - \frac{(l+1/2)^2}{R'^2} - \frac{2M}{\hbar^2}V^J(jl\,|\,jl\,|R')\right]^{1/2} dR' - k_jR + \frac{(l+1/2)\pi}{2}\right\} \qquad (203)$$

with

$$\frac{\hbar^2k_j^2}{2M} = E - B_e j(j+1) \qquad (204)$$

and R_t the outermost classical turning point for channel j, l. Note that it is the *full* diagonal part of the potential matrix, $V^J(jl\,|\,jl\,|R)$ which enters and not just the spherical part. The distorted-wave integral $D^J_{j'l',jl}$ is given by[140]

$$D^J_{j'l',jl} = \frac{2(1-\delta_{jj'}\delta_{ll'})}{(k_jk_{j'})^{1/2}}\int_0^\infty dR\,R^2W_{j'l'}(R)\,V^J(j'l'|jl\,|R)\,W_{jl}(R) \qquad (205)$$

where the $W_{jl}(R)$ are the WKB approximations to the solution of the Schrödinger equation

$$\left[\frac{1}{R}\frac{\partial^2}{\partial R^2}R + k_j^2 - \frac{l(l+1)}{R^2} - \frac{2M}{\hbar^2}V^J(jl\,|\,jl\,|R)\right]W_{jl}(R) = 0 \tag{206}$$

We remark that while more sophisticated versions of the approximation have been developed, it is our feeling that unless the present version is accurate, the method should not be employed. We indicate later, in Section 8.5, when the method may be used.

7. Semiclassical Approximation

There are a great variety of semiclassical approximations which have been reported in the past few years. Excellent discussions of these can be found in the reviews by Miller,[141] Clark *et al.*,[142] and Balint-Kurti.[140] The methods which seem most promising are those that do not require multidimensional root searches.[143-145] A particularly attractive approach is based on the time-dependent form of Schrödinger's equation in which the dependence on the scattering vector has been approximated by a classical trajectory approximation. Then the atom–diatom interaction becomes time dependent, and use of an expansion in internal states of the target yields the infinite set of first-order differential equations,[142]

$$i\hbar\frac{\partial S}{\partial t}(\mathbf{n}'|\mathbf{n}|t) = \sum_{\mathbf{m}}\langle\mathbf{n}'|V|\mathbf{m}\rangle\, S(\mathbf{m}|\mathbf{n}|t)\exp(i[E_{n'} - E_m]\,t/\hbar) \tag{207}$$

where

$$\lim_{t\to-\infty} S(\mathbf{n}'|\mathbf{n}|t) = \delta_{\mathbf{n},\mathbf{n}'} \tag{208}$$

and $\mathbf{n}$ represents the collection of quantum numbers characterizing the internal state of the target. The cross section for the transition $\mathbf{n}\to\mathbf{n}'$ is then

$$\sigma(\mathbf{n}\to\mathbf{n}') = 2\pi\int_0^\infty db\,b\,|S(\mathbf{n}'|\mathbf{n})|^2 \tag{209}$$

where

$$S(\mathbf{n}'|\mathbf{n}) = \lim_{t\to\infty} S(\mathbf{n}'|\mathbf{n}|t) \tag{210}$$

and b is the impact parameter associated with the trajectory used in expressing $\mathbf{R}$ as a function of time. The semiclassical methods of interest basically make approximations such that equation (207) can be solved in a simple fashion. The choice of trajectory for use in expressing the potential (through its dependence on $\mathbf{R}$) as a function of time is important. Usually, the collision energy is chosen to be the average of the initial and final energy.[142] In addition, the simplification is often made of using only the spherical part of the potential to determine the trajectory. [However, the resulting $\mathbf{R}(t)$ is then used in the *full* potential in order to solve equation (207).] The strong coupling correspondence principle (SCCP) method of Percival and Richards[146,147] and Levine and Johnson[148] then assumes that all strongly coupled levels are equally spaced (best for small B_e and large quantum numbers). The Heisenberg correspondence principle can then be used to write

$$\langle \mathbf{n}' | F(\mathbf{q}) | \mathbf{n} \rangle \cong F_{\mathbf{s}}(\mathbf{I}) \tag{211}$$

where

$$\mathbf{s} = \mathbf{n}' - \mathbf{n} \tag{212}$$

and the quantity $F_{\mathbf{s}}(\mathbf{I})$ is the Fourier component of $F(\mathbf{q})$ evaluated along a classical trajectory with action variables $\mathbf{I}$. Thus

$$F_{\mathbf{s}}(\mathbf{I}) = \left(\frac{1}{2\pi}\right)^N \int d\boldsymbol{\theta} F(\mathbf{q}(\boldsymbol{\theta}, \mathbf{I})) \exp(i\mathbf{s} \cdot \boldsymbol{\theta}) \tag{213}$$

Again, an average trajectory is assumed such that, e.g.,

$$\mathbf{I} = (\mathbf{n} + \mathbf{n}')\hbar/2 \tag{214}$$

In addition, the assumption is made that the purely classical motion of the collision does not deviate strongly from the unperturbed motion. Then the resulting solution of equation (207) can be shown to be[142]

$$s(\mathbf{n}'|\mathbf{n}) = \left(\frac{1}{2\pi}\right)^N \int_0^{2\pi} d\boldsymbol{\theta} \exp\left\{ i \left[\boldsymbol{\theta} \cdot (\mathbf{n} - \mathbf{n}') - \frac{1}{\hbar} \int_{-\infty}^{\infty} dt V^c(\mathbf{I}, \boldsymbol{\theta} + \boldsymbol{\omega}^c t; t) \right] \right\} \tag{215}$$

where

$$\frac{1}{\hbar} \int_{-\infty}^{\infty} dt V^c(\mathbf{I}, \boldsymbol{\theta} + \boldsymbol{\omega}^c t; t)$$

is the change in action of the target occurring during the average classical trajectory. By use of classical perturbation theory, a multipole expansion of the potential and angular momentum algebra, the change in action for a rotor becomes explicitly

$$A(\alpha, \beta, \gamma; E, b) = 4\pi \sum_{\lambda=1}^{N} (2\lambda + 1)^{-1} \sum_{\mu,\nu=-\lambda}^{\lambda} N_{\lambda\mu} N_{\lambda\nu} D^{\lambda}_{\nu\mu}(\gamma, \beta, \alpha) V^{\lambda}_{\nu\mu}(E, b) \tag{216}$$

where γ and α are the angle variables conjugate to j and m, respectively, β is the angle between the angular momentum of the rotor and of the atom, and $D^{\lambda}_{\nu\mu}(\gamma, \beta, \alpha)$ is the usual representation coefficient for finite rotations. $V^{\lambda}_{\nu\mu}(E, b)$ is a Fourier transform coefficient for the potential, so that if

$$V(R, \theta) = \sum_{\lambda=0}^{N} F_{\lambda}(R) P_{\lambda}(\cos\theta) \tag{217}$$

then

$$V^{\lambda}_{\nu\mu}(E, b) = \frac{2}{\hbar} \int_{0}^{\infty} dt F_{\lambda}(R) \cos(\nu\omega^{c}t - \mu\phi) \tag{218}$$

where $R, \phi + \pi/2$ are the plane polar coordinates of the atom trajectory under the influence of the spherical part of the potential $F_0(R)$, and ω^c is the classical frequency of the rotor. Finally, $N_{\lambda\mu}$ is given by

$$N_{\lambda\mu} = Y_{\lambda\mu}(\pi/2, 0) \tag{219}$$

An important point of simplification in the above is that the dependence on the angles of the rotor appears solely in the $D^{\lambda}_{\nu\mu}$ and the dependence on the trajectory is contained in the Fourier coefficients $V^{\lambda}_{\nu\mu}$. This makes the evaluation of the target action change for any orientation quite simple once the orientation-independent $V^{\lambda}_{\nu\mu}$ are evaluated. The changes in the rotor action variables j and m [i.e., the components of $(\mathbf{n} - \mathbf{n}')$] are then given by[142]

$$\Delta j = -\partial A/\partial\gamma \tag{220}$$

and

$$\Delta m = -\partial A/\partial\alpha \tag{221}$$

Then the probability amplitude S becomes

$$S(\Delta j, \Delta m; \beta, E, b) = \frac{1}{4\pi^2} \int_{0}^{2\pi} d\alpha \int_{0}^{2\pi} d\gamma \exp\{i[\gamma\Delta j + \alpha\Delta m - A(\alpha, \beta, \gamma; E, b)]\} \tag{222}$$

We note that related simpler expressions may also be derived in a CS version of the SCCP method.[147] In addition, analytic results can be gotten for certain potential forms.[142] The relation of the SCCP method to a variety of other semiclassical methods is discussed in detail in the review by Clark *et al.*[142] and the reader is referred to that work for more details and references to the literature.

8. Method Selection

We now turn to the important question of selecting the approximation (if any) which is most appropriate for treating a particular collision system. The situation is not simple because there are a variety of parameters which can significantly affect the physics of molecular collisions. It is therefore, unfortunately, not easy to write down inequalities that are generally applicable and simultaneously highly precise and quantitative. Instead, we shall content ourselves with a somewhat qualitative discussion (although inequalities will be given in certain cases). We begin by discussing the conditions under which the various sudden approximations will be appropriate. Next, we discuss the criteria for selecting other methods to apply when the sudden approximations are not expected to be valid. In addition, we briefly indicate when semiclassical and perturbative approximations may be employed.

8.1. Energy Sudden Approximation

In the energy sudden approximation (which is employed in obtaining both the IOS and the l_z-conserving approximations), one replaces the various wave numbers k_j by an effective wave number k [see the discussion preceding equation (80)]. The primary condition for the validity of this replacement is that the velocity of the projectile relative to the target center of mass must be reasonably larger than the velocities of the atoms in the molecule. For a given transition $j_0 \to j$, this must be true for all states j' of the diatom which are strongly coupled to the transition. It is also reasonable that the difference in the velocity in states j_0 and j of the atoms comprising the diatom be small compared to the projectile's average velocity. This may be expressed as follows. The average relative kinetic energy of the projectile is

$$\{E - B_e[j_0(j_0 + 1) + j(j + 1)]\}/2$$

The average (asymptotic) velocity of the projectile is

$$(\{2E - B_e[j_0(j_0 + 1) + j(j + 1)]\}/M)^{1/2}$$

where we emphasize that M is the reduced mass of the projectile relative to the diatom center of mass. Now the kinetic energy of the rotor in state j is $B_e j(j + 1)$, which also equals $(1/2)\, I\omega^2$, where I is the moment of inertia of the diatom and ω is the angular velocity. Since

$$I = \mu r_e^2 \tag{223}$$

where μ is the reduced mass of the atoms comprising the rotor and r_e is the equilibrium separation of the diatom (which may depend on the state j), we find that the characteristic velocity of the rotational state j is $[2B_e j(j+1)/\mu]^{1/2}$. Then our condition that the collision be sudden in the energy is that the velocity ratio t_{ES}, defined by Top and Kouri[126] as

$$t_{ES} = \left(\frac{2MB_e}{\mu}\right)^{1/2} \frac{|[j_0(j_0+1)]^{1/2} - [j(j+1)]^{1/2}|}{\{2E - B_e[j_0(j_0+1) + j(j+1)]\}^{1/2}} \tag{224}$$

must satisfy

$$t_{ES} < 1 \tag{225}$$

This is an extremely useful criterion but it is not without fault. The principal flaw is that it does not incorporate the influence of the closed (or almost closed) rotational states. If the well depth is large compared to the relevant kinetic energies and the anisotropy is strong, then coupling to states which are asymptotically closed or almost closed can be important. These virtual excitations violate the energy sudden condition equation (225) and lead to error. However, as the collision energy increases so that the kinetic energy of strongly coupled states is comparable to or greater than the well depth of the potential, the closed channels become less important even in strongly coupled systems and the energy sudden approximation becomes quite good even for these systems. Of course, if one is considering collision energies and potentials where Feshbach resonances are important, the energy sudden approximation will fail.

A slightly simpler version of the criterion equation (225) is also useful when B_e is very small. Then one may compare $B_e\tilde{j}(\tilde{j}+1)/\mu$ with E/M, where $\tilde{j}$ is the maximum of j_0 and j. One requires that

$$\left[\frac{B_e\tilde{j}(\tilde{j}+1)M}{\mu E}\right]^{1/2} \ll 1 \tag{226}$$

If the ratio in equation (226) is not much less than one, one should use equation (225).

When equation (225) or (226) is satisfied, it is important to realize that the cross section factorizations, equations (103)–(105), (120), (125), (127), (130)–(132), and (135), due to Goldflam *et al.*[119] are valid and should always be employed. This is true even if one suspects the validity of the centrifugal sudden approximation (which in the IOS is made in addition to the energy sudden approximation). The justification for this is that Khare's work[136] on the l_z-conserving approximation has shown that such factorizations depend *only* on the energy sudden approximation and not on the centrifugal sudden approximation. We now consider the conditions for accuracy of the centrifugal sudden approximation.

8.2. Centrifugal Sudden Approximation

We now discuss conditions under which equations (31)–(32), (45), (50)–(53), (56), (59)–(61), and (77) for the various CS quantities are applicable. Quantitative criteria for the CS approximation are much harder to develop than for the energy sudden approximation, and at least part of those in use are the result of empirical studies rather than rigorous theoretical derivations.[4,28,32,34-36,39,42,44,46,47,54,60,61] While we do not object to this, we do comment that more work would be valuable. A reason for the difficulty in developing simple inequalities such as equation (225) or (226) for the centrifugal sudden approximation is due to the variation with R of the centrifugal potential. As a result, the accuracy of the approximation depends strongly on which region of R one is considering. Discussions of the validity of the centrifugal sudden approximation have been given by Walker and Light,[4] McGuire,[39] Kouri *et al.*[63] (KHS), and by Dickinson and Richards.[147] Qualitatively, one can compare the relative radial velocity of the projectile to the velocity of its rotation with respect to the diatom center of mass in the various l states. It is then reasonable to expect that the region where the major breakdown of the approximation occurs is in the vicinity of classical turning points (and, of course, within nonclassical regions). KHS[63] have argued, on the basis of the behavior of the exact body frame Lippmann–Schwinger equation, that for impacts in which the classical turning point R_t occurs further out than the impact parameter X, so that $R_t > X$, the CS is valid. They show that this is true for scattering by the hard core of the potential so that the CS may always be used in such cases. This also implies that for systems where the short-range part of the potential dominates the scattering, the CS will be accurate. McGuire[39] has given an empirical rule for potentials expressed in the form

$$V(R,\theta) = \sum_{\lambda=0} [a_S^\lambda V_S(R) + a_L^\lambda V_L(R)]\, P_\lambda(\cos\theta) \tag{227}$$

where a_S^λ is the coefficient of the short-range potential $V_S(R)$ and a_L^λ is the coefficient of the long-range potential $V_L(R)$ contributing to the λth multipole of the potential. Then the criterion is that $a_L^\lambda < a_S^\lambda$. Arguments given by McGuire[39] were based on a physical picture of intermultiplet transitions occurring in the long-range region and not in the short-range region. Although the empirical rule is useful as a sufficient condition, McGuire's physical picture of the intermultiplet transitions occurring far from turning points is not correct. Rather, as stated earlier, the primary region where breakdown of the CS occurs is in the vicinity of the turning points. Then whether the λ transitions occur in the long- or short-range region is affected by when the turning points occur. For low l, the λ transitions occur in the small-R region, and for large l they occur in the large-R region. In the region of the hard core and well of a potential, the effect on the turning point positions of approximating all l's by the effective L is relatively small (unless perhaps one is dealing with the range of orbital angular momenta leading to a three-turning-point problem). As discussed by KHS,[63] the breakdown of the CS in the vicinity of the turning points does continue to propagate into the long-range region so that an extremely long-range anisotropy can magnify the effects arising from the turning point defect. As the total angular momentum is increased so that the long-range part of the potential dominates, the effect on the turning points due to replacing all l's by L is larger. This is because the potential is quite flat in the region away from the hard core and well, and one must have a large change in R in order for the potential to change enough to compensate for the different centrifugal potential arising from replacing all l's by L. There is another important effect which must be mentioned. In order for λ to be conserved in a given transition $j_0 \to j$, where $j \neq j_0$, it is necessary that the orientation of the plane of the rotor change (as noted by McGuire and Kouri[1]). Thus the orientation angle is given by

$$\cos\theta = \frac{\lambda}{[j(j+1)]^{1/2}} \tag{228}$$

and since λ is being approximately conserved in the CS, one concludes that for $j_0 \neq j$, the orientation of the rotor *must change*. However, if $j_0 = j$, then the rotor orientation remains unchanged if λ is conserved. In the large l domain, one is primarily sampling the long-range part of the potential so that relatively weak forces are acting on the rotor. As a result, we expect that it is difficult to change the orientation of the rotor in such glancing collisions. This suggests that the CS approximation will be quite accurate for rotationally elastic $j = j_0$ collisions in the large l region. This was first noted theoretically and verified computationally by Shimoni and Kouri.[6-9] It also implies that the accuracy of the CS for rotationally inelastic transitions deteriorates as l gets so large that all the scattering is occurring due to the long-range part of the potential. Then how good the CS is for a $j \neq j_0$ transition is basically determined by what per cent of the overall $j_0 \to j$ transition cross section is contributed by this long-range region. In the case where $a_L^\lambda < a_S^\lambda$, this per cent contribution to rotationally inelastic scattering is small (especially as the relative kinetic energy increases) so that good accuracy for the CS results. We also expect that the case of a $j \neq j_0$ transition whose cross section is dominated by the large l region should be treated well either by the DLD method, or if equation (225) holds, by the l_z approximation. An example is the $j_0 = 0 \to j = 2$ transition in the Li^+–H_2 system.[137]

So far we have contrasted the effects of the long- and short-range anisotropies on the CS approximation. The analysis of KHS[63] also predicts that as the energy of the collision increases, the CS improves. This behavior has been observed computationally in a number of studies,[1,25,35,36,46,54] and it partially reflects the shifting importance of the long- and short-range parts of the potential to the scattering. In particular, as the energy increases, the long-range tail of the potential becomes less important. In the analysis of KHS,[63] they give as estimates for the impact parameter and turning point $X_j \sim (l + 1/2)/k_j$ and

$$R_t \sim \left\{ \frac{l(l+1)\hbar^2}{2M[E_j - V(R_t)]} \right\}^{1/2}$$

Here, X_j is the impact parameter in state j and E_j is the *asymptotic* relative kinetic energy in state j. As E_j becomes large, and l is also large, $V(R_t)$ is generally negligible, and then R_t becomes approximately equal to X_j. This then implies that the turning point defects will be small and the CS should become accurate. However, when E_j is small, and in particular close to threshold, we see that X_j is large, while the turning point R_t can still be small (for low l). In such cases, the system penetrates into a region of strong λ transitions and the CS will fail. This failure of the CS close to threshold has been observed computationally in comparisons with CC results for He + H_2 and He + CO.[1,25] A useful energy criterion for *neutral* systems has been given by KHS.[63] It is strictly a sufficiency condition and should not be used as a necessary condition. If, for a transition of interest, the relative kinetic energy of both the initial and final states are greater than the well depth, the CS approximation is expected to be valid. The condition ensures that one does not encounter difficulties with three-turning-point regions, and also that one is far enough from threshold. [We do note that there are computational cases where this condition was *not* satisfied and yet the CS worked well. An interesting one is the model study of Walker and Light[4] where E_j was less than the well depth for all open channels. However, only the results for $j = 8$ (the highest open channel) were poor, and even then, primarily for the $0 \rightarrow 8$ transition. We note that the failure of the CS in that work for elastic $j_0 \rightarrow j_0$ transitions was due to their having been labeled by J rather than by l.[8,9]]

In considering the CS, it is also important to note that the method can be used in calculations of magnetic transitions. In some of the early literature, the CS was interpreted as conserving the R (or P) helicity. It was demonstrated rather conclusively by Shimoni and Kouri[9,12] that this was not the case and that in fact, the CS approximations to the helicity amplitudes $T^J(j\lambda|j_0m_0)$ and $f(j\lambda|j_0m_0|\hat{\mathbf{R}})$ were not diagonal in λ, m_0. Secrest[5] gave CS expressions for the space-oriented magnetic transition amplitude $f(jm|j_0m_0|\hat{\mathbf{R}})$ which were not diagonal in m, m_0. In fact, depending on the choice of the CS orbital angular momentum parameter L, there can be a frame in which j_z is conserved. It has been shown that choosing L equal to the final l_f yields the diagonal McGuire–Kouri[1] result for the space-fixed Arthurs–Dalgarno[73] amplitude $f(jm|j_0m_0|\hat{\mathbf{R}})$, while choosing L equal to the initial l_i leads to a reinterpretation of the diagonal McGuire–Kouri amplitude as $f(j\lambda|j_0\lambda_0|\hat{\mathbf{R}})$. It is clear from equation (54) that if a diagonal amplitude occurs with respect to one set of axes, in general the amplitude is nondiagonal with respect to other choices of axes. Additionally, it was shown by Kouri and Shimoni,[12] Khare,[14] and Khare *et al.*[16] that all of the various degeneracy-averaged cross sections were independent of the choice of quantization axes. Thus the previous success of the McGuire–Kouri[1] amplitude, the Shimoni–Kouri[6-9] analysis of opacities and forward and backward scattering, and the Goldflam–Kouri[13] CS theory of general relaxation processes cannot be used to choose between the various possible ways of choosing the axis along which to conserve j_z. At present, computational results of Monchick and Kouri[61] indicate that semiquantitative agreement between CC and CS results for $d\sigma(j_0m_0 \to jm)/d\Omega$ occurs when the choice $L = l_i$ is made. CC studies of He + H_2 by Shimoni *et al.*[149] indicate that CS and CC will agree *qualitatively* regardless of whether $L = l_i$ or l_f is chosen. These results indicate that transitions of the form $j_0m_0 \to j_0m$ are diagonally dominant, while for $j_0m_0 \to jm$, $j \neq j_0$, significant $\Delta m \neq 0$ transitions occur as l gets larger. Pfeffer and Secrest[150] have suggested using $L = (l_i + l_f)/2$, and computational results of Fitz[127] for Ar + N_2 are significantly better than those for $L = l_i$ or l_f. This is likely a reflection of the fact that $L = l_i$ or l_f leads to T matrices which do not have the proper time reversal symmetry (until one degeneracy averages). Shimoni and Kouri[11] have presented an alternative solution to this problem which has only been tested for He + H_2 at $J = 4$ and 36. However, their method also leads to significant improvement in the description at the T-matrix level. Khare *et al.*[16] showed that l_i and l_f labeling gives the *same* cross sections for the standard experimental quantization axis. We suggest that one use $L = l_i$ to obtain $f(jm|j_0m_0|\hat{\mathbf{R}})$, $L = l_f$ to get $f(j\lambda|j_0\lambda_0|\hat{\mathbf{R}})$, and either to get the experimental axis result. Otherwise, one can use the slightly more complex $L = (l_i + l_f)/2$ choice of the CS parameter. $L = l_i$ and $L = l_f$ correspond to using equations (45) and (54). Finally, in concluding the discussion of the CS, we encourage the reader to consult the literature cited above for additional details and also remark that more work is needed to further quantify the conditions for validity of the CS.

8.3. Infinite-Order Sudden Approximation

In the case of the IOS, one is making *both* the CS and the energy sudden approximations. Therefore both the two preceding discussions and the concomitant sufficiency criteria also apply in order for equations (82), (83), (99), (101), (103), (105), (120), (125)–(132), (134), and (135) to be employed. However, it is important to emphasize again that the Goldflam *et al.*[119,120] factorizations apply provided *only* that the energy sudden condition equation (225) is met. *We also note that this approximation should be seriously considered even in cases where the validity conditions are not well satisfied provided one requires only qualitatively accurate results.* The reason is that the approximation is extremely simple and therefore can yield results with very little computational effort (especially since only the cross sections out of the $j = 0$ state are required using the factorizations). We also note that Pack[2] has suggested the use of the WKB approximation in solving equations (85) and (86) for the orientation-dependent $T_{l_0}(\cos\theta)$. Noting that the orientation-dependent S matrix $S_{l_0}(\cos\theta)$ is related to $T_{l_0}(\cos\theta)$ via

$$T_{l_0}(\cos\theta) = 1 - S_{l_0}(\cos\theta) \tag{229}$$

it is easy to show that $T_{l_0}(\cos\theta)$ can be written in terms of an orientation-dependent phase shift $\eta_{l_0}(\theta)$ according to

$$T_{l_0}(\cos\theta) = -2i \sin\eta_{l_0} \exp(i\eta_{l_0}) \tag{230}$$

The WKB approximation to $\eta_{l_0}(\theta)$ is expected to apply when the distance ΔR over which the potential varies is large compared to the projectile de Broglie wavelength $h/(2ME_j)^{1/2}$. Thus our WKB test ratio t_{WKB}, defined as

$$t_{\text{WKB}} = h/\Delta R(2ME_j)^{1/2} \tag{231}$$

must satisfy

$$t_{\text{WKB}} \ll 1 \tag{232}$$

Then the WKB expression for $\eta_{l_0}(\theta)$ is given by[2]

$$\eta_{l_0}(\theta) = \lim_{R\to\infty} \left\{ \int_{R_i}^{R} \left[k^2 - \frac{(l_0 + 1/2)^2}{R'^2} - \frac{2M}{\hbar^2} V(R', \theta) \right]^{1/2} dR' - kR + \frac{(l_0 + 1/2)\pi}{2} \right\} \tag{233}$$

where R_i is the outermost classical turning point and Langer's replacement of $l_0(l_0 + 1)$ by $(l_0 + 1/2)^2$ is employed. This approximation, along with the factorization of Goldflam *et al.*,[119,120] makes the IOS calculations exceedingly simple. It is clear that equation (232) is satisfied as the projectile reduced mass becomes large, or as E_j gets large, or as the potential varies more slowly so that ΔR becomes large. It is also clear that if the effective potential involves three turning points and the de Broglie wavelength is such that significant tunneling is expected, then precise solutions of equations (85) and (86) should be obtained.

8.4. l_z-Conserving and DLD Approximations

The l_z-conserving approximation, equations (171), (172), (185), (186), (190), (193), (198), and (199), applies when the energy sudden criterion equation (225) is valid. It does not approximate the centrifugal potential and therefore can be applied to collisions for which the CS and IOS are expected to fail. It should be useful for treating collisions of molecules with large moments of inertia at energies where many rotational states are energetically accessible. It will not be suitable for transitions in which the rotor velocity is comparable to the projectile velocity. Thus, as the projectile mass increases, the l_z method (like the IOS) becomes less accurate. Also, transitions involving very high rotor states will not be well treated by this method.

The DLD method, equation (200), is also complementary to the CS and IOS since it is appropriate for treating inelastic scattering dominated by larger orbital angular momenta.[137] Conditions for its validity have been considered by De Pristo and Alexander.[138] Like the l_z-conserving method, no approximation is made to the centrifugal potential in the DLD. In addition, the full rotor energy operator $\mathbf{j}^2/2I$ is included. Instead, only coupling involving orbital angular momenta satisfying $l_i + j_i = l_f + j_f$ is retained. Thus the potential is modified in a manner which includes the orbital angular momenta which make the greatest contribution to the scattering at large impacts. The method is expected to be best for treating collisions in which all turning points occur in the region outside the potential well. Since the number of coupled equations encountered is equal to the number of rotor states coupled, the method is most useful for molecules with small moments of inertia. The most impressive application to date has been to the $j_i = 0 \rightarrow j_f = 2$ transition in the Li^+–H_2 system.[137] The DLD may be used when both the CS and equation (204) are invalid so that the l_z-conserving method is not appropriate.

8.5. *Exponential Distorted-Wave Approximation*

In addition to the l_z-conserving and DLD methods, there is yet another class of approximations which can be quite useful in complementing the CS and IOS methods. These are the infinite-order perturbation theory methods, and in particular, the exponential distorted-wave (EDW) approximation[140,151–155] [see equations (202)–(206)]. The approach is expected to work well in the case where the classical turning points do not permit the system to penetrate into the region of the minimum (or repulsive hard core) of the potential. In addition, it should be well suited to situations where the energy sudden condition, equation (225), is not valid (i.e., the adiabatic regime). Unlike the ordinary distorted-wave procedure, the exponential distorted-wave method retains unitarity and has been found to be accurate for describing collisions which are dominated by the long-range part of the potential. A potentially useful procedure would be to use the CS for the partial waves dominated by the hard core and short-range anisotropy and then use the EDW for the higher partial waves. A procedure that should significantly decrease the amount of labor in calculating the EDW approximation is to employ the WKB approximation for the distorted waves.[155] This eliminates the necessity of having to numerically compute the distorted waves. In addition, it is expected to be very important that the full diagonal part of the potential be included in determining the distorted waves instead of using just the spherical part of the potential. It is felt that using distorted waves that only incorporate the spherical part of the potential will lead to serious error in situations where the initial state is other than the $j = 0$ state and when the anisotropy is large. An excellent summary of the method can be found in the review by Balint-Kurti.[140]

8.6. Semiclassical Approximations

We have seen that the CS, IOS, l_z-conserving, DLD, and EDW methods provide valuable tools for treating a wide range of collision systems. However, there is an important class of systems (or transitions) that is not conveniently dealt with by these methods. For example, if one is considering a system where the projectile reduced mass M is large, the rotor constant B_e is small, and the quantum numbers of the transition are large, then the WKB condition, equation (232), holds but the energy sudden condition, equation (225), does not hold. That is, due to the large projectile reduced mass, the relative motion is expected to be semiclassical. Also, due to the high rotational quantum state, and small B_e, the rotor motion is expected to be semiclassical. In this situation, one would expect that a classical S-matrix approach should be appropriate.[141-148] Of the (large) variety of procedures available, it is felt that methods not requiring a multidimensional root-search[143-145] technique are preferred. This is primarily because of the rather substantial amount of computational labor required in the methods necessitating a root search. The most attractive classical S-matrix approach seems to be that based on the use of the correspondence principle. A particularly simple form has been presented by Percival and Richards[146] and Levine and Johnson[148]; see also Dickinson and Richards.[156] Thus in the situation where both the projectile and target may be treated semiclassically, it is felt that equations (216)–(222) are the most appropriate. For an excellent detailed discussion of these methods, the reader is referred to the review article by Clark *et al.*[142]

Finally, we point out that for the calculation of quantities involving extensive averages over quantum states, the Miller–Marcus type of classical S-matrix methods becomes much easier computationally. This is due to the reduction in the dimensionality of the root searches. Also, if highly averaged quantities are the ultimate objective, serious consideration must be given to the use of classical trajectory methods (for systems where the semiclassical conditions are met). Classical trajectory techniques are discussed elsewhere in this volume so we will not pursue them here.

8.7. *Full Close Coupling*

Although the methods discussed above cover most of the possible atom–diatom collision systems and scattering conditions, there are still some situations where none of the methods (CS, IOS, l_z-conserving, DLD, EDW, and semiclassical) is expected to apply. The most glaring situations are (a) when one desires transitions very close to threshold and (b) transitions which are strongly coupled to closed channels, e.g., Feshbach resonances. In these cases, it appears that recourse must be made to the full close-coupling method. Further work on developing efficient approximations for treating these situations is needed.

ACKNOWLEDGMENTS

Supported under National Science Foundation Grant CHE76-18970, Petroleum Research Fund Grant 10162-AC6-C, and R. A. Welch Foundation Grant E-608.

References

1. P. McGuire and D.J. Kouri, Quantum mechanical close coupling approach to molecular collisions: j_z-conserving coupled states approximation, *J. Chem. Phys.* **60**, 2488–2499 (1974).
2. R.T Pack, Space-fixed vs. body-fixed axes in atom–diatomic molecule scattering: Sudden approximations, *J. Chem. Phys.* **60**, 633–639 (1974).
3. M. Tamir and M. Shapiro, The approximate conservation of *P*-helicity in rotational excitation: A new decoupling scheme, *Chem. Phys. Lett.* **31**, 166–171 (1975).
4. R.B. Walker and J.C. Light, Body-fixed equations for atom–molecule scattering: Exact and centrifugal decoupling methods, *Chem. Phys.* **7**, 84–93 (1975).
5. D. Secrest, Theory of angular momentum decoupling approximations for rotational transitions in scattering, *J. Chem. Phys.* **62**, 710–719 (1975).
6. Y. Shimoni and D.J. Kouri, Accurate opacity functions for atom–diatom scattering: Average definite parity j_z-conserving approximation, *J. Chem. Phys.* **65**, 3372–3373 (1976).
7. Y. Shimoni and D.J. Kouri, On representative orbital angular momentum quantum numbers in the definite parity j_zCCS approximation, *J. Chem. Phys.* **65**, 3958–3967 (1976).
8. Y. Shimoni and D.J. Kouri, Quantum mechanical close coupling approach to molecular collisions: Averaged definite parity j_z-conserving coupled states approximation, *J. Chem. Phys.* **66**, 675–688 (1977).
9. Y. Shimoni and D.J. Kouri, Quantum mechanical close coupling approach to molecular collisions: Averaged definite parity j_z approximation with Clebsch–Gordan weights, *J. Chem. Phys.* **66**, 2841–2849 (1977).

10. G.A. Parker and R.T Pack, Identification of the partial wave parameter and simplification of the differential cross section in the j_zCCS approximation in molecular scattering, *J. Chem. Phys.* **66,** 2850–2853 (1977).
11. D.J. Kouri and Y. Shimoni, A new method for accurately approximating individual T-matrix elements in atom–diatom collisions, *J. Chem. Phys.* **65,** 5021–5023 (1976).
12. D.J. Kouri and Y. Shimoni, On the j_z-conserving coupled states approximation: Magnetic transitions and angular distributions in rotating and fixed frames, *J. Chem. Phys.* **67,** 86–98 (1977).
13. R. Goldflam and D.J. Kouri, On accurate quantum mechanical approximations for molecular relaxation phenomena: Averaged j_z-conserving coupled states approximation, *J. Chem. Phys.* **66,** 542–547 (1977).
14. V. Khare, On the equivalence between the space fixed and the body fixed formulations of the j_z-conserving approximation, *J. Chem. Phys.* **67,** 3897–3900 (1977).
15. D.J. Kouri, R. Goldflam, and Y. Shimoni, On the coupled states and infinite order sudden approximations to the body frame wavefunction, *J. Chem. Phys.* **67,** 4534–4544 (1977).
16. V. Khare, D.J. Kouri, and R.T Pack, On magnetic transitions and the interpretation of the partial wave parameter in the CS and IOS approximations in molecular scattering theory, *J. Chem. Phys.* **69,** 4419–4430 (1978).
17. K.P. Lawley and J. Ross, Semiclassical theory of rotational excitation of a diatomic molecule by an atom, *J. Chem. Phys.* **43,** 2930–2942 (1965).
18. H. Klar, Theory of collision induced rotational energy transfer in the Π-state of diatomic molecules, *J. Phys. B* **6,** 2139–2149 (1973).
19. A.E. De Pristo and M. Alexander, Relationships among the coupled states, P-helicity decoupling and effective potential methods, *Chem. Phys.* **19,** 181–192 (1977).
20. D.J. Kouri, On the decoupling of angular momenta in molecular collisions, *Chem. Phys. Lett.* **31,** 599–601 (1975).
21. H. Rabitz, Effective Hamiltonians in molecular collisions, in *Modern Theoretical Chemistry,* W.H. Miller, editor, Vol. 2, Plenum Press, New York (1976), pp. 33–80, and references therein.
22. B.H. Choi and K.T. Tang, On j_z-conserving coupled-states approximation for molecular collisions, *J. Chem. Phys.* **62,** 4568–4569 (1975).
23. P. McGuire, Elastic and inelastic angular distributions in the j_z-conserving coupled states approximation for molecular collisions, *Chem. Phys. Lett.* **23,** 575–578 (1973).
24. D.J. Kouri and P. McGuire, A coupled states approximation study of Li^+–H_2 collisions, *Chem. Phys. Lett.* **29,** 414–417 (1974).
25. S.I. Chu and A. Dalgarno, The rotational excitation of carbon monoxide by hydrogen atom impact, *Proc. R. Soc. London A* **342,** 191–207 (1975).
26. P. McGuire and J.P. Toennies, *A priori* low temperature vibrational relaxation rates for He–H_2, *J. Chem. Phys.* **62,** 4623–4627 (1975).
27. P. McGuire and H. Kruger, Elastic and rotationally inelastic H–H_2 and H–D_2 collisions, *J. Chem. Phys.* **63,** 1090–1094 (1975).
28. S.-I. Chu and A. Dalgarno, Approximations for the rotational excitation of molecules by atoms, *J. Chem. Phys.* **63,** 2115–2118 (1975). Erratum: Approximations for the rotational excitation of molecules by atoms [*J. Chem. Phys.* **63,** 2115 (1975)], *J. Chem. Phys.* **64,** 3085 (1976).
29. M. Tamir and M. Shapiro, Application of the PHD approximation to the high rotational transitions in the Ar + N_2 system, *Chem. Phys. Lett.* **39,** 79–81 (1976).
30. M. Alexander and P. McGuire, Cross sections and rate constants for low temperature ^{4}He–H_2 vibrational relaxation, *J. Chem. Phys.* **64,** 452–459 (1976).
31. M. Shapiro and M. Tamir, Rotational excitation in the Ar + N_2 system, *Chem. Phys.* **13,** 215–225 (1976).
32. M. Alexander and P. McGuire, Inelastic collisions between ^{4}He and H_2: Study of dynamical approximations, *Chem. Phys. Lett.* **12,** 31–36 (1976).

33. J.M. Launay, Body-fixed formulation of rotational excitation: Exact and centrifugal decoupling results for CO–He, *J. Phys. B* **9**, 1823–1838 (1976).
34. L. Monchick and S. Green, Validity of approximate methods in molecular scattering. III. Effective potential and coupled states approximations for differential and gas kinetic cross sections, *J. Chem. Phys.* **66**, 3085–3093 (1976).
35. S. Green and L. Monchick, Validity of approximate methods in molecular scattering: Thermal HCl–He collisions, *J. Chem. Phys.* **63**, 4198–4205 (1975).
36. S. Green, Accuracy of decoupling approximations for rotational excitation: Low energy CO–He collisions, *Chem. Phys. Lett.* **38**, 293–296 (1976).
37. T.G. Heil, Atom–molecule scattering: Close coupled and coupled states calculations for argon–methane scattering, Ph.D. Thesis, Department of Chemistry, University of Illinois, 1977.
38. T.G. Heil and D. Secrest, private communication.
39. P. McGuire, Validity of the coupled states approximation for molecular collisions, *Chem. Phys.* **13**, 81–94 (1976).
40. A.B. Elkowitz and R.E. Wyatt, J_z-conserving approximation for the hydrogen exchange reaction, *Mol. Phys.* **31**, 189–201 (1976).
41. A. Kuppermann, G.C. Schatz, and J.P. Dwyer, Angular momentum decoupling approximations in the quantum dynamics of reactive systems, *Chem. Phys. Lett.* **45**, 71–73 (1977).
42. L.A. Collins and N.F. Lane, Quantum mechanical calculations of rotational excitation cross sections for HCl, DCl, HF and DF in slow collisions with He, *Phys. Rev. A* **14**, 1358–1367 (1976).
43. R. Goldflam and D.J. Kouri, Quantum scattering theory of relaxation phenomena: Coupled states calculations of rotational relaxation and spectral line shapes in He–H_2 gas mixtures, *J. Chem. Phys.* **66**, 2452–2456 (1977).
44. P. McGuire, Convergence of the elastic and the inelastic cross sections in the coupled-states and the close coupling methods, *Chem. Phys.* **8**, 231–238 (1975).
45. T.G. Heil and D.J. Kouri, Diatom–diatom collisions: j_z-conserving coupled states calculations for para-H_2 + para-H_2 scattering, *Chem. Phys. Lett.* **40**, 375–389 (1976).
46. S. Green, L. Monchick, R. Goldflam, and D.J. Kouri, Computational tests of angular momentum decoupling approximations for pressure broadening cross sections, *J. Chem. Phys.* **66**, 1409–1412 (1977).
47. D.E. Fitz and P. McGuire, Comparison of the coupled states, infinite order sudden, and exponential Born methods for Ar–CsF collisions, *Chem. Phys. Lett.* **44**, 503–506 (1976).
48. S. Green, On the accuracy of the "decoupled l-dominant" approximation for atom-molecule scattering, *J. Chem. Phys.* **65**, 68–70 (1976).
49. U. Buck and P. McGuire, Rotational excitation of HCl by Ar impact, *Chem. Phys.* **16**, 101–108 (1976).
50. R. Goldflam and D.J. Kouri, On predicting quantal cross sections by interpolation: Surprisal analysis of j_zCCS and statistical results, *J. Chem. Phys.* **65**, 4218–4227 (1976).
51. G.D. Billing, The semiclassical coupled states method, *J. Chem. Phys.* **65**, 1–6 (1976).
52. M.H. Alexander, Cross sections and rate constants for the vibrational relaxation of $D_2(v = 1, j = 0)$ in collisions with ^{4}He, *Chem. Phys. Lett.* **40**, 101–105 (1976).
53. P. McGuire, Coupled-states approach for elastic and for rotationally and vibrationally inelastic atom–molecule collisions, *J. Chem. Phys.* **62**, 525–534 (1975).
54. P. McGuire, Coupled states approximation study of inelastic H^+–H_2 collisions at 3.7 eV, *J. Chem. Phys.* **65**, 3275–3279 (1976).
55. P. McGuire, K. Rudolph, and J.P. Toennies, Comparison of quantum and experimental rotational transition probabilities for H^+–H_2 collisions at 3.7 eV. *J. Chem. Phys.* **65**, 5522–5523 (1976).
56. B.J. Garrison and W.A. Lester, Jr., Coupled states cross sections for rotational ex-

citation of H_2CO by He impact at interstellar temperatures, *J. Chem. Phys.* **66**, 531–536 (1977).

57. S. Green, Rotational excitation of symmetric top molecules by collisions with atoms: Close coupling, coupled states and effective potential calculations for NH_3-He, *J. Chem. Phys.* **64**, 3463–3473 (1976).
58. T.G. Heil, S. Green, and D.J. Kouri, The coupled states approximation for scattering of two diatoms, *J. Chem. Phys.* **68**, 2562–2583 (1978).
59. P. McGuire, H. Schmidt, V. Hermann, and F. Linder, Quantum oscillations in the transition probability for rotational excitation of H_2 by H^+ impact at 4.67 eV, *J. Chem. Phys.* **66**, 4243–4244 (1977).
60. S.M. Tarr, H. Rabitz, D.E. Fitz, and R.A. Marcus, Classical and quantum centrifugal decoupling approximations for HCl–Ar, *J. Chem. Phys.* **66**, 2854–2859 (1977).
61. L. Monchick and D.J. Kouri, Magnetic transitions in the initial-labeled CS approximation: Computations for He + HCl, *J. Chem. Phys.* **69**, 3262–3267 (1978).
62. D.J. Kouri, T.G. Heil, and Y. Shimoni, On the Lippmann–Schwinger equation for atom–diatom collisions: A rotating frame treatment, *J. Chem. Phys.* **65**, 226–235 (1976).
63. D.J. Kouri, T.G. Heil, and Y. Shimoni, Sufficiency conditions for the validity of the j_z-conserving coupled states approximation, *J. Chem. Phys.* **65**, 1462–1473 (1976).
64. B.A. Lippmann and J. Schwinger, Variational principles for scattering processes. I, *Phys. Rev.* **79**, 469–480 (1950).
65. R.D. Levine, *Quantum Mechanics of Molecular Rate Processes*, Clarendon Press, Oxford (1969).
66. R. Goldflam, Formal and computational studies of molecular collision dynamics, Ph.D. Thesis, Department of Physics, University of Houston, May 1977.
67. W.N. Sams and D.J. Kouri, Noniterative solutions of integral equations for scattering. I. Single channels, *J. Chem. Phys.* **51**, 4809–4814 (1969).
68. W.N. Sams and D.J. Kouri, Noniterative solutions of integral equations for scattering. II. Coupled channels, *J. Chem. Phys.* **51**, 4815–4819 (1969).
69. D. Secrest, Amplitude densities in molecular scattering, in *Methods in Computational Physics*, B. Alder, S. Fernbach, and M. Rotenberg, editors, Vol. 10, Academic Press, New York (1971), pp. 243–286.
70. R.A. White and E.F. Hayes, A model potential for vibrational excitation of diatomic molecules, *Chem. Phys. Lett.* **14**, 98–102 (1972).
71. R.A. White and E.F. Hayes, Quantum mechanical studies of the vibrational excitation of H_2 by Li^+, *J. Chem. Phys.* **57**, 2985–2993 (1972).
72. E.R. Smith and R.J.W. Henry, Noniterative integral-equation approach to scattering problems, *Phys. Rev. A* **7**, 1585–1590 (1973).
73. A.M. Arthurs and A. Dalgarno, The theory of scattering by a rigid rotator, *Proc. R. Soc. London A* **256**, 540–551 (1960).
74. M.H. Alexander, P.J. Dagdigian, and A.E. De Pristo, Quantum interpretation of fully state selected rotationally inelastic collision experiments, *J. Chem. Phys.* **66**, 59–66 (1977).
75. A.R. Edmonds, *Angular Momentum in Quantum Mechanics*, Princeton University Press, Princeton, New Jersey (1957).
76. M. Jacob and G.C. Wick, On the general theory of collisions of particles with spin, *Ann. Phys. (N.Y.)* **7**, 404–428 (1959).
77. J.O. Hirschfelder, C.F. Curtiss, and R.B. Bird, *Molecular Theory of Gases and Liquids*, Wiley, New York (1965).
78. U. Fano, Pressure broadening as a prototype of relaxation, *Phys. Rev.* **131**, 259–268 (1963).
79. A. Ben-Reuven, Impact broadening of microwave spectra, *Phys. Rev.* **145**, 7–22 (1966).
80. R. Shafer and R.G. Gordon, Quantum scattering theory of rotational relaxation and spectral line shapes in H_2-He gas mixtures, *J. Chem. Phys.* **58**, 5422–5443 (1973).

81. W.K. Liu and R.A. Marcus, On the theory of the relaxation matrix and its application to microwave transient phenomena, *J. Chem. Phys.* **63**, 272–289 (1975).
82. A.E. De Pristo and M. Alexander, Rotationally inelastic scattering of two HF molecules, *J. Chem. Phys.* **66**, 1334–1342 (1977).
83. M.H. Alexander and A.E. De Pristo, Symmetry considerations in the quantum treatment of collisions of two diatomic molecules, *J. Chem. Phys.* **66**, 2166–2172 (1977).
84. S.I. Drozdov, The scattering of fast neutrons by non-spherical nuclei, *JETP* **1**, 591–592 (1955).
85. D.M. Chase, Adiabatic approximation for scattering processes, *Phys. Rev.* **104**, 838–842 (1956).
86. K. Adler and A. Winther, On the theory of multiple Coulomb excitation with heavy ions, *Mat. Fys. Medd. Dan. Vid. Selsk.* **32**, 1–72 (1960).
87. L. Monchick and E.A. Mason, Transport properties of polar gases, *J. Chem. Phys.* **35**, 1676–1697 (1961).
88. K. Takayanagi, Vibrational and rotational transitions in molecular collisions, *Prog. Theor. Phys. Suppl.* **25**, 1–98 (1963).
89. K.H. Kramer and R.B. Bernstein, Sudden approximation applied to rotational excitation of molecules by atoms. I, *J. Chem. Phys.* **40**, 200–203 (1964).
90. R.B. Bernstein and K.H. Kramer, Sudden approximation applied to rotational excition of molecules by atoms. II. Scattering of polar diatomics, *J. Chem. Phys.* **44**, 4473–4485 (1966).
91. R.J. Cross and D.R. Herschbach, Classical scattering of an atom from a diatomic rigid rotor, *J. Chem. Phys.* **43**, 3530–3540 (1965).
92. R.J. Cross, Jr., Classical small-angle scattering from anisotropic potentials, *J. Chem. Phys.* **46**, 609–618 (1967).
93. C.F. Curtiss, Molecular collisions. VI. Diagrammatic methods, *J. Chem. Phys.* **48**, 1725–1731 (1968).
94. C.F. Curtiss, Molecular collisions. VIII, *J. Chem. Phys.* **49**, 1952–1957 (1968).
95. R.J. Cross, Quenching of glory undulations in the total cross section by anisotropy in the intermolecular potential, *J. Chem. Phys.* **49**, 1976–1977 (1968).
96. R.D. Levine, Opacity analysis of inelastic molecular collisions: I. The quantum mechanical sudden approximation, *Chem. Phys. Lett.* **4**, 211–213 (1969).
97. R.D. Levine, Opacity analysis of inelastic molecular collisions. VI. The sudden approximation for rotational excitation, *J. Chem. Phys.* **54**, 997–1004 (1971).
98. R.J. Cross, Jr., Effects of anisotropy in the potential on molecular scattering, *J. Chem. Phys.* **52**, 5703–5707 (1970).
99. T.P. Tsien and R.T Pack, Rotational excitation in molecular collisions: A strong coupling approximation, *Chem. Phys. Lett.* **6**, 54–56 (1970).
100. C.F. Curtiss, Molecular collisions. XII. Generalized phase shifts, *J. Chem. Phys.* **52**, 4832–4841 (1970).
101. M.A. Wartell and R.J. Cross, Jr., Semiclassical theory of vibrationally inelastic scattering in three dimensions, *J. Chem. Phys.* **55**, 4983–4991 (1971).
102. T.P. Tsien and R.T Pack, Rotational excitation in molecular collisions: A many-state test of the strong coupling approximation, *Chem. Phys. Lett.* **8**, 579–581 (1971).
103. E.S. Chang and U. Fano, Theory of electron–molecule collisions by frame transformations, *Phys. Rev. A* **6**, 173–185 (1972).
104. P.G. Burke and N. Chandra, Electron–molecule interactions. III. A pseudo-potential method for $e^- - N_2$ scattering, *J. Phys. B* **5**, 1696–1711 (1972).
105. D.E. Golden, N.F. Lane, A. Temkin, and E. Gerjuoy, Low energy electron–molecule scattering experiments and the theory of rotational excitation, *Rev. Mod. Phys.* **43**, 642–678 (1971).
106. M.A. Brandt and D.G. Truhlar, Importance of long-range forces and short-range forces

in electron scattering: Elastic scattering by N_2 at 10 and 30 eV, *Chem. Phys. Lett.* **23**, 48–52 (1973).
107. T.P. Tsien, G.A. Parker, and R.T Pack, Rotational inelastic molecular scattering: Computational tests of some simple solutions of the strong coupling problem, *J. Chem. Phys.* **59**, 5373–5381 (1973).
108. L.W. Hunter, On infinite order sudden approximations for an arbitrary potential energy, *J. Chem. Phys.* **62**, 2855–2859 (1975).
109. C.F. Curtiss, Semiclassical theory of rotational excitation in collisions of diatomic molecules, *J. Chem. Phys.* **63**, 2738–2744 (1975).
110. F.S. Collins and R.J. Cross, Jr., Vibrationally inelastic scattering at high energies: $H^+ - H_2$, *J. Chem. Phys.* **65**, 644–652 (1976).
111. C.F. Curtiss, Rotational and vibrational excitation in the scattering of atoms by diatomic molecules, preprint.
112. D.G. Truhlar, R.E. Poling, and M.A. Brandt, Infinite order sudden approximation for rotational excitation of hydrogen molecules by electrons in the energy range 10–40 eV, *J. Chem. Phys.* **64**, 826–829 (1976).
113. J.M. Bowman and J. Arruda, Static rotation and quasiclassical total cross sections for atom-rigid rotor scattering: $^4He + I_2^*(j = 12) \rightarrow {}^4He + I_2^*(j')$, *Chem. Phys. Lett.* **41**, 43–45 (1976).
114. J.M. Bowman and S.C. Leasure, Sudden rotation calculations of atom–molecule scattering, *J. Chem. Phys.* **66**, 288–295 (1977). Erratum: Sudden rotation calculations of atom–molecule scattering [*J. Chem. Phys.* **66**, 288 (1977)], *J. Chem. Phys.* **66**, 4724 (1977).
115. J.M. Bowman, Sudden rotation total rotational inelastic cross sections for $^4He + I_2$, *J. Chem. Phys.* **66**, 296–297 (1977).
116. S.C. Leasure and J.M. Bowman, Fully converged sudden rotation total integral cross sections for $^4He + H_2(v = 0, j = 0) \rightarrow {}^4He + H_2(v' = 1, j')$, *Chem. Phys. Lett.* **48**, 179–183 (1977).
117. R.T Pack, Effect of partial wave parameter identification on IOS opacities and integral cross sections for rotationally inelastic collisions, *J. Chem. Phys.* **66**, 1557–1561 (1977).
118. C.F. Curtiss, Semiclassical theory of rotational transitions in collisions of diatomic molecules, *J. Chem. Phys.* **67**, 5770–5774 (1977).
119. R. Goldflam, S. Green, and D.J. Kouri, Infinite order sudden approximation for rotational energy transfer in gaseous mixtures, *J. Chem. Phys.* **67**, 4149–4161 (1977).
120. R. Goldflam, D.J. Kouri, and S. Green, On the factorization and fitting of molecular scattering information, *J. Chem. Phys.* **67**, 5661–5675 (1977).
121. G.A. Parker and R.T Pack, Rotationally and vibrationally inelastic scattering in the rotational IOS approximation: Ultrasimple calculation of total (differential, integral and transport) cross sections for nonspherical molecules, *J. Chem. Phys.* **68**, 1585–1601 (1978).
122. R. Schinke and P. McGuire, On the sudden centrifugal potential in the space-fixed system: Dependence of proton–H_2 cross sections on the choice of angular momentum parameter, *Chem. Phys.* **28**, 129–145 (1978).
123. D.A. Varshalovich and V.K. Khersonsky, Collisional excitations of interstellar molecules, *Astrophys. Lett.* **18**, 167–172 (1977).
124. A.S. Dickinson and D. Richards, Sudden atom–molecule collisions: CsF scattered by Ne, Ar, Kr and Xe, *J. Phys. B* **11**, 1085–1103 (1978).
125. D.E. Fitz, A study of reorientation effects in CsF–Ar using approximate quantum methods, *Chem. Phys.* **24**, 133–142 (1977).
126. Z.H. Top and D.J. Kouri, Numerical tests of the factorization of cross section data, *Chem. Phys.* (in press).
127. D.E. Fitz, On the choice of partial wave parameter for IOS calculations of *m*-dependent rotationally inelastic cross sections, *Chem. Phys. Lett.* **55**, 202–205 (1978).

128. S. Green, On the relation between microwave transient T_1 and T_2 relaxation times, *J. Chem. Phys.* **69**, 4076–4082 (1978).
129. R. Schinke and P. McGuire, Combined rotationally sudden and vibrationally exact quantum treatment of proton–H_2 collisions, *Chem. Phys.* **31**, 391–412 (1978).
130. I.C. Percival and M.J. Seaton, The partial wave theory of electron–hydrogen atom collisions, *Proc. Camb. Philos. Soc.* **53**, 654–662 (1957).
131. R.P. Srivastava and H.R. Zaidi, Intermolecular forces revealed by Raman scattering, in *Contemporary Raman Spectroscopy, Topics in Current Physics Series*, Springer-Verlag, New York (1977).
132. H. Rabitz, Rotation and rotation–vibration pressure-broadened spectral lineshapes, in *Ann. Rev. Phys. Chem.*, H. Eyring, C.J. Christensen, and H.S. Johnston, editors, Vol. 25, Annual Reviews, Inc., Palo Alto (1974), pp. 155–177.
133. L. Monchick, K.S. Yun, and E.A. Mason, Formal kinetic theory of transport phenomena in polyatomic gas mixtures, *J. Chem. Phys.* **39**, 654–669 (1963).
134. L. Monchick and S. Green, Validity of central field approximations in molecular scattering: Low energy CO–He collisions, *J. Chem. Phys.* **63**, 2000–2009 (1975).
135. A. Messiah, *Quantum Mechanics*, Wiley, New York (1961).
136. V. Khare, On the l_z-conserving energy sudden approximation for atom–diatom scattering, *J. Chem. Phys.* **68**, 4631–4640 (1978).
137. A.E. De Pristo and M.H. Alexander, A decoupled l-dominant approximation for ion–molecule and atom–molecule collisions, *J. Chem. Phys.* **64**, 3009–3013 (1976).
138. A.E. De Pristo and M.H. Alexander, Limits of validity of the decoupled l-dominant approximation, *Chem. Phys. Lett.* **44**, 214–218 (1976).
139. A.E. De Pristo and M.H. Alexander, Decoupling approximations for rotationally inelastic collisions between ions and polar molecules: H^+–CN, *J. Phys. B* **9**, 2713–2721 (1976).
140. G.G. Balint-Kurti, The theory of rotationally inelastic molecular collisions, in *Int. Rev. of Science, Physical Chemistry*, A.D. Buckingham and C.A. Coulson, editors, Series Two, Butterworths, Boston (1975), pp. 283–326.
141. W.H. Miller, Classical-limit quantum mechanics and the theory of molecular collisions, in *Advances in Chemical Physics*, I. Prigogine and S.A. Rice, editors, Vol. 25, Interscience, New York (1974), pp. 69–177.
142. A.P. Clark, A.S. Dickinson, and D. Richards, The correspondence principle in heavy-particle collisions, in *Advances in Chemical Physics*, I. Prigogine and S.A. Rice, editors, Vol. 36, Interscience, New York (1977), pp. 63–139.
143. W.H. Miller, Semiclassical theory of atom–diatom collisions: Path integrals and the classical S matrix, *J. Chem. Phys.* **53**, 1949–1959 (1970).
144. R.A. Marcus, Extension of the WKB method to wave functions and transition probability amplitudes (S-matrix) for inelastic or reactive collisions, *Chem. Phys. Lett.* **7**, 525–532 (1970).
145. R.A. Marcus, Theory of semiclassical transition probabilities (S matrix) for inelastic and reactive collisions, *J. Chem. Phys.* **54**, 3965–3979 (1971).
146. I.C. Percival and D. Richards, A correspondence principle for strongly coupled states, *J. Phys. B* **3**, 1035–1046 (1970).
147. A.S. Dickinson and D. Richards, A semiclassical study of the body-fixed approximation for rotational excitation in atom–molecule collisions, *J. Phys. B* **10**, 323–343 (1977).
148. R.D. Levine and B.R. Johnson, On the classical and semi-classical limits in collision theory, *Chem. Phys. Lett.* **7**, 404–408 (1970).
149. Y. Shimoni, D.J. Kouri, and A. Kumar, Close coupling calculations of magnetic transitions for He + H_2 in an uncoupled space frame, *Chem. Phys. Lett.* **52**, 299–304 (1977).
150. G. Pfeffer and D. Secrest, The rotational sudden approximation at low energies, *J. Chem. Phys.* **67**, 1394–1404 (1977).

151. I.C. Percival, A variational principle for scattering phases, *Proc. Phys. Soc. (London)* **76**, 206–216 (1960).
152. M.J. Seaton, Strong coupling in optically allowed atomic transitions produced by electron impact, *Proc. Phys. Soc. (London)* **77**, 174–183 (1961).
153. G.G. Balint-Kurti and R.D. Levine, Opacity analysis of inelastic molecular collisions: Computational studies of the exponential Born approximation for rotational excitation, *Chem. Phys. Lett.* **7**, 107–111 (1970).
154. R.D. Levine, Exponential approximations in collision theory, *Mol. Phys.* **22**, 497–523 (1971).
155. S. Bosanac and G.G. Balint-Kurti, Cross sections for the rotationally inelastic scattering of Ne + N_2: Application of the exponential semi-classical distorted wave approximation, *Mol. Phys.* **29**, 1797–1811 (1975).
156. A.S. Dickinson and D. Richards, Rotational excitation of diatomic molecules by atoms, *J. Phys. B* **7**, 1916–1936 (1974).

10
Rotational Excitation III: Classical Trajectory Methods

M. D. PATTENGILL

1. Introduction

The use of classical trajectory (CT) methods in the solution of molecular scattering problems has been the subject of several recent review articles.[1-5] Here we restrict attention to nonreactive, rotationally inelastic collisions and focus on the practical details involved in executing a trajectory calculation. Before proceeding, a few comments are in order on the accuracy and utility of CT methods as applied to rotational excitation.

Within the assumption of classical mechanics, CT studies provide exact solutions for model scattering problems. The question of CT validity thus hinges on the importance of quantum effects. As discussed by Miller,[6] the origin of quantum effects is the principle of superposition, absent from purely classical treatments, which gives rise to interference, tunneling, quantization, and selection rules. In order for the CT method to be accurate, these effects either must be unimportant or must be accounted for in some satisfactory fashion.

Unfortunately, as discussed in Reference 7, no simple criterion presently exists for gauging the validity of classical mechanics as applied to inelastic scattering. However, if we restrict attention to fundamental physical observables for rotational excitation, viz., integral and differential cross sections for state-to-state transitions, some useful insights are available. These derive from studies[7-20] in which direct quantum vs. classical cross section comparisons have been made for specific model systems. As dis-

M.D. PATTENGILL • Department of Chemistry, University of Kentucky, Lexington, Kentucky 40506.

cussed in detail later, in these studies state-to-state cross sections were extracted from the classical results by a procedure in which (a) rotational quantum numbers j_c were assigned to classical rotational energies by use of a correspondence rule, and (b) cross sections were calculated, usually by binning trajectories on the basis of final rotational quantum numbers, taking account, if applicable, of the parity selection rule.* Thus the CT calculations ignore interference and tunneling, while accounting for quantization in an approximate manner and selection rules in a somewhat *ad hoc* manner.

Results of the comparative cross-section studies are summarized in Table 1 (similar to Table I of Reference 18). Overall, the results indicate that, used with caution, the CT method can provide a useful approximation at collision energies somewhat above the threshold for rotational excitation. This finding is in accord with the argument of Miller[21] that for

* According to the parity selection rule, for potentials symmetric with respect to inversion of the rotor through its center of mass (i.e., homonuclear diatomics), the change in rotational quantum number, Δj, is restricted to even values.

Table 1. Comparative Studies of Classical Trajectory and Quantum-Mechanical Cross Sections for Rotationally Inelastic Scattering[a]

System	Ref.	j	E_{rel} (eV)	E_{th}(eV)	Quality of CT results
Integral cross section comparisons					
$He + H_2$	12, 16, 19	0	0–0.81	0.05	CT underestimates near threshold, overestimates well above threshold. At higher energies, agreement to within 25–50%
$Li^+ + H_2$	8, 10	0	0–0.60	0.05	Similar to $He + H_2$. At higher
		1	0.60	0.08	energies CT overestimates by about 20–50%.
$Ar + N_2$	14, 15	0	0.066	0.001	Within statistics of CT calcu-
		2	0.065	0.003	lations, agreement essential-
		4	0.061	0.005	ly quantitative.
		6	0.056	0.007	
		8	0.048	0.009	
		10	0.039	0.011	
$H + CO$	13	0	0.01, 0.05	0.0005	Good overall agreement.
		2	0.01, 0.05	0.001	CT tends to overestimate
		5	0.01, 0.05	0.003	small Δj and underestimate large Δj transitions.
$He + CO$	12, 20	0	0–0.02	0.0005	Fair agreement overall. CT does not reproduce quantum preference for even Δj transitions. $\Delta j = 1$ overestimated if (14) used.

Table 1 *(cont'd.)*

System	Ref.	j	E_{rel} (eV)	E_{th} (eV)	Quality of CT results
"H_2" + CS[b]	20	0	0–0.012	0.0002	CT seriously overestimates $\|\Delta j\| = 1$ transitions. Others well represented.
		2	0–0.012	0.0006	
He + HCl	20	0	0.025, 0.062	0.003	Fair overall agreement. CT tends to overestimate small Δj and underestimate large Δj transitions.
		1	0.025, 0.062	0.005	
		4	0.025, 0.062	0.012	
Ar + HCl	17, 18, 20	0	0.025–0.077	0.003	CT seriously overestimates large Δj transitions from $j = 0$. For $j \neq 0$ overall agreement good. CT may tend to overestimate. Some $\|\Delta j\| = 1$ transitions.
		1	0.025–0.077	0.005	
		3	0.013–0.062	0.012	
He + HCN	13	0	0.007	0.0004	Poor agreement. CT fails to reproduce quantum preference for even Δj transitions.
		3	0.007	0.001	
		5	0.007	0.002	
"H_2" + OCS[b]	20	0	0–0.012	0.0001	Similar to He + HCN.
Differential cross section comparisons					
He + H_2	9	0	1.09	0.005	CT underestimates at small angles. Good agreement for large angles.
Li^+ + H_2	7, 10, 11	0	0.6, 1.2	0.05	Good agreement for large angles beyond rainbow angle.
		1	0.6, 1.2	0.08	

[a] j is the initial rotor state, E_{rel} is the initial relative translational energy, and E_{th} is the energy threshold for the lowest energy rotational excitation.

[b] "H_2" represents H_2 rigorously constrained to remain in the state $j = 0$. With this constraint, "H_2" behaves formally as a spherical atom.

classically allowed processes,[6] interference and tunneling effects are rapidly quenched in averaged collision properties such as cross sections.

Above threshold, two notable failings of the CT method occur in situations where quantum effects might be expected to be important. Thus for He + CO, He + HCN, and "H_2" + OCS, the CT integral cross sections fail to reproduce the quantum-mechanical preference for even Δj transitions. This failure can be explained[12,13] in terms of a near symmetry in the atom–molecule potential function which leads to interference effects similar to those responsible for the parity selection rule in scattering on symmetric potentials. In addition, for He + H_2 and Li^+ + H_2, the CT differential cross sections, while good at large angles, are poor in the small-angle rainbow region where interference effects are strong.[7,9]

Despite the caveats discussed above and others apparent from Table 1, it is clear that the CT method merits serious consideration as an approximation technique for the calculation of rotationally inelastic cross sections. When justified, its use is often attractive from a cost point of view, particularly for collisions involving heavy particles (i.e., many accessible states) and/or strong coupling. In this connection, it should be noted that for a given level of relative statistical error, the efficiency of the CT method decreases with decreasing cross section, Thus even when tunneling and interference effects are unimportant, the CT method is of limited utility for the accurate determination of small cross sections.

For the remainder of this chapter we concentrate on the details of performing a trajectory calculation. We specialize from the outset to the case of atom–diatomic collisions occurring at fixed initial translational and rotational energies and focus on obtaining rotational energy transfer cross sections. For simplicity, and since it is frequently a time-saving and accurate approximation,[22] we treat the diatomic as a rigid rotor.

2. *Ingredients of a Trajectory Calculation*

2.1. *Equations of Motion*

A convenient coordinate system for A + BC scattering is one which employs the Cartesian coordinates of atom A relative to the center of mass of BC (given by vector **R**) and the Cartesian coordinates of atom C relative to atom B (given by vector **r**). With these, after separation of the ABC center-of-mass motion, the kinetic energy T of the A + BC system is given by[23,24]

$$T = \tfrac{1}{2}\mu_{\mathrm{A,BC}}\dot{\mathbf{R}}\cdot\dot{\mathbf{R}} + \tfrac{1}{2}\mu_{\mathrm{BC}}\dot{\mathbf{r}}\cdot\dot{\mathbf{r}} \tag{1}$$

where $\mu_{\mathrm{A,BC}}$ is the reduced mass of the A,BC pair, μ_{BC} is the reduced mass of BC, and a dot over a vector denotes differentiation of each element with respect to time.

A decision must be made on whether to integrate Lagrange's or Hamilton's equations of motion. Although either set of equations can be (and have been) used, here we consider only Hamilton's equations. Introducing the momenta

$$\mathbf{P} = \mu_{\mathrm{A,BC}}\dot{\mathbf{R}}, \qquad \mathbf{p} = \mu_{\mathrm{BC}}\dot{\mathbf{r}} \tag{2}$$

conjugate to **R** and **r** into (1) yields

$$T = (\mathbf{P}\cdot\mathbf{P}/2\mu_{\mathrm{A,BC}}) + (\mathbf{p}\cdot\mathbf{p}/2\mu_{\mathrm{BC}}) \tag{3}$$

and Hamiltonian

$$H(\mathbf{P},\mathbf{p},\mathbf{R},\mathbf{r}) = T(\mathbf{P},\mathbf{p}) + V(\mathbf{R},\mathbf{r}) \tag{4}$$

where $V(\mathbf{R},\mathbf{r})$ is the A + BC interaction potential.*

* Throughout, we assume $V(\mathbf{R},\mathbf{r})$ is a known function. For rotational excitation studies, $V(\mathbf{R},\mathbf{r})$ can often be conveniently represented as a Legendre expansion in R (the magnitude of **R**) and the angle γ between **R** and **r** (i.e., $\mathbf{R}\cdot\mathbf{r} = Rr\cos\gamma$).

With (3) and (4), Hamilton's equations can be obtained. In doing this, it is necessary to take into account the rigid-rotor constraint that the magnitude of $\mathbf{r}$ be constant. This constraint can be introduced by the method of Lagrange multipliers,[25] to yield the set of 12 first-order coupled equations[24]:

$$\begin{aligned} \dot{\mathbf{R}} &= \mathbf{P}/\mu_{\mathrm{A,BC}}, & \dot{\mathbf{P}} &= -\nabla_{\mathbf{R}} V \\ \dot{\mathbf{r}} &= \mathbf{p}/\mu_{\mathrm{BC}}, & \dot{\mathbf{p}} &= -\nabla_{\mathbf{r}} V + \lambda \mathbf{r} \end{aligned} \tag{5}$$

with

$$\lambda = (1/r^2)[\mathbf{r}\cdot\nabla_{\mathbf{r}}V - (\mathbf{p}\cdot\mathbf{p}/\mu_{\mathrm{BC}})] \tag{6}$$

where r is the magnitude of $\mathbf{r}$, and $\nabla_{\mathbf{R}}$, $\nabla_{\mathbf{r}}$ are del operators for $\mathbf{R}$ and $\mathbf{r}$, respectively.

2.2. Selection of Initial Conditions

Generation of initial values for $\mathbf{P}$, $\mathbf{p}$, $\mathbf{R}$, and $\mathbf{r}$ can be accomplished in a variety of ways. Here we follow closely the prescription of Karplus *et al.*[23] Since the initial relative translational energy E_T is assumed to be fixed, the initial relative motion is characterized by a fixed relative velocity, v_R $[= (2E_T/\mu_{\mathrm{A,BC}})^{1/2}]$. Choosing the z axis as the initial direction of v_R and placing atom A initially in the yz plane yields for the initial components of $\mathbf{R}$ and $\mathbf{P}$,

$$X = 0, \qquad Y = b, \qquad Z = -(\rho^2 - b^2)^{1/2} \tag{7}$$

$$P_X = 0, \qquad P_Y = 0, \qquad P_Z = \mu_{\mathrm{A,BC}} v_R \tag{8}$$

where b is the impact parameter and ρ is the initial separation between A and the BC center of mass [chosen large enough that at ρ, $V(\mathbf{R},\mathbf{r})$ is negligible]. The BC orientation can be specified in terms of spherical polar coordinates defined by r_e, the (constant) magnitude of $\mathbf{r}$, and angles θ and ϕ, so that initially,

$$x = r_e \sin\theta\cos\phi, \qquad y = r_e \sin\theta \sin\phi, \qquad z = r_e\cos\theta \tag{9}$$

The magnitude of the initial BC rotational angular momentum is given in terms of the (fixed) initial BC rotational energy E_R by

$$L = (2\mu_{\mathrm{BC}} E_R)^{1/2} r_e \tag{10}$$

and, since

$$E_R = (\mathbf{p}\cdot\mathbf{p})/2\mu_{\mathrm{BC}} \tag{11}$$

the magnitude of $\mathbf{p}$ is given by

$$p = L/r_e \tag{12}$$

If η is defined as the angle of **p** with the vector $\mathbf{r} \times \mathbf{k}$ (**k** is the unit vector along the z axis) which is perpendicular to **r**, it follows that initially,

$$\begin{aligned} p_x &= -L(\sin\phi\cos\eta + \cos\phi\cos\theta\sin\eta)/r_e \\ p_y &= L(\cos\phi\cos\eta - \sin\phi\cos\theta\sin\eta)/r_e \\ p_z &= L(\sin\theta\sin\eta)/r_e \end{aligned} \tag{13}$$

Thus initial values of **P**, **p**, **R**, and **r** are completely specified by the values of E_T (or v_R), E_R, b, ρ, r_e, θ, ϕ, and η.

The value of E_T (or v_R) is fixed, and since BC is a rigid rotor, so is r_e. E_R is also fixed by placing BC in a particular rotational state* j and using a correspondence rule to relate j to L. The most widely used correspondence relationships have been

$$L = [j(j+1)]^{1/2}\hbar \tag{14}$$

and

$$L = (j + \tfrac{1}{2})\hbar \tag{15}$$

which differ significantly only for very low j values. Studies[8,18,20] which have compared cross sections obtained using both (14) and (15) for $j = 0$, have not established a clear preference for one over the other.† Here we use (14), which leads to

$$E_R = j(j+1)B_e \tag{16}$$

where $B_e = \hbar^2/2\mu_{BC}r_e^2$. In addition, as noted above, ρ is fixed at a value large enough so that $V(\mathbf{R},\mathbf{r})$ is initially negligible.

The remaining four quantities, b, θ, ϕ and η, are varied from trajectory to trajectory in such a way that their possible values occur with proper statistical weight. Several methods for performing these variations are possible (cf. Section 4 below). Here, we use the most straightforward, in which parameter values are chosen at random from appropriate probability distributions.

Since ϕ and η are uniformly distributed over the interval 0–2π rad,

$$\phi = 2\pi\xi_1, \qquad \eta = 2\pi\xi_2 \tag{17}$$

where ξ_i is a random number chosen from a uniform distribution between

* Another possibility which has been investigated for at least one system[26] is initial state boxing. In this, noninteger initial j values are chosen from an interval centered at the quantum-mechanical (integer) initial j value.

† For He + CO, use of (15) rather than (14) leads to much better classical vs. quantum agreement for the $j = 0 \rightarrow j' = 1$ transition.[20] For Ar + HCl ($j = 0$), use of (15) leads to better agreement for large Δj transitions, but worse agreement for small Δj transitions.[18]

0 and 1. θ and b have normalized probability distributions*

$$P(\theta) = \tfrac{1}{2}\sin\theta, \qquad P(b) = 2b/b_{\text{max}}^2 \tag{18}$$

over the ranges 0–π and 0–b_{max}, respectively. Equating the cumulative probabilities

$$C(\theta) = \int_0^\theta P(\theta)\,d\theta, \qquad C(b) = \int_0^b P(b)\,db \tag{19}$$

to random numbers ξ_3 and ξ_4 yields

$$\cos\theta = 1 - 2\xi_3, \qquad b = (\xi_4)^{1/2} b_{\text{max}} \tag{20}$$

The value of b_{max} is chosen large enough so that all interesting scattering events are included. In general, for a particular calculation, this value must be determined by running trial trajectories. Since $P(b)$ weights large b trajectories more heavily, it is critical to the efficiency of the calculation that b_{max} be no larger than necessary.

Chains of pseudorandom numbers can be generated conveniently by a congruence method.[1] Thus starting with a seed integer random number $I_0 < 2^{N-1}$ (N is the number of binary digits in the computer's word length) supplied by the user,

$$I_{i+1} = 5^x I_i \tag{21}$$

where x is the largest odd number such that $5^x < 2^{N-1}$. ξ_i is obtained from I_i by performing

$$\xi_i = I_i \bmod 2^N \tag{22}$$

2.3. *Integration of Equations of Motion*

Given a complete set of trajectory initial conditions, integration of the equations of motion (5) can be accomplished by a variety of methods. A convenient procedure which has been used in several trajectory studies of rotational excitation[15,16,22] makes use of a combined fourth-order Runge–Kutta–Gill, fourth-order Adams–Moulton algorithm. The integration, which uses a constant time increment Δt is begun by executing five cycles of Runge–Kutta–Gill integration which prepares a table of coordinates and momenta at the first five multiples of Δt. This table is then used to continue the integration by the more efficient, but non-self-starting Adams–Moulton method. The integration scheme has been described by

* If desired, b can be restricted to quantum-mechanically allowed values given by either $b_l = [l(l+1)]^{1/2}\,\hbar/P$ or $(l+\tfrac{1}{2})\,\hbar/P$, where P is the initial magnitude of $\mathbf{P}$, and l is the orbital angular momentum quantum number. For Li^+–H_2 scattering,[7] discrete vs. continuous b selection leads to negligible differences in computed cross sections.

Bunker.[1] For completeness, we reproduce it here in somewhat different notation and with a misprint in Reference 1 corrected.

Denoting the elements of **P**, **p**, **R**, and **r** by x_i's, the equations of motion can be written

$$\dot{x}_i(t) = f_i(\{x_i(t)\}), \qquad i = 1,2,\ldots,12 \tag{23}$$

with initial conditions

$$x_i(t_0) = x_{i0}, \qquad i = 1,2,\ldots,12 \tag{24}$$

In the Runge–Kutta–Gill procedure, values of the x_i at time $t_0 + \Delta t$ are given by

$$\left.\begin{aligned}
k_{i1} &= f_i(\{x_{i0}\})(\Delta t)\\
x_{i1} &= x_{i0} + \tfrac{1}{2}k_{i1}\\
q_{i1} &= k_{i1}\\
k_{i2} &= f_i(\{x_{i1}\})(\Delta t)\\
x_{i2} &= x_{i1} + (1 - \tfrac{1}{2}2^{1/2})(k_{i2} - q_{i1})\\
q_{i2} &= (2 - 2^{1/2})k_{i2} - (2 - \tfrac{3}{2}2^{1/2})q_{i1}\\
k_{i3} &= f_i(\{x_{i2}\})(\Delta t)\\
x_{i3} &= x_{i2} + (1 + \tfrac{1}{2}2^{1/2})(k_{i3} - q_{i2})\\
q_{i3} &= (2 + 2^{1/2})k_{i3} - (2 + \tfrac{3}{2}2^{1/2})q_{i2}\\
k_{i4} &= f_i(\{x_{i3}\})(\Delta t)\\
x_{i4} &= x_{i3} + \tfrac{1}{6}(k_{i4} - 2q_{i3})\\
x_i(t_0 + \Delta t) &= x_{i4}
\end{aligned}\right\} \quad i = 1,2,\ldots,12 \tag{25}$$

In the Adams–Moulton method, previous values of the x_i are used to evaluate predictors:

$$\begin{aligned}
x_i(t_n + \Delta t) = x_i(t_n) + (\Delta t/720)\,[&1901f_i(\{x_i(t_n)\}) - 2774f_i(\{x_i(t_n - \Delta t)\})\\
&+ 2616f_i(\{x_i(t_n - 2\Delta t)\}) - 1274f_i(\{x_i(t_n - 3\Delta t)\})\\
&+ 251f_i(\{x_i(t_n - 4\Delta t)\})], \qquad i = 1,2,\ldots,12
\end{aligned} \tag{26}$$

and then correctors

$$\begin{aligned}
x_i(t_n + \Delta t) &= (\Delta t/1440)\,[475f_i(\{x_i(t_n + \Delta t)\}) + 1427f_i(\{x_i(t_n)\})\\
&- 798f_i(\{x_i(t_n - \Delta t)\}) + 482f_i(\{x_i(t_n - 2\Delta t)\})\\
&- 173f_i(\{x_i(t_n - 3\Delta t)\}) + 27f_i(\{x_i(t_n - 4\Delta t)\})], i = 1,2,\ldots,12
\end{aligned} \tag{27}$$

In general, the value of Δt used in the integration is determined by running sample trajectories. The accuracy of the integration for a particular trajectory can be tested either by (a) back integration, i.e., integrating from final to initial conditions with time increment $-\Delta t$, or (b) repeating the trajectory for smaller values of Δt. If Δt is adequate, either (a) the initial conditions will be regenerated, or (b) the final conditions will not change within the desired level of accuracy (usually several significant figures). It is extremely important to note that total energy and total angular momentum conservation are necessary, but not sufficient, conditions for accurate trajectory integration.

2.4. *Analysis of Final Conditions*

Individual trajectories are terminated when the A to BC center-of-mass separation attains a value (usually equal to ρ) large enough that $V(\mathbf{R}, \mathbf{r})$ is again negligible. At this point, a complete set of final momenta and coordinates, $\mathbf{P}'$, $\mathbf{p}'$, $\mathbf{R}'$, $\mathbf{r}'$, are available. Although other final-state properties may be of interest in particular situations, we restrict attention here to the extraction of integral and differential cross sections for rotational excitation. Needed to obtain these are the magnitude of the final BC rotational angular momentum

$$L' = |\mathbf{r}' \times \mathbf{p}'| \tag{28}$$

and the atomic scattering angle* χ, given by

$$\cos\chi = \mathbf{P} \cdot \mathbf{P}'/|\mathbf{P}|\,|\mathbf{P}'| \tag{29}$$

The most frequently used method† for calculating cross sections first assigns (noninteger) final BC rotational quantum numbers j_c'. This is done using the same correspondence rule (14) or (15), as used in the initial condition selection. For the choice used here,

$$L' = [j_c'(j_c' + 1)]^{1/2}\hbar \tag{30}$$

j_c' is then assigned to the nearest (integer) quantum-mechanically allowed BC rotational state j'. Thus trajectories are binned on the basis of j_c' so that (a) for heteronuclear BC (no parity selection rule), trajectories with j_c' in the range $j' - \frac{1}{2}$ to $j' + \frac{1}{2}$ are ascribed to j', and (b) for homonuclear BC (parity selection rule), trajectories with j_c' in the range $j' - 1$ to $j' + 1$ are ascribed to j'.

* The molecular scattering angle is given by $\chi_{mol} = 2\pi - \chi$.
† For examples of other methods, see References 8 and 13.

State-to-state cross sections and their associated statistical errors can be calculated* as follows.[4] If $N_{j'}$ is the number of trajectories assigned to j' and N_T is the total number of trajectories in the sample, the probability of scattering into state j' is

$$P_{j'} = N_{j'}/N_T \tag{31}$$

An estimate of the variance of $P_{j'}$ is given by

$$\mathrm{var}(P_{j'}) = P_{j'}(1 - P_{j'})/N_T \tag{32}$$

The state-to-state integral cross section $\sigma_{j'j}$ for scattering from j to j' is thus

$$\sigma_{j'j} = \pi b_{\max}^2 P_{j'} \tag{33}$$

with variance

$$\mathrm{var}(\sigma_{j'j}) = \pi^2 b_{\max}^4 [P_{j'}(1 - P_{j'})/N_T] \tag{34}$$

The differential cross section (per unit solid angle) for BC scattered into state j' is given by

$$\frac{d^2\sigma_{j'j}(\chi)}{d^2\omega} = \frac{\pi b_{\max}^2 P_{j'}(\chi)}{2\pi \sin\chi} \tag{35}$$

where $P_{j'}(\chi)$ is the probability of scattering into j' at angle χ (i.e., the angular distribution for j'). Using the histogrammic method, the differential cross section can be approximately represented by

$$\frac{d^2\sigma_{j'j}(\chi_m)}{d^2\omega} = \frac{\pi b_{\max}^2}{2\pi N_T} \sum_{i=1}^{N_{j'}(\Delta\chi)} \frac{1}{\sin\chi_i} \tag{36}$$

where the sum is over the $N_{j'}(\Delta\chi)$ trajectories assigned to j' having χ in the range $\Delta\chi$, which defines the width of the histogram interval, and χ_m is the midpoint of $\Delta\chi$. An estimate of the variance of $d^2\sigma_{j'j}(\chi)/d^2\omega$ can be obtained from†

$$\mathrm{var}\left[\frac{d^2\sigma_{j'j}(\chi_m)}{d^2\omega}\right] = \frac{\pi^2 b_{\max}^4}{4\pi^2}\left\{\mathrm{var}[P_{j'}(\chi_m)]\,\mathrm{var}\left[\frac{1}{\sin\chi_m}\right] + P_{j'}^2(\chi_m)\,\mathrm{var}\left[\frac{1}{\sin\chi_m}\right] + \left(\frac{1}{\sin\chi_m}\right)^2 \mathrm{var}[P_{j'}(\chi_m)]\right\} \tag{37}$$

* See also Reference 27.

† If a is a constant, and x and y are statistically independent variables with means $\bar{x}$ and $\bar{y}$, respectively, it follows that[28] $\mathrm{var}(axy) = a^2[\mathrm{var}(x)\,\mathrm{var}(y) + \bar{x}^2\mathrm{var}(y) + \bar{y}^2\mathrm{var}(x)]$.

where

$$
\begin{aligned}
P_{j'}(\chi_m) &= N_{j'}(\Delta\chi)/N_T \\
\operatorname{var}[P_{j'}(\chi_m)] &= P_{j'}(\chi_m)[1 - P_{j'}(\chi_m)]/N_T \\
\frac{1}{\sin\chi_m} &= \frac{1}{N_{j'}(\Delta\chi)} \sum_{i=1}^{N_{j'}(\Delta\chi)} \frac{1}{\sin\chi_i}
\end{aligned}
\tag{38}
$$

$$
\operatorname{var}\left[\frac{1}{\sin\chi_m}\right] = \frac{1}{[N_{j'}(\Delta\chi)]^2} \times \left\{\left[\sum_{i=1}^{N_{j'}(\Delta\chi)} \frac{1}{\sin^2\chi_i}\right] - \frac{1}{N_{j'}(\Delta\chi)}\left[\sum_{i=1}^{N_{j'}(\Delta\chi)} \frac{1}{\sin\chi_i}\right]^2\right\}
$$

For $N_{j'}(\Delta\chi)$ small, this estimate must be used with caution.

3. Construction of a Trajectory Program

There are, of course, any number of ways of writing an operable trajectory program. An often used and convenient method employs a modular approach in which the program is composed of a number of more or less independent subroutines. Besides permitting individual subroutine debugging, such construction facilitates future program modification. We illustrate here a sample organization based on this approach.

A convenient set of units for trajectory calculations is that in which the unit of mass is the amu, the unit of length is the Å, and the unit of time is 10^{-13} sec. The resultant unit of energy is just the kJ/mole.

Basic storage requirements are four 1×6 arrays, P, Q, PDOT, and QDOT, which contain current values of **P** and **p**, **R** and **r**, $\dot{\mathbf{P}}$ and $\dot{\mathbf{p}}$, and $\dot{\mathbf{R}}$ and $\dot{\mathbf{r}}$, respectively, and a 5×12 array, TABLE, which contains five previous sets of **P**, **p**, **R**, and **r** required by the Adams–Moulton integrator. With these, the calculation can be conveniently organized in terms of the following subroutines:

MAIN—reads input parameters and converts units if necessary; calls HEAD to output batch information; for each trajectory: calls SELECT, which generates initial coordinates and momenta; saves initial values of **P** for scattering-angle calculation; calls EAMCAL, which calculates initial values of the total energy and total angular momentum; calls WRIN to output trajectory initial conditions; calls CNS, which schedules trajectory integration, and, upon completion, returns a set of final coordinates and momenta; calculates scattering angle, final rotational energy, final rotational "quantum number," j'_c, and any other final properties of interest; calls EAMCAL, which calculates final values of the total energy and total angular momentum; calls WROU to output trajectory final conditions.

HEAD—called from MAIN; outputs desired batch information.

SELECT—called from MAIN; generates a set of initial coordinates and momenta: calls RAND, which generates random numbers as needed.

EAMCAL—called from MAIN; uses current values in P and Q to calculate total energy and total angular momentum.

WRIN—called from MAIN; outputs desired trajectory initial conditions.

CNS—called from MAIN; schedules trajectory integration; on entry, calls RKG, which performs 5 cycles at Runge–Kutta–Gill integration, thereafter, calls AMS which performs 1 cycle of Adams–Moulton integration; after each AMS cycle, tests for end of trajectory, and terminates integration either when end test is satisfied or when the user-specified maximum number of integration cycles has been exceeded.

RAND—called from SELECT; generates a random number by a congruence method; on initial call, uses random number input by user; on subsequent calls, uses the previously generated random number.

RKG—called from CNS; uses initial coordinates and momenta to perform 5 cycles of Runge–Kutta–Gill integration; prepares TABLE for use by AMS; calls TDS and VDS to generate current values for PDOT and QDOT as needed.

AMS—called from CNS; uses TABLE to perform 1 cycle of Adams–Moulton integration and updates TABLE for use by next cycle; calls TDS and VDS as needed.

TDS—called from RKS and AMS; uses current values in P to calculate current values for QDOT.

VDS—called from RKS and AMS; uses current values in P and Q to calculate current values for PDOT.

WROU—called from MAIN; outputs desired trajectory final conditions, including checks on total energy and total angular momentum conservation.

To facilitate cross section calculation, it is convenient to have WROU output j'_c and χ values either on cards or tape. These can then be input to a separate data analysis program.

The adequacy of single precision arithmetic for trajectory calculations is dependent on both the computer word length and the characteristics of the collision system being investigated. For studies of the type considered here, in which the trajectories are usually direct and simple, single precision is usually sufficient, although double precision, if affordable, is desirable.

4. Efficiency-Improving Techniques

4.1. Alternative Sampling Schemes

The sampling procedure described above for the generation of trajectory initial conditions is straightforward, leads to simple data analysis, and is adequate for most rotational excitation studies. However, in some instances, other sampling techniques can be used with greater efficiency. Here, we briefly illustrate two, uniform sampling and stratified sampling, as applied to the selection of impact parameters.

In uniform sampling, a nonuniformly distributed variable is sampled from a uniform distribution. For example, as is apparent from (20), random sampling from the natural distribution for b produces b^2 values which are uniformly distributed between 0 and b_{max}^2. Thus large b values are preferentially weighted. Use of too large a value of b_{max} thus generates many uninteresting (i.e., "elastic") trajectories. To attempt to avoid this situation we might sample b (instead of b^2) from a uniform distribution between 0 and b_{max}, i.e.,

$$b = \xi_4 b_{max} \tag{39}$$

Generation of the natural distribution of b values can be accomplished by assigning to each trajectory a statistical weight equal to its impact parameter, i.e.,

$$w = b \tag{40}$$

during the data analysis. Uniform sampling of b has been used in a recent study[20] of rotational excitation.

The application of stratified sampling to CT studies has been described in Reference 29. It offers another possible solution to efficiency problems encountered in impact parameter selection. Stratified sampling applies the method of (20) to restricted intervals (strata) in b. This allows one to probe outward stratum-by-stratum from $b = 0$, stopping when the cross section per stratum becomes negligible. During the data analysis, trajectories in a particular stratum are each assigned a weight corresponding to the probability of a collision occurring in that stratum. This probability is proportional to the area of the stratum; thus*

$$w = (b_{j+1}^2 - b_j^2)/b_{max}^2 \tag{41}$$

where b_j and b_{j+1} define the inner and outer boundaries of the outermost stratum. Stratified sampling on b has been used in recent CT studies of Ar–HCl scattering.[17,18]

* The area of the stratum between b_j and b_{j+1} is $A_{j,j+1} = \pi b_{j+1}^2 - \pi b_j^2$, and the total area for the region between 0 and b_{max} is $A_{tot} = \pi b_{max}^2$. Thus $w = A_{j,j+1}/A_{tot}$.

4.2. Moment Methods

Readily accessible from a CT calculation are the moments of the distribution of the rotational energy change, i.e.,

$$\langle (\Delta E_R)^n \rangle = \int (\Delta E_R)^n P(\Delta E_R) dE_{R'} \tag{42}$$

where $\Delta E_R = E_{R'} - E_R$. Thus as an alternative to the direct procedure for integral cross-section calculation described above, it is possible to directly calculate the first N moments of $P(\Delta E_R)$ and then obtain the cross sections by a moment inversion procedure. If $P(\Delta E_R)$ is accurately characterized by the values of the first few moments, moment inversion can be economical. Of the several methods proposed and investigated(8, 20, 27, 30-36) for extracting state-to-state cross sections from a set of moments, we consider only the simple case in which a continuous $P(\Delta E_R)$ curve is obtained, and cross sections are then calculated by evaluating

$$\sigma_{j'j} = \pi b_{\max}^2 \int_{\Delta E_1}^{\Delta E_2} P(\Delta E_R)\, dE_{R'} \tag{43}$$

where ΔE_1 and ΔE_2 are ΔE_R values corresponding to the bin boundaries associated with j' in the direct calculation.

For random sampling of initial trajectory variables from their natural distributions, CT values of the moments of $P(\Delta E_R)$ are given by

$$M_n = \langle (\Delta E_R)^n \rangle = \sum_{i=1}^{N_T} [(E_{R'})_i - E_R]^n / N_T \tag{44}$$

where $(E_{R'})_i$ is $E_{R'}$ for trajectory i and N_T is the total number of trajectories in the sample. To obtain a continuous analytical expression for $P(\Delta E_R)$ from the computed M_n, a variety of techniques might be used. If $P(\Delta E_R)$ is characterized by the first moment defined in terms of $|\Delta E_R|$, the information-theoretic approach of Levine, Bernstein, and their co-workers(32-34) can be used effectively. For the more general case, an adaptation of a method developed(8,35,36) to construct $P(\Delta E_R, b)$ curves [i.e., $P(\Delta E_R)$ curves at a particular impact parameter] should be useful. According to this, denoting ΔE_R by x, $P(\Delta E_R)$ is expanded as

$$P(x) = [(2\pi)^{1/2}\sigma]^{-1} \exp\left(-\frac{(x-\mu)^2}{2\sigma^2}\right) \sum_{n=0}^{N} a_n \mathrm{He}_n\left(\frac{x-\mu}{\sigma}\right) \tag{45}$$

where μ is the average value of x,

$$\mu = \int x P(x)\, dx \tag{46}$$

σ^2 is the second moment of $P(x)$ about μ,

$$\sigma^2 = \int (x-\mu)^2 P(x)\, dx \tag{47}$$

and the $\text{He}_n(y)$ are Hermite polynomials with weight factor $\exp(-y^2/2)$. From (45) and the orthogonality of the $\text{He}_n(y)$, it follows that the expansion coefficients are given by

$$a_m = \frac{1}{m!} \int \text{He}_m\left(\frac{x-\mu}{\sigma}\right) P(x) dx \tag{48}$$

Inserting explicit forms for the $\text{He}_n(y)$ into (48) yields[35] the following expressions for the first 11 expansion coefficients in terms of the computed moments M_n:

$$\begin{aligned}
a_0 &= 1 \qquad a_1 = a_2 = 0 \\
a_3 &= (3!)^{-1}[\mu_3/\sigma^3] \\
a_4 &= (4!)^{-1}[(\mu_4/\sigma^4) - 3] \\
a_5 &= (5!)^{-1}[(\mu_5/\sigma^5) - 10(\mu_3/\sigma^3)] \\
a_6 &= (6!)^{-1}[(\mu_6/\sigma^6) - 15(\mu_4/\sigma^4) + 30] \\
a_7 &= (7!)[(\mu_7/\sigma^7) - 21(\mu_5/\sigma^5) + 105(\mu_3/\sigma^3)] \\
a_8 &= (8!)^{-1}[(\mu_8/\sigma^8) - 28(\mu_6/\sigma^6) + 210(\mu_4/\sigma^4) - 315] \\
a_9 &= (9!)^{-1}[(\mu_9/\sigma^9) - 36(\mu_7/\sigma^7) + 378(\mu_5/\sigma^5) - 1260(\mu_3/\sigma^3] \\
a_{10} &= (10!)^{-1}[(\mu_{10}/\sigma^{10}) - 45(\mu_8/\sigma^8) + 630(\mu_6/\sigma^6) - 3150(\mu_4/\sigma^4) + 3780]
\end{aligned} \tag{49}$$

where $\sigma^2 = \mu_2$, and

$$\mu_n = \sum_{r=0}^{n} \binom{n}{r} (-M_1)^{n-r} M_r \tag{50}$$

Thus generation of moments up to $n = N$ yields values of a_n up to a_N. In practice, inversions can be performed for varying N until convergence in $P(\Delta E_R)$ is (hopefully) obtained.

Legendre moment methods have been investigated[37,38] for the calculation of differential cross sections in CT studies of reactive scattering.

5. *Concluding Remarks*

The preceding describes an efficient and workable method for the CT calculation of rotationally inelastic cross sections for atom–diatomic collisions occurring at fixed translational and rotational energies. Besides values for the atomic masses, for E_T and for E_R, the prerequisite input datum is an analytical representation of the atom–molecule interaction potential $V(\mathbf{R}, \mathbf{r})$. With these, the most straightforward calculation proceeds from initial condition selection, (7)–(22), to integration of the equations of motion, (5) and (23)–(27), to final condition analysis and cross section calculation, (28)–(38).

Although the fixed E_T and E_R assumptions and the rigid-rotor constraint have been invoked, these do not represent inherent limitations of the CT method. Descriptions of the techniques used for variable E_T and/or E_R and for vibrating diatomics can be found in a variety of sources, e.g., References 1, 4, 5, 7, 22, and 23.

References

1. D.L. Bunker, Classical trajectory methods, *Methods Comput. Phys.* **10,** 287–325 (1971).
2. J.C. Keck, Monte Carlo trajectory calculations of atomic and molecular excitation in thermal systems, *Adv. At. Mol. Phys.* **8,** 39–69 (1972).
3. R.N. Porter, Molecular trajectory calculations, *Ann. Rev. Phys. Chem.* **25,** 317–355 (1974).
4. J.C. Polanyi and J.L. Schreiber, The dynamics of bimolecular reactions, in *Physical Chemistry—An Advanced Treatise,* H. Eyring, W. Jost, and D. Henderson, editors, Vol. 6A, Kinetics of Gas Reactions, Academic Press, New York (1974),Chap. 6, pp. 383–487.
5. R.N. Porter and L.M. Raff, Classical trajectory methods in molecular collisions, in *Dynamics of Molecular Collisions, Part B,* W.H.Miller, editor, Plenum Press, New York (1976), Chap. 1, pp. 1–52.
6. W.H. Miller, Classical-limit quantum mechanics and the theory of molecular collisions, *Adv. Chem. Phys.* **25,** 69–179 (1974).
7. G.D. Barg, G.M. Kendall, and J.P. Toennies, Quasi-classical calculations of elastic and rotationally and vibrationally inelastic differential cross sections for $Li^+ + H_2$, *Chem. Phys.* **16,** 243–268 (1976).
8. R.A. LaBudde and R.B. Bernstein, Classical study of rotational excitation of a rigid rotor: $Li^+ + H_1$. II. Correspondence with quantal results, *J. Chem. Phys.* **59,** 3687–3691 (1973).
9. G.D. Barg, H. Fremery, and J.P. Toennies, *Faraday Discuss. Chem. Soc.* **55,** 59 (1973).
10. G. M. Kendall and J. P. Toennies, *Faraday Discuss. Chem. Soc.* **55,** 227–229 (1973).
11. P. McGuire, Close-coupling elastic and inelastic integral and differential cross sections for $Li^+ + H_2$ collisions, *Chem. Phys.* **4,** 249–258 (1974).
12. S.D. Augustin and W.H. Miller, Classical trajectory study of rotational excitation in low energy He-CO and He-H_2 collisions, *Chem. Phys. Lett.* **28,** 149–152 (1974).
13. P. Brumer, Rotational excitation and interference effects in atom-rotor collisions, *Chem. Phys. Lett.* **28,** 345–351 (1974).
14. R.T Pack, Close coupling test of classical and semiclassical cross sections for rotationally inelastic Ar–N_2 collisions, *J. Chem. Phys.* **62,** 3143–3148 (1975).
15. M.D. Pattengill, A comparison of classical trajectory and exact quantal cross sections for rotationally inelastic Ar–N_2 collisions, *Chem. Phys. Lett.* **36,** 25–28 (1975).
16. M.D. Pattengill, A classical trajectory study of inelastic scattering in the system He + H_2, *J. Chem. Phys.* **65,** 5033–5036 (1976).
17. S.M. Tarr, H. Rabitz, D.E. Fitz, and R.A. Marcus, Classical and quantum centrifugal decoupling approximations for HCl–Ar, *J. Chem. Phys.* **66,** 2854–2859 (1977).

18. J.C. Polanyi, N. Sathyamurthy, and J.L. Schreiber, Rotational energy transfer (theory). I. Comparison of quasiclassical and quantum mechanical results for elastic and rotationally inelastic HCl–Ar collisions, *Chem. Phys.* **24**, 105–110 (1977).
19. C.S. Lin and D. Secrest, Rotational excitation of H_2 in collision with He, *J. Chem. Phys.* **67**, 1291–1292 (1977).
20. S. Chapman and S. Green, Rotational excitation of linear molecules by collisions with atoms: Comparison of classical and quantum methods, *J. Chem. Phys.* **67**, 2317–2332 (1977).
21. W.H. Miller, Classical S matrix for rotational excitation; quenching of quantum effects in molecular collisions, *J. Chem. Phys.* **54**, 5386–5397 (1971).
22. M.D. Pattengill, A classical mechanical study of the effects of vibration in inelastic He–H_2 scattering, *J. Chem. Phys.* **66**, 1449–1451 (1977).
23. M. Karplus, R.N. Porter, and R.D. Sharma, Exchange reactions with activation energy. I. Simple barrier potential for (H, H_2), *J. Chem. Phys.* **43**, 3259–3287 (1965).
24. R.A. LaBudde and R.B. Bernstein, Classical study of rotational excitation of a rigid rotor: $Li^+ + H_2$, *J. Chem. Phys.* **55**, 5499–5516 (1971).
25. H. Goldstein, *Classical Mechanics*, Addison-Wesley, Reading, Mass. (1950), p. 42ff.
26. A. Gelb, Classical trajectory study of rotational excitation of HD by collisions with He, *Chem. Phys.* **27**, 245–249 (1978).
27. J.M. Bowman and S.C. Leasure, An improved quasiclassical histogram method, *J. Chem. Phys.* **66**, 1756–1757 (1977); Erratum, *J. Chem. Phys.* **67**, 4784 (1977).
28. W.C. Hamilton, *Statistics in Physical Science*, Ronald Press, New York (1964), p. 30ff.
29. C.A. Parr, J.C. Polanyi, and W.H. Wong, Distribution of reaction products (theory), VIII. Cl + HI, Cl + DI, *J. Chem. Phys.* **58**, 5–20 (1973).
30. D.G. Truhlar and J.W. Duff, Classical probability matrix: Prediction of quantum state distributions by a moment analysis of classical trajectories, *Chem. Phys. Lett.* **36**, 551–554 (1975).
31. D.G. Truhlar, Quasiclassical predictions of final vibrational state distributions in reactive and nonreactive collisions, *Int. J. Quantum Chem. Symp.* **10**, 239–250 (1976).
32. R.D. Levine, R.B. Bernstein, P. Kahana, I. Procaccia, and E.T. Upchurch, Surprisal analysis and probability matrices for rotational energy transfer, *J. Chem. Phys.* **64**, 796–807 (1976).
33. I. Procaccia and R.D. Levine, Cross sections for rotational energy transfer: An information theoretic synthesis, *J. Chem. Phys.* **64**, 808–817 (1976).
34. M.D. Pattengill and R.B. Bernstein, Surprisal analysis of classical trajectory calculations of rotationally inelastic cross sections for the Ar–N_2 system; influence of the potential energy surface, *J. Chem. Phys.* **65**, 4007–4015 (1976).
35. M.D. Pattengill, C.F. Curtiss, and R.B. Bernstein, Molecular collisions. XV. Classical limit of the generalized phase shift treatment of rotational excitation: Atom–rigid rotor, *J. Chem. Phys.* **55**, 3682–3693 (1971).
36. M.D. Pattengill, R.A. LaBudde, R.B. Bernstein, and C.F. Curtiss, Molecular collisions. XVI. Comparison of GPS with classical trajectory calculations of rotational inelasticity for the Ar–N_2 system, *J. Chem. Phys.* **55**, 5517–5522 (1971).
37. D.G. Truhlar and N.C. Blais, Legendre moment method for calculating differential scattering cross sections from classical trajectories with Monte Carlo initial conditions, *J. Chem. Phys.* **67**, 1532–1540 (1977).
38. N.C. Blais and D.G. Truhlar, Monte Carlo trajectories: Alignment of HBr rotational angular momentum as a function of scattering angle for the reaction $H + Br_2 \rightarrow HBr + Br$, *J. Chem. Phys.* **67**, 1540–1547 (1977).

11
Vibrational Excitation I: The Quantal Treatment

DON SECREST

1. Introduction

The close-coupling treatment of excitation of vibration by collision is formally the same as that for rotational excitation described in Chapter 8. The expansion of the wave function contains, in addition to the rotational wave functions, the vibrational wave functions of the noninteracting systems. The potential integral becomes more complicated and has more subscripts. The radial wave function has more subscripts and there are a larger number of coupled equations to consider. Still the form of the system of coupled equations is the same as it was for pure rotational excitation, and the methods described in Chapter 8 may be used for solving these systems also. There are, however, differences which arise in vibrational transition problems under certain circumstances and these will be discussed; and the methods described in Chapter 8 may be used for solving these problems will be pointed out later.

The complete close-coupling formalism for vibrational energy transfer has not proven very useful. It has been applied to collisions of atoms and ions with H_2.[1] These are tedious and time-consuming calculations. At energies high enough to excite vibrational transitions to any appreciable extent, there are in general large numbers of open rotational channels. The hydrogen molecule has very widely spaced rotational energy levels, and there are few rotational levels between the vibrational levels. They nonethe-

DON SECREST • School of Chemical Sciences, University of Illinois, Urbana, Illinois 61801.

less lead to an exceedingly large number of channels in close coupling at an energy at which vibration could be excited. For molecules other than H_2 or its isotopes, the rotational spacings are extremely small, leading to unmanageably large coupled systems. This can be nicely illustrated by an example. Let us consider a collision of ground-state N_2 with an atom at high enough energy to excite the first vibrational level. Since N_2 is a homonuclear diatomic, we cannot easily cause transitions from an even rotational state to an odd one. Since initially N_2 is in the $j = 0$ state, we need consider only even rotational states, meaning only half of the rotation states in this simple example. There is another advantage to choosing the initial state to be the ground state. In the total angular momentum representation, the coupled equations break up into smaller sets of equations, one for each total angular momentum state J. The relative angular momentum quantum number of the atom–molecule system, l, associated with each rotation state j runs from $l = |J - j|$ to $J + j$. These equations break up further into two blocks, one of even l and one of odd l. For $j = 0$ there is only one l equal to J. If we are interested in collisions with N_2 in the $j = 0$ state, for each J we need consider only the coupled equations with l quantum numbers of a single parity. Those of the other parity do not couple with the ground state. Thus for this example we may again throw out half of the remaining equations. For N_2 there are more than 30 j states below the first excited vibration state. If we carry only $j = 0, 2, 4, \ldots, 30$ in the calculation, the l quantum number associated with $j = 0$ is $l = J$. With $j = 2$, we must carry $l = J - 2$, J, and $J + 2$, or three channels. With $j = 4$ there are five channels. The total number of rotational channels associated with the ground vibrational state is $16^2 = 256$. This would lead to 256 coupled equations, not including any rotational channels with the excited vibrational state. Even with this extremely simple example, the number of coupled equations becomes prohibitive. Clearly, vibrational excitation calculations are practical only by the use of approximations.

One of the most popular approximations used in studying vibrational transitions is the collinear approximation. In this approximation one considers a model of a linear molecule confined to a straight line and an atom colliding with the molecule on the same straight line. This avoids the question of rotational transitions all together. Vibration-to-vibration energy transfer may also be studied in a collision of two linear molecules constrained to remain on the same straight line.

The collinear model has many of the features of a real collision process, and useful information has been obtained from such studies. Close-coupling calculations on such systems are also useful for evaluating approximations. The collinear atom–diatomic-molecule collision is the simplest of all inelastic quantum-mechanical scattering problems to solve. It requires only a few seconds' computing time on a modern computer to obtain close-coupled solutions correct to three or four figures using the method of Sams and Kouri. For this reason the model is ideally suited for testing new numerical methods or approximations which can be applied to the model. It is often also used to test new ideas and theories. If a reasonable model for a collision process to be studied can be made in the collinear collision mold, then this is an ideal first model to use.

A number of attempts have been made to relate the results from collinear models to real three-dimensional situations. Though many examples can be given where such studies have yielded useful results, for the most part collinear collisions serve only as a qualitative model.

A somewhat more realistic treatment of vibration is provided by the breathing sphere model. This model can be visualized in two different ways. From one point of view, a breathing sphere calculation could be considered a model in which one eliminates the angle dependence of the molecule by averaging the interaction potential over all orientations, and then solves the resulting close-coupling equations for the spherically averaged potential. Another way of looking at the problem is to start the molecule in the ground rotation state and include no other rotation states in the problem, i.e., completely ignoring rotational excitation. In either case, the close-coupled equations are the same. There is one channel for each vibrational state carried, and the number of coupled equations for each partial wave is equal to the number of vibrational states included. The total angular momentum is just the relative angular momentum of the atom–molecule system.

Though the breathing sphere approximation is capable of giving physically useful results, it is clear that the rotational degrees of freedom are important in vibrational energy transfer. In recent years a number of approximations have been proposed which treat the rotational channels with acceptable accuracy. These approximations developed out of the early work of Rabitz[(2)] and are discussed in Chapter 9. The coupled-states and effective-potential approximations give accurate rotational transitions, while the infinite-order sudden approximation, though somewhat less accurate than the other two, completely decouples the rotational transitions and makes the calculation extremely simple. These approximations are valuable for handling the rotational channels in vibrational calculations. While they do not give extreme accuracy at the low energies at which they have been tested, they do treat the rotational channels properly, as attested by the somewhat better than qualitative results of their use. Furthermore, their accuracy improves as the energy of collision increases, and it is expected to be very good indeed at the higher energies at which vibrational excitation becomes important. These approximations are powerful tools for accurate calculation of vibrational transitions.

2. Angular Momentum Decoupling Approximations

The effective-potential approximations,[3] of which the coupled-states approximation* is a special case, affect only the rotational channels in a collision calculation. Though vibrational degrees of freedom are usually ignored when these approximations are derived, the presence of one or more vibrational degrees of freedom in no way interferes with the derivation or application of these approximations, and computational savings of several orders of magnitude can be realized by their use. The approximations are new and their application to vibrational energy transfer problems are just beginning. A number of calculations have been performed using these techniques.[3-5] The presence of a vibrational degree of freedom simply adds another quantum number to the equations. To illustrate this we give the equation for the coupled-states approximation [equation (32), Chapter 9] for a system with a degree of vibrational freedom:

$$\left[\frac{\hbar^2}{2M}\frac{1}{R}\frac{\partial^2}{\partial R^2}R - \frac{l_0(l_0+1)\hbar^2}{2MR^2} + \frac{\hbar^2 k_{jv}^2}{2M} - V_\lambda(jv|jv|R)\right]\psi_{l_0\lambda}(jv|j_0v_0|R)$$

$$= \sum_{j'\neq j} V_\lambda(jv|j'v'|R)\,\psi_{l_0\lambda}(j'v'|j_0v_0|R) \qquad (1)$$

The only difference between this equation and equation (32) of Chapter 9 is the appearance of the vibrational subscripts v, on k_{jv}, the potential V, and the wave function. Here,

$$k_{jv}^2 = E - E_{jv} \qquad (2)$$

where E_{jv} is the rotation–vibration eigenenergy of the rotating–vibrating molecule in this atom–diatom collision. The potential is given by

$$V_\lambda(jv|j'v'|R) = \sum_l (-1)^\lambda\,[(2j+1)(2j'+1)]^{1/2}$$

$$\times \begin{pmatrix} j & l & j' \\ 0 & 0 & 0 \end{pmatrix}\begin{pmatrix} j & l & j' \\ -\lambda & 0 & \lambda \end{pmatrix}\int H_{vj}^*(r)\,V_l(R,r)\,H_{v'j'}(r)\,dr \qquad (3)$$

Here $H_{vj}(r)$ is the vibrational wave function of the diatom in rotation state j. The expansion coefficients $V_l(R,r)$ describe the interaction potential in the form

$$V(R,r,\theta) = \sum_l V_l(R,r)\,P_l(\cos\theta) \qquad (4)$$

where θ is the angle between the atom and the internuclear axis of the diatomic molecule.

Equation (1) is easily solved by the numerical methods of Chapter 8.

* See Chapter 9 and references therein for a discussion of this approximation.

In the notation of that chapter, equation (1) becomes

$$\left[-\frac{\partial^2}{\partial R^2}+\frac{l_0(l_0+1)}{R^2}-k_{jv}^2\right]f_{jvj_0v_0}^{l_0\lambda}(R)=-\sum_{j'v'}V_{jvj'v'}^{\lambda}(R)\,f_{j'v'j_0v_0}^{l_0\lambda}(R) \quad (5)$$

where

$$f_{j'v'j_0v_0}^{l_0\lambda}(R)=\psi_{l_0\lambda}(j'v'|j_0v_0|R)/R \quad (6)$$

and

$$V_{jvj'v'}(R)=\frac{2M}{\hbar^2}V_\lambda(jv|j'v'|R) \quad (7)$$

The choice of the method of solution depends on the number of energies for which the solution is sought, the accuracy required, and the detailed form of the interaction potential, as described in Chapter 8.

Since the l subscript does not appear on the potential, equation (7), in this approximation or in any of the effective-potential approximations, the number of *coupled* equations has been reduced. The total number of equations has not been reduced in the present case, but they are decoupled. There is still a system of equations for each l_0 which must be solved, and there are the same number of l_0 quantum numbers to be considered as J states in the uncoupled equations. Each coupled system is smaller however, but more systems must be solved. We must solve equation (5) for every λ, and λ runs from $-j_{max}$ to j_{max} giving the same number of equations as for the coupled system. The saving comes from the fact that the effort in solving a coupled system by any of the methods described in Chapter 8 goes as N^3, where N is the number of equations in the coupled system. Thus it is easier to solve more systems with less coupling than it is to solve the same size system when the equations are all coupled together.

It is not entirely accurate to say that there are the same number of equations to solve in the coupled-states approximation as in the coupled system. The equations for negative λ are identical to those for positive λ of the same magnitude. Accordingly, there are approximately half as many equations to solve as in the coupled system. In practice a further saving is realized. In a typical calculation one wishes to compute a set of cross sections $\sigma_{j_fv_f\leftarrow j_Iv_I}$ for transitions from an initial j, j_I to a final j, j_f. There will be a maximum j_f and j_I in which one is interested. In the calculation it is necessary to carry j states greater than either of these maximum interesting rotational states. In the coupled systems, the number of equations depends on the largest j state carried, and we are forced to compute complete information for all states carried. In the coupled-states approximation, one needs to consider only those values of λ needed to compute the desired cross sections. These are the values of λ, $\lambda = 0, 1, \ldots, j_{max}$, where j_{max} is the largest interesting j_f or j_I, whichever is smaller. Thus if one is only interested in transitions out of the rotational ground state, $j_I = 0$ to five or six final states, only $\lambda = 0$ needs to be considered. When the full close-coupled equations are solved, one gets information on all transitions whether desired or not.

In many of the effective-potential approximations[3] there is no quantum number corresponding to λ, and one has only one set of coupled equations to solve for each l_0, no matter what transitions are interesting.

If one now considers the problem of scattering an atom from N_2, which was discussed in the Introduction, one sees that only 16 coupled equations are associated with the ground vibrational state instead of the 256 for the close-coupled state. To solve the problem of vibrational transitions, however, one must consider more than the ground vibrational state. At least three vibrational states with their associated rotational states would need to be included. Including fewer rotational states in the excited vibrational states than in the ground state, one might need on the order of 25 or 30 states for a reasonable approximation to a vibrational transition. This is certainly a tractable number of states, but it is by no means certain that this is enough states to represent the collision accurately. Furthermore, though such a calculation is possible, it is not an easy calculation. For a collision of two diatomic molecules, one would, even in this approximation, be faced with an unmanageable number of coupled equations.

The infinite-order sudden approximation completely eliminates the rotational channel coupling. This approximation is discussed in Chapter 9 for pure rotational transitions. When only rotations are considered, the approximation completely decouples the equations and one needs to solve only the ordinary second-order uncoupled differential equations given by equation (85) in Chapter 9. When vibration is included, the equations one must solve in the infinite-order sudden approximation are given by

$$\left[-\frac{\partial^2}{\partial R^2} + \frac{l_0(l_0+1)}{R^2} - k_\upsilon^2\right] f_{l_0\upsilon}(R,\theta) = \sum_{\upsilon'} V_{\upsilon\upsilon'}(R,\theta)\, f_{l_0\upsilon'}(R,\theta) \tag{8}$$

where

$$k_\upsilon^2 = \frac{2M}{\hbar^2}(E - E_\upsilon) \tag{9}$$

E_υ is the υth vibrational energy level of the molecule, and

$$V_{\upsilon\upsilon'}(R,\theta) = \frac{2M}{\hbar^2}\int H_\upsilon^*(r)\, V(R,r,\theta)\, H_{\upsilon'}(r)\, d\upsilon \tag{10}$$

Here $H_\upsilon(r)$ is the vibrational wave function of the diatomic molecule. The radial wave function $f_{l_0\upsilon}$ is related to the ψ_{l_0} of Chapter 9 by

$$f_{l_0\upsilon}(R,\theta) = \frac{1}{R}\psi_{l_0\upsilon}(R,\theta) \tag{11}$$

As in the pure rotational case, when the boundary conditions of equation (8) are computed, one obtains an S matrix which depends parametrically on θ. To obtain the S matrix (or T matrix) for the rotational transition, one must perform the integral given by equation (95) of Chapter 9. This is usually accomplished by Gaussian quadrature, by computing solutions to equation (8) for values of θ for which $\cos\theta$ are Gaussian quadrature evaluation points. Then the integral in equation (95) of Chapter 9 becomes a simple quadrature sum. This technique has worked well in tests of this approximation at low energies where comparisons were made with exact calculations for pure rotational transitions. At higher energies where vibrational transitions become important, it has been found that the phase of $S_{l_0}(\theta)$ is a rather steep function of molecule orientation.

In a calculation of Li^+–N_2 scattering[4] at 4.23 eV, the phases of $S_{l_0}(\theta)$ for a range of l_0, which contribute significantly to the cross section, were found to vary over a large range. As a result, the real and imaginary parts of the S matrix were found to oscillate rapidly as the angle θ varied from 0 to π.

To compute the integrals corresponding to equation (95) of Chapter 9 by Gaussian integration would require a great number of Gaussian points. A technique was developed for handling this problem. The infinite-order sudden S-matrix element may be written

$$S_{l_0}(\theta) = |S_{l_0}(\theta)| \exp[i\eta_{l_0}(\theta)] \tag{12}$$

The magnitude of $S_{l_0}(\theta)$ is a slowly varying function of θ, and $\eta_{l_0}(\theta)$ is nearly linear. Thus we may write

$$\eta_{l_0}(\theta) = r_{l_0}(\theta) + K_{l_0}\theta \tag{13}$$

Here $r_{l_0}(\theta)$ is relatively slowly varying. Now the oscillating part of the integrand is removed and $S_{l_0}(\theta)$ may be written

$$S_{l_0}(\theta) = |S_{l_0}(\theta)| \exp[ir_{l_0}(\theta)] (\cos K_{l_0}\theta + i \sin K_{l_0}\theta) \tag{14}$$

where all terms are slowly varying except the sine and cosine term. This brings the integral into a form which is well suited to Filon quadrature.[6] A far fewer number of points is required to perform the integral using Filon quadrature. Of course, in computing $S_{l_0}(\theta)$ the absolute phase is unknown and unimportant, but in order to arrive at equation (13) it is necessary to compute $S_{l_0}(\theta)$ at a large number of values of θ so as to follow the phase change. With this large number of values available, one could use the trapezoid rule to do the integration. It is not necessary, however, to compute S_{l_0} at such a large number of θ values for every l_0. One may compute $S_{l_0}(\theta)$ on a fine mesh at one value of l_0, $\bar{l}_0$. At the next l_0, $l_0 = \bar{l}_0 + 1$, one needs compute S_{l_0} only at the points needed for the Filon quadrature. The phases at the computed points are within π of those at the next higher l_0. The relative phase may be followed in this way over large regions in the

l_0 space. When for some l_0 and θ the magnitude of $S_{l_0}(\theta)$ becomes very small, the phase varies unusually rapidly and it is necessary to relocate the phase by doing another fine-mesh calculation. In general, however, a much sparser mesh of points is required with Filon quadrature than with other quadrature methods.

Though the phase does not vary greatly from one l_0 value to the next, over several l values it can vary considerably. If a function could be found which followed the phase closely, this function could be used to aid interpolation of the phase over a large range of l_0 values, and one could compute solutions to the differential equations only at selected l_0 values and interpolate for the rest. For large l_0 values, so large that inelastic transitions are no longer present, the WKB method gives an accurate *absolute* phase for the elastic S-matrix elements. It has been found in one case at least[(4)] that, as lower l_0 values are calculated, the elastic S-matrix phase creeps away from the WKB phase slowly. One may compute every nth l_0 exactly and keep track of the difference between the exact phase and the WKB phase. This difference was a slowly varying function in the case studied and easily interpolated. One needs only to compute the WKB phase for each l_0, and interpolate the difference in phase, and the magnitude of the S matrix for the elastic S-matrix elements. Due to the unitarity of the S matrix, the inelastic phases can be followed except when a matrix element becomes small. This usually happens only for particular orientations of the diatomic, and the actual phase can be established from a fine mesh of orientations in this case.

The infinite-order sudden approximation treats the dynamics of the collision properly and is an approximation to the coupled-states approximation, which disturbs the energetics of the collision only. This can be seen from one of the derivations of the infinite-order sudden approximation. The IOS can be obtained from the coupled-states approximation[(7)] by replacing the wave vector k_j (or k_{jv} in the rotation–vibration case) by $\bar{k}$, which no longer depends on j. This has the effect, for the pure rotation situation, of making all energy levels degenerate. The potential is unchanged, however, and the equations solved in the IOS approximation are obtained by an exact transformation of the energy-modified coupled-states equations. Thus the only approximation beyond coupled states is to ignore the difference in energy between the various rotational levels. For a system with closely spaced energy levels and a high collision energy, this may well be a justifiable approximation. This may also be a good approximation for the rotation states when one is interested in vibration-to-translation energy transfer. It would very likely be a bad approximation in off-resonant vibration–vibration energy transfer where one might hope that a concurrent rotational transition would take up the energy defect.

In vibrational collision calculations using the IOS approximation one is not limited to a single wave vector. In each vibrational state there is but one wave vector. For the inelastic transitions one may use a different j in each vibrational state. This of course requires solving the equations for each set of initial and final rotational states. It is nonetheless possible in this way to account for rotational energy in vibrationally *inelastic* transitions. How accurate such a technique is has not yet been determined.

3. *Asymptotic Expansion Technique for Handling Long-Range Potentials*

For very-long-range potentials such as those going as $1/R^3$, it is necessary to integrate quite far before the potential has decayed sufficiently that the wave function can be fit to the Riccati Bessel functions. If all of the long-range terms in the potential are of the form $1/R^n$, where n is an integer as is usually the case,* it is possible to find a complete set of asymptotic solutions for the long-range potential and match these to the solution to obtain the R matrix.† The asymptotic solution for the long-range potential in scattering was first used in electron scattering.[9] It was adapted to molecular collisions by Brandt and Truhlar.[10,11]

For large R, for which the short-range terms in the potential have become negligible, the system of coupled equations we wish to solve takes on the form

$$\left(-\frac{\partial^2}{\partial R^2} - k_i^2\right) f_{il}(R) = -\sum_{i'}\sum_{n} V^n_{ii'} \frac{1}{R^n} f_{i'l}(R), \qquad R \to \infty \tag{15}$$

Here i is a composite subscript consisting of whatever quantum numbers apply, rotational or vibrational or both. The method we are describing applies to any scattering problem. The $V^n_{ii'}$ are the constant coefficients of the $1/R^n$ term in the potential. We have included $l(l+1)/R^2$ in the potential expansion.

When R is so large that all of the short-range parts of the potential have become negligible, the solution must have approached a solution to equation (15). The most general solution to this equation may be expanded in an asymptotic series of the form

$$F_i(R) = \sum_{rs}\left[\alpha^i_{rs}\frac{1}{R^s}\sin(k_r R - l_r\pi/2) + \beta^i_{rs}\frac{1}{R^s}\cos(k_r R - l_r\pi/2) + \gamma^{-i}_{rs} e^{-\kappa_r R}\frac{1}{R^s} + \gamma^{+i}_{rs} e^{\kappa_r R}\frac{1}{R^s}\right] \tag{16}$$

* This is not always the case. Occasionally the results of an *a priori* potential calculation are fit to terms with a $1/R^\alpha$ dependence (see Ref. 8). In such a case the method discussed here does not work.

† See Chapter 8 for a discussion of the R matrix.

Here k_r is the wave number for open channels, α^i_{rs} and β^i_{rs} are taken to be zero when r is the index of a closed channel, and κ_r is the absolute value of the imaginary wave number for the closed channel corresponding to the indices r; $\gamma^{\pm i}_{rs} = 0$ when the indices represented by r correspond to an open channel. All of the coefficients, $\alpha^i_{rs} = \beta^i_{rs} = \gamma^{\pm i}_{rs} = 0$ when s is less than 0.

Substituting equation (16) into (15) and equating coefficients of the powers of R and of similar sine, cosine, and exponential functions, one obtains recursion relations for the coefficients in equation (16):

$$(k_i^2 - k_r^2)\,\alpha^i_{rs} + (s-1)(s-2)\,\alpha^i_{rs-2} + 2k_r(s-1)\,\beta^i_{rs-1} = \sum_{i'}\sum_{n} V^n_{ii'}\alpha^{i'}_{rs-n} \tag{17}$$

$$(k_i^2 - k_r^2)\,\beta^i_{rs} + (s-1)(s-2)\,\beta^i_{rs-2} - 2k_r(s-1)\,\alpha^i_{rs-1} = \sum_{i'}\sum_{n} V^n_{ii'}\beta^{i'}_{rs-n} \tag{18}$$

$$(k_i^2 + \kappa_r^2)\,\gamma^{\pm i}_{rs} + (s-1)(s-2)\,\gamma^{\pm i}_{rs-2} + 2\kappa_r(s-1)\,\gamma^{\pm i}_{rs-1} = \sum_{i'}\sum_{n} V^n_{ii'}\gamma^{\pm i'}_{rs-n} \tag{19}$$

The above recursion relations give a unique asymptotic solution to equation (15) when a set of coefficients α^i_{r0}, β^i_{r0}, and $\gamma^{\pm i}_{r0}$ are chosen. A useful choice when i is an open channel,

$$\begin{aligned} \alpha^i_{r0} &= \delta_{ir} \\ \beta^i_{r0} &= \gamma^{\pm i}_{r0} = 0 \end{aligned} \tag{20}$$

With this choice,

$$F^i(R) \sim \sin(k_i R - l_i\pi/2), \qquad R \to \infty \tag{21}$$

For the closed channels, one chooses

$$\begin{aligned} \alpha^i_{r0} &= \beta^i_{r0} = \gamma^{-i}_{r0} = 0 \\ \gamma^{+i}_{r0} &= \delta_{ri} \end{aligned} \tag{22}$$

These correspond to incoming waves. Similarly, outgoing waves are obtained by choosing β^i_{r0} and γ^{-i}_{r0} as Kronecker deltas.

Now the technique for matching the numerical solution of the full coupled system to this solution, equation (16), of the asymptotic equations becomes very simple. If we let $\mathbf{F}(R)$ be the matrix of solutions to the full coupled system given by one of the methods of Chapter 8, and let $\mathbf{u}(R)$ and $\mathbf{v}(R)$ be matrices of incoming and outgoing waves obtained from equation (16) and the recursion relations, equations (17)–(19), then we may write

$$\mathbf{F}(R) = \mathbf{u}(R)\cdot\mathbf{A} + \mathbf{v}(R)\cdot\mathbf{B} \tag{23}$$

If we do this at two different values of R, we can solve the two resulting systems for the unknown matrices $\mathbf{A}$ and $\mathbf{B}$. The two values of R, for which the matrices **F**, **u**, and **v** are evaluated, must each be large enough that equation (15) is valid and they should not be too close together. It is not a good idea to use two successive steps in the integration for this purpose. I usually choose ΔR, the distance between the two R values, large enough that the dimensionless product $k_i \Delta R$ is of the order of one. Here k_i is the smallest wave number.

Once **A** and **B** are found, the reactance matrix **R** is given by solving the system

$$\mathbf{R} \cdot \mathbf{A} = \mathbf{B} \tag{24}$$

for the R matrix.

It is important to note that the series given by equation (16) is asymptotic[12]; that is to say, the series does not converge. As more and more terms are added to the series, the added terms become smaller and the series approaches the correct solution up to a point. Beyond that point the added terms begin to get larger and the series diverges. Thus it is important to quit computing terms in the series at the proper place. As R gets larger the series solution becomes better. A good approach is to stop adding terms to the series when the terms start getting larger. It is not certain how accurate the solution is at this point. When equation (24) is solved, one obtains an R matrix of questionable accuracy. One then integrates the full equations farther to a new R and uses the last two R values to compute new matrices **A** and **B**, which should be the same as those computed the first time. Then equation (24) gives a new **R**. If it agrees well enough with that previously computed, one may stop.

This technique can yield considerable savings. In a calculation for which the potential had $1/R^3$, $1/R^4$, and $1/R^5$ terms, we were able, using this technique, to stop the integration after integrating only 1/10 of the range required to arrive at the region in which we could terminate using Riccati Bessel functions. Furthermore, using this technique, we found that we could use a larger step size to achieve the same accuracy obtained by integrating the full range. The necessity of a smaller step size for integrating the full range is very likely due to rounding-error buildup during the long integration. A small step size was required to keep the rounding error within bounds.

Near a threshold, Burke and Schey[9] tell us that this termination method has trouble. They describe a modification for this situation. We have never used this technique close enough to a threshold to have difficulty. They claim trouble when one of the $k_i^2 < 0.005$.

4. *Effects of the Dissociative Continuum*

In the study of vibrational energy transfer one may encounter a problem which does not occur in rotational excitation problems. As described in

Chapter 8, in deriving the coupled system of equations, one expands the wave function in a complete set of functions for the internal states of the separated systems. In practice, one cannot use the entire set of functions since the complete set is usually infinite. Only a finite number of states of the system are energetically accessible. It is often necessary to retain some states which are not energetically accessible (closed channels) in order to obtain a converged scattering result. This is necessary because during the collision the internal states of the molecule are perturbed by the interaction, and the higher states are a mathematical necessity for representing the new perturbed eigenstates of the interacting system. These closed channels are necessary only in the strongly interacting region of the calculation and may be dropped as the calculation moves into the more weakly interacting region of the potential.

The use of the internal states of the noninteracting system as an expansion set for the scattering wave function causes no trouble for a vibrational problem if the vibrator is treated as a harmonic oscillator. Often in vibration problems, a Morse potential or some similar possibly more realistic potential with a dissociative continuum is used. The bound states of such potentials are not complete without the continuum. This may appear to be unimportant since a complete set is never used in practice, and indeed if the potential has many bound states and only states near the bottom of the well are accessible, a finite expansion in bound states converges satisfactorily. If one is interested in vibrational deexcitation of vibrational states on the order of halfway between the bottom of the well and the dissociation limit, however, one finds that even when perturbed by a weak collision, the perturbed states of a Morse potential cannot be represented well even when the entire set of bound states is used.[13] Though often there is an appearance of convergence, if some other types of functions are added to the expansion, it is seen to be a false limit that was attained. This is similar to the problem of computing the energy levels of He using only the bound hydrogenic functions. In this case, there are an infinite number of bound hydrogenic functions, and the helium energy levels converge nicely, as states are added, to the wrong result. The continuum functions lie in an orthogonal space to the bound set and the convergence is to a projection on the space of bound hydrogenic functions. The problem is not difficult to handle in vibrational scattering. One expands in the noninteracting internal states only because it is convenient to do so. In practice any set of functions could be used for this purpose.

Let us consider, for illustration purposes, a collinear collision of a Morse oscillator with a structureless particle interacting with a repulsive exponential. The Schrödinger equation can be written in reduced center-of-mass units:

$$\left[-\frac{\partial^2}{\partial x^2}-\frac{\partial^2}{\partial y^2}+D(e^{-\alpha y}-2e^{-\alpha y/2})+Ae^{-\alpha(x-\lambda y)}\right]\psi(x,y)=E\psi(x,y) \tag{25}$$

As in Chapter 8, one starts the integration of this system at small $x = x_0$, well into the nonclassical region of the potential. At this point the potential that y, the molecule internal coordinate, sees is

$$V_{x_0}(y) = D(e^{-\alpha y} - 2e^{-\alpha y/2}) + Ae^{-\alpha x_0}e^{\alpha\lambda y} \tag{26}$$

Thus a good basis set to use for expanding the internal states of the oscillator is the eigenfunction of the Schrödinger equation:

$$H_{x_0}(y)\,\psi_i(y) = \left[-\frac{\partial^2}{\partial y^2} + V_{x_0}(y)\right]\psi_i(y) = E_i\psi_i(y) \tag{27}$$

One can find the eigenfunctions for this system by the variation method, using some expansion set of functions such as harmonic oscillator functions or sines and cosines. Then equation (25) can be rewritten:

$$\left[-\frac{\partial^2}{\partial x^2} + H_{x_0}(y) + A(e^{-\alpha x} - e^{-\alpha x_0})\,e^{+\alpha y} - E\right]\psi(x, y) = 0 \tag{28}$$

Now expanding the wave function in this set,

$$\psi(x, y) = \sum_i \psi_i(y)\, f_i(x) \tag{29}$$

one obtains the coupled system,

$$\left(-\frac{\partial^2}{\partial x^2} - k_i^2\right) f_i(x) = -\sum_{i'} \int_{-\infty}^{\infty} \psi_i^*(y)\, A(e^{-\alpha x} - e^{-\alpha x_0})\, e^{\alpha y}\psi_{i'}(y)\, dy f_{i'}(x) \tag{30}$$

This equation may be solved in a large region around x_0 using any of the methods of Chapter 8. As one gets far from x_0 this is no longer a reasonable expansion and one may reexpand around a new point x_1 and continue in this way into the asymptotic region, where the bound Morse oscillator states are a reasonable expansion set.

Whenever a new expansion point is encountered it is necessary to match the wave function and its derivative at the point where the connection is made. This requires computing the overlap of the expansion set for x_i and x_{i+1}, in order to make the connection.

This is standard procedure in the methods of Light[14] and the R-matrix method of Light and Walker.[15] They developed their method for reactive scattering for which they were integrating from the space of reactants to the space of products and this gave a natural method of changing the coordinate system. It is no extra work for their methods as well as that of Gordon, since the Hamiltonian must be diagonalized at every step in any case. If a simple set of expansion functions, such as harmonic oscillator functions, is used, the calculation of the Hamiltonian matrix is trivial and the diagonalization must be done in any case. In using the Sams and Kouri method or the log-derivative method, one would not make a new expansion at every step, but only five or six times during the calculation. Creating a new expansion set is not a time-consuming process, however, and for potentials with only a few bound states it is necessary to do so. If a number of collision energies are to be considered, one needs to perform the diagonalization and the overlap calculation only with the first energy and save the overlap matrices in core. One finds by this method that far fewer states are necessary for a converged result than when one uses the asymptotic states, even in a case where the asymptotic states are sufficient, as in the harmonic oscillator collision problem. This is true because a large number of asymptotic states are needed to represent the perturbed oscillator, whereas by the present method most of the perturbation is included in the basis set, and the interaction potential is always small.

In fact, in the methods of Light[14] and the R-matrix method,[15] the interaction potential is approximated by a constant with its value at the expansion point. Thus by both of these methods, the interaction potential is always zero and between matching points the wave functions are freely propagating waves.

References

1. J. Schaefer and W.A. Lester, Jr., Theoretical study of inelastic scattering of H_2 by Li^+ on SCF and CI potential energy surfaces, *J. Chem. Phys.* **62,** 1913–1924 (1975).
2. H. Rabitz, Effective potentials in molecular collisions, *J. Chem. Phys.* **57,** 1718–1725 (1972).
3. H. Rabitz, Effective Hamiltonians in molecular collisions, in *Modern Theoretical Chemistry, Part A,* W.H. Miller, editor, Plenum Press, New York (1976), pp. 33–80.
4. G. Pfeffer, The calculation of the scattering of lithium ions from nitrogen molecules, Ph.D. Thesis, University of Illinois, Urbana, Illinois (1977).
5. J.M. Bowman and S.C. Leasure, Sudden rotational calculations of atom–molecule scattering, *J. Chem. Phys.* **66,** 288–295 (1977); Erratum: *J. Chem. Phys.* **66,** 4724 (1977).
6. R.W. Hamming, *Numerical Methods for Scientists and Engineers,* McGraw-Hill, New York (1962).
7. D. Secrest, Theory of angular momentum decoupling approximations for rotational transitions in scattering, *J. Chem. Phys.* **62,** 710–719 (1975).
8. L. Monchick and S. Green, Validity of central field approximations in molecular scattering: Low energy CO–He collisions, *J. Chem. Phys.* **63,** 2000–2009 (1975).
9. P.G. Burke and H.M. Schey, Elastic scattering of low-energy electrons by atomic hydrogen, *Phys. Rev.* **126,** 147–162 (1962).
10. M.A. Brandt and D.G. Truhlar, Importance of long-range forces and short-range forces in electron scattering: Elastic scattering by N_2 at 10 and 30 eV, *Chem. Phys. Lett.* **23,** 48–52 (1973).
11. M.A. Brandt, Topics in quantum mechanical scattering theory with application to electron–nitrogen-molecule scattering, Masters Thesis, University of Minnesota, Minneapolis, Minnesota (1975).
12. A. Erdély, *Asymptotic Expansions,* Dover, New York (1956).
13. A.L. Scherzinger, Inelastic scattering of dissociable molecules, Ph.D. Thesis, University of Illinois, Urbana, Illinois (1977).
14. J.C. Light, Quantum calculations in chemically reactive systems, in *Methods of Computational Physics,* B. Alder, S. Fernbach, and M. Rotenberg, editors, Vol. 10, Academic Press, New York (1971), pp. 111–142.
15. J.C. Light and R.B. Walker, An R matrix approach to the solution of coupled equations for atom–molecule reactive scattering, *J. Chem. Phys.* **65,** 4272–4282 (1976).

12

Vibrational Excitation II: Classical and Semiclassical Methods

W. RONALD GENTRY

1. Introduction

In this chapter we consider methods for calculating cross sections or rates for vibrational energy transfer in which at least one degree of freedom in the collision system is treated classically. We ignore cases in which electronic transitions or electronically nonadiabatic effects play an important role, but we do include methods for treating simultaneous vibrational and rotational transitions, since the two are often coupled strongly in real systems. The literature on classical and semiclassical treatments of vibrational energy transfer is immense. Because of the special nature of this volume as a "consumers handbook" designed to be of practical rather than of pedagogical value, no attempt has been made here to be comprehensive in scope or historical in perspective.* Instead we focus on those aspects of the theory which at present appear to be the most useful in the interpretation of modern experimental data. Our chief criterion for usefulness is the degree of success with which a given method has actually been applied to the interpretation of experimental results. However, in some cases the lack of appropriate data precludes comparison with experiment, and the value of a given approach must be judged by the available comparisons with other theoretical results.

* There are, however, several reviews[1-3] which provide some historical perspective and relatively comprehensive bibliographies of earlier work.

W. RONALD GENTRY • Chemical Dynamics Laboratory, Department of Chemistry, University of Minnesota, Minneapolis, Minnesota 55455.

In choosing between classical, semiclassical, and quantal methods the primary consideration is the degree of quantum detail which is required in the description. In general, at least a partially quantal calculation will be necessary in order to reproduce features of the data which contain quantal phase information (interference structure), such as quantum oscillations in the differential cross section or glory undulations in the integral cross section.* Such quantum structure disappears very quickly in averaging over initial or final state distributions[4] or distributions in collision energy and scattering angle, and therefore appears in few experimental data. In the absence of features in the data which are *directly* a consequence of the wave nature of the colliding species, a classical treatment may be appropriate. At total system energies corresponding to even a single vibrational quantum the number of angular momentum states in three spatial dimensions is so large as to render a converged coupled-channel calculation unfeasible except in special cases,[5] hence some sort of dynamical approximation or reduction in the dimensionality of the problem must be sought. The alternatives are to introduce approximations into the quantal equations (see Chapter 11) or to replace all or part of the quantal equations with the corresponding classical description. The penalty one pays in computational effort for including all degrees of freedom in a dynamically exact classical calculation is relatively small. A classical or semiclassical calculation is to be preferred whenever a fully quantal treatment would require approximations which are poorer than that of substituting classical equations of motion for part of the quantum scattering problem. Fortunately for the experimentalist investigating vibrational energy transfer, some classical and semiclassical methods are found to be remarkably accurate for vibrational transitions. A bonus associated with treating all or part of the problem classically is that it is relatively easy to understand those features of the collision dynamics which influence the discrete classical motion. By examining individual classical trajectories it is sometimes possible to obtain insights into the "mechanism" of the collision process which are helpful in analyzing the effects of approximations or in generalizing the results to other systems. Such insights are more difficult to obtain from fully quantal calculations, in which the details of the wave motion are much more complicated and are typically buried in numerical code understandable only to a computer.

The degree to which all or part of the quantal equations of motion may be safely replaced by classical mechanics depends not only on the collision parameters (masses, kinetic energy, initial and final states, scattering angle) but also on the extent to which individual dynamical calculations are to be averaged over collision parameters prior to comparison with experiment. In general, a classical description will be more accurate for a

* An exception to this rule is classical S-matrix theory (see below), which in its most fundamental form does retain phase information.

more highly averaged quantity. It is relatively easy to determine *sufficiency* conditions for a classical description to be adequate,[6] but to determine under what conditions a quantal description is actually *necessary* is much more difficult. For example, one certainly expects classical mechanics to be valid in the limit of large quantum numbers, and this argument is often invoked to justify treating the translational and rotational degrees of freedom classically while retaining a quantum description of the vibrational mode. However, in some methods the important features of the vibrational motion may be described well classically even for small quantum numbers.[7]

Apart from the selection of a general classical or semiclassical approach to the collision problem, there is also the question of whether additional approximations can be introduced without significant loss of accuracy in the final result. These approximations are of three types: dynamical approximations which simplify the equations of motion in particular degrees of freedom, dimensional approximations which reduce the number of degrees of freedom, and approximations to the potential energy surface which can sometimes simplify the numerical calculations by permitting part of the problem to be solved analytically. If enough approximations of all three types are made simultaneously, then the *entire* problem can be solved analytically. Shin[8] has presented a compilation of analytic methods which have been useful for calculating thermal vibrational relaxation rates. However, often the approximations which are necessary for analytic solutions are individually of dubious accuracy, and great care must be exercised in interpreting the result.[9] The methods considered here will usually require numerical solutions.

In Sections 2 and 3 we outline several general theoretical approaches to the vibrational excitation problem more or less independently of the optional approximations which may be imposed on the actual numerical calculations. In Section 4 we consider various useful approximations and their ranges of validity.

We will adopt the notation T, R, V for general references to the translational, rotational, and vibrational degrees of freedom, respectively. Other notation is given below.

Notation

b = impact parameter

θ, ϕ = polar, azimuthal orientation angles of rotational angular momentum

ω = vibrational frequency

γ, β = vibrational, rotational phase

$\mathbf{r}$ = diatom internuclear separation vector

$\mathbf{R}$ = diatom center of mass–atom separation vector

μ = reduced mass of oscillator

m = reduced mass of collision system

t = time

$\mathbf{i}, \mathbf{f}$ = initial, final internal quantum states

E_i, E_f = initial, final internal energy

T_i, T_f = initial, final translational energy

v_i, v_f = initial, final vibrational quantum number

j_i, j_f = initial, final rotational quantum number

m_i, m_f = initial, final rotational projection quantum number

P_{if} = transition probability for $\mathbf{i} \rightarrow \mathbf{f}$

σ_{if} = cross section for $\mathbf{i} \rightarrow \mathbf{f}$

χ = center-of-mass scattering angle

$I_{if}(\chi)$ = differential cross section for $\mathbf{i} \rightarrow \mathbf{f}$

Boldface type indicates a vector quantity. A dot over a symbol indicates the time derivative.

2. *Quasiclassical Methods*

Given a potential energy surface with analytic derivatives and a well-defined set of initial positions and momenta for all the nuclei, the classical equations of motion for a simple molecular collision may be solved exactly* using a digital computer of very modest capability (by 1978 standards). If the initial conditions are chosen to be only those which correspond to the quantally allowed values of internal energy and rotational angular momentum, then the method is termed "quasiclassical." For comparison with an experimental result on a VRT (vibrational–rotational–translational) energy transfer process, the quasiclassical method consists of three steps; (i) selection of initial conditions from the weighted set appropriate to the experiment, (ii) numerical solution of the classical equations of motion between time t_1 (before the collision) and time t_2 (after the collision), and (iii) analyzing the distribution of the results among states of the products. Alternative means of carrying out these steps have been discussed in reviews[10-12] giving bibliographies of the original work, and will be mentioned only briefly here. Many details of the numerical methods common to both RT and VRT energy transfer are given explicitly in Chapter 10.

Once a protocol has been adopted for assigning the final quantum state $\mathbf{f}$ to the products of a specific collision trajectory, the calculation of the cross section σ_{if} for the transition $\mathbf{i} \rightarrow \mathbf{f}$ at fixed collision energy amounts to performing a multidimensional integral over all initial conditions subject to the constraints that the final state be $\mathbf{f}$. In principle the integration could be carried out by scanning (varying in small regular increments) all of the initial conditions to generate the entire hypersurface of any specified final property as a function of all initial parameters. In practice, because of the complexity of this hypersurface this approach is seldom feasible, and

* We bow to the common use of the term "exact" to mean "numerically accurate to a few significant figures, within the context of a certain set of fundamental assumptions."

the actual calculation is usually done by Monte Carlo (random) selection of most or all initial conditions[10,13] The Monte Carlo procedure is especially appealing to experimentalists because it is closely analogous to the way in which scattering experiments are carried out—with control of certain collision parameters (e.g., collision energy) and a random selection of others (e.g., impact parameter). Usually the cyclical variables such as the initial phase and orientation angles are sampled over their entire ranges, but a limit must be placed on the sampled range of impact parameter $0 \leq b \leq b_{\max}$ such that no collisions with $b > b_{\max}$ contribute to **f**. If all initial classical variables $(b, \theta, \phi, \gamma, \beta)$ are sampled randomly with the correct statistical weights[10] then the operational definition of the total cross section for the transition $\mathbf{i} \rightarrow \mathbf{f}$ at a specified initial kinetic energy T_i is simply

$$\sigma_{if}(T_i) = \pi b_{\max}^2 \lim_{N \to \infty} \frac{n_{if}}{N} \tag{1}$$

where n_{if} is the number of trajectories connecting **i** and **f**, and N is the total number of trajectories calculated. In the Monte Carlo method, the uncertainty in σ_{if} for finite N can easily be estimated statistically from the rate of convergence as N is increased.* The differential cross section $I_{if}(\chi)$ (averaged over the angular width $\Delta\chi$ centered at χ) is calculated similarly by further distributing n_{if} over χ:

$$I_{if}(T_i, \chi) = \frac{\pi b_{\max}^2}{\Delta\Omega(\chi, \Delta\chi)} \lim_{N \to \infty} \frac{n_{if}(\chi, \Delta\chi)}{N} \tag{2}$$

where $\Delta\Omega$ is the finite increment of solid angle included in $\Delta\chi$. For interpretation of the vibrational excitation mechanism it is often of interest to scan b and observe how the transition probability

$$P_{if}(T_i, b) = \lim_{N \to \infty} \frac{n_{if}(b)}{N} \tag{3}$$

varies with impact parameter. In this case the cross section is

$$\sigma_{if}(T_i) = 2\pi \int_0^{b_{\max}} b P_{if}(T_i, b) db \tag{4}$$

In some cases[6] the values of b have been chosen to correspond to quantized values of the collisional orbital angular momentum. If a more highly averaged quantity, such as a rate coefficient, is desired then the Monte Carlo selection of initial conditions can also cover a weighted

* Other (nonrandom) sampling techniques have been proposed, which converge more rapidly for small N.[14] However, such methods have not been popular because of difficulty in estimating the uncertainty of the results.[15]

random selection of T_i and of internal quantum states, thus avoiding the expense of calculating unneeded microscopic quantities.

The actual trajectory calculation is performed by direct numerical integration of the coupled classical equations of motion, usually in Hamiltonian form,[10,13] and usually in Cartesian coordinates to avoid numerical problems near poles. Many numerical methods have been used. Two of of the most popular are the Runge–Kutta–Gill and Adams–Moulton routines.[10]* Tests of numerical accuracy include convergence of the result with decreasing step size Δt, back integration from final to initial conditions, and conservation of energy and angular momentum.

For some purposes, such as calculating the rate coefficient for excitation or relaxation in bulk experiments, it may be sufficient to determine only the *average* vibrational energy gained or lost in the collision[16-18]; however, in general one will want to calculate the (discrete) distribution of final vibrational and rotational quantum numbers from the (continuous) distribution of final classical internal energy E and angular momentum L. All procedures for doing this are in some sense arbitrary, and if the results depend strongly on which of several different but physically reasonable methods is used, then the calculation is probably not to be trusted. One widely used protocol[6,19] is to assign the rotational quantum number to be the integer nearest to the classical parameter j_c, where

$$j_c(j_c + 1)\hbar^2 = L^2 \tag{5}$$

This procedure must be modified for homonuclear molecules in which only $\Delta j = 2$ transitions are permitted by symmetry.[20] The classical vibrational energy E_v can be separated approximately from the rotational energy by the formula

$$E_v = E - \frac{L^2}{2\mu r_m} \tag{6}$$

where r_m is the minimum in the effective potential for radial motion. The final vibrational quantum number (for a harmonic oscillator) is the integer nearest to the classical parameter v_c, where

$$(v_c + \tfrac{1}{2})\hbar\omega = E_v \tag{7}$$

An alternative is to transform the classical equations into action-angle variables and to assign the vibrational and rotational quantum numbers by rounding off the radial and angular actions, respectively, to the nearest integers.[21] These and similar methods are known as histogram methods; each trajectory contributes to only one final state $(v_f j_f)$. An entirely different approach, which represents the discrete set of final vibrational quantum numbers in terms of the lowest-order discrete moments of the

* In equation 41 of Ref. 10, (-1475) should be replaced by $(+1427)$.

continuous classical vibrational energy distribution, has recently been suggested by Truhlar and Duff[22] and tested for collinear vibrational excitation collisions.[22-24] The moment methods appear to give generally better results than histogram methods for small transition probabilities. The results depend on the number of moments used, however, and it is not yet clear how one should pick the number. Bowman and Leasure[25] have shown how a histogram distribution can be modified so that its first moment agrees with that of the continuous classical distribution. Moment methods have also been tried for rotational state distributions, but were found to be inferior to histogram methods.[26]

In cases where the vibrational transition probabilities are large, the quasiclassical histogram method has given generally fair (about a factor of 2) agreement with exact quantal calculations for collinear atom–diatom collisions,[22,27] and, in at least one case, good agreement with an experimental state-resolved differential cross section.[6] The histogram method is not valid for small probabilities, and indeed will predict zero transition probability for those transitions for which no set of initial conditions results in transfer of at least half a quantum of classical vibrational energy. For threshold processes, which are typical of thermal bulk experiments, the histogram method is sometimes found to fail badly.[24] However, three-dimensional calculations of thermal rates for vibrational deexcitation have often yielded good qualitative agreement with the experimental rates.[28-31]

3. *Semiclassical Methods*

The remaining methods considered in this section are "semiclassical" in the sense that they each contain elements of both classical and quantal dynamics, in various admixtures. All but a small fraction of the semiclassical treatments of vibrational energy transfer which have appeared to date have been confined to the case of a collinear collision between an atom and a diatomic molecule. These studies have been very valuable in the conceptual development of theories and for testing approximations by comparison with exact fully quantal results. Despite some early claims to the contrary, however, collinear calculations generally do not provide a reasonable basis for the interpretation of experimental data on vibrational energy transfer. Here we will focus our attention on those approaches for which three-dimensional calculations are currently practical.

3.1. *Quantal Internal Modes Coupled through the Interaction Potential to Classical Translational Motion*

The earliest and perhaps conceptually the most straightforward semiclassical approach to the inelastic scattering problem is to view the internal motion of the molecule quantally while treating the translational motion classically. Molecular transitions are induced through the interaction potential* $V_{int}(\mathbf{R},\mathbf{r})$, which is parametrically time dependent through the classical translational trajectory $\mathbf{R}(t)$. From time-dependent quantum mechanics the (complex) amplitudes c_l of internal states are given by

$$i\hbar\dot{c}_l = \sum_k c_k(t)\exp[-it(E_l - E_k)/\hbar]\langle l|V_{int}(\mathbf{R},\mathbf{r})|k\rangle \tag{8}$$

where $|i\rangle$ represents the eigenstate $|v_i j_i m_i\rangle$ of the isolated molecule. A numerically exact semiclassical treatment would consist of simultaneous integration of the coupled equations (8) and Hamilton's equations for $\mathbf{R}(t)$ to obtain the set of final-state amplitudes $c_l(+\infty)$ at each value of classical impact parameter b and for each initial-state amplitude $c_i(-\infty) = \delta_{ik}$. As with the quantal coupled-channel method, the number of states included in the basis set must be sufficient to ensure the desired numerical accuracy. For three-dimensional collisions such a calculation would be only slightly (if at all) less laborious than an exact fully quantal calculation. Except for some collinear calculations,[32-34] virtually all semiclassical treatments of this general type introduce two further approximations into equation (8): (*i*) decoupling of the translational motion $\mathbf{R}(t)$ from part or all of the internal motion, and (*ii*) neglect of some or all of the off-diagonal rotational matrix elements of V_{int}. Often, in addition to these dynamical approximations, numerical approximations are introduced as well.

Approximations of type (*i*) include the following.

(*i-a*) *The Sudden Approximation* (*SA*). The SA assumes that some or all internal degrees of freedom are "frozen" during the encounter, and that the forces acting in these modes may therefore be treated as impulses. The SA may be applied separately to the rotational[35-40] or vibrational[35,36,38,41] degrees of freedom, and will be most valid at high collision speeds. A conceptually similar and complementary approximation assumes that the vibrational coordinate r responds adiabatically to the interaction and adjusts so as to remain at the minimum of potential energy throughout the encounter.[37,42-44] The adiabatic approximation is appropriate only for collisions at very low kinetic energies and has not been much used.

(*i-b*) *Impact Parameter* (*IP*) *Approximation.* In the IP approach the translational motion is taken to be uniform and rectilinear, so that $\mathbf{R}(t)$ reduces to

$$R(t) = (b^2 + t^2u_0^2)^{1/2} \tag{9}$$

where u_0 is the translational speed and $t = 0$ is the time corresponding to the minimum radial separation $R = b$. The IP approximation is valid only for small scattering angles and small $\Delta T/T$, and is sometimes useful

* The interaction potential is defined as the total Born–Oppenheimer potential energy of the three-body system, $V(\mathbf{R},\mathbf{r})$ minus the potential energy of the isolated molecule $V(\infty,r)$.

for T–V energy transfer at high kinetic energies[38-40,45] or for nearly resonant V–V or V–R processes at lower energies,[46,47] provided that the process of interest be dominated by collisions at large b.

(*i-c*) *The Spherical Interaction* (*SI*) *Approximation.* A less stringent approximation than the IP treatment is to solve the radial equations of motion for only the spherical term $P_0(\cos\theta)$ in the Legendre expansion of $V_{\text{int}}(\mathbf{R}, \mathbf{r})$.[48] Usually, however, the SI approximation is combined with either the SA[35-38] or IP[38,47] treatments.

(*i-d*) *Effective-Potential* (*EP*) *Approximation.* A more sophisticated alternative to the SI approximation is to use a translational trajectory determined from an "effective potential,"[49-50] which is the interaction potential averaged over some, but not all, of the internal molecular states (usually m and sometimes both j and m).[47,51-55] Thus, depending on other approximations, one would have a different effective translational trajectory for each initial state (v_i) or $(v_i j_i)$ or for each pair of initial and final states $(v_i|v_f)$ or $(v_i j_i|v_f j_f)$.

Approximations of type (*ii*) include the following.

(*ii-a*) *Breathing Sphere* (*BS*) *Approximation.* The most drastic method for dealing with rotational degrees of freedom is to remove them from the problem completely. The BS approximation assumes $V_{\text{int}}(\mathbf{R}, \mathbf{r}) = V_{\text{int}}(R, r)$ and likewise drops all molecular orientation variables from the wave function.[38,41,56]

(*ii-b*) *Classical Rotation* (*CR*). Rotational degrees of freedom may be removed from the quantal part of the problem, the $\langle l|V_{\text{int}}|k\rangle$ matrix elements, and incorporated into the classical part of the problem along with translational degrees of freedom.[51,53,54] This leaves only vibrational matrix elements of V_{int} to be evaluated, but these will now be functions not only of b, but also of θ, ϕ, and β. The additional classical variables may be averaged over by the Monte Carlo method just as in the quasiclassical approach, with the final rotational state distribution determined by the histogram technique. This method has the advantage of including *all* rotational degrees of freedom, albeit classically. The quasiclassical histogram method is often satisfactory for rotational transitions,[6,20,26,57] particularly for the large-j states which are sometimes involved in vibrational transitions.[6,30,58-61]

(*ii-c*) *Coupled-States* (*CS*) *Approximation.* An alternative procedure which does not retain all rotational degrees of freedom, but which does treat rotation quantally, is the CS method developed originally as an approximation to the fully quantal coupled-channel problem (see Chapters 7, 9, and 11). The semiclassical CS method reduces the number of coupled equations in equation (8) by neglecting all Δm transitions. Billing[55] has formulated a semiclassical CS model which also incorporates an effective spherical potential for the translational motion. The quantal CS and EP approximations have recently been reviewed by Toennies.[3] The semiclassical CS method of Billing might be expected to have a domain of validity similar to that shared by the corresponding fully quantal EP and CS approximations,[62,63] but this has not yet been verified.

Space does not permit enumeration of all the computational formulas which result from the various combinations of approximations (*i-a*)–(*i-d*) and (*ii-a*)–(*ii-c*), together with the useful analytic approximations which are helpful in special cases. Also, no single combination of approximations in the context of equation (8) has yet emerged which is at once *both* sufficiently economical and sufficiently general to distinguish it clearly as an optimum choice for a large set of cases. The reader is best advised to consult the original literature on particular methods, guided perhaps by the general discussion of approximations in Section 4. Below we mention briefly some representative cases where theories of this type have been compared with experiment or fully quantal calculations.

Billing has combined the EP and CR approximations to study VRT transitions in He + H_2[53] and Li^+ + H_2.[54] Agreement with the He + H_2 vibrational relaxation time (a highly averaged quantity) is very good for temperatures above 250 K, although it must be appreciated that there may be considerable variation in the result with choice of potential energy surface. Agreement with the quantal EP calculations* of Alexander[64] and Rabitz and Zarur[65] is less satisfactory, with total cross sections for individual state-to-state transitions differing by factors of 2–10. For Li^+ + H_2, the rotationally averaged, vibrationally inelastic total cross sections are larger by about a factor of 2 than the corresponding coupled-channel results of Schaefer and Lester.[66] Billing has also applied similar treatments to diatom–diatom collisions.[67-69]

Krüger and Schinke[39] and Schinke *et al.* [40] have used a semiclassical IP treatment and the rotational SA to examine H^+ + H_2 vibrational transitions at high collision energies ($\geq$ 10 eV) and small scattering angles. Provided that a sufficiently large vibrational basis set is employed in order to take account of the distortion of the H_2 vibrational potential during the collision,[60,70] the agreement with the experimental vibrationally resolved relative differential cross sections[40,70,71] is excellent.

Collins and Cross[38] also studied H^+ + H_2 at high energies by an IP method, but in addition they used the vibrational SA, which results in great simplification of the calculation, but limits the range of validity in the H^+ + H_2 system to energies higher than several hundred eV.[72] At these energies the agreement of the vibrationally resolved total cross section with experiment[73] is within the (large) experimental uncertainty. These authors also examined the BS approximation and found that this additional approximation changed the results by only about 15% typically. It should be pointed out, however, that the H^+ + H_2 system is unusual

* The quantal calculations use a different effective potential.

in that the vibrational distortion of the molecule at large separations R has a rather weak dependence on molecular orientation.[60]

3.2. Classical S-Matrix Theory

One of the most important conceptual developments in semiclassical scattering theory is the classical limit of quantum mechanics known as "classical S-matrix theory," which was developed from different points of view by Miller and Marcus. The theory and its applications in many areas have been reviewed recently by Miller.[74,75] In classical S-matrix theory the dynamical part of the scattering problem is done purely classically, as in the semiclassical method; however, the classical trajectories which are used are only those which satisfy double-ended quantal boundary conditions for both the initial and final states. Although the numerical trajectory calculation may be carried out in the usual position-momentum variables,[10] the results are formulated in the action–angle variables,[21,74] in which the vibrational action n is the classical (continuous) analog of the vibrational quantum number v and the conjugate angle variable q is the vibrational phase. During the collision n and q are both functions of time to be evaluated numerically, but asymptotically, at $t \to \pm\infty$; n is a constant of the motion, and q is linear in t.[76] For a collinear atom–diatom collision at kinetic energy T_i, the probability of the vibrational transition $v_i \to v_f$ is calculated by first finding all classical trajectories for which both conditions $n(-\infty) = v_i$ and $n(\infty) = v_f$ are satisfied. The first condition, $n_i = v_i$, is taken as an initial classical boundary condition for the trajectory, but the second, $n_f = v_f$, must be found by searching (iteratively) the space of the initial phase shift

$$\bar{q}_i = q_i - R(m/2T_i)^{1/2} \tag{10}$$

for the result

$$n_f(\bar{q}_i) = v_f \tag{11}$$

The S-matrix elements are

$$S_{v_i v_f} = \sum \left[2\pi \frac{\partial n_f(\bar{q}_i, n_i)}{\partial \bar{q}_i} \right]^{-1/2} \exp[i\hbar^{-1}\xi(\bar{q}_i, n_i)] \tag{12}$$

where the phase ξ is given by the integral

$$\xi(\bar{q}_i, n_i) = -\int_{t_i}^{t_f} (mR\dot{R} + q\dot{n})\, dt \tag{13}$$

and the sum is over all trajectories which satisfy equation (11).[76] The transition probability is then

$$P_{v_i v_f} = |S_{v_i v_f}|^2 \tag{14}$$

Equation (12) is the "primitive" semiclassical result, which fails near an extremum in $n_f(\bar{q}_i, n_i)$. This problem can be remedied in many cases by "uniformization,[76-78] which maintains the proper coherence near an extremum. The S-matrix formulation is valuable partly because the S-matrix *amplitudes* are combined coherently, permitting the semiclassical description of quantal interference phenomena, which is not possible in the quasiclassical method. Another important feature of the classical S-matrix approach is that it can be extended to classically forbidden processes, i.e., those in which no real classical trajectory connects v_i and v_f, as is often the case in vibrational energy transfer. Classically forbidden transitions are handled in the same way as classically allowed transitions, except that the trajectories are calculated in complex coordinates and complex time. Miller[75] has discussed the numerical aspects and philosophy of this procedure. With the introduction of imaginary components in n, q, and R, the exponential factor in equation (12) becomes a damping factor which accounts for the exponential attenuation of the wave function with depth of penetration (tunneling) into classically forbidden phase space. When only one complex trajectory contributes to $S_{v_i v_f}$,[75] the resulting probability is

$$P_{v_i v_f} = \left[2\pi \left| \frac{\partial n_f(\bar{q}_i, n_i)}{\partial \bar{q}_i} \right| \right]^{-1} \exp[-2\hbar^{-1} \operatorname{Im} \xi(\bar{q}_i, n_i)] \tag{15}$$

The uniform classical S-matrix representation has been shown to be remarkably accurate for calculating vibrational transition probabilities for collinear atom–diatom collisions.[27] However, the complex-trajectory and iterative root-searching procedures which are required are computationally laborious, even for the collinear case.* In three dimensions, the root-searching problems become essentially prohibitive, since the double-ended boundary conditions must be applied to all rotational degrees of freedom as well as the vibrational mode. Furthermore, the integral representations which have been proposed to alleviate the root-searching difficulties sometimes result in large errors, particularly for small transition probabilities.[27,80]

At present it appears that the only practical method for applying classical S-matrix theory to vibrational excitation processes in three dimensions is to eliminate or average over the rotational degrees of freedom. Cross[81] has suggested that the sudden approximation can be helpful in some cases. A more generally useful method is the "partial-averaging" technique,[4,74,75,82-84] in which interference effects are ignored and the rotational phase boundary conditions are averaged over by the Monte Carlo method rather than constrained to satisfy double-ended boundary

* One exception is for the special case of a hard-sphere interaction potential, where the classical S-matrix result can be obtained analytically.[79]

conditions. For rotational transitions, this method is similar to the standard quasiclassical procedure. For classically forbidden vibrational transitions, however, the partially-averaged S-matrix method incorporates the additional features of complex-valued trajectories and the exponential damping factor, which permit classical mechanics in this context to describe small-probability processes. The probability is still given by equation (15), but $P_{v_i v_f}$ becomes parametrically a function of all the initial conditions other than v_i and γ. These can be averaged over and the final rotational states assigned by the histogram method to yield the appropriate averaged value $\langle P_{v_i v_f} \rangle$. Raczkowski and Miller[84] calculated integral vibrational–rotational transition cross sections for $Li^+ + H_2$, using double-ended boundary conditions for the vibrational action but a Monte Carlo average over b, θ, ϕ, and β. The agreement with nonconverged coupled-channel calculations was encouraging, but not conclusive.

3.3. *Classical–Quantal Correspondence Methods*

In some sense all semiclassical methods are based on classical–quantal correspondences. However, the two methods summarized below are ones which make explicit use of correspondence relations in order to take account of some uniquely quantal features of the dynamics (e.g., discrete vibrational energy levels) while avoiding the numerical complication of a wave function representation of the vibrational motion.

3.3.1. *The* DECENT *and* INDECENT *Methods*

The harmonic oscillator belongs to a special class of problems in which the classical and quantal equations of motion are related very intimately. For the motion of a particle of mass M in one dimension r the classical equation of motion can be written

$$F(r, t) = \frac{-\partial V(r, t)}{\partial r} = M \frac{d^2 r}{dt^2} \tag{16}$$

The Ehrenfest theorem relates the corresponding quantal expectation values:

$$\langle F(r, t) \rangle = M \frac{d^2 \langle r \rangle}{dt^2} \tag{17}$$

Provided only that the potential be of the general form

$$V(r, t) = a(t) + b(t)r + \tfrac{1}{2}c(t)r^2 \tag{18}$$

where a, b, and c are arbitrary functions of time, then

$$\langle F(r, t) \rangle = -b(t) - c(t)\langle r \rangle = F(\langle r \rangle, t) \tag{19}$$

Therefore, for this special class of potentials, equation (17) shows that the *quantal* expectation value $\langle r \rangle$ obeys exactly the *classical* equation of motion, equation (16). For bound systems, equation (18) permits a time-dependent shift in the location of the potential minimum through the coefficient $b(t)$ and in the oscillator force constant through the coefficient $c(t)$. The quantal equation of motion for this one-dimensional problem is given by equation (8), where now $|l\rangle$ and $|k\rangle$ are harmonic oscillator eigenfunctions. It is easy to see in terms of the ladder operators α_+ and α_- [where r is proportional to $(\alpha_+ - \alpha_-)$] that the term $a(t)$ can induce no transitions, while the term $b(t)r$ couples states with $\Delta v = \pm 1$ and the term $c(t)r^2$ couples states with $\Delta v = \pm 2$. Except for extremely violent collisions the effect of direct $\Delta v = \pm 2$ coupling is negligible compared to that of two successive $\Delta v = \pm 1$ transitions. Under these conditions a harmonic oscillator subjected to a general time-dependent perturbation can be well-described by the potential

$$V(r, t) = b(t)r + \tfrac{1}{2}\mu\omega^2(r - r_e)^2 \tag{20}$$

in which $-b(t)$ represents the externally applied force, which has completely arbitrary time dependence. The exact solution to the time-dependent Schrödinger equation for the potential in equation (20) gives the transition probabilities[85] in terms of a single parameter ε:

$$P_{v_i v_f} = v_i!\,v_f!\,e^{-\varepsilon}\varepsilon^{(v_i+v_f)}(S_{v_i v_f})^2 \tag{21}$$

where

$$S_{v_i v_f} = \sum_{k=0}^{\min(v_i, v_f)} \frac{(-1)^k \varepsilon^{-k}}{(v_i - k)!\,k!\,(v_f - k)!} \tag{22}$$

As a consequence of the close classical–quantal correspondence for $\langle r \rangle$,[86,87] the parameter ε is simply

$$\varepsilon = \mathscr{E}/\hbar\omega \tag{23}$$

where $\mathscr{E}$ is the final vibrational energy of a *classical* oscillator initially at rest $[r(-\infty) = r_e, \dot{r}(-\infty) = 0)]$ which is driven by the same time-dependent force $-b(t)$. Even though $\mathscr{E}$ is always positive, the initially stationary classical oscillator correctly corresponds to any stationary quantum state $(\langle r \rangle = r_e)$ before the collision, and therefore equation (21) applies to both excitation and deexcitation processes. All of the quantal information for this problem is contained in the classical motion and in the correspondence relation of equation (21). This correspondence does *not* depend on the quantum numbers v_i and v_f being large.

Even though the classical–quantal correspondence is *exact* only for a harmonic oscillator driven by a time-dependent potential which is linear in r, it will be valid *approximately* for any interaction potential which is approximately linear over the actual extent of the quantum oscillator wave packet at any instant during the collision.* If this condition is satisfied, then regardless of the actual time dependence of the driving force $-b(t)$ the wave packet will not spread and will remain centered on the classical value of $r(t)$. Therefore for an arbitrary potential energy surface, $\mathscr{E}$ can be calculated from the exact classical trajectory, and the corresponding approximate quantal transition probabilities are then specified by equation (21). This approximation forms the basis for the DECENT and INDECENT methods† suggested by Gentry and Giese.[60,61,91,92] To apply these methods to three-dimensional collisions, one calculates the exact classical trajectory numerically for selected initial values of b, θ, ϕ, and β, just as in the quasiclassical method, except that the initial vibrational boundary condition appropriate to all transitions is $r(-\infty) = r_e$, $\dot{r}(-\infty) = 0$. Thus there is no averaging over the initial vibrational phase. The final rotational state can be assigned by any acceptable quasiclassical procedure. These probabilities are then averaged over θ, ϕ, and β to calculate the total transition cross section

$$\sigma_{if} = 2\pi \int_0^\infty b\langle P_{if}(b)\rangle\, db \tag{24}$$

or further distributed in χ to obtain the differential cross section.

The difference between the DECENT and INDECENT models lies in the choice of the initial translational boundary condition $\dot{R}(-\infty)$. Quantum-mechanically the initial and final kinetic energies T_i and T_f are separated by the discrete energy defect $E_f - E_i$, whereas the classical energy defect $\mathscr{E}$ is continuous and is often small compared to $\hbar\omega$. In the DECENT model the initial kinetic energy T used to calculate the classical trajectory is taken to be

$$T = T_i \tag{25}$$

and the single value of $\mathscr{E}$ found for this trajectory is used to evaluate the entire matrix of transition probabilities $P_{v_i v_f}$ corresponding to a given value of T_i. Detailed balancing is not obeyed, since the probability for the reverse transition $P_{v_f v_i}$ would be calculated from a trajectory with $T = T_f$. The INDECENT model, however, uses an initial kinetic energy corresponding to the arithmetic mean of the speeds in the initial and final channels,

$$\bar{T} = \tfrac{1}{4}(T_i^{1/2} + T_f^{1/2})^2 \tag{26}$$

* The exact correspondence is broken if the quadratic term in equation (18) is also time dependent.[86] Even though the quantal wave packet still remains centered on the classical value of r, a time-dependent oscillator frequency induces spreading of the wave packet. Heller[88-90] has developed elegant schemes for dealing with wave packet dynamics for locally quadratic potentials.

† DECENT = Distribution of Exact Classical ENergy Transfer; INDECENT, INtermediate speed DECENT approximation.

thus enforcing the detailed balance condition

$$P_{v_i v_f} = P_{v_f v_i} \tag{27}$$

but requiring the calculation of a different trajectory for each different value of the quantum level spacing $|E_f - E_i|$. This choice of trajectory usually results in excellent agreement between the semiclassical and quantal transition probabilities for collinear collisions, even at energies very close to threshold.[92] At higher colliison energies, $T \gtrsim 10|E_f - E_i|$, the computationally more efficient DECENT model may be used with little loss of accuracy.

The INDECENT model has been compared with exact quantal calculations for collinear A + BC collisions over a large range of kinetic energy and for several interaction potentials.[27,80,92,93] For harmonic BC molecules the error found for transitions among the lower vibrational states is typically of the order of 10%. As one should expect, the best results are obtained for small values of the dimensionless mass parameter[92]

$$M = M_A M_C / M_B (M_A + M_B + M_C) \tag{28}$$

which correlates with small vibrational distortion of the BC molecule. The error is somewhat greater if BC is a Morse oscillator. Taking the quantum level spacing $\hbar\omega$ in equation (23) to be the smallest quantum level spacing involved in the transition,[92] one finds an error for transitions among the lower vibrational states of the order of 20%.* Even for high vibrational states of a Morse oscillator, the INDECENT model is fairly accurate for single-quantum transitions, provided that $\hbar\omega$ in equation (29) is replaced by the actual energy defect of the transition.[92] The INDECENT model works well both for classically allowed transitions and for those which are classically forbidden. Transition probabilities as small as 10^{-10} are predicted about as accurately as by the classical S-matrix theory, but without the use of complex trajectories or double-ended boundary conditions. Very importantly, the INDECENT model can reproduce the sometimes very large qualitative differences in the transition probabilities due to relatively subtle differences in various model potential energy surfaces for a given chemical system.[92,93,95]

The DECENT model has been compared with experimental differential cross sections for state-resolved $H^+ + H_2$[60,70,71,91,96] and $Li^+ + H_2$[61] collisions. The potential energy surfaces and the mechanisms of vibrational excitation are very different for these two cases. Nevertheless the DECENT

* The reason that the INDECENT model does fairly well for a Morse oscillator, despite the spreading of the time-dependent wave packet which occurs in a Morse potential, can be seen from the calculations of Walker and Preston.[94] A minimum-uncertainty Morse wave packet typically begins to break up only after several oscillations, which is usually less than the duration of the collision. Even after the wave packet has lost coherence it still follows the classical oscillatory motion for many oscillations.

model predictions for the excitation probability at fixed scattering angle are almost within the estimated experimental error (~10% for $v_i = 0 \rightarrow v_f = 1$). The predicted rotational excitation for $H^+ + H_2$ was also in good qualitative agreement with subsequent rotationally resolved experiments[96] and a quantal coupled-states calculation,[97] carried out at lower collision energies.

The INDECENT model has recently been extended to diatom–diatom collisions.*[95,99] The agreement with exact quantal results for both VV and TV energy transfer is about the same as for atom–diatom collisions.

3.3.2. The Strong-Coupling Correspondence Principle

The strong-coupling correspondence principle (SCCP) was proposed by Percival and Richards[100] as a means for efficiently treating transitions among highly excited electronic states of atoms. Clark and Dickinson[101] applied the same methods successfully to the problems of rotational and vibrational transitions in molecules. The method is based on Heisenberg's form of the classical–quantal correspondence principle,[102] which relates the matrix elements $\langle l|U|k\rangle$ of an operator U to the Fourier component of the operator at the frequency $|E_l - E_k|/\hbar$, evaluated along an "average" classical path. The theory neglects differences in the transition frequencies for all strongly coupled states, and takes for the classical path the trajectory defined by first-order classical perturbation theory (CPT), in which the internal motion of the molecule is assumed to be unaffected by the interaction, while the time dependence of the radial atom–molecule separation is determined by the spherically averaged interaction potential.

The SCCP method can be derived as the CPT limit of classical S-matrix theory.[102] For the simultaneous transition $v_i j_i \rightarrow v_f j_f$, the SCCP approximation to the S-matrix is given by[103]

$$\bar{S}_{if}(b,\theta,\phi,\bar{T}) = \frac{1}{4\pi^2}\int_0^{2\pi}\int_0^{2\pi} \exp\{i[|v_f - v_i|\gamma + |j_f - j_i|\beta - \Delta n(\gamma,\beta)]\}\,d\beta\,d\gamma \qquad (29)$$

where $\Delta n(\gamma,\beta)$ is the change in the classical action of the molecule from CPT. The particular CPT trajectory chosen is the one corresponding to the arithmetic means of the vibrational frequencies, the rotational frequencies, and the translational speeds in the initial and final states. The $\bar{S}$ matrix is therefore symmetric in the interchange of initial and final states, $\bar{S}_{if} = \bar{S}_{fi}$. To enforce detailed balancing for the transition cross

* The formalism for a closely related treatment of diatom–diatom collisions was given previously by Kelley.[98]

section,

$$(2j_i + 1)T_i\sigma_{if} = (2j_f + 1)T_f\sigma_{fi} \tag{30}$$

the weighted probability averaged over molecular orientation is taken to be[103]

$$\langle P_{if}(b, T_i)\rangle = \frac{1}{4\pi}\frac{\bar{T}}{T_i}\left(\frac{2j_f + 1}{2j_i + 1}\right)^{1/2}\int_0^{2\pi}\int_0^{\pi}|\bar{S}_{if}(b,\theta,\phi,\bar{T})|^2 \sin\theta\, d\theta\, d\phi \tag{31}$$

The total cross section for the transition is then obtained by integrating over b [equation (24)].

For realistic potentials the action difference Δn in equation (29) must be obtained by numerical integration. However, the computational effort for CPT trajectories is significantly less than for the exact trajectories because of the smaller number of degrees of freedom. This is especially true for pure rotational transitions, for which the techniques of angular momentum algebra can be used to obtain analytically the orientation dependence of Δn.[63,102] Although the CPT dynamical approximation will be valid only for small $\Delta T/T$, it will usually be much better than the sudden approximation. The SCCP method is most accurate for small Δv transitions among highly excited states. Just as in the INDECENT method, the CPT vibrational energy transfer will be more accurate for small values of M [equation (28)], because the vibrational distortion of the molecule during the collision increases with increasing M.[92] At the expense of additional numerical computation, the SCCP method can be modified to include an approximate second-order correction to CPT,[104] which can improve the result substantially for small v. In extreme cases, such as the collision of an ion with a polar molecule in its ground rotational state, the CPT approximation may fail badly for individual trajectories at energies up to even several hundred eV.[105,106] However, CPT will be much better for results averaged over orientation,[105] and for shorter-range potentials.

There has not yet been a comparison of SCCP predictions with experimental results for vibrational excitation, but Clark[103] has compared the three-dimensional SCCP results with those of the quantal CS and EP approximations for simultaneous rotational and vibrational excitation in $He + H_2$ collisions. The SCCP vibrationally inelastic $0 \rightarrow 1$ cross sections agree with the approximate quantal cross sections to within about a factor of 2. The agreement seems to be better for pure rotational transitions.

3.4. *Models for Special Cases*

If dynamical and dimensional approximations are combined with a drastic simplification of the potential energy surface, the validity of the result becomes very hard to assess. Nevertheless, simple models may be useful qualitative guides for some special cases. We summarize several models of this type below.

3.4.1. ITFITS *Models*

The ITFITS model of Heidrich *et al.*[107] uses the forced-oscillator correspondence relation, equation (21), but combined with an approximate trajectory for which $\mathscr{E}$ can be calculated analytically. ITFITS assumes a collinear collision of atom A with molecule BC, where the interaction potential is

$$V_{\text{int}} = \text{const} \cdot e^{-z/L} \tag{32}$$

Here z is the A–B separation, and the results are independent of the scaling constant. With the additional assumption of an impulsive collision between A and B,[108] the classical energy transfer is given by

$$\varepsilon = 2\{\pi\alpha^{-1}(1 + M^{-1})^{-1} \operatorname{csch}[\pi\alpha^{-1}(M/2\bar{T})^{1/2}]\}^2 \tag{33}$$

where

$$\alpha = L^{-1}[\hbar M_C/\omega M_B(M_B + M_C)]^{1/2} \tag{34}$$

The impulse approximation gives an accurate representation of the exact energy transfer for this model for $M \leq 0.5$.[92,109] Morse [110] has used the breathing sphere model to extend the ITFITS approach to three-dimensional collisions, and Morse and LaBreque[111]* developed a technique for treating the excitation of a Morse oscillator within the same framework. Zelechow *et al.*[112] extended the ITFITS model to collinear diatom–diatom collisions. The same approach has also been used by Shobatake *et al.*[113] to formulate a model for atom–polyatomic-molecule energy transfer. Rubinson *et al.*[28] were able to use a modified form of the model for the qualitative interpretation of data on the rate of I_2 vibrational relaxation. Bass[114] proposed a similar model based on a hard-sphere interaction potential.

3.4.2. *Angular Dependence of Impulsive Energy Transfer*

Cheng *et al.*[115] have given a formula which can be used to estimate the dependence of the total inelastic energy transfer on scattering angle for high-energy collisions. The model assumes an impulsive (hard-sphere) collision between A and B, from which the energy transfer and scattering angle can be obtained by vector addition. The result is

$$\frac{\Delta T}{T_i} = \frac{2[M^{-1} + \sin^2\chi - \cos^2\chi(M^{-2} - \sin^2\chi)^{1/2}]}{(1 + M^{-1})^2} \tag{35}$$

* See also comment in Ref. 131.

The predicted energy transfer agrees qualitatively with the measured average translational energy loss in O_2^+ + He collisions at energies of about 10 eV.[115]

3.4.3. Near-Resonant V–V and V–R Transitions Induced by Long-Range Forces

Sharma[116,117] and Sharma and Brau[118] proposed a model for near-resonant transitions induced by the long-range forces which depend only on the multipole expansions of the charge distributions in the isolated molecules. The model is based on the first-order perturbation theory approximation to equation (8), which gives the transition probability as

$$P_{kl} = \frac{1}{\hbar^2} \left| \int_{-\infty}^{\infty} \exp[it(E_l - E_k)/\hbar] \langle l|V_{\text{int}}(t)|k\rangle \right|^2 \tag{36}$$

The interaction potential $V_{\text{int}}(t)$ is decomposed into a sum of terms, each of which is proportional to $Q_{s_1}(1)Q_{s_2}(2)R^{-s_1-s_2-1}$, where $Q_{s_i}(i)$ is the the sth multipole moment of molecule i. Rotational transitions are induced by the angular dependence of the Q_{s_i}, the matrix elements for which obey simple selection rules. Vibrational transitions are induced by $(\partial Q_{s_i}/\partial r_i)$. Since the potential is accurate only at large separations, where the two charge distributions do not overlap, the model is useful only for transitions dominated by large impact parameter collisions, and relatively weak interactions, i.e., ones for which the energy defect $|E_l - E_k|$ is very small compared to T_i. For such interactions an impact parameter treatment of the translational motion is appropriate, and only the longest-range potential terms (smallest values of $s_1 + s_2$) will contribute significantly. The model is applicable only when a particular VV or VR process is so nearly resonant that it can be presumed to dominate over all other channels involving that particular vibrational state. To apply the model, one selects the one or two lowest-order potential terms which connect the two states involved, discarding all matrix elements of V_{int} from higher-order potential terms as well as the ones which are eliminated by selection rules. In the IP approximation, useful analytic approximations can sometimes be found for the resulting expression, greatly reducing the computational effort in averaging over impact parameter and a thermal speed distribution. Comparison of the thermally averaged transition probability with experimental relaxation data has been carried out for processes such as

$$CO_2(01^10) + H_2(v_i, j_i = 0, 1) \rightarrow CO_2(000) + H_2(0, 3) + 80.5\,\text{cm}^{-1} \tag{37}$$

for which the Sharma model correctly predicts the negative temperature dependence of the relaxation rate.[116] The theory has been extended by Chiu[119] to the use of the Born approximation for the translational motion. Recently, a similar model treatment by Krutein and Linder[120] was used successfully to interpret vibrationally resolved data on $H^+ + CO_2$ small-angle scattering.

4. Approximations

Approximations can sometimes result in great savings in computational effort. It is incumbent on the user, of course, to demonstrate that any approximation which is adopted is in fact valid for the particular case to which it is applied. Parameters which determine the range of validity for a given approximation include the mass combination of the collision system, and the kinetic energy range and the range of scattering angle in which comparisons between experiment and theory are to be made. Another important consideration is the nature of the potential energy surface, particularly the range of the interaction, the shape of the interaction potential around the oscillator minimum, and the anisotropy.

4.1. Dynamical Approximations

4.1.1. Sudden Approximation (SA)

The SA neglects the internal motion of the molecule during the collision by taking the time derivatives of all internal normal coordinates equal to zero. The SA is therefore more restrictive than CPT, which assumes merely that these time derivatives are *constant* during the encounter. Use of the SA has sometimes been justified on the basis of a comparison between the collision duration τ and the period λ_i of oscillation in a particular degree of freedom of the initial state, the criterion for validity being $\tau \ll \lambda_i$. This criterion is not sufficient, however, if the collision couples states with significantly different frequencies, such as high and low rotational states. A better criterion is

$$\tau \ll \min(\lambda_i, \lambda_f) \tag{38}$$

The SA may be applied to quantal[121,122] as well as to semiclassical collision theory. Generally the SA will be much better for rotation than for vibration.[35-40] Schinke *et al.*[40] showed the rotational SA to be very good for small angle $H^+ + H_2$ T$\rightarrow$ VR energy transfer at kinetic energies of a few eV. Collins and Cross[38] applied the vibrational SA to the same system, but Skodje *et al.*[72] showed that convergence of the SA cross sections to within 20% of the dynamically exact cross sections is not obtained until the kinetic energy exceeds several hundred eV. For neutral systems where the interaction is of much shorter range the vibrational SA may be applicable at much lower kinetic energies.[36] In cases where strong long-range forces are present, such as in collisions of ions with polar molecules, even the rotational SA may not be valid at low energies.[105] The validity of the vibrational SA (or impulse approximation[108]) has been tested for collinear atom–diatom collisions as a function of the mass parameter M from equation (28). For a repulsive exponential interaction and $M \leq 0.5$, the SA agrees well with the exact classical results for vibrational energy transfer at kinetic energies as low as $0.01(\hbar\omega/\alpha^2)$.[92] For $M \geq 2/3$ the SA has essentially no range of validity,[92] although an *ad hoc* correction factor[109] has been devised which applies over a limited energy range. These conclusions may not apply to three-dimensional collisions or to other potentials.

4.1.2. Impact Parameter (IP) Approximation

The IP approximation assumes uniform rectilinear translational motion [equation (9)]. Although its use is obviously restricted to small $\Delta T/T$ and to small scattering angles, it may still be helpful for many problems. Gentry[105] showed that most of the error in the rotational energy transfer from CPT for high-energy $H^+ + HF$ ($j = 0$) collisions is due to the SA rather than the IP approximation. The IP approximation has been used very successfully for the semiclassical interpretation of data on high-energy $H^+ + H_2$ vibrational excitation.[39,40] Serious errors can occur for thermal-energy collisions, however,[123] where the requirement that $\Delta T \ll T$ is not usually satisfied.

4.1.3. Neglect of ΔT

A less restrictive approximation is the use of an "elastic trajectory," which assumes $\Delta T = 0$ but permits the translational trajectory to be curved. However, unless the additional assumption is made that the internal molecular motion is uncoupled from the translational motion (as in the CPT method of Clark *et al.*[102]), the use of an elastic trajectory results in only a small decrease in computational effort.

4.2. Dimensional Approximations

4.2.1. Collinear Models

It was rather casually assumed in some early treatments of vibrational energy transfer that collinear collisions would be more effective than any others in inducing vibrational transitions, and that therefore the vibrational transition cross section could be written as the product of the collinear transition probability and the gas-kinetic cross section, multiplied by a "steric factor"[124] between 0 and 1 to account for noncollinear collisions. McKenzie[47] has recently shown, within the context of a semiclassical IP treatment, that a collinear model can under some circumstances agree reasonably well with a three-dimensional calculation, provided that the ratio of vibrational to rotational frequency is sufficiently large. However, in cases where the collinear and three-dimensional calculations have been compared for realistic potentials and exact dynamics, the correlation has usually been poor. For example, Gentry and Giese[61] found for the large-angle scattering of Li^+ from H_2 at $T = 3.65$ eV that the orientation-averaged exact classical vibrational energy transfer in a three-dimensional calculation was about 40 times *larger* than the collinear energy transfer. In this case the discrepancy is due to strong rotational–vibrational coupling in slightly eccentric collisions. Such effects will often be important in strongly repulsive collisions where the anisotropy of the potential is large and the interaction time is approximately equal to or greater than the vibrational period.[61,125] Strong VR coupling has also been seen often in fully quantal calculations in three dimensions.[59,64,65,122,126-128] Furthermore, Faubel and Toennies[129] found that in collisions of a Morse oscillator with a structureless atom, the collinear geometry usually results in less vibrational excitation than does a collision along the perpendicular bisector of the molecular bond. Although collinear calculations will remain useful for testing approximate theories, they are generally not to be recommended for comparison with experiment.

4.2.2. *The Breathing Sphere (BS) Model*

The breathing sphere model fails for the same reason as collinear models—because it neglects VR coupling and the orientation dependence of $(\partial V_{int}/\partial r)$. The BS model is more realistic, however, in that it gives the proper weighting to large impact parameter trajectories, which explore the weaker part of the interaction. In some systems, such as $H^+ + H_2$, the driving force $(\partial V_{int}/\partial r)$ is not a strong function of orientation at large separations ($R > 2.5$ Å) and a BS model may work well for sufficiently small scattering angles.[60] As with other approximations one expects that the BS results may be more accurate for a highly averaged quantity than for a microscopic one, because of cancellation of errors. Verter and Rabitz[130] used a BS model and obtained good agreement with experimental data on the thermal vibrational relaxation rate in CO + He mixtures. However, they were not able to demonstrate the separability of the average vibrational and rotational transition rates, which would be sufficient to validate the BS model.

4.3. *Influence of the Potential Energy Surface*

In many cases the greatest obstacle to definitive theoretical interpretation of experimental data will be the accuracy of the potential energy surface used for the dynamical calculation. "Exact" dynamical calculations are feasible either in classical mechanics (with ease) or in quantum mechanics (with difficulty for simultaneous Δv Δj transitions); however, there has never been, in the same sense, an "exact" potential energy surface for an inelastic collision, which would require convergence of the dynamical calculation with respect to the electronic basis set used for calculating the potential energy. If one carries out a dynamically exact quantal calculation, then comparison with experiment will provide a test of the potential energy surface. However, with any approximate dynamical treatment one must consider the possibility that cancellation of errors between approximations to the dynamics and approximations to the potential energy surface may result in better (or worse) agreement with experiment than is justified. If the dynamical calculation is not too expensive, this possibility may be examined by repeating the calculation with other potential surfaces that lie within the estimated uncertainty of the "best" available potential.

4.3.1. Diatomic Molecule Potential

If the interaction potential is poorly known, then the use of an accurate potential function for the isolated molecule may not add much realism. One should be aware, however, that the form of the diatom potential sometimes makes a large difference in the vibrational excitation probability. Clark and Dickinson[131] compared harmonic oscillator (HO) and Morse oscillator (MO) transition probabilities in an exact quantal treatment of collinear collisions with a repulsive exponential interaction. For small M the differences were small, but for $M = 2/3$ the HO single-quantum transition probabilities were about a factor of 2 larger than the MO probabilities. The INDECENT model [27,80,92] reproduces the qualitative differences well, indicating that the difference correlates with the classical energy transfer and is not attributable to a subtle difference between the HO and MO wave functions. Systems with small M correspond to the collision of a light atom with a molecule in which both atoms are heavy. The difference between the MO and HO transition probabilities for such cases is small because the energy transfer is nearly impulsive.[92] Drake and Lin[132] showed that for He + H_2 collinear collisions neither the HO nor the MO potential reproduces the results obtained with an accurate *ab initio* H_2 potential. As previously noted, the collinear geometry is not necessarily representative. Schinke and Toennies[133] demonstrated that HO transition probabilities are usually greater than the corresponding MO probabilities in collinear collisions, but that in perpendicular collisions the reverse is true. This result encourages one to hope that subtle differences will tend to cancel in averaging over molecular orientations. The three-dimensional unconverged coupled-channel calculations of Kouri and Wells[126] on Li^+ + H_2 collisions indicated little difference in the vibrational excitation cross sections between HO and MO potentials for H_2 other than that which is attributable to the difference in energy level spacing.

4.3.2. Interaction Potential

If a constant force $\pm F$ is applied to an oscillator for a time interval τ, the oscillator gains energy at times during τ when $\pm\dot{r} > 0$ and loses energy when $\pm\dot{r} < 0$. If τ is equal to the vibrational period, then the net energy transfer is zero regardless of the magnitude of F. Similar cancellation of strong driving forces can occur in real atom–molecule collision systems whenever the interaction time is comparable to or greater than the vibrational period of the molecule. In such cases the average energy transfer can change greatly with a relatively small change in the strength of the driving force ($\partial V_{\text{int}}/\partial r$) or especially with a shift in its timing relative to the oscillator motion.[60,61,72,91,93] Great care should therefore be given to the selection of the potential energy surface for any calculation of vibrational energy transfer.

Quantum-mechanical calculations yield points on the potential energy surface one at a time. For an A + BC interaction there are three independent internuclear separations, so that even a fairly coarse grid of 10 points for each coordinate would require 10^3 separate calculations of the potential energy. Even for simple few-electron systems this would be very expensive if high accuracy is required at each point. To do dynamical calculations one must use many more values of the potential energy than can be calculated accurately by *ab initio* means; therefore if *ab initio* potential points are to be used, the potential must be represented either analytically or by interpolation[30] on the grid of points. The most popular scheme for using *ab initio* potential points has been to fit the points with an analytic function containing several adjustable parameters. This method generally requires fewer points than the interpolation method. An excellent demonstration of the pitfalls which may be encountered has been provided by Alexander and Berard,[134] who performed exact quantal calculations of the vibrational excitation probabilities for fixed-orientation He + H_2 collisions, using five different analytic "best" fits to a small set of interaction potential points. Even though the standard deviations (from the calculated points) of all five surfaces were about 1 kcal/mole or less, four of the surfaces showed systematic differences over a range of about a factor of 2 in the $0 \rightarrow 1$ transition probability, and the fifth differed from the others by orders of magnitude at some orientations. Thus one must face the sobering possibility that a potential energy surface having the traditional "chemical accuracy" of 1 kcal/mole may not be adequate to describe vibrational excitation processes in some cases. Apparently, however, high accuracy is not required everywhere on the surface. Duff and Truhlar[93] showed that the results of Alexander and Berard could be reproduced semiquantitatively using classical trajectory methods. Thus the critical region of the potential surface, where the highest accuracy is required, is the relatively small region explored by the classical trajectories. One could imagine an iterative approach to improving the calculated potential surface, in which, first, enough points are calculated to generate a reasonable analytically fitted surface, then a classical trajectory method such as the INDECENT model is used to discover for what coordinates additional potential points need to be calculated to refine the surface in critical regions.

The extreme sensitivity of the vibrational transition probability to slight alterations in the interaction potential may to some extent be an artifact of fixed-orientation dynamical calculations. Averaging over molecular orientations (or angular momentum projections) and collision orbital angular momentum can help to wash out the interference structure due to idiosyncratic phase relationships which can greatly influence results calculated in reduced dimensionality. White and Hayes,[135] in a fixed-orientation study of $Li^+ + H_2$ collisions, found that the results were very sensitive to a small change in a model potential. However, Schaefer and Lester[66] carried out three-dimensional coupled-channel calculations for the same system using three different *ab initio* potentials, and generally found very little difference in the inelastic differential cross sections, although some of the partial cross sections differed substantially.

In most cases of interest to the experimentalist there will not exist a highly accurate set of calculated potential energy points, and one must rely on a semiempirical description[16,17,28,29,31,58,119,136,137] or a completely *ad hoc* model.[114,115,138] There are many examples in which potential parameters have been varied in order to fit data on vibrationally inelastic processes,[18,73,81,139-141] but only in cases where the dynamical approximations are known to be trustworthy can the resulting potential be considered reliable. Methods exist (see Chapters 2 and 3) for estimating the potential energy surface appropriate to any atom–molecule collision process. For vibrational excitation calculations, particular attention should be given to the shape of the interaction potential around the minimum in the oscillator coordinate. The success of the INDECENT model, which is based on the classical–quantal correspondence for a *linearly* forced harmonic oscillator, suggests strongly that the most important feature of the surface for vibrational excitation is $r_e(\mathbf{R})$, the variation in the location of the oscillator minimum as a function of the atom–molecule separation vector. Any available information about this feature should be used in selecting the form of the potential energy surface. Relatively large changes in r_e are to be expected for reactive systems, in which strong valence forces are involved.[16,17,29,60,142] In such cases it may be possible to obtain information on the bond distances in the intermediate from spectroscopic data. Another very important feature is the range in $\mathbf{R}$ of the oscillator forcing function $\partial V_{\text{int}}(r, \mathbf{R})/\partial r$. The longer the range of the driving force, the less vibrationally impulsive the collision will be, and the more critical will be the detailed shape of the potential. In systems where strong long-range forces exist, e.g., charge–dipole or dipole–dipole interactions, it may be very important to include the dependence of the coefficients of the long-range terms on the oscillator coordinate. This information is sometimes available from data on infrared absorption intensities or from molecular beam electric resonance experiments.

5. *Conclusions and Recommendations*

There are several classical and semiclassical methods that will, within limited ranges of validity, yield theoretical results on vibrational energy transfer of sufficient accuracy to be useful in the interpretation of experimental data. In selecting a method for doing the dynamical calculation, the first important considerations are: (a) Does the data contain highly microscopic information? and (b) Is there accurate and extensive information available on the potential energy surface for the interaction? If the answer to both (a) and (b) is yes, then clearly one should employ the most accurate practical dynamical method in order to extract the most information by comparison with experiment. If the answer to both (a) and (b) is no, then the best one can hope for is a qualitative interpretation of the data and perhaps a model treatment is justified. One should then consider using whatever dynamical approximations seem reasonable, and also using a simple model potential to reduce the computational effort. Usually the answer to (b) will be that some information about the potential exists which is not as extensive as one would like and is of unknown accuracy. In this case, if an accurate dynamical calculation is to be undertaken, the first step is to generate as accurate a potential energy surface as possible from whatever data is available. On the basis of the features of the potential surface selected and the range of collision parameters appropriate to the experiments a more-or-less optimum choice of method can be made.

The methods that use exact classical trajectories, i.e., the quasiclassical methods, the DECENT and INDECENT methods, and the classical S-matrix methods, have the advantage that they make no approximations regarding the dynamical coupling between the translational, rotational, and vibrational degrees of freedom and can therefore be applied to any type of potential energy surface without concern that some (classically) important dynamical feature implicit in the surface will be ignored. For strongly allowed transitions among high vibrational and rotational states, the quasiclassical histogram method should be very accurate. Vibrational and rotational transitions are treated in a consistent way by the quasiclassical method and, in cases where it is expected to be accurate, the accuracy of the results should not be very sensitive to the influence of VR coupling and vibrational anharmonicity. The INDECENT method (or, at high energies, the DECENT method) is to be preferred for transitions among low vibrational states, since it is more accurate than quasiclassical methods, can treat both classically allowed and classically forbidden transitions equally well, and is computationally more efficient. Those forms of classical S-matrix theory which are simplest computationally do not appear to be as reliable as the INDECENT method, while the uniform representations, which one might expect to be more accurate, are at present much less efficient computationally for three-dimensional collision problems.

The other useful semiclassical methods involve dynamical approximations (usually the decoupling of the internal and translational motions), the validity of which should be carefully examined before such a method is selected. The CPT approximation used in the SCCP method of Clark,[103] will be valid for high rotational energies and moments of inertia and for potentials in which the long-range distortion of the molecule is small both rotationally and vibrationally. When these conditions are met the SCCP method will probably yield better results than the INDECENT method for transitions between highly excited vibrational states, but poorer results for low vibrational states. The semiclassical coupled-states approach of Billing[55] will be less sensitive than the SCCP method to errors caused by neglect of coupling of the translational and internal modes, because an effective potential is used. The m_l-conservation approximation, however, may make this method more sensitive to error caused by VR coupling in the molecule. The CS method has been found to work best for short-range weakly anisotropic potentials.[3,143] The advantage that both of these methods have is that they permit a quantal description of rotational as well as vibrational transitions, while the rotational motion is treated classically in the INDECENT method. This feature may be important for those experiments which resolve both rotational and vibrational states. It is possible that for low rotational states it is more important to retain a quantal description of rotation than it is to perform the dynamical calculation exactly.[144] For the special case of small-angle scattering and small $\Delta T/T$, a semiclassical IP method such as that of Krüger and Schinke[39] or Collins and Cross[38] may be both accurate and economical, provided that the potential contains no strong long-range anisotropy (which could invalidate the rotational sudden approximation) or strong long-range driving force $\partial V_{\text{int}}/\partial r$ (which could invalidate the vibrational sudden approximation).

A final note of encouragement can be taken from the fact that most of the references given herein are to work carried out during the last few years. The current rapid pace of both theoretical and experimental research on vibrational energy transfer will certainly result soon in the further clarification of issues surrounding the use of approximations and the application of new approaches to realistic energy transfer problems.

ACKNOWLEDGMENTS

The author's research in this area has been supported in part by the National Science Foundation, and by an Alfred P. Sloan Research Fellowship. He is grateful to Professor D. G. Truhlar for his comments on the manuscript.

References

1. D. Rapp and T. Kassal, The theory of vibrational energy transfer between simple molecules in nonreactive collisions, *Chem. Rev.* **69,** 61–102 (1969).
2. D. Secrest, Theory of rotational and vibrational energy transfer in molecules, *Ann. Rev. Phys. Chem.* **24,** 379–406 (1973).
3. J.P. Toennies, The calculation and measurement of cross sections for rotational and vibrational excitation, *Ann. Rev. Phys. Chem.* **27,** 225–260 (1976).
4. W.H. Miller, Classical *S*-matrix for rotational excitation; quenching of quantum effects in molecular collisions, *J. Chem. Phys.* **54,** 5386–5397 (1971).
5. R.G. Gordon, Rational selection of methods for molecular scattering calculations, *Faraday Discuss. Chem. Soc.* **55,** 22–29 (1973).
6. G.D. Barg, G.M. Kendall, and J.P. Toennies, Quasi-classical calculations of elastic and rotationally and vibrationally inelastic differential cross sections for Li^+–H_2, *Chem. Phys.* **16,** 243–268 (1976).
7. P. Pechukas and M.S. Child, Is semiclassical scattering theory accurate for transitions from low-lying vibrational states?, *Mol. Phys.* **31,** 973–987 (1976).
8. H.K. Shin, in *Dynamics of Molecular Collisions, Part A,* W.H. Miller, editor, Plenum Press, New York (1976), pp. 131–210.
9. M.H. Alexander, Inaccuracy of an approximate dynamical treatment of He–H_2 vibrational energy transfer, *Chem. Phys. Lett.* **39,** 485–487 (1976).
10. D.L. Bunker, Classical trajectory methods, *Methods Comput. Phys.* **10,** 287–325 (1971).
11. J.N.L. Connor, Theory of molecular collisions and reactive scattering, *Ann. Rep. A. Chem. Soc.* **70,** 5–30 (1973).
12. R.N. Porter, Molecular trajectory calculations, *Ann. Rev. Phys. Chem.* **25,** 317–355 (1974).
13. R.N. Porter and L.M. Raff, in *Dynamics of Molecular Collisions, Part B,* W.H. Miller, editor, Plenum Press, New York (1976), pp. 1–52.
14. V.B. Cheng, H.H. Suzukawa, and M. Wolfsberg, Investigations of a nonrandom numerical method for multidimensional integration, *J. Chem. Phys.* **59,** 3992–3999 (1973).
15. H.H. Suzukawa, D.L. Thompson, V.B. Cheng, and M. Wolfsberg, Empirical testing of the suitability of a nonrandom integration method for classical trajectory calculations: Comparison with Monte Carlo techniques, *J. Chem. Phys.* **59,** 4000–4008 (1973).
16. D.L. Thompson, Monte Carlo trajectory calculation of the rates of F-atom vibrational relaxation of HF and DF, *J. Chem. Phys.* **57,** 4164–4169 (1972).
17. D.L. Thompson, Monte Carlo trajectory calculation of the rates of H- and D-atom vibrational relaxation of HF and DF, *J. Chem. Phys.* **57,** 4170–4173 (1972).
18. M.K. Matzen and G.A. Fisk, A classical trajectory study of inelastic collisions between highly vibrationally excited KBr and Ar, *J. Chem. Phys.* **66,** 1514–1522 (1977).
19. J.D. Kelley and R.L. Thommarson, Vibrational deactivation and atom exchange in $O(^3P) + CO(X^1\Sigma^+)$ collisions, *J. Chem. Phys.* **66,** 1953–1959 (1977).
20. R.A. LaBudde and R.B. Bernstein, Classical study of rotational excitation of a rigid rotor: $Li^+ + H_2$. II. Correspondence with quantal results, *J. Chem. Phys.* **59,** 3687–3691 (1973).

21. N.C. Blais and D.G. Truhlar, Monte Carlo trajectory study of $Ar + H_2$ collisions. I. Potential energy surface and cross sections for dissociation, recombination and inelastic scattering, *J. Chem. Phys.* **65**, 5335–5356 (1976).
22. D.G. Truhlar and J.W. Duff, Classical probability matrix: prediction of quantum state distributions by a moment analysis of classical trajectories, *Chem. Phys. Lett.* **36**, 551–554 (1975).
23. D.G. Truhlar, Quasiclassical predictions of final vibrational state distributions in reactive and nonreactive collisions, *Int. J. Quantum Chem. Symp.* **10**, 239–250 (1976).
24. R.J. Gordon, A comparison of exact classical and quantum mechanical calculations of vibrational energy transfer, *J. Chem. Phys.* **65**, 4945–4957 (1976).
25. J.M. Bowman and S.C. Leasure, An improved quasiclassical histogram method, *J. Chem. Phys.* **66**, 1756–1757 (1977).
26. S. Chapman and S. Green, Rotational excitation of linear molecules by collisions with atoms: Comparison of classical and quantum methods, *J. Chem. Phys.* **67**, 2317–2331 (1977):
27. J.W. Duff and D.G. Truhlar, Tests of semiclassical treatments of vibrational–translational energy transfer in collinear collisions of helium with hydrogen molecules, *Chem. Phys.* **9**, 243–273 (1975).
28. M. Rubinson, B. Garetz, and J.I. Steinfeld, Energy transfer processes in monochromatically excited iodine. IX. Classical trajectory and semiclassical calculations of vibrationally and rotationally inelastic cross sections, *J. Chem. Phys.* **60**, 3082–3097 (1974).
29. D.L. Thompson, Quasiclassical trajectory studies of reactive energy transfer, *Accts. Chem. Res.* **9**, 338–344 (1976).
30. M. Sathyamurthy and L.M. Raff, Inelastic scattering calculations in polyatomic systems using an *ab initio* potential energy surface: The $CO_2^*(0,0,1,0) + H_2(D_2) \rightarrow CO_2(0,0,0,0) + H_2(D_2)$ systems, *J. Chem. Phys.* **66**, 2191–2211 (1977).
31. H.E. Bass and D.L. Thompson, Vibrational relaxation of Cl_2 by HCl and DCl and self-relaxation of HCl and DCl: A Monte Carlo quasiclassical trajectory study, *J. Chem. Phys.* **66**, 2545–2553 (1977).
32. D.J. Locker and D.J. Wilson, Exact semiclassical transition probabilities for collinear collisions, *J. Chem. Phys.* **52**, 271–278 (1970).
33. R.L. McKenzie, Vibration–translation energy transfer in anharmonic diatomic molecules. I. A comparative evaluation of the semiclassical approximation, *J. Chem. Phys.* **63**, 1655–1662 (1975).
34. R.L. McKenzie, Vibration–translation energy transfer in anharmonic diatomic molecules. II. The vibrational quantum number dependence, *J. Chem. Phys.* **64**, 1498–1508 (1976).
35. M.A. Wartell and R.J. Cross, Semiclassical theory of vibrationally inelastic collisions, *J. Chem. Phys.* **55**, 4983–4991 (1971).
36. M.A. Wartell, Sudden approximation applied to vibrational inelastic scattering: He–HD, *J. Chem. Phys.* **58**, 4700–4701 (1973).
37. J.R. Stallcop, Inelastic scattering in atom–diatomic molecule collisions. II. Effect of rotation on vibrational transitions, *J. Chem. Phys.* **62**, 690–695 (1975).
38. F.S. Collins and R.J. Cross, Vibrationally inelastic scattering at high energies: $H^+ + H_2$, *J. Chem. Phys.* **65**, 644–652 (1976).
39. H. Krüger and R. Schinke, Impact parameter approach to three-dimensional vibrationally inelastic ion(atom)–molecule collisions: Application to $H^+ + H_2$, *J. Chem. Phys.* **66**, 5087–5092 (1977).
40. R. Schinke, H. Krüger, V. Hermann, H. Schmidt, and F. Linder, Vibrational excitation in $H^+ + H_2$ collisions: Comparison between experiment and rotationally sudden impact parameter calculations, *J. Chem. Phys.* **67**, 1187–1190 (1977).
41. R.J. Gordon and A. Kuppermann, String-plucking model for vibrational excitation of molecules, *J. Chem. Phys.* **58**, 5776–5785 (1973).

42. D. Storm and E. Thiele, Collision of an atom and a diatomic: A semiadiabatic approximation, *J. Chem. Phys.* **59,** 5102–5108 (1973).
43. W.A. Cady, Semiclassical treatment of vibrational–translational energy transfer in the near-adiabatic approximation, *J. Chem. Phys.* **61,** 1214–1220 (1974).
44. W.A. Cady, A.P. Clark, and A.S. Dickinson, Comment on "Semiclassical treatment of vibrational–translational energy transfer in the near-adiabatic approximation," *J. Chem. Phys.* **63,** 5505–5506 (1975).
45. B. Ritchie, Theory of small-energy-transfer collisions in dominant long-ranged forces: $H^+ + H_2$ and $e^- - H_2$ vibrational excitation, *Phys. Rev. A* **6,** 1456–1460 (1972); Erratum: *Phys. Rev. A* **7,** 828 (1973).
46. T.A. Dillon and J.C. Stephenson, Multiquantum vibrational-energy exchange, *Phys. Rev. A* **6,** 1460–1468 (1972).
47. R.L. McKenzie, The influence of molecular rotation on vibration–translation energy transfer, *J. Chem. Phys.* **66,** 1457–1474 (1977).
48. G.D. Billing Sorenson, Semiclassical three-dimensional model for vibrational relaxation, *J. Chem. Phys.* **57,** 5241–5250 (1972).
49. H. Rabitz, Effective potentials in molecular collisions, *J. Chem. Phys.* **57,** 1718–1725 (1972).
50. H. Rabitz, in *Dynamics of Molecular Collisions, Part A,* W.H. Miller, editor, Plenum, New York (1976), pp. 33–80.
51. G. Billing Sorenson, Semiclassical three-dimensional scattering theory, *J. Chem. Phys.* **61,** 3340–3343 (1974).
52. G.D. Billing, Semiclassical three-dimensional model for vibrational relaxation. II, *J. Chem. Phys.* **62,** 1480–1487 (1975).
53. G.D. Billing, Semi-classical calculations of rotational/vibrational transitions in He–H_2, *Chem. Phys.* **9,** 359–369 (1975).
54. G.D. Billing, Semiclassical calculations of rotational–vibrational transitions in $Li^+ - H_2$, *Chem. Phys.* **14,** 267–273 (1976).
55. G.D. Billing, The semiclassical coupled states method, *J. Chem. Phys.* **65,** 1–6 (1976).
56. D.J. Wilson and D.J. Locker, Quantum vibrational transition probabilities in atom–diatomic molecule collisions. A tractable three-dimensional model, *J. Chem. Phys.* **57,** 5393–5402 (1972).
57. R.T. Pack, Close coupling test of classical and semiclassical cross sections for rotationally inelastic Ar + N_2 collisions, *J. Chem. Phys.* **62,** 3143–3148 (1975).
58. T.A. Dillon and J.C. Stephenson, Calculation of vibrational and rotational energy transfer between HF, DF, HCl and CO_2, *J. Chem. Phys.* **58,** 2056–2064 (1973).
59. J. Schaefer and W.A. Lester, Effect of rotation on vibrational excitation of H_2 by Li^+ impact, *Chem. Phys. Lett.* **20,** 575–580 (1973).
60. C.F. Giese and W.R. Gentry, Classical trajectory treatment of inelastic scattering in collisions of H^+ with H_2, HD and D_2, *Phys. Rev. A* **10,** 2156–2173 (1974).
61. W.R. Gentry and C.F. Giese, Vibrational excitation in the DECENT approximation: The large-angle scattering of Li^+ from H_2, *J. Chem. Phys.* **62,** 1364–1371 (1975).
62. L.A. Collins and N.F. Lane, Quantum-mechanical calculations of rotational excitation cross sections for HCl, DCl, DF and HF in slow collisions with He, *Phys. Rev. A* **14,** 1358–1367 (1976).
63. A.S. Dickinson and D. Richards, A semiclassical study of the body-fixed approximation for rotational excitation in atom–molecule collisions, *J. Phys. B* **10,** 323–343 (1977).
64. M.H. Alexander, Effective potential study of rotationally–vibrationally inelastic collisions between He and H_2, *J. Chem. Phys.* **61,** 5167–5181 (1974).
65. H. Rabitz and G. Zarur, Vibration–rotation inelasticity in He – H_2, *J. Chem. Phys.* **61,** 5076–5084 (1974).
66. J. Schaefer and W.A. Lester, Theoretical study of inelastic scattering of H_2 by Li^+ on SCF and CI potential energy surfaces, *J. Chem. Phys.* **62,** 1913–1924 (1975).
67. G. Billing Sorenson, Semiclassical three-dimensional model for vibrational energy transfer in diatomic molecules, *Chem. Phys.* **5,** 244–254 (1974).

68. G.D. Billing and E.R. Fischer, VV and VT rate coefficients in H_2 by a quantum-classical model, *Chem. Phys.* **18**, 225–232 (1976).
69. G.D. Billing, Rotational and vibrational relaxation of hydrogen and deuterium, *Chem. Phys.* **20**, 35–42 (1977).
70. H. Udseth, C.F. Giese, and W.R. Gentry, Transition probabilities and differential cross sections for vibrational excitation in collisions of H^+ with H_2, HD and D_2, *Phys. Rev. A* **8**, 2483–2493 (1973).
71. H. Schmidt, V. Hermann, and F. Linder, Spectroscopy of low-energy $H^+ + H_2$ collisions: Rotational and vibrational excitation of H_2, *Chem. Phys. Lett.* **41**, 365–369 (1976).
72. R.T. Skodje, W.R. Gentry, and C.F. Giese, On the use of the sudden approximation for vibrational excitation in high-energy collisions, and the sensitivity of the result to the choice of potential energy surface, *J. Chem. Phys.* **65**, 5532–5533 (1976).
73. F.A. Herrero and J.P. Doering, Vibrational excitation of H_2 by proton impact, *Phys. Rev. A* **5**, 702–712 (1972).
74. W.H. Miller, Classical-limit quantum mechanics and the theory of molecular collisions, *Adv. Chem. Phys.* **25**, 69–177 (1974).
75. W.H. Miller, The classical *S*-matrix in molecular collisions, *Adv. Chem. Phys.* **30**, 77–136 (1975).
76. W.H. Miller, Classical *S*-matrix: Numerical application to inelastic collisions, *J. Chem. Phys.* **53**, 3578–3587 (1970).
77. R.A. Marcus, Theory of semiclassical transition probabilities for inelastic and reactive collisions. V. Uniform approximation in multidimensional systems, *J. Chem. Phys.* **57**, 4903–4909 (1972).
78. J.N.L. Connor and H.R. Mayne, Semiclassical theory of molecular collisions: Collinear atom harmonic oscillator collisions with attractive interactions, *Mol. Phys.* **32**, 1123–1135 (1976).
79. W. Eastes and J.D. Doll, Semiclassical calculation of the harmonic oscillator transition probability for a collinear hard sphere collision, *J. Chem. Phys.* **60**, 297–302 (1974).
80. J.W. Duff and D.G. Truhlar, Classical *S*-matrix: Application to classically forbidden vibrational excitation for He + HBr and H + Br_2, *Chem. Phys.* **17**, 249–254 (1976).
81. R.J. Cross, Sudden and semiclassical approximations in inelastic and reactive scattering, *J. Chem. Phys.* **58**, 5178–5179 (1973).
82. J.D. Doll and W.H. Miller, Classical *S*-matrix for vibrational excitation of H_2 by collision with He in three dimensions, *J. Chem. Phys.* **57**, 5019–5026 (1972).
83. W.H. Miller and A.W. Raczkowski, Partial averaging in classical *S*-matrix theory, *Faraday Discuss. Chem. Soc.* **55**, 45–50 (1973).
84. A.W. Raczdowski and W.H. Miller, Classical *S*-matrix calculation for vibrationally inelastic transitions in three-dimensional collisions of Li^+ with H_2, *J. Chem. Phys.* **61**, 5413–5420 (1974).
85. D. Rapp, *Quantum Mechanics*, Holt, Rinehart, and Winston, New York (1971).
86. L.P. Presnyakov and A.M. Urnov, Quantum transitions between highly excited atomic levels induced by external time-dependent forces, *J. Phys. B* **3**, 1267–1271 (1970).
87. A.P. Clark and I.C. Percival, Vibrational excitation and the Feynman correspondence identity, *J. Phys. B* **8**, 1939–1952 (1975).
88. E.J. Heller, Time-dependent approach to semiclassical dynamics, *J. Chem. Phys.* **62**, 1544–1555 (1975).
89. E.J. Heller, Time-dependent variational approach to semiclassical dynamics, *J. Chem. Phys.* **64**, 63–73 (1976).
90. E.J. Heller, Classical *S*-matrix limit of semiclassical dynamics, *J. Chem. Phys.* **65**, 4979–4989 (1976).
91. W.R. Gentry and C.F. Giese, Application of a classical trajectory model to vibrational excitation in high-energy $H^+ + H_2$ collisions, *Phys. Rev. A* **11**, 90–96 (1975).

92. W.R. Gentry and C.F. Giese, Quantum vibrational transition probabilities from real classical trajectories: Collinear atom–diatom collisions, *J. Chem. Phys.* **63**, 3144–3155 (1975).
93. J.W. Duff and D.G. Truhlar, Use of semiclassical collision theory to compare analytic fits to the interaction potential for vibrational excitation of H_2 by He, *J. Chem. Phys.* **63**, 4418–4429 (1975).
94. R.B. Walker and R.K. Preston, Quantum versus classical dynamics in the treatment of multiple photon excitation of the anharmonic oscillator, *J. Chem. Phys.* **67**, 2017–2028 (1977).
95. R.T. Skodje, W.R. Gentry, and C.F. Giese, Quantum vibrational transition probabilities from real classical trajectories: Symmetric diatom–diatom collisions, *J. Chem. Phys.* **66**, 160–168 (1977).
96. P. McGuire, K. Rudolph, and J.P. Toennies, Comparison of quantum and experimental rotational transition probabilities for $H^+ + H_2$ collisions at 3.7. eV, *J. Chem. Phys.* **65**, 5522–5523 (1976).
97. P. McGuire, H. Schmidt, V. Hermann, and F. Linder, Quantum oscillations in the transition probability for rotational excitation of H_2 by H^+ impact at 4.67 eV, *J. Chem. Phys.* **66**, 4243–4244 (1977).
98. J.D. Kelley, Vibrational energy transfer processes in collisions between diatomic molecules, *J. Chem. Phys.* **56**, 6108–6117 (1972).
99. R.T. Skodje, W.R. Gentry, and C.F. Giese, Quantum vibrational transition probabilities from real classical trajectories: Asymmetric diatom–diatom collisions, *J. Chem. Phys.* in press (1979).
100. I.C. Percival and D. Richards, A correspondence principle for strongly coupled states, *J. Phys. B* **3**, 1035–1046 (1970).
101. A.P. Clark and A.S. Dickinson, Correspondence principle methods applied to a forced harmonic oscillator, *J. Phys. B* **4**, L112–L116 (1971).
102. A.P. Clark, A.S. Dickinson, and D. Richards, The correspondence principle in heavy-particle collisions, *Adv. Chem. Phys.* **36**, 63–139 (1977).
103. A.P. Clark, Simultaneous rotational and vibrational transitions in $He + H_2$ collisions, *J. Phys. B* **10**, L389–L394 (1977).
104. A.P. Clark, A modified correspondence principle for strongly coupled states, *J. Phys. B* **6**, 1153–1164 (1973).
105. W.R. Gentry, Ion–dipole scattering in classical perturbation theory, *J. Chem. Phys.* **60**, 2547–2553 (1974).
106. H. Udseth, C.F. Giese, and W.R. Gentry, Rotational excitation in the small-angle scattering of protons from diatomic molecules, *J. Chem. Phys.* **60**, 3051–3056 (1974).
107. F.H. Heidrich, K.R. Wilson, and D. Rapp, Collinear collisions of an atom and harmonic oscillator, *J. Chem. Phys.* **54**, 3885–3897 (1971).
108. B.H. Mahan, Refined impulse approximation for the collisional excitation of the classical anharmonic oscillator, *J. Chem. Phys.* **52**, 5221–5225 (1970).
109. J.D. Kelley and M. Wolfsberg, Comparison of approximate translational–vibrational energy-transfer formulas with exact classical calculations, *J. Chem. Phys.* **44**, 324–338 (1966).
110. R.I. Morse, Collisions of an atom with a harmonic oscillator, *J. Chem. Phys.* **54**, 4138–4139 (1971).
111. R.I. Morse and R.J. LaBreque, Collinear collisions of an atom and a Morse oscillator: An approximate semiclassical approach, *J. Chem. Phys.* **55**, 1522–1530 (1971).
112. A. Zelechow, D. Rapp, and T.E. Sharp, Vibrational–vibrational–translational energy transfer between two diatomic molecules, *J. Chem. Phys.* **49**, 286–299 (1968).
113. K. Shobatake, S.A. Rice, and Y.T. Lee, ITFITS model for vibration–translation energy partitioning in atom–polyatomic molecule collisions, *J. Chem. Phys.* **59**, 2483–2489 (1973).

114. J.N. Bass, Translation to vibration energy transfer in O + NH_3 and O + CO_2 collisions, *J. Chem. Phys.* **60**, 2913–2921 (1974).
115. M.H. Cheng, M.H. Chiang, E.A. Gislason, B.H. Mahan, C.W. Tsao, and A.S. Werner, Collisional excitation of small molecular ions, *J. Chem. Phys.* **52**, 6150–6156 (1970).
116. R.D. Sharma, Deactivation of bending mode of CO_2 by hydrogen and deuterium, *J. Chem. Phys.* **50**, 919–923 (1969).
117. R.D. Sharma, Transfer of vibrational energy from asymmetric stretch of CO_2 to ν_3 of N_2O, *Phys. Rev. A* **2**, 173–187 (1970).
118. R.D. Sharma and C.A. Brau, Energy transfer in near-resonant molecular collisions due to long-range forces with application to transfer of vibrational energy from ν_3 mode of CO_2 to N_2, *J. Chem. Phys.* **50**, 924–930 (1969).
119. L.-Y. C. Chiu, Vibrational and rotational energy transfer upon molecular collisions, *Chem. Phys.* **16**, 269–279 (1976).
120. J. Krutein and F. Linder, Differential scattering experiments on vibrational excitation in low-energy H^+ + CO_2 collisions, *J. Phys. B* **10**, 1363–1375 (1977).
121. P. Eckelt, H.J. Korsch, and V. Phillip, Energy loss spectra for vibrorotational molecular excitation in an impulsive spectator moldel, *J. Phys. B* **7**, 1649–1665 (1974).
122. H.J. Korsch and V. Phillip, Quantum mechanical impulse approximation for vibrorotational excitation in atom–diatom collisions, *Phys. Rev. A* **13**, 497–499 (1976).
123. T.A. Dillon and J.C. Stephenson, Effect of the straight path approximation and exchange forces on vibrational energy transfer, *J. Chem. Phys.* **58**, 3849–3854 (1973).
124. J.D. Kelley and M. Wolfsberg, Simple expression for "steric factor" in translational–vibrational energy transfer, *J. Chem. Phys.* **50**, 1894–1896 (1969).
125. J.D. Kelley and M. Wolfsberg, Exact classical calculations on collisional energy transfer to diatomic molecules with a rotational and a vibrational degree of freedom, *J. Chem. Phys.* **53**, 2967–2977 (1970).
126. D.J. Kouri and C.A. Wells, Comparisons of Morse and harmonic oscillator models for vibration–rotation excitation of H_2 by Li^+, *J. Chem. Phys.* 60, 2296–2304 (1974).
127. P. McGuire and J.P. Toennies, *A priori* low temperature vibrational relaxation rates for He–H_2, *J. Chem. Phys.* **62**, 4623–4627 (1975).
128. H.J. Korsch and V. Phillip, Collision-induced molecular excitation: Comparison of the collinear model with three-dimensional calculations, *Chem. Phys. Lett.* **31**, 296–300 (1975).
129. M. Faubel and J.P. Toennies, Energy transfer in classical collinear and perpendicular collisions of a structureless atom with a Morse oscillator, *Chem. Phys.* **4**, 36–44 (1974).
130. M.R. Verter and H. Rabitz, Theoretical evaluation of vibrational transition rates and relaxation in CO – He, *J. Chem. Phys.* **64**, 2939–2952 (1976).
131. A.P. Clark and A.S. Dickinson, Collinear collision of an atom and a Morse oscillator: Exact quantum mechanical results, *J. Phys. B* **6**, 164–180 (1973).
132. G.W.F. Drake and C.S. Lin, Semiclassical study of the vibrational excitation of H_2 in collision with He, *J. Phys. B* **7**, 398–405 (1974); Corrigendum: *J. Phys. B* **7**, 2440 (1974).
133. R. Schinke and J.P. Toennies, Semiclassical calculations of vibrational energy transfer: Comparison of the harmonic and the Morse oscillator in collinear and perpendicular collisions with a structureless atom, *J. Chem. Phys.* **62**, 4871–4879 (1975).
134. M.H. Alexander and E.V. Berard, Potential energy surface dependence of vibrationally inelastic collisions between He and H_2, *J. Chem. Phys.* **60**, 3950–3957 (1974).
135. R.A. White and E.F. Hayes, A model potential for vibrational excitation of diatomic molecules, *Chem. Phys. Lett.* **14**, 98–102 (1972).
136. H.K. Shin, Vibrational transitions in atom–diatomic systems. Use of the Lennard-Jones potential, *J. Phys. Chem.* **77**, 1666–1673 (1973).
137. H.K. Shin, Vibrational relaxation of hydrogen fluorides: HF($\nu = 1$) + F→ HF($\nu = 0$) + F, *Chem. Phys. Lett.* **43**, 4–10 (1976).
138. A. Bogan, Impulse approximation for three-dimensional inelastic atom–diatomic molecule collisions, *Phys. Rev. A* **9**, 1230–1241 (1974).

139. R.L. Thommarson, G.C. Berend, and S.W. Benson, Molecular relaxation processes in triatomic molecules. I. T–V transition probabilities in the CO_2–He syste, *J. Chem. Phys.* **54,** 1313–1316 (1971).
140. J.W. Ioup and A. Russek, Vibrational–rotational excitation in atom–diatomic molecule collisions, *Phys. Rev. A* **8,** 2898–2914 (1973).
141. W.L. Dimpfl and B.H. Mahan, Large-angle inelastic scattering of Na^+ by D_2, *J. Chem. Phys.* **60,** 3238–3250 (1974).
142. W.R. Gentry, H. Udseth, and C.F. Giese, Evidence for vibrational excitation in H^+ + CH_4, CD_4 collisions by means of a surface-hopping mechanism, *Chem. Phys. Lett.* **36,** 671–673 (1975).
143. D.J. Kouri, T.G. Heil, and Y. Shimoni, Sufficiency conditions for the validity of the j_z-conserving coupled states approximation, *J. Chem. Phys.* **65,** 1462–1473 (1976).
144. L.L. Poulsen, P.L. Houston, and J.I. Steinfeld, Near-resonant vibration ↔ rotation energy transfer, *J. Chem. Phys.* **58,** 3381–3388 (1973).

13

Electronic Excitation: Nonadiabatic Transitions

M. S. CHILD

1. Introduction

1.1. Physical Considerations

Nonadiabatic transitions are, strictly, those arising from a breakdown of the Born–Oppenheimer approximation, the term adiabatic being reserved for the limit in which the electronic problem is solved with the nuclei fixed. This corresponds to a situation in which, in classical terms, the electrons would pass through many cycles of their periodic or quasiperiodic[1] motion in a time short compared with that required to achieve a significant perturbation to this motion by changing the nuclear geometry.

Similar considerations ultimately apply to any inelastic scattering process, but the special features of the electronic, as distinct from the rotational or vibrational, case are on the one hand the typically high frequencies of the electronic motion favoring weakly nonadiabatic transitions, and on the other hand the strong sensitivity of the electronic wave functions to changes in nuclear geometry, which acts in the opposite sense. Any student of the molecular structure literature[2] is familiar with the number of electronic configurations required to accommodate the changes involved in mapping out even a triatomic potential energy surface.

Two situations are of particular importance. The first arises from an intersection between potential energy curves or surfaces, as typified by the systems $M + X$,[3,4] $M + X_2$[4-6] (X = F, Cl, Br, I), and $A^+ + H_2$[7-10] (A = H, He, Ar). Here the existence of an intersection may be interpreted

---

M. S. CHILD • Theoretical Chemistry Department, University of Oxford, 1 South Parks Road, Oxford OX1 3TG, England.

as arising from a change in the classical electronic motion, such that one or more of its frequency components falls to zero, so that, at that point, any change in nuclear position will be fast compared with this component of the electronic motion. The rate of divergence of the curves or surfaces will therefore govern the range of nuclear separations over which this anomaly in the relative electronic and nuclear velocities persists, and hence over which the nonadiabatic coupling mechanism applies. Clearly, in view of the competition with the electronic frequency, the higher the nuclear velocity, the wider will be the nonadiabatic transition zone. These remarks are placed on a more quantitative footing in Section 3.1.1.

The second important situation arises from a switch from one dominant interaction scheme to another as the nuclear position changes. The prototype in this case would be that of a diatomic molecule showing a transition from Hund's case (e) at large internuclear separations, where the atomic spin–orbit coupling is dominant, to case (a) or (b) at smaller separations where the electronic orbital angular momentum is quenched by the dominance of two-center electron–nuclear and electron–electron interactions. The most striking consequences of this Demkov[11] or perturbed symmetric resonance[12,13] situation are seen when the initial asymptotic separation is small, and when the rate of transition from one coupling scheme to another is high. Examples discussed in the recent literature include intermultiplet transitions in the systems F + Xe,[14] X + H_2 (X = F, Br).[15,16]

Before considering the detailed solution of the above curve-crossing and Demkov coupling problems and the practical application of the results to atom–atom and atom–molecule scattering, it is important to outline the overall formulation of the theory. The first step in Section 2.1 is to write down the Hamiltonian in a body-fixed reference frame, because the electronic coordinates used to describe the wave function are always described with respect to the nuclear positions, rather than to a space-fixed system. This leads naturally to a distinction between angular (or Coriolis) and radial coupling terms, of which the latter are normally of dominant importance, although some well-established Coriolis-dominated processes may be recognized, particularly at energies above those encountered on the molecular scale. Given this division of the Hamiltonian, a distinction is drawn in Section 2.2 in the treatment of the radial coupling, between the adiabatic and so-called diabatic formulations according to the choice of the electronic basis. The adiabatic basis, as the name implies, is the set of Born–Oppenheimer wave functions, parametrically dependent on the nuclear coordinates $\{Q_i\}$; these are coupled by the nuclear kinetic energy terms in the total Hamiltonian. The strictly diabatic basis, on the other hand, would be independent of the nuclear coordinates, and coupled by the electronic Hamiltonian itself; a natural choice, which will be termed the asymptotic diabatic basis, would be the set of separated atom wave functions. At this level the two representations are equivalent in the sense that the scattering problem may be solved with equal accuracy in either, the choice being normally one of computational convenience. The adiabatic formulation has, however, a marked advantage whenever the basis is to be truncated to simplify the computation, because in cases of strong configuration interaction the two lowest adiabatic states may contain significant admixtures of many more than two diabatic states. On the other hand, the diabatic coupling scheme, via the electronic Hamiltonian matrix rather than the nuclear kinetic energy, may be computationally more convenient, particularly in atom–molecule scattering problems (see Section 2.2.4). It is also more readily related to the standard analytical results described for the two-state curve-crossing and Demkov coupling cases in Section 3.1. Smith[17] and Baer[18] have therefore developed an algorithm for transformation from the truncated two-state adiabatic basis to an exactly equivalent locally diabatic two-state basis—in the sense of local to the relevant initial and final states. The connections between these representations, and some implications relevant to any two-state model, are discussed in Sections 2.2.3 and 2.2.4. The concluding section, 2.3, of this part of the chapter gives a brief summary of approximation methods for elements of the electronic Hamiltonian matrix, with particular emphasis on charge transfer processes between neutral species, which have received concentrated attention in the recent past.[4]

The final part contains in Section 3.1 an outline of the essential analytical results applicable to a two-state model with one nuclear degree of freedom, followed in Section 3.2 by an account of their immediate implications for the forms of the inelastic atom–atom differential and total cross sections. Finally, the generalization of these results to the more complicated atom–molecule situation is discussed in Section 3.3. Here an important distinction is drawn between what may be termed "state-based" and "trajectory-based" philosophies. The former, which include close-coupling methods,[19] and the Bauer–Fischer–Gilmore (BFG)[20] and Franck–Condon[21] approximations, start from a conventional vibrational-state expansion for the total vibrational–translational wave function, while the latter, surface-hopping trajectory[22-24] (SHT), methods rely on the addition or quantum superposition[23,24] of electronic state-to-state transition probabilities from a spectrum of relevant surface-hopping trajectories. One point, which has received little attention in the literature, is argued quite strongly; namely, that the optimum choice of approximation method should depend on the ratio of initial translational energy to the relevant vibrational quantum, the reason being an elaboration of the arguments in the first paragraph of this Introduction, except that one is now interested in periods of the vibrational rather than the electronic motion. Thus the high-vibrational-frequency–low-transitional-velocity regime results in a highly oscillatory classical trajectory, giving many periods of the vibrational motion in the time required for any significant change in the translational coordinate, and hence in a set of almost sharply defined vibrational states. In the opposite, low-frequency, high-translational-velocity limit, on the other hand, the translational coordinate may take the system to a vastly different part of the potential surface during one vibrational period, so that the concept of vibrational state loses any meaning. This argues for the use of a state-based method in the former regime and a trajectory-based method in the latter, although the precise connection between the "state" and "trajectory" formulations remains to be investigated.

Those seeking a more detailed mathematical description of the theory are referred to important reviews by Nikitin[25,26] and the papers of Crothers[12,13] and Delos and Thorson.[27] Related material is also given elsewhere by the present author.[28,29] Other recent reviews that may be used to supplement the present discussion have been given by Tully,[30] Janev,[31] and Baede.[4]

1.2. Notation

The following notational conventions will be adopted.

Coordinates

Electronic: $\mathbf{q}$.

Nuclear: global $\mathbf{Q}$; translational $\mathbf{R}$; vibrational $\mathbf{r}$.

Body-fixed coordinates for the atom–diatom system are shown in Figure 1, with z along the translational coordinate $\mathbf{R}$, and xz defining the triatomic plane.

Reduced Mass

Diatom BC:

$$m = m_B m_c/(m_b + m_c)$$

Atom–diatom A, BC:

$$m = m_a(m_b + m_c)/(m_a + m_b + m_c)$$

$$\mu = m_b m_c/(m_b + m_c)$$

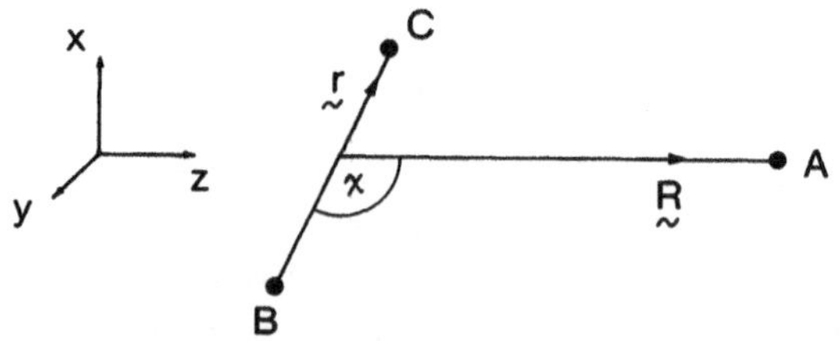

Figure 1. The coordinate system.

Wave Function

Total: $\Psi(\mathbf{q}, \mathbf{Q})$, $\Psi(\mathbf{q}, R)$, $\Psi(\mathbf{q}, R, r)$, ..., etc.
Electronic: adiabatic, $\chi_v(\mathbf{q}; \mathbf{Q})$, $\chi_v(\mathbf{q}, R)$, ..., etc.
diabatic, $\xi_i(\mathbf{q})$
Nuclear: adiabatic, $\phi_v(R)$ or $\eta_{vn}(r)\,\phi_{vn}(R)$
diabatic, $\psi_i(R)$ or $\eta_{in}(r)\,\psi_{in}(R)$

Thus

$$\Psi(\mathbf{q}, R) = \sum_v \chi_v(\mathbf{q}; R)\,\phi_v(R)$$

$$= \sum_i \xi_i(\mathbf{q})\,\psi_i(R)$$

$$\Psi(\mathbf{q}, R, r) = \sum_{v,n} \chi_v(\mathbf{q}; R, r)\,\eta_{vn}(r)\,\phi_{vn}(R)$$

$$= \sum_{i,n} \xi_i(\mathbf{q}; R, r)\,\eta_{in}(r)\,\psi_{in}(R)$$

Potential Surface

$W_v(\mathbf{Q})$, $W_v(R)$, $W_v(R, r)$, etc.

2. *Equations of Motion*

2.1. *The Body-Fixed Hamiltonian*

After separating off the center-of-mass motion, the total Hamiltonian for a diatomic system may be expressed with respect to a rotating coordinate system such that z lies along the internuclear axis, in the form[32]

$$\hat{H}(\mathbf{q}, \mathbf{R}) = \hat{H}_{\text{el}}(\mathbf{q}, R) - \frac{\hbar^2}{2mR^2}\frac{\partial}{\partial R}\left(R^2\frac{\partial}{\partial R}\right)$$
$$+ \frac{1}{2mR^2}\{\hat{J}^2 + \hat{L}^2 - 2\hat{J}_z\hat{L}_z - \hat{J}_+\hat{L}_- - \hat{J}_-\hat{L}_+\} \qquad (1)$$

where $\hat{H}_{\text{el}}(\mathbf{q}, R)$ is the Born–Oppenheimer electronic Hamiltonian, and $\hat{\mathbf{J}}$ and $\hat{\mathbf{L}}$ denote the total and electronic orbital angular momentum (in the presence of unpaired electron spin $\hat{\mathbf{L}}$ is replaced by $\hat{\mathbf{L}} + \hat{\mathbf{S}}$).

The corresponding form for an atom–diatom interaction is more complicated because there are now three contributions to the total angular momentum, namely, those produced by the electronic ($\hat{\mathbf{L}}$), diatomic rotational ($\hat{\mathbf{j}}$) and atom–diatom orbital ($\hat{\mathbf{l}}$) motions. (As in the diatomic case, a fourth contribution $\hat{\mathbf{S}}$ must be added to $\hat{\mathbf{L}}$ in the equations which follow, in which spin effects are significant.) It is also necessary to specify the orientation of the coordinate axes within the triatomic system:

$$\hat{H}(\mathbf{q}, \mathbf{R}, \mathbf{r}) = \hat{H}_{el}(\mathbf{q}, \mathbf{R}, \mathbf{r}) - \frac{\hbar^2}{2mR^2}\frac{\partial}{\partial R}\left(R^2\frac{\partial}{\partial R}\right) - \frac{\hbar^2}{2\mu r^2}\frac{\partial}{\partial r}\left(r^2\frac{\partial}{\partial r}\right)$$
$$+ \frac{\hat{j}^2}{2\mu r^2} + \frac{1}{2mR^2}\{\hat{K}^2 + \hat{j}^2 - 2\hat{K}_z\hat{j}_z - \hat{K}_+\hat{j}_- - \hat{K}_-\hat{j}_+\} \tag{2}$$

where

$$\hat{\mathbf{K}} = \hat{\mathbf{l}} + \hat{\mathbf{j}} + \hat{\mathbf{L}} \tag{3}$$

The general molecule–molecule cáse may be covered by extension of the work of Vezzetti and Rubinow,(33) Miller,(34) Pack,(35) and Baer,(18) but the details will not be included here.

The structure of equations (1) and (2) allows a useful distinction between radial and angular (or Coriolis) nonadiabatic coupling terms, the latter being represented by $(\hat{J}_+\hat{L}_- + \hat{J}_-\hat{L}_+)$ or $(\hat{K}_+\hat{j}_- + \hat{K}_-\hat{j}_+)$, as the case may be. Of these, the Coriolis coupling which acts by causing the electron cloud to drift behind the nuclei as the molecule rotates, is relatively insignificant for scattering from the ground electronic state, at energies on the molecular scale (0–10 eV). It is, however, of major importance in causing transitions between diatomic states differing by substitution of an $np\sigma$ for an $np\pi$ orbital at energies sufficient to penetrate close to the united atom limit. The process

$$H^+ + H(1s) \rightarrow H^+ + H(2s, 2p) \tag{4}$$

involving a transition from the $2p\sigma$, $^2\Sigma_u^+$ to the $2p\pi$, $^2\Pi_u$ state is a classic example.(36,37,38) Thorson(38) has attributed an appreciable nonadiabatic correction ($\sim$2 eV at 0.5 Å) to the effective He_2 potential observed by scattering at 2 keV, again dominated by Coriolis coupling between the $2p\sigma$ and $2p\pi$ orbitals. One other situation in which Coriolis coupling cannot be neglected would arise in the asymptotic description of a system which dissociates into atoms with nonzero angular momentum, because then the resulting states would be strictly quantized with respect to a space-fixed rather than a body-fixed frame. The spectroscopic analog, termed L or S uncoupling(39,40) would correspond with a transition as one increased the internuclear separation from Hund's case (a) or (b), first to case (c) or (d) and then to case (e).

These angular coupling terms are also responsible in first order for heterogeneous ($\Delta\Lambda = \pm 1$, or $\Delta\Omega = \pm 1$) perturbations and predissociations of diatomic spectra, and in second-order for λ-type doubling in the spectrum.(40) Computational techniques specifically designed for these spectroscopic applications, but also applicable to scattering problems, have been developed by Hinkley *et al.*(41)

Attention will be concentrated below on the radial-coupling terms, working unless otherwise stated at $J = 0$ (or $l = 0, j = 0$) in order to avoid the complexities of the angular momentum coupling problem (see Chapters 6–10). It is also convenient to include any purely electronic angular momentum terms, such as $\hat{L}^2/2\mu R^2$ in equation (1), in the electronic Hamiltonian $\hat{H}_{el}$, and to adopt the following forms for the total wave function $\phi(\mathbf{q}, \mathbf{Q})$ in order to remove any first derivative terms from the Schrödinger equation. Thus for the atom–atom case, with $J = 0$,

$$\phi(\mathbf{q}, \mathbf{R}) = R^{-1}(\mathbf{q}, R) \tag{5}$$

with $\Psi(\mathbf{q}, R)$ given by

$$\begin{aligned} \hat{H}_{\text{eff}}\,\Psi(\mathbf{q}, R) &= E(\mathbf{q}, R)\Psi(\mathbf{q}, R) \\ \hat{H}_{\text{eff}} &= \hat{H}_{el}(\mathbf{q}, R) - \frac{\hbar^2}{2m}\frac{\partial^2}{\partial R^2} \end{aligned} \tag{6}$$

while in the atom–diatom case, with $j = l = 0$,

$$\Phi(\mathbf{q}, \mathbf{R}, \mathbf{r}) = (rR)^{-1}\,\Psi(\mathbf{q}, R, r) \tag{7}$$

with $\Psi(\mathbf{q}, R, r)$ taken as an eigenfunction of

$$\hat{H}_{\text{eff}}(\mathbf{q}, \mathbf{R}, \mathbf{r}) = \hat{H}_{el}(\mathbf{q}, \mathbf{R}, \mathbf{r}) - \frac{\hbar^2}{2m}\frac{\partial^2}{\partial R^2} - \frac{\hbar^2}{2\mu}\frac{\partial^2}{\partial r^2} \tag{8}$$

2.2. *Adiabatic and Diabatic Representations*

2.2.1. *Diabatic Representation*

The diabatic representation is obtained by expanding the total wave function in terms of a set of real electronic functions $\{\xi_i(\mathbf{q})\}$ most conveniently defined in the atom–atom case as the eigenfunctions of $\hat{H}_{el}(\mathbf{q}, R)$ at $R \to \infty$, and in the atom–diatom case at $R \to \infty$, $r = r_0$, where r_0 has a fixed value close to the potential energy minima of the relevant diatomic potential energy curves:

$$\Psi(\mathbf{q}, \mathbf{Q}) = \sum_i \xi_i(\mathbf{q})\,\psi_i(\mathbf{Q}) \tag{9}$$

so that on applying $\hat{H}_{\text{eff}}(\mathbf{q}, \mathbf{Q})$, defined by equation (6) in the atom–atom case, for example, and performing the conventional manipulations,

$$\left(\sum_\alpha \frac{\hbar}{2m_\alpha}\frac{\partial}{\partial Q_\alpha^2} + E\right)\psi_i(\mathbf{Q}) = \sum_j H_{ij}(\mathbf{Q})\,\psi_j(\mathbf{Q}) \tag{10}$$

where

$$H_{ij}(\mathbf{Q}) = \langle \xi_i | \hat{H}_{el} | \xi_j \rangle \tag{11}$$

and Q_α denotes one of the nuclear coordinates, with corresponding reduced mass m_α. [$Q_1 = R, m_1 = \mu$ in the atom–atom case, and $(Q_1, Q_2) = (R, r)$, $(m_1, m_2) = (m, \mu)$ in the atom–diatom case, for example.]

By the above definition of the $\{\xi_i(\mathbf{q})\}$,

$$\lim_{R\to\infty} H_{ij}(R) = E_i\,\delta_{ij} \tag{12}$$

where E_i are the sums of atomic energy levels. The atom–diatom equivalent of equation (10) is complicated only to the extent of replacing $(\hbar^2/2\mu)(\partial^2/\partial R^2)$ by $(\hbar^2/2m)(\partial^2\partial R^2) + (\hbar^2/2\mu)(\partial^2/\partial r^2)$, but the limit $R \to \infty$ is less simple, because the matrix $\mathbf{H}(R, r)$ is necessarily diagonal only at $R \to \infty$ and $r = r_0$. The problems raised by this are discussed in Section 2.4.6.

2.2.2. *Adiabatic Representation*

The adiabatic basis, on the other hand, is the set of Born–Oppenheimer electronic functions $\{\chi_\nu(\mathbf{q}; \mathbf{Q})\}$ parametrically dependent on the nuclear coordinates $\mathbf{Q}$, determined by the equation

$$\hat{H}_{\text{el}}(\mathbf{q}, \mathbf{Q})\,\chi_\nu(\mathbf{q}; \mathbf{Q}) = W_\nu(\mathbf{Q})\,\chi_\nu(\mathbf{q}; \mathbf{Q}) \tag{13}$$

where $W_\nu(\mathbf{Q})$ are the electronic potential energy surfaces. It is readily

$$\Psi(\mathbf{q}, \mathbf{Q}) = \sum_\nu \chi_\nu(\mathbf{q}; \mathbf{Q})\,\phi_\nu(\mathbf{Q}) \tag{14}$$

that the adiabatic translational factors must satisfy the following equations:

$$\left[-\sum_\alpha \frac{\hbar^2}{2m_\alpha}\frac{\partial^2}{\partial Q_\alpha^2} + W_\nu(\mathbf{Q}) - E\right]\phi_\nu(\mathbf{Q}) = \sum_\mu\left[\sum_\alpha \tau^\alpha_{\nu\mu}(\mathbf{Q})\frac{\partial}{\partial Q_\alpha} + \rho^\alpha_{\nu\mu}(\mathbf{Q})\right]\phi_\mu(\mathbf{Q}) \tag{15}$$

The coupling terms $\tau^\alpha_{\nu\mu}(\mathbf{Q})$ and $\rho^\alpha_{\nu\mu}(\mathbf{Q})$ depend on the rate of change of the $\chi_\nu(\mathbf{q}; \mathbf{Q})$ with changes in the nuclear geometry:

$$\tau^\alpha_{\nu\mu}(\mathbf{Q}) = \frac{\hbar^2}{m_\alpha}\left\langle \chi_\nu \left| \frac{\partial}{\partial Q_\alpha} \right| \chi_\mu \right\rangle \tag{16}$$

$$\rho^\alpha_{\nu\mu}(\mathbf{Q}) = \frac{\hbar^2}{2m_\alpha}\left\langle \chi_\nu \left| \frac{\partial}{\partial Q_\alpha^2} \right| \chi_\mu \right\rangle \tag{17}$$

These are the nuclear equations of motion in the adiabatic representation, all coupling terms being due to the nuclear kinetic energy.

The asymptotic limits of the $W_\nu(\mathbf{Q})$ as $R \to \infty$ are now sums of atomic energy levels in the atom–atom case, or of atomic energies and diatomic potential curves in the atom–diatom case, but the forms of the coupling terms $\tau^{\alpha}_{\mu\nu}(\mathbf{Q})$ and $\rho^{\alpha}_{\mu\nu}(\mathbf{Q})$ may give rise to difficulties in cases involving internal degrees of freedom, as discussed below.

Of the two types of coupling terms in equation (15), those labeled $\tau^{\alpha}_{\nu\mu}(\mathbf{Q})$ are normally the most important, particularly as the energy increases, because the derivative $\partial/\partial Q_{\alpha}$ generates a term proportional to the nuclear momentum P_{α}. The diagonal terms necessarily vanish in any real electronic basis,[28] and computationally convenient alternative expressions for the off-diagonal elements are given in equations (28) and (29) below.

The physical significance of these nonadiabatic terms may now be considered, specializing to the atom–atom case for simplicity. One is concerned with the relative magnitudes of $|\tau_{\nu\mu}(R)\,\partial/\partial R|$ and the level spacing $|W_{\nu}(R) - W_{\mu}(R)|$, recognizing that the term $\partial/\partial R$ may be assumed under semiclassical conditions to extract the local reduced nuclear momentum $[\mu v(R)/\hbar]$. Hence the condition for neglect of the terms on the right-hand side of equation (15) is that

$$\left| \frac{\langle \chi_{\nu} | \partial/\partial R | \chi_{\mu} \rangle \, \hbar v(R)}{W_{\nu}(R) - W_{\mu}(R)} \right| \ll 1 \tag{18}$$

Now the matrix element $\langle \chi_{\nu} | \partial/\partial R | \chi_{\mu} \rangle$ has the magnitude of the inverse of a distance δR related to the rate of change of the $\chi_{\mu}(\mathbf{q}; R)$ with R. Secondly, the level spacing $|W_{\nu}(R) - W_{\mu}(R)|$ may be used to define an R-dependent frequency component of the electronic motion $\omega_{\nu\mu}(R)$:

$$|W_{\nu}(R) - W_{\mu}(R)| = \hbar\omega_{\nu\mu}(R) \tag{19}$$

Equation (18) therefore reduces to

$$[\delta R \, \omega_{\nu\mu}(R)/v(R)] \gg 1 \tag{20}$$

showing that the time required to travel distance δR at velocity $v(R)$ must be sufficient to allow many periods of the electronic motion. This is a sensible criterion for the definition of a sharply defined electronic state.

Deviations from the adiabatic limit are expected to increase in importance (i) for small δR, implying a rapid change in composition of the electronic wave functions, (ii) for small $\omega_{\nu\mu}(R)$, implying a small level spacing $|W_{\nu}(R) - W_{\mu}(R)|$, and (iii) for large $v(R)$, a high nuclear velocity. The two most important cases in practice are those of a weakly avoided curve crossing, and of Demkov coupling, discussed in detail in Sections 3.1.1 and 3.1.2, both being characterized by a combination of small $\omega_{\nu\mu}(R)$ and small δR.

2.2.3. Diabatic-to-Adiabatic Transformation

The following elements of the standard transformation between diabatic and adiabatic representations may be useful for practical applications. It is assumed that solutions of the Born–Oppenheimer equation, (13), are sought in the form

$$\chi_\nu(\mathbf{q};\mathbf{Q}) = \sum_j \xi_j(\mathbf{q})\, U_{j\nu}(\mathbf{Q}) \tag{21}$$

so that the components $U_{j\nu}$ define an eigenvector of the matrix **H** defined by equation (11):

$$\mathbf{H}(\mathbf{Q})\,\mathbf{U}_\nu(\mathbf{Q}) = \mathbf{U}_\nu(\mathbf{Q})\, W_\nu(\mathbf{Q}) \tag{22}$$

These vectors $\mathbf{U}_\nu$ provide the columns of the orthogonal transformation matrix **U** between the two representations, in terms of which equations (21) and (22) may be written

$$\boldsymbol{\chi} = \tilde{U}\boldsymbol{\xi} \qquad \boldsymbol{\xi} = U\boldsymbol{\chi} \tag{23}$$

$$\tilde{\mathbf{U}}\mathbf{H}\mathbf{U} = \mathbf{W} \tag{24}$$

where $\tilde{\mathbf{A}}$ denotes the transpose of any matrix **A**, $\tilde{\boldsymbol{\xi}}$ and $\tilde{\boldsymbol{\chi}}$ are row vectors, and **W** is a diagonal matrix. Similar transformations between the translational components $\psi_i(\mathbf{Q})$ and $\phi_\nu(\mathbf{Q})$ are implied by the identity

$$\begin{aligned}\Psi(\mathbf{q},\mathbf{Q}) &= \sum_i \xi_i(\mathbf{q})\,\psi_i(\mathbf{Q}) = \tilde{\boldsymbol{\xi}}\boldsymbol{\psi} \\ &= \sum_\nu \chi_\nu(\mathbf{q}_i;\mathbf{Q})\,\phi_\nu(\mathbf{Q}) = \tilde{\boldsymbol{\chi}}\boldsymbol{\phi}\end{aligned} \tag{25}$$

Thus

$$\boldsymbol{\psi} = \mathbf{U}\boldsymbol{\phi}, \qquad \boldsymbol{\phi} = \tilde{\mathbf{U}}\boldsymbol{\psi} \tag{26}$$

or in expanded form

$$\begin{aligned}\psi_i(\mathbf{Q}) &= \sum_\nu \mathbf{U}_{i\nu}(\mathbf{Q})\,\phi_\nu(\mathbf{Q}) \\ \phi_\nu(\mathbf{Q}) &= \sum_\nu \psi_i(\mathbf{Q})\, U_{i\nu}(\mathbf{Q})\end{aligned} \tag{27}$$

Finally the dominant nonadiabatic coupling terms $\tau^\alpha_{\nu\mu}(\mathbf{Q})$ defined by equation (16) may be written

$$\tau^\alpha_{\mu\nu}(\mathbf{Q}) = \frac{\hbar^2}{m_\alpha}\sum_i U_{i\mu}\frac{\partial \tilde{\mathbf{U}}_{\nu i}}{\partial R} \tag{28}$$

This assumes that the Born–Oppenheimer electronic wave functions have been obtained by configuration interaction. It could not be used in conjunction with a Hartree–Fock calculation, for example, because the orbitals used to construct the basis would then vary with the nuclear coordinates, but the following alternative Hellmann–Feynman form would be applicable:

$$\tau^\alpha_{\mu\nu}(Q) = \frac{\hbar^2}{m_\alpha}\frac{\langle \chi_\mu | \partial \tilde{H}_{\mathrm{el}}/\partial Q_\alpha | \chi_\nu\rangle}{|W_\nu(\mathbf{Q}) - W_\mu(\mathbf{Q})|} \tag{29}$$

2.2.4. *Two-State Model*

In the particular case of a two-state model, which is of importance for the following discussion, the matrix $\mathbf{U}(\mathbf{Q})$ may be expressed in terms of a mixing angle $\gamma(\mathbf{Q})$:

$$U(\mathbf{Q}) = \begin{pmatrix} \cos\gamma(\mathbf{Q}) & \sin\gamma(\mathbf{Q}) \\ -\sin\gamma(\mathbf{Q}) & \cos\gamma(\mathbf{Q}) \end{pmatrix} \tag{30}$$

where

$$\tan 2\gamma(\mathbf{Q}) = 2H_{12}(\mathbf{Q})/[H_{11}(\mathbf{Q}) - H_{22}(\mathbf{Q})]$$

and the first and second columns of $\mathbf{U}(\mathbf{Q})$ are associated with the upper and lower adiabatic potential surfaces, respectively:

$$W_\pm(\mathbf{Q}) = \tfrac{1}{2}[H_{11}(\mathbf{Q}) + H_{22}(\mathbf{Q})] \pm \tfrac{1}{2}\{[H_{11}(\mathbf{Q}) - H_{22}(\mathbf{Q})]^2 + 4H_{12}^2(\mathbf{Q})\}^{1/2} \tag{31}$$

The nonadiabatic coupling matrices take the form

$$\boldsymbol{\tau}^\alpha(\mathbf{Q}) = \frac{\hbar^2}{m_\alpha}\frac{\partial\gamma}{\partial Q_\alpha}\begin{pmatrix} 0 & 1 \\ -1 & 0 \end{pmatrix}$$

and

$$\boldsymbol{\rho}^\alpha(\mathbf{Q}) = \frac{\hbar^2}{2m_\alpha}\frac{\partial^2\gamma}{\partial Q_\alpha^2}\begin{pmatrix} 0 & 1 \\ -1 & 0 \end{pmatrix} - \left(\frac{\partial\gamma}{\partial Q_\alpha}\right)^2\begin{pmatrix} 1 & 0 \\ 0 & 1 \end{pmatrix} \tag{32}$$

Knowledge of the dependence of the mixing angle $\gamma(\mathbf{Q})$ on the nuclear coordinates $\{Q_\alpha\}$ is therefore sufficient to determine the magnitude of the coupling.

Two special cases giving rise to a peak in $(\partial\gamma/\partial Q_\alpha)$ are (i) the existence of a curve crossing between the curves obtained by cutting sections along the coordinate Q_α through the diabatic surfaces $H_{11}(\mathbf{Q})$ and $H_{22}(\mathbf{Q})$, and (ii) a sharp increase in $H_{12}(\mathbf{Q})$ due to a change in Q_α, with $H_{11}(\mathbf{Q}) - H_{22}(\mathbf{Q})$ small but roughly constant. The latter is the Demkov coupling case[11] or perturbed symmetric resonance.[12,13] Analytical results for models containing one nuclear degree of freedom, which illustrate this behavior, are outlined in Section 3.1 below.

2.2.5. *Reduced Adiabatic and Diabatic Representations*

It was seen in Section 2.2.1 that the occurrence of strong nonadiabatic coupling is normally associated with a region of close proximity between the two adiabatic potential curves or surfaces. Hence it is frequently sufficient to reduce equations (12) to the relevant pair of coupled equations in the adiabatic representation. This will be termed the reduced adiabatic representation, the adequacy of which could be tested by using equation (18) to estimate the strength of coupling to other surfaces, provided the coupling terms $\langle \chi_\nu | \partial/\partial R | \chi_\mu \rangle$ are available.

Truncation of the diabatic representation is more complicated because retention of only two terms from the original diabatic basis would neglect polarization and configuration interaction contributions, which could markedly affect the composition of the total wave function and the location of the equivalent adiabatic potential surfaces. Smith[17] has therefore devised a procedure for constructing a diabatic formulation equivalent to the reduced adiabatic representation. The information required includes knowledge of the adiabatic potentials $W_\pm(R)$ and the nonadiabatic coupling term $\tau_{+-}(R)$. The algorithm is to specify a reference point R^0, taken in the asymptotic region where the adiabatic and diabatic descriptions coincide, and to invert equation (31) to deduce the variation of the mixing angle from that of $\tau_{+-}(R)$:

$$\gamma(R) = \int_{R^0}^{R} \tau_{+-}(R)\, dR \tag{33}$$

The diabatic potential matrix is then given by

$$\mathbf{H}(R) = \mathbf{U}(R)\,\mathbf{W}(R)\,\tilde{\mathbf{U}}(R) \tag{34}$$

with $\mathbf{U}(R)$ defined in terms of $\gamma(R)$ by equation (29).

The generalization of this procedure to several nuclear degrees of freedom, due to Baer,[18] requires knowledge of the elements $\tau^{\alpha}_{+-}(\mathbf{Q})$ for all the nuclear coordinates Q_α. The function $\gamma(\mathbf{Q})$ is then taken as a line integral along any convenient path from the reference point $\mathbf{Q}^0$ to the point $\mathbf{Q}$ of interest:

$$\gamma(\mathbf{Q}) = \sum_{\alpha} \int_{Q^0_\alpha}^{Q_\alpha} \tau^{\alpha}_{+-}(\mathbf{Q})\, dQ_\alpha \tag{35}$$

2.2.6. *Choice of Representation*

Formally the diabatic and adiabatic representations are completely equivalent, in both their full and reduced forms, but a number of practical considerations may weigh in favor of one or the other.

It is convenient first to consider cases involving only one nuclear degree of freedom, for which the diabatic representation has a special place in that the parameters required for the model analytical descriptions given in Section 3.1 below are diabatic quantities. The absence of first-derivative terms in equation (10) may also make the diabatic representation more convenient. On the other hand, the results of any *ab initio* calculation are normally adiabatic wave functions, but this is largely a formal advantage for the adiabatic representation because the equivalent diabatic form is readily recovered, as described in Sections 2.2.3 and 2.2.5. Furthermore, the adiabatic formulation is subject to a serious computational disadvantage whenever the level separation $|W_\nu(R) - W_\mu(R)|$ becomes very small, as at a narrowly avoided curve crossing, because the coupling term $\tau_{\mu\nu}(R)$ develops a near singularity. Overall there is therefore a preference for the diabatic representation, except possibly when the functions $\tau_{\mu\nu}(R)$ are known to vary slowly with R, compared with the de Broglie wavelength.

The more general problem, involving internal as well as translational nuclear degrees of freedom, is complicated by the possibility that the electronic Hamiltonian $\mathbf{H}(r, R)$ may be diagonal only at a particular point $r = r_0$ in the asymptotic region $R \to \infty$. This argues in favor of the adiabatic representation in order to ensure separability between the relevant initial (and final) states. Situations may occur, however, when only the presence of an approaching atom can cause any mixing between the diatomic diabatic states. In thise cases the off-diagonal element $H_{ij}(r, R)$ vanishes accidentally as $R \to \infty$, and the nonadiabatic coupling terms $\tau_{ij}^{\alpha}(\mathbf{Q})$ will diverge as $R \to \infty$.[42] The diabatic representation is therefore more convenient. Further complications could arise in any general multistate treatment if some but not all of the off-diagonal elements H_{ij} vanished as $R \to \infty$, but these remain to be investigated.

Another complication, peculiar to the reduced diabatic representation in the reactive case, is that the definition of $\gamma(\mathbf{Q})$ given by equation (35) does not ensure any particular value of this mixing angle in the products region. Hence it is necessary to apply a further transformation based on the final angle $\gamma(\mathbf{Q}^P)$ in order to transform back to the uncoupled adiabatic representation in the products asymptotic region. Similar transformations at intermediate reference points may also be computationally convenient in order to remove embarrassingly large off-diagonal diabatic coupling terms $H_{12}(\mathbf{Q})$.[42]

2.3. Nonadiabatic Coupling Mechanisms

Three important nonadiabatic coupling mechanisms may be recognized: configuration interaction, spin–orbit coupling, and angular momentum decoupling. In each case one is interested more in the rate of change of the electronic wave function than in the absolute magnitude of the coupling involved. In other words, the separations between the diagonal elements of the diabatic electronic Hamiltonian matrix are at least as important as the magnitudes of the off-diagonal terms.

2.3.1. Configuration Interaction

The occurrence of configuration interaction is a familiar feature of *ab initio* electronic structure calculations,[2] but application of the full *ab initio* method over a useful region of configuration space will rarely be feasible except for systems involving two or three light atoms. Recourse must therefore be made to one or the other of the semiempirical methods discussed in Chapter 3 of this volume. Of these the diatomics-in-molecules method,[10,43,44] which is readily parametrized to cover the entire energetically accessible region of configuration space, has been applied to the calculation of nonadiabatic transition probabilities in the systems $H^+ + H_2$,[42] $H^+ + D_2$,[22,24] $Ar^+ + H_2$,[45,46] $F + H_2$, and $F + Br_2$.[15] This requires knowledge of the diatomic potential energy curves correlating with all relevant pairs of atomic states, and the extent of any configuration interaction within each diatomic. The curves themselves may frequently be available from diatomic spectroscopy, but an *ab initio* valence bond calculation may be required for any diatomic requiring the inclusion of more than one state of a given symmetry, in order to assess the configuration interaction. The formulation given in Chapter 3 leads naturally [in equation (75)] to the diatomic curves, but an *ab initio* valence bond calculation may be required to determine coupling matrix elements.

One other situation, which has received concentrated attention,[47-53] is the interaction between covalent and ionic configurations leading to the charge transfer process

$$M + X \rightarrow M^+ + X^- \tag{36}$$

Smirnov[47,48] has argued, in the atom–atom case, that at the separation at which the covalent, $H_{11}(R)$, and ionic, $H_{22}(R)$, potential curves intersect, the off-diagonal coupling term $H_{12}(R)$ must depend only on the exponential tails of the donor and acceptor orbitals, the forms of which may be deduced from the ionization potential I_M and electron affinity E_X. This has led Olson *et al.*[49] to suggest the following empirical expression:

$$H_{12}(R) = \tfrac{1}{2}\nu\gamma R^* \exp(-0.86R^*) \tag{37}$$

where

$$\nu = (2I_M)^{1/2} \qquad \gamma = (2E_X)^{1/2} \qquad R^* = 0.5(\nu + \gamma)R \tag{38}$$

all distances and energies being given in atomic units. The exponent 0.86 was obtained by a least-squares fit to a semilog plot of 99 measured or calculated values covering a range of 11 orders of magnitude in $H_{12}(R)$, although many of the calculated values were obtained from wave functions without the correct asymptotic form. Equation (37) may therefore be regarded as semiquantitatively reliable. A straightforward procedure for obtaining a more accurate estimate has been described by Grice and Herschbach[50] and Adelman and Herschbach.[51]

Finally, Los and co-workers[52,53] have modified the form of equation (37) for use in atom–diatom calculations, by introducing the following dependence on the diatomic bond length r and orientation χ (see section 1.2):

$$H_{12}(R, r) = \frac{c_1}{2}\, \nu\gamma(r)\, R^*(r) \exp[-c_2 R^*(r)] \cos\chi \tag{39}$$

where $\gamma(r)$ and $R^*(r)$ are defined by (38), but with the vertical electron affinity $E_{X_2}(r)$ in place of E_X. This results in a decrease in $H_{12}(R, r)$ as r increases, due to the increase in $E_{X_2}(r)$.

2.3.2. *Spin–Orbit Coupling*

The calculation of off-diagonal spin–orbit coupling matrix elements has been extensively developed for applications to diatomic spectroscopy.[54-56] As a cruder alternative methods have been suggested[57-59] for incorporating knowledge of the atomic spin–orbit coupling strength within the diatomics-in-molecules method, but this inevitably neglects any two-center contributions to the spin–orbit coupling.

2.3.3. *Angular Momentum Decoupling*

The term angular momentum decoupling is used here to describe the transition from one angular momentum scheme to another, typified for example by the quenching of spin–orbit coupling due to increasing electrostatic interactions as the internuclear separation decreases. The onset of such decoupling would be marked by the point at which the Σ–Π separation between states correlating with the same separated atom states becomes comparable with the spin–orbit coupling. A simple procedure for the calculation of such $\Delta\Lambda$ separations remains to be developed, however. It is normally assumed to vary exponentially with the particle separation.

3. *Nonadiabatic Transition Probabilities and Cross Sections*

3.1. *Two-State Models in One Degree of Freedom*

The two-state problem in one degree of freedom is most conveniently formulated from the diabatic viewpoint, using the equations

$$\left[\frac{\hbar^2}{2m}\frac{d^2}{dR^2} + E - H_{11}(R)\right]\psi_1(R) = H_{12}(R)\psi_2(R)$$
$$\left[\frac{\hbar^2}{2m}\frac{d^2}{dR^2} + E - H_{22}(R)\right]\psi_2(R) = H_{21}(R)\psi_1(R) \qquad (40)$$

where the diagonal terms $H_{ii}(R)$ are assumed (when relevant) to include the centrifugal term $l(l+1)\hbar^2/2mR^2$. It is also assumed in order to avoid ambiguity, that $H_{11}(R) < H_{22}(R)$ in the limit $R \to \infty$ (see Figure 2).

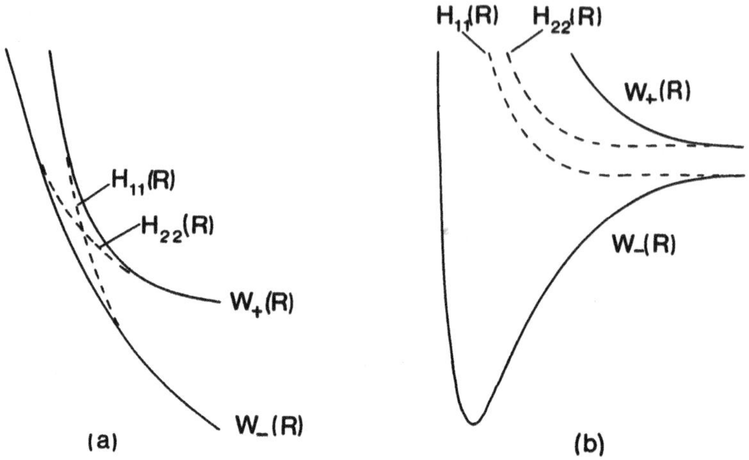

Figure 2. Adiabatic (solid line) and diabatic (dashed line) potential curves for (a) the curve-crossing model and (b) the Demkov (perturbed symmetric resonance) model.

3.1.1. Curve Crossing

The simplest curve-crossing model assumed by Landau,(60) Zener,(61) and Stückelberg,(62) and subsequently analyzed by a variety of authors,(12,25,26,27,63-69) is based on the equations

$$H_{11}(R) - H_{22}(R) = -(F_1 - F_2)(R - R_x)$$
$$H_{12}(R) = H_{12}(R_x) = \text{const} \qquad (41)$$

with the additional assumption that the radial velocity may be treated as constant over the relevant transition region, with the value

$$v = \{2[E - H_{ii}(R_x)]/m\}^{1/2} \qquad (42)$$

the limits of this transition region being set according to equation (20) by the distance ΔR from the crossing point R_x traveled by the particle during one period of the internal motion. But in this linear potential model,

$$\hbar\omega(R) = H_{22}(R) - H_{11}(R) = (F_1 - F_2)\Delta R \qquad (43)$$

Hence the equation for boundaries of the transition region

$$\omega(R)\,\Delta R/v \gg 1$$

becomes

$$\Delta R \gg [\hbar v/(F_1 - F_2)]^{1/2} \qquad (44)$$

Bates(63) first emphasized the implications of this velocity dependence, which must lead to a breakdown of the model at sufficiently high energies, in any practical case.

To the extent, however, that equations (41) and (42) are valid over this region, equations (40) may be solved by a variety of techniques to yield

$$S_{12} = 2ie^{-\pi\delta}(1 - e^{-2\pi\delta})^{1/2}\sin(\Gamma + \chi)\exp(i\eta_+ + i\eta_-) \tag{45}$$

where

$$\delta = H_{12}^2(R_x)/\hbar v(F_1 - F_2) \tag{46}$$

$$\Gamma = \int_{a_+}^{R_x} k_+(R)\,dR - \int_{a_-}^{R_x} k_-(R)\,dR \tag{47}$$

$$\chi = \arg\Gamma(i\delta) + \delta - \delta\ln\delta + \pi/4 \tag{48}$$

$$\eta_\pm = \lim_{R\to\infty}\int_{a_\pm}^{R} k_\pm(R)\,dR - k_\pm(\infty)R + (l + \tfrac{1}{2})\pi/2 \tag{49}$$

with $k_\pm(R)$ dependent on the adiabatic curves $W_\pm(R)$ defined by equation (31):

$$k_\pm^2(R) = 2m[E - W_\pm(R)]/\hbar^2 \tag{50}$$

and with $a_\pm$ used to denote the turning points on $W_\pm(R)$. The phase correction term χ varies monotonically between $\chi = -\pi/4$ for $\delta = 0$ and $\chi = -(1/12\delta)$ for $\delta \gg 1$. Expressed in terms of Γ and δ, the validity of equations (45)–(49) requires that[25]

$$\Gamma \gg 1 \quad \text{and} \quad (\Gamma/\delta) \gg 5 \tag{51}$$

These conditions break down even for the Landau–Zener model at energies close to the curve-crossing because $\Gamma \to 0$, but the following perturbation formula valid for $\delta \ll 1$ may be employed in this region[70,71]:

$$S_{12} = -2i(2\delta)^{1/2}\zeta^{1/4}\,\mathrm{Ai}(-\zeta)\exp(i\eta_+ + i\eta_-) \tag{52}$$

where

$$\zeta = (-\tfrac{3}{2}\Gamma)^{2/3} \tag{53}$$

By this definition ζ depends via Γ (which is a negative quantity) on the adiabatic potentials, although the normal derivation[71] results in an expression for ζ dependent on the diabatic potential curves to which (53) reduces when $H_{12}(R_x) = 0$. This small phase correction, which ensures continuity between equation (45) and the asymptotic form of equation (52), is expected to give a significantly more accurate representation of the phase variation of S_{12}.

More accurate but computationally less convenient extensions of the theory have also been developed to handle curve-crossing situations involving nonlinear potentials and a variable interaction $H_{12}(R)$.

The major correction is to replace equations (46) and (47) by the

Stückelberg forms[12,62]

$$\delta = \frac{1}{2\pi} \operatorname{Im} \int_{R_c}^{R_c^*} [k_-(R) - k_+(R)]\, dR \simeq \frac{1}{2\pi} \operatorname{Im} \frac{1}{\hbar v} \int_{R_c}^{R_c^*} [W_+(R) - W_-(R)]\, dR \tag{54}$$

$$\Gamma = \operatorname{Re}\left[\int_{a_+}^{R_c} k_+(R)\, dR - \int_{a_-}^{R_c} k_-(R)\, dR \right] \tag{55}$$

where R_c and R_c^* are the complex conjugate roots of the equation

$$[H_{11}(R) - H_{22}(R)]^2 + 4H_{12}^2(R) = 0 \tag{56}$$

These may be recognized according to equations (31) and (32) as points at which the adiabatic terms $W_\pm(R)$ intersect and the nonadiabatic coupling $\tau_{+-}(R)$ diverges, because $(d\gamma/dR) \to 0$, where $\gamma(R)$ is the angle used to represent the transformation between diabatic and adiabatic electronic states.

It is readily verified by use of equations (41) that in the Landau–Zener model

$$R_c = R_x + 2iH_{12}(R_x)/(F_1 - F_2) \tag{57}$$

and hence that equation (54) reduces to equation (46). This also shows that $R_x = \operatorname{Re}(R_c)$ in the simple model, thereby reconciling equations (47) and (55), but there is evidence from a recent computation by Faist and Levine[72] that equation (47) for the Stückelberg phase Γ may be justified in general only by replacing the upper integration limit R_x by the point at which the nonadiabatic coupling function $\tau_{12}(R)$ reaches its maximum. This prescription must be approximately equivalent to the use of equation (55), without the necessity to perform an analytical continuation of the elements $H_{ij}(R)$ to locate the complex roots of equation (56). This discrepancy between R_x and $\operatorname{Re}(R_c)$ may also account for a small error in the phase of the single-passage diagonal Landau–Zener S matrix elements in the diabatic representation, discussed by Grosser and de Vries.[73]

3.1.2. Demkov Coupling

The Demkov coupling[11] or perturbed symmetric resonance[12,13,26] case is characterized by a sharp increase in the off-diagonal coupling $H_{12}(R)$ between two diabatic terms $H_{11}(R)$ and $H_{22}(R)$ separated by a small and roughly constant energy. This results in a transition in the composition of the adiabatic wave functions from pure to mixed diabatic states as R decreases from infinity. In other words the mixing angle $\gamma(R)$ characterizing the transformation between the diabatic and adiabatic states changes from $\gamma = 0$ at $R \to \infty$ to a limiting value $\gamma(0) = \pi/4$, as distinct from the change from $\gamma = 0$ to $\gamma = \pi/2$ in the curve-crossing case. This alters the treatment of the boundary conditions in seeking solutions to equation (40), but it may still lead to a sharp peak in $(d\gamma/dR)$ and hence to strong nonadiabatic coupling.

The theory has been developed in its most general form by Crothers[12,13] and Nikitin.[26] Crothers[12,13] considers a general model, with no restriction on the Hamiltonian matrix $\mathbf{H}(R)$ except (i) that there should be only two significant complex roots, (R_c, R_c^*) of equation (56), at which the two adiabatic surfaces intersect, and (ii) that $H_{12}(R) \gg |H_{11}(R) - H_{22}(R)|$ as $R \to 0$. In other words, $\gamma(0) = \pi/4$. The real parts of R_c and R_c^*, which fix the location of the transition zone, therefore lie roughly at the point at which $H_{12}(R) = |(H_{11}(R) - H_{22}(R)|$. In this model[12;13]

$$S_{12} = i \operatorname{sech} \pi\delta \sin\Gamma \exp(i\eta_+ + i\eta_-) \tag{58}$$

with δ and Γ again given in terms of R_c and R_c^* by equations (54) and (55).

Nikitin,[26] on the other hand, considers the model

$$H_{12}(R) = \tfrac{1}{2} A \sin\theta e^{-\alpha R}, \qquad H_{22}(R) - H_{11}(R) = \Delta - A\cos\theta e^{-\alpha R} \tag{59}$$

allowing for variation in the limiting mixing angle, $\gamma(0) = \theta/2$, but with a restriction to the exponential forms. This yields

$$S_{12} = \frac{2\sin\Gamma\{\sinh[\pi\xi(1-\cos\theta)]\sinh[\pi\xi(1+\cos\theta)]\}^{1/2}\exp(i\eta_+ + i\eta_-)}{\exp(-\pi\xi\cos\theta)\sinh 2\pi\xi} \tag{60}$$

where

$$\xi = \Delta/2\alpha\hbar v \tag{61}$$

v being the velocity in the transition region.

Finally both equations (58) and (60) may be shown[13,26] to reduce to the Demkov[11] form:

$$S_{12} = i \operatorname{sech}(\pi\Delta/2\alpha\hbar v) \sin\Gamma \exp(i\eta_+ + i\eta_-) \tag{62}$$

in the simpler model

$$H_{12}(R) = Ae^{-\alpha R}, \qquad H_{22}(R) - H_{11}(R) = \Delta = \text{const} \tag{63}$$

The argument from equation (60) merely involves setting $\theta = \pi/2$, but in going from equation (58) to (62) it is necessary to perform the integral specified by equation (54).

3.2. *Inelastic Atom–Atom Scattering*

The analysis in Section 3.1 allows a relatively simple interpretation of the main features of the experimental inelastic scattering cross sections provided they fall into one or the other of the curve-crossing or Demkov categories. More complicated situations can arise,[63] however, showing possibly both types of contribution to the scattering matrix, and these must be handled by special methods.

3.2.1. *Differential Cross Section*

The main features of the inelastic differential cross section

$$\frac{d\sigma_{12}}{d\Omega} = |f_{12}(\theta)|^2 \tag{64}$$

may be understood by performing a stationary phase[74-77] reduction of the scattering amplitude

$$f_{12}(\theta) = (2ik_1)^{-1} \sum_{l=0}^{\infty} (2l+1)[S_{12}(l) - 1] P_l(\cos\theta) \tag{65}$$

The analysis is simplified by introducing the half-odd angular momentum label

$$\lambda = l + \tfrac{1}{2} \tag{66}$$

and by recognizing that both the Landau–Zener and Demkov expressions for $S_{12}(l)$ [see equations (45) and (58)] take the characteristic form

$$S_{12}(l) = S_{12}^0(\lambda) \sin\tau(\lambda) \exp[i\eta_+(\lambda) + i\eta_-(\lambda)] \tag{67}$$

where $S_{12}^0(\lambda)$ is a relatively slowly varying function of λ, with an exponential cutoff at

$$\lambda_{\max} = [2m(E - E_0)]^{1/2} R_0/\hbar \tag{68}$$

where R_0 is the center of the transition zone, close to R_x or to $\mathrm{Re}(R_c)$ in the Landau–Zener or Demkov cases, respectively, and E_0 is the mean potential energy at this point:

$$E_0 = \tfrac{1}{2}[H_{11}(R_0) + H_{22}(R_0)] \tag{69}$$

The presence of two complex exponential contributions to $S_{12}(l)$ implies two branches to the scattering amplitude, corresponding, following the classic work of Ford and Wheeler,[74,75] to the existence of two inelastic deflection functions $\chi_a(\lambda)$ and $\chi_b(\lambda)$ defined by the equations[76,77]

$$\chi_{a,b}(\lambda) = \frac{\partial}{\partial\lambda}[\eta_+(\lambda) + \eta_-(\lambda) \pm \tau(\lambda)] \tag{70}$$

with a and b used to denote upper and lower signs in equation (70), and in the equations which follow. The physical interpretation is that $\chi_a(\lambda)$ may be seen in the light of equations (49) and (55) to arise from a switch from motion under the lower adiabatic potential $W_-(R)$ to the upper potential $W_+(R)$ on the inward part of the trajectory, while the second function $\chi_b(\lambda)$ arises from a similar switch during outward motion.

To the extent that $\chi_a(\lambda)$ and $\chi_b(\lambda)$ both decrease monotonically with λ, the differential cross section shows the simple two-branch Stückelberg interference pattern illustrated in Figure 3, which may be described at the stationary phase level by the equations[76,77]

$$f_{12}(\theta) = f_a(\theta)\, e^{i\gamma_a(\theta)} - f_b(\theta)\, e^{i\gamma_b(\theta)} \tag{71}$$

where, with $v = a, b$,

$$\begin{aligned} f_v(\theta) &= \frac{S^0_{12}(\lambda_v)}{k_1}\left(\frac{\lambda_v}{\sin\theta\,|d\chi_v/d\lambda|}\right)^{1/2} \\ \gamma_v(\theta) &= \eta_+(\lambda_v) + \eta_-(\lambda) \pm \tau(\lambda_v) - \lambda_v\theta \end{aligned} \tag{72}$$

the value of λ_v being the angular momentum at which $\chi_v(\lambda)$ coincides with the scattering angle θ of interest:

$$\chi_v(\lambda_v) = \theta \tag{73}$$

A more sophisticated uniform version of equation (71) has been given by Child and Gerber,[78] on the assumption that $S_{12}(\lambda)$ may be expressed in

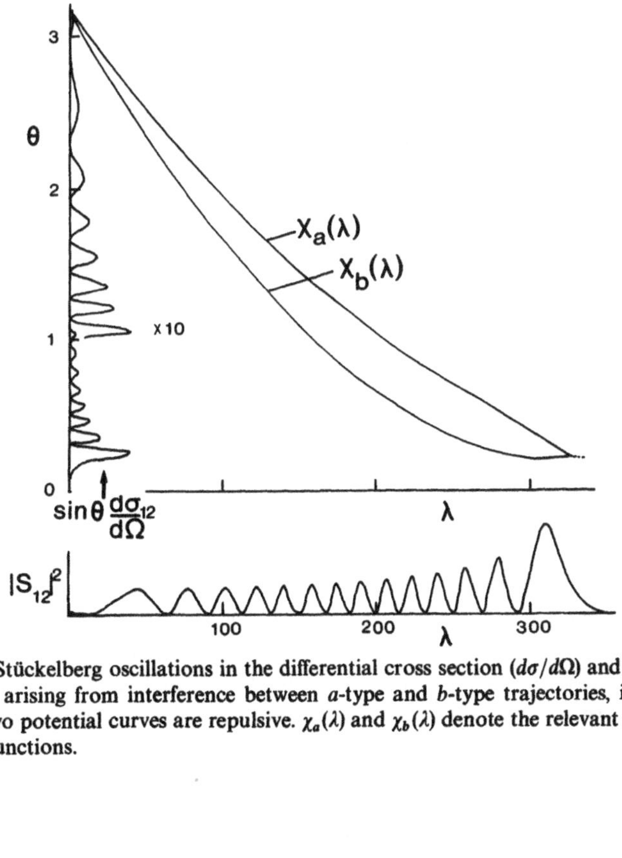

Figure 3. Stückelberg oscillations in the differential cross section ($d\sigma/d\Omega$) and in the S matrix $|S_{12}|^2$ arising from interference between a-type and b-type trajectories, in the case that the two potential curves are repulsive. $\chi_a(\lambda)$ and $\chi_b(\lambda)$ denote the relevant classical deflection functions.

the form

$$S_{12}(\lambda) = A(\lambda)\,\mathrm{Ai}[-\zeta(\lambda)]\exp[i\eta_+(\lambda) + i\eta_-(\lambda)] \tag{74}$$

where, consistent with equation (67),

$$\zeta(\lambda) = [3\tau(\lambda)/2]^{\frac{2}{3}} \tag{75}$$

The result may be written

$$f_{12}(\theta) = \frac{e^{i\bar{\gamma}(\theta)}}{2k_1(\sin\theta)^{1/2}}[(\pi_a^{1/2} + \pi_b^{1/2})\,\xi^{1/4}\,\mathrm{Ai}(-\xi) - i(\pi_a^{1/2} - \pi_b^{1/2})\,\xi^{-1/4}\,\mathrm{Ai}'(-\xi)] \tag{76}$$

where in the notation of equations (72) and (73),

$$\begin{aligned} \bar{\gamma}(\theta) &= \tfrac{1}{2}[\gamma_a(\theta) + \gamma_b(\theta)] \\ \xi(\theta) &= \{\tfrac{3}{4}[\gamma_a(\theta) - \gamma_b(\theta)]\}^{2/3} \end{aligned} \tag{77}$$

and

$$\pi_v(\theta) = \lambda_v A(\lambda_v)\,\zeta^{1/2}(\lambda_v)/\chi_v'(\lambda_v) \tag{78}$$

with $\chi_v'(\lambda)$ used to denote $d\chi_v/d\lambda$.

An interesting phase conservation theorem arises from the total enclosure of the region in Figure 3 between the two curves $\chi_a(\lambda)$ and $\chi_b(\lambda)$. In the first place the phase difference $2\tau(\lambda)$ between the two branches of $S_{12}(l)$ given by equation (67) must vanish at the cutoff angular momentum $\lambda = \lambda_{\max}$. Hence on integrating equation (70)

$$\tau(\lambda) = \frac{1}{2}\int_{\lambda}^{\lambda_{\max}} [\chi_a(\lambda) - \chi_b(\lambda)]\,d\lambda \tag{79}$$

with the interpretation that the phase of oscillations of $|S_{12}(l)|^2$ is obtained by integrating the vertical separation $\chi_a(\lambda) - \chi_b(\lambda)$ in Figure 3. Equally, however, the phase difference governing the interference structure of $|f_{12}(\theta)|^2$ may be expressed as

$$\gamma_a(\theta) - \gamma_b(\theta) = \int_{\lambda_a}^{\lambda_{\max}} \chi_a(\lambda)\,d\lambda - \int_{\lambda_b}^{\lambda_{\max}} \chi_b(\lambda)\,d\lambda - (\lambda_a - \lambda_b)\,\theta \tag{80}$$

which may be seen in light of equation (73) as arising from integration over the horizontal displacement between $\chi_a(\lambda)$ and $\chi_b(\lambda)$. It follows on taking the limit $\lambda \to \lambda_a \to \lambda_b \to 0$ and $\theta \to \pi$ in equations (68) and (69) that $\tau(0)$ and $\gamma_a(\pi) - \gamma_b(\pi)$, which fix the total number of oscillations in $|S_{12}(\lambda)|^2$ and $|f_{12}(\theta)|^2$, respectively, are both given by the total area enclosed between the curves $\chi_a(\lambda)$ and $\chi_b(\lambda)$ in Figure 2. This equality is readily confirmed by counting the maxima in $|S_{12}(\lambda)|$ and $|f_{12}(\theta)|^2$ in Figure 2.

The general case is more complicated, however, because one or both of the deflection functions $\chi_v(\lambda)$ may pass through a maximum or minimum,

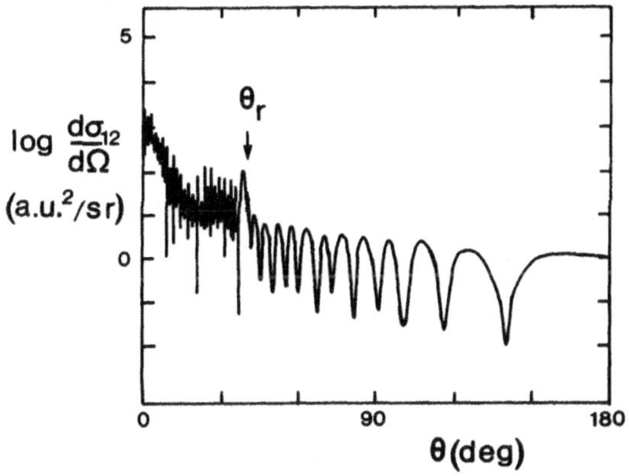

Figure 4. Computed interference structure in the differential cross section, for the charge neutralization, $Li^+ + I^- \rightarrow Li + I$, at $E = 7.292$ eV, adapted from Faist and Levine.[72] Note the appearance of Stückelberg oscillations for angles greater than the "ionic" rainbow angle θ_r. Similar results for other systems are reported by Delvigne and Los,[77] and Faist and Bernstein.[80]

leading to a mixture of Stückelberg and rainbow (see Chapters 4 and 5) interference patterns.[72,76,77,79,80] The simple two-branch Stückelberg pattern emerges, however, from the resulting complicated structure at angles greater than the rainbow angle, as shown in Figure 4.

3.2.2. *Total Cross Section*

The energy dependence of the total cross section

$$\sigma_{12}(E) = \frac{2\pi}{k_1^2} \sum_{l=0}^{\infty} (2l + 1) |S_{12}(l)|^2 \simeq \frac{2\pi}{k_1^2} \int_0^{\infty} |S_{12}(\lambda)|^2 \, d\lambda^2 \tag{81}$$

may be obtained in both the Landau–Zener and Demkov cases, by applying a random-phase approxmation according to which the term $\sin^2\tau(\lambda)$ derived from equation (67) is replaced by an average value of 1/2, and by noting that according to equations (46) and (61) the λ dependence of $S_{12}^0(\lambda)$ arises only from the λ dependence of $\delta(l)$, which may be parametrized in the form

$$2\pi\delta(l) = (v^*/v)\,[1 - (\lambda/\lambda_{\max})^2]^{-1/2} \tag{82}$$

where v is the initial collision velocity and v^* is a characteristic parameter given in the Landau–Zener and Demkov cases, respectively, by

$$v_{LZ}^* = 2\pi H_{12}^2(R_x)/\hbar(F_1 - F_2)\,[1 - (E/E_0)]^{1/2} \tag{83}$$

and

$$v_D^* = \pi\Delta/\alpha\hbar\,[1 - (E/E_0)]^{1/2} \tag{84}$$

It follows on introducing

$$x = 2\pi[1 - (\lambda/\lambda_{max})^2]^{-1/2} \tag{85}$$

as the integration variable in equation (81), and using the forms for $|S_{12}^0(\lambda)|^2$ implied by equations (45) and (58), that

$$\sigma_{12}(E) = \pi R_0^2[1 - (E_0/E)]\, G(v/v^*) \tag{86}$$

with the function $G(v/v^*)$ given in the Landau–Zener and Demkov cases by

$$G_{LZ}(\alpha) = 4\int_1^\infty e^{-x/\alpha}(1 - e^{-x/\alpha})\, x^{-3}\, dx \tag{87}$$

$$G_D(\alpha) = \int_1^\infty \operatorname{sech}^2(x/2\alpha)\, x^{-3}\, dx \tag{88}$$

The characteristic behavior of these two functions is illustrated in Figure 5. They are identical for sufficiently small reduced velocities, $\alpha = v/v^*$, such that over the entire integration region

$$4e^{-x/\alpha}[1 - e^{-x/\alpha}] \simeq \operatorname{sech}^2(x/2\alpha) \simeq 4e^{-x/\alpha} \tag{89}$$

but $G_{LZ}(\alpha)$ vanishes for large α while $G_D(\alpha)$ tends to 1/2 because the integrand in (87) vanishes for small x/α, while $\operatorname{sech}^2(x/2\alpha) \to 1$ as $x/\alpha \to 0$.

The practical consequences of equation (86) and Figure 4, are that the magnitude of $\sigma_{12}(E)$ depends on the radius R_0 of the transition region, from which it may be possible to estimate E_0 by use of equation (69). The residual behavior $G(v/v^*)$ may be used to determine the parameter v^*, and hence according to equations (83) and (84), the coupling strength $H_{12}(R_x)$ and the range α of the coupling term in the Landau–Zener and Demkov

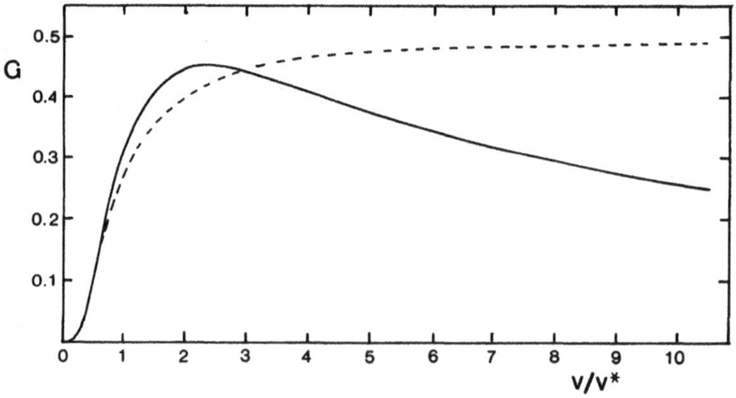

Figure 5. Reduced total cross sections $G(v/v^*)$ for the Landau–Zener (solid line) and Demkov (dashed line) models.

cases, respectively. Convenient properties of the functions $G(v/v^*)$ for this purpose are that $G_{LZ}(v/v^*)$ takes its maximum value 0.452 at $v/v^* = 2.358$, and that $G_D(v/v^*)$ reaches half its limiting value at $v/v^* = 0.922$.

It should be noted that the above conclusions were deduced on the assumption that $|d\tau/d\lambda| \gg 1$ in order to justify the random-phase approximation. The presence of a turning point in $\tau(\lambda)$ would lead to oscillations around the background defined by the functions $G(v/v^*)$.[81]

3.3. Inelastic and Reactive Collinear Atom–Diatom Scattering

3.3.1. Vibronic Network Formulation

The collinear atom–diatom problem is most readily formulated in the diabatic representation by dividing the electron matrix elements into diatomic and interaction parts:

$$H_{ij}(r,R) = H_{ij}^0(r) + V_{ij}(r, R) \tag{90}$$

and expanding the nuclear wave functions in terms of the diatomic eigenfunctions defined by

$$\left[-\frac{\hbar^2}{2\mu}\frac{d^2}{dr^2} + H_{ii}^0(r) \right] \eta_{in}(r) = E_{in}\eta_{in}(r) \tag{91}$$

Thus

$$\psi_i(r, R) = \sum \eta_{in}(r)\, \psi_{in}(R) \tag{92}$$

leading to the close-coupled equations

$$\left[\frac{\hbar^2}{2m}\frac{d^2}{dR^2} + E - E_{in} \right] \psi_{in}(R) = \sum_{jm} V_{in,jm}(R)\, \psi_{jm}(R) \tag{93}$$

where

$$V_{in,jm}(R) = \int_0^\infty \eta_{in}(r)\, V_{ij}(r, R)\, \eta_{jm}(r)\, dr \tag{94}$$

Here it has been assumed that the relevant translational variable is the interparticle separation R, as would be the case for an inelastic collision. Reactive situations may be handled in a similar way by applying a series of coordinate transformations designed to pass smoothly from reactants to products, and joining solutions of the close-coupled equations similar to (93) at the boundaries of each integration region.[19,42]

The manifold of diabatic vibronic terms

$$W_{in}(R) = E_{1n} + V_{in,in}(R) \tag{95}$$

coupled by the interaction terms $V_{in,jm}(R)$, results in the network of adiabatic curves shown in Figure 6. Coupling terms off-diagonal in i and j induce the

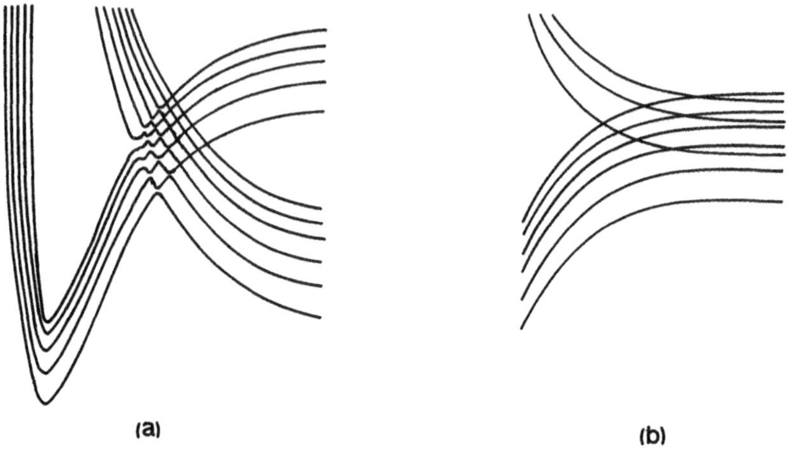

Figure 6. Adiabatic vibronic networks for (a) surface crossing, and (b) perturbed symmetric resonance (Demkov) models.

electronic transitions; those diagonal in i or j and off-diagonal in n and m are responsible for purely vibrational transitions.

In assessing the extent to which the coupling between the two vibrational manifolds covers the vibronic network, the major consideration is the displacement between the valleys on the two potential energy surfaces, because this determines the number of channels on one surface with significant Franck–Condon overlap with any given channel on the other. A secondary consideration is the dependence of the interaction function $H_{12}(R, r)$ on the internal coordinate r, but this will be of minor importance unless the two valleys are almost exactly superimposed.

The magnitude of this important displacement clearly depends on the chemical change induced by the collision. Thus a case involving charge exchange may be expected to cause a large change in the equilibrium diatomic bond length, and hence to strong coupling over a large part of the vibronic network. A fine-structure transition in the systems $F + H_2$ or $Br + H_2$, on the other hand, involves no change in the asymptotic diatomic bond length, and hence would be subject to the vibrational selection rule $\Delta v = 0$, except in cases of very close resonance.[82] The network will therefore be extensively coupled in charge transfer situations, but may break up into a set of two-state curve crossings between levels with a common vibrational quantum number when the transition primarily affects only the incoming atom.

A variety of propagation methods[83-88] are available for exact numerical solution of the resulting close-coupled equations. Of these the methods of Gordon[85] and Johnson[88] appear to have been most widely applied to nonadiabatic transitions.

3.3.2. Multi-Curve-Crossing Approximations

Transitions arising from the presence of avoided intersections in Figure 5 have been treated at the simplest level by Bauer *et al.*[20] in the approximation that each intersection is treated in isolation, with the transition probabilities between the relevant diabatic states $|1n\rangle, |2m\rangle$, say, given by the Landau–Zener matrix

$$\mathbf{P}_{nm} = \begin{pmatrix} \exp(-2\pi\delta_{nm}) & 1 - \exp(-2\pi\delta_{nm}) \\ 1 - \exp(-2\pi\delta_{nm}) & \exp(-2\pi\delta_{nm}) \end{pmatrix} \tag{96}$$

with δ_{nm} given by equation (46) in terms of the interaction element $H_{1n,2m}(R)$, the radial velocity v, and the difference between the derivatives of the curves $W_{in}(R)$ defined by equation (95), all quantities being evaluated at the intersection between $W_{1n}(R)$ and $W_{2n}(R)$. The final probability distribution matrix is then obtained as a product of elementary matrices containing $\mathbf{P}_{nm}$ in the 2×2 nm position.

One obvious weakness of this procedure is that any semiclassical interference terms are ignored. This is almost certainly unimportant in estimating the total cross section, because even in a three-state model any interference effects, other than those attributable to Feshbach resonances, have been found to average out.[89] The calculation of the internal-state distribution, the differential cross section, or of individual S-matrix elements at a given orbital angular momentum poses a severer test. The first improvement would be to replace equation (96) by the transition matrix[90]

$$T_{nm} = \begin{pmatrix} \exp(-\pi\delta_{nm}) & -[1 - \exp(-2\pi\delta_{nm})]^{1/2} \exp(-i\chi_{nm}) \\ [1 - \exp(-2\pi\delta_{nm})]^{1/2} \exp(i\chi_{nm}) & \exp(-\pi\delta_{nm}) \end{pmatrix} \tag{97}$$

with χ_{nm} given in terms of δ_{nm} by equation (48), and where it is assumed that the first row and column apply to the state with the larger slope, $F_{1n} = -\partial W_{1n}/dR$. These matrices would then be linked by diagonal phase evolution matrices with elements

$$L_{vv} = \exp\left| i \int_{R_x}^{R'_x} k_v(R)\, dR \right| \tag{98}$$

obtained by integrating the appropriate adiabatic reduced momentum $k_v(R)$ between the diabatic curve-crossing points R_x and R'_x, the sign of $k_v(R)$ being that appropriate to the inward or outward part of the trajectory as the case may be. A recent calculation in this spirit, but with δ_{nm} given by the Stückelberg formula, equation (54), and with S_{12} described by equation (63), has been reported by Laing *et al.*[82]

A second more fundamental difficulty is associated with the localization of each transition point, because the width of the transition zone defined by equation (44) may well exceed the spacing between successive transition points. Physically this would correspond, according to the discussion in Section 2.2.2, to a situation in which the time spent in the transition zone is small compared to a period of the internal motion. Hence the concept of an internal-state expansion loses any physical meaning.

The optimum conditions for application of either of the above localized approximations are therefore that the velocity should be sufficiently low, and the internal frequency sufficiently high that the spacing in time between successive curve crossings should be small compared with the internal time period.

3.3.3. Franck–Condon Approximation

One proposal, due to Kendall and Grice,[21] intended to meet the objections to the above localized picture, is to apply a single Landau–Zener surface-hopping probability, multiplied by a Franck–Condon factor, $|S_{1n2m}|^2$, to determine the distribution over final internal states. The transition amplitudes assigned to a single passage through the crossing zone may therefore be written

$$\begin{aligned} T_{1n\to 2m} &= (1 - p_{1n})^{1/2}\, S_{1n2m} \\ T_{1n\to 1n'} &= p_{1n}{}^{1/2}\, \delta_{nn'} \end{aligned} \tag{99}$$

where p_{1n} is the Landau–Zener (or Stückelberg) probability of remaining on the same diabatic curve

$$p_{1n} = \exp(-2\pi\delta) \tag{100}$$

with δ evaluated by equations (46) or (54) at the intersection between $W_{1n}(R)$ and the Franck–Condon weighted curve[91]:

$$\overline{W}_{2n}(R) = \sum_m |S_{1n2m}|^2\, W_{2m}(R) \tag{101}$$

The original formulations of the theory[21] apply only to a single passage, but the obvious extension would be to assign the following classical probabilities to the overall change incurred during a double passage through the transition zone:

$$\begin{aligned} p_{1n\to 2m} &= [(1 - p_{1n})\, p_{2m} + p_{1n}(1 - p_{2m})]\, S_{1n2m}^2 \\ p_{1n\to 1n'} &= p_{1n}^2 \delta_{nn'} + (1 - p_{1n}) \sum_m (1 - p_{2m})\, S_{1n2m} S_{2m1n'} \end{aligned} \tag{102}$$

p_{2m} is calculated at the intersection between $W_{2m}(R)$ and the weighted mean of the family $W_{1n}(R)$. Semiclassical versions of the theory, including Stückelberg interference effects, could also be envisaged, but these remain to be developed.

The validity of this picture has been examined by Child.[91] It is found to apply in all perturbation situations ($p_{in} \ll 1$ for all i, n) to the extent that the location of the different effective crossing points does not markedly affect the transition probabilities. It is also trivially valid if the two families of vibrational functions overlap exactly ($S_{1n2m} = \delta_{nm}$). In cases of more extensive overlap, and of intermediate or strong coupling ($0.1 < p_{in} < 1.0$), the validity criterion implies that the time spent in crossing the full transition region should be short compared with a vibrational period. It is therefore a very high velocity model.

3.3.4. Surface-Hopping Trajectory Approximations

The analytical results contained in Section 3.1 have been used not only in conjunction with an internal-state expansion as described in Section 3.3.2, but also as an extension of the classical and semiclassical trajectory methods discussed in Chapters 10, 12, 16, 18. The underlying argument is that any trajectory cuts a one-dimensional section through the potential surface manifold; hence the Landau–Zener and Demkov formulas may be used to calculate the transition probability from one potential surface to another while the distribution over internal states is deduced from the end points of an appropriate family of trajectories. Two methods have been developed, termed below the classical and semiclassical models.

3.3.4.1. Classical Model. The classical model due to Tully and Preston,[22] and recently reviewed by Tully,[30] was developed to handle transitions arising at an avoided seam between the two adiabatic surfaces. The procedure is to integrate Hamilton's equations on the initial diabatic (or adiabatic) surface, to the vicinity of the potential seam. The component of motion perpendicular to the seam is then assumed to induce the transition, all other momentum components being conserved. The surface-hopping transition probability is obtained either by means of a local numerical solution of the quantum-mechanical equations,[22] or by use of the Landau–Zener matrix given by equation (96), with the velocity and potential difference derivative required for the evaluation of δ taken normal to the potential seam. The classical trajectory is then continued on the new surface after an energy-conserving adjustment to the momentum normal to the seam, if trajectories over the adiabatic surfaces are employed. Any subsequent encounters with the seam are handled in a similar way, and appropriately weighted histograms are constructed to determine the final internal-state distributions on the two surfaces.

The strengths of this procedure are those applicable to any classical calculation, namely, that one requires knowledge only of the potential surfaces and that the coordinate transformation required in the reactive case may be applied at the end of the calculation. The major practical difficulty is that at thermal energies and high internal vibrational frequencies, a given trajectory may cross the potential seam many times. For example, at a collision velocity of 600 msec^{-1} and a vibrational frequency of 1000 cm^{-1} one could encounter a double crossing through the seam during every 0.2 Å of the translational motion. Although an exponential variation in the interaction $H_{12}(r, R)$ frequently ensures that the range of R over which the diabatic transition probability differs from zero (at large R) or unity (at small R) may be restricted to 2–3 Å, this means that it is difficult to follow every branch of every trajectory.

The problem does not arise, however, at higher energies because the number of oscillations through the transition region is substantially reduced. An interesting example of the physical insight offered by the classical trajectory picture is the interpretation given to the charge transfer collisions

$$M + X_2 \rightarrow \begin{cases} M^+ + X_2^- \\ M^+ + X + X^- \end{cases}$$

studied in the keV range by Los and co-workers.[52,92-95] The presence of two maxima in the total ionization cross section for Na + Br_2, shown in Figure 7 for example, is attributed[92] to trajectories which make a double passage across the seam between covalent and ionic surfaces, represented by the dashed line in Figure 8. The amplitude of Br_2 vibrational motion on the covalent surface is sufficiently small that every trajectory reaches the seam at the same point, but the subsequent motion on the ionic surface and

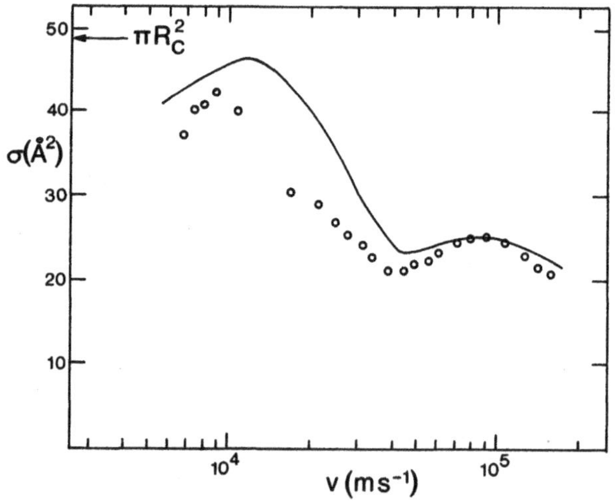

Figure 7. The total charge exchange cross section, Na + Br_2 → Na^+ + Br_2^- as a function of collision velocity, adapted from Hubers *et al.*[52] The solid line was computed in a classical surface-hopping trajectory approximation. The experimental points were scaled to the calculated ones at the position of the (Landau–Zener) maximum of the corresponding covalent cross section. πR_c^2 denotes the cross section implied by 100% ionization for trajectories reaching the point R_c in Figure 8.

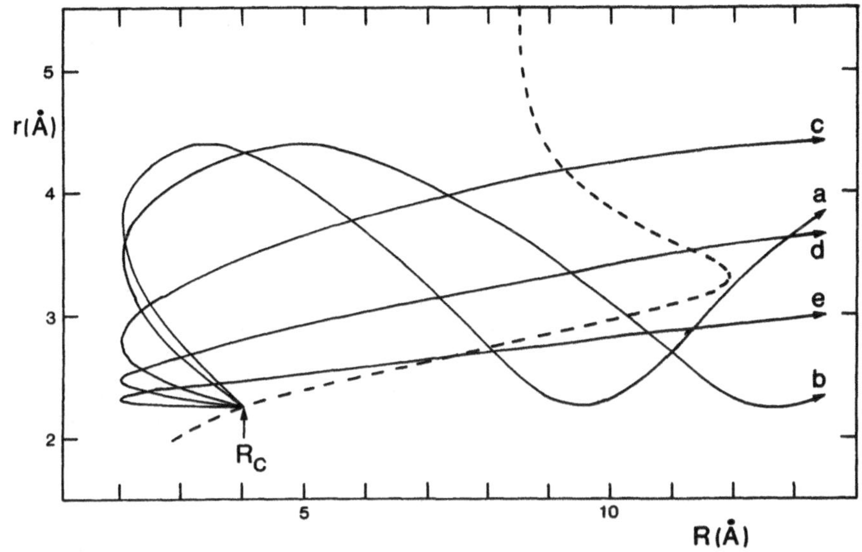

Figure 8. Bond-stretching trajectories on the ionic surface for the $Na^+ + Br_2^-$ system at impact parameter $b = 2$ Å, and collision velocities (a) 4×10^3 msec^{-1}, (b) 5×10^3 msec^{-1}, (c) 10^4 msec^{-1}, (d) 2×10^4 msec^{-1}, and (e) 4×10^4 msec^{-1}. The dashed line denotes the seam between the covalent and ionic surfaces. All trajectories are assumed to switch to the ionic surface at $R = R_c$ and at the equilibrium bond length, $r = r_e$, of the Br_2 molecule, but the probability of return to the covalent surface is a decreasing function of the coordinate R at the point at which the trajectory recrosses the seam.

the ultimate fate of the system depend on the collision velocity. At low velocities the magnitude of the interaction $H_{12}(r_x, R_x)$ is such that there is unit probability of transition to the ionic surface at the first encounter with the seam, but at $v \simeq 0.4 \times 10^4$ msec^{-1} the trajectory recrosses the seam at roughly the same point and the system switches back to the covalent state. At higher velocities the location of the second crossing first moves to larger R values, leading to smaller H_{12} and hence to a higher probability of remaining in the ionic state, but as the velocity increases further the time available for ionic "bond stretching" is reduced and the location of the second crossing point moves back to that of the first, and the overall ionization probability decreases. It is this bond-stretching mechanism which leads after summation over all impact parameters to the first maximum in Figure 6. The second is attributed[92] to the Landau–Zener effect (see Figure 5) due to an overall transition probability $P_1(1 - P_1)$ with P_1 decreasing from unity to zero as $v \to \infty$.

Similar but more striking behavior has been observed for $Cs + O_2$,[95] for which the total ionization cross section shows several bond-stretching maxima, attributed to the higher O_2^- vibrational frequency and hence to several critical energies at which the trajectory finally crosses the seam adiabatically after one, two, or three oscillations in the inner region.

3.3.4.2. Semiclassical Model. A more sophisticated semiclassical version of the theory has been suggested by Miller and George[23] and applied to a variety of problems by George and co-workers[24,96–98]; the most complete description is given by Lin *et al.*[24] The major differences between this and the classical model of Tully and Preston are the assignment of a probability amplitude (rather than a simple probability) to each surface-hopping transition, and the calculation of semiclassical phases, obtained by accumulation of the classical action along each branch of the trajectory. Semiclassical interference effects are therefore automatically taken into account.

In its fullest form the theory is a direct generalization of the Stückelberg treatment of the one-dimensional models discussed in Section 3.1. Thus the emphasis is placed on the location of the complex intersections between the adiabatic potential energy surfaces $W_+(\mathbf{Q})$ and $W_-(\mathbf{Q})$, which themselves define a $2(N-1)$-dimensional surface in the full complex coordinate space. This intersection reduces to a point in the one-dimensional case, but to a two-dimensional complex hypersurface around the potential seam in a surface-crossing collinear atom–diatom problem. Given this emphasis on the complex transition regions the theory is equally applicable to generalized Demkov situations ($F + H_2$[98] as well as $F + Xe$[99]) as to surface-crossing cases.

The original computational procedure is, however, quite complicated because one must seek all surface-hopping trajectories from a given internal state n, on one adiabatic surface, to the desired final quantum state m, on the second surface, with the transition between the surfaces achieved by taking carefully chosen complex time steps in the transition region designed to pass round the branch point in time, t_* say, at which the two adiabatic surfaces intersect. The scattering matrix for a transition from the lower to the upper surface is then given in the standard semiclassical form, as a sum over all trajectories from n_- to m_+:

$$S_{m_+n_-} = \sum_i J_i^{1/2} \exp[i\Phi_i(m_+, n_-)] \tag{103}$$

where J_i is the usual Jacobian factor necessary for normalization,[100,101] and

$$\Phi_i = \int_{t_1}^{t_2} T\,dt - \int_{t_1}^{t_*} W_-\,dt - \int_{t_*}^{t_2} W_+\,dt \tag{104}$$

the line integral being taken along the ith trajectory. Thus the real part of Φ_i gives rise to semiclassical interference effects, governing the distribution of final internal states, and the imaginary part of Φ_i acquired during the complex time part of the trajectory determines the transition amplitude from one surface to the other. Trajectories passing several times through the transition region would obviously acquire more than one imaginary part.

More recently, Komornicki *et al.*[98] have proposed a simpler "decoupling" procedure, whereby one follows only real trajectories, accumulating the usual semiclassical phase, but the transition amplitude is deduced (as in the classical Tully and Preston model) from knowledge of the variation of the two surfaces in a direction Q normal to the seam between them. Komornicki *et al.*[98] suggest the Stückelberg form for the transition amplitude

$$T = \exp(-\pi\delta) \tag{105}$$

where

$$2\pi\delta = \mathrm{Im} \int_{Q_c}^{Q_c^*} [k_+(Q) - k_-(Q)]\, dQ \tag{106}$$

which necessitates a search for the complex intersections (Q_c and Q_c^*) between W_+ and W_- in the direction Q, but the Landau–Zener formula [equation (46)] would be equally applicable provided the linear diabatic potential and constant interaction approximations were valid.

Equation (103) has the advantage that it also applies to the perturbed symmetric resonance situation, except that in this case one would anticipate from the discussion in Section 3.1.2 that equation (105) should be replaced by

$$T = \tfrac{1}{2}\,\mathrm{sech}\,\pi\delta \tag{107}$$

for small values of $\pi\delta$, or more generally by the form implied by equation (60). This point has not been examined in the literature.

3.3.5. *Comparison and Conclusions*

The alternatives available for the interpretation of nonadiabatic transitions in atom–molecule collisions may be listed as (a) close-coupled calculations (Section 3.3.1); (b) classical or semiclassical propagation through a vibronic network based, as appropriate, on Landau–Zener [equations (45) and (46)], Stückelberg [(45) and (54)], Demkov [(58) and (61)], or Nikitin [(61) and (63)] transition probabilities or transition amplitudes; (c) the Franck–Condon approximation; and (d) a classical or semiclassical surface-hopping trajectory formulation again based on Landau–Zener, Stückelberg, Demkov, or Nikitin transition probabilities. Since the available exact numerical calculations have been limited to the systems $H^+ + H_2$,[42] $Ar^+ + H_2$,[45] $F + H_2$,[15] $Br + H_2$,[15] and $C^+ + H_2$,[102] often over a limited energy range, and since there has been no systematic investigation of the relative validity of the above approximation even for these systems. it is impossible to make any firm recommendation concerning the available approximations. The following general criteria may however be applied, for which it is convenient to draw a distinction between the surface-crossing and perturbed symmetric resonance or Demkov situations, although this separation is not always clear cut, particularly in the vibronic network formulation.

3.3.5.1. Surface Crossing. The typical crossing problem will be characterized by a range ΔR over which the Landau–Zener (or Stückelberg) transition probabilities, in either the vibronic network or the classical trajectory formulation, lie between 0.1 and 0.9, say. The vibronic network approach is to be preferred if the number of intersections suffered by a given curve over this region is small (less than five, say), provided the velocity is sufficiently low that the dynamic range given by equation (44) is less than the spacing between successive intersections. This condition, which may be expressed in the form

$$\hbar\omega/(F_1 - F_2) \gg [\hbar v/(F_1 - F_2)]^{1/2} \tag{105}$$

shows that the network formulation is favored at low[102] velocities v, and high internal frequencies ω, both taken in the transition region. The surface-hopping trajectory formulation is favored in the opposite extreme because any given trajectory would cross the potential seam a relatively small number of times.

The choice between a classical and a semiclassical formulation is more difficult. Any semiclassical interference effects will almost certainly average out (as in the atom–atom case) in the calculation of the total nonadiabatic transition cross section, but the situation is less clear with regard to the product internal-state distribution. It has recently been argued[103] that semiclassical considerations may lead to an interference structure in such distributions arising from vibrationally excited reactants, but it is not clear to what extent these conclusions apply to nonadiabatic processes.

Another question concerns the use of the Landau–Zener or Stückelberg formulas. Since the latter reduces to the former, in the linear potential, constant interaction model, it is always to be preferred on grounds of accuracy, but it is very much less convenient in practice. Hence except in situations requiring the highest accuracy, the Landau–Zener formula will be preferred in most surface-crossing situations. Care must be employed, however, in treating threshold processes by any method, particularly those involving tunneling through a lower adiabatic barrier. These can only be handled by a local numerical solution of the two-state equations generated by either the vibronic network or the trajectory formulations.

Finally, the relevance of the Franck–Condon approximation may be considered. The situation identified in Section 3.3.3 in which the seam is normal to the two superimposed entrance valleys is of little interest because the Franck–Condon and vibronic network models both reduce to a set of isolated curve-crossing problems which may be solved immediately.

The high-energy limit in which the time in the transition region is short compared with the internal frequency is, however, one in which the classical trajectory approach has been recommended above. The similarity between the two approximations is readily recognized, because at energies sufficiently high that the trajectories are effectively linear, any surface-hopping transitions occur without any change in the vibrational coordinate, as required by the Franck–Condon principle. It would be extremely interesting to explore this connection further, both in the classical and in the semiclassical approximation. If complete equivalence were established, the Franck–Condon formulation would be preferable to the trajectory method in leading directly to quantized internal-state distributions.

3.3.5.2. Demkov Transitions. The only atom–molecule Demkov-type processes which have been investigated in detail are the fine-structure transitions in the $F + H_2$ and $Br + H_2$ systems.[15,82,98] Here the negligible displacement between the entrance valleys on the two surfaces results in considerable simplifications in both the vibronic network and the trajectory formulations, in the former case because the number of significantly coupled levels is small, and in the latter because the transition region runs normal to the entrance valley, so that any trajectory crosses this region only twice. In both cases the transition probability could be deduced from equations (54) and (58) which would necessitate a search for the relevant complex transition points.[82,98] The alternative would be to employ equation (60) provided it were possible to make the exponential approximations given by equation (59) to the diabatic vibronic Hamiltonian matrix, or to the section of the electronic Hamiltonian matrix selected by the classical trajectory.

Of the two methods the vibronic network formulation would appear to be simpler, but Komornicki *et al.*[98] have obtained excellent agreement with exact numerical results for the $F + H_2$ system by both the full complex trajectory method and the simplified decoupled version described at the end of Section 3.3.4.

References

1. H. Goldstein, *Classical Mechanics*, Addison-Wesley, Reading, Massachusetts (1959).
2. H.F. Shaefer III, *The Electronic Structure of Atoms and Molecules*, Addison-Wesley, Reading, Massachusetts (1972).
3. G.A.L. Delvigne and J. Los, The differential cross-sections for chemi-ionization in alkali atom–halogen molecule collisions: Classical interpretation, *Physica (Utrecht)* **59**, 61–76 (1972).
4. A.P.M. Baede, Charge transfer between neutrals at hyperthermal energies, *Adv. Chem. Phys.* **30**, 463–535 (1975).
5. G.G. Balint-Kurti, Potential energy surfaces for simple chemical reactions: $Li + F_2 \rightarrow LiF + F$, *Mol. Phys.* **25**, 393–435 (1973).
6. C. Nyeland and J. Ross, Estimate of potential surface for K-Cl-Cl, *J. Chem. Phys.* **54**, 1665–1670 (1971).
7. R.K. Preston and J.C. Tully, Effects of surface crossing in chemical reactions: The H_3^+ system, *J. Chem. Phys.* **54**, 4297–4303 (1971).

8. J. Krenos, R.K. Preston, R. Wolfgang, and J.C. Tully, Reaction of hydrogen atomic ions with hydrogen molecules, *ab initio* theory and a conceptual model, *Chem. Phys. Lett.* **10,** 17–21 (1971).
9. C.W. Bauschlicher, Jr., S.V. O'Neil, P.K. Preston, and H.F. Shaefer III, Avoided intersection of potential energy surfaces: The ($H^+ + H_2$, $H + H_2^+$) system, *J. Chem. Phys.* **59,** 1286–1291 (1973).
10. P.J. Kuntz and A.C. Roach, Ion–molecule reactions of rare gases with hydrogen, *J. Chem. Soc. Faraday II* **68,** 259–280 (1972); Erratum: **69,** 926 (1973).
11. Yu.N. Demkov, Charge transfer at small resonance defects, *Sov. Phys. JETP* **18,** 138–142 (1964).
12. D.S.F. Crothers, A critique of Zwaan–Stückelberg phase integral techniques, *Adv. Phys.* **20,** 405–451 (1971).
13. D.S.F. Crothers, Perturbed symmetric resonance: An exact formula, *J. Phys. B* **6,** 1418–1425 (1973).
14. R.K. Preston, C. Sloane, and W.H. Miller, Semi-classical theory of collisionally induced fine-structure transitions in fluorine atoms, *J. Chem. Phys.* **60,** 4961–4969 (1974).
15. I.H. Zimmerman and T.F. George, Quantum mechanical study of electronic transitions in collinear atom–molecule collisions, *Chem. Phys.* **7,** 323–335 (1975).
16. J.C. Tully, Collisions of $F(^2P_{1/2})$ with H_2, *J. Chem. Phys.* **60,** 3042–3050 (1974).
17. F.T. Smith, Diabatic and adiabatic representations for atomic collision problems, *Phys. Rev.* **179,** 111–123 (1969).
18. M. Baer, Adiabatic and diabatic representations for atom–molecule collisions: Treatment of the three dimensional case, *Chem. Phys.* **15,** 49–57 (1976).
19. Z.H. Top and M. Baer, Incorporation of electronically non adiabatic effects into bimolecular reactive systems I. Theory, *J. Chem. Phys.* **66,** 1363–1371 (1975).
20. E. Bauer, E.R. Fisher, and F.R. Gilmore, De-excitation of electronically excited sodium by nitrogen, *J. Chem. Phys.* **51,** 4173–4181 (1969).
21. G.M. Kendall and R. Grice, Vibrational coordinates in the electron jump model, *Mol. Phys.* **24,** 1373–1382 (1972).
22. J.C. Tully and R.K. Preston, Trajectory surface hopping approach to non-adiabatic molecular collisions, *J. Chem. Phys.* **55,** 562–572 (1971).
23. W.H. Miller and T.F. George, Semiclassical theory of electronic transitions in low energy atomic and molecular collisions involving several nuclear degrees of freedom, *J. Chem. Phys.* **56,** 5637–5652 (1972).
24. Y.W. Lin, T.F. George, and K. Morokuma, Semiclassical treatment of electronic transitions in molecular collisions: $H^+ + D_2 \rightarrow HD^+ + D$, *J. Chem. Phys.* **60,** 4311–4322 (1972).
25. E.E. Nikitin, Theory of non-adiabatic transitions. Recent development of the Landau–Zener–(Linear) model, in *Chemische Elementarprozesse,* H. Hartmann, editor, Springer-Verlag, Berlin (1968), pp. 43–77.
26. E.E. Nikitin, Theory of non-adiabatic transitions: Recent development with exponential models, *Adv. Quantum Chem.* **5,** 135–184 (1970).
27. J.B. Delos and W.R. Thorson, *Phys. Rev. A* **6,** 728–745 (1972); Erratum: **9,** 1026 (1974).
28. M.S. Child, *Molecular Collision Theory,* Academic Press, New York (1973).
29. M.S. Child, Semiclassical effects in heavy particle collisions, *Adv. Atom. Mol. Phys.* (to be published).
30. J.C. Tully, Nonadiabatic processes in molecular collisions, in *Dynamics of Molecular Collisions, Part B,* W.H. Miller, editor, Plenum Press, New York (1976), pp. 217–267.
31. P.K. Janev, Non-adiabatic transitions between ionic and covalent states, *Adv. Atom. Mol. Phys.* **12,** 1–38 (1976).
32. W.R. Thorson, Inelastic scattering of atoms I, *J. Chem. Phys.* **34,** 1744–1757 (1961).
33. D.J. Vezzetti and S.I. Rubinov, Asymptotic solution of the Schrödinger equation for the three body problem, *Ann. Phys. (New York)* **35,** 373–395 (1965).

34. W.H. Miller, Distorted-wave theory for collisions of an atom and a diatomic molecule, *J. Chem. Phys.* **49**, 2373–2381 (1968).
35. R.T. Pack, Space-fixed vs body-fixed axes in atom–diatomic molecule scattering. Sudden approximations, *J. Chem. Phys.* **60**, 633–639 (1974).
36. D.R. Bates and D.A. Williams, Low energy collisions between hydrogen atoms and protons, *Proc. Phys. Soc. London* **83**, 425–433 (1964).
37. S.K. Knudson and W.R. Thorson, Lyman-α excitation and resonant charge exchange in slow H^+–H(1*s*) collisions, *Can. J. Phys.* **48**, 313–329 (1970).
38. W.R. Thorson, Nonadiabatic effects in the high energy scattering of normal helium atoms, *J. Chem. Phys.* **39**, 1431–1441 (1963).
39. I. Kovács, *Rotational Structure of the Spectra of Diatomic Molecules* (trans. L. Nemes), Hilger, London (1969).
40. G. Herzberg, *Spectra of Diatomic Molecules*, Van Nostrand, New York (1950).
41. R.K. Hinkley, J.A. Hall, T.E.H. Walker, and W.G. Richards, Λ doubling in $^2\Pi$ states of molecules, *J. Phys. B.* **5**, 204–212 (1972).
42. Z.H. Top and M. Baer, Incorporation of electronically non-adiabatic effects into reactive systems II. The collinear ($H_2 + H^+, H_2^+ + H$) system, *Chem. Phys.* **25**, 1–18 (1977).
43. F.O. Ellison, A method of diatomics in molecules I. General theory and application to H_2O, *J. Am. Chem. Soc.* **85**, 3540–3544 (1963).
44. J.C. Tully, Diatomics-in-molecules potential energy surfaces I. First-row triatomic hydrides, *J. Chem. Phys.* **58**, 1396–1410 (1973).
45. M. Baer and J.A. Beswick, Electronic non-adiabatic transitions in the reactive ($Ar^+ + H_2$, $Ar + H_2^+$, $ArH^+ + H$) system: Numerical results for the collinear configuration, *Chem. Phys. Lett.* **51**, 360–364 (1977).
46. M. Baer, Electronic non-adiabatic transitions in the reaction $Ar^+ + H_2(v_i = 0) \rightarrow ArH^+ + H$: A comparison between exact collinear results and a two state model calculation, *Mol. Phys.* **35**, 1637–1648 (1978).
47. B.M. Smirnov, Formation and decay of negative ions, *Sov. Phys. Dokl.* **10**, 218–221 (1965).
48. B.M. Smirnov, Negative ion gas laser, *Sov. Phys. Dokl.* **12**, 242–244 (1967).
49. R.E. Olson, F.T. Smith, and E. Bauer, Estimation of the coupling matrix elements for one electron transfer systems, *Appl. Opt.* **10**, 1848–1855 (1971).
50. R. Grice and D.R. Herschbach, Long-range configuration interaction of ionic and covalent states, *Mol. Phys.* **27**, 159–175 (1974).
51. S.A. Adelman and D.R. Herschbach, Asymptotic approximation for ionic-covalent configuration mixing in hydrogen and alkali hydrides, *Mol. Phys.* **33**, 793–809 (1977).
52. M.M. Hubers, A.W. Kleyn, and J. Los, Ion pair formation in alkali–halogen collision at high velocities, *Chem. Phys.* **17**, 303–325 (1976).
53. J.A. Aten, G.E.H. Lanting, and J. Los, The energy dependence of differential cross-sections for ion-pair formation in Na, K, Cs + I_2 collisions, *Chem. Phys.* **19**, 241–251 (1977).
54. T.E.H. Walker and W.G. Richards, Molecular spin–orbit coupling constants: The role of core polarization, *J. Chem. Phys.* **52**, 1311–1314 (1970).
55. W.H. Moores and R. McWeeny, The calculation of spin–orbit splitting and *g* tensors for small molecules and radicals, *Proc. R. Soc. London Ser. A* **332**, 365–384 (1973).
56. J.A. Hall, T.E.H. Walker, and W.G. Richards, One-centre matrix elements of the spin–other orbit operator in linear molecules: Diagonal terms, *Mol. Phys.* **20**, 753–759 (1971).
57. J.C. Tully, Diatomics in molecules potential energy surfaces II. Nonadiabatic and spin–orbit interactions, *J. Chem. Phys.* **59**, 5122–5134 (1973).
58. F.H. Mies, Molecular theory of atomic collisions: Fine-structure transitions, *Phys. Rev. A* **7**, 942–951 (1973).

59. J.S. Cohen and B. Schneider, Ground and excited states of Ne_2 and Ne_2^+ I. Potential curves with and without spin–orbit coupling, *J. Chem. Phys.* **61**, 3230–3239 (1974).
60. L.D. Landau, Zur Theorie der Energieübertragung II, *Phys. Z. Sowjetunion* **2**, 46–51 (1932).
61. C. Zener, Non-adiabatic crossing of energy levels, *Proc. R. Soc. London Ser. A* **137**, 696–702 (1932).
62. E.C.G. Stückelberg, Theorie der unelastischen Stösse zwischen Atomen, *Helv. Phys. Acta* **5**, 369–422 (1932).
63. D.R. Bates, Collisions involving the crossing of potential energy curves, *Proc. R. Soc. London Ser. A* **257**, 22–31 (1960).
64. L.P. Kotova, Angular distribution in inelastic atomic collisions, *Sov. Phys. JETP* **28**, 719–722 (1969).
65. M.S. Child, Thermal energy scattering of alkali atoms from halogen atoms and molecules: The effect of curve crossing, *Mol. Phys.* **16**, 313–327 (1969).
66. A.D. Bandrauk and M.S. Child, Analytical predissociation line widths from scattering theory, *Mol. Phys.* **19**, 95–111 (1970).
67. M.S. Child, Curve crossing and the WKB approximation, *Mol. Phys.* **20**, 171–184 (1971).
68. M.S. Child, On the Stückelberg formula for non-adiabatic transitions, *Mol. Phys.* **28**, 495–501 (1974).
69. D.S.F. Crothers, Stückelberg curve-crossing phases: The parabolic model, *J. Phys. B.* **8**, L442–L446 (1975).
70. W.H. Miller, Uniform semiclassical approximations for elastic scattering and eigenvalue problems, *J. Chem. Phys.* **48**, 464–467 (1968).
71. M.S. Child, A uniform approximation for one-dimensional matrix elements, *Mol. Phys.* **29**, 1421–1429 (1975).
72. M.B. Faist and R.D. Levine, Collisional ionization and elastic scattering in alkali–halogen atom collisions, *J. Chem. Phys.* **64**, 2953–2970 (1976).
73. J. Grosser and A.E. de Vries, Single path transition probabilities in curve-crossing phenomena II. Phase shift, *Chem. Phys.* **10**, 229–233 (1975).
74. K.W. Ford and J.A. Wheeler, Semiclassical description of scattering, *Ann. Phys. (N.Y.)* **7**, 259–286 (1959).
75. K.W. Ford and J.A. Wheeler, Application of semiclassical analysis, *Ann. Phys. (N.Y.)* **7**, 286–322 (1959).
76. R.E. Olson and F.T. Smith, Collision spectroscopy IV. Semiclassical theory of inelastic scattering with application to He^+ + Ne, *Phys. Rev. A* **3**, 1607–1617 (1971).
77. G.A.L. Delvigne and J. Los, Rainbow, Stückelberg oscillations and rotational coupling on the differential cross-section of Na + I → Na^+ + I^-, *Physica (Utrecht)* **67**, 166–196 (1973).
78. M.S. Child and R.B. Gerber, Inversion of the Stückelberg oscillations in atom–atom scattering (to be published).
79. J.B. Delos, Studies of the potential-curve-crossing problem III. Collisional spectroscopy of close crossings, *Phys. Rev. A* **9**, 1626–1634 (1974).
80. M.B. Faist and R.B. Bernstein, Computational study of elastic and electronically inelastic scattering of Br by ground state I atoms: Role of potential curve crossing, *J. Chem. Phys.* **64**, 2971–2984 (1976).
81. R.E. Olson, Two-state Stückelberg–Landau–Zener theory applied to oscillatory inelastic total cross-sections, *Phys. Rev. A* **2**, 121–126 (1970).
82. J.R. Laing, T.F. George, I.H. Zimmerman, and Y.W. Lin, The vibronic representation for collinear atom–diatom collisions: Two state semiclassical model, *J. Chem. Phys.* **63**, 842–851 (1975).
83. A.C. Allison, The numerical solution of coupled differential equations arising from the Schrödinger equation, *J. Comput. Phys.* **6**, 378–391 (1970).

84. W.A. Lester Jr. and R.B. Bernstein, Computational procedure for the close coupled rotational excitation problem, *J. Chem. Phys.* **48**, 4896–4904 (1968).
85. R.G. Gordon, Quantum scattering using piecewise analytic solutions, *Methods Comput. Phys.* **10**, 81–110 (1971).
86. W.N. Sams and D.J. Kouri, Noniterative solutions of integral equations of scattering II. Coupled channels, *J. Chem. Phys.* **51**, 4815–4819 (1969).
87. B.R. Johnson and D. Secrest, The solution of the non-relativistic quantum scattering problem without exchange, *J. Math. Phys. (N.Y.)* **7**, 2187–2195 (1966).
88. B.R. Johnson, The multichannel log-derivative method for scattering calculations, *J. Comput. Phys.* **13**, 445–449 (1973).
89. A.M. Wooley, Semiclassical scattering theory and total cross sections for systems with many crossing points, *Mol. Phys.* **22**, 607–618 (1971).
90. M.S. Child, Semi-classical theory of tunnelling and curve-crossing problems: A diagrammatic approach, *J. Mol. Spectrosc.* **53**, 280–296 (1974).
91. M.S. Child, Franck–Condon transitions in multi-curve crossing processes, *Faraday Discuss. Chem. Soc.* **55**, 30–33 (1973).
92. A.P.M. Baede and J. Los, Total cross-sections for charge transfer and production of free electrons by collisions between alkali atoms and some molecules, *Physica (Utrecht)* **52**, 422–440 (1971).
93. A.M.C. Moutinho, A.P.M. Baede, and J. Los, Charge transfer between alkali atoms and oxygen molecules, *Physica (Utrecht)* **51**, 432–445 (1970).
94. K. Lacmann and D.R. Herschbach, Collisional excitation and ionization of K atoms by diatomic molecules: Role of ion pair states, *Chem. Phys. Lett.* **6**, 106–110 (1970).
95. A.W. Kleyn, M.M. Hubers, J.A. Aten, and J. Los (to be published).
96. Y.W. Lin, T.F. George, and K. Morokuma, Semiclassical treatment of charge exchange at low energies: Collinear $H^+ + H_2$ collisions, *J. Phys. B* **8**, 265–272 (1975).
97. Y.W. Lin, T.F. George, and K. Morokuma, Semiclassical treatment of electronic transitions in molecular collisions: Three-dimensional $H^+ + D_2 \rightarrow HD^+ + D$, *Chem. Phys. Lett.* **30**, 49–53 (1975).
98. A. Komornicki, T.F. George, and K. Morokuma, Decoupling scheme for a semi-classical treatment of electronic transitions in atom–diatom collisions: Real-valued trajectories and local analytic continuation, *J. Chem. Phys.* **65**, 48–54 (1976).
99. R. K. Preston, C. Sloane, and W.H. Miller, Semiclassical theory of collisionally induced fine-structure transitions in fluorine atoms, *J. Chem. Phys.* **60**, 4961–4969 (1974).
100. W.H. Miller, Classical limit quantum mechanics and the theory of molecular collisions, *Adv. Chem. Phys.* **25**, 69–177 (1974).
101. W.H. Miller, The classical *S* matrix in molecular collisions, *Adv. Chem. Phys.* **30**, 77–136 (1975).
102. S.I. Chu and A. Dalgarno, Fine structure transitions of C^+ in collisions with H_2, *J. Chem. Phys.* **62**, 4009–4015 (1975).
103. M.S. Child, The Condon reflection principle in collision dynamics, *Mol. Phys.* **35**, 759–770 (1978).

14

Reactive Scattering Cross Sections I: General Quantal Theory

JOHN C. LIGHT

1. Introduction

Reactive scattering events are those in which the "chemical identity" of at least some of the particles changes. Included in this would be chemical exchange reactions, dissociation, ionization, and charge transfer, or a combination of these. Since the "chemical identity" of the species in a scattering process is a persistent asymptotic property, the theory of such processes must include all relevant asymptotic states which may occur or be measured. Being chemists, on the other hand, we may not want to include, even in the theory at first, such often unnecessary complications as exchange of core electrons, electronic excitations in adiabatic chemical reactions, etc. Thus we will want theories which deal effectively with those (and only those) changes of "chemical identity" which are likely to occur for collision energies and reactants of interest.

Although the mathematical questions associated with the development of such theories have received considerable attention,[1-3] and a number of formal approaches have been (and are being) developed,[4-8] their presentation in this section will be relatively limited for several reasons. First, they lie outside my personal experience in terms of actual calculations; second, they have not provided the formalism within which most "exact" quantal approaches to calculation of reactive scattering have been developed; and

JOHN C. LIGHT • The James Franck Institute and The Department of Chemistry, The University of Chicago, Chicago, Illinois 60637.

lastly, the R-matrix approach emphasized here is, perhaps, the most elementary formal approach which is readily translated into practical computational schemes. The R-matrix approach does, however, have one formal defect, which accounts for its relative simplicity. It cannot be used currently to describe dissociative reactions such as either atomic fragmentations or ionization. Finally, it should be mentioned that the major current "exact" and practical approaches to the calculation of reactive scattering events are strictly based neither on formal reactive scattering theory nor on formal R-matrix theory, but on approximate (but accurate, we believe) representations of the problem in terms of local properties.

With this rather extended caveat in mind, this chapter will present a definition of the reactive scattering process, a brief review of the formal scattering theory approaches, the R-matrix approach, and finally a discussion of the approaches currently implemented in a relatively exact fashion. The formal theory has provided the primary base for approximate (perturbative) approaches discussed in Chapter 15; the R-matrix approach has provided the theoretical basis for both L^2 computations and a numerical integration method, and the current direct approaches have provided, via the detailed methods discussed in the following chapter, the most accurate quantal reactive scattering data to date.

2. *Quantal Reactive Scattering Processes*

Although on a particular electronically adiabatic potential energy surface the Hamiltonian for nuclear motion is unique, in systems for which reactive scattering is possible there will always exist more than one mathematically convenient way of writing the Hamiltonian for large separations of reactants or products. If we denote the various reactant and product channels by α, β, $\gamma, \ldots$, then we can write the Hamiltonian as

$$H = T_\alpha + H_\alpha + V_\alpha = T_\beta + H_\beta + V_\beta = \cdots \tag{1}$$

where T_λ $(\lambda = \alpha, \beta, \gamma, \ldots)$ is the relative kinetic energy operator of the fragments in the chemical channel λ, H_λ is the internal Hamiltonian of the atoms, molecules and molecular fragments in channel λ, and V_λ is the interaction between the separating molecular fragments in channel λ with the property that $V_\lambda \to 0$ as $R_\lambda \to \infty$. $\mathbf{R}_\lambda$ is the relative (center-of-mass) vector between the separating chemical fragments in channel λ. We denote all the internal coordinates in channel λ by $\{u_\lambda\}$ and will assume that no dissociative processes can occur. Let the orthonormal internal channel eigenfunctions $\phi_i^\lambda(\{u_\lambda\})$ satisfy the internal channel Hamiltonian

$$H_\lambda \varphi_i^\lambda(\{u_\lambda\}) = \varepsilon_i^\lambda \varphi_i^\lambda(\{u_\lambda\}) \tag{2}$$

We can then write the asymptotic form of the entire wave function ($R_\lambda \to \infty$) for scattering from initial channel λ' and state i to all other states as

$$\psi_i^{\lambda'} \to \sum_{\lambda,j} \{ \delta_{ij}\delta_{\lambda\lambda'}\varphi_j^\lambda(\{u_\lambda\})\exp(-i\mathbf{k}_j^\lambda \cdot \mathbf{R}_\lambda) - [\exp(ik_j^\lambda R_\lambda)/R_\lambda]\,\varphi_j^\lambda(\{u_\lambda\})\, f_{ji}^{\lambda\lambda'}(\hat{\mathbf{R}}_\lambda)\} \tag{3}$$

where $|k_j^\lambda|^2 = |(E - \varepsilon_j^\lambda)|\,(2\mu_\lambda/\hbar^2)$ is the square of the relative wave vector. As the reactants and products separate into the asymptotic region ($|R_\lambda| \to \infty$) all the internal eigenfunctions, being bounded, are orthogonal, independent of channel. The coefficients of each φ_j^λ, therefore, represent the amplitude of scattering into that state as a function of angle, given the initial state and wave vector, $\mathbf{k}_i^{\lambda'}$. They are related to the cross sections as for inelastic scattering:

$$d\sigma(\lambda, j, \Omega; \lambda', i, E)\frac{k_j^\lambda}{k_i^{\lambda'}} = |f_{ji}^{\lambda\lambda'}(\Omega)|^2\, d\Omega \tag{4}$$

The wave function, throughout space, must also satisfy the Schrödinger equation.

Within the brief description above are the roots of the mathematical and physical problems associated with quantal reactive scattering. The wave function must satisfy the Schrödinger equation throughout space, yet in order to satisfy the boundary conditions it must be decomposed in the different asymptotic regions corresponding to different chemical channels in different ways. In particular, the different asymptotic coordinate systems (R_λ, $\{u_\lambda\}$) are not orthogonal to each other, and the different basis sets, $\{\varphi_j^\lambda(\{u_\lambda\})\}$, are, for practical purposes, incomplete, and orthogonal to each other only asymptotically. The use of different coordinate systems and basis sets is inherent to the mathematical description of the experimental or physical processes occurring, since the flux of products in particular internal states is desired.

3. *Formal Scattering Theory for Reactive Collisions*

Since there exist a number of discussions of this topic in the literature,[1-8] we shall not attempt a complete discussion here. Rather we shall give a brief review of the problems and present those results which seem most appropriate. If the exact Green's function for the entire Hamiltonian were known, then there would be no problem in solving for the reactive scattering. One does not, however, know the exact Green's function, and, following the normal procedure for inelastic scattering, one can attempt to partition the Hamiltonian as in equation (1):

$$H = H_\alpha^0 + V_\alpha = H_\beta^0 + V_\beta = \cdots \tag{5}$$

(H_λ^0 now contains the entire kinetic energy operator, $H_\lambda^0 = T_\lambda + H_\lambda$), and write the integral equation for the Green's function as

$$G^+ = G_\alpha^{0+} + G_\alpha^{0+} V_\alpha G^+ = G_\beta^{0+} + G_\beta^{0+} V_\beta G^+ = \cdots \tag{6}$$

where the + refers to outgoing boundary conditions and G_λ^{0+} is defined as

$$G_\lambda^{0+} = \lim_{\varepsilon \to 0} (E + i\varepsilon - H_\lambda^0)^{-1} \tag{7}$$

The T matrix, which gives directly the transition matrix elements, is related to the S matrix by

$$S = I - 2\pi i T \tag{8}$$

and, for *inelastic* scattering problems is given by the integral equation

$$T = V + V G_0^+ T \tag{9}$$

where $H = H_0 + V$ and G_0^+ is the scattering Green's function for H_0. Using equation (6) for inelastic scattering only (one channel), this can be rearranged to yield the T matrix in terms of the full scattering Green's function G^+:

$$T = V + V G^+ V \tag{10}$$

For inelastic scattering equation (9) can be used to generate explicit coupled integral equations for the scattering, as in Chapter 6.

For reactive scattering, however, formal problems arise with the analogous procedure.[1-5] The T matrix is blocked into nonreactive and reactive portions with indices indicating channels, and the analog to (10) is

$$T_{\lambda\lambda'} = V_\lambda + V_\lambda G^+ V_{\lambda'} \tag{11}$$

Since one cannot, in general, work with the full Green's function, one wants to use the analogs of equation (9), which are

$$T_{\lambda\lambda'} = V_\lambda + V_\lambda G_\lambda^{0+} T_{\lambda\lambda'} \tag{12}$$

These equations, however, suffer from formal difficulties in that the kernel is not compact. A nonmathematical way of looking at this problem is that since each channel formulation of the Green's function in equation (6) is exact and each integral equation produces a solution valid throughout space (if solved), one is attempting, with a Green's function appropriate to channel λ, to construct the bound-state solutions in the other channels. Aside from the formal mathematical difficulties which it causes it would seem that the uncoupled Lippmann–Schwinger (L–S)-type equations (12) are not convenient for reactive scattering. Before proceeding, however, it is worthwhile noting that the L–S equations are useful for perturbative theories of reactive scattering in low orders, such as perturbed stationary state theory. In first order only V enters, and the problems with the kernel do not arise.

Although there are several approaches which avoid the formal mathematical problems, the relatively recent work of Kouri and co-workers[6] appears to be the most straightforward. This work has recently been extended by Rabitz and Rabitz,[8] who give an excellent analysis. We follow a portion of their analysis in the brief review here. Since the problems are associated with the uncoupled form of the L–S equations, it is not surprising that their resolution lies in a proper coupling of the separate channel equations. This is easily done by introducing a square-channel coupling matrix **W** (of the dimensionality of the number of channels) which couples the channels. The matrix which does this is required to satisfy

$$\sum_{\lambda} W_{\lambda\lambda'} = \sum_{\lambda'} W_{\lambda\lambda'} = 1 \tag{13}$$

A satisfactory, but nonunique, choice is to set one element of each row and column to one, the other elements to zero, and to have the ones cyclic ($W_{N,1} = W_{1,2} = \cdots W_{N-1,N} = 1$), i.e., to make **W** a channel permutation matrix.

Using **W**, we can write the Green's function, equation (6), more symmetrically as

$$G^{+} = \sum_{\lambda'} W_{\lambda\lambda'}(G_{\lambda'}^{0+} + G_{\lambda'}^{0+} V_{\lambda'} G^{+}) \tag{14}$$

Using the fact that equation (11) can also be written

$$T_{\lambda\lambda'} = V_{\lambda'} + V_{\lambda} G^{+} V_{\lambda'} \tag{15}$$

and substituting (14) in (11), the coupled equations for reactive scattering are found to be

$$T_{\lambda\lambda'} = V_{\lambda} W_{\lambda\lambda'} + V_{\lambda} \sum_{\lambda''} W_{\lambda\lambda''} G_{\lambda''}^{0+} T_{\lambda''\lambda'} \tag{16}$$

These equations do not suffer from the mathematical problems associated with the L–S equations. The exact solution will be unique although lower-order solutions may depend on the choice of **W**,[8] since, for more than two channels different **W**'s "match" solutions in different channels in different cyclic order. As yet this approach has been tried only on the "exchange" reaction of e^{-}, H(1s), but with good results.[6,9] It should be noted that with this approach one should be able to avoid some of the problems of coordinate systems and basis sets mentioned earlier, although this has not yet been seriously tested. The freedom in the choice of **W**, however, may cause considerable ambiguity in truncated calculations. To date no exact calculations of a chemical exchange problem in three dimensions has been attempted utilizing this formal approach.

4. R-Matrix Theory

The origins and early development of R-matrix theory are well described in the exhaustive review by Lane and Thomas,[10] and a brief review of R-matrix theory as applied to inelastic scattering was given earlier in Chapter 6 of this volume. Since R-matrix theory as a square integrable (L^2) or algebraic technique has been applied only to a few model chemical reactions,[11-14] it cannot be regarded as a panacea. It is of interest, however, for several reasons: In the nondissociative energy region R-matrix theory is a formal exact approach which avoids the mathematical pitfalls of the L–S equations. It is also extremely flexible, allowing one, for example, to combine L^2 and close-coupled differential equation solutions in the same calculation, using each where convenient.

The application of R-matrix techniques to chemically reactive systems differs little in theory from the application to inelastic scattering. The primary difference, as might be expected, is that the surface $\mathbf{s}$ on which the "internal" and external wave functions are matched must cut through regions of space corresponding to each chemical channel. The R matrix may be considered an operator which, for a specified surface, relates the wave function and derivative

$$\psi = \mathscr{R}_{\mathbf{s}} \mathbf{n} \cdot \nabla \psi \tag{17}$$

The $\mathscr{R}$ operator is Hermitian so that when it is evaluated in a real basis on $\mathbf{s}$ it yields a symmetric R matrix and thus a unitary $\mathbf{S}$ matrix. Since the wave function can be written asymptotically as

$$\psi = \sum_{\lambda,j} f_j^\lambda(\mathbf{R}_\lambda)\varphi_j^\lambda(\{u_\lambda\}) \tag{18}$$

the $\mathscr{R}$ operator evaluated on a surface $\mathbf{s}$ on which each asymptotic vector $\mathbf{R}_\lambda$ is perpendicular in the appropriate channel region of space will satisfy the following:

$$\mathbf{f}\,|_{\mathbf{s}} = \mathbf{R}_{\mathbf{s}}\mathbf{f}'\,|_{\mathbf{s}} \tag{19}$$

where $\mathbf{f}$ is the vector of channel coefficients, $f_j^\lambda(R_\lambda)$ evaluated on $\mathbf{s}$ and $\mathbf{f}'|_{\mathbf{s}}$ is the vector of their normal derivatives. Since the open-channel $\mathbf{S}$ matrix can be obtained from the open-channel R matrix[10] with high accuracy,[13] and the surface is placed where there is no overlap between the open internal functions in different channels, the R matrix, properly used, can avoid the problems often associated with the different coordinates and nonorthogonal basis sets for different channels. We note that for energies below the dissociation limit the R matrix need *not* be evaluated on those portions of the surface $\mathbf{s}$ on which the bound reactant or product internal functions are zero. Thus in the three-atom collinear case, the R matrix may be evaluated on the surface shown schematically in Figure 1. As one

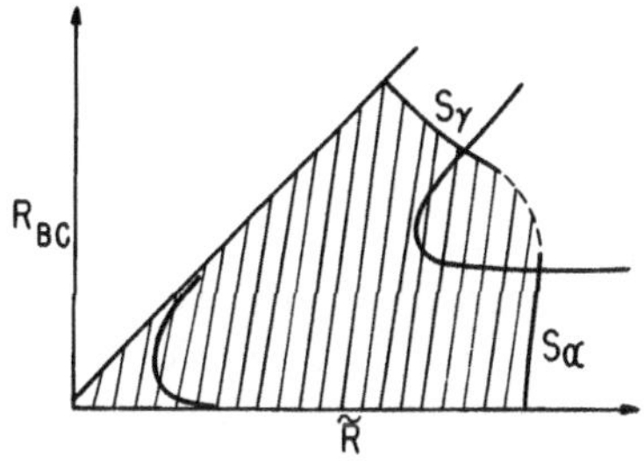

Figure 1. *R*-matrix "box" for collinear reactive scattering, Shaded area is "volume" in which the *R* matrix is evaluated. The representation $\mathcal{R}$ on the channel surfaces S_α, S_γ only is required if $E < D$. The dissociation energy contour D is shown schematically by the heavy line.

would expect, the *R* matrix evaluated as in equation (19) is blocked by channel. For three chemical channels, it would be

$$\mathbf{R} = \begin{pmatrix} \mathbf{R}^{\alpha\alpha} & \mathbf{R}^{\alpha\beta} & \mathbf{R}^{\alpha\gamma} \\ \mathbf{R}^{\beta\alpha} & \mathbf{R}^{\beta\beta} & \mathbf{R}^{\beta\gamma} \\ \mathbf{R}^{\gamma\alpha} & \mathbf{R}^{\gamma\beta} & \mathbf{R}^{\gamma\gamma} \end{pmatrix} \tag{20}$$

As mentioned above, the evaluation of the *R* matrix can be approached in a number of ways. As originally envisaged,[10] an L^2 basis throughout the entire *R*-matrix "box" is used, the Bloch Hamiltonian (see Chapter 6) is diagonalized, and the internal solutions are projected onto the external asymptotic solutions at the boundaries. In the few chemically reactive cases examined so far, this has required a large basis in translational functions for each internal function in order to obtain convergence. Since one of the primary practical problems in quantal reactive scattering theory is the large number of internal basis functions required, no one has yet had the temerity to attempt a solution of a three-dimensional reactive scattering problem by this direct approach.

An alternative approach is to subdivide the *R*-matrix box into smaller ones, approximate the true potential in each box by one analytically soluble in the translation, and use these solutions to construct, within each smaller box, an *R* matrix. These "step" *R* matrices are then combined recursively[10,15] into the overall *R* matrix, equation (20). The recursion relations combine two *R* matrices which share a common boundary into one for the combined volume. Even in the reactive scattering case where the internal basis must change between at least some *R*-matrix "boxes," it has been shown[16] that the symmetry of the *R* matrix is preserved, thus producing a (desirable) unitary **S** matrix.

In the past *R*-matrix theory has been used primarily as a formalism within which to examine resonances,[10] statistical theories,[14,17] etc. However, the desirable formal properties, the flexibility, and the relatively stable numerical properties[15] would seem to place the *R*-matrix approach high among the candidates for further development as a practical approach to three-dimensional reactive scattering.

5. *Coupled-Equations Approach*

This approach is described in considerable detail in Chapter 17, and only a few comments on it will be made here. The current coupled equations for quantal reactive scattering are based on the utilization of reaction coordinates[18] of one form or another. With the Hamiltonian written in terms of these coordinates, a set of coupled equations (differential or integral) is derived for the coefficients (functions of the translational coordinates) of the internal basis functions. These coupled equations are then solved in given regions of space, and the wave functions and derivatives are matched to each other at any common boundaries and to the proper asymptotic forms at the "external" (asymptotic) boundaries in each chemical channel. As shown in Chapter 17, the theoretical and computational methods for this have been worked out, and realistic calculations have been done.

The only major formal restriction on such calculations lies in the utilization of reaction coordinates. Inherent to such coordinate systems are regions of space which can be defined equally well in terms of "reactant" or "product" coordinates. Since these regions are by construction only in regions of very high potential energy, the wave function is taken to be zero there. This appears to be quite proper for energies below the dissociative limit.

An alternative choice of using a fixed, *nonorthogonal* coordinate system and a wave function expansion in (nonorthogonal) basis sets for *both* reactants and products was used by Diestler[19] in a collinear reactive scattering calculation. The coupled equations are then equivalent to an inelastic scattering problem in nonorthogonal coordinates. Asymptotically the basis becomes orthogonal, but the application of scattering boundary conditions is somewhat modified due to the nonorthogonal coordinates. This approach yielded very good results[19] in the collinear calculations. Even though some complications are introduced into the coupled equations (first derivative terms and a nonsymmetric effective interaction matrix), the extension of this method to three-dimensional reactive scattering would seem to be feasible.

In all of the approaches above the exact quantal treatment appears to result in calculational techniques that are quite time consuming on a computer. This is a direct result of the large numbers of channels required to obtain convergence with respect to truncation of the internal basis sets. Even in the case of $H + H_2$ reactive scattering, at $\sim$ 1-eV total energy about 30–50 coupled equations are required for some symmetries, total angular momenta, and parities to obtain convergence.[20-22] (When the channel calculations are properly combined, these yield **S** matrices coupling more than 100 states.) For other reactions the requisite numbers of states will be even higher, and thus great emphasis is now being placed on determining relatively accurate approximate approaches to the problem. These are described in following chapters.

Finally, Knapp and Diestler[23] have recently presented a quantal approach to collision-induced dissociation. Since this is described in detail in Chapter 20 it will not be discussed here.

6. *Summary*

The problem of accurate quantal calculations of three-dimensional reactive scattering has been treated formally by three distinct approaches: coupled-channel integral equations of the Lippmann–Schwinger type, a square integrable, R-matrix approach, and coupled differential equations, usually based on a collision coordinate representation. Of these, only the last has been developed to the point of actual application to realistic systems, as is described in detail in Chapter 17. Although the theory extending these techniques to asymmetric reactive systems is known,[24] the application of exact theory to many reactions will be inhibited until a rather large reduction in computation costs takes place. This may occur either because of better algorithms or cheaper computation.

All of the formal approaches mentioned here, however, have found use in the development of approximate quantal theories of chemical reaction. These are reviewed in Chapter 15.

ACKNOWLEDGMENT

This work was supported in part by the National Science Foundation.

References

1. R.G. Newton, *Scattering Theory of Waves and Particles,* McGraw-Hill, New York (1966).
2. R.D. Levine, *Quantum Mechanics of Molecular Rate Processes,* Oxford University Press, London (1969).
3. L.S. Rodberg and R.M. Thaler, *Introduction to the Quantum Theory of Scattering,* Academic Press, New York (1967).
4. J.C.Y. Chen, Coupled integral equations approach to nonrelativistic three-body systems with applications to atomic problems, *Case Stud. At. Phys.* **3,** 305 (1973).
5. L.D. Faddeev, *Mathematical Aspects of the Three-Body Problem in Quantum Scattering Theory,* Israel Program for Scientific Translations, Jerusalem, 1965.
6. M. Baer and D.J. Kouri, Coupled channel operator approach to e–H scattering, *J. Math. Phys.* **14,** 1637 (1973); D.J. Kouri and F.S. Levin, A new method in the theory of many-body scattering, *Phys. Lett. B* **50,** 421 (1974); D.J. Kouri, F.S. Levin, M. Craigie, and D. Secrest, On a general relationship between solutions of alternative sets of equations for rearrangement channel T operators, *J. Chem. Phys.* **61,** 17 (1974).
7. W. Tobocman, New integral equations for the transition operators of many body systems, *Phys. Rev. C* **9,** 2466 (1974).

8. S. Rabitz and H. Rabitz, Decomposition theory of chemical reactions, *J. Chem. Phys.* **67**, 2964 (1977).
9. D.J. Kouri, M. Craigie, and D. Secrest, Coupled channel operators and rearrangement scattering I. Comparative study of one state *S*-wave *e–H* scattering, *J. Chem. Phys.* **60**, 1851 (1974).
10. A. Lane and R. Thomas, *R*-matrix theory of nuclear reactions, *Rev. Mod. Phys.* **30**, 257 (1958).
11. O. Crawford, Calculation of chemical reaction rates by the *R*-matrix method, *J. Chem. Phys.* **55**, 2563 (1971).
12. R. Der, O. Gebhardt, and R. Haberlandt, *R*-matrix theory of reactive atom–diatom collisions, *Chem. Phys. Lett.* **27**, 107 (1974).
13. D. Zvijac and J. Light, *R*-matrix theory for collinear chemical reactions, *Chem. Phys.* **12**, 237 (1976).
14. D. Zvijac and J. Light, Reactions of large molecules proceeding through an intermediate complex I. Theory, *Chem. Phys.* **21**, 411 (1977).
15. J. Light and R. Walker, An *R*-matrix approach to the solution of coupled equations for atom–molecule reactive scattering, *J. Chem. Phys.* **65**, 4272 (1976).
16. E. Stechel, R. Walker, and J. Light, R-matrix propagation methods II. Inelastic scattering, *J. Chem. Phys.* **69**, 3518 (1978).
17. B.C. Eu and J. Ross, Some applications of the *R*-matrix theory to reactive and elastic molecular scattering, *J. Chem. Phys.* **44**, 2467 (1966).
18. R.A. Marcus, Analytical mechanics of chemical reactions III. Natural collision coordinates, *J. Chem. Phys.* **49**, 2610 (1968).
19. D.J. Diestler, Close-coupling technique for chemical exchange reaction of the type $A + BC \rightarrow AB + C, H + H_2 \rightarrow H_2 + H$, *J. Chem. Phys.* **54**, 4547 (1971).
20. G. Schatz and A. Kuppermann, Quantum mechanical reactive scattering for three-dimensional atom plus diatom systems II. Accurate cross sections for $H + H_2$, *J. Chem. Phys.* **65**, 4668 (1976).
21. A. Elkowitz and R.E. Wyatt, Quantum mechanical reaction cross sections for the three-dimensional hydrogen exchange reaction, *J. Chem. Phys.* **62**, 2504 (1975).
22. E. Stechel, R. Walker, and J. Light, private communication.
23. E.-W. Knapp and D. Diestler, Quantum mechanical treatment of collision-induced dissociation, *J. Chem. Phys.* **67**, 4969 (1977).
24. R. Walker, J. Light, and A. Altenberger-Sizcek, Chemical reaction theory for asymmetric atom–molecule collisions, *J. Chem. Phys.* **64**, 1166 (1976).

15
Reactive Scattering Cross Sections II: Approximate Quantal Treatments

ROBERT E. WYATT

1. Introduction

In this chapter approximate quantal methods for predicting reaction probabilities and cross sections will be discussed. Emphasis is placed upon practical methods which employ quantum dynamics for the reactive scattering of neutral, electronically adiabatic A + BC three-dimensional collisions. Some of the terminology and techniques used in this chapter will be considered in somewhat more detail in Chapter 17, where "accurate" methods for atom–diatom reactions are discussed.

All of the approximate methods to be reviewed here have undergone extensive, rapid development during the past few years. In most cases, not many chemical reactions have been studied in detail. Because of this, the range of applicability of the methods is not fully understood. However, the three major methods to be discussed all look very promising, and each has already been applied to at least several different reactions. The first method, *J_z conserving*, has been used to reduce the number of coupled equations which are simultaneously integrated in the close-coupling scheme. The second method, *Born approximations*, actually includes a number of free-wave and distorted-wave modifications. The Born approximations, for selected reactions, lead to reasonable predictions of product vibration–rotation distributions, and may, in the distorted-wave version, yield accurate

ROBERT E. WYATT • Department of Chemistry, The University of Texas at Austin, Austin, Texas 78712.

absolute cross sections. The third method, *overlap models*, also includes a number of submodels which employ different physical and mathematical approximations to the product–reactant nuclear wave function overlap integral. As mentioned above, emphasis here is placed upon methods which have already been applied to three-dimensional reactions. We of course are deeply indebted to the hundreds of calculations on model collinear and coplanar reactions which make this extension to three "real" dimensions possible.

2. Angular Momentum Decoupling: J_z Conserving

The J_z-conserving approximation, formulated originally by McGuire and Kouri[1] and by Pack[2] for atom–diatom inelastic scattering, has been recently extended to diatom–diatom inelastic[3] and to three-dimensional atom–diatom reactive scattering, with application to the hydrogen exchange reaction.[4,5] A body-fixed, BF, (rotating) coordinate system is first introduced, followed by neglect within each arrangement channel of those kinetic energy coupling terms (referred to as kinematic, as opposed to potential coupling) which mix BF wave functions of different K, the component of $\mathbf{J}$ along the rotating z axis. Many fewer coupled channels need be simultaneously integrated (compared to full close coupling), so the computer time is only about 1/30th of the time required for full close-coupling calculations. The two formulations of J_z conserving for reactive scattering will be reviewed separately, since they involve different choices of rotating axes and different rotational basis sets.

In the formulation of Elkowitz and Wyatt,[4] natural collision coordinates[6,7] are used, with the BF z axis changing alignment in the interaction region from the BC to A relative vector in reactants to the C to AB vector in products. This realignment is accomplished through use of a switching angle $\alpha(s)$, whose value depends upon the extent of evolution (measured by s) along the reference curve leading from reactants to products. The total Hamiltonian is then partitioned into terms diagonal in K, and those producing $K \pm 1$ coupling (these are asymmetry terms coming from off-diagonal elements of the inertia tensor in addition to Coriolis interaction terms between internal bending and triangle tumbling). It is important to note that K in this formulation does not mean the same thing in the reaction zone as the value used for example by Schatz and Kuppermann, since the orientations of the tumbling z axes relative to the three nuclei are different.

At any stage of reaction, the rotational dynamics of the system is that of an (unstable) asymmetric top, freely tumbling, but which has an internal (bending) degree of freedom characterized by the angle γ, which determines the deviation from collinearity. Adiabatic rotational states of the asymmetric top Hamiltonian may be expanded in normalized symmetric top functions $N^J_{MK}(\theta, \phi, \chi)$ times hindered internal rotor functions $R^{JK}_{jl}(\gamma; s)$,

$$\Omega^{JM}_{jl}(\theta, \phi, \chi, \gamma ; s) = \sum_{K=-J}^{+J} N^J_{MK}(\theta, \phi, \chi)\, R^{JK}_{jl}(\gamma ; s) \tag{1}$$

The asymmetric top rotational energies W^J_{jl} asymptotically reduce to (in the rigid-rotor approximation)

$$W^J_{jl} \rightarrow \frac{j(j+1)\hbar^2}{2\mu r_0^2} + \frac{l(l+1)\hbar^2}{2\mu Z^2} \tag{2}$$

where Z is the scaled atom–diatom distance and r_0 is the scaled diatom distance. At each value of s, asymmetric top states form a basis for expansion of the rotational part of the scattering wave function; the rotational energies $W^J_{jl}(s)$ appear as effective potentials in the resultant close-coupling equations (see Chapter 17).

First Elkowitz and Wyatt examined in detail the $J = 0$ rotational energy correlation diagram for $H + H_2$ on a model fit to the Yates–Lester–Liu potential surface.[8,9] For this value of J, only one value of l is allowed for each j ($l = j$), and in addition K is exactly conserved, $K = 0$. In order to scatter when $J > 0$, an approximate correlation diagram may be obtained from the $J = 0$ case by modifying W^0_{jj} in order to (a) account for the increased centrifugal energy of the complex near $s = 0$, and (b) adjust the asymptotic energy from W^0_{jj} to W^J_{jl} given in equation (2). Assuming that the $J = 0$ rotational problem has been solved (the effect of the rotational potential has been included exactly), the new levels in the $J > 0$ correlation diagram in the reactant arrangement region may be approximately determined from the "interpolation formula"[4]

$$W^J_{jl} = W^0_{jj}(s) + [J(J+1)(1+f) + fj(j+1) - fl(l+1)]\,\frac{A\hbar^2}{2} \tag{3}$$

where A is an inertia coefficient evaluated on the reaction path, and the function f has the limiting values $f \rightarrow -1$ for separated reactants and $f = 0$ at $s = 0$ (transition state). A single value of l for each value of j and J can be determined by examination of the hindered rotor wave functions in equation (1). The fact that only one l is used for each Jj (instead of all values allowed by the triangle inequality $l = |J - j|, |J - j| + 1, \ldots, J + j$) is the reason that this method is so much faster than full close-coupling calculations. From equation (1), the probability of finding angular momentum component K for different values of l when Jj are fixed is given by

$$P_K(l; s) = \left| \int_0^{\pi} R^{JK*}_{jl}(\gamma ; s)\, R^{JK}_{jl}(\gamma ; s) \sin\gamma\, d\gamma \right|^2 \tag{4}$$

with

$$\sum_{K=-J}^{+J} P_K(l;s) = 1 \tag{5}$$

Examination of distributions $\{P_K(l;s), l \text{ ranging}, Jj \text{ fixed}\}$ near the transition state showed that each l in reactants correlated with one value of $|K|$ near $s = 0$. The special values of l which lead to $K = 0$ concentration near the the transition state are given by the algorithm

$$\begin{aligned}
&J(\text{even}) \begin{cases} j(\text{odd}) & l = |J - j| \\ j(\text{even}) & l = \begin{cases} J & \text{if } j \leq J \\ j & \text{if } j > J \end{cases} \end{cases} \\
&J(\text{odd}) \begin{cases} j(\text{odd}) & l = |J - j| \\ j(\text{even}) & l = \begin{cases} J & \text{if } j < J \\ j+1 & \text{if } j > J \end{cases} \end{cases}
\end{aligned} \tag{6}$$

The one-to-one $l \to |K|$ correlation between reactants or products and the transition state tends to break down as J or j increases (each l then leads to a distribution over $|K|$ near $s = 0$). Reactive scattering with $K = 0$ dominates at low J, since (a) in a semiclassical picture, the molecular angular momentum j would be perpendicular to the rotating z axis and more time would be spent by the three atoms in near-linear geometries, which favors reaction, and (b) only the spherical harmonics $Y_j^K(\gamma, 0)$ with $K = 0$ allow for appreciable density accumulation near $\gamma = 0$ or $\gamma = \pi$ (the collinear geometries). This dominance of $K = 0$ had been observed in earlier $H + H_2$ calculations, and is discussed in a review article by Truhlar and Wyatt.[10]

The J_z-conserving method in the natural coordinate-asymmetric top formulation was used for rotational excitation calculations on the reaction $H + H_2(v = 0, j = 0) \to H_2(v' = 0, j') + H$. Reaction probabilities vs. J (for distinguishable atoms) are shown in Figure 1a at $E_{tr} = 0.528$ eV for both 50-channel close coupling and 18-channel J_z-conserving calculations. Good agreement was obtained, particularly for $J \lesssim 10$. Total reaction cross sections from the J_z model in the post-threshold energy region shown in Figure 1b are within 8% of accurate close-coupling results on the same potential.

Kuppermann *et al.*[5] considered a different J_z formulation, again with numerical application to the $H + H_2$ reaction. Their method was not based upon properties of adiabatic asymmetric top states in natural coordinates. Within each arrangement tube, the rotational part of the scattering wave function was expanded in a $D_{MK}^J Y_j^K$ basis, with translation–vibration expansion coefficients:

$$\Psi^{JM}(\mathbf{r}, \mathbf{R}) = \sum_{K=-J}^{+J} \sum_{j=|K|}^{\infty} D_{MK}^J(\theta, \phi, \chi)\, Y_j^K(\beta; 0)\, w_j^{JK}(r, R) \tag{7}$$

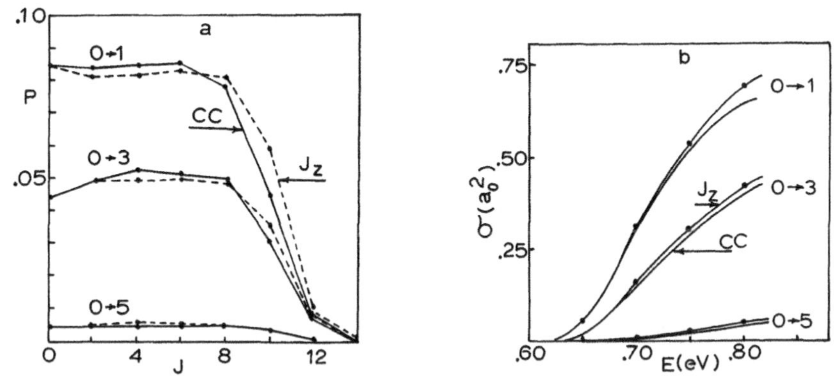

Figure 1. Close-coupling (CC) and J_z-conserving (J_z) results for the $H + H_2$ reaction.[4] (a) Para → ortho reaction probabilities vs. J from the ground-state H_2 reactant molecule at $E_{total} = 0.80$ eV. Probabilities are for reaction with both ends of the molecule. (b) Total reaction cross sections vs. total energy ($E_{total} = E_{trans} + 0.27$ eV) for $j = 0 \to j' = 1, 3, 5$. J_z results are shown as solid dots at four energies. Cross sections are for reaction with both ends of the molecule, without antisymmetrization for identical particles.

where β is the **r**, **R** angle (not the "natural" bending angle as in Elkowitz and Wyatt). This type of expansion is performed separately, in the reactant and in each product arrangement region, where $\phi, \theta, \chi, \beta, R, r$ or $\phi', \theta', \chi', \beta', R', r'$, respectively, are the six coordinates defining the orientation, size, and shape of the three-body triangle. In reactants, K is defined along **R**, while K' in products is defined along **R**$'$. Substitution of equation (7) into the Schrödinger equation, followed by projection onto the rotational basis functions, yields a set of coupled equations for the translation–vibration functions:

$$(\hat{h}_j^{JK} - E)\, w_j^{JK} + \sum_{j'=|K|}^{\infty} V_{jj'}^{K} w_{j'}^{JK} + \hat{h}_j^{J,K+1} w_j^{J,K+1} + \hat{h}_j^{J,K-1} w_j^{J,K-1} = 0 \quad (8)$$

where the K to $K \pm 1$ coupling terms arise from the off-diagonal matrix elements of the orbital angular momentum operator $\hat{\mathbf{l}}^2 = (\hat{\mathbf{J}} - \hat{\mathbf{j}})^2$. In equation (8), the kinematic coupling operators have a long-range R^{-2} dependence:

$$\hat{h}_j^{J,K\pm 1} = \frac{-\hbar^2}{2\mu R^2}[J(J+1) - K(K \pm 1)]^{1/2}[j(j+1) - K(K \pm 1)]^{1/2} \quad (9)$$

The other operators in equation (8) are defined later in Chapter 17; the important point is that the potential is off-diagonal in j, but diagonal in K, while the K-coupling terms are diagonal in j. In the "tumbling-decoupling" approximation employed by Kuppermann *et al.*, the $K \pm 1$ coupling terms in equation (8) were neglected within each arrangement tube, but the remaining potential coupling terms (diagonal in K) were retained. The functions $w_j^{JK}(R, r)$ were in turn expanded in a local vibrational basis, and close-coupling equations were solved for the expansion coefficients (translational wave functions).

At some point in the reaction zone, it is necessary to change angular bases, since the coordinates used θ, ϕ, χ, β for reactant geometries are different from the product set $\theta', \phi', \chi', \beta'$. As discussed in more detail in Chapter 17, these coordinates are changed on matching surfaces which separate the A + BC, B + AC, and C + AB arrangement regions of configuration space. The exact matching of the scattering wave function and its normal derivative involves K mixing, since $\mathbf{J} \cdot \hat{\mathbf{R}} \neq \mathbf{J} \cdot \hat{\mathbf{R}}'$ if $\hat{\mathbf{R}}$ and $\hat{\mathbf{R}}'$ are noncollinear. The result is that the wave function $w_{j'}^{JK'}(R', r')$ on the matching surface in the product arrangement region is "contaminated" with various K components of the reactant channel wave functions $w_j^{JK}(R, r)$.

Kuppermann *et al.*[5] reported probabilities for several reactive and nonreactive transitions from their calculations on the Porter–Karplus H_3 surface.[11] The intense individual transitions and the degeneracy-averaged ones are in good agreement with full close-coupling results. With matching surface $K \to K'$ mixing included, the J_z approximation agrees with close-coupling results to within about 7% over the energy range $E_{tr} = 0.18$–0.38 eV, where the total cross section increases by three orders of magnitude. However, for weak transitions, where the initial-channel and final-channel K values differ (these probabilities are an order of magnitude or more smaller than the $K = K' = 0$ processes), considerable error may result. For example, for the $v = 0, j = 0, m_j = 0\,(K = 0) \to v' = 0, j' = 1, m_{j'} = 1$ $(K' = 1)$ transition, the probability from the decoupling method is only about 4% of the exact value (5.7×10^{-4}). Many individual nonreactive transitions are in error by similar amounts. In spite of these errors, it should be recalled that only degeneracy-averaged total cross sections are needed for predicting the type of thermal rate constants usually used in chemical kinetics. The J_z-conserving method predicts total cross sections quite well.

The J_z-conserving approximation has also been applied by Redmon and Wyatt[12] to the $F + H_2(v = 0, j = 0) \to FH(v'j') + H$ reaction for $E_{tr} <$ 0.20 eV. As many as 80 coupled vibration–rotation channels were used for each J, for $J \lesssim 20$. The J_z approximation for this reactive scattering problem should work rather well; like $H + H_2$, the preferred transition state is linear and $K = 0$ scattering is expected to dominate. Because of the large number of open channels (>100 for $J > 0$), "exact" close-coupling calculations are not currently feasible on exoergic systems of this type. The J_z approximation is the significant feature that made the FH_2 calculations feasible. It is likely that this approximation will soon be applied to several other reactive systems, including $Cl + H_2$, $H + DT$, and $F + HD$.

3. Born-Type Approximations

3.1. T-Matrix Elements and Distortion Potentials

The Born approximation (BA) and the distorted-wave Born approximation (DWBA) have provided qualitatively and sometimes quantitatively accurate information on differential cross sections, total cross sections, and product state distributions in chemical reactions. Approximate transition amplitudes for individual state-to-state processes may be computed one at a time, so that unlike close-coupling techniques, only those processes of interest need be focused upon. (The collection of approximate S-matrix elements will not form a unitary S matrix, but unitarization is possible.) A readable description of these approximations is in the paper by Tang and Karplus.[13] Formal scattering theory[14] provides *exact* expressions for the T-matrix elements in a reaction leading from $A + BC \rightarrow AB + C$, symbolized $\alpha \rightarrow \beta$ (where α or β temporarily denote both arrangement channel and internal quantum numbers):

$$T_{\alpha\beta} = \langle \Phi_\beta | \hat{V}_\beta | \Psi_\alpha^{(+)} \rangle = \langle \Psi_\beta^{(-)} | \hat{V}_\alpha | \Phi_\alpha \rangle \tag{10}$$

If the total nuclear Hamiltonian is decomposed

$$\begin{aligned} \hat{H} &= \hat{K} + \hat{H}^0_{BC} + \hat{V}_\alpha(\mathbf{R}, \mathbf{r}) = \hat{H}^0_\alpha + \hat{V}_\alpha \\ &= \hat{K} + \hat{H}^0_{AB} + \hat{V}_\beta(\mathbf{R}', \mathbf{r}') = \hat{H}^0_\beta + \hat{V}_\beta \end{aligned} \tag{11}$$

where $\hat{K}$ is the kinetic energy operator for atom–molecule relative motion, and $\hat{H}^0_{BC}$ ($\hat{H}^0_{AB}$) is the molecular Hamiltonian for the initial (final) state, then $\hat{V}_\alpha$ ($\hat{V}_\beta$) is defined as the interaction potential in initial (final)-state form. As a result, $\hat{V}_\alpha$ provides both for elastic and inelastic processes and for rearrangement leading to $AB + C$. The functions Φ_α and Φ_β are solutions to the noninteracting atom–molecule problem at total energy E:

$$(\hat{H}^0_\alpha - E)\,\Phi_\alpha = 0, \qquad (\hat{H}^0_\beta - E)\,\Phi_\beta = 0 \tag{12}$$

These solutions are usually written as products of plane waves times molecular vibration–rotation wave functions, for example,

$$\Phi_\alpha(\mathbf{R}, \mathbf{r}) = e^{i\mathbf{k}_{vj}\cdot\mathbf{R}} H_{vj}(r)\, Y_j^{m_j}(\theta\phi) \tag{13}$$

where $\mathbf{k}_{vj}$ (or $\mathbf{k}_{v'j'}$) specifies the initial (or final) direction for relative motion. In addition, $\Psi^{(\pm)}_{\alpha(\beta)}$ are exact scattering wave functions for Φ_α (Φ_β) as the "initial" state, with the ($\pm$) superscript denoting outgoing or incoming boundary conditions:

$$(\hat{H} - E)\,\Psi_\alpha^{(\pm)} = (\hat{H} - E)\,\Psi_\beta^{(\pm)} = 0 \tag{14}$$

These exact wave functions obey Lippmann–Schwinger integral equations, with boundary conditions explicitly included:

$$\Psi_\alpha^{(+)} = \Phi_\alpha + (E + i\varepsilon - \hat{H}_\alpha^0)^{-1} \hat{V}_\alpha \Psi_\alpha^{(+)}$$
$$\Psi_\beta^{(-)} = \Phi_\beta + (E - i\varepsilon - \hat{H}_\beta^0)^{-1} \hat{V}_\beta \Psi_\beta^{(-)} \tag{15}$$

Because of the virtual impossibility of solving equation (15) exactly for most chemical systems, numerous approximate solutions have been suggested. If only the first term on the right side of equation (15) is retained in equation (10), the (first) Born approximation for the T-matrix element is obtained:

$$T_{\alpha\beta}^{\mathrm{BA}} = \langle \Phi_\beta | \hat{V}_\beta | \Phi_\alpha \rangle = \langle \Phi_\beta | \hat{V}_\alpha | \Phi_\alpha \rangle \tag{16}$$

This approximation is not expected to provide accurate cross sections for low-energy chemical reactions, but in the next section it is demonstrated that (for some processes) propensity rules for final-state distributions and the energy dependence of total cross sections are correctly predicted.

The distorted-wave Born approximation is expected to provide more accurate (absolute) cross sections, particularly at low energy near the reaction threshold. In the DWBA, the initial and final interaction potentials are each decomposed into a sum of two terms. Usually, the first term is chosen to account for the strong repulsive interaction between the atom and molecule. It is denoted $\hat{V}_\alpha^0(R)$ or $\hat{V}_\beta^0(R')$ and produces elastic scattering only, since it contains no molecular geometry coordinates (vibration–rotation dependence). The DWBA partitioning for the interaction potentials is thus

$$\hat{V}_\beta(R, r, \gamma) = \hat{V}_\alpha^0(R) + \hat{V}'_\alpha(R, r, \gamma)$$
$$\hat{V}_\beta(R', r', \gamma') = \hat{V}_\beta^0(R') + \hat{V}'_\beta(R', r', \gamma') \tag{17}$$

where the distortion potentials $\hat{V}'_\alpha$ and $\hat{V}'_\beta$ are responsible for both inelastic processes and reaction. It is assumed that distorted-wave solutions can be found from the equations

$$(\hat{H}_\alpha + \hat{V}_\alpha^0 - E)\,\chi_\alpha^{(\pm)} = 0$$
$$(\hat{H}_\beta + \hat{V}_\beta^0 - E)\,\chi_\beta^{(\pm)} = 0 \tag{18}$$

Formal scattering theory[14] provides another form for the *exact* T-matrix element:

$$T_{\alpha\beta} = \langle \chi_\beta^{(-)} | \hat{V}'_\beta | \Psi_\alpha^{(+)} \rangle = \langle \Psi_\beta^{(-)} | \hat{V}'_\alpha | \chi_\alpha^{(+)} \rangle \tag{19}$$

The DWBA then arises when the exact scattering wave function is approximated by distorted-wave solutions:

$$T_{\alpha\beta}^{\mathrm{DW}} = \langle \chi_\beta^{(-)} | \hat{V}'_\beta | \chi_\alpha^{(+)} \rangle = \langle \chi_\beta^{(-)} | \hat{V}'_\alpha | \chi_\alpha^{(+)} \rangle \tag{20}$$

Once the T-matrix elements have been estimated with the BA or DWBA, the differential reaction cross section is computed from the standard formula

$$\frac{d\sigma_{\alpha\beta}(\theta)}{d\Omega} = \frac{\mu_\alpha \mu_\beta}{\hbar^4} \frac{k_\beta}{k_\alpha} |T_{\alpha\beta}(\mathbf{k}_\alpha, \mathbf{k}_\beta)|^2 \tag{21}$$

where $\mathbf{k}_\beta$ is oriented relative to $\mathbf{k}_\alpha$ by the scattering angle θ. Also, $\mu_\alpha(\mu_\beta)$ is the reduced mass for relative motion in the entrance (exit) channel. Specific applications of these approximations will now be considered.

3.2. *Born Approximation: Energy Dependence of the Total Reaction Cross Section*

Accurate analytic formulas for the energy dependence of reaction cross sections are difficult to derive, so it is interesting to examine Eu's recent derivation, which started with the Born approximation.[15] This formula is not expected to be valid near threshold, but applies to the post-maximum energy range of the cross section for an exoergic reaction. Specific application was made to the $K + CH_3I \rightarrow KI + CH_3$ reaction, with CH_3 treated as a structureless particle. Eu started with the Raff–Karplus[16] parameterization of the potential surface and wrote the final-state interaction potential in the form (for $A + BC \rightarrow AB + C$)

$$\hat{V}_\beta(R', r') = V_{BC}(r) + V_{AC}(|r - r'|) + \eta_{AB}(r')\, S_{BC}(r) + \eta_{BC}(r)\, S_{AB}(r') \tag{22}$$

where V_{BC} is a Morse potential, V_{AC} is a Yukawa potential for AC repulsion $[V_{AC}(x) \sim x^{-1} \exp(-\beta x)]$, and η_{ij} is a switching function. The two terms involving η_{ij} lead to higher-order terms in the inverse energy expansion of the cross section given below and are neglected.

The Born approximation for the $\alpha \rightarrow \beta$ T-matrix element is then the six-dimensional integral

$$T^{BA}_{\alpha\beta} = \langle e^{i\mathbf{k}_\beta \cdot \mathbf{R}'} \phi_\beta(\mathbf{r}') \,|\, \hat{V}_\beta \,|\, e^{i\mathbf{k}_\alpha \cdot \mathbf{R}} \phi_\alpha(\mathbf{r}) \rangle \tag{23}$$

Each plane wave is next expanded in spherical Bessel functions. In order to avoid numerical integrations, the method of stationary phase is used to evaluate the integral, the integrand of which is highly oscillatory at sufficiently high energy. The state-to-state cross sections are found to decrease as E_{tr}^{-1} as E_{tr} increases. A total (average) cross section is defined by summing over all final product states:

$$\bar{\sigma} = \sum_{v'j'} \left\{ (2j+1)^{-1} \sum_{mjm_j'} \sigma_{vjm_j v'j'm_j'} \right\} \tag{24}$$

The final formula for the average cross section is then

$$\bar{\sigma}(E_{tr}) = \sigma_0 E_{tr} \left(1 + \frac{Q}{E_{tr}} \right)^{5/2} \tag{25}$$

where Q is the ground-state-to-ground-state exoergicity ($Q > 0$). An explicit formula for σ_0 [equation (3-16) in Reference 15] involves Morse parameters, moments of inertia, and reduced masses for AB and BC.

The Born average cross section formula was applied by Eu to the $K + CH_3I \rightarrow KI + CH_3$ reaction. Evaluation of σ_0 from spectroscopic parameters led to cross sections about four times the experimental values of Gersh and Bernstein.[17] However, if the cross section is fit to the experimental value at $E_{tr} = 0.94$ eV (the reaction cross section peaks near 0.2 eV translational energy and declines smoothly until at least 1.2 eV), the energy dependence of the cross section in equation (25) provides an excellent fit to the data. In addition, when the experimental data above 0.2 eV is plotted in the form $\ln(\sigma_{exp}/E_{tr})$ vs. $\ln(1 + QE_{tr}^{-1})$, the slope of the linear plot is 5/2! Further studies of the energy dependence of state-to-state and average cross sections as provided by Born formulas are clearly of great interest.

3.3. Born Approximation: Product State Internal Energy Distributions

The Born approximation has also been used to predict product state vibration–rotation distributions in atom–diatom reactions. For the A + BC → AB + C reaction, Kafri *et al.*[18] neglected end-atom AC repulsion and approximated the total potential as $\hat{V} = \hat{V}_{BC}(r) + \hat{V}_{AB}(r')$. This gives for the final-state interaction potential just the binding potential of the reactant diatom (a harmonic oscillator potential) $\hat{V}_\beta = \hat{V} - \hat{V}_{AB}(r') = \hat{V}_{BC}(r)$. When inserted into equation (23) for the T-matrix element, the resultant integral breaks down into a product of three factors:

$$T^{BA}_{vjm_jv'j'm_j'} = \left[E_{vj} + \frac{\hbar^2 q_C^2}{2\mu_{BC}} \right] g^*_{vjm_j}(\mathbf{q}_A)\, g_{v'j'm_j'}(\mathbf{q}_C) \tag{26}$$

where E_{vj} is the BC diatomic internal energy. This important relation can also be more rigorously derived from the formal reaction theory of three-body systems (see Section 3.4). The form factors g are Fourier transforms of the internal wave functions:

$$g_{vjm_j}(\mathbf{q}) = \int d\mathbf{r} e^{-i\mathbf{q}\cdot\mathbf{r}} \phi_{vjm_j}(\mathbf{r}) \tag{27}$$

The quantities $\mathbf{q}_C$ and $\mathbf{q}_A$ are momentum transfers (in units of $\hbar$), and are defined by noting the following momentum expressions (in the cm frame).

$$\left.\begin{aligned} \mathbf{k} &= \text{momentum of A before reaction} \\ -\mathbf{k} &= \text{momentum of BC} \\ -\frac{m_C}{m_B + m_C}\mathbf{k} &= \text{momentum of C} \end{aligned}\right\} \text{reactants}$$

$$\left.\begin{aligned} \mathbf{k}' &= \text{momentum of C after reaction} \\ -\mathbf{k}' &= \text{momentum of AB} \\ -\frac{m_A}{m_A + m_B}\mathbf{k}' &= \text{momentum of A} \end{aligned}\right\} \text{products}$$

The momentum changes of A and C are then the arguments of the form factors in equation (26):

$$\mathbf{q}_A = -\mathbf{k} - \frac{m_A}{m_A + m_B}\mathbf{k}'$$

$$\mathbf{q}_C = +\mathbf{k}' + \frac{m_C}{m_B + m_C}\mathbf{k} \tag{28}$$

In order to further evaluate the form factors, the plane waves are expanded in spherical Bessel functions, and the angular integrals are done explicitly; this gives

$$g_{vjm_j}(\mathbf{q}) = 4\pi(-i)^j\, Y_j^{m_j}(\hat{q})\, g_{vj}(q) \tag{29}$$

where the vibrational form factor is

$$g_{vj}(q) = \int_0^\infty r\, dr\, j_i(qr)\, \phi_v(r) \tag{30}$$

[In an appendix to their paper, Kafri *et al.*[18] consider the evaluation of $g_v(q)$ for harmonic oscillator vibrational functions.] For nonpolarized reactant molecules, and for product molecules where the polarization is not measured, the differential reaction cross section is obtained by averaging over m_j and summing over m'_j; this yields

$$\sigma_{vjv'j'}(\theta) = \frac{k_{v'j'}}{k_{vj}}(2j' + 1)\left[E_{vj} + \frac{\hbar^2 q_C^2}{2\mu_{BC}}\right] g_{vj}^2(q_A)\, g_{v'j'}^2(q_C) \tag{31}$$

Since this result is not expected to provide reliable estimates of the magnitudes of total cross sections, emphasis is placed upon predictions of the vj and $v'j'$ dependence of the normalized probabilities,

$$P_{vjv'j'} = \sigma_{vjv'j'} \Big/ \sum_{v'j'} \sigma_{vjv'j'}$$

Specific applications to the Cl + HI(DI) → I + HCl(DCl), F + H_2 → FH + H, and F + HD → FH + D, FD + H reactions were considered by Kafri *et al.*[18] For F + H_2, product distributions $P_{00v'j'}$ were computed and a surprisal analysis was performed on the resultant vibrational distributions. The surprisal slope parameter came out indicating a higher tendency to produce inverted products relative to the experimental results[19] [$\lambda_v(\text{Born}) \approx -10$, $\lambda_v(\text{exp}) \approx -6$]. In addition, the inversion ratio $P(v' = 2)/P(v' = 1)$ (where sums have been performed over product rotational levels) was correctly higher for para-H_2 than for the even-j–odd-j H_2 mixture referred to as normal-H_2. The theoretical inversion ratio was only slightly larger than the experimental value.[20] For the F + HD reaction the HF/DF branching ratio at $E_{tr} = 0.09$ eV was 0.6, 1.25, 1.37, and 1.57 for j values of 0, 1, 2, and 3, respectively. The increase of this branching ratio is more pronounced than in trajectory computations.[21]

Kimura[22] has recently applied the Born approximation to the I + HI(v) → IH(v') + I reaction, with $v = 0$ and $v' = 0, 1,$ and 2. He emphasized that equation (26) for the cross section depends upon the product of two Franck–Condon factors describing the overlap of bound vibrational functions with "free" plane waves. This viewpoint is also utilized when the spectator-stripping limit is considered in the following section.

3.4. Spectator-Stripping Model

In a recent series of papers, Micha and collaborators have developed a multiple-collision expansion of the Faddeev coupled integral equations. The atom–diatom reaction is viewed as proceeding through a succession of atom–atom encounters, with higher terms in the expansion of the T matrix describing multiple-pair collisions.[23] The theory is based on a total interaction potential V written as a sum of pair potentials $\hat{V}_{AB} + \hat{V}_{BC} + \hat{V}_{CA}$. This was shown to lead to realistic potential surfaces for reactive systems, provided the pair potentials were allowed to depend on atomic spins. Using the potential $\hat{V}$ in the formalism of three-body rearrangements, and iterating the resulting set of coupled equations, one arrives at the multiple-collision expansion for the transition amplitude:

$$T_{\beta\alpha} = \langle \Phi_\beta | \hat{G}_0^{-1} + \hat{t}_{AC} + \hat{t}_{BC}\hat{G}_0\hat{t}_{AC} + \hat{t}_{BC}\hat{G}_0\hat{t}_{AB} + \hat{t}_{AC}\hat{G}_0\hat{t}_{AB} + \cdots | \Phi_\alpha \rangle$$

where $\hat{G}_0$ is the propagator for free motion of three atoms, and the $\hat{t}_{IJ}$ are transition operators for collisions between atoms I and J inside the system. The lowest-order term automatically gives the "spectator-stripping"[24,25] matrix element of equation (26), in which the rearrangement A + BC → AB + C is seen to result from a matching of atomic momenta that allows for a smooth reactants → products transition. The exact Faddeev theory has been developed by Micha,[23] and the multiple-collision expansion has been described by Micha and Yuan.[26] Recent applications have been made to the following reactions: T + H_2,[27] Ar^+ + H_2,[26,28] K + I_2,[28,29] K + Br_2,[29] and K + IBr.[29] Because of its complexity, the K + Br_2 reactions will be discussed in more detail, although the K + I_2 reaction has been more thoroughly studied.

Yuan and Micha[28,29] considered the K + Br_2 reaction to occur through the following three-step sequence.

(i) Electron transfer (vertical) harpooning model[30]:
$K + Br_2(v = 0, j = 0) \rightarrow K^+ + Br_2^-(v'', j'')$

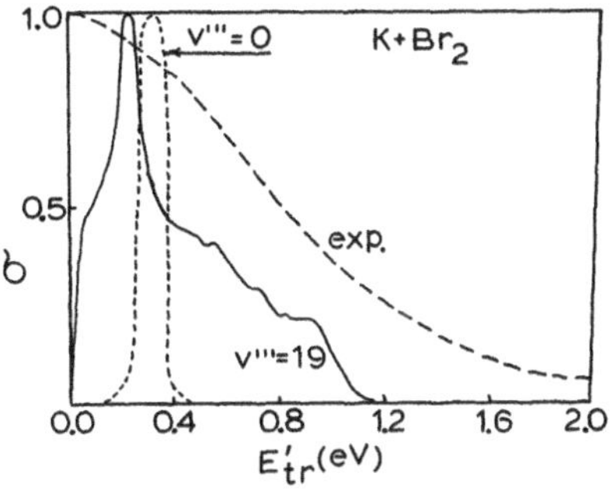

Figure 2. Product translational energy distribution for the $K + Br_2(v=0) \rightarrow KBr + Br$ reaction at $E_{tr} = 0.14$ eV. Dotted and solid curves are spectator stripping results[29] for $v''' = 0$ and 19, respectively. Experimental data[31] are shown as the dashed curve. The maximum value of σ is set to unity in all cases.

(ii) Partial relaxation of Br_2^- in the Coulomb field of K^+:
$K^+ + Br_2^-(v'', j'') \rightarrow K^+ + Br_2^-(v''', j''')$

(iii) Spectator-stripping reaction of relaxed molecule:
$K^+ + Br_2^-(v''', j''') \rightarrow KBr(v', j') + Br$

A stochastic model was used to estimate the electron transfer probability in step (i); using Franck–Condon vibrational overlap factors, the most probable value of v'' was 19, with a spread from 12 to 26. As in equation (31), the final step proceeds in the single-collision limit with an amplitude proportional to the product of the molecular vibrotor states in momentum space. Harmonic oscillator vibrational states were used to evaluate equation (31). Experimental[31] and computed relative reaction cross sections for reactions proceeding through specific v''' intermediate levels vs. product translational energy are shown in Figure 2. The higher vibrational states v''' contribute greatly to the broadening of the energy distribution. The computed maximum does not occur at $E'_{tr} = 0$, in contrast to the experimental results. However, it is very significant that at least preliminary quantum studies have begun on these complex reactions.

3.5. *Distorted Waves: Methodology*

The DWBA calculations begun by Micha[32] and by Karplus and Tang[13,33] in the early to mid-1960's on the $H + H_2$ (and isotopic) reactions were the first extensive quantal calculations on atom–diatom reactions. These studies and their relation to other (classical and statistical) reaction theories, as applied to $H + H_2$, have been reviewed by Truhlar and Wyatt.[10] We will focus here on a recent reformulation of the DWBA method by Choi and Tang[34] which led to a significant reduction in computer time requirements (compared to the 1965 calculations) and has resulted in rather *accurate* absolute reaction cross sections (unlike Born approximation results).

In the adiabatic perturbed model,[13,33] the entrance channel distorted wave function is approximated as the product of a deformed molecular wave function times a translational function:

$$\chi_\alpha^{(+)}(\mathbf{R}, \mathbf{r}) = T_\alpha^{(+)}(\mathbf{R})\, \phi_\alpha(\mathbf{r}\,; R) \tag{32}$$

where the molecular function is in turn approximated by the product of a deformed vibrational factor and a deformed (hindered) rotor function:

$$\phi_\alpha(\mathbf{r}\,; R) = H_{v_j}(r\,; R)\Lambda_j(\gamma\,; R) \tag{33}$$

At each value of R, the deformed molecular functions are defined as approximate eigenfunctions of the operator $(\hat{H}^0_{\mathrm{BC}} + \hat{V})$, where $\hat{H}^0_{\mathrm{BC}}$ is the asymptotic (free) molecular Hamiltonian. Note that the full interaction potential is used for deformation. The translational functions are then obtained as distorted-wave solutions of the atom–molecule *central-field* problem defined by the potential

$$V_\alpha(R) = \int d^3\mathbf{r}\phi_\alpha^*(\mathbf{r}\,; R)\, \hat{V}_{\mathrm{total}}\,(r, R, \gamma)\, \phi_\alpha(\mathbf{r}\,; R) \tag{34}$$

As Karplus has pointed out,[35] the best two-body (angular-averaged) distortion potential for atom–molecule relative motion is between the extremes of free or adiabatic molecular deformations and should depend upon the molecular state and the relative energy. There has apparently been no implementation of a procedure like this. Adiabatic molecular deformation in $\chi_\alpha^{(+)}$ leads to much more accurate threshold behavior than the free-molecule approximation.

For $\chi_\beta^{(-)}$, Choi and Tang used the unperturbed model, which employs a free molecular function and a distorted translational function

$$\chi_\beta^{(-)}(\mathbf{R}', \mathbf{r}') = T_\beta^{(-)}(\mathbf{R}')\, \phi_\beta(\mathbf{r}') \tag{35}$$

The exit channel distortion potential in this case is the average over the ground-state asymptotic vibrational wave function of the spherical part of the interaction potential.

The adiabatic deformation model was pioneered by Karplus and Tang,[13,33] and its use is not the reason for the enhanced efficiency of the recent DWBA calculations. Rather, the improvement arises because the angular parts of the reactant and product distorted functions that refer to overall triangle orientation—the Euler angles—are referred to a common axis system. The integrations over the Euler angles are then done analytically. This leads to T-matrix elements of the form

$$T_{vjm_jv'j'm'_j} = \sum_{l=0,1,\ldots} (2l + 1)\, A_l P_l^{|m_j - m'_j|}(\cos\theta) \tag{36}$$

where θ is the scattering angle. This sum of three-dimensional integrals A_l can be evaluated much more rapidly (in about 1% of the time required previously[36]) than the five-dimensional variety tackled by Karplus and Tang.

Another formulation of DWBA for atom–diatom reactions was presented by Miller.[37] In this case, the helicity representation was used: molecular angular momentum projections were referred to either **R** in reactants, or **R**′ in products. On the other hand, Choi and Tang referred all molecular angular momentum projections to the direction of **k**, the reactant wave vector. There are clearly a number of choices available for definitions of angular coordinates and axes of quantization. In addition, Suck[38] has emphasized the usefulness of an expansion of the T-matrix elements in terms of the rotational angular momentum transferred during the reaction.

3.6. Distorted Waves: Numerical Results

Choi and Tang have calculated DWBA cross sections within the adiabatic model for rotational transitions in the $H + H_2$ and $D + H_2$ three-dimensional reactions. On the Porter–Karplus H_3 surface[11] (collinear barrier height 0.396 eV), the total indistinguishable atom reactive cross section[39] for $j = 0 \rightarrow j' = 1$ is $1.8a_0^2$, compared to $1.6a_0^2$ from close-coupling calculations on the same potential surface.[40] The differential reaction cross section is backpeaked, with a half-width of 60°, and smoothly decreases from the value at 180°. At $E_{tr} = 0.278$ eV, $\sigma_{0\rightarrow 1}$ (DWBA) $= 0.26a_0^2$, which is in good agreement with the close-coupling result[41] of $0.04a_0^2$. In Figure 3a, the distorted-wave cross section to all open-product rotational channels (for indistinguishable atoms) is compared to recent close-coupling results. This figure illustrates that the DWBA results for this reaction are in excellent agreement with close-coupling cross sections over the translational energy range from 0.27 to 0.50 eV.

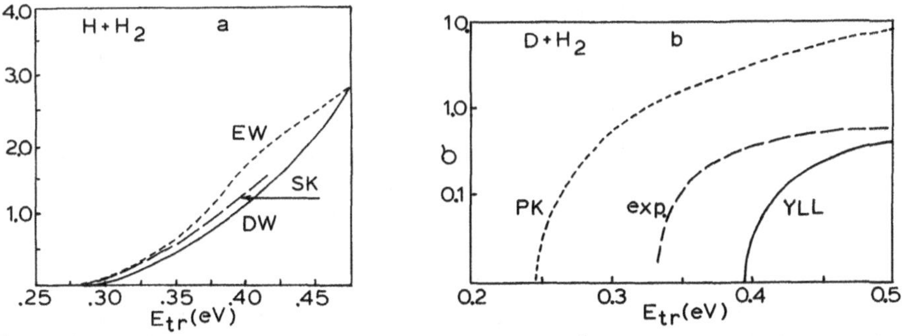

Figure 3. $H + H_2$ and $D + H_2$ reaction cross sections (a_0^2). (a) Comparison between the DWBA total reaction cross section[39] summed over all final states ($j' = 0, 1, 2, \ldots$) with the CC results of Schatz and Kuppermann[41] and Elkowitz and Wyatt.[40] These calculations are on the PK H_3 potential surface.[11] (b) DWBA[43] and experimental total reaction cross sections for $D + H_2$. (——) DWBA calculation with Yates–Lester–Liu potential surface.[8,9] (- - - -) DWBA results with PK surface. (– – –) renormalized experimental result.[45]

DWBA results were also obtained for the $D + H_2 \rightarrow DH + H$ reaction[42,43] on both the PK (Porter–Karplus) and YLL (Yates–Lester–Liu)[8,9] potential surfaces. The total reaction cross sections $\sigma(0 \rightarrow \Sigma j')$ for both surfaces are plotted vs. energy in Figure 3b. The PK surface has a threshold (where $\sigma \approx 0.01\, a_0^2$) which is about 0.15 eV lower than the YLL value. This is much lower than one would expect on the basis of the difference in collinear barrier heights (0.03 eV). The lowered reactivity of the YLL surface is due to the increased repulsiveness of this surface for *non*collinear configurations. The $D + H_2$ reaction cross section from the crossed-beam experiments of Geddes *et al.*,[44] renormalized by Gengenbach *et al.*,[45] is also shown in Figure 3b. (The estimated experimental error is $\pm 50\%$.) The experimentally estimated threshold energy[46] of $E_{tr} = 0.33$ eV was used to fit an energy dependence to the beam results, which were done at about 0.48 eV. It is believed that YLL surface leads to too high a threshold energy for this reaction.

The DWBA results for the $H + H_2$ reaction are remarkably accurate when compared to close-coupling results. Extension of the method to exoergic chemical reactions would be very useful in providing a severe test of the method, and in providing new theoretical results.

4. *Overlap Models (Franck–Condon Factors)*

4.1. *Introduction*

There has been considerable recent interest in developing simple analytic models for product state distributions, particularly in exoergic atom–diatom reactions. Although a number of different derivations are available, the main result is that the relative probabilities of populating various final states β from an initial state α are proportional to the squares of generalized (i.e., multidimensional) Franck–Condon factors, $\langle \chi_\beta^{(-)} | \chi_\alpha^{(+)} \rangle$, where $\chi_\alpha^{(+)}(\chi_\beta^{(-)})$ is a distorted wave in the entrance (exit) channel. With additional approximations, the multidimensional integral is usually reduced to products of one-dimensional overlap factors. The simplicity of the final formulas has made it possible to study the role of mass effects and various potential surface parameters in determining the product state distributions in selected reactions.

Overlap factors have a long history of success in molecular spectroscopy and it is only recently that they have been used to generate chemical reaction amplitudes. Pre-1975 uses of overlap factors include O'Malley's[47] study of product vibrational excitation in the $O^+ + N_2$ reaction, Herschbach's[48] study of translational energy distributions in the $H + Cl_2$ and $K + CH_3I$ reactions, Berry's[49] study of population inversion in the collinear $F + H_2$ and $H + Cl_2$ reactions, and Baer's[50] qualitative use of overlap factors in explaining multimodal HCl vibrational distributions in the $H + Cl_2$

reaction. More recently, Eu[51] presented a derivation (starting from the DWBA, and using excited electronic potential curves as distortion potentials) of the translational energy distribution formula previously used by Herschbach.[48]

4.2. Collinear Reactions: Vibrational Distributions

Extensive development and testing of the overlap model for collinear reactions was recently reported by Halavee and Shapiro.[52] They derived a closed expression for the v' distribution in collinear triatomic reactions

$$T_{vv'} = \iint dR\, dr\, \chi_{v'}^{(-)*}(R', r')\, \chi_v^{(+)}(R, r) \tag{37}$$

where $\chi_{v'}^{(-)}$ is a product channel distorted wave and $\chi_v^{(+)}$ is a reactant channel distorted wave. The distortion potential for reactants A + BC is chosen to prevent product formation, while the distortion potential for products AB + C is chosen to prevent "dereaction" to A + BC. The overall reaction amplitude is thus proportional to the two-dimensional overlap of wave functions for two "half-collisions." The distorted wave for each half-collision is in turn approximated by the product of a translational factor and an adiabatically deformed vibrational wave function,

$$\chi_{v'}^{(+)}(R, r) = f_v(R)\, H_v(r; R) \tag{38}$$

Here H_v, a local adiabatic vibrational wave function, may in turn be expanded in asymptotic vibrational wave functions appropriate for the free molecule:

$$H_v(r; R) = \sum_p b_{vp}(R)\, H_p(r) \tag{39}$$

Because of the form of the distortion potentials, the integrand in equation (37) is expected to be localized in a small region of the interaction zone.[53] Since the translational wave functions tend to maximize near the translational turning points, the two-dimensional integral may be approximately reduced to a (sum of) one-dimensional overlap integrals by replacing each translational wave function by its value at the turning point (R_v or $R_{v'}$):

$$T_{vv'} \approx \sum_{p,q} b_{v'p}^*(R_{v'})\, b_{vp}(R_v)\, f_{v'}^{(-)}(R_{v'})^*\, f_v^{(+)}(R_v)\, W_{pq}(R_v) \tag{40}$$

The one-dimensional vibration overlap integrals are

$$W_{pq}(R_v) = \int dr\, H_p(R_v - ar)\, H_q(r) \tag{41}$$

in which $(R_v - ar)$ is the product vibrational coordinate expressed in reactant integration variables. The quantities W_{pq} are overlap factors between two shifted harmonic oscillators with different force constants.

Further approximations are then made to evaluate the quantities b and f in equation (40). Near the classical turning point, the translational function is approximated by an Airy function, for example,

$$f_v^{(+)}(R_v) = \frac{\mu_\alpha^{1/2}}{[-\mu_\alpha D(R_v)]^{1/2}} Ai(0) \tag{42}$$

where $D(R_v)$ is the slope of the adiabatic channel energy $\varepsilon_v(R)$ at the classical turning point $\varepsilon_v(R) = \varepsilon_v(R_v) + D(R_v)(R - R_v)$. In addition, the b_{vp}, which are inelastic transition amplitudes on the distorted potential surface, are evaluated by a semiclassical method based upon a perturbed harmonic oscillator. For harmonic oscillator functions, all factors in equation (40) can thus be evaluated analytically.

The net result is that equation (40) for the $v \to v'$ transition amplitude may be written

$$T_{vv'} = (\mathbf{B}^b \mathbf{W} \mathbf{B}^a)_{vv'} \tag{43}$$

where $\mathbf{B}^a$ contains factors telling how the A + BC state v deforms on the approach to the reactant channel turning point, while $\mathbf{B}^b$ tells how AB + C states prepared near the product channel turning points relax as C accelerates. The overlap matrix $\mathbf{W}$ tells how much of each product vibrational state "originated" with each prepared reactant vibrational state.

Of course, it should be recognized that the form for $T_{vv'}$ in equation (37) follows from the distorted-wave formula in equation (20) by removing the potential from the integral as a constant factor. Even if this is *not* done, an equation similar to equation (43) may be derived, with the W_{pq} overlap factors replaced by similar terms containing the distortion potential $\hat{V}'_\beta$ in the integrand. However, for simplicity we will emphasize here the analysis based upon the direct overlap factors in equation (37). In applications to the reaction $H + Cl_2 \to HCl(v') + Cl$, Halavee and Shapiro[52] found for $v = 0$ that (at low energy), $v' = 4$ is most populated, with $v' = 3$ more populated than $v' = 5$. As shown in Figure 4a, these results are in good qualitative agreement with exact (close-coupling) results.[50] For reaction

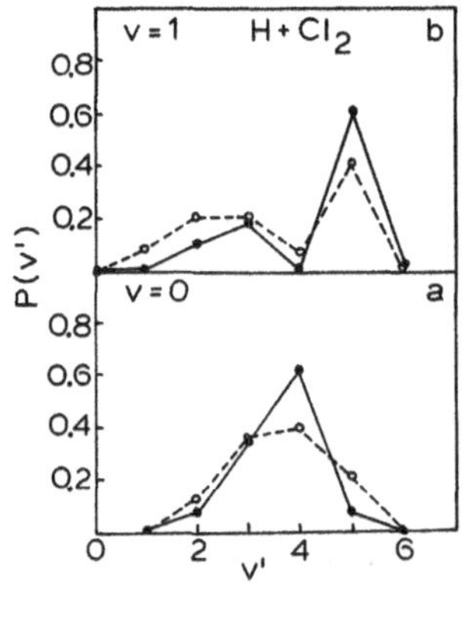

Figure 4. Close-coupling[50] and overlap model[52] product vibrational distributions in the $H + Cl_2$ reactions: (a) $H + Cl_2(v = 0) \to HCl(v') + Cl$ at 0.18 eV total energy; (b) $H + Cl_2(v = 1) \to HCl(v') + Cl$ at 0.25 eV. ●, overlap model, ♀, close-coupling results.

from $v = 1$, they found a bimodal vibrational distribution, with a lower peak at $v' = 3$, almost no $v' = 4$, and a higher peak at $v' = 5$. Figure 4b shows that good agreement was again obtained with the exact quantum collinear studies. In a later study,[54] the same authors studied the collinear $F + H_2$, $H + I_2$, and $F + HCl$ reactions. They pointed out that if the London formula is adequate to describe the ground electronic state potential surface, then all parameters related to the interaction region (translational turning points and vibrational force constants) may be obtained from spectroscopic information (including excited electronic state potential curves) on the separated reactants and products.

Schatz and Ross[55] have also derived and applied a collinear analytic model for vibrational energy distributions in collinear exoergic atom–diatom reactions. Starting with the exact T-matrix element written in terms of a transition between two "quasiadiabatic" potential surfaces (the diagonal elements of an exact but nondiagonal representation of the electronic Hamiltonian in each arrangement channel, which are analogous to the distortion potentials mentioned above), they reduced it to a two-dimensional nuclear overlap factor as in equation (37) above. The vibrationally adiabatic reactant nuclear function was then written as (R_c is the "critical" value of R near the region of maximum overlap between $\chi_\alpha^{(+)}$ and $\chi_\beta^{(-)}$)

$$\chi_v^{(+)}(R, r) = f_v^{(+)}(R)\, H_v(r; R_c) \tag{44}$$

with the translation function approximated by the form appropriate for a hard-wall potential:

$$f_v^{(+)}(R) \approx \sin k_v(R - R_v) \approx k_v(R - R_v) \tag{45}$$

(R_v is the translation turning point in channel v). In the product channel, a separable wave function of the form (an asymptotic vibrational function is assumed)

$$\chi_{v'}^{(-)}(R', r') = f_v^{(-)}(R')\, H_{v'}(r') \tag{46}$$

was chosen, in which the translational function was approximated by a step function:

$$f_{v'}^{(-)}(R') = \begin{cases} 0, & R' < R'_{v'} \\ \text{const}, & R' \geq R'_{v'} \end{cases} \tag{47}$$

where $R'_{v'}$ is the translational turning point in product channel v' on the quasiadiabatic potential surface. The forms chosen by Schatz and Ross for the translational functions $f_v^{(+)}$ and $f_v^{(-)}$ are clearly much simpler (and cruder) than the Airy functions used by Halavee and Shapiro. In addition, Schatz and Ross neglected vibration–translation coupling on each surface. However, the localized nature of the overlap may lead to a lack of sensitivity to these approximations. In equations (44) and (46), H_v and $H_{v'}$ were approximated by both harmonic or Morse oscillator functions. Analytic formulas for $T_{vv'}$ in both the harmonic [equation (3.18) in Reference 55] and Morse oscillator [equation (3.21) in Reference 55] cases were derived.

Product vibrational distributions were then predicted for the collinear $F + H_2(D_2)$ and the $D(D) + Cl_2$ reactions. For $F + H_2$, slow growth of the $0 \rightarrow 3$ relative to the $0 \rightarrow 2$ reaction probability was predicted (a "dynamic threshold" effect), in qualitative agreement with exact close-coupling calculations. However, the smaller reaction probabilities are almost always underestimated, although relative orderings are correct. The predicted probability distributions for $H + Cl_2$ are not as accurate as the FH_2 results, but qualitative trends are again correctly predicted.

Fischer and Venzl[56] have recently derived an overlap model from DWBA theory for collinear reactions. Very good agreement was obtained between predictions from their vibrational distribution formula and exact quantum results for the $X + F_2$ (X = Mu, H, D, T) reactions. In addition, Mukamel and Ross[57] have shown how to variationally construct quasi-adiabatic surfaces which optimize the accuracy of the Franck–Condon approximation for reactions. The overlap model is clearly capable, at least for collinear reactions, of providing qualitatively correct trends with minimal computer effort.

4.3. Three-Dimensional Reactions: Vibration–Rotation Distributions

Schatz and Ross have also extended the overlap model to three-dimensional reactions,[58] with numerical application to the H and F reactions with H_2. Several different formulas were derived for probabilities $P_{vjv'j'}$, depending upon the estimated relative strengths of potential or "kinematic" (K to $K \pm 1$) coupling. The T-matrix element for the reactive transition at total angular momentum J is approximated by the six-dimensional overlap factor (integration is over θ, ϕ, χ, R, r, and γ)

$$T^J_{vjm_j v'j'm'_j} = \langle \chi^{J(-)}_{v'j'm'_j} | \chi^{J(+)}_{vjm_j} \rangle \tag{48}$$

Expansion of each distorted wave in normalized symmetric top functions [i.e., a rotating coordinate system is introduced for reactants, with z (rotating) parallel to $\mathbf{R}$; another rotating coordinate system is introduced for products, with z' (products) parallel to $\mathbf{R}'$] leads to

$$\begin{aligned} \chi^{J(+)}_{vjm_j}(R, r, \gamma, \theta, \phi, \chi) &= \sum_K N^J_{MK}(\theta, \phi, \chi) \Psi^{JK}_{vjm_j}(R, r, \gamma) \\ \chi^{J(-)}_{v'j'm'_j}(R', r', \gamma', \theta', \phi', \chi') &= \sum_{K'} N^J_{MK'}(\theta', \phi', \chi') \Psi^{JK'}_{v'j'm'_j}(R', r', \gamma') \end{aligned} \tag{49}$$

Then, integration over the Euler angles yields a sum of integrals over internal coordinates:

$$T^J_{vjm_jv'j'm'_j} = \sum_{KK'} \int R^2\,dR r^2\,dr \sin\gamma\,d\gamma\,\Psi^{JK'*}_{v'j'm'_j}(R', r', \gamma')\,d^J_{K'K}(\Delta)\,\Psi^{JK}_{vjm_j}(R, r, \gamma) \tag{50}$$

In deriving this expression, the relation between (unnormalized) symmetric top functions referred to different rotating z axes was used:

$$D^J_{MK'}(\theta', \phi', \chi') = \sum_{K=-J}^{+J} d^J_{K'K}(\Delta)\,D^J_{MK}(\theta, \phi, \chi) \tag{51}$$

where Δ is the angle between $\mathbf{R}$ and $\mathbf{R}'$ ($0 \leq \Delta \leq \pi$). Quite analogous to equation (8), the distorted wave functions in each arrangement channel are solutions of the $(2J + 1)$ coupled equations (where $\alpha = vjm_j$)

$$(\hat{H} - E)\,\Psi^{J,K}_\alpha + \hat{H}^{J,K+1}\Psi^{J,K+1}_\alpha + \hat{H}^{J,K-1}\Psi^{J,K-1}_\alpha = 0 \tag{52}$$

The coupling operators will be considered in more detail in Chapter 17. Note once more that the kinematic coupling terms (which arise from the kinetic energy operator expressed in rotating coordinates) link functions of different K, but are diagonal in vjm_j.

Schatz and Ross considered four subcases in estimating solutions Ψ^{JK}_α to the body-fixed coupled equations for input into the integrand in equation (50). For weak kinematic coupling, the terms involving $\hat{H}^{JK\pm 1}$ are neglected in equation (52), but they are retained for strong coupling. In addition, deformed or asymptotic internal rotation and vibrational wave functions may be used depending on the strength of the potential coupling near the "critical" configuration. Only one of these four cases, strong potential-weak kinematic coupling, will be illustrated here.

Near the critical configuration, each internal function is approximated in separable-adiabatic form (translation–vibration factor times rotation factor)

$$\Psi^{JK}_{vjm_j}(R, r, \gamma) = \Phi^{JK}_{vj}(R, r)\,C^{JK}_{jn}\xi^K_n(\gamma\,; R_c) \tag{53}$$

where ξ^K_n is a normalized wave function for bending state n and C^{JK}_{jn} is an amplitude which tells how much of the asymptotic rotor jK is present at the critical configuration in bending state n. The bending functions, assuming a harmonic bending potential centered at $\gamma = 0$, are solutions to the equation

$$\left[\frac{-\hbar^2}{2I^*}\left(\frac{1}{\sin\gamma}\frac{\partial}{\partial\gamma}\sin\gamma\frac{\partial}{\partial\gamma} - \frac{K^2}{\sin^2\gamma}\right) + \frac{1}{2}k_b\gamma^2 - E_{nK}\right]\xi^K_n = 0 \tag{54}$$

where I^* is the bending moment of inertia. For the lowest $K = 0$ bending state (the only case considered by Schatz and Ross), $\xi^0_0(\gamma\,; R_c) \approx \exp\{-\beta_b\gamma^2/2\}$, where $\beta_b = (k_b I^*)^{1/2}/\hbar$. Note that k_b determines the potential anisotropy near R_c. The coefficients C^{JK}_{jn} are then the rotational overlap integrals ($n = K = 0$ is assumed)

$$C_{jn}^{JK} = \int_0^\pi \sin\gamma \xi_n^K(\gamma; R_c) P_j^K(\sin\gamma)\, d\gamma$$
$$\approx \delta_{K,0}(2j+1)^{1/2} \exp\{-j(j+1)/2\beta_b\} \quad (55)$$

where ξ_n^K has been expanded in a series in β_b^{-1} to obtain an integral to first order in β_b^{-1}.

With the separable wave function in equation (53), the T-matrix element is also the product of an angular (bend) factor times a translation–vibration factor,

$$T^J_{v\,jm_jv'j'm_j'} = C_{jn}^{JK} C_{j'n'}^{JK'} I^J_{nKn'K'} L^J_{v\,jKv'j'K'} \quad (56)$$

where the angular and translation–vibration overlap factors are

$$I^J_{nKn'K'} = \int_0^\pi \sin\gamma d\gamma \xi_n^K(\gamma)\, d^J_{K'K}(\Delta)\, \xi_{n'}^{K'}(\gamma)$$
$$L^J_{vjKv'j'k} = \int R^2\, dR \int r^2\, dr\, \Phi_{vj}^{JK}(R,r)\, \Phi_{v'j'}^{JK'}(R',r') \quad (57)$$

In the simplest case, $n = n' = K = K' = 0$, equation (56) yields the relative transition probabilities

$$|T_{vjm_jv'j'm_j'}|^2 \approx (2j+1)(2j'+1)\, e^{-j(j+1)/\beta_b} e^{-j'(j'+1)/\beta_b'} |L^J_{vjKv'j'K'}|^2\, |I^J_{nKn'K'}|^2 \quad (58)$$

In applications to the $H + H_2$ reaction, the jK dependence was neglected in the translation–vibration wave functions Φ_{vj}^{JK} that appear in equation (53). This gives simplified translation–vibration overlap factors denoted $K^J_{vv'}$. In this case, the predicted rotational distribution arises entirely from the angular coordinate (γ) part of the Franck–Condon factor. However, for exoergic reactions, the jK dependence is retained, so that the overlap depends upon J, j, and K (because the positions of the translational turning points on the reactant and product distortion surfaces are dependent upon J, j, and K).

For the $H + H_2$ reaction at total angular momentum J, the relative rotational distribution is predicted to have the form

$$P^J_{jj'} \approx (2j+1)(2j'+1)\, e^{-E_j/\beta_b} e^{-E_{j'}/\beta_b} \quad (59)$$

A plot of $\ln[P^J_{jj'}/(2j'+1)]$ vs. $E_{j'}$ should thus be linear, with a slope that is independent of J. Accurate reaction probabilities from close-coupling calculations agree with this prediction, at least for $J = 0$ and 4 (Figure 3 in Reference 58). In addition, equation (59) predicts that the reagent and product rotational distributions are uncorrelated: the j' distribution for a given j is independent of j. This is also substantiated by accurate calculations. In a related study, Pollak[60] has shown that the transfer of momentum constraint in information theory leads to $H + H_2$ reaction cross sections (linear surprisal plots are obtained for quantum-mechanical reaction cross sections for para → ortho conversion in this reaction[59]) which depend only upon the sum of the reactant and product rotational energies, $E_j + E_{j'}$; this also implies a lack of correlation between j and j'. Even though the quantal close-coupling results are in substantial agreement with the above prediction, the value of β_b evaluated by fitting the close-coupling results to the predicted Franck–Condon distributions is "much higher" than the value calculated directly from the bending potential of the Porter–Karplus H_3 surface.[58]

The strong potential-weak kinematic model was also applied to the three-dimensional $F + H_2$ reaction at $E_{tr} = 0.1$ eV. Because the reaction probabilities fall off rapidly with increasing J, centrifugal effects were neglected in determining product translational turning points. Morse oscillator functions were again used in computing translation–vibration factors $K_{vv'}$. As shown in Figure 5a, the vibrational distribution $P_{vv'}$ (the net probability obtained after summing over j' in each v' manifold) peaks at the correct ($v' = 2$) level and gives $f_{v'} = 0.65$, in agreement with experiment,[61,62] where $f_{v'} = 0.66$. However, the predicted v' distribution is much narrower than the experimental result, for example, $P_{03}/P_{02} \approx 0.45$ from experiment, whereas the overlap model leads to a value of about 0.05 for this ratio. The product rotational distribution $P_{jj'}$ (this time a sum is performed over all v') has a shape that is in reasonable agreement with

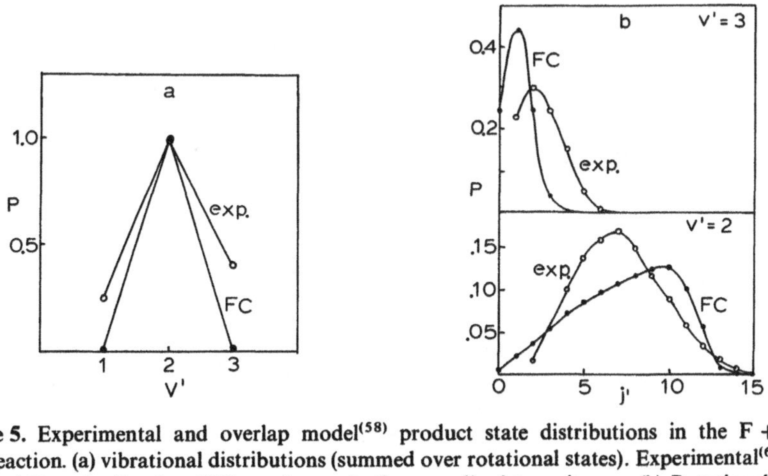

Figure 5. Experimental and overlap model[58] product state distributions in the $F + H_2$ reaction. (a) vibrational distributions (summed over rotational states). Experimental[61,62] and Franck–Condon (FC) results are both normalized to unit sum. (b) Rotational distributions for $v' = 2$ and 3.

experiment: $j' = 6$ is the peak of the experimental distribution in the $v' = 2$ manifold, whereas the computed distribution peaks at $j' = 10$. These rotational distributions are shown in Figure 5b.

The Franck–Condon model has also been related to the information–theoretic analysis of product distributions,[63] and to the role of symmetry in both concerted[64] and nonconcerted reactions.[65] Photochemical dissociation of polyatomic molecules has recently been treated with overlap models.[66,67]

5. Other Approaches, and Conclusions

In addition to the approaches discussed in the preceding sections, mention should be made of several other methods of generating approximate quantal reaction probabilities for three-dimensional reactions. These methods are: (i) the use of optical potentials and optical models,[68] (ii) the use of collinear[69–71] or coplanar[72] reaction probabilities to generate cross sections for three-dimensional reactions, (iii), "detailed-quantum transition state theory"[73] to produce the net approximate reaction probability from selected reagent states, and (iv), application of the infinite-order sudden approximation to reactive collisions.[74] The collection of techniques reviewed in this chapter provides an enormously more sophisticated array than was available just a few years ago for theory to predict and interpret modern state-to-state experiments on chemical reactions.

ACKNOWLEDGMENTS

This work was supported in part by research grants from the Robert A. Welch Foundation of Houston, Texas, and by the National Science Foundation.

For constructive comments on an easy draft of this chapter, I am very grateful to John Ross, George C. Schatz, David A. Micha, Robert B. Walker, Moshe Shapiro, K. T. Tang, and B. C. Eu.

References

1. P. McGuire and D.J. Kouri, Quantum mechanical close coupling approach to molecular collisions: J_z-conserving coupled states approximation, *J. Chem. Phys.* **60**, 2488–2499 (1974).
2. R.T. Pack, Space-fixed vs. body-fixed axes in atom–diatomic molecule scattering: Sudden approximations, *J. Chem. Phys.* **60**, 633–639 (1974).
3. T.G. Heil, S. Green, and D.J. Kouri, The coupled states approximation for scattering of two diatoms, *J. Chem. Phys.* **68**, 2562–2583 (1968).

4. A.B. Elkowitz and R.E. Wyatt, J_z-conserving approximation for the hydrogen exchange reaction, *Mol. Phys.* **31**, 189–201 (1976).
5. A. Kuppermann, G.C. Schatz, and J.P. Dwyer, Angular momentum decoupling approximations in the quantum dynamics of reactive systems, *Chem. Phys. Lett.* **45**, 71–73 (1977).
6. R.A. Marcus, Analytical mechanics of chemical reactions III. Natural collision coordinates, *J. Chem. Phys.* **49**, 2610–2616 (1968).
7. R.E. Wyatt, Theory of three-dimensional reactive collisions using natural collision coordinates, *J. Chem. Phys.* **56**, 390–400 (1972).
8. A.C. Yates and W.A. Lester, A new H_3 potential energy surface and its implication for chemical reaction, *Chem. Phys. Lett.* **24**, 305–309 (1974).
9. B. Liu, *Ab initio* potential energy surface for H_3, *J. Chem. Phys.* **58**, 1925–1937 (1973).
10. D. Truhlar and R.E. Wyatt, History of H_3 kinetics, *Ann. Rev. Phys. Chem.* **27**, 1–43 (1976).
11. R.N. Porter and M. Karplus, Potential energy surface for H_3, *J. Chem. Phys.* **40**, 1105–1115 (1964).
12. M.J. Redmon and R.E. Wyatt, Computational methods for reactive scattering, *Int. J. Quantum Chem. Symp.* **11**, 343–351 (1977).
13. K.T. Tang and M. Karplus, Quantum theory of (H, H_2) scattering: Approximate treatments of reactive scattering, *Phys. Rev. A* **4**, 1844–1858 (1971).
14. P. Roman, *Advanced Quantum Theory,* Addison-Wesley, Reading, Mass. (1965), pp. 281–308, 370–377.
15. B.C. Eu, Energy dependence of reaction cross sections at high energy, *Chem. Phys.* **5**, 95–106 (1974).
16. L.M. Raff and M. Karplus, Theoretical investigations of reactive scattering in molecular beams, K + CH_3I and related systems, *J. Chem. Phys.* **44**, 1212–1229 (1966).
17. M.E. Gersh and R.B. Bernstein, Measurement of the energy dependence of the cross section for the reaction $K + CH_3I \rightarrow KI + CH_3$ from 0.1 to 1.0 eV, *J. Chem. Phys.* **55**, 4661–4663 (1971).
18. A. Kafri, Y. Shimoni, R.D. Levine, and S. Alexander, The Born approximation as a simple diagnostic method for direct molecular collisions with applications to Cl + HI and F + H_2 reactions, *Chem. Phys.* **13**, 323–337 (1976).
19. M.J. Berry, F + H_2, D_2 and HD reactions: Chemical laser determination of the product vibrational state populations, *J. Chem. Phys.* **59**, 6229–6253 (1973).
20. R.D. Coombe and G.C. Pimentel, Effects of rotation on the vibrational energy distributions in the reaction F + H_2, *J. Chem. Phys.* **59**, 1535–1536 (1973).
21. J.T. Muckerman, Classical dynamics of the reaction of fluorine atoms with hydrogen molecules II. Dependence on the potential energy surface, *J. Chem. Phys.* **56**, 2997–3906 (1972).
22. M. Kimura, Theoretical consideration of the role of the Franck–Condon factor in chemical reactions, *Chem. Phys. Lett.* **45**, 489–491 (1977).
23. D.A. Micha, Collision dynamics of three interacting atoms: The Faddeev equations, *J. Chem. Phys.* **57**, 2184–2192 (1972).
24. S. Datz and R.E. Minturn, Study of the reaction Cs with Br_2 in crossed molecular beams, *J. Chem. Phys.* **41**, 1153–1154 (1964).
25. A. Henglein, Kinematics of ion–molecule reactions, in *Molecular Beams and Reaction Kinetics,* International School of Physics XLIV, Ch. Schlier, editor, Academic Press, New York (1970), pp. 139–183.
26. D.A. Micha and J. Yuan, Collision dynamics of three interacting atoms: The multiple collision expansion, *J. Chem. Phys.* **63**, 5462–5469 (1975).
27. D.A. Micha and P. McGuire, Collision dynamics of three interacting atoms: Foundations of the spectator-stripping model, *Chem. Phys. Lett.* **17**, 207–209 (1972).
28. J. Yuan and D.A. Micha, Collision dynamics of three interacting atoms: Stripping reactions of $Ar^+ + H_2$ and K + I_2, *J. Chem. Phys.* **64**, 1032–1041 (1976).

29. J. Yuan and D.A. Micha, Collision dynamics of three interacting atoms: Electron transfer in the reactions of K + Br_2, BrI, and I_2, *J. Chem. Phys.* **65**, 4876–4884 (1976).
30. D.H. Herschbach, Reactive scattering in molecular beams, *Adv. Chem. Phys.* **10**, 319–393 (1966).
31. T.T. Warnock, R.B. Bernstein, and A.E. Grosser, Internal energy of reaction products by velocity analysis III. Product excitation function for K + Br_2, *J. Chem. Phys.* **46**, 1685–1693 (1967).
32. D.A. Micha, The exchange reaction of H and H_2, *Ark. Fysik* **20**, 425–436 (1965).
33. M. Karplus and K.T. Tang, Quantum mechanical study of H + H_2 reactive scattering, *Discuss. Faraday Soc.* **44**, 56–67 (1967).
34. B.H. Choi and K.T. Tang, Theory of distorted-wave Born approximation for reactive scattering of an atom and a diatomic molecule, *J. Chem. Phys.* **61**, 5147–5157 (1974).
35. M. Karplus, Special results of theory: Distorted waves, in *Molecular Beams and Reaction Kinetics*, International School of Physics XLIV, Ch. Schlier, editor, Academic Press, New York (1970), pp. 406–426.
36. K.T. Tang and M. Karplus, Quantum theory of (H, H_2) scattering: Two-body potential and elastic scattering, *J. Chem. Phys.* **49**, 1676–1692 (1968).
37. W.H. Miller, Distorted wave theory for collisions of an atom and a diatomic molecule, *J. Chem. Phys.* **49**, 2373–2381 (1968).
38. S.H. Suck, Theory of atom–diatom reactive scattering based on the distorted-wave Born approximation, *Phys. Rev. A* **15**, 1893–1899 (1977).
39. B.H. Choi and K.T. Tang, Three-dimensional quantum mechanical studies of the H + H_2 reactive scattering, *J. Chem. Phys.* **65**, 5161–5180 (1976).
40. A.B. Elkowitz and R.E. Wyatt, Quantum mechanical reaction cross sections for the three-dimensional hydrogen exchange reaction, *J. Chem. Phys.* **62**, 2504–2506 (1975).
41. G.C. Schatz and A. Kuppermann, Quantum mechanical reactive scattering for three-dimensional atom-diatom systems II. Accurate cross sections for H + H_2, *J. Chem. Phys.* **65**, 4668–4682 (1976).
42. K.T. Tang and B.H. Choi, Three-dimensional quantum mechanical studies of D + $H_2 \rightarrow$ DH + H reactive scattering, *J. Chem. Phys.* **62**, 3642–3658 (1975).
43. B.H. Choi and K.T. Tang, Three-dimensional quantum mechanical studies of D + $H_2 \rightarrow$ DH + H reactive scattering II, *J. Chem. Phys.* **63**, 2854–2860 (1975).
44. J. Geddes, H.F. Krause, and W.L. Fite, Atom–molecule reaction D + $H_2 \rightarrow$ DH + H studied by molecular beams, *J. Chem. Phys.* **56**, 3298–3307 (1972).
45. R. Gengenbach, Ch. Hahn, and J.P. Toennies, Molecular beam measurements of the D + H_2 potential and recalibration of the reactive cross section, *J. Chem. Phys.* **62**, 3620–3630 (1975).
46. A. Kuppermann and J.M. White, Energy threshold for the D + $H_2 \rightarrow$ DH + H reaction, *J. Chem. Phys.* **44**, 4352–4353 (1966).
47. T.F. O'Malley, Simple model for the high energy reaction of O^+ ions with N_2, *J. Chem. Phys.* **52**, 3269–3277 (1970).
48. D. Herschbach, Reactive scattering, *Discuss. Faraday Soc.* **55**, 233–251 (1973).
49. M.J. Berry, Golden rule calculation of reaction product vibronic state distributions, *Chem. Phys. Lett.* **27**, 73–77 (1974).
50. M. Baer, Isotopic reactive systems H + Cl_2 and D + Cl_2: A quantum mechanical treatment of the collinear arrangement, *J. Chem. Phys.* **60**, 1058–1063 (1974).
51. B.C. Eu, Franck–Condon transitions and chemical reaction, *Mol. Phys.* **31**, 1261–1276 (1976).
52. U. Halavee and M. Shapiro, A collinear analytic model for atom–diatom chemical reactions, *J. Chem. Phys.* **64**, 2826–2839 (1976).
53. R.B. Walker and R.E. Wyatt, DWBA: Study of the collinear H + H_2 reaction, *Chem. Phys. Lett.* **16**, 52–56 (1972).
54. M. Shapiro and U. Halavee, The London formula and the product vibrational state

distributions in A + BC → AB + C chemical reactions, *Chem. Phys. Lett.* **40,** 381–386 (1976).

55. G.C. Schatz and J. Ross, Franck–Condon factors in studies of dynamics of chemical reactions I. General theory, application to collinear atom–diatom reactions, *J. Chem. Phys.* **66,** 1021–1037 (1977).
56. S. Fischer and G. Venzl, On the dynamics of exothermic triatomic exchange reaction, *J. Chem. Phys.* **67,** 1335–1343 (1977).
57. S. Mukamel and J. Ross, On the quasiadabatic description of the dynamics of electronically adiabatic chemical reactions, *J. Chem. Phys.* **66,** 3759–3786 (1977).
58. G.C. Schatz and J. Ross, Franck–Condon factors in studies of dynamics of chemical reactions II. Vibration–rotation distributions in atom–diatom reactions, *J. Chem. Phys.* **66,** 1037–1053 (1977).
59. R.E. Wyatt, Information theoretic analysis of quantum mechanical reaction cross sections, *Chem. Phys. Lett.* **34,** 167–169 (1975).
60. E. Pollak, Rotational surprisal plots for H + H_2: A simple theory, *Chem. Phys. Lett.* **47,** 513–516 (1977).
61. J.C. Polanyi and K.B. Woodall, Energy distribution among reaction products VI. F + H_2, D_2, *J. Chem. Phys.* **57,** 1574–1586 (1972).
62. M.J. Berry, F + H_2, D_2, HD reactions: chemical laser determination of the product vibrational state populations and the F + HD isotope effect, *J. Chem. Phys.* **59,** 6229–6253 (1973).
63. G.C. Schatz and J. Ross, Franck–Condon factors in studies of the dynamics of chemical reactions III. Analysis of information theory for vibration rotation distributions and isotopic branching ratios, *J. Chem. Phys.* **66,** 2943–2958 (1977).
64. T.F. George and J. Ross, Analysis of symmetry in chemical reactions, *J. Chem. Phys.* **55,** 3851–3866 (1971).
65. H. Metiu, J. Ross, and T.F. George, Symmetry rules for nonconcerted reactions, *Chem. Phys.* **11,** 259–272 (1975).
66. Y.B. Band and K.F. Freed, Energy distribution in selected fragment vibrations in dissociation processes in polyatomic molecules, *J. Chem. Phys.* **67,** 1462–1472 (1977).
67. K.F. Freed and Y.B. Band, Product energy distributions in the dissociation of polyatomic molecules, in *Excited States*, Vol. 3, E.C. Lim, editor, Academic Press, New York (1977), pp. 110–203.
68. D.A. Micha, Optical models in molecular collision theory, in *Dynamics of Molecular Collisions, Part A*, W.H. Miller, editor, Plenum Press, New York, (1976), p. 81.
69. J.N.L. Connor, W. Jakubetz, and J. Manz, Information theoretic analysis and collinear to three-dimensional transformation of reaction probabilities for F + H_2, H + F_2, H + Cl_2 and D + Cl_2, *Chem. Phys.* **17,** 451–469 (1976).
70. R.B. Bernstein and R.D. Levine, The relation between collinear and 3D dynamical calculations of reactive molecular collisions, *Chem. Phys. Lett.* **29,** 314–318 (1974).
71. M. Baer, Weak and strong interactions in chemical reactions, *J. Chem. Phys.* **62,** 4545–4550 (1975).
72. R.B. Walker and R.E. Wyatt, Three-dimensional reaction cross sections from planar scattering data, *Mol. Phys.* **28,** 101–111 (1974).
73. J.C. Light and A. Altenberger-Siczek, Practical approximation to reaction probabilities from selected initial quantum states, *Chem. Phys. Lett.* **30,** 195–199 (1975).
74. J.M. Bowman and K.T. Lee, Sudden approximation calculations of reactive scattering: The H + H_2 reaction, *J. Chem. Phys.* **68,** 3940–3941 (1978).

16

Reactive Scattering Cross Sections III: Quasiclassical and Semiclassical Methods

DONALD G. TRUHLAR and JAMES T. MUCKERMAN

1. Introduction

Classical trajectory methods for calculating inelastic scattering cross sections are covered in earlier chapters of this book, especially Chapters 10 and 12. This chapter covers the extension of this technique to treat reactive scattering. The first question which must be answered in a classical trajectory study of a reactive system is whether one should be using this method at all. Classical trajectory studies are useful not just because they yield reaction cross sections, angular distributions, reactivity as a function of initial and final energy distribution, and other observable reaction attributes, but also for the insight they may offer into the actual reaction event. One may look at the atomic motions in representative trajectories, and one may calculate such nonobservables as opacity functions (probability of reaction as a function of impact parameter) and dependence on features of the potential energy surface. But one must be careful not to overinterpret the reaction by a trajectory study. Because many reaction attributes depend sensitively on quantitative and qualitative features of the potential energy surfaces which are not quantitatively understood, one must be cautious about believing that the dynamical details of a particular trajectory calculation are in general accord with reality. Trajectory calculations are discussed from this point of view in Chapter 18 of this book.

DONALD G. TRUHLAR • Chemical Dynamics Laboratory, Department of Chemistry, University of Minnesota, Minneapolis, Minnesota 55455. JAMES T. MUCKERMAN • Chemistry Department, Brookhaven National Laboratory, Upton, New York 11973.

Chapter 18 also discusses some simple theories which sometimes give as much reliable information as a full trajectory calculation and should be considered as alternative or complementary theoretical approaches.

Another consideration in deciding to use trajectory methods is the accuracy of classical mechanics. It is well known that quantum effects (e.g., tunneling and interference) play a role, and in some instances an extremely important one, in molecular collisions, so it is worthwhile to provide a few guidelines about when to expect a good first approximation to the nuclear motion problem from the quasiclassical trajectory method. These are as follows:

(i) The quasiclassical trajectory method becomes more reliable the larger the masses of the nuclei and the higher the energy in each degree of freedom of the system (within the validity of the Born–Oppenheimer approximation). This follows from the fact that the de Broglie wavelength becomes smaller as the momentum is increased. When the de Broglie wavelengths for motion along all generalized coordinates become smaller than the distances over which the potential energy changes rapidly as a function of those coordinates, the wave (quantum) nature of the internuclear motion need not be considered and it may be treated classically.

(ii) The quasiclassical method yields more reliable results when the desired information is highly averaged. This property is related to the Ehrenfest relations.[1] These relations show that the quantum-mechanical expectation values of position $\langle \mathbf{x} \rangle$ and momentum $\langle \mathbf{p} \rangle$ satisfy the classical equations of motion for the average potential $\langle V(\mathbf{x}) \rangle$. But classical mechanics assumes they are governed by the potential $V(\langle \mathbf{x} \rangle)$ at the average position. Thus the validity of classical mechanics requires that the spread $\Delta\mathbf{x}$ of coordinates covered by a representative wave packet be small. Unfortunately, wave packets representing atoms and molecules at chemical energies may have considerable spread. But classical mechanics may still be accurate for sufficiently averaged reaction attributes.

(iii) When the total energy of reactants is comparable to or below the minimum potential energy barrier separating reactants from products (i.e., low-energy collisions of internally "cold" molecules), trajectory methods may seriously underestimate reaction probabilities. The use of the method (without semiclassical corrections for classically forbidden paths as discussed in Section 3.4) in such threshold regions where tunneling or its multidimensional analogs are important should be avoided.

(iv) The quasiclassical trajectory method, when used to calculate state-to-state reaction probabilities, will misestimate probabilities which are "classically forbidden" in the classical S matrix sense[2] (see Section 3.4). The method is not reliable for predicting small transition probabilities.

(v) The quasiclassical trajectory method is not appropriate for predicting resonance features or other interference phenomena.

Once it has been determined that the quasiclassical trajectory method is applicable and useful for a given system, one must determine the number of relevant potential surfaces as discussed in Chapter 18 and obtain analytic representations for them by, e.g., the methods of Chapters 2 and 3. If more than one surface is important, one must also characterize the nonadiabatic coupling terms. Trajectory techniques, like other theories of chemical reactions, have been applied most extensively to single-surface reactions, and these are emphasized in these chapters. The next questions which arise are: (i) How does one correspond the initial and final conditions of the trajectory to real (quantized) initial and final collision conditions? (ii) How does one minimize the errors in the classical treatment? Question (ii) may be restated: How does one incorporate semiclassical corrections? The answers to questions (i) and (ii) are closely related. The most straightforward way to incorporate quantized initial conditions is called the quasiclassical trajectory technique. This is explained in Section 2 of this chapter. This is the largest section of this chapter. It presents working equations for the quasiclassical trajectory method as applied to single-surface reactions of atoms with diatomic molecules.* The literature already contains several descriptions (mostly incomplete) of various techniques and equations for classical or quasiclassical forward trajectory calculations on three-dimensional atom–diatom reactions.[3-29] Rather than review all the computational schemes which have been used for various aspects of such calculations or compare the techniques used by various workers, we mainly present one set of approaches which we recommend as efficient and more-or-less optimal. However, we do present more than one approach to the calculation of product distributions since the optimum method may depend on the problem of interest.

The methods discussed in Section 2 are the ones we recommend in general for practical applications, but there are some problems where alternative correspondence methods for initial and final quantized degrees of freedom may be required or desirable. Section 3 presents a brief discussion of such alternative approaches to this correspondence.

Trajectory methods for studying multisurface reactions are discussed briefly in Chapters 13 and 18. We discuss some computational aspects of these methods in Section 4.

2. *Quasiclassical Trajectory Method*

The quasiclassical trajectory method assumes that each of the nuclei comprising a chemical system moves according to the laws of classical mechanics in the force field arising from the adiabatic electronic energy of

* Many of the techniques in this chapter are also applicable to reactions of atoms with polyatomic molecules and to molecule–molecule reactions. Nevertheless, we restrict our explicit considerations to atom–diatom reactions.

the system. The method may also be extended (see Section 4) to cases where the Born–Oppenheimer separation of electronic and nuclear motion breaks down provided that such breakdown is confined to localized regions of configuration space. The term "quasiclassical" is used to denote the manner in which molecules are prepared before collision (i.e., the *initial conditions*). In the quasiclassical trajectory method, molecules are prepared in discrete internal energy states corresponding to the quantum states of the molecule. Once the trajectory is begun, this quantum restriction is relaxed so that the time evolution of the system is governed solely by classical mechanics. A similar "quantization" is often employed in the analysis of *product molecule* internal energy states.

In accordance with the above, this section first considers the calculation of a single trajectory for an A + BC collision corresponding to specified initial quantized rotational and vibrational energies and a fixed initial relative translational energy. Then we consider the calculation of reaction cross sections and rate constants from a batch or batches of such trajectories. Finally, we consider the calculation of other reaction attributes, such as distributions of final conditions of the products and reactivity as a function of initial conditions.

2.1. Equations of Motion

It is assumed in the present discussion that the potential energy function (hypersurface) is an analytic function of the three internuclear distances, i.e., $V \equiv V(R_1, R_2, R_3)$, where R_1, R_2, and R_3 are the AB, BC, and AC distances, respectively, during the A + BC collision.

The first step in setting up a trajectory calculation is to define a coordinate system in which to express the Hamiltonian for the system. It is convenient to choose as a *reference* coordinate system a space-fixed Cartesian system consisting of the nine coordinates $\mathbf{x} \equiv \{x_i; i = 1, \ldots, 9\}$ of the three nuclei, denoted A, B, and C, respectively, and the nine momenta $\mathbf{p}_x \equiv \{p_{x_i}; i = 1, \ldots, 9\}$ conjugate to these coordinates. The Hamiltonian in this reference coordinate system is[5]

$$H(\mathbf{x}, \mathbf{p}_x) = T(\mathbf{p}_x) + V[R_1(\mathbf{x}), R_2(\mathbf{x}), R_3(\mathbf{x})] \tag{1a}$$

where

$$T(\mathbf{p}_x) = \sum_{i=1}^{3} \left(\frac{1}{2m_A} p_{x_i}^2 + \frac{1}{2m_B} p_{x_{i+3}}^2 + \frac{1}{2m_C} p_{x_{i+6}}^2 \right) \tag{1b}$$

Hamilton's equations of motion[30] are

$$\dot{x}_i \equiv \frac{dx_i}{dt} = \frac{\partial H}{\partial p_{x_i}} = \frac{\partial T}{\partial p_{x_i}} \qquad (i = 1, \ldots, 9) \tag{2a}$$

$$\dot{p}_{x_i} \equiv \frac{dp_{x_i}}{dt} = -\frac{\partial H}{\partial x_i} = -\frac{\partial V}{\partial x_i} = -\sum_{k=1}^{3} \frac{\partial V}{\partial R_k} \frac{\partial R_k}{\partial x_i} \qquad (i = 1, \ldots, 9) \tag{2b}$$

The three internuclear distances in terms of the reference coordinates are

$$R_1 \equiv R_{AB} = \left[\sum_{i=1}^{3} (x_i - x_{i+3})^2 \right]^{1/2} \tag{3a}$$

$$R_2 \equiv R_{BC} = \left[\sum_{i=1}^{3} (x_{i+3} - x_{i+6})^2 \right]^{1/2} \tag{3b}$$

$$R_3 \equiv R_{AC} = \left[\sum_{i=1}^{3} (x_i - x_{i+6})^2 \right]^{1/2} \tag{3c}$$

from which the required $\partial R_k/\partial x_i$ terms can readily be obtained. There are, of course, a total of 18 coupled first-order differential equations implied by equations (2a) and (2b).

It is indeed possible to carry out the entire trajectory calculation, including the specification of initial conditions, in this reference set of Cartesian coordinates. While it is usually convenient to transform the coordinate system as described below, it should be noted that it is not necessary to do so.

From equation (2b) it is seen that if the potential energy function is independent of some coordinate (i.e., the coordinate is *cyclic*), the momentum conjugate to that coordinate is a *constant of the motion* (independent of time). The elimination of any cyclic coordinate reduces the number of coupled equations which must be integrated to obtain each trajectory.

Anticipating that the coordinates of the center of mass of the entire system will prove to be cyclic, we wish to devise a new set of coordinates in which the three components of the position of the center of mass are three of the nine new coordinates. It would further prove convenient in the specification of initial conditions in A + BC to have internal coordinates corresponding to the displacement of A from the center of mass of the BC molecule. Hence we define[5]

$$q_i = x_{i+6} - x_{i+3} \qquad (i = 1, 2, 3) \tag{4a}$$

$$Q_i = x_i - \frac{1}{(m_B + m_C)}[m_B x_{i+3} + m_C x_{i+6}] \qquad (i = 1, 2, 3) \tag{4b}$$

$$S_i = \frac{1}{M}[m_A q_i + m_B q_{i+3} + m_C q_{i+6}] \qquad (i = 1, 2, 3) \tag{4c}$$

where $M \equiv m_A + m_B + m_C$, as our new system of *generalized coordinates*. These particular generalized coordinates for a three-body system are often called Jacobi coordinates. The coordinates in the three equations in equation (4) are called internal, relative, and center-of-mass coordinates, respectively.

The Hamiltonian can be obtained in terms of the generalized coordinates $\{q_i, p_i, Q_i, P_i, S_i, P_{S_i};\ i = 1, 2, 3\}$, where p_i, P_i, and P_{S_i} are the momenta conjugate to q_i, Q_i, and S_i, respectively by employing the F_2-type generator function[30]:

$$F_2(\mathbf{p},\mathbf{P},\mathbf{P}_S,\mathbf{q},\mathbf{Q},\mathbf{S}) = \sum_{j=1}^{3} [p_j q_j(\mathbf{x}) + P_j Q_j(\mathbf{x}) + P_{S_j} S_j(\mathbf{x})] \tag{5}$$

Using equation (4) and the relation $p_{x_i} = (\partial F_2/\partial x_i)$, one derives expressions for $p_{x_i}(\mathbf{p},\mathbf{P},\mathbf{P}_S)$ which are substituted into equation (1). The result is

$$H(\mathbf{q},\mathbf{Q},\mathbf{p},\mathbf{P},\mathbf{P}_S) = T(\mathbf{p},\mathbf{P},\mathbf{P}_S) + V[R_1(\mathbf{q},\mathbf{Q}), R_2(\mathbf{q},\mathbf{Q}), R_3(\mathbf{q},\mathbf{Q})] \tag{6a}$$

$$T(\mathbf{p},\mathbf{P},\mathbf{P}_S) = \sum_{i=1}^{3} \left(\frac{1}{2\mu_{\mathrm{BC}}} p_i^2 + \frac{1}{2\mu_{\mathrm{A,BC}}} P_i^2 + \frac{1}{2M} P_{S_i}^2 \right) \tag{6b}$$

Here

$$\mu_{\mathrm{BC}} \equiv m_{\mathrm{B}} m_{\mathrm{C}}/(m_{\mathrm{B}} + m_{\mathrm{C}}) \tag{7a}$$

and

$$\mu_{\mathrm{A,BC}} \equiv m_{\mathrm{A}}(m_{\mathrm{B}} + m_{\mathrm{C}})/M \tag{7b}$$

are the *reduced masses* corresponding to internal and relative motion, respectively, and $\{R_i(\mathbf{q},\mathbf{Q}); i = 1, 2, 3\}$ are given by[5]

$$R_1 = \left[\sum_{i=1}^{3} \left(\frac{m_{\mathrm{C}}}{m_{\mathrm{B}} + m_{\mathrm{C}}} q_i + Q_i \right)^2 \right]^{1/2} \tag{8a}$$

$$R_2 = \left[\sum_{i=1}^{3} Q_i^2 \right]^{1/2} \tag{8b}$$

$$R_3 = \left[\sum_{i=1}^{3} \left(\frac{m_{\mathrm{B}}}{m_{\mathrm{B}} + m_{\mathrm{C}}} q_i - Q_i \right)^2 \right]^{1/2} \tag{8c}$$

While the generator function method has the advantage of generality, the transformation of the kinetic energy to conjugate momenta of Jacobi coordinates can be achieved more simply by the mobile method.[31]

As was anticipated, the R_i are independent of S_1, S_2, and S_3, the coordinates of the center of mass. These coordinates are therefore cyclic and the conjugate momenta P_{S_1}, P_{S_2}, and P_{S_3} are constants of the motion. We may without loss of generality define the center of mass to be stationary (or, equivalently, define our coordinate system to move with the uniform motion of the center of mass) and delete the terms involving P_{S_1}, P_{S_2}, and P_{S_3} from the Hamiltonian. The resulting set of twelve Hamilton's equations is

$$\dot{q}_i = \frac{\partial H}{\partial p_i} = \frac{\partial T}{\partial p_i} \qquad (i = 1, 2, 3) \tag{9a}$$

$$\dot{Q}_i = \frac{\partial H}{\partial P_i} = \frac{\partial T}{\partial P_i} \qquad (i = 1, 2, 3) \tag{9b}$$

$$\dot{p}_i = -\frac{\partial H}{\partial q_i} = -\frac{\partial V}{\partial q_i} = -\sum_{k=1}^{3} \frac{\partial V}{\partial R_k}\frac{\partial R_k}{\partial q_i} \qquad (i = 1, 2, 3) \tag{9c}$$

$$\dot{P}_i = -\frac{\partial H}{\partial Q_i} = -\frac{\partial V}{\partial Q_i} = -\sum_{k=1}^{3} \frac{\partial V}{\partial R_k}\frac{\partial R_k}{\partial Q_i} \qquad (i = 1, 2, 3) \tag{9d}$$

where $\{R_i(\mathbf{q}, \mathbf{Q}); i = 1, 2, 3\}$ is given by equations (8). This is the usual form in which the equations of motion are integrated, although some workers(32,33) have found it convenient to use the equivalent second-order form obtained from Newton's or Lagrange's instead of Hamilton's equations of motion.

It is, of course, possible to reduce the number of equations even further by recognizing the total energy and total angular momentum (three components) as constants of the motion. This would reduce the number of equations from 12 to 8, but the form of the resulting equations would be much more complicated and will not be discussed here. Moreover, the conservation of these quantities can serve as an accuracy criterion when integrating equations (9) numerically.

2.2. *Initial Conditions*

Before Hamilton's equations of motion, equations (9), can be integrated to obtain a trajectory, initial values of the coordinates and momenta, $\{q_i^0, Q_i^0, p_i^0, P_i^0; i = 1, 2, 3\}$, must be specified. These initial values depend on the chosen set of *collision parameters*, which are those geometric parameters which characterize a collision and make one collision differ from another. For collisions of an atom A with some selected vibrational–rotational state (n, j) of the molecule BC at some fixed center-of-mass collisions energy E_{rel}, there are *five* collision parameters. Defining (without loss of generality) the atom A and the center of mass of the BC molecule to lie initially in the yz plane on the $-z$ side of the origin of the reference set of Cartesian coordinates with the initial relative velocity vector $\mathbf{v}_{\text{rel}}$ directed along the $+z$ axis, the five collision parameters are as follows:

- b the impact parameter (the y component of the initial relative coordinate, i.e., Q_2^0)
- θ the initial azimuthal orientation angle of the BC internuclear axis (the angle between $\mathbf{q}$ and the $+z$ axis)
- ϕ the initial polar orientation angle of the BC internuclear axis (the angle between the projection of $\mathbf{q}$ onto the xy plane and the $+x$ axis)
- η the initial orientation of the BC angular momentum (the angle between $\mathbf{q} \times \mathbf{p}$ and some reference vector normal to the BC internuclear axis)
- ξ the initial phase angle of the BC vibration (defined to be zero at the inner turning point, π at the outer turning point, and to vary linearly with time)

It now remains to express the initial values of the generalized coordinates and momenta $\{q_i^0, Q_i^0, p_i^0, P_i^0; i = 1, 2, 3\}$ in terms of these collision parameters. If ρ is the initial separation between A and the center of mass of BC and b is the impact parameter of the collision, then

$$Q_1^0 = 0 \tag{10a}$$

$$Q_2^0 = b \tag{10b}$$

$$Q_3^0 = -(\rho^2 - b^2)^{1/2} \tag{10c}$$

$$P_1^0 = 0 \tag{10d}$$

$$P_2^0 = 0 \tag{10e}$$

$$P_3^0 = (2\mu_{A,BC}E_{rel})^{1/2} = P^0 \tag{10f}$$

Let $r \equiv |\mathbf{q}|$ and let θ and ϕ be defined as above so that

$$q_1^0 = r^0 \sin\theta \cos\phi \tag{11a}$$

$$q_2^0 = r^0 \sin\theta \sin\phi \tag{11b}$$

$$q_3^0 = r^0 \cos\theta \tag{11c}$$

One convenient method for specifying the remaining variables $\mathbf{p}^0$ and r^0 (the as yet undetermined initial internuclear separation) in terms of η, ξ, θ, and ϕ is to place r^0 at the inner turning point and to define ρ, the initial separation of A and BC, as

$$\rho = \rho_0 + \frac{\xi}{2\pi} v_{rel} \tau_{BC}^{n,j} \tag{12a}$$

$$= \rho_0 + \frac{\xi}{2\pi} \frac{P^0 \tau_{BC}^{n,j}}{\mu_{A,BC}} \tag{12b}$$

so that when along the trajectory the separation of A and BC decreases to ρ_0, the BC vibrational phase angle has the value ξ. Here $\tau_{BC}^{n,j}$ is the vibrational period of BC in the state (n, j) and ρ_0 is the "collision shell" radius, which must be sufficiently large so that the interaction between A and BC is negligible. For very long-range interaction potentials corrections for nonvanishing interaction can be made but these will not be discussed here. To minimize the distance beyond ρ_0 that the trajectory is started, it is more efficient to use a random number to choose r between r_- and r_+, the inner and outer turning points of BC, respectively, in each trajectory (or, alternatively, select either r_- and r_+ randomly) and replace the vibrational period in equation (12) by the half-period*

$$\frac{1}{2}\tau_{\mathrm{BC}}^{n,j} = \left(\frac{\mu_{\mathrm{BC}}}{2}\right)^{1/2} \int_{r_-}^{r_+} \left[\varepsilon_{n,j} - V_{\mathrm{BC}}(r) - \frac{j(j+1)\hbar^2}{2\mu_{\mathrm{BC}} r^2}\right]^{-1/2} dr \tag{13}$$

where $\varepsilon_{n,j}$ is the internal energy of the state (n, j) of the BC molecule and $V_{\mathrm{BC}}(r)$ is defined as

$$V_{\mathrm{BC}}(r) \equiv V_{\mathrm{BC}}(R_2) = \lim_{R_1, R_3 \to \infty} V(R_1, R_2, R_3) \tag{14}$$

The calculation of $\varepsilon_{n,j}$ and of $(1/2)\,\tau_{\mathrm{BC}}^{n,j}$ will be discussed below.

An even more efficient method[27] is always to choose $\rho = \rho_0$, but before integrating the three-body equations of motion to integrate the diatomic equation of motion from $r = r_-$ for a time $(\xi/2\pi)\,\tau_{\mathrm{BC}}^{n,j}$. If $\xi \geq \pi$, one can do this more efficiently by integrating from $r = r_+$ for a time $[(\xi - \pi)/2\pi]\,\tau_{\mathrm{BC}}^{n,j}$.

With the selection of $r = r_\pm$, all the initial momentum of the BC molecule is angular momentum, i.e., at a turning point there is no radial component of $\mathbf{J}_r$ where $\mathbf{J}_r$ is the BC angular momentum vector which has the magnitude $J_r = [j(j+1)]^{1/2}\,\hbar$ or $(j + 1/2)\,\hbar$, depending on the correspondence scheme chosen. We will use the former in the rest of this chapter, although in many cases there is not much difference between the two choices. We may therefore write

$$\mathbf{q}^0 \cdot \mathbf{J}_r = 0 \tag{15}$$

and

$$J_r^2 = j(j+1)\,\hbar^2 = (r_\pm p^0)^2 \tag{16}$$

Let $\boldsymbol{\kappa}$ be a reference vector normal to the BC internuclear axis defined by

$$\boldsymbol{\kappa} = \mathbf{q}^0 \times \hat{\mathbf{e}}_z = q_2^0 \hat{\mathbf{e}}_x + q_1^0 \hat{\mathbf{e}}_y \tag{17}$$

where $\hat{\mathbf{e}}_x$, $\hat{\mathbf{e}}_y$, $\hat{\mathbf{e}}_z$ are the unit vectors along Cartesian axes so that

$$\mathbf{p}^0 \cdot \boldsymbol{\kappa} = q_2^0 p_1^0 - q_1^0 p_2^0 = J[(q_1^0)^2 + (q_2^0)^2]^{1/2} \cos\eta \tag{18}$$

where the second equality in equation (18) follows from the definition of η as the angle between the angular momentum of BC and the reference vector $\boldsymbol{\kappa}$. Combining equations (11), (15), (16), and (18) leads to the follow-

* Alternatively one could alternate r^0 between r_- and r_+ in successive trajectories. If one does not pick new values for all the other random variables in the alternate trajectories this is a form of cluster sampling and is less efficient than the procedure recommended in the text. If one does pick new values for all the other random variables this is a form of stratified sampling which slightly complicates the error analysis without any expected significant increase in accuracy.

ing initial values of internal (BC) momenta

$$p_1^0 = J_r(\sin\phi\cos\eta - \cos\theta\cos\phi\sin\eta)/r_\pm \tag{19a}$$

$$p_2^0 = -J_r(\cos\phi\cos\eta + \cos\theta\sin\phi\sin\eta)/r_\pm \tag{19b}$$

$$p_3^0 = J_r(\sin\theta\sin\eta)/r_\pm \tag{19c}$$

Given some set of values for the collision parameters, these values are used in equations (10), (11), (12), and (19) with $r^0 = r_\pm$ and $J_r = [j(j+1)]^{1/2}\hbar$. The constant ρ_0 must be determined for the particular system studied. The calculation of $r_\pm$, $\varepsilon_{n,j}$, and $(1/2)\,\tau_{\mathrm{BC}}^{n,j}$ will now be considered.

In the semiclassical theory of bound states, the internal energy of the BC molecule, $\varepsilon_{n,j}$, is defined by the implicit equation[34]

$$\tfrac{1}{2}J_v - \pi\hbar(n + \tfrac{1}{2}) = 0 \tag{20a}$$

where

$$J_v = 2\int_{r_-}^{r_+}\left\{2\mu_{\mathrm{BC}}\left[\varepsilon_{n,j} - V_{\mathrm{BC}}(r) - \frac{J_r^2}{2\mu_{\mathrm{BC}}r^2}\right]\right\}^{1/2} dr \tag{20b}$$

The integral in equation (20b) is readily evaluated by numerical integration. A particularly efficient scheme is obtained by defining a new variable

$$y \equiv \frac{2r - r_+ - r_-}{r_+ - r_-} \tag{21a}$$

and using a Chebyshev quadrature of the second kind[35]:

$$\int_{-1}^{1} f(y)(1 - y^2)^{1/2}\,dy \cong \sum_{i=1}^{k} w_i f(y_i) \tag{21b}$$

where

$$w_i = \frac{\pi}{k+1}\sin^2\frac{i\pi}{k+1} \tag{21c}$$

and

$$y_i = \cos\frac{i\pi}{k+1} \tag{21d}$$

After an initial estimate of $\varepsilon_{n,j}$, the Newton–Raphson method[36] may be used to solve iteratively for the correct value of $\varepsilon_{n,j}$. Very few points ($k \leq 15$) are required to evaluate the integral in equation (20b). The integral

$$\frac{\partial J_v}{\partial \varepsilon_{n,j}} = \mu_{\mathrm{BC}}\int_{r_-}^{r_+}\left\{2\mu_{\mathrm{BC}}\left[\varepsilon_{n,j} - V_{\mathrm{BC}}(r) - \frac{J_r^2}{2\mu_{\mathrm{BC}}r^2}\right]\right\}^{-1/2} dr \tag{22}$$

which also arises in the application of the Newton–Raphson method to find the zero of equation (20a), is simply the half-period as defined in equation (13) and is readily evaluated using numerical differentiation of J_v (see

Reference 27) or by another numerical integration. If the latter procedure is chosen it is convenient again to make the substitution of equation (21a) and to use a Chebyshev quadrature of the first kind[35]:

$$\int_{-1}^{1} \frac{f(y)}{(1-y^2)^{1/2}}\,dy = \sum_{i=1}^{k} w_i f(y_i) \tag{23a}$$

where

$$w_i = \frac{\pi}{k} \tag{23b}$$

and

$$y_i = \cos\frac{(2i-1)\pi}{2k} \tag{23c}$$

Chebyshev quadratures are also called Gauss–Mehler quadratures[37] and are widely used for this kind of semiclassical integral.[38,39] In evaluating the integrals in equations (20) and (13), one must first solve for the appropriate turning points $r_\pm$ at each trial energy using, for example, the Newton–Raphson method as described in Reference 5.

2.3. Calculation of a Trajectory

Having obtained the equations of motion and the initial conditions, the next problem is to calculate a trajectory. This is accomplished by numerical integration of the equations of motion for the given initial values of the $\{q_i, Q_i, p_i, P_i;\ i = 1, 2, 3\}$. There are a large number of well-tested algorithms for integrating coupled sets of differential equations, and these are described sufficiently well in the numerical-analysis literature.[40] The most popular integrators for trajectory studies of chemical reactions have been the fourth-order Runge–Kutta–Gill and the Adams–Moulton method. The latter is a predictor–corrector scheme and is most often used with fourth-order predictor and fifth-order corrector, although higher orders are used sometimes. Many problems have been solved with a fixed-size time step, but variable-step-size Runge–Kutta and variable-step-size predictor–corrector schemes have also been widely used. Predictor–corrector schemes are more efficient but more complicated than Runge–Kutta schemes for this kind of calculation if a fixed step size is used. Variable-step-size algorithms are more problematic. For many problems, they are actually less efficient than fixed-step-size schemes because of the extra "overhead" involved in the variable-step-size routines. But for some problems significant savings in computer time were achieved by using variable-step-size predictor–corrector schemes. A good program library will include a choice of more than one of these kinds of algorithms. If the computer costs for the project being undertaken are small, then it is probably most efficient just to use any one of these which is handy. If the computer costs will be significant, one should test several algorithms for cost efficiency on the particular problem of interest. The algorithm which is most efficient for one problem (with a given potential energy function, set of atomic masses, range of energies and initial conditions, and number of significant figures of precision required on given trajectories) is not necessarily the most efficient for another problem.

There has not been much work reported in systematically comparing the efficiencies of various numerical integration schemes specifically as applied to chemical dynamics problems. In general it is difficult to tell how efficient an integration scheme will be for practical work on the basis of formal properties like order of the finite difference error, stability regions, or number of function evaluations per step. Matzen and Riley[(41)] made an interesting study of two $F + H_2$ trajectories (with H_2 initially in the ground state, E_{rel} equal to about 2 kcal/mol, and the potential given by an extended LEPS function). For each trajectory they systematically varied the numerical parameters for each of five integration schemes and determined the computer time requirements for various accuracies. The most efficient integration scheme was the Bulirsch–Stoer extrapolation method[(42)] with automatic error control applied to the second-order (Newtonian) equations of motion.[(33)] The other schemes evaluated, in order of decreasing efficiency for five-significant-figure accuracy were: the Bulirsch–Stoer extrapolation method with automatic error control applied to the first-order (Hamiltonian) equations of motion, the fixed-step-size, fourth-order Adams–Moulton predictor–corrector method used in QCPE program No. 273,* the fixed-step-size eleventh-order Adams–Mouton predictor–corrector routine used in QCPE program No. 229, and the variable-order, variable-step-size, automatic-error-control Adams method of Shampine and Gordon.[(43)] Their results are summarized in more detail in Table 1. Of course these results are not indicative of the efficiencies of the various schemes in general, but they do show the kinds of differences in computing times that may occur. And they show that higher-order and variable-step-size methods are not always more efficient than lower-order, fixed-step-size methods.

Other comparisons of the efficiencies of various integration schemes have been made by Parr[(44)] and Brumer.[(45,46)] Parr tested an Adams–Moulton eleventh-order predictor/eleventh-order corrector (AM11), an Adams–Moulton fifth-order predictor/sixth-order corrector (AM6), and a Runga–Kutta–Gill fourth-order integrator (RKG4) for the unimolecular dissociation of a triatomic molecule with a pairwise Morse potential. The first two integrators gave comparable results in his tests and both were superior to the RKG4 scheme. Brumer was particularly concerned with trajectories involving long-lived collision complexes. He found that Nord-

* The QCPE programs are discussed in Section 2.7.

Table 1. Relative Computing Times for Two F + H_2 Trajectories[a]

Integration scheme	Computing time		
	3 s.f.[b]	5 s.f.	6 s.f.
Bulirsch–Stoer, second-order equations	1.0	1.3	1.5
Bulirsch–Stoer, first-order equations	1.4	2.0	2.4
QCPE 273	1.1	2.0[c]	2.4[c]
QCPE 229	2.2	2.3	2.4
Shampine–Gordon (ODE)	1.9	3.2	3.8[c]

[a] The results in this table are based on unpublished calculations by M. K. Matzen and M. E. Riley. They are the average for two trajectories of the minimum computing time necessary to get a specified number of significant figures correct in all of the following final-state properties: positions, momenta, translational energy, vibrational energy, and rotational energy.
[b] s.f. = significant figures.
[c] Extrapolated.

sieck's sixth-order variable-step-size Adams method led to a serious error buildup for long-lived trajectories. Hamming's fourth-order variable-step-size predictor–corrector method suffered from considerable roundoff error propagation and unreliability of the step-size changing algorithm. He was able to integrate long-lived trajectories accurately with the RKG4 scheme and with Gear's sixth-order hybrid method (G6).[(47)] Since Parr[(44)] was able to integrate long-lived trajectories using the AM6 scheme, Brumer tested the latter three schemes by integrating a NaCl oscillator for 125 periods. The G6 method was the most efficient. In the range of five to eight significant figures of accuracy, the AM6 scheme gave errors six times larger for the same number of function evaluations. The RKG4 method gave errors 4500 times larger than the G6 method with twice as many function evaluations as the G6 method used to obtain eight significant figures. For problems involving short-lived trajectories, Brumer expected that the Gear method will be competitive with other variable-step-size predictor–corrector schemes of the same order but that the improvement should not be significant.

Once a method has been chosen one must find a step size (for methods with fixed step sizes) or a value of the error control parameter (for methods with variable step sizes and/or variable orders) which yields results of sufficient accuracy for the given problem. This should always be determined carefully for a new problem and rechecked when the problem, energy range, or initial conditions are varied significantly. There are several ways to determine what step size (or step-size control parameter) is needed. One is to integrate a few trajectories with very small step sizes and then increase the step size until the error in the quantities of interest (see Section 2.5) reaches its maximum allowable limit. This is the safest method but it is not always the most convenient. Necessary but not sufficient checks for accuracy are conservation of total energy and angular momentum. Sometimes these are conserved to more significant figures than the accuracy of the trajectory. Another check is back integration of trajectories, i.e., starting a new integration with final coordinates and with reversed final momenta of the original integration to try to recover initial conditions. This is one of the best checks, but sometimes the initial conditions are recovered to more significant figures than the accuracy of the original forward integration.[44] Another danger in performing checks is that some trajectories, especially those for collisions of long duration, are harder to integrate than others. Thus one must be sure that accuracy checks are really performed on a representative sample. It is a good practice to monitor energy and angular momentum conservation on all trajectories and routinely to back-integrate a reasonable fraction of trajectories on all computer runs. If any change in the parameters of the problem causes the errors in these checks to increase, then one should again recheck the results against trajectories integrated with smaller step-sizes.

Each trajectory is integrated in time until the chemical species produced by the collision, either the reactants (A + BC) or some set of products, have separated beyond the "collision shell" distance ρ_0'. As soon as one or another such end test is satisfied, the integration may be stopped and the identity of the products determined. Care must be taken not to stop the trajectory as the reactants are approaching one another, but otherwise a good end test consists of finding two of the R_i larger than ρ_0'. This allows a tentative assignment of the products. The pair of atoms with the smallest internuclear separation may be a stable molecule or a quasibound molecule, but might also have dissociated if the trajectory had been integrated for a longer time. Analysis of the tentative diatomic molecule's internal energy and angular momentum allows a definitive assignment.

The end test procedure just described has the advantages that (i) comparing the various R_i to ρ_0' is very fast computationally and (ii) there is no possible misassignment of products. The end tests used in Reference 5 (comparing to ρ_0' the relative separation of each atom from the center of mass of the other two atoms) can for certain mass combinations (e.g., F + HD) incorrectly identify the product diatomic molecule.

Of course one must be sure that ρ_0' is large enough. It must be large enough not only to insure that the trajectory can be judged reactive or not but also to insure that properties of the reactive collision, such as discussed in Section 2.5, are accurate to the desired number of significant figures. This, like step size, has to be determined numerically.

2.4. Sampling and Averaging over the Initial Conditions

If one can identify some maximum impact parameter such that collisions with impact parameter b greater than b_{max} cannot possibly react according to classical mechanics, then the reaction cross section can be expressed classically as the product of a geometrical cross section πb_{max}^2 and an average probability of reaction over all possible collisions with $b < b_{max}$, i.e.,

$$\sigma_r = \pi b_{max}^2 \langle \mathscr{P}_r \rangle \tag{24}$$

where $\mathscr{P}_r$ is the reaction probability as a function of all the collision parameters as defined in Section 2.2. The average of $\mathscr{P}_r$ is taken over all possible collisions with the impact parameter $b < b_{max}$ and all possible values of all other variables. If we seek a reactive scattering cross section at some fixed center-of-mass collision energy for atom A with some selected state (n, j) of the molecule BC, then

$$\sigma_r(E_{rel}, n, j) = \pi b_{max}^2 \langle \mathscr{P}_r(E_{rel}, n, j) \rangle \tag{25a}$$

and

$$\langle \mathscr{P}_r(E_{rel}, n, j) \rangle = \frac{1}{(2\pi)^3 b_{max}^2} \times \int_{b=0}^{b_{max}} \int_{\theta=0}^{\pi} \int_{\phi=0}^{2\pi} \int_{\eta=0}^{2\pi} \int_{\xi=0}^{2\pi} \mathscr{P}_r(b, \theta, \phi, \eta, \xi; E_{rel}, n, j)\, b \times \sin\theta \, db \, d\theta \, d\phi \, d\eta \, d\xi \tag{25b}$$

where the normalization constant is determined by the definite integral of the volume element. Should we desire to average $\mathscr{P}_r$ over some distribution of collision energies, equation (25) can be appropriately modified. For example, a state-selected thermal rate constant for the reaction of A with the state (n, j) of BC is defined as

$$k_r^{n,j}(T) = \langle v_{rel} \sigma_r(E_{rel}, n, j) \rangle_T \tag{26a}$$

$$= \left(\frac{8kT}{\pi \mu_{A,BC}} \right)^{1/2} \pi b_{max}^2 [(2\pi)^3 b_{max}^2 (kT)^2]^{-1} \times \int_{E_{rel}=0}^{\infty} \int_{b=0}^{b_{max}} \int_{\theta=0}^{\pi} \int_{\phi=0}^{2\pi} \int_{\eta=0}^{2\pi} \int_{\xi=0}^{2\pi} \mathscr{P}_r(E_{rel}, b, \theta, \phi, \eta, \xi; n, j) \times E_{rel}\, e^{-E_{rel}/kT}\, b \sin\theta \, dE_{rel} \, db \, d\theta \, d\phi \, d\eta \, d\xi \tag{26b}$$

where k is the Boltzmann constant, T is the temperature, and v_{rel} is the initial relative speed given by

$$v_{rel} = P^0 / \mu_{A,BC} \tag{26c}$$

One may further obtain the completely thermally averaged rate constant $k_r(T)$ by averaging over Boltzmann distributions of n and j.

The function $\mathscr{P}_r$ in equations (25b) and (26b) is the probability of reaction for a collision specified by some set of collision parameters. In equation (25) it is specified by $(b, \theta, \phi, \eta, \xi)$ at fixed E_{rel}, n, and j. As all classical trajectories with the same set of collision parameters (initial conditions) are identical and either do or do not lead to reaction, the function $\mathscr{P}_r$ is a Boolean function which takes on one of two possible values: zero for no reaction, one for reaction. To determine whether $\mathscr{P}_r$ is zero or one for some particular set of collision parameters, one must compute a trajectory and examine the collision products to see if a reaction has occurred. Selecting initial values of the collision parameters for each trajectory is nothing more or less than selecting points in the collision parameter space for the numerical evaluation of the multidimensional integral in equation (25) or (26). The values of initial collision parameters, therefore, depend upon the numerical integration method chosen.

The two methods in current use in trajectory calculations for evaluation of multidimensional integrals of the form

$$I = \int_0^1 \int_0^1 \cdots \int_0^1 f(\boldsymbol{\beta})\, d\boldsymbol{\beta} \tag{27}$$

are the *Monte Carlo method*[48-50] and the method of "fixed lattice points" (a so-called *diophantine method*).[17] In the former the selection of points in the unit hypercube is completely random, while in the latter the selection is predetermined by a set of "good" or "optimum" lattice points which depend upon the dimensionality of the hypercube and the total number of points at which the function is to be evaluated. Although there is evidence that the latter method may often converge more rapidly than the Monte Carlo method as the number of points is increased[17] and that practical error estimates are possible, the method allows no rigorous estimate of the error, and none of the previously computed function values can be used to obtain an improved estimate of the integral using a larger number of points. We recommend the use of the Monte Carlo method because it allows a rigorous estimate of the error and because *all previously computed function values are used along with additional points computed to obtain convergence.* We will accordingly restrict our discussion of selection of initial values of collision parameters to that method.

In the Monte Carlo method the integral in equation (27) is approximated by the average value of the function $f(\boldsymbol{\beta})$ over N randomly selected values $\boldsymbol{\beta}^{[i]}$ from a uniform distribution in the unit hypercube, i.e.,

$$I \approx \varepsilon(I) = \frac{1}{N} \sum_{i=1}^{N} f(\boldsymbol{\beta}^{[i]}) \tag{28}$$

where $\varepsilon(I)$ is the Monte Carlo estimator of the integral I.[48-50] In the limit of a sufficiently large number of points, N, the variance of the estimator $\varepsilon(I)$ about the value I is

$$\operatorname{var}[\varepsilon(I)] \underset{N\to\infty}{\sim} \frac{\operatorname{var}[f]}{N} \tag{29}$$

where the var$[f]$ is the variance of the function $f(\boldsymbol{\beta})$ over the unit hypercube and is some constant. The standard deviation of the estimator is then

$$s = \{\mathrm{var}[\varepsilon(I)]\}^{1/2} \underset{N\to\infty}{\sim} \frac{\{\mathrm{var}[f]\}^{1/2}}{N^{1/2}} \tag{30}$$

The estimated error in $\varepsilon(I)$ is usually taken to be some multiple of the standard deviation, the exact multiple being determined by the desired confidence limits, so that for a sufficiently large number of points

$$\Delta\varepsilon(I) \propto \frac{1}{N^{1/2}} \tag{31}$$

The probability that the estimator $\varepsilon(I)$ is within $\Delta\varepsilon(I)$ of I is 0.6826 if $\Delta\varepsilon(I)$ is taken as one standard deviation and 0.9554 if it is taken as two standard deviations. Confidence limits for the case when N is not very large can be obtained from standard statistical tables.[51]

Now the variance of the function $f(\boldsymbol{\beta})$ is defined as

$$\mathrm{var}[f] = \langle f^2 \rangle - \langle f \rangle^2 \tag{32}$$

where the average values are taken over the unit hypercube. In the *special case* in which f is a Boolean function (see above), $\langle f^2 \rangle = \langle f \rangle$ so that

$$\mathrm{var}[f_{\mathrm{Boolean}}] = \langle f_{\mathrm{Boolean}} \rangle (1 - \langle f_{\mathrm{Boolean}} \rangle) \tag{33}$$

Further, $\langle f \rangle = I$ so that $\varepsilon(I)$ is an estimator of $\langle f \rangle$. For a Boolean function

$$\varepsilon(I) = \frac{N_r}{N} \tag{34}$$

where N_r is the number of nonzero values of f_{Boolean} sampled (i.e., the number of reactive trajectories or the number of different trajectories for which $f_{\mathrm{Boolean}} = 1$). Therefore

$$\mathrm{var}[f_{\mathrm{Boolean}}] \cong \frac{N_r}{N}\left(1 - \frac{N_r}{N}\right) = \frac{N_r(N - N_r)}{N^2} \tag{35}$$

and from equation (29)

$$\mathrm{var}[\varepsilon(I)] \approx \frac{N_r(N - N_r)}{N^3} = \frac{N_r^2}{N^2}\,\frac{N - N_r}{N_r N} \tag{36}$$

Using a one standard deviation rule (68% confidence level), we obtain the result for the error in $\varepsilon(I)$,[5]

$$\Delta\varepsilon(I) \approx \frac{N_r}{N}\left(\frac{N - N_r}{N_r N}\right)^{1/2} \tag{37a}$$

for the case of a Boolean function. In many practical applications one has $N_r \ll N$. In this case equations (34) and (35) may be combined to yield the useful limiting form

$$\frac{\Delta\varepsilon(I)}{\varepsilon(I)} \underset{\substack{N\to\infty \\ N>>N_r}}{\sim} \frac{1}{N_r^{1/2}} \tag{37b}$$

It is apparent from the $N^{-1/2}$ dependence of the estimated error in equation (31) that the Monte Carlo method converges very rapidly at first but then very slowly as the number of points N is increased. In other words not too many points are required to obtain a moderately accurate estimate of the integral, but very many points are required to obtain high accuracy.

To put the integral in equation (25) in the form of equation (27) we must transform variables according to

$$d\beta_1 = \frac{1}{c_1} b\, db \tag{38a}$$

$$d\beta_2 = \frac{1}{c_2} \sin\theta\, d\theta \tag{38b}$$

$$d\beta_3 = \frac{1}{c_3} d\phi \tag{38c}$$

$$d\beta_4 = \frac{1}{c_4} d\eta \tag{38d}$$

$$d\beta_5 = \frac{1}{c_5} d\xi \tag{38e}$$

This implies that

$$\langle \mathscr{P}_r \rangle \equiv \frac{c_1 c_2 c_3 c_4 c_5}{(2\pi)^3 b_{\max}^2} \times \int_{\beta_1=0}^{1} \int_{\beta_2=0}^{1} \int_{\beta_3=0}^{1} \int_{\beta_4=0}^{1} \int_{\beta_5=0}^{1} \mathscr{P}_r(\boldsymbol{\beta})\, d\beta_1\, d\beta_2\, d\beta_3\, d\beta_4\, d\beta_5 \tag{38f}$$

The relationships of the new variables to the old are then obtained straightforwardly by integration:

$$\beta_1 = \int_0^{\beta_1} d\beta_1 = \frac{1}{c_1} \int_0^b b\, db = \frac{1}{c_1} \frac{b^2}{2} \tag{39a}$$

$$\beta_2 = \int_0^{\beta_2} d\beta_2 = \frac{1}{c_2} \int_0^\theta \sin\theta\, d\phi = \frac{1}{c_2}(1 - \cos\theta) \tag{39b}$$

$$\beta_3 = \int_0^{\beta_3} d\beta_3 = \frac{1}{c_3} \int_0^\phi d\phi = \frac{1}{c_3} \phi \tag{39c}$$

$$\beta_4 = \int_0^{\beta_4} d\beta_4 = \frac{1}{c_4}\int_0^{\eta} d\eta = \frac{1}{c_4}\eta \tag{39d}$$

$$\beta_5 = \int_0^{\beta_5} d\beta_5 = \frac{1}{c_5}\int_0^{\xi} d\xi = \frac{1}{c_5}\xi \tag{39e}$$

The normalization constants c_i are determined from the conditions $\beta_i = 1$ at the upper limit of the corresponding variable. The result is

$$\beta_1 = \left(\frac{b}{b_{\max}}\right)^2 \tag{40a}$$

$$\beta_2 = \tfrac{1}{2}(1 - \cos\theta) \tag{40b}$$

$$\beta_3 = \frac{1}{2\pi}\phi \tag{40c}$$

$$\beta_4 = \frac{1}{2\pi}\eta \tag{40d}$$

$$\beta_5 = \frac{1}{2\pi}\xi \tag{40e}$$

and the expression for $\langle\mathscr{P}_r\rangle$ becomes a special case of equation (27) as follows:

$$\langle\mathscr{P}_r\rangle = \int_0^1\int_0^1\int_0^1\int_0^1\int_0^1 \mathscr{P}_r(\boldsymbol{\beta})\, d\boldsymbol{\beta} \tag{41}$$

The procedure for selecting appropriate values of the collision parameters over which to average the reaction probability $\mathscr{P}_r$ is as follows.

(i) Select a set of five random numbers $\beta_1, \beta_2, \ldots, \beta_5$ uniformly distributed in the interval (0, 1). Most modern computing centers now have system libraries which include programs for random number or pseudorandom number generation. These are always or almost always good enough for our purposes so this step will not be discussed in detail.

(ii) Obtain the corresponding values of the collision parameters by inverting equations (39). This gives

$$b = \beta_1^{1/2} b_{\max} \tag{42a}$$

$$\theta = \cos^{-1}(1 - 2\beta_2) \tag{42b}$$

$$\phi = 2\pi\beta_3 \tag{42c}$$

$$\eta = 2\pi\beta_4 \tag{42d}$$

$$\xi = 2\pi\beta_5 \tag{42e}$$

These values of the collision parameters are then used in equations (10), (11), (12), and (19) to define the initial values of the generalized coordinates and momenta for a trajectory, and the trajectory is computed to determine whether $\mathscr{P}_r(b(\beta_1), \theta(\beta_2), \phi(\beta_3), \eta(\beta_4), \xi(\beta_5); E_{\text{rel}}, n, j)$ is one or zero. If this procedure is repeated N times and N_r nonzero values of $\mathscr{P}_r$ are obtained, then from equations (25), (28), (34), (37), and (41) we have

$$\sigma_r(E_{\text{rel}}, n, j) = \pi b_{\text{max}} \langle \mathscr{P}_r \rangle \tag{43a}$$

$$\cong \pi b_{\text{max}}^2 N_r/N \tag{43b}$$

and

$$\Delta\sigma_r \approx \pi b_{\text{max}}^2 \frac{N_r}{N}\left(\frac{N - N_r}{N_r N}\right)^{1/2} \tag{44}$$

Similarly, an expression for the state-selected rate constant $k_r^{n,j}$ is obtained by the Monte Carlo integration of equation (26). The result is

$$k_r^{n,j} \cong \left(\frac{8kT}{\pi\mu_{\text{A,BC}}}\right)^{1/2} \pi b_{\text{max}}^2 \frac{N_r}{N} \tag{45}$$

and

$$\Delta k_r^{n,j} \approx \left(\frac{8kT}{\pi\mu_{\text{A,BC}}}\right)^{1/2} \pi b_{\text{max}}^2 \frac{N_r}{N}\left(\frac{N - N_r}{N_r N}\right)^{1/2} \tag{46}$$

The collision energy may be selected in the following manner. A new transformed variable

$$d\beta_6 = ue^{-u}\,du \tag{47}$$

where $u = E_{\text{rel}}/kT$ is defined analogous to those in equation (38). Integrating equation (47) to obtain $\beta_6(u)$ we have

$$\beta_6 = \int_0^{\beta_6} d\beta_6 = \int_0^u ue^{-u}\,du = 1 - (u + 1)\,e^{-u} \tag{48}$$

which, unfortunately, cannot be inverted to obtain a closed-form expression for $u(\beta_6)$. Successive approximations to u, however, may be obtained using the relation

$$u_{i+1} = u_i + \frac{u_i + 1 - (1 - \beta_6)\,e^{u_i}}{u_i} \tag{49}$$

which is derived using the Newton–Raphson iteration method. The sequence of partial sums, u_i, is absolutely convergent for all values of β_6 in the (open) interval (0, 1) when $u_1 = 1$, the inflection point of $\beta_6(u)$, is used as the initial guess. Once a converged value $u(\beta_6)$ is obtained using equation (49), the collision energy is given by $E_{\text{rel}} = ukT$. Alternatively the initial relative translational energy may be selected by a rejection method.[8]

To compute the completely thermally averaged rate constant one must also average over discrete distributions of n and j. When the number of internal states is not too large this can be done in the following manner.[27] Pick a number $N_{\text{select}}^{\text{total}}$ much larger than the number of trajectories to be run. For example, let $N_{\text{select}}^{\text{total}}$ be 10^7. Compute the energies of all states with a probability of occurrence greater than $1/N_{\text{select}}^{\text{total}}$ by using equation (20) and compute the number $N_{\text{select}}(n, j)$ of times each such state should be sampled in 10^7 collisions. If the total probability of occurrence of states with probabilities less than $1/N_{\text{select}}^{\text{total}}$ is nonnegligible, then $N_{\text{select}}^{\text{total}}$ is too small. Arrange the states in order and store the subtotals of the $N_{\text{select}}(n, j)$. Then, for each trajectory, pick a random number in the range (0, 1), multiply by $N_{\text{select}}^{\text{total}}$, and compare efficiently to the table of subtotals to decide what n and j should be used. If the number of states of the reactant is too large for this scheme to be practical it can be modified as necessary, e.g., by analytic approximations for $\varepsilon_{n,j}$.

The transformed variables defined by equations (38) and (47) are those appropriate for so-called *crude Monte Carlo sampling*. Crude Monte Carlo sampling is a completely "safe" and unbiased sampling technique (so long as $b_{\max}$ is chosen properly), but it is not necessarily optimum in the sense of yielding the most precise results for a particular reaction attribute for a given number of trajectories computed. The goal of the alternative sampling methods is variance reduction, i.e., by biasing the selection process for initial conditions we attempt to accelerate the convergence of the Monte Carlo evaluation of unbiased reaction attributes. This is the most important area for research in the theory of Monte Carlo integration, and the newcomer to the field of Monte Carlo trajectory calculations of chemical dynamics might be surprised to find how little such variance-reducing techniques have been used in this area. This, however, is not without a good reason. Simply stated, a nonuniform selection procedure which reduces the variance of one computed reaction attribute will generally increase the variance of one or more other interesting attributes. Since we are typically interested in several such reaction attributes (as discussed in later sections), we often adopt the compromise of using no nonuniformity at all. As an example, some workers prefer to skew their sample in favor of low impact parameters. This does not produce biased estimates of reaction attributes if trajectories with high impact parameters are weighted relatively more highly than trajectories with low impact parameters. And it may lead to a smaller Monte Carlo uncertainty (variance) in the estimated reaction cross section if, as is often the case, the probability of reaction is a decreasing function of impact parameter. But it will also generally lead to much poorer statistics for the differential cross section in the forward direction if, as is usually also the case, the forward-scattered trajectories tend to come from large impact parameters. However, if one is really interested in only one or a few reaction attributes, e.g., in only the thermal rate coefficient, it may be possible to use variance-reducing techniques quite profitably. In general the use of any variance-reducing technique requires knowledge of the dependence of the trajectory results (which are the integrand of an integral being evaluated by the Monte Carlo method) on the independent variable whose distribution is to be sampled nonuniformly. Thus we usually have no grounds for biasing the selection of initial variables like the vibrational phase or the orientation of the BC angular momentum, but we may be able to bias the selection of initial energy quantities or impact parameters. Two alternative sampling methods—importance sampling and stratified sampling—will now be discussed and illustrated by applying each to the selection of the impact parameter.

Consider the *opacity function* (this is one of the *reactivity functions* discussed in Section 2.6.1),

$$\mathscr{P}_r(b; E_{\text{rel}}, n, j) = \frac{1}{2(2\pi)^3} \times \int_{\theta=0}^{\pi} \int_{\phi=0}^{2\pi} \int_{\eta=0}^{2\pi} \int_{\xi=0}^{2\pi} \mathscr{P}_r(b, \theta, \phi, \eta, \xi; E_{\text{rel}}, n, j) \times \sin\theta \, d\theta \, d\phi \, d\eta \, d\xi \tag{50}$$

denoted simply by $\mathscr{P}_r(b)$ in the following discussion, which has the properties

$$\langle \mathscr{P}_r \rangle = \frac{2}{b_{\max}^2} \int_0^{b_{\max}} \mathscr{P}_r(b) \, b \, db \tag{51}$$

$$\sigma_r = 2\pi \int_0^{b_{\max}} \mathscr{P}_r(b) \, b \, db \tag{52}$$

and

$$\frac{d\sigma_r}{db} = 2\pi b \mathscr{P}_r(b) \tag{53}$$

where

$$\sigma_r = \int_0^{b_{\max}} \frac{d\sigma_r}{db} \, db \tag{54}$$

If an *expected* opacity function $\mathscr{P}_r^0(b)$ is defined, then equation (51) may always be rewritten as[52]

$$\langle \mathscr{P}_r \rangle = \frac{2}{b_{\max}^2} \int_0^{b_{\max}} \frac{\mathscr{P}_r(b)}{\mathscr{P}_r^0(b)} \mathscr{P}_r^0(b) \, b \, db \tag{55}$$

$$= \int_0^1 \frac{\mathscr{P}_r(\beta_1)}{\mathscr{P}_r^0(\beta_1)} \, d\beta_1 \tag{56}$$

where

$$d\beta_1 = \frac{2}{b_{max}^2} \mathscr{P}_r^0(b)\, b\, db \tag{57}$$

If the *actual* opacity function resembles the *expected* function, the ratio $\mathscr{P}_r(b)/\mathscr{P}_r^0(b)$ should vary quite slowly with impact parameter (or equivalently, with β_1). The more slowly varying this ratio can be made, the smaller the variance of the integrand becomes, and the smaller the estimated error for a given number of trajectories computed. The function $\mathscr{P}_r^0(b)$ is an *importance-sampling* function.

It is clear from equation (57) that use of the crude Monte Carlo selection of the impact parameter defined by equations (38a) or (40a) corresponds to taking the expected opacity function (importance-sampling function) as a step function, i.e.,

$$\mathscr{P}_r^0(b) = \begin{cases} 1, & b \leq b_{max} \\ 0, & b > b_{max} \end{cases} \tag{58}$$

If the step function in equation (58) is not expected to have the same shape as the actual $\mathscr{P}_r(b)$, then crude Monte Carlo sampling might not be the method of choice for the calculation of the reaction cross section or reaction rate constant. One alternative is to use the following importance-sampling function[52]:

$$\mathscr{P}_r^0(b) = \begin{cases} 3\left(1 - \dfrac{b}{b_{max}}\right), & b \leq b_{max} \\ 0, & b > b_{max} \end{cases} \tag{59}$$

in which case

$$d\beta_1 = \frac{6}{b_{max}^2}\left(1 - \frac{b}{b_{max}}\right) b\, db \tag{60}$$

$$\beta_1 = \int_0^{\beta_1} d\beta_1 = 3\left(\frac{b}{b_{max}}\right)^2 - 2\left(\frac{b}{b_{max}}\right)^3 \tag{61}$$

$$\langle \mathscr{P}_r \rangle = \int_0^1 \frac{\mathscr{P}_r(\beta_1)}{\mathscr{P}_r^0(\beta_1)} d\beta_1 \tag{62}$$

and

$$b = b_{max}\{\cos[\tfrac{1}{3}\cos^{-1}(1 - 2\beta_1) + \tfrac{4}{3}\pi] + \tfrac{1}{2}\} \tag{63}$$

Notice that the factor of 3 in equation (59) was chosen so that $\beta_1(b = b_{max}) = 1$, and in deriving equation (61) we used the fact that $\beta_1(b = 0) = 0$. When using an importance-sampling function for the selection of the impact parameter, equation (41) becomes

$$\langle \mathscr{P}_r \rangle = \int_0^1 \int_0^1 \int_0^1 \int_0^1 \int_0^1 \frac{\mathscr{P}_r(\boldsymbol{\beta})}{\mathscr{P}_r^0(\beta_1)} d\beta_1\, d\beta_2\, d\beta_3\, d\beta_4\, d\beta_5 \tag{64}$$

and its Monte Carlo estimator becomes

$$\langle \mathscr{P}_r \rangle \cong \frac{1}{N} \sum_{i=1}^{N} \frac{\mathscr{P}_r(\boldsymbol{\beta}^{[i]})}{\mathscr{P}_r^0(\beta_1^{[i]})} = \frac{1}{N} \sum_{i=1}^{N_r} \frac{1}{\mathscr{P}_r^0(\beta_1^{[i]})} \tag{65}$$

where the N computed trajectories are for β_1 selected at random uniformly on the interval (0, 1). It is evident from equation (65) that each trajectory is *weighted* by

$$w_i \equiv \frac{1}{\mathscr{P}_r^0(\beta_1^{[i]})} \tag{66a}$$

when importance sampling is used. Note that $w_i = 1$ for each trajectory when $\mathscr{P}_r^0$ is given by equation (58). Because we have required that $\mathscr{P}_r^0(b)$ be normalized so that $\beta_1(b = b_{\max}) - \beta_1(b = 0)$ is unity, one will find in general that

$$\sum_{i=1}^{N} w_i \underset{\text{large } N}{\sim} N \tag{66b}$$

The importance-sampling estimator for the reaction cross section is

$$\sigma_r \cong \pi b_{\max}^2 \frac{1}{N} \sum_{i=1}^{N_r} w_i \tag{67}$$

with estimated error (for sufficiently large N)

$$\Delta\sigma_r \approx \pi b_{\max}^2 \frac{1}{N} \left[\sum_{i=1}^{N_r} w_i^2 - \frac{1}{N} \left(\sum_{i=1}^{N_r} w_i \right)^2 \right]^{1/2} \tag{68}$$

It should be pointed out that the above procedure for importance sampling always results in an unbiased estimator for the cross section but is advantageous only if $\mathscr{P}_r/\mathscr{P}_r^0$ has a smaller variance than $\mathscr{P}_r$ itself. For example, the importance-sampling function $\mathscr{P}_r^0(b) = b_{\max}/2b$ is a poor choice because of its unphysical behavior near $b = 0$. This leads to extremely large weights for some trajectories and thus to a very uneven approach of the calculated results to convergence. This sampling function leads to a flat b distribution, i.e., equal probability of sampling any b rather than the probability being proportional to b. In practice, when flat b sampling has been employed it has usually been combined[12,13,19,20,22] with a discrete selection of the corresponding angular momentum J_{rel}, given by $\mu_{\text{A,BC}} v_{\text{rel}} b/\hbar$, but this does not eliminate the problems. We recommend that equation (59) is generally a better way to give higher weights to small-b collisions when that is appropriate.

Equations (64)–(66) are easily generalized for the importance sampling of variables other than (or in addition to) the impact parameter.

Another type of sampling which is often quite useful is *stratified sampling.* In stratified sampling the unit hypercube is divided into subvolumes (usually by hyperplanes perpendicular to the edges). Sampling is carried out within each subvolume exactly as in crude Monte Carlo sampling, but all subvolumes are not sampled equally densely.

As an example of stratified sampling consider partitioning the impact parameter interval $(0, b_{max})$ into k_{strat} subintervals (b_{k-1}, b_k) where k goes from 1 to k_{strat} and $b_0 = 0$. This partition corresponds to a division of the unit hypercube by the $(k_{strat} - 1)$ hyperplanes $\beta_1 = \beta_{1,k} \equiv (b_k/b_{max})^{1/2}$. Let subvolume k, denoted by $\mathscr{V}_k$, be defined as the subvolume characterized by β_1 in the interval $\beta_{1,k-1} \leq \beta_1 \leq \beta_{1,k}$. By design,

$$\sum_{k=1}^{k_{strat}} \mathscr{V}_k = 1 \tag{69}$$

The stratified-sampling analog of equation (41) is

$$\langle \mathscr{P}_r \rangle = \sum_{k=1}^{k_{strat}} \mathscr{V}_k \langle \mathscr{P}_r \rangle_k \tag{70a}$$

$$\cong \sum_{k=1}^{k_{strat}} \mathscr{V}_k \frac{N_{rk}}{N_k} \tag{70b}$$

with estimated error

$$\Delta \langle \mathscr{P}_r \rangle \approx \left\{ \sum_{k=1}^{k_{strat}} \mathscr{V}_k^2 \frac{N_{rk}(N_k - N_{rk})}{N_k^3} \right\}^{1/2} \tag{71}$$

where N_{rk} and N_k are subtotals of N_r and N for the kth subvolume. Any number of trajectories N_k may be computed at random in the kth stratum, but the estimated error in equation (71) is minimized for a fixed total number of trajectories

$$N = \sum_{k=1}^{k_{strat}} N_k$$

when for each stratum k,[(53)]*

$$N_k \propto \mathscr{V}_k \frac{N_{rk}}{N_k} \left(\frac{N_k - N_{rk}}{N_{rk} N_k} \right)^{1/2} \tag{72}$$

Alternatively we may ask: what is the optimum allocation of stratification boundaries in the sense that all N_k are equal for the optimum boundaries? Clearly, for impact parameter strata, the answer is that the bounda-

* This is a consequence of the Tschuprow–Neyman theorem.

ries should correspond to an equipartition of the variable $\beta_1(b)$ as computed by integrating equation (57) with the optimum $\mathscr{P}_r^0(b)$. The generalization to boundaries for the other β_i is straightforward. Allocating trajectories to the various strata according to equation (72) or using the strata which are optimum for $\langle \mathscr{P}_r \rangle$ may, however, defeat one of the very useful applications of stratified sampling. As will be discussed in Section 2.6, stratified sampling can be used to obtain greatly improved statistics on cross sections for producing *minor products* or final-state distributions of any products in regions of their final-state parameter space that are lightly populated if the region of initial conditions leading to that product or that region of product space can be identified. Equation (72) applies only to minimizing the error of $\langle \mathscr{P}_r \rangle$ or of the cross section related to it by equation (25).

We conclude our discussion of stratified sampling by noting that from equation (70)

$$\langle \mathscr{P}_r \rangle \cong \frac{1}{N} \sum_{k=1}^{k_{\text{strat}}} \frac{\mathscr{V}_k N}{N_k} N_{rk} \tag{73a}$$

$$= \frac{1}{N} \sum_{k=1}^{k_{\text{strat}}} \sum_{i=1}^{N_k} \frac{\mathscr{V}_k N}{N_k} \mathscr{P}_{rk}(\boldsymbol{\beta}^{[i]}) \tag{73b}$$

This implies that the statistical weight of each trajectory in the kth stratum is $\mathscr{V}_k N/N_k$.

2.5. *Product Analysis*

Once a trajectory is complete and a product diatomic molecule tentatively identified, it is most convenient to transform to a new set of generalized coordinates and momenta in which the Hamiltonian is asymptotically separable, i.e.,

$$H(\mathbf{q}', \mathbf{Q}', \mathbf{p}', \mathbf{P}') \sim T_{\text{rel}}(\mathbf{P}') + H_{\text{int}}(\mathbf{q}', \mathbf{p}') \tag{74}$$

Here $\mathbf{q}'$ and $\mathbf{p}'$ are the coordinates and conjugate momenta for the *product diatomic molecule* and $\mathbf{Q}'$ and $\mathbf{P}'$ are those for the *relative motion of the products.* As the trajectory is integrated in the generalized coordinates appropriate for reactants, $\{q_i, Q_i, p_i, P_i; i = 1, 2, 3\}$, it is necessary to derive expressions for the product coordinates $\{q'_i, Q'_i, p'_i, P'_i; i = 1, 2, 3\}$ in terms of those for reactants. (Of course, if the product molecule is BC no transformation is necessary.) This is accomplished in two steps:

(i) Derive expressions analogous to equation (4) for $\mathbf{q}'$ and $\mathbf{Q}'$ in terms of the reference Cartesian coordinates, $\{x_i; i = 1, 9\}$. This defines the transformation matrix $\mathbf{T}'$ in the transformation

$$\begin{bmatrix} \mathbf{q}' \\ \mathbf{Q}' \end{bmatrix} = \mathbf{T}'\mathbf{x} \tag{75}$$

The matrix **T** is defined by equation (4) and

$$\begin{bmatrix} \mathbf{q} \\ \mathbf{Q} \\ \mathbf{S} \end{bmatrix} = \mathbf{T}\mathbf{x} \tag{76}$$

Note **T**′ is 6 × 9 but **T** is 9 × 9. It should be pointed out that the convention for the *scattering angle* depends upon whether the relative motion coordinate is taken to be the atom with respect to the center of mass of the diatomic, or vice versa. If, as is often the case, the scattering angle is to be defined as the angle between the final relative velocity of the species which contains the (initially free) atom A and the incident direction of A, then the sign of the right-hand side of (4b) should be changed for **T**′ but not for **T**.

(ii) Find the inverse of the transformation defined by equation (76), i.e., find the inverse of the square matrix **T**; then

$$\mathbf{x} = \mathbf{T}^{-1}\begin{bmatrix} \mathbf{p} \\ \mathbf{Q} \\ \mathbf{S} \end{bmatrix} \tag{77a}$$

which upon substitution in equation (75) gives

$$\begin{bmatrix} \mathbf{q}' \\ \mathbf{Q}' \end{bmatrix} = \mathbf{T}'\mathbf{T}^{-1}\begin{bmatrix} \mathbf{q} \\ \mathbf{Q} \\ \mathbf{S} \end{bmatrix} \tag{77b}$$

the desired result. It does not matter than **T**′ is not a square matrix as its inverse is not required. The new conjugate generalized momenta are obtained from

$$p_i' = m\dot{q}_i' \qquad (i = 1, 2, 3) \tag{78a}$$

$$P_i' = \mu\dot{Q}_i' \qquad (i = 1, 2, 3) \tag{78b}$$

where

$$\begin{bmatrix} \dot{\mathbf{q}}' \\ \dot{\mathbf{Q}}' \end{bmatrix} = \mathbf{T}'\mathbf{T}^{-1}\begin{bmatrix} \dot{\mathbf{q}} \\ \dot{\mathbf{Q}} \\ \dot{\mathbf{S}} \end{bmatrix} \tag{79}$$

and

$$\dot{q}_i = \frac{1}{\mu_{\mathrm{BC}}} p_i \qquad (i = 1, 2, 3) \tag{80a}$$

$$\dot{Q}_i = \frac{1}{\mu_{\mathrm{A,BC}}} P_i \qquad (i = 1, 2, 3) \tag{80b}$$

$$\dot{S}_i = \frac{1}{M} P_{S_i} \qquad (i = 1, 2, 3) \tag{80c}$$

For the products AB + C, m is μ_{AB} and μ is $\mu_{C,AB}$. For the products AC + B, m is μ_{AC} and μ is $\mu_{B,AC}$. The (6 × 9) matrix product $\mathbf{T'T}^{-1}$ appropriate for the transformation to collision products other than reactants is defined by

$$q'_i = -\frac{m_C}{m_B + m_C} q_i - Q_i \qquad (i = 1, 2, 3) \quad (81a)$$

$$Q'_i = -\frac{m_B M}{(m_A + m_B)(m_B + m_C)} q_i + \frac{m_A}{m_A + m_B} Q_i \qquad (i = 1, 2, 3) \quad (81b)$$

for the products AB + C, and by

$$q'_i = \frac{m_B}{m_B + m_C} q_i - Q_i \qquad (i = 1, 2, 3) \quad (82a)$$

$$Q'_i = \frac{m_C M}{(m_A + m_C)(m_B + m_C)} q_i + \frac{m_A}{m_A + m_C} Q_i \qquad (i = 1, 2, 3) \quad (82b)$$

for the products AC + B. Dissociative collisions producing A + B + C may (for convenience) be analyzed in the *reactant* generalized coordinate system.

Expressions for various quantities associated with collision products in terms of the generalized coordinates and momenta appropriate for reactants (the final values of which are known from the calculation of the trajectory) are obtained from the relations given below by substitution of the proper transformation [equations (4) for A + BC or A + B + C, equations (81) for AB + C, or equations (82) for AC + B]. The scalar diatomic internuclear separation is denoted by r ($\equiv [\mathbf{q}' \cdot \mathbf{q}']^{1/2}$) and the diatomic potential energy, defined analogous to equation (14), is represented by $V_D(r)$. A tilde is used to denote a quantity which is continuous according to the expression given, but would be discrete according to quantum mechanics. Expressions derived for the AB + C products may be compared to those given in Reference 5:

Total energy (conserved):

$$H' = \frac{1}{2\mu} \sum_{i=1}^{3} P_i'^2 + \frac{1}{2m} \sum_{i=1}^{3} p_i'^2 + V(\mathbf{R}(\mathbf{Q})) \quad (83)$$

Internal energy:

$$\mathscr{E}'_{\text{int}} = \frac{1}{2m} \sum_{i=1}^{3} p_i'^2 + V_D(r) \quad (84)$$

Internal angular momentum:

$$\tilde{\mathbf{J}}'_r = \mathbf{q}' \times \mathbf{p}' \tag{85a}$$

$$= (q'_2 p'_3 - q'_3 p'_2)\hat{\mathbf{e}}_x + (q'_3 p'_1 - q'_1 p'_3)\hat{\mathbf{e}}_y + (q'_1 p'_2 - q'_2 p'_1)\hat{\mathbf{e}}_z \tag{85b}$$

Relative angular momentum:

$$\tilde{\mathbf{J}}'_{\text{rel}} = \mathbf{Q}' \times \mathbf{P}' \tag{86}$$

Total angular momentum (conserved):

$$\tilde{\mathbf{J}}'_{\text{tot}} = \tilde{\mathbf{J}}'_r + \tilde{\mathbf{J}}'_{\text{rel}} \tag{87a}$$

$$= \mathbf{q}' \times \mathbf{p}' + \mathbf{Q}' \times \mathbf{P}' \tag{87b}$$

Relative velocity and relative speed:

$$\mathbf{v}'_{\text{rel}} = \frac{1}{\mu}\mathbf{P}' \tag{88a}$$

$$v'_{\text{rel}} = \frac{1}{\mu}(\mathbf{P}' \cdot \mathbf{P}')^{1/2} = \frac{1}{\mu}\left(\sum_{i=1}^{3} P_i'^2\right)^{1/2} \tag{88b}$$

Scattering angle:

$$\theta = \cos^{-1}\frac{\mathbf{v}_{\text{rel}} \cdot \mathbf{v}'_{\text{rel}}}{v_{\text{rel}} v'_{\text{rel}}} \tag{89a}$$

where

$$\mathbf{v}_{\text{rel}} = \frac{P^0}{\mu_{\text{A,BC}}}\hat{\mathbf{e}}_z \tag{89b}$$

and $\mathbf{v}_{\text{rel}}$ is the initial relative velocity.

Diatomic vibrational and rotational energies:

$$\tilde{\mathscr{E}}'_{\text{rot}} = \min\left\{V_D(r) + \frac{\tilde{\mathbf{J}}'_r \cdot \tilde{\mathbf{J}}'_r}{2mr^2}\right\} - V_D(r_e) \tag{90a}$$

$$\tilde{\mathscr{E}}'_{\text{vib}} = \tilde{\mathscr{E}}'_{\text{int}} - \tilde{\mathscr{E}}'_{\text{rot}} \tag{90b}$$

where r_e is the internuclear separation at the minimum of V_D, $V_D(r_e)$ is the classical zero of internal energy for the product molecule, and the minimum of the effective potential $V_D(r) + \tilde{\mathbf{J}}'_r \cdot \tilde{\mathbf{J}}'_r/2mr^2$ may be determined, for example, using the Newton–Raphson iteration technique.

Diatomic rotational "quantum number" (continuous):

$$\tilde{j}' = -\tfrac{1}{2} + \tfrac{1}{2}[1 + 4\tilde{\mathbf{J}}'_r \cdot \tilde{\mathbf{J}}'_r/\hbar^2]^{1/2} \tag{91}$$

Diatomic vibrational "quantum number" (continuous):

$$\tilde{n}' = -\frac{1}{2} + \frac{1}{\pi\hbar}\int_{r_-}^{r_+}\left\{2m\left[\varepsilon'_{\text{int}} - V_D(r) - \frac{J'_r \cdot J'_r}{2mr^2}\right]\right\}^{1/2} dr \tag{92}$$

where $r_\pm$ are the diatomic turning points of the effective potential at the energy $\varepsilon'_{\text{int}}$ and are determined using the Newton–Raphson iteration technique as in equations (13) and (20). The integral is evaluated using equations (21).

The definitions of the quantities in equations (83)–(88) are unambiguous. We have already mentioned the arbitrary choice of direction for $\mathbf{v}'_{\text{rel}}$ which affects the definition of scattering angle. Diatomic vibrational and rotational energies cannot be unambiguously defined. Various definitions that make them sum to $\tilde{\varepsilon}'_{\text{int}}$ are acceptable in various circumstances; the choice given in equations (90) seems reasonable for many purposes.* The correspondence principle for rotational quantum number is also not unique; the choice given in equation (91) is equivalent to $\tilde{J}_r'^2 = j(j+1)\hbar^2$ as opposed to $\tilde{J}_r'^2 = (j+1/2)^2\hbar^2$. The continuous "quantum numbers" j' and $\tilde{n}'$ are often rounded to the nearest integer value in making the correspondence between the result of a trajectory calculation and the discrete nature of the quantum-mechanical states of the product molecule. While this is an extremely powerful procedure it should be exercised with caution, especially in cases where rounding to a higher value places the product in a "closed" (energetically forbidden) state.

After calculating each trajectory and analyzing the collision products, it is quite useful to store certain information about that trajectory on some mass storage device (tape, disk, etc.). Having the pertinent information about each trajectory in such a form greatly facilitates the calculation of reaction attributes (see the following section, 2.6) through the analysis of a large number of trajectories. In fact, it is sometimes more convenient to create *two data files*: one to be used in most of the data analysis which contains only the more commonly used information (e.g., channel designation, impact parameter, scattering angle, product internal state, final relative velocity, etc.) and another which contains the initial and final coordinates of each trajectory from which any additional information may be computed at a later time if desired.

* Often it is best to define the rotational and vibrational energies in terms of the quantum-mechanical eigenenergies of the states with discrete quantum numbers n and j as assigned by the histogram method. Then $\tilde{\varepsilon}_{\text{int}}$ will be one of the allowed discrete product internal energies. Optionally, v'_{rel} can then be adjusted for energy conservation but it is difficult and almost always unnecessary to readjust all product variables in a consistent fashion this way.

2.6. Calculation of Reaction Attributes

Having completed the calculation of a statistically significant sample of trajectories and calculated a reactive scattering cross section using equations (43) and (44), or a state-selected rate constant using equations (45) and (46), one is faced with the realization of having calculated far more information than is represented by $\sigma_r \pm \Delta\sigma_r$, or $k_r \pm \Delta k_r$. The following discussion focuses on extracting this additional information, calculated for individual trajectories using the above formulas, from the trajectory data in a form amenable to comparison with experiment.

2.6.1. Reactivity Functions

We use the term *reactivity function* to denote the probability of reaction, or any other quantity derived from the reaction probability, as a function of some *reactant* collision parameter. The *opacity function*, or probability of reaction as a function of impact parameter, defined by equation (50), is one example of a reactivity function. Another (closely related) example is $d\sigma_r/db$, defined by equation (53). Still another example is the *excitation function* $\sigma_r(E_{rel})$. We will limit our discussion to a few of the more commonly used reactivity functions involving the impact parameter and collision energy. Other reactivity functions (e.g., the dependence of reactivity upon reactant orientation or vibrational phase) may be obtained in an analogous manner.

In constructing *histogrammic* representations of reactivity functions, the proper specification of the "bins" into which the trajectory results are to be sorted is intimately coupled to the manner in which initial values of the collision parameters were selected. For reactivity as a function of initial collision parameters, there is an *a priori* "best method" for specifying the bins assuming the initial conditions were selected optimally. We note that the choice of sampling method implies some knowledge about the reaction probability function $\mathscr{P}_r$. This same knowledge should be brought to bear on the presentation of results; reactivity function histograms should be constructed in a manner consistent with the sampling method.[52]

Consider the histogrammic representation of the opacity function $\mathscr{P}_r(b)$. The initial selection of $b(\beta_1)$ according to equation (42a) or (63) implies some *expected* opacity function $\mathscr{P}_r^0$, either equation (58) or (59), respectively. To the extent that the choice of $\mathscr{P}_r^0(b)$ was optimal, the bins of the histogrammic representation of $\mathscr{P}_r(b)$ should correspond to an equipartition of the variable $\beta_1(b)$.[52] The following considerations will make the meaning of this statement more clear. If $\mathscr{P}_r^0(b)$ is defined by equation (58), the values of the impact parameter corresponding to the boundaries of the k_{max} bins are given by

$$b_k = (k/k_{max})^{1/2}\, b_{max} \qquad (k = 0, \ldots, k_{max}) \tag{93a}$$

In general, they are given by

$$b_k = b(\beta_1 = \beta_{1,k}) = b\left(\beta_1 = \frac{k}{k_{\text{max}}}\right) \qquad (k = 0, \ldots, k_{\text{max}}) \qquad \text{(93b)}$$

The value of the opacity function in each bin may be calculated more than one way. First one may use

$$\mathscr{P}_{rk} \cong \sum_{i=1}^{N_{rk}} w_i \Big/ \sum_{i=1}^{N_k} w_i \qquad \text{(94a)}$$

where $\mathscr{P}_{rk}$ is the average value of $\mathscr{P}_{rk}(b)$ for b in the kth bin (i.e., $b_{k-1} < b \le b_k$) when impact parameters in this range are properly sampled at random from a distribution uniform in b^2, and where N_{rk} and N_k are the number of reactive and total collisions, respectively, with impact parameter b in the kth bin. The estimated error for each $\mathscr{P}_{rk}$ is

$$\Delta\mathscr{P}_{rk} \approx \left[\frac{1}{N_k}(\mathscr{P}_{rk} - \mathscr{P}_{rk}^2)\right]^{1/2} \qquad [\text{with } \mathscr{P}_{rk} \text{ from (94a)}] \qquad \text{(94b)}$$

Note that if the expected opacity function is within a constant factor of the true opacity function, each of the numerators on the right-hand side of equation (94a) is approximately equal to $N\langle\mathscr{P}_r\rangle/k_{\text{max}}$. In this case each bin accounts for roughly an equal fraction of the reactive cross section. This is the sense in which the equipartition of the variable $\beta_1(b)$ was optimum. Furthermore, we see that the optimum allocation of bins for reactivity functions of collision parameters is closely related to the optimum allocation of strata for stratified sampling of collision parameters.

A second expression for $\mathscr{P}_{rk}$ may be obtained by replacing the denominator of equation (94a) by its converged value (which is *not* generally N_k when the weights are not all unity):

$$\sum_{i=1}^{N_k} w_i \underset{N\,\text{large}}{\sim} N_k \frac{\beta_{1,k}^{\text{original}} - \beta_{1,k-1}^{\text{original}}}{\beta_{1,k}^{\text{new}} - \beta_{1,k-1}^{\text{new}}} \qquad \text{(95a)}$$

$$\underset{N\,\text{large}}{\sim} N(\beta_{1,k}^{\text{original}} - \beta_{1,k-1}^{\text{original}}) \qquad \text{(95b)}$$

where $\beta_{1,k}^{\text{original}}$ and $\beta_{1,k}^{\text{new}}$ are defined by equations (38a) and (57), respectively. Thus

$$\mathscr{P}_{rk} \cong \frac{b_{\text{max}}^2 \sum_{i=1}^{N_{rk}} w_i}{N(b_k^2 - b_{k-1}^2)} \qquad \text{(96a)}$$

with error estimate

$$\Delta\mathscr{P}_{rk} \approx \frac{b_{\text{max}}^2}{N(b_k^2 - b_{k-1}^2)}\left[\sum_{i=1}^{N_{rk}} w_i^2 - \frac{1}{N}\left(\sum_{i=1}^{N_{rk}} w_i\right)^2\right]^{1/2} \qquad \text{(96b)}$$

Although equations (94a) and (96a) give the same results for infinite N, one or the other may be more useful or more accurate in different calculations with finite N. Notice that if we take $k_{\max} = 1$, then $\mathscr{P}_{rk}$ becomes $\langle \mathscr{P}_r \rangle$. In this case equations (96a) and (96b) reduce to equations (67) and (68), but equations (94a) and (94b) provide an alternative method of calculation. In general it requires some judgment or a comparison of error estimates to decide which formula to use in a specific case.

Consider now the histogrammic representation of the related function $d\sigma_r/db$, as defined by equation (54). We may define the expected function [cf. equation (53)]

$$\left(\frac{d\sigma_r}{db}\right)^0 = 2\pi b \mathscr{P}_r^0(b) \tag{97}$$

which, for $\mathscr{P}_r^0$ defined by equation (58), is

$$\left(\frac{d\sigma_r}{db}\right)^0 = \begin{cases} 2\pi b, & b \le b_{\max} \\ 0, & b > b_{\max} \end{cases} \tag{98}$$

Equation (54) may always be rewritten as

$$\sigma_r = \int_0^{b_{\max}} \frac{(d\sigma_r/db)}{(d\sigma_r/db)^0} \left(\frac{d\sigma_r}{db}\right)^0 db \tag{99}$$

which upon substitution of equations (57) and (97) becomes

$$\sigma_r = \pi b_{\max}^2 \int_0^1 \frac{d\sigma_r/db}{(d\sigma_r/db)^0} d\beta_1 \tag{100}$$

Again it is clear that, if the expected function $(d\sigma_r/db)^0$ is an accurate one, the "best" impact parameter bins, in the sense that each bin contains an equal fraction of the cross section, correspond to an equipartition of the variable β_1. The histogrammic representation of $d\sigma_r/db$ is derived from equation (54) as follows. We write

$$\sigma_r = \sum_{k=1}^{k_{\max}} \int_{b_{k-1}}^{b_k} \frac{d\sigma_r}{db} db \tag{101a}$$

$$= \sum_{k=1}^{k_{\max}} (b_k - b_{k-1}) \left(\frac{d\sigma_r}{db}\right)_k \tag{101b}$$

$$\cong \pi b_{\max}^2 \frac{1}{N} \sum_{k=1}^{k_{\max}} \sum_{i=1}^{N_{rk}} w_i \tag{102}$$

where $(d\sigma_r/db)_k$ is the average value of $d\sigma_r/db$ in the kth interval of b (i.e., in $[b_{k-1}, b_k]$). Equating equations (101) and (102) term by term gives[52,54]

$$\left(\frac{d\sigma_r}{db}\right)_k \cong \frac{\pi b_{\max}^2}{b_k - b_{k-1}} \frac{1}{N} \sum_{i=1}^{N_{rk}} w_i \tag{103}$$

Clearly, if one substitutes each $(d\sigma_r/db)_k$ into equation (101a) for

$b_{k-1} < b \leq b_k$ and integrates, one obtains the correct value of the integral as given by equation (67). The estimated error in each bin is given by

$$\Delta\left(\frac{d\sigma_r}{db}\right) \approx \frac{\pi b_{\max}^2}{(b_k - b_{k-1})} \frac{1}{N}\left[\sum_{i=1}^{N_{rk}} w_i^2 - \frac{1}{N}\left(\sum_{i=1}^{N_{rk}} w_i\right)^2\right]^{1/2} \tag{104}$$

Other histogrammic distributions as functions of initial collision parameters can be constructed in a similar fashion. Of course in some cases one does not necessarily desire an equal contribution to the cross section from each bin of a reactivity function, and then one should adjust the boundaries according to some alternative criterion.[52]

Another reactivity parameter of interest when calculating $k_r^{n,j}$ is the *activation energy*

$$E_a^{n,j}(T) = \frac{d \log k_r^{n,j}}{d(-1/RT)} \tag{105}$$

This is the activation energy for a state-selected rate constant. By analogy to the derivations for completely thermally averaged rate constants,[55-57] it can be shown to be[54,58]

$$E_a^{n,j}(T) = \langle E_{\text{rel}} \rangle_{\text{rc};T} - \langle E_{\text{rel}} \rangle_{\text{ap};T} \tag{106a}$$

$$= \langle E_{\text{rel}} \rangle_{\text{rc};T} - \langle E_{\text{rel}} \rangle_{\text{ac};T} + \tfrac{1}{2}kT \tag{106b}$$

where the subscripts rc, ap, and ac mean reactive collisions, all pairs, and all collisions, respectively. The use of this definition of the activation energy will be called the Tolman method. It yields a far more precise estimate of that quantity than any algorithm involving differences in $k_a^{n,j}$ at different temperatures.[56-58] To apply equation (106) to a set of trajectory data at some fixed temperature, the average values are evaluated by the Monte Carlo method,

$$E_a^{n,j} \cong \frac{\sum_{i=1}^{N_r} w_i E_{\text{rel},i}}{\sum_{i=1}^{N_r} w_i} - \frac{\sum_{i=1}^{N} w_i E_{\text{rel},i}}{\sum_{i=1}^{N} w_i} + \tfrac{1}{2}kT \tag{107a}$$

$$\cong \frac{\sum_{i=1}^{N_r} w_i E_{\text{rel},i}}{\sum_{i=1}^{N_r} w_i} - \frac{\sum_{i=1}^{N} w_i E_{\text{rel},i}}{N} + \tfrac{1}{2}kT \tag{107b}$$

For a sufficiently large number N of trajectories, the second term in both equations (107a) and (107b) becomes $2kT$. For the activation energy of completely thermally averaged rate constants one must also compute the average value of the internal energy for all reactions minus that for all collisions. For N sufficiently large one will obtain the same results for the averages over all collisions using $\sum_{i=1}^{N} w_i$ as using N. But if importance sampling is being used there may be a significant difference. This illustrates the general principle that if importance sampling is used to accelerate the convergence of one reaction attribute, e.g., $\langle \mathscr{P}_r \rangle$, the convergence of others, in these examples $\langle E_{rel}\rangle_{ac:T}$ or $\langle \varepsilon_{int}\rangle_{ac:T}$, may be decelerated. The user of these algorithms must therefore use discretion in deciding which results are reliable and which are not.

2.6.2. *Final-State Distributions*

We subdivide the discussion of the analysis of final-state distributions according to the number and type of variables involved, i.e., according to whether the distribution is a function of one or two variables, and whether the variables take on discrete values quantum mechanically or are really continuous. More than two variables may be considered but such distributions are difficult to plot and usually suffer from poor statistics.

2.6.2.1. One Variable—Discrete. The discrete variables of interest in product distributions are those corresponding to the quantized internal states of the product molecule. It is of interest to calculate a reactive scattering cross section for collisions of some fixed vibrational–rotational state (n, j) of the BC molecule producing some specific product state (n', j') or group of product states (e.g., all states with a specific n' irrespective of j'). For fixed E_{rel}, this is called a *state-to-state cross section* and for a thermal distribution of E_{rel} it is called a *state-to-state rate constant.* We will explicitly discuss state-to-state cross sections but the methods are also applicable to state-to-state rate constants and to final-state distributions when the reactant states are completely thermal. The total reaction cross section for producing arrangement α' may be decomposed as

$$\sigma_{r\alpha'}(E_{rel}, n, j) = \sum_{n'} \sum_{j'} \sigma_r(n, j \rightarrow \alpha', n', j'; E_{rel}) \tag{108a}$$

$$= \sum_{n'} \sigma_r(n, j \rightarrow \alpha', n'; E_{rel}) \tag{108b}$$

$$= \sum_{j'} \sigma_r(n, j \rightarrow \alpha', j'; E_{rel}) \tag{108c}$$

where the index α' refers to the chemically distinct product molecules (AB or AC). To calculate a histogrammic representation of $\sigma_r(n, j \rightarrow n', j')$, $\sigma_r(n, j \rightarrow n')$, or $\sigma_r(n, j \rightarrow j')$ for a given product, the continuous values of $\tilde{n}'$ and $\tilde{j}'$ from equations (91) and (92) for each trajectory are first rounded to the nearest integer value (as discussed above). The Boolean reaction probability function $\mathscr{P}_r$ is then redefined (for each trajectory) such that

$$\mathscr{P}_r = \begin{cases} 1, & \text{for } (\alpha', n', j'),\ (\alpha', n'), \text{ or } (\alpha', j') \text{ equal to some} \\ & \text{preselected values} \\ 0, & \text{otherwise} \end{cases} \tag{109}$$

The state-to-state cross section and estimated error are then computed using the same equations as for the total reaction cross section [e.g., using equations (43) and (44) for crude Monte Carlo sampling] except that N_r now represents the number of trajectories with $\mathscr{P}_r = 1$ as defined by equation (109). N is still the total number of trajectories calculated. Estimated relative errors, $\Delta\sigma_r(n, j \to \alpha', n', j')/\sigma_r(n, j \to \alpha', n', j')$, for state-to-state cross sections are generally much larger than those for the total reactive scattering cross section because N_r is generally much smaller.

As an example, if we wish to compute the way in which the products are distributed in their various vibrational states, we would "bin" the product states generated from a set of N trajectories according to the identity of the products (α'), and within each of these arrangement bins we would further "bin" the trajectories according to the value of $\tilde{n}'$ rounded to the nearest integer. The result would be a one-variable (n') histogram for each distinct product molecule.

Another method for calculating final-state distributions of variables which quantum mechanically take on discrete values is classical P matrix theory.[59-64] Because quantal rotational energies are more closely spaced than vibrational energies, the histogram method is more justified for assigning j' than for assigning n'. It may be useful to use the histogram method for assigning j' and classical P matrix theory to assign n'. To do this we write the state-to-state cross section as (suppressing the index α')

$$\sigma_r(n, j \to n', j'; E_{\text{rel}}) = \sigma_r(n, j \to j'; E_{\text{rel}})\, P_{n'}(n, j \to j'; E_{\text{rel}}) \tag{110}$$

where $\sigma_r(n, j \to j'; E_{\text{rel}})$ is computed by the histogram method as described above and $P_{n'}(n, j \to j'; E_{\text{rel}})$ is the probability that collisions with initial conditions n, j, and E_{rel} and final rotational quantum number j' should be assigned final vibrational quantum number n'. Define moments of the final vibrational quantum number n' for such collisions by

$$\langle (n')^m \rangle (n, j \to j'; E_{\text{rel}}) = \frac{\sum_{n'=0}^{n'_{\max}} (n')^m\, \sigma_r(n, j \to n', j'; E_{\text{rel}})}{\sum_{n'=0}^{n'_{\max}} \sigma_r(n, j \to n', j'; E_{\text{rel}})} \tag{111}$$

$$= \frac{\sum_{n'=0}^{n'_{\max}} (n')^m\, P_{n'}(n, j \to j'; E_{\text{rel}})}{\sum_{n'=0}^{n'_{\max}} P_{n'}(n, j \to j'; E_{\text{rel}})} \tag{112}$$

where $n'_{\max}$ is the maximum value allowed by conservation of total energy. These moments may be approximated from the trajectories as an average over the $N_{rj'}$ trajectories which are assigned final rotational quantum number j' as follows:

$$\langle (\tilde{n}')^m \rangle (n, j \to j'; E_{\text{rel}}) \cong \frac{\sum_{i=0}^{N_{rj'}} (\tilde{n}'_i)^m\, w_i}{\sum_{i=0}^{N_{rj'}} w_i} \tag{113}$$

Then the classical P matrix theory prescribes that the quantities $P_{n'}(n, j \to j'; E_{\text{rel}})$ be given by the least-biased estimate in the information-theory sense consistent with the constraints that the probabilities sum to one and that the first N_{mom} moments ($m = 1, 2, \ldots, N_{\text{mom}}$) of equation (112) are equal to the first N_{mom} moments of equation (113). A least-biased estimate in the information-theory sense is one that maximizes the information-theory entropy of the distribution. For a probability distribution $P_{n'}$ the information-theory entropy is[65]

$$I = -K \sum_{n'=0}^{n'_{\max}} P_{n'} \ln\left(\frac{P_{n'}}{P_{n'}^0}\right) \tag{114}$$

where $P_{n'}^0$ is an *a priori* probability and where K is a constant. For the present problem, maximization of the information-theory entropy subject to the constraints yields

$$P_{n'} = \frac{P_{n'}^0 \exp\left[-\sum_{m=1}^{N_{\text{mom}}} \lambda_m (n')^m\right]}{\sum_{n'=0}^{n'_{\max}} P_{n'}^0 \exp\left[-\sum_{m=1}^{N_{\text{mom}}} \lambda_m (n')^m\right]} \tag{115}$$

where the arguments $(n, j \rightarrow j'; E_{\text{rel}})$ have been suppressed on $P_{n'}$, $P_{n'}^0$, and λ_m. The quantities λ_m are Lagrange multipliers whose values are found by using the trajectory results [equation (113)] on the left side of equation (112) and substituting equation (115) on the right side of equation (112). This gives N_{mom} simultaneous nonlinear equations for N_{mom} Lagrange multipliers.

The above description shows that a complete specification of the algorithm requires specifying N_{mom} and the prior probabilities $P_{n'}^0$. There are three approaches to each of these choices. For N_{mom} one might try to use (i) the value that would make the classical P matrix predictions be most accurate as compared to a quantum-mechanical calculation,[59,60] (ii) the minimum value required to reach a region of stability of the information-theory entropy with respect to increasing N_{mom},[62,63] or (iii) some fixed small value such as 1, 2, or 3.[61,64] Choice (i) requires theoretical considerations of how many classical moments might be accurate as compared to quantum-mechanical ones.[60] The other choices have the disadvantage that inaccurate classical moments might be used or accurate classical moments might be unused. The choice of N_{mom} is not independent of the choice of *a priori* probabilities. There are four approaches to this choice. The original one was to use the statistical phase-space theory[66,67] predictions as the *a priori* probability.[59,60] This has the advantage that every individual quantum state allowed by energy and angular momentum conservation has the same weight. Including angular momentum conservation requires a model so it might be neglected. Then for the present problem $P_{n'}^0$ is proportional to the sum of the degeneracies of all energetically allowed rotational states of vibrational level n'. The second choice is $P_{n'}^0 = 1$,[61,63] which coincides with the first choice for collinear collisions. The third choice of *a priori* probabilities is the microcanonical statistical thermodynamic one[68,69] which allows information-theory entropy to be treated like statistical mechanical entropy.[62] Fourthly, we might use a model for $P_{n'}^0$.[62,63] Atabek and Lefebvre used N_{mom} in the range 1–3 and two sets of $P_{n'}^0$ to predict the final vibrational distribution of CN in the photodissociation of

ICN. They found for that problem that for $N_{mom} \geq 2$, the results are fairly insensitive to either the *a priori* probabilities or N_{mom}; the information-theory entropy and the larger probabilities become invariant to varying either. The classical P matrix approach is recommended for consideration because it is capable of much better accuracy than the histogram method for final vibrational state distributions. But the optimal choices for N_{mom} and the *a priori* probabilities are not clear at this time and are subjects for further research.

2.6.2.2. One Variable—Continuous. As the choice of sampling method for initial conditions does not necessarily imply any knowledge of distributions of product state variables (e.g., scattering angle, final relative velocity, etc.), there is no *a priori* method for binning the final values of these variables to achieve the maximum resolution of their distributions. However, one extremely useful technique for obtaining histogrammic representations of product state distributions is to construct "bins" such that each bin contains an equal volume of phase space.[70] This uniform phase space (UPS) method for constructing histograms is applicable to any number of discrete and/or continuous variables. The basic premise of the UPS method is that having no *a priori* knowledge of a distribution is equivalent to "expecting" the distribution to be uniformly spread over all the available phase space, and that in the absence of such knowledge the histogrammic representation should be constructed according to this expectation.[70] Of course as soon as some trajectories have been computed one can make a more knowledgeable estimate of the shape of the distribution, and one may find that it differs significantly from the uniform distribution. If so, then the UPS prescription for binning may be inappropriate. For example, because of vibrational population inversion in the products, the distribution of final relative velocities for an exoergic reaction is often peaked at a value corresponding to a low fraction of the total available energy being in translation; but a uniform phase-space distribution is peaked at high final relative velocity. Thus the UPS prescription corresponds to large bins at low v'_{rel} and small bins at high v'_{rel}, but a more optimum presentation of the results would involve small bins at low v'_{rel} and large bins at high v'_{rel}. The systematics of choosing bin sizes so that all bins contain the same number of reactive trajectories or so that all bins contain the same fraction of the total cross section are presented elsewhere.[52] In this chapter we present a detailed treatment only for the UPS prescription for binning trajectories. This illustrates the general techniques, but as just discussed it is not always the preferred method for binning.

Consider the distribution $g(\gamma)$, where γ is some continuous product variable, which is related to the total reaction cross section by

$$g(\gamma) = \frac{d\sigma_r}{d\gamma} \tag{116a}$$

i.e.,

$$\sigma_r = \int_{\gamma_{\min}}^{\gamma_{\max}} g(\gamma)\, d\gamma \tag{116b}$$

$$= \int_{\gamma_{\min}}^{\gamma_{\max}} \frac{g(\gamma)}{\rho(\gamma)} \rho(\gamma)\, d\gamma \tag{116c}$$

Here $\rho(\gamma)$ is the density of phase space along the γ coordinate[70] and is assumed to have the normalization

$$\int_{\gamma_{\min}}^{\gamma_{\max}} \rho(\gamma)\, d\gamma = 1 \tag{117}$$

Equation (116c) may be rewritten as

$$\sigma_r = \int_0^1 g(u)\, du \tag{118}$$

where

$$g(u) \equiv \frac{g(\gamma)}{\rho(\gamma)} \tag{119}$$

and

$$du = \rho(\gamma)\, d\gamma \tag{120}$$

Equal increments of the transformed variable

$$u = \int_0^u du = \int_{\gamma_{\min}}^{\gamma} \rho(\gamma)\, d\gamma \tag{121}$$

contain equal volumes of phase space. If the density of products were uniformly distributed over all allowed phase space, the distribution $g(u)$ would be a constant. In such a case, if the variable $\gamma(u)$ were binned in equal increments of u, then $g_k \Delta\gamma_k$, where $g_k \equiv \langle g(\gamma) \rangle_k$, would be identical for each (in general, nonuniform) interval $\Delta\gamma_k$. In fact, any deviation from a uniform (flat) distribution $g(u)$, or equal volumes (areas) in the bars of the histogram g_k vs. γ, reflects a dynamical preference for certain regions of phase space and is directly related to the dynamical information contained in the distribution.[70,71] The UPS method will now be used to derive expressions for histogrammic representations of several common product state distributions.

The solid-angle differential reaction cross section $d^2\sigma_r/d\Omega$ is related to the integral reaction cross section by

$$\sigma_r = \int_0^{4\pi} \frac{d^2\sigma_r}{d\Omega}\, d\Omega \tag{122a}$$

$$= \int_0^{4\pi} \frac{(d^2\sigma_r/d\Omega)}{\rho(\Omega)}\, \rho(\Omega)\, d\Omega \tag{122b}$$

Taking $\rho(\Omega) = 1/4\pi$ yields

$$\sigma_r = 4\pi \int_0^1 \frac{d^2\sigma_r}{d\Omega}\, du \tag{123}$$

where (using the cylindrically symmetric $d\Omega = 2\pi \sin\theta\, d\theta$)

$$du = \rho(\Omega)\, d\Omega = \rho(\theta)\, d\theta \tag{124}$$

$$= \tfrac{1}{2} \sin\theta\, d\theta \tag{125}$$

Therefore,

$$u = \tfrac{1}{2}(1 - \cos\theta) \tag{126}$$

The UPS partition of θ is then given by

$$\theta_k = \cos^{-1} \frac{k_{\max} - 2k}{k_{\max}} \qquad (k = 0, \ldots, k_{\max}) \tag{127}$$

An expression for $d^2\sigma_r/d\Omega_k$ is derived as follows. Equation (123) is rewritten over the partition of θ given by equation (127) and equated to the Monte Carlo expression for the integral reaction cross section over the same partition,

$$\sigma_r = 4\pi \sum_{k=1}^{k_{\max}} \int_{(k-1)/k_{\max}}^{k/k_{\max}} \frac{d^2\sigma_r}{d\Omega}\, du \tag{128a}$$

$$= \frac{4\pi}{k_{\max}} \sum_{k=1}^{k_{\max}} \left\langle \frac{d^2\sigma_r}{d\Omega} \right\rangle_k \tag{128b}$$

$$\cong \pi b_{\max}^2 \frac{1}{N} \sum_{k=1}^{k_{\max}} \sum_{i=1}^{N_{rk}} w_i \tag{128c}$$

where $\langle d^2\sigma_r/d\Omega \rangle_k$ is the average value of $(d^2\sigma_r/d\Omega)$ in the kth interval of u. Equating equations (128b) and (128c) term by term gives the desired result[52,72]

$$\left\langle \frac{d^2\sigma_r}{d\Omega} \right\rangle_k \cong \frac{k_{\max} b_{\max}^2}{4} \frac{1}{N} \sum_{i=1}^{N_{rk}} w_i \tag{129}$$

The estimated error is given by

$$\Delta\left\langle \frac{d^2\sigma_r}{d\Omega} \right\rangle_k \approx \frac{k_{\max} b_{\max}^2}{4} \frac{1}{N} \left[\sum_{i=1}^{N_{rk}} w_i^2 - \frac{1}{N} \left(\sum_{i=1}^{N_{rk}} w_i \right)^2 \right]^{1/2} \tag{130}$$

The polar differential reactive scattering cross section, $(d\sigma_r/d\theta)$, is related to σ_r and $d^2\sigma_r/d\Omega$ by

$$\sigma_r = \int_0^{\pi} \frac{d\sigma_r}{d\theta} \, d\theta \tag{131}$$

and

$$\frac{d\sigma_r}{d\theta} = 2\pi \sin\theta \frac{d^2\sigma_r}{d\Omega} \tag{132}$$

As indicated by equation (122b), the density of phase space is not uniform along the θ coordinate. If the integral in equation (131) is partitioned according to equation (127), then each interval of θ will contain the same volume of phase space. Hence

$$\sigma_r = \sum_{k=1}^{k_{\max}} \int_{\theta_{k-1}}^{\theta_k} \frac{d\sigma_r}{d\theta} \, d\theta \tag{133}$$

$$= \sum_{k=1}^{k_{\max}} (\theta_k - \theta_{k-1}) \left\langle \frac{d\sigma_r}{d\theta} \right\rangle_k \tag{134}$$

$$\cong \pi b_{\max}^2 \sum_{k=1}^{k_{\max}} \sum_{i=1}^{N_{rk}} w_i \tag{135}$$

where $\langle d\sigma_r/d\theta \rangle_k$ is the average value of $d\sigma_r/d\theta$ in the interval $[\theta_{k-1}, \theta_k]$. Equating equations (134) and (135) term by term yields[52]

$$\left\langle \frac{d\sigma_r}{d\theta} \right\rangle_k \cong \frac{\pi b_{\max}^2}{\theta_k - \theta_{k-1}} \frac{1}{N} \sum_{i=1}^{N_{rk}} w_i \tag{136}$$

The estimated error is given by

$$\Delta\left\langle \frac{d\sigma_r}{d\theta} \right\rangle_k \approx \frac{\pi b_{\max}^2}{\theta_k - \theta_{k-1}} \frac{1}{N} \left[\sum_{i=1}^{N_{rk}} w_i^2 - \frac{1}{N} \left(\sum_{i=1}^{N_{rk}} w_i \right)^2 \right]^{1/2} \tag{137}$$

For differential scattering cross sections we have just seen that the UPS criterion suggests that histogrammic boundaries should correspond to equal intervals of $\cos\theta$ rather than equal intervals of θ. The use of equal intervals of $\cos\theta$ rather than θ has also been suggested[72] on the grounds that such bins contain equal fractions of the available 4π sr of solid angle. But the argument involving equal fractions of the available phase space is more general and can also be applied to other kinds of product distributions. For example, consider the relative translational degree of free-

dom of the products. The continuous variable under consideration here could be the final relative translational energy E'_{rel} or final relative speed v'_{rel}. We may define $f_{E'} \equiv E'_{\text{rel}}/E'_{\max}$ and $f_{v'} \equiv v'_{\text{rel}}/v'_{\max}$ as the fractions of
could be the final relative translational energy E'_{rel} or final relative speed
respectively, which are related by $f_{E'} = f^2_{v'}$. The densities of phase space along the coordinates $f_{E'}$ and $f_{v'}$ are rather complicated to specify exactly, but in the limit of classical (continuous) vibrational and rotational energies for a rigid-rotor harmonic oscillator (RRHO), which should resemble a real system well enough for the purpose of constructing bins of a histogram, the phase-space densities are[65,70]

$$\rho(f_{E'}) = \tfrac{15}{4} f_{E'}^{1/2}(1 - f_{E'}) \tag{138}$$

$$\rho(f_{v'}) = \tfrac{15}{2} f_{v'}^{2}(1 - f_{v'}^{2}) \tag{139}$$

It is easily shown that[70]

$$z_{E'} = z_{v'} \equiv z \tag{140}$$

where

$$z_{E'} = \int_0^{f_{E'}} \rho(f_{E'})\, df_{E'} \tag{141}$$

$$z_{v'} = \int_0^{f_{v'}} \rho(f_{v'})\, df_{v'} \tag{142}$$

This unique translational variable, along which equal increments represent equal volumes of phase space, may be written in terms of either $f_{E'}$ or $f_{v'}$ as

$$z = \tfrac{5}{2} f_{E'}^{3/2}(1 - \tfrac{3}{5} f_{E'}) \tag{143}$$

or

$$z = \tfrac{5}{2} f_{v'}^{3}(1 - \tfrac{3}{5} f_{v'}^{2}) \tag{144}$$

To calculate a UPS histogrammic representation of, say, $d\sigma_r/dv'_{\text{rel}}$ one constructs an equipartition of the uniform variable z and finds by interpolation of equation (144) the value of $f_{v',k}$ corresponding to each z_k. As the function $z(f_{v'})$ has an inflection point at $f_{v'} = 1/2^{1/2}$, this point is a good initial guess when using the Newton–Raphson method for the interpolation. The final-relative-speed bins are then defined by $v'_{\text{rel},k} = v'_{\max} f_{v',k}$. The desired result is then obtained by analyzing

$$\sigma_r = \int_0^{v'_{\max}} \frac{d\sigma_r}{dv'_{\text{rel}}}\, dv'_{\text{rel}} \tag{145}$$

analogous to the way equations (134) and (135) were analyzed. The result is[52]

$$\left(\frac{d\sigma_r}{dv'_{\text{rel}}}\right)_k \cong \frac{\pi b^2_{\max}}{v'_{\text{rel},k} - v'_{\text{rel},k-1}} \frac{1}{N} \sum_{i=1}^{N_{rk}} w_i \tag{146}$$

with estimated error

$$\Delta\left(\frac{d\sigma_r}{dv'_{\text{rel}}}\right)_k \approx \frac{\pi b_{\max}^2}{v'_{\text{rel},k} - v'_{\text{rel},k-1}} \frac{1}{N}\left[\sum_{i=1}^{N_{rk}} w_i^2 - \frac{1}{N}\left(\sum_{i=1}^{N_{rk}} w_i\right)^2\right]^{1/2} \tag{147}$$

It is also possible (and, in some cases, more appropriate) to treat v'_{rel} as a discrete variable resulting from the quantized (n', j') internal states of the product diatomic molecule.(70,71)

Other histogrammic distributions as functions of a single continuous variable may be derived in a similar fashion. In addition, if one has some knowledge of the product distribution (on general grounds or by having already calculated some trajectories) one may generalize the treatment given above by replacing $\rho(\gamma)$ by a more appropriate function.

An alternative way to calculate continuous distributions of product variables is to expand them in a set of functions and determine the expansion coefficients by moment methods.(29,72,73) The most convenient functions are the natural orthogonal functions for the interval involved. We present here the details only for final angular distributions; extension of the method to other continuous distributions proceeds analogously. In addition the extension to recoil velocity distributions will be clear from the treatment of joint distributions.

Just as for histogram methods, it is recommended first to transform to an appropriate variable u such that, e.g., equal increments of u correspond to equal volumes of phase space. For angular distributions the variable u is given by equation (126). Angular distributions are defined on a finite interval so the convenient orthogonal polynomials are the Legendre polynomials. They are usually defined on the interval $-1 < \tilde{u} < 1$ so we make a linear transformation

$$\tilde{u} = Au + B \tag{148}$$

to a variable $\tilde{u}$ which has this range and is therefore a more convenient argument for Legendre polynomials but, since the transformation is linear, still provides a UPS representation. This yields

$$\tilde{u} = \cos\theta \tag{149}$$

We recall the following important properties of the Legendre polynomials:

$$P_0(\tilde{u}) = 1 \tag{150}$$

$$P_1(\tilde{u}) = \tilde{u} \tag{151}$$

$$P_{l+1}(\tilde{u}) = (2l + 1)\,\tilde{u}P_l(\tilde{u}) - lP_{l-1}(\tilde{u}) \tag{152}$$

$$\int_{-1}^{1} P_l(\tilde{u})\,P_{l'}(\tilde{u}) = \frac{2\delta_{ll'}}{2l + 1} \tag{153}$$

So it is convenient to define normalized Legendre polynomials by

$$\tilde{P}_l(\tilde{u}) = [(2l + 1)/2]^{1/2} P_l(\tilde{u}) \tag{154}$$

Then the differential reaction cross section per unit solid angle is expanded as

$$\frac{d^2\sigma_r}{d\Omega} = \frac{\sigma_r}{2\pi}\sum_l d_{rl}\tilde{P}_l(\tilde{u}) \tag{155}$$

and the expansion coefficients are determined by the usual method of multiplying both sides by one normalized Legendre polynomial and integrating. This gives

$$d_{rl} = \frac{2\pi}{\sigma_r}\int_{-1}^{1} \frac{d^2\sigma_r}{d\Omega}\tilde{P}_l(\tilde{u})\,d\tilde{u} \tag{156}$$

$$= \frac{1}{\sigma_r}\int_{-1}^{1} \frac{d\sigma_r}{d\tilde{u}}\tilde{P}_l(\tilde{u})\,d\tilde{u} \tag{157}$$

$$= \frac{1}{\sigma_r}\int \tilde{P}_l(\tilde{u})\,d\sigma_r \tag{158}$$

If σ_r and the last integral are both evaluated from the same sample of trajectories, this becomes

$$d_{rl} \cong \sum_{i=1}^{N_r} P_l(\tilde{u}_i)\,w_i \bigg/ \sum_{i=1}^{N_r} w_i \tag{159}$$

where the sums are over the N_r trajectories which yield the product of interest. Substitution of equation (159) into equation (155) yields the desired result except that one must decide what upper limit to use in equation (155). For exact values of the d_{rl}, the limits should be 0 to ∞. But for the Monte Carlo values of d_{rl} given by equation (159) it is best not to include the highest l values in the sum. If the values of l used are too high, the computed differential cross section will exhibit meaningless statistical noise. It is recommended for working purposes to use only those terms for which d_{rl} is reasonably well converged with respect to increasing the number N of trajectories run. This provides an appropriately smoothed analytic representation of $d^2\sigma_r/d\Omega$. If desired, the differential reaction cross section per unit polar angle may then be obtained from it by equation (132) as before.

When using the moment method one does not get as convenient an error estimate for the differential cross section as one gets for the differential cross section averaged over a bin width in the histogram method. But one does get an analytic smoothed result as a compensating advantage. The differential cross section computed by the moment method must be checked both with respect to increasing $l_{\max}$ and increasing N. If $l_{\max}$ is increased indefinitely for fixed N, the result will tend to a sequence of delta functions at the set of N_r scattering angles. But for large enough N, there should be a region of stability where the result is insensitive to increasing $l_{\max}$. This result is the actual differential cross section. Distributions computed with smaller l_{max} and smaller N are approximate smoothed versions of this converged result. It is not possible to give a universal recommendation for how far the result should be converged with respect to increasing l_{max} (always including only terms reasonably well converged with respect to increasing N) and increasing N (to converge better the individual terms). These decisions depend on the use to which the calculation will be put.

2.6.2.3. Two Variables—Discrete, Continuous, and Mixed. Provided the two variables are separable, the construction of histogrammic representations of distributions of two variables is an obvious generalization of the one-variable cases discussed above.* Only one example need be given here. Consider the distribution $d^3\sigma_r/d\Omega\, dv'_{rel}$ in velocity–angle space. The defining relation is

$$\sigma_r = 2\pi \int_0^{\pi} \int_0^{v'_{max}} \frac{d^3\sigma_r}{d\Omega\, dv'_{rel}} \sin\theta\, d\theta\, dv'_{rel} \tag{160}$$

which may be rewritten in terms of the UPS angle variable u as

$$\sigma_r = 4\pi \int_0^{1} \int_0^{v'_{max}} \frac{d^3\sigma_r}{d\Omega\, dv'_{rel}} du\, dv'_{rel} \tag{161}$$

If boundaries $\{u_j\}$ are defined by an equipartition of u, and boundaries $\{v'_{rel,k}\}$ are defined by an equipartition of the UPS translational variable z, then

$$\sigma_r = 4\pi \sum_{j=1}^{j_{max}} \sum_{k=1}^{k_{max}} \int_{u_{j-1}}^{u_j} \int_{v'_{rel,k-1}}^{v'_{rel,k}} \frac{d^3\sigma_r}{d\Omega\, dv'_{rel}} du\, dv'_{rel} \tag{162}$$

$$= \frac{4\pi}{j_{max}} \sum_{j=1}^{j_{max}} \sum_{k=1}^{k_{max}} \left\langle \frac{d^3\sigma_r}{d\Omega\, dv'_{rel}} \right\rangle_{j,k} (v'_{rel,k} - v'_{rel,k-1}) \tag{163}$$

$$\cong \frac{\pi b_{max}^2}{N} \sum_{j=1}^{j_{max}} \sum_{k=1}^{k_{max}} \sum_{i=1}^{N_{rjk}} w_i \tag{164}$$

where N_{rjk} is the number of reactive trajectories which produced a product in the jth angle bin *and* the kth velocity bin. The final expresion is[52]

$$\left(\frac{d^3\sigma_r}{d\Omega\, dv'_{rel}} \right)_{j,k} \cong \frac{j_{max} b_{max}^2}{4(v'_{rel,k} - v'_{rel,k-1})} \frac{1}{N} \sum_{i=1}^{N_{rjk}} w_i \tag{165}$$

with estimated error

$$\Delta \frac{d^3\sigma_r}{d\Omega\, dv'_{rel}} \approx \frac{j_{max} b_{max}^2}{4(v'_{rel,k} - v'_{rel,k-1})} \frac{1}{N} \left[\sum_{i=1}^{N_{rjk}} w_i^2 - \frac{1}{N} \left(\sum_{i=1}^{N_{rjk}} w_i \right)^2 \right]^{1/2} \tag{166}$$

Unless a very large sample of trajectories has been computed, the N_{rjk} for

* The case of nonseparable variables is treated elsewhere.[70]

most (j, k) bins will generally be quite small, resulting in large relative errors.

Distributions which are functions of two or more variables can also be calculated by moment methods. We will illustrate this only for one important distribution—molecular-beam, velocity–angle maps. The extension to other cases proceeds analogously. The quantity to be calculated is the double differential reaction cross section $d^3\sigma_r / d\Omega\, dv'_{\text{rel}}$, where v'_{rel} is final relative speed. For a fixed initial energy, v'_{rel} is confined to a finite interval from zero to a maximum value determined by conservation of total energy. Just as for histograms of final relative velocity we first transform to the variable which provides a UPS representation. Then, since the conventional finite interval for orthogonal polynomials is $-1 < \tilde{z} < 1$, we transform to a new UPS variable defined on this range. This yields

$$\tilde{z} = 5f_{v'}^3(1 - \tfrac{3}{5}f_{v'}^2) - 1 \tag{167}$$

Then the triple differential reaction cross section is expanded as

$$\frac{d^3\sigma_r}{d\Omega\, dv'_{\text{rel}}} = \frac{\sigma_r}{2\pi} \sum_k \sum_l d_{rkl}\tilde{P}_k(\tilde{z})\, \tilde{P}_l(\tilde{u}) \tag{168}$$

where $\tilde{u}$ is $\cos\theta$ as in Section 2.6.2.2. Multiplying both sides of this equation by a product of normalized Legendre polynomials and integrating yields

$$d_{rkl} = \frac{2\pi}{\sigma_r} \int_{-1}^{1} \int_{-1}^{1} \frac{d^3\sigma_r}{d\Omega\, dv'_{\text{rel}}} \tilde{P}_k(\tilde{z})\, \tilde{P}_l(\tilde{u})\, d\tilde{z}\, d\tilde{u} \tag{169}$$

$$= \frac{1}{\sigma_r} \int_{-1}^{1} \int_{-1}^{1} \frac{d^2\sigma_r}{d\tilde{u}\, d\tilde{z}} \tilde{P}_k(\tilde{z})\, \tilde{P}_l(\tilde{u}) \frac{d\tilde{z}}{dv'_{\text{rel}}}\, d\tilde{z}\, d\tilde{u} \tag{170}$$

$$= \frac{1}{\sigma_r} \int \tilde{P}_k(\tilde{z})\, \tilde{P}_l(\tilde{u}) \frac{d\tilde{z}}{dv'_{\text{rel}}}\, d\sigma \tag{171}$$

If σ_r and the last integral are both evaluated from the same sample of trajectories, this becomes

$$d_{rkl} \cong \frac{\sum_{i=1}^{N_r} \tilde{P}_k(\tilde{z}_i)\, \tilde{P}_l(\tilde{u}_i) \left.\dfrac{d\tilde{z}}{dv'_{\text{rel}}}\right|_{v'_{\text{rel}} = v'_{\text{rel},i}} w_i}{\sum_{i=1}^{N_r} w_i} \tag{172}$$

where the Jacobian will be considered below and the sums are over the N_r trajectories which yield the product of interest. Substitution of equation (172) into equation (168) yields the desired result. As for angular distributions one must decide what upper limits to use in the sums. Again we recommend that only converged coefficients be used in the sums. This automatically smooths out statistical noise.

Equation (172) involves a Jacobian which can be evaluated analytically for each trajectory. It equals

$$\frac{d\tilde{z}}{dv'_{rel}} = \frac{15}{v'_{max}} f_{v'}^2 (1 - f_{v'}^2) \tag{173}$$

where we have used equation (141) and the definitions preceding equation (138).

Finally, we mention that moment methods are also very convenient for calculating quantities such as the mean final relative velocity as a function of scattering angle or the mean component of product rotational angular momentum along some axis as a function of scattering angle. Details are in the literature.[29]

We have explicitly discussed only the case of two continuous final variables. One could also consider two discrete final variables, e.g., n' and j', or a mixed pair. e.g., j' and θ. The application of the above techniques to such cases is straightforward.

2.6.3. *Initial-State–Final-State Correlations*

Another application of the above techniques is to calculate initial-state–final-state correlations. An example is the average product rotational angular momentum as a function of reactant orbital angular momentum l.[26] Another example is the average value of l as a function of scattering angle.[29] More complicated examples would be the distribution of product rotational angular momentum as a function of l, the distribution of l as a function of scattering angle, or the distribution of scattering angle as a function of b.[27] These kinds of results can all be obtained efficiently by various combinations of the techniques explained above.

2.7. *Available Programs*

The Quantum Chemistry Program Exchange (QCPE) has four computer programs (QCPE Nos. 229, 248, 273, and 316) for carrying out trajectory calculations on reactive atom–diatom collisions. Although they do not incorporate all the most recommended techniques described in this section, they may be convenient for some trajectory studies where the use of the optimal techniques is not of the utmost importance.

All four programs are written in single-precision FORTRAN IV. Program 248 is an expanded version of program 229. Program 229 is the program used for one of the author's trajectory studies[9,14,74] on the $F + H_2$, HD, and D_2 systems. Program 316 is a version of program 273 that is suitable for use on minicomputers; thus it is less efficient than 273 when used on large computers. Program 273 is, in the words of the manual, a program "expressly intended for easy transferability between computers and for use by people who may be unfamiliar with the numerical techniques involved."

Table 2. QCPE Trajectory Codes

Number	Authors	Name	Number of symbolic cards
229	J. T. Muckerman	CLASTR: Monte Carlo quasiclassical trajectory program	2000
248	D. G. Hopper	CTAMYM: Modification of the atomic–diatomic quasiclassical trajectory program CLASTR	3000
273	S. Chapman, D.L. Bunker, and A. Gelb	A + BC: General trajectory program	1225
316	S. Chapman, K.R. Wright, D. L. Bunker, A. Gelb, and J. Santamaria	MINI A + BC: Minicomputer adapted version of the QCPE # 273 general trajectory program	1500

It includes several options and is probably to be most recommended for general purposes. The program writers and the names and sizes of the programs are given in Table 2. Further documentation and/or the programs themselves may be ordered for small fees from QCPE (Chemistry Department, Indiana University, Bloomington, Indiana).

3. *Other Trajectory Methods for Single-Surface Reactions*

The quasiclassical trajectory method discussed in Section 2 is unsymmetrical in the way it treats initial and final states. Initial vibrational and rotational energies or actions are quantized exactly,* but final states are quantized by histogram or moment methods, if at all. Thus cross sections and rate constants calculated by the quasiclassical trajectory method do not satisfy detailed balance and time-reversal requirements.[75] It has been suggested that more accurate quasiclassical histogram results can be obtained for some state-to-state cross sections by quantizing the actions in the final state rather than the initial one.[76] This involves running trajectories on the reverse reaction using the procedures of Section 2 to calculate the reverse state-to-state cross section. The state-to-state cross section of original interest is calculated from the detailed balance relation which relates the exact forward and reverse state-to-state cross sections. This is called the quasiclassical trajectory reverse histogram (QCTRH) method, and the original method may be called the quasiclassical trajectory forward histogram (QCTFH) method. The QCTRH region may be useful in some cases near thresholds for a given state-to-state process, but

* Some workers also quantize the initial orbital angular momentum l of relative translational motion, but there is usually no good reason to do this.

it is hard to decide for a given case which method is more accurate.[60,77] We recommend that if QCTRH calculations are done they not be considered, without further evidence, as probably more accurate than the QCTFH calculations. Rather any large difference of the QCTRH results from the QCTFH results should be considered as a warning that perhaps neither is reliable.

There are also some techniques which do treat initial and final states more symmetrically. These are discussed in the rest of this section.

3.1. Unquantized Initial and Final Conditions

When one is interested in thermally averaged reaction rates rather than state-to-state rates or cross sections, it may not be necessary to use the quasiclassical quantization procedures of Section 2. In some cases, for example, for all the atoms heavy, equivalent results may be obtained if one treats the initial rotational and vibrational energies purely classically and, for thermal averaging, samples from a classical continuous Boltzmann distribution instead of a quantum discrete one. This may be called the forward classical trajectory method. If the problem of interest allows one to forego the quasiclassical quantization, then one can use the Liouville theorem to make the calculation more efficient by beginning the trajectories in a transition region instead of the reactant region. Liouville's theorem states that probability density is conserved along phase-space trajectories.[78] Therefore, in the absence of transition-region trajectories which never reach reactants or products, a classical equilibrium distribution of reactant and product states implies an equilibrium distribution at the transition region. This is rather obvious, but it has caused some confusion. Some workers have argued that this cannot be true because if a batch of trajectories sampled from an equilibrium distribution in the reactant region (where the reactant atom is not interacting with the molecule) is followed to a dividing surface in the region where all the atoms are interacting, a nonequilibrium distribution is obtained there. But this should be expected because the batch sampled does not contain those interaction-region trajectories which began in the product region. Further, if the trajectories are stopped when they first reach the dividing surface, one will be missing those interaction-region trajectories which started as reactants, have reached the dividing surface at least once previously, and are returning to it. Thus to obtain the thermal reaction rate by sampling in the interaction region, one can sample from an equilibrium distribution along the dividing surface, but one must count only those trajectories which originated from reactants and which proceed directly to products without recrossing the dividing surface. Alternatively, one could count only those trajectories which arrived directly from reactants (without previously crossing the dividing surface) and which eventually proceed to products. But there may be more computational advantage in one formulation than the other. In the formulation given second one must integrate every sampled trajectory forward until it reaches one of the asymptotic regions, but one needs to integrate trajectories backward only until they recross the dividing surface or reach the reactant region. In the formulation given first one needs to integrate every trajectory backwards until it reaches an asymptotic region, but one can stop the forward integration if a trajectory recrosses the dividing surface. This method is called the combined phase–space–trajectory (CPST) method. It was developed and applied by Keck,[79-81] has been applied most extensively to atom–molecule reactions by Anderson and co-workers,[82-84] and has also been discussed by Mayer[85] and Miller.[86] In this method the equilibrium reaction rate constant $k_r(T)$ is given by the product $k_e(S, T)\,\xi(S, T)$, where $k_e(S, T)$ is the one-way equilibrium rate constant at temperature T for flow of phase points through the dividing surface S which separates reactants and products, and $\xi(S, T)$ is the fraction of trajectories, sampled from a Boltzmann distribution along the surface S at temperature T, which initiated as reactants and proceed directly to products.* If S is placed in the reactant region this yields the classical trajectory method mentioned in the first two sentences of this paragraph. This has the advantage that $k_e(S, T)$ is very easy to calculate and unweighted sampling may be done by general methods, as in Section 2. If S is in the interaction region, however, one may gain the computational advantage that $\xi(S, T)$ is much larger. But $k_e(S, T)$ and the distributions to be sampled need to be recalculated for every new choice of S. For processes with small rates the calculation may be economically prohibited if trajectories are sampled in the reactant region. But the higher efficiency of the CPST method opens these systems to the possibility of trajectory analysis.

Since the converged results of the CPST method are necessarily identical to the converged results of a classical trajectory calculation with sampling in the reactant region, one may calculate product distributions, using the methods of Section 1. By applying "product analysis" to the reactant end of the trajectory too, one may also calculate reactivity functions and reactant–product correlations. If state-to-state cross sections or rates are not needed and effects of initial- or final-state quantization are not important, the CPST method should be considered whenever the inefficiency of sampling in the reactant region is a serious problem.

3.2. *Symmetrically Averaged Initial and Final Conditions*

It was mentioned at the end of Section 3.1 that one could sample trajectories in the interaction region, integrate both forward and backward, and apply "product analysis" techniques to both ends of the trajectories.

* Alternatively, one may sample from a surface dividing reactants and products, integrate in both directions, and sum the quantity

$$(1/2)(N_c^{[i]} + 1)^{-1}[1 + (-1)^{N_c^{[i]}}$$

where $N_c^{[i]}$ is the number of additional crossings of the $[i]$th trajectory as time runs in both directions.[79,86]

For example, one could calculate state-to-state rate constants by applying the histogram method to both ends of the trajectory. One could calculate the same kind of approximation to the state-to-state rate constants by starting the trajectories at one end and averaging the initial conditions over a histogram interval instead of fixing them at the center as in the quasiclassical trajectory method. By sampling in the reactant region this way one can also calculate state-to-state cross sections. This is called the symmetrically averaged classical trajectory method.[75,86-88] It has not been applied widely, and the few results have not been encouraging. While this method has the advantage over the quasiclassical trajectory method that it can be carried out in such a way as to satisfy detailed balance, it has the disadvantage that it appears to be less accurate. Apparently quasiclassical quantization of at least one end of the trajectories is an important element in the correspondence of trajectory results with state-to-state cross sections. The symmetrically averaged classical trajectory method is not recommended.

3.3. Exactly Quantized Initial and Final Conditions

Another way to eliminate the asymmetry of the methods of Section 2 is to calculate all results using only trajectories which satisfy the double-ended quantized boundary conditions. If this is done without including any interference effect,† it yields what has variously been called the classical limit or the classical–semiclassical method.[2,60,89-92] The quasiclassical trajectories already satisfy the quantized boundary conditions in the reactant channel, but imposition of such boundary conditions on the products requires a root search. Having found these root trajectories, the state-to-state cross sections may be calculated from the Jacobian determinant D of the final quantized action variables with respect to the initial values of the angle variables conjugate to the corresponding reactant action variables.[2,91,92] However, it has been shown both on general grounds[90] and by computational examples[60] that this method is not more accurate than the quasiclassical trajectory histogram method. Since (because of the root search) it is also more difficult computationally, it is not recommended.

3.4. Classical S Matrix Theory and Other Semiclassical Methods That Include Interference Effects

In this section we discuss a few methods for incorporating more quantum effects into trajectory calculations. In particular the methods discussed here all incorporate some effects due to the phase of the quantum-mechanical wave function, i.e., interference effects, and they also allow for the treatment of multidimensional tunneling effects. They have not been demonstrated to be practical for three-dimensional reactive scattering cal-

† Interference effects are discussed in the following section, 3.4.

culations, and we are not recommending them for such. But they provide valuable insight into the reasons for the successes and failures of simpler trajectory techniques, and they provide a language for discussion of semiclassical and quantum corrections to these techniques. For these reasons we recommend the study of these methods.

The first method is the classical S matrix theory,[2,89-92] also called exact semiclassical theory.[93] Using the root trajectories of Section 3.3, which satisfy double-ended quantized boundary conditions, one calculates a phase along each trajectory. The phase is an action integral computed from the local momenta and is basically the number of de Broglie wavelenghts in the trajectory divided by 2π. Each root trajectory j contributes an amplitude $D_j^{1/2}e^{i\phi_j}$, where D_j is the Jacobian determinant and ϕ_j the phase. Adding amplitudes of all root trajectories for a given-state-to-state transition and then squaring gives what is called the primitive semiclassical approximation of classical S matrix theory. Because amplitudes are added, interference effects are included. If there are no root trajectories for a given state-to-state process then the state-to-state transition is said to be classically forbidden or classically inaccessible. A state-to-state process for which the quasiclassical trajectory histogram methods, either forward or reverse, yield zero is always classically forbidden. But the converse is not true. If a state-to-state process is classically forbidden, one may still always find solutions to the classical equations of motion which satisfy the double-ended quantized boundary conditions by letting the time step be a complex number in the numerical integration. This causes the coordinates, momenta, and phase to become complex valued (i.e., have a nonzero imaginary part) during the trajectory although, to satisfy physical boundary conditions at the end of the trajectory, the positions and coordinates must all become real valued again at the end of the trajectory. This requirement is sometimes associated with computational instabilities. The instabilities must be relieved by judicious choice of path through the complex t plane. This freedom of choice exists because the final values of coordinates, momenta, and phases are independent of the path chosen as long as no branch lines of the trajectory are crossed.[91] For complex time paths, the phase is complex and the physically interesting solutions have $|e^{i\phi_j}| < 1$, often $\ll 1$. Thus the contribution of complex-valued trajectories to the cross section for a state-to-state process is exponentially small. This procedure is called the analytic continuation of classical mechanics and it provides a multidimensional generalization for classically forbidden transitions of the tunneling picture for one-dimensional barrier penetrations. When any real-valued root trajectories contribute to a particular state-to-state cross section, their contributions will often but not always dominate the contributions of complex-valued trajectories. For practical applications to real problems it is difficult to know whether or not one has found all the significant root trajectories, especially the complex ones. Another serious problem is that usually, and especially in threshold regions for the state-to-state process of interest, accurate numerical results can be obtained only by using a uniformized expression (rather than a simple sum) to add the amplitudes of the root trajectories.[60,90,94-97] The appropriate uniformized expression may be special to each new case, and at this time no straightforward general way to derive the appropriate uniformized expressions for new cases is known. An additional difficulty in applying uniformized expressions is that they require absolute phase differences, including the phase differences coming from $D_j^{1/2}$, but phase differences are easier to calculate modulo π.[96]

Although, as mentioned above, it is generally impractical to apply the classical S matrix theory to real problems, the theory does help one to understand the limitations of the quasiclassical trajectory histogram method. The quasiclassical trajectory histogram method can be derived as a suitable average of the primitive semiclassical method with real-valued trajectories when applied to an observable which does not correspond to a completely specified (including a component of rotational angular momentum) final state or which involves some averaging over initial conditions.[90] But for classically forbidden processes or threshold regions, the analytically continued trajectories and the phase information are crucial aspects of the problem which must be included for quantitative results.

An alternative approach to semiclassical dynamics is the Gaussian wave packet method of Heller.[98-100] In this method the initial states are decomposed into time-dependent localized wave packets, each of which is separately propagated using the assumption that the potential energy may be expanded quadratically around the instantaneous center of the wave packet. Because of this assumption the center of the wave packet in both coordinate and momentum space follows the classical equations of motion. The wave packets carry phase information and must be added coherently in the product region. This method has the advantages that it requires no root searches and in a sense uniformization is automatic. In addition the visualization of the wave packets as quantum trajectories has appeal for intuitive understanding of quantum effects. But the method has the disadvantage that effects of barrier penetration, rapidly varying potential surfaces, and long-lived collisions require wave packet bifurcation or multifurcation to attain good accuracy. There is no simple criterion for how often to bifurcate the wave packet. If the reexpansion of the wave packets is repeated often enough, the method yields the exact quantum-mechanical result in the form of a discrete phase-space path integral. Many of the practical details of application of this new method to semiclassical dynamics have not been thoroughly explored, but the method seems well suited in principle to computational exploitation, and further study is clearly warranted.

Another approach to semiclassical dynamics involves the use of Wigner phase-space densities.[101] It has been suggested that Wigner phase-space densities might provide semiclassical initial conditions for a trajectory study; the semiclassical phase-space distributions might be more accurate than the quasiclassical ones of Section 1.[102] The Wigner phase-space densities have a high-energy tail which can provide a semiclassical description of multidimensional tunneling effects. Further, they have negative weights in some regions of phase space; this suggests that interference effects might be included. Heller has analyzed some of the errors in classically propagating such phase-space distributions, and he has suggested a revised classical-like dynamics for ameliorating the difficulties.[102] General practical solutions to all the computational difficulties have not yet been obtained. But comparison of the quasiclassical initial phase-space distributions to the Wigner ones does help visualize some of the effects missing in the former approach.

4. *Trajectory Methods for Multisurface Reactions*

So far in this chapter we have only considered the case where the electronic motion is treated completely adiabatically. Then there are no electronic transitions, and the electronic energy plus internuclear electrostatic interactions provide a potential energy surface for internuclear motion. This is the correct limit if electronic motion is treated by quantum mechanics and nuclear motion is treated classically, i.e., if the wave function is expanded in exact fixed-nuclei electronic states at any time, but the limit $\hbar \rightarrow 0$ is taken when considering the changes in the system's wave function due to nuclear motion.[2,103,104] Such a limit is most valid when the spacing between electronic energy states (for fixed nuclei) is large compared to the kinetic energy of internuclear motion. For some chemical reactions, e.g., $H + H_2$ at low energies, this is true for all points on all trajectories. For some other problems, e.g., low-energy collisions of systems with isolated surface crossings or isolated weakly avoided surface crossings, this is true for all except a few isolated regions of configuration space. These regions may be considered as narrow subspaces in the configuration space, and collisions may be considered as electronically adiabatic except when the trajectories enter these subspaces. Other problems of interest involve weak interaction of electronic states with large energy differences, interaction of electronic states which are close in energy over wide regions of configuration space, and interactions of dense (in number of electronic states per unit energy) sets of electronic states where successive surface crossings or avoided surface crossings are not well isolated. For these classes of problems, electronic and internuclear motion are not well separated dynamically. Since electronic motion in bound states must almost always be treated quantum mechanically, its nonseparation from internuclear motion means that internuclear motion must be treated quantum mechanically and the trajectory concept must be discarded as inapplicable for such collisions. For the former class of problems, involving only narrow regions of nonadiabaticity, we can use the model of trajectory surface hopping. In this model one defines a seam in the nonadiabatic region. The internuclear motion is treated by trajectories everywhere else, and transitions are assumed to be able to occur only when a seam is crossed. At seams it is assumed that there is a certain probability for an electronic transition to occur. If it does the internuclear motion is continued as a new trajectory, starting at the seam and governed by the potential surface corresponding to the new electronic state. If it does not then the original trajectory may be continued as if a single-surface problem were being considered. This is called the trajectory surface-hopping (TSH) model.

The account just given indicates that there are three new elements in a TSH calculation: (i) defining the seam, (ii) calculating the probability of a surface hop, and (iii) determining the momenta to start the trajectory on the new surface. All the other aspects of the calculation are the same as for single-surface calculations discussed in Sections 2 and 3. Element (i) depends on electronic structure considerations and is discussed briefly in Chapter 18. Here we discuss elements (ii) and (iii).

The probability of a surface hop is governed by the nonadiabatic coupling terms in the quantum-mechanical equations[105,106] for coupled electronic and internuclear motion. In an adiabatic representation, transitions are caused by internuclear kinetic energy operators acting on the internuclear parameters in the electronic fixed-nuclei wave functions. The simplest theory of nonadiabatic transitions is the Landau–Zener semiclassical model (see Chapter 13) for transitions caused by the radial kinetic energy in atom–atom collisions. It is natural to try to extend this to treat nonadiabatic transitions in atom–molecule systems as caused by the velocity component perpendicular to the seam. Tully and Preston have shown by classical path calculations (see Chapter 18) that this velocity component is the effective one in causing nonadiabatic transition in some cases.[107] It is clear from its derivation, however, that the Landau–Zener model has serious shortcomings. We recommend instead that the surface-hop probability be calculated by the original method of Tully and Preston,[107] which is to perform classical path semiclassical calculations for the time-dependent probability amplitudes of the coupled electronic states and parametrize an analytic expression to the results for probabilities of a surface hop to be used in the TSH calculations. However, if the conditions for validity of the Landau–Zener approximation are to be assumed to apply, then the parameters of the Landau–Zener expression should be evaluated by the prescription of Stine and one of the authors.[108] This problem is discussed further in Chapters 13 and 18.

The prescription for changing momenta when an electronic transition occurs at a seam can be obtained by solving the classical equations of motion for the internuclear motion on a potential surface given by the old surface $V_1(R_1, R_2, R_3)$ on one side of the seam and the new surface $V_2(R_1, R_2, R_3)$ on the other side of the seam.[103] In general the location of the seam is given by a function $w(\mathbf{q}, \mathbf{Q})$ of the generalized coordinates of equation (9).* Since $w(\mathbf{q}, \mathbf{Q})$ is independent of time and depends only on relative coordinates $(q_1, q_2, q_3, Q_1, Q_2, Q_3)$ the solution to the equations of motion automatically conserves total energy and total angular momentum. The explicit solution to the equations of motion indicates that the change in the generalized momenta is proportional to the gradient of w evaluated at the point where the trajectory crosses the seam. The final expression is[103,108]

$$p_i^{(2)} = p_i^{(1)} - \left(\frac{\partial w}{\partial q_i}\right)_{\text{seam}} \frac{A}{B}\left[1 - \left(1 - 2\Delta W_{\text{seam}} \frac{B}{A^2}\right)^{1/2}\right] \tag{174}$$

$$P_i^{(2)} = P_i^{(1)} - \left(\frac{\partial w}{\partial Q_i}\right)_{\text{seam}} \frac{A}{B}\left[1 - \left(1 - 2\Delta W_{\text{seam}} \frac{B}{A^2}\right)^{1/2}\right] \tag{175}$$

where

$$A = \sum_{i=1}^{3}\left[\left(\frac{\partial w}{\partial q_i}\right)_{\text{seam}} \frac{1}{\mu_{\text{BC}}} p_i^{(1)} + \left(\frac{\partial w}{\partial Q_i}\right)_{\text{seam}} \frac{1}{\mu_{\text{A,BC}}} P_i^{(1)}\right] \tag{176}$$

and

$$B = \sum_{i=1}^{3}\left[\frac{1}{\mu_{\text{BC}}}\left(\frac{\partial w}{\partial q_i}\right)_{\text{seam}}^2 + \frac{1}{\mu_{\text{A,BC}}}\left(\frac{\partial w}{\partial Q_i}\right)_{\text{seam}}^2\right] \tag{177}$$

where the superscripts (1) and (2) indicate values just before and just after crossing the seam, respectively. In equations (174) and (175)

$$\Delta W_{\text{seam}} \equiv (V_2 - V_1)_{\text{seam}} \tag{178}$$

One final point should be mentioned. It is computationally more efficient to let the trajectory branch rather than hop.[107] In other words, instead of letting the trajectory change surfaces or not based on the probability P, one can continue integrating two trajectories, one on the original surface with weight $(1 - P)$ and one on the final surface with weight P. As each branch reaches additional crossing points it branches further and each branch acquires new reduced weights. This complicates the error analysis but is more efficient. Tully and Preston have named these two techniques the anteater and the ants technique. They found the most efficient procedure was a compromise between the two in which trajectories were allowed to branch at first but when the number of branches became too great they were forced to hop.[107]

* For cases where the seam is the locus of intersection of two diabatic surfaces, $w(R_1, R_2, R_3)$ is the energy difference of the two diabatic surfaces.

Just as semiclassical methods are useful for understanding the errors in trajectory studies of single-surface reactions, they are helpful in understanding the errors of trajectory surface-hopping calculations. The classical path method[107] and classical S matrix theory[2,103,109-114] are most useful in this regard. The former shows clearly how TSH calculations must fail when the nonadiabatic regions become too nonlocalized. The latter shows clearly how TSH calculations must fail when successive nonadiabatic regions occur too close to each other or when nonadiabatic regions are too close to classical turning points or in classically forbidden regions of configuration space. The latter regions are regions which are not visited by trajectories starting with the initial conditions of interest.

5. Concluding Remarks

In this chapter we have tried to recommend computationally efficient techniques for carrying out molecular trajectory calculations for atom–diatom reactive collisions. We have discussed different ways of using the trajectories to predict final distributions of both discrete and continuous properties of the products. We have also discussed semiclassical theories which shed some light on the errors due to using classical mechanics.* Despite these errors the trajectory method is one of the most powerful computational techniques for studying the details of reactive molecular collisions and we envision its increasing use in the years to come.

ACKNOWLEDGMENTS

This work was supported in part by a National Science Foundation grant to Donald G. Truhlar and was partly carried out by the other author at Brookhaven National Laboratory under the auspices of the U. S. Department of Energy. The authors are grateful to their collaborators Dr. Normand C. Blais and Dr. Michael B. Faist for many helpful discussions and to Dr. Merle E. Riley and Dr. M. Keith Matzen for providing unpublished results.

References

1. J.L. Powell and B. Craseman, *Quantum Mechanics,* Addison-Wesley, Reading, Massachusetts (1961), pp. 98–100.
2. W.H. Miller, Classical-limit quantum mechanics and the theory of molecular collisions, *Adv. Chem. Phys.* **25**, 69–177 (1974).

* The errors in using classical mechanics for reactive collisions are discussed again in Chapter 18 where some comparisons of classical and quantum calculations are referenced and discussed.

3. M. Karplus and L.M. Raff, Theoretical investigations of reactive collisions in molecular beams: K + CH_3I, *J. Chem. Phys.* **41,** 1267–1277 (1964).
4. D.L. Bunker and N.C. Blais, Monte Carlo calculations. V. Three-dimensional study of a general bimolecular interaction potential, *J. Chem. Phys.* **41,** 2377–2386 (1964).
5. M. Karplus, R.N. Porter, and R.D. Sharma, Exchange reactions with activation energy. I. Simple barrier potential for (H, H_2), *J. Chem. Phys.* **43,** 3259–3287 (1965).
6. P.J. Kuntz, M.H. Mok, and J.C. Polanyi, Distribution of reaction products. V. Reactions forming an ionic bond, M + XC(3D), *J. Chem. Phys.* **50,** 4623–4652 (1969).
7. J.C. Polanyi and W.H. Wong, Location of energy barriers. I. Effect on the dynamics of reactions A + BC, *J. Chem. Phys.* **51,** 1439–1450 (1969).
8. D.L. Bunker, Classical trajectory methods, *Methods Comput. Phys.* **10,** 287–325 (1971).
9. J.T. Muckerman, Monte Carlo calculations of energy partitioning and isotope effects in reactions of fluorine atoms with H_2, HD, and D_2, *J. Chem. Phys.* **54,** 1155–1164 (1971).
10. R.L. Jaffe and J.B. Anderson, Classical trajectory analysis of the reaction F + H_2 $\rightarrow$ HF + H, *J. Chem. Phys.* **54,** 2224–2236 (1971); Erratum: **56,** 682 (1972).
11. A.G. Clarke and G. Burns, Trajectory studies of atomic recombination reactions, *J. Chem. Phys.* **55,** 4717–4730 (1971).
12. D.L. Thompson, On a classical trajectory study of energy transfer in some atom–diatomic molecule systems, *J. Chem. Phys.* **56,** 3570 (1972); Erratum: **65,** 5031–5032 (1976).
13. L.M. Raff, D.L. Thompson, L.B. Sims, and R.N. Porter, Dynamics of the molecular and atomic mechanisms for the hydrogen–iodine exchange reaction, *J. Chem. Phys.* **56,** 5998–6027 (1972); Erratum: **65,** 4338 (1976).
14. J.T. Muckerman, Classical dynamics of the reaction of fluorine atoms with hydrogen molecules. III. The hot-atom reactions of ^{18}F with HD, *J. Chem. Phys.* **57,** 3388–3396 (1972).
15. C.A. Parr, J.C. Polanyi, and W.H. Wong, Distribution of reaction products (theory). VIII. Cl + HI, Cl + DI, *J. Chem. Phys.* **58,** 5–20 (1973).
16. N.C. Blais and D.G. Truhlar, Monte Carlo trajectories: Dynamics of the reaction F + D_2 on a semiempirical valence-bond potential energy surface, *J. Chem. Phys.* **58,** 1090–1108 (1973).
17. H.H. Suzukawa, Jr., D.L. Thompson, V.B. Cheng, and M. Wolfsberg, Empirical testing of the suitability of a nonrandom integration method for classical trajectory calculations: Comparisons with Monte Carlo techniques, *J. Chem. Phys.* **59,** 4000–4008 (1973).
18. J.C. Polanyi and J.L. Schreiber, The dynamics of bimolecular reactions, in *Physical Chemistry: An Advanced Treatise,* Vol. VIA, *Kinetics of Gas Reactions,* W. Jost, editor, Academic Press, New York (1974), pp. 383–487.
19. J.M. White and D.L. Thompson, Monte Carlo quasiclassical trajectory study of Br + HBr and H + HBr: Effect of reactant vibration and rotation on reaction rates and energy transfer, *J. Chem. Phys.* **61,** 719–732 (1974).
20. N.C. Blais and D.G. Truhlar, Monte Carlo trajectories: The reactions H + Br_2 $\rightarrow$ HBr + Br, *J. Chem. Phys.* **61,** 4186–4203 (1974); Erratum: **65,** 3803 (1976).
21. P.J. Kuntz, The classical trajectory method, in *Interaction between Ions and Molecules,* P. Ausloos, editor, Plenum Press, New York (1975), pp. 123–137.
22. R.N. Porter, D.L. Thompson, L.M. Raff, and J.M. White, Comparison of the combined phase-space/trajectory and quasiclassical trajectory methods in the study of reaction dynamics: H + I_2 and H + Br_2, *J. Chem. Phys.* **62,** 2429–2445 (1975).
23. D.J. Malcome-Lawes, Dynamics of some hydrogen isotopic exchange reactions at high energies, *J. Chem. Soc. Faraday Trans. II* **71,** 1183–1199 (1975).
24. R.N. Porter, L.M. Raff, and W.H. Miller, Quasiclassical selecton of initial coordinates and momenta for a rotating Morse oscillator, *J. Chem. Phys.* **63,** 2214–2218 (1975).

25. J.C. Polanyi, J.L. Schreiber, and J.J. Sloan, Distribution of reaction products, *Chem. Phys.* **9,** 403–412 (1975).
26. N.H. Hijazi and J.C. Polanyi, Magnitude and orientation of rotation in exchange reactions $A + BC \rightarrow AB + C$. II, *Chem. Phys.* **11,** 1–16 (1975).
27. N.C. Blais and D.G. Truhlar, Monte Carlo trajectory study of $Ar + H_2$ collisions. I. Potential energy surface and cross sections for dissociation, recombination, and inelastic scattering, *J. Chem. Phys.* **65,** 5335–5356 (1976).
28. R.N. Porter and L.M. Raff, Classical trajectory methods in molecular collisions, in *Modern Theoretical Chemistry,* Vol. 2, *Dynamics of Molecular Collisions: Part B,* W.H. Miller, editor, Plenum Press, New York (1976), pp. 1–52.
29. N.C. Blais and D.G. Truhlar, Monte Carlo trajectories: Alignment of HBr rotational angular momentum as a function of scattering angle for the reaction $H + Br_2 \rightarrow HBr + Br$, *J. Chem. Phys.* **67,** 1540–1546 (1977).
30. H. Goldstein, *Classical Mechanics,* Addison-Wesley, Reading, Massachusetts (1950).
31. J.O. Hirschfelder, Coordinates which diagonalize the kinetic energy of relative motion, *Int. J. Quantum Chem. Symp.* **3,** 17–31 (1969).
32. R.A. LaBudde and R.B. Bernstein, Classical study of rotational excitation of a rigid rotor: $Li^+ + H_2$, *J. Chem. Phys.* **55,** 5499–5516 (1971).
33. M.E. Riley and M.K. Matzen, Non-Maxwellian H and F velocity distributions in in an $H_2 - F_2$ reaction, *J. Chem. Phys.* **63,** 4787–4799 (1975).
34. I.C. Percival, Semiclassical theory of bound states, *Adv. Chem. Phys.* **36,** 1–61 (1977).
35. P.J. Davis and I. Polonsky, Numerical interpolation, differentiation, and integration, in *Handbook of Mathematical Functions with Formulas, Graphs, and Mathematical Tables,* M. Abramowitz and I.A. Stegun, editors, U.S. Government Printing Office, Washington, D.C. (1964), pp. 875–924, especially pp. 888–889.
36. J.B. Scarborough, *Numerical Mathematical Analysis,* 4th ed., Johns Hopkins Press, Baltimore, Maryland (1958), pp. 192–198.
37. Z. Kopal, *Numerical Analysis,* Chapman and Hall, London (1955), p. 381.
38. F.J. Smith, The numerical evaluation of the classical angle of deflection and of the J.W.K.B. phase shift, *Physica (Utrecht)* **30,** 497–504 (1964).
39. R.T. Pack, Space-fixed vs. body-fixed axes in atom–diatomic molecule scattering: Sudden approximations, *J. Chem. Phys.* **60,** 633–639 (1974).
40. L. Lapidus and J.H. Seinfeld, *Numerical Solution of Ordinary Differential Equations,* Academic Press, New York (1971).
41. M.K. Matzen and M.E. Riley, private communication.
42. R. Bulirsch and J. Stoer, Numerical treatment of ordinary differential equations by extrapolation methods, *Numer. Math.* **8,** 1–13 (1966).
43. L.F. Shampine and M.K. Gordon, *Computer Solution of Ordinary Differential Equations: The Initial Value Problem,* W.H. Freeman, San Francisco (1975).
44. C.A. Parr, Classical dynamics of triatomic systems, Ph.D. Thesis, California Institute of Technology, Pasadena, California (1968), pp. 47–55, 156–159, 168–176.
45. P.W. Brumer, Structure and collision complex dynamics of alkali halide dimers, Ph.D. Thesis, Harvard University, Cambridge, Massachusetts (1972), pp. IV-39 to IV-61.
46. P. Brumer and M. Karplus, Collision complex dynamics in alkali halide exchange reactions, *Discuss. Faraday Soc.* **55,** 80–92 (1973).
47. C.W. Gear, Hybrid methods for initial value problems in ordinary differential equations, *J. SIAM Numer. Anal. Ser. B* **2,** 69–86 (1964).
48. *Symposium on Monte Carlo Methods Held at the University of Florida,* H.A. Meyer, editor, John Wiley and Sons, New York (1956); see especially the contributions of A.W. Marshall, pp. 1–14, H. Kahn, pp. 146–190, and J.W. Butler, pp. 249–264.
49. J.M. Hammersley and D.C. Handscomb, *Monte Carlo Methods,* Methuen & Co., London (1964, corrected printing 1965).

50. *The Monte Carlo Method: The Method of Statistical Trials,* Yu. A. Schreiber, editor, Pergamon Press, Oxford (1966); see especially the first chapters by Yu. A. Schreiber, pp. 1–90, and I.M. Sobol', pp. 91–130.
51. M. Zelen and N.C. Severo, in *Handbook of Mathematical Functions,* M. Abramowitz and I.A. Stegun, editors, National Bureau of Standards, Washington, D.C. (1964), pp. 925–995; see Table 26.10.
52. M.B. Faist, J.T. Muckerman, and F.E. Schubert, Importance sampling and histogrammic representations of reactivity functions and product distributions in Monte Carlo quasiclassical trajectory calculations *J. Chem. Phys.* **69,** 4087–4096 (1978).
53. M.H. Hansen, W.N. Hurwitz, and W.G. Madow, *Sample Survey Methods and Theory,* Vols. 1, 2, John Wiley and Sons, New York (1953), p. 209 (Vol. 1) and pp. 132–134 (Vol. 2.)
54. N.C. Blais and D.G. Truhlar, Monte Carlo trajectory study of Ar + H_2: Vibrational selectivity of dissociative collisions at 4500°K, *J. Chem. Phys.*, in press.
55. R.C. Tolman, *Statistical Mechanics with Applications to Physics and Chemistry,* Chemical Catalog Company, New York (1927), pp. 260–270.
56. D.G. Truhlar, Interpretation of the activation energy, *J. Chem. Educ.* **55,** 309–311 (1978).
57. D.G. Truhlar and N.C. Blais, Direct calculation of the equilibrium value of the energy of activation for dissociation of H_2 by Ar and evidence for the important contribution of collisional dissociation from low vibrational quantum numbers at shock-tube temperatures, *J. Am. Chem. Soc.* **99,** 8108–8109 (1977).
58. D.G. Truhlar and J.C. Gray, Interpretation and temperature dependence of the energy of activation for the reactions H + Cl_2, H_2 + I, H + H_2, and isotopic analogues, *Chem. Phys. Lett.* **57,** 93–99 (1978).
59. D.G. Truhlar and J.W. Duff, Classical probability matrix: Prediction of quantum-state distributions by a moment analysis of classical trajectories, *Chem. Phys. Lett.* **36,** 551–554 (1975).
60. D.G. Truhlar, Quasiclassical predictions of final vibrational state distributions in reactive and nonreactive collisions, *Int. J. Quantum Chem. Symp.* **10,** 239–250 (1976).
61. R.J. Gordon, A comparison of exact classical and quantum mechanical calculations of vibrational energy transfer, *J. Chem. Phys.* **65,** 4945–4957 (1976).
62. U. Halavee and R.D. Levine, Quantal transition probabilities from classical trajectories: Application to the collinear reactive H + Cl_2 system, *Chem. Phys. Lett.* **46,** 35–41 (1977).
63. O. Atabek and R. Lefebvre, Moment analysis of dynamics of photodissociation of linear triatomics, *Chem. Phys. Lett.* **52,** 29–33 (1977).
64. R.J. Gordon, A comparison of exact classical and quantum mechanical calculations of vibrational energy transfer. II. The effect of long-lived collision complexes, *J. Chem. Phys.* **67,** 5923–5929 (1977).
65. A. Ben-Shaul, R.D. Levine, and R.B. Bernstein, Entropy and chemical change. II. Analysis of product energy distributions: Temperature and entropy deficiency, *J. Chem. Phys.* **57,** 5427–5447 (1972).
66. J.C. Light, Statistical theory of bimolecular exchange reactions, *Discuss. Faraday Soc.* **44,** 14–29 (1968).
67. D.G. Truhlar and A. Kuppermann, Exact and approximate quantum mechanical reaction probabilities and rate constants for the collinear H + H_2 reaction, *J. Chem. Phys.* **56,** 2232–2252 (1972).
68. A. Ben-Shaul, R.D. Levine, and R.B. Bernstein, Prior-expectation distribution functions for energy disposal and energy consuption in reactive molecular collisions, *J. Chem. Phys.* **61,** 4937–4938 (1974).
69. A. Ben-Shaul, On statistical models and prior distributions in the theory of chemical reactions, *Chem. Phys.* **22,** 341–366 (1977).
70. M.B. Faist, The uniform phase space representation of product state distributions of elementary chemical reactions, *J. Chem. Phys.* **65,** 5427–5435 (1976).

71. R.B. Bernstein and M.B. Faist, Information content of detailed differential cross sections (molecular beam scattering velocity-angle maps) vs. resolved product state distributions, *J. Chem. Phys.* **65**, 5436–5444 (1976).
72. D.G. Truhlar and N.C. Blais, Legendre moment method for calculating differential scattering cross sections from classical trajectories with Monte Carlo initial conditions, *J. Chem. Phys.* **67**, 1532–1539 (1977).
73. E.A. Gislason and J.G. Sachs, Expansion of differential cross sections determined from classical trajectory studies in a series of Legendre polynomials, *Chem. Phys. Lett.* **52**, 270–275 (1977).
74. J.T. Muckerman, Classical dynamics of the reaction of fluorine atoms with hydrogen molecules. II. Dependence on the potential surface, *J. Chem. Phys.* **56**, 2997–3006 (1972).
75. J.L. Schreiber, Discussion, *Faraday Discuss. Chem. Soc.* **55**, 72–73 (1973).
76. S.C. Leasure and J.M. Bowman, A dynamic threshold in the $F + H_2$ reaction as determined from three-dimensional quasiclassical-reverse trajectory calculations, *Chem. Phys. Lett.* **39**, 462–467 (1976).
77. D.G. Truhlar, J.A. Merrick, and J.W. Duff, Comparison of trajectory calculations, transition state theory, quantum mechanical reaction probabilities, and rate constants for the collinear reaction $H + Cl_2 \rightarrow HCl + Cl$, *J. Am. Chem. Soc.* **98**, 6771–6783 (1976).
78. R.C. Tolman, *The Principles of Statistical Mechanics*, Oxford University Press, London (1938), pp. 48–52.
79. J. Keck, Statistical investigation of dissociation cross-sections for diatoms, *Discuss. Faraday Soc.* **33**, 173–182 (1962).
80. J. Keck, Variational theory of reaction rates, *Adv. Chem. Phys.* **13**, 85–121 (1967).
81. J.C. Keck, Monte Carlo trajectory calculations of atomic and molecular excitation in thermal systems, *Adv. At. Molec. Phys.* **8**, 39–69 (1972).
82. J.B. Anderson, Statistical theories of chemical reactions: Distributions in the transition region, *J. Chem. Phys.* **58**, 4684–4692 (1973).
83. R.L. Jaffe, J.M. Henry, and J.B. Anderson, Variation theory of reaction rates: Application to $F + H_2 \leftrightarrows HF + H$, *J. Chem. Phys.* **59**, 1128–1141 (1973).
84. J.B. Anderson, A test of the validity of the combined phase-space/trajectory method, *J. Chem. Phys.* **62**, 2446–2453 (1975).
85. I. Mayer, Some remarks on "Statistical theories of chemical reactions. Distribution in the transition region," *J. Chem. Phys.* **60**, 2564–2565 (1974).
86. W.H. Miller, Quantum mechanical transition state theory and a new semiclassical model for reaction rate constants, *J. Chem. Phys.* **61**, 1823–1834 (1974).
87. W.H. Miller, Discussion, *Faraday Discuss. Chem. Soc.* **55**, 71–72 (1973).
88. H. Essén, G.D. Billing, and M. Baer, Comparison of quantum mechanical and quasiclassical calculations of collinear reaction rate constants for the $H + Cl_2$ and $D + Cl_2$ systems, *Chem. Phys.* **17**, 443–449 (1976).
89. W.H. Miller, The semiclassical nature of atomic and molecular collisions, *Acc. Chem. Res.* **4**, 161–167 (1971).
90. W.H. Miller, Classical *S* matrix for rotational excitation: Quenching of quantum effects in molecular collisions, *J. Chem. Phys.* **54**, 5386–5397 (1971).
91. J.D. Doll, T.F. George, and W.H. Miller, Complex-valued classical trajectories for reactive tunneling in three-dimensional collisions of H with H_2, *J. Chem. Phys.* **58**, 1343–1351 (1973).
92. A.W. Raczkowski and W.H. Miller, Classical *S* matrix calculation for vibrationally inelastic transitions in three-dimensional collisions of Li^+ with H_2, *J. Chem. Phys.* **61**, 5413–5420 (1974).
93. R.A. Marcus, The theoretical approach, *Faraday Discuss. Chem. Soc.* **55**, 9–21 (1973).
94. H. Kreek, R.L. Ellis, and R.A. Marcus, Semiclassical collision theory: Multidimensional Bessel uniform approximation, *J. Chem. Phys.* **61**, 4540–4544 (1974).

95. H. Kreek, R.L. Ellis, and R.A. Marcus, Semiclassical collision theory: Application of multidimensional uniform approximations to atom–rigid rotor system, *J. Chem. Phys.* **62**, 913–926 (1975).
96. J.W. Duff and D.G. Truhlar, Classical *S* matrix: Numerical applications to classically allowed chemical reactions, *Chem. Phys.* **4**, 1–23 (1974).
97. J.W. Duff and D.G. Truhlar, Classical *S* matrix: Application of the Bessel uniform approximation to a chemical reaction, *Chem. Phys. Lett.* **40**, 251–256 (1976).
98. E.J. Heller, Time dependent approach to semiclassical dynamics, *J. Chem. Phys.* **62**, 1544–1555 (1975).
99. E.J. Heller, Wavepacket path integral formulation of semiclassical dynamics, *Chem. Phys. Lett.* **34**, 321–325 (1975).
100. E.J. Heller, Classical *S* matrix limit of wavepacket dynamics, *J. Chem. Phys.* **65**, 4979–4989 (1976).
101. E. Wigner, On the quantum correction for thermodynamic equilibrium, *Phys. Rev.* **40**, 749–759 (1932).
102. E.J. Heller, Wigner phase space method: Analysis for semiclassical applications, *J. Chem. Phys.* **65**, 1289–1298 (1976).
103. W.H. Miller and T.F. George, Semiclassical theory of electronic transitions in low energy atomic and molecular collisions involving several nuclear degrees of freedom, *J. Chem. Phys.* **56**, 5637–5652 (1972).
104. J.-T. Hwang and P. Pechukas, The adiabatic theorem in the complex plane and the semiclassical calculation of nonadiabatic transition amplitudes, *J. Chem. Phys.* **67**, 4640–4653 (1977).
105. J.O. Hirschfelder and W.J. Meath, The nature of intermolecular forces, *Adv. Chem. Phys.* **12**, 3–106 (1967).
106. H. Eyring and S.H. Lin, Potential energy surfaces, in *Physical Chemistry: An Advanced Treatise,* Vol. VIA, *Kinetics of Gas Reactions,* W. Jost, editor, Academic Press, New York (1974), pp. 121–186.
107. J.C. Tully and R.K. Preston, Trajectory surface hopping approach to nonadiabatic molecular collisions: The reaction of H^+ with D_2, *J. Chem. Phys.* **55**, 562–572 (1971).
108. J.R. Stine and J.T. Muckerman, On the multidimensional surface intersection problem and classical trajectory "surface hopping," *J. Chem. Phys.* **65**, 3975–3984 (1976).
109. T.F. George and Y.-W. Lin, Multiple transition points in a semiclassical treatment of electronic transitions in atom(ion)–diatom collisions, *J. Chem. Phys.* **60**, 2340–2349 (1974).
110. Y.-W. Lin, T.F. George, and K. Morokuma, Semiclassical treatment of electronic transitions in molecular collisions: Collinear $H^+ + D_2 \rightarrow HD^+ + D$, *J. Chem. Phys.* **60**, 4311–4322 (1974).
111. Y.-W. Lin, T.F. George, and K. Morokuma, Semiclassical treatment of electronic transitions in molecular collisions: Three-dimensional $H^+ + D_2 \rightarrow HD^+ + D$, *Chem. Phys. Lett.* **30**, 49–53 (1975).
112. A. Komornicki, T.F. George, and K. Morokuma, Decoupling scheme for a semiclassical treatment of electronic transitions in atom–diatom collisions: Real-valued trajectories and local analytic continuation, *J. Chem. Phys.* **65**, 48–54 (1976).
113. J.R. Stine and J.T. Muckerman, Conditions for the equivalence of the "surface hopping" and complex-valued-time methods to effect nonadiabatic transitions, *Chem. Phys. Lett.* **44**, 46–49 (1976).
114. G.L. Bendazzolli, M. Raimondi, B.A. Garetz, T.F. George, and K. Morokuma, Semiclassical study of collision-induced predissociation: Comparison of the Landau–Zener model with the method of analytic continuation, *Theor. Chim. Acta* **44**, 341–350 (1977).

17

Direct-Mode Chemical Reactions I: Methodology for Accurate Quantal Calculations

ROBERT E. WYATT

1. Introduction

Accurate quantum-mechanical state-to-state probabilities and cross sections for chemical reactions in three physical dimensions are just beginning to become available for selected neutral electronically adiabatic atom–diatom collision systems. Intense activity in a number of research groups, particularly since 1970, has led to the development of theoretical methods and computational algorithms which have been applied to several dozen model collinear reactions,[1-5] to several coplanar reactions,[6-10] and to the three-dimensional $H + H_2$[11-14] and (to a lesser extent) $F + H_2$[15] reactions. These results provide a body of knowledge against which approximate methods (particularly quantum approximations, Chapter 15, and classical mechanics, Chapter 18) may be compared. In addition, analysis of the quantum results, including state populations, probability density, and flux distributions in the transition state region, will surely provide us with a deeper understanding of "what happens during a chemical reaction."

For historical reasons, and because of its relative simplicity, the hydrogen exchange reaction has been subjected to extensive computational study via the quantum-mechanical close-coupling (CC) technique. In the first

ROBERT E. WYATT • Department of Chemistry, The University of Texas at Austin, Austin, Texas 78712.

CC study of a three-dimensional reaction, Wolken and Karplus[11] employed an integrodifferential equation formalism with relatively small rotational basis sets and one vibrational basis function. Much more extensive (and expensive!) calculations based upon integration of systems of coupled differential equations have recently been presented by Elkowitz and Wyatt[12,13] and by Schatz and Kuppermann.[14,16] Using a fast new numerical integration procedure,[17] Stechel *et al.** have obtained accurate converged reaction probabilities from the threshold energy region to energies beyond the opening of the $v = 1$ product channel. The very significant theoretical development represented by all of these calculations will find application in the next several years to other three-dimensional reactions, including $F + H_2$, $H + O_2$, Li + HF, H + DT, and possibly $O + H_2$, F + HD, and $H + Cl_2$. The F + HD and H + DT reactions are significant in that they are asymmetric A + BC ($B \neq C$) systems, with two reaction paths leading to both AB and AC product molecules. In addition, the important class of electronic nonadiabatic reactions have begun to receive computational study[18-22] through integration of the CC equations.

In this chapter, the currently used methodology for the accurate study of atom–diatom reactions will be outlined. The six coordinates required to specify the orientation, size, and shape of the three-body triangle are discussed in Section 2. Body-fixed (rotating) internal coordinates are used. The division of the internal coordinate space into three arrangement zones and the matching surfaces between them are also considered. Internal basis sets for vibration and bending on each arrangement tube are defined in Section 3. Expansion of the scattering wave function in the internal bases, and numerical integration of the close-coupled differential equations for a set of translational wave functions are presented in Section 4. Joining the primitive solutions on the arrangement channel matching surfaces, and satisfying asymptotic boundary conditions for the S matrix are considered in Section 5. A small set of selected numerical results are presented in Section 6. Other quantum methods, and discussion of numerical results for collinear and coplanar reactions have appeared in several review articles.[1-5]

2. Coordinate Geometry and Hamiltonian

2.1. Space-Fixed and Body-Fixed Coordinates

The three research groups that are responsible for developing much of the theory in this chapter have not used a uniform notation; hence Table 1 is presented as a dictionary to aid in comparing the local coordinate definitions. For an atom–diatom reaction at subdissociative

* Private communication.

Table 1. Coordinate Notation[a]

Coordinates		SK		WLA	EW
Euler angles[b]		θ		θ	θ
		ϕ		ϕ	ϕ
		Ψ		γ	χ
Internal coordinates[c]	I, II	III	IV		
Translation[d]	R	ϕ	η	u	s
Vibration	r	ρ	ζ	v	ρ
Bend	γ	γ	γ	χ	γ

[a] SK = Schatz and Kuppermann,[14] WLA = Walker, Light, and Altenberger-Siczek,[25] EW = Elkowitz and Wyatt.[12]
[b] In SK and WLA, the BF y axis is perpendicular to the nuclear triangle, but in EW, the BF x axis is the perpendicular axis.
[c] I, ..., IV refer to different regions in Figure 1b.
[d] $u = 0$ on match surface and $u \to \infty$ asymptotically; $s = 0$ on match surface and $s \to -\infty$ asymptotically in reactants.

energies, there are three arrangement channels which lead to separated A + BC (channel α), B + AC (β), or C + AB (γ). After separating off center-of-mass motion, we introduce mass-scaled vectors, which for channel α are (analogous definitions follow for the other two channels; λ denotes any arrangement channel)

$$\mathbf{R}_\alpha = c_\alpha^{-1}\left[\mathbf{r}_\mathrm{A} - \frac{m_\mathrm{B}\mathbf{r}_\mathrm{B} + m_\mathrm{C}\mathbf{r}_\mathrm{C}}{m_\mathrm{B} + m_\mathrm{C}}\right]$$
$$\mathbf{r}_\alpha = c_\alpha[\mathbf{r}_\mathrm{B} - \mathbf{r}_\mathrm{C}] \tag{1}$$

where the dimensionless scale factor is $c_\alpha = (\mu_{\mathrm{BC}}/\mu_{\mathrm{A,BC}})^{1/4}$. Scaled coordinates have the advantage that the reduced mass is the same for all nuclear degrees of freedom in any arrangement:

$$\mu = (m_\mathrm{A} m_\mathrm{B} m_\mathrm{C})/(m_\mathrm{A} + m_\mathrm{B} + m_\mathrm{C}).$$

It is convenient at this stage to introduce body-fixed (BF) coordinates since the adiabatic electronic potential within each channel is a function of the three triangle shape–size coordinates, independent of overall orientation. In addition, as discussed in Chapter 15, useful models for reactive scattering are conveniently formulated in the BF coordinate frame. The BF coordinate axes x, y, and z are oriented relative to space-fixed (SF) axes by the three Euler angles θ_λ, ϕ_λ, and χ_λ, where χ_λ denotes twist about the rotating z axis. BF coordinates and three Euler angles are introduced separately for each atom–diatom arrangement, each of which is characterized by a set of arrangement vectors, $\mathbf{R}_\lambda$ and $\mathbf{r}_\lambda$. The orientation of the BF z axis is chosen to either (a) point parallel to $\mathbf{R}_\lambda$ for geometries within the λ arrangement zone (see below), or (b) point parallel (or antiparallel) to $\mathbf{R}_\lambda$ in the asymptotic region of channel λ (large R_λ) but to change alignment smoothly during the course of reaction. In the latter case, when natural collision coordinates[23] (NCC) are used for the $\alpha \to \beta$ reaction (for example), the BF z axis rotates smoothly from the direction $\mathbf{R}_\alpha$ in reactants to the direction $-\mathbf{R}_\beta$ in products. [The switching mechanism is the switching angle $\alpha(s)$, whose value depends upon the extent of reaction (measured by s).] The reason that the BF z axis in products is chosen to point in the $-\mathbf{R}_\beta$ direction, instead of along $+\mathbf{R}_\beta$, is so that the z axis will not have to undergo a sudden flip through about an angle of π as progression through the transition state region occurs.

In order to specify the instantaneous size and shape of the nuclear triangle within the BF frame, three internal coordinates are required. These may be R_λ, r_λ, and β_λ (where β_λ is the geometric angle between $\mathbf{R}_\lambda$ and $\mathbf{r}_\lambda$; $\beta_\lambda = 0$ or π denote collinear geometries), or the NCC set s, ρ, γ or some other set of orthogonal curvilinear coordinates which are introduced at a specific stage of the reaction.

Since $[\hat{H}, \hat{J}^2] = [\hat{H}, \hat{J}_z(\mathrm{SF})] = 0$, J and M are good quantum numbers throughout the reactive collision, provided that $\hat{H}$ is free of external-field terms. In each arrangement region, the total wave function may be expanded in normalized symmetric top functions, which form a complete set in $\theta_\lambda, \phi_\lambda, \chi_\lambda$:

$$\Psi_{vjl}^{JM\lambda} = \sum_{K=-J}^{+J} N_{MK}^{J}(\theta_\lambda, \phi_\lambda, \chi_\lambda)\, \Psi_{vjl}^{JK\lambda}(R_\lambda, r_\lambda, \beta_\lambda) \tag{2}$$

The internal translation–vibration–bend functions $\Psi_{vjl}^{JK\lambda}$ for the collision starting in channel vjl are determined by solving a system of close-coupled (CC) equations, subject to asymptotic boundary conditions ($R_\lambda \to \infty$) in each channel.

An important point concerns the definition of K in equation (2); it is the component of total angular momentum J along the BF z axis. In NCC, if the switching angle α is employed to reorient the z axis during reaction, then K is still the component of J along the instantaneous z axis. However, in the reaction zone, this value of K is not the same as the projection of J along the $\mathbf{R}_\lambda$ axis. This distinction is important in the wave function matching procedure (Section 5) and in developing reactive scattering models (Chapter 15).

2.2. Natural Translation–Vibration Coordinates

Orthogonal curvilinear translation–vibration coordinates now will be introduced in each arrangement region; the "natural" bending angle γ will

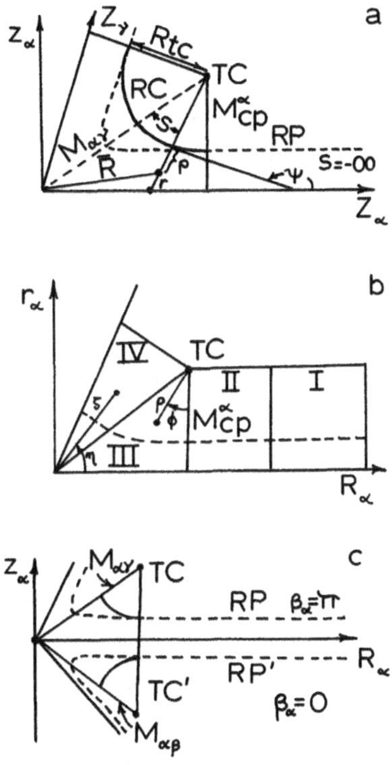

Figure 1. Internal translation–vibration coordinates. (a) Natural collision coordinates[23] used by Elkowitz and Wyatt,[12] drawn in the first quadrant only. On the match surface $M_{\alpha\gamma}$, $s = 0$, and on the approach to $M_{\alpha\gamma}$ from reactants, s increases toward zero. (b) Segmentation of coordinate space used by Schatz and Kuppermann,[24] showing local polar coordinates in the strong interaction (III) and matching (IV) regions. (c) Coordinates for asymmetric A + BC reaction, introduced by Walker, Light, and Altenberger-Siczek.[25] In the region between $M_{\alpha\gamma}$ and $M_{\alpha\beta}$ about the R_α axis, collision coordinates u_α and ρ are introduced relative to the reference curve RC.

be introduced in Section 2.3. In A + B_2 arrangement tubes, Elkowitz and Wyatt[12] used s and ρ defined in the first quadrant of the collinear plane in Figure 1a. The vibrational coordinate ρ is defined in cuts perpendicular to the reference curve (RC), along which the arc length s is measured. A Cartesian–polar coordinate transformation occurs along match line M^{α}_{CP}. Schatz and Kuppermann[24] divided the collinear subspace of each arrangement tube into an asymptotic region (I), a weak-interaction region (II), a strong-interaction region (III), and a matching region (IV). The propagation–vibration coordinates (Figure 1b) in these four zones are (R_λ, r_λ), (R_λ, r_λ), (ϕ, ρ), and (η, ζ), respectively. A Cartesian–polar transformation again occurs at M^{α}_{CP}. For asymmetric A + BC reactions, Walker, Light, and Altenberger-Siczek[25] (WLA) used reference curves RC and RC′ symmetrically displaced about R_α (Figure 1c), with a progression variable $u(s)$ defined along RC. The vibrational coordinate ρ again measures displacements from RC. It is important to note that these three coordinate systems share the common feature that they are not defined relative to the reaction path (RP). In addition, the turning centers (TC) or origins for the polar coordinates are in regions of high potential where there is little wave function penetration to the coordinate singularity which occurs there.

2.3. Natural Bending Angle

The geometric angle β_α between the α-channel vectors $\mathbf{R}_\alpha$ and $\mathbf{r}_\alpha$ does not describe internal rotation in the product channels, since $\beta_\lambda \neq \beta_\beta \neq \beta_\alpha$. The natural bending angle γ for the $\alpha \to \gamma$ reaction, for example, converts smoothly from β_α in reactants to β_γ in products. In the reaction zone, γ has the disadvantage that it is not the angle between any pair of arrangement vectors $\mathbf{R}_\lambda$ and $\mathbf{r}_\lambda$. The bending angle γ and the switching angle $\alpha(s)$, which realigns the rotating z axis during reaction, may be introduced simultaneously in the following manner.[23] If the BF x axis is always forced to be perpendicular to the nuclear plane, then $\mathbf{r}_\alpha$ and $\mathbf{R}_\alpha$ have BF components $\{0, y_\alpha, z_\alpha\}$ and $\{0, Y_\alpha, Z_\alpha\}$. We now replace $\{y_\alpha, Y_\alpha\}$ with $\{m, \alpha\}$, where $m = r\sin\gamma$ in Figure 2a: $y_\alpha = -m\cos\alpha$, $Y_\alpha = m\sin\alpha$. In Figure 2a the m axis is perpendicular to the collinear subspace ($m = 0$), with $\{r, \gamma\}$ locating point P. During reaction, the switching angle α, Figure 2b, inverses smoothly from 0 in reactants to $\xi_{\alpha\gamma}$ in products ($\xi_{\alpha\gamma}$ is the kinematic skewing angle between Z_α and Z_γ in Figure 2a). As a result, $Y_\alpha = 0$ in asymptotic reactant regions and $Y_\gamma = 0$ in separated product regions. In this way, the BF z axis realigns during reaction and γ is used to specify deviations from collinearity at all values of s. This argument has focused upon single-path reactions, but extension to $A + B_2$ arrangement regions is possible.[12]

2.4. Matching Surfaces and Arrangement Tubes

Three internal coordinates, for example s, ρ, and γ, have now been introduced within each arrangement tube. Arrangement tube α extends

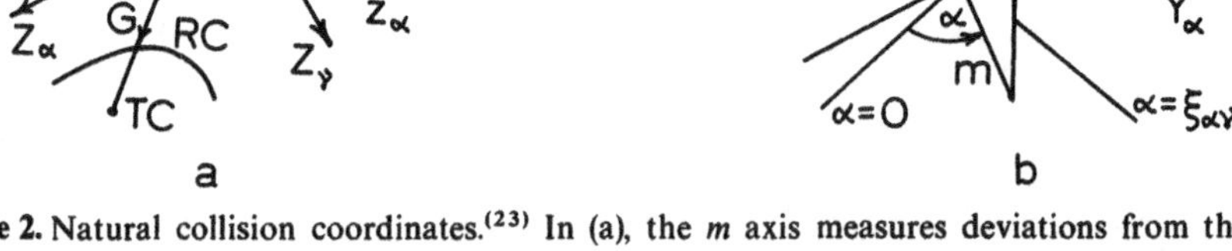

Figure 2. Natural collision coordinates.[23] In (a), the m axis measures deviations from the collinear ($Z_\alpha z_\alpha$) plane. At one value of s, polar coordinates $r\gamma$ locate point P. In (b), the G axis of (a) is drawn perpendicular to the plane defined by the y components of R_α and r_α (or R_γ and r_γ). In (b), point P is located by coordinates r, γ, and α. As s increases toward products in (a), α in (b) increases from 0 (where $Y_\alpha = 0$) to $\zeta_{\alpha\gamma}$ (where $Y_\gamma = 0$). In either (a) or (b), the product components (γ arrangement) are obtained from the reactant components (α) by rotation through the kinematic skewing angle $\zeta_{\alpha\gamma}$.

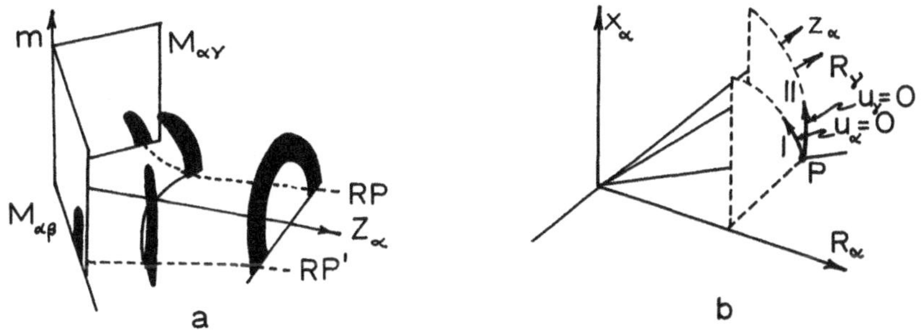

Figure 3. (a) Reactant tube for α arrangement, showing reaction paths RP and RP′, and match surfaces $M_{\alpha\gamma}$ and $M_{\alpha\beta}$. The dark regions locate geometries where $V < 1$ eV, for example, for a reaction passing through two near-linear intermediates. (b) Rotation out of the collinear plane (at P) about the R_α or R_γ (with $u_\alpha = 0$ and $u_\gamma = 0$, respectively) does not define the $\alpha\gamma$ match surface. $(x_\alpha z_\alpha)$ are the components of r_α on the BF axes.

from the match surfaces $M_{\alpha\gamma}$ and $M_{\beta\gamma}$ out to large values of R_α. For example, in Figure 3a, the half-plane $M_{\alpha\gamma}$ is perpendicular to the collinear Z_α, z_α plane and is defined by the condition $s = 0$. Within this plane, ρ and γ range freely. For the H_3 system in the $M_{\alpha\gamma}$ plane, all isosceles triangle configurations with atom B (the transferred particle) at the apex are generated as ρ and γ vary. In the $M_{\alpha\beta}$ matching plane, atom C forms the apex of a similar sequence of isosceles triangles.

Kuppermann has introduced a novel coordinate system[26] in terms of which the matching surfaces and arrangement tubes are easily visualized. First, spherical polar coordinates ζ, ω, and β are defined from R_α, r_α, and β_α, with $\zeta = (R_\alpha^2 + r_\alpha^2)^{1/2}$, $\omega = 2\cos^{-1}(R_\alpha/\zeta)$, and $\beta = \beta_\alpha$, or in terms of Cartesian coordinates, $X_\alpha = \zeta \sin\omega \cos\beta$, $Y_\alpha = \zeta \sin\omega \sin\beta$, and $Z_\alpha = \zeta \cos\omega$. For H_3, surfaces of constant potential, say 1 eV, define three arrangement tubes, with their axes skewed by 120° relative to one another. The collinear subspace corresponds to the plane $\beta_\alpha = 0, \pi$. The three match surfaces (each a half-plane) are defined to meet on the OY_α axis, two of them making angles of $\pm 60°$ with the positive $OY_\alpha Z_\alpha$ plane, and the third lying in the negative $OY_\alpha Z_\alpha$ plane. Within the $M_{\alpha\gamma}$ match plane, for example, $r_\alpha = r_\gamma$ and $R_\alpha = R_\gamma$, and isosceles triangles are again formed.

For the asymmetric A + BC reaction, WLA define matching surfaces by requiring two of the internuclear distances to be constant as the nuclei are bent out of the collinear plane.[25] For the $\alpha \to \gamma$ reaction, $M_{\alpha\gamma}$ is defined by the condition $r_\gamma/r_\alpha =$ constant (the constant may be chosen as a function of the three masses). In this case, for R_λ, r_λ, β_λ internal coordinates, the surface $M_{\alpha\gamma}$ is not obtained by setting the propagation variable s equal to a constant and letting the vibration-bend coordinates range freely. For example, in Figure 3b, starting at point P and $M_{\alpha\gamma}$ in the collinear plane (with $s_\alpha = s_\gamma = 0$), as β_α increases, curve I is traced out, but curve II is traced as β_γ increases; the surface $s_\alpha = 0$ is tangent to $s_\gamma = 0$ only in collinear geometries. In order to remain on $M_{\alpha\gamma}$, the propagation variable must increase as β_α increases. This necessitates complicated matching transformations in order to smoothly join the wave function and its normal derivative across $M_{\alpha\gamma}$ (for the $\alpha \to \gamma$ reaction). These transformations have now been derived and are being used in reactive scattering calculations. This point will be treated in more detail in Section 5 below.

3. *Internal Basis Sets*

3.1. *Partitioning of Hamiltonian*

In NCC, the nuclear kinetic energy operator may be partitioned into contributions from translation (s), vibration (ρ), and rotation-bend ($\theta, \phi, \chi, \gamma$) motions, plus higher-order coupling terms which vanish for collinear geometries, but which become important for bent intermediates[12]:

$$\begin{aligned}
\hat{T} &= \hat{T}_{\text{tr}} + \hat{T}_{\text{vib}} + \hat{T}_{\text{rot}} + \hat{T}_{\text{coupl}} \\
\hat{T}_{\text{tr}} &= \frac{-\hbar^2}{2\mu}\frac{1}{\eta g}\frac{\partial}{\partial s}\frac{g}{\eta}\frac{\partial}{\partial s} \\
\hat{T}_{\text{vib}} &= \frac{-\hbar^2}{2\mu}\frac{1}{(r-\rho)^2}\frac{\partial}{\partial \rho}(r-\rho)^2\frac{\partial}{\partial \rho} \\
\hat{T}_{\text{rot}} &= \hat{T}_{\text{st}} + \hat{T}_{\text{asym}} + \hat{T}_{\text{cor}} + \hat{T}_{\text{bend}}
\end{aligned} \tag{3}$$

where the rotational kinetic energy operator has been partitioned into a symmetric top term, an asymmetry term (which arises because the BF axes are not principal axes, except at $s = 0$), a Coriolis term which couples bending (γ) motion to overall tumbling, and finally, the bending kinetic energy. These rotational terms are

$$\begin{aligned}
\hat{T}_{\text{st}} &= \frac{A}{2}[\hat{J}^2 - (1 + f^2)\hat{J}_z^2] \\
\hat{T}_{\text{asym}} &= \frac{fA\cot\gamma}{2}(\hat{J}_z\hat{J}_y + \hat{J}_y\hat{J}_z) \\
\hat{T}_{\text{cor}} &= \frac{-ifA\hat{J}_x}{2\sin\gamma}\left(\frac{\partial}{\partial\gamma}\sin\gamma + \sin\gamma\frac{\partial}{\partial\gamma}\right) \\
\hat{T}_{\text{bend}} &= \frac{-\hbar^2}{2I^*}\left(\frac{1}{\sin\gamma}\frac{\partial}{\partial\gamma}\sin\gamma\frac{\partial}{\partial\gamma} - \frac{\hat{J}_z^2}{\hbar^2\sin^2\gamma}\right)
\end{aligned} \tag{4}$$

The various geometric coefficients (η, g, r, etc.) in these equations are defined in Table 2. The kinetic energy operator defined in equations (3),

Table 2. Geometric Parameters and Inertia Coefficients

$$g(s) = \bar{R}^2 - r^2 f^2; \; \bar{R}, r \text{ defined in Figure 1a}$$
$$f(s) = \cos(\alpha - \Psi); \; \Psi \text{ defined in Figure 1a}$$
$$\alpha(s) = \text{switching angle defined in Section 2.3}$$
$$A(s) = (\mu g)^{-1}$$
$$(I^*)^{-1} = A(\bar{R}/r)^2$$
$$\eta(s, s) = 1 + K\rho$$
$$K = \text{curvature of reference curve} = \begin{cases} 0, & \text{Cartesian region} \\ R_{tc}^{-1}, & \text{polar region} \end{cases}$$

with $\hat{T}_{\text{coupl}} = 0$, was used by Elkowitz and Wyatt[12,13] in close-coupling studies on the $H + H_2$ reaction. In addition, they chose $\alpha(s)$ so that several terms in $\hat{T}_{\text{coupl}}$ were forced to vanish.

When γ is chosen as the geometric bending angle ($\gamma \equiv \beta$), and a circular arc is used for the reference curve within the polar region of Figure 1a, the kinetic energy components assume simpler forms, in which $\hat{T}_{\text{coupl}} = 0$, and $f = -1$, $A = (\mu R^2)^{-1}$, and $(I^*)^{-1} = (R^{-2} + r^{-2})/\mu$. These expressions for the kinetic energy components are then identical to those in WLA[25] [their equation (17)], when it is recognized that the BF x axis is assumed to be perpendicular to the nuclear plane in the above expressions, whereas WLA oriented the BF y axis perpendicular to the nuclear plane. The kinetic energy operator used by Schatz and Kuppermann (see Appendix A of Reference 24) is equivalent to the one used by WLA, when allowance is made for different location of the TC in regions III and IV of Figure 1b.

For $A + A_2$ or $A + B_2$ arrangement tubes, the electronic hypersurface may be decomposed into translational, vibrational, bending, and vibration-bend coupling terms:

$$\hat{V}(s, \rho, \gamma) = \hat{V}_{\text{tr}}(s) + \hat{V}_{\text{vib}}(\rho; s) + \hat{V}_{\text{bend}}(\gamma; s) + \hat{V}_{\text{vb}}(s, \rho, \gamma) \tag{5}$$

where $\hat{V}_{\text{tr}}$ is the potential on the reaction path when s is the position on the reference curve, $\hat{V}_{\text{vib}}$ is the increment in the potential (at fixed s) for vibrational motion (usually in the collinear plane), $\hat{V}_{\text{bend}}$ is the increment in the potential as γ increases (starting at the value of ρ on the reaction path), and $\hat{V}_{\text{vb}}$ is whatever remains after the sum of the three preceding components is subtracted from the actual potential. For small excursions from the reaction path, $\hat{V}_{\text{vb}} \approx 0$. This decomposition of the kinetic energy operator and potential is very useful in defining local vibration–rotation bases for expansion of the scattering wave function.

The potential may also be decomposed in other ways, for example,

$$\hat{V}(s, \rho, \gamma) = \sum_{l,n} a_{ln}(s) [\rho - \rho_0(s)]^n P_l^0(\cos\gamma)$$

where ρ_0 is the local distance from the RC to the RP in Figure 1(a). This form is useful in evaluating matrix elements of $\hat{V}$ in the spherical-harmonic harmonic-oscillator basis.

3.2. Vibrational Basis

Local adiabatic vibrational eigenvalues and basis functions are defined relative to the vibrational Hamiltonian in equation (3),

$$\left[\frac{-\hbar^2}{2\mu}\frac{d^2}{d\rho^2} + \hat{V}_{\text{vib}}(\rho; s) - E_v(s)\right] H_v(\rho; s) = 0 \tag{6}$$

where V_{vib} is usually chosen to be either a local harmonic or Morse oscillator fit to the collinear vibrational potential. This presents no problems for symmetric $A + A_2$ or $A + B_2$ type reactions, but for a general asymmetric A + BC reaction passing, for example, through two near-linear intermediates (ABC, ACB), the vibrational basis generated in the $\beta_\alpha = 0$ collinear subspace may be a poor model for the vibrational functions near the other reaction path, where $\beta_\alpha = \pi$. For this reason, it is advantageous to define two vibrational bases,[25] $\{H_v^0\}$ for expansion of the wave function near the $\beta_\alpha = 0$ subspace, and $\{H_v^\pi\}$ for similar expansions near $\beta_\alpha = \pi$. This technique would only be used for $s > s_B$, where s_B denotes a position on the reaction path where the reactant tube wave function is well bifurcated (with respect to bending motion) into segments ψ_α^0 and ψ_α^π each of which is localized near one of the two reaction paths. For $s < s_B$, if the bending motion is relatively unhindered, then either vibrational basis may be used for expanding ψ_α^{JM}. The two vibrational potentials which generate these bases are defined by partitioning the total potential in two ways:

$$\hat{V} = \begin{cases} \hat{V}_{\text{tr}}^0(s) + \hat{V}_{\text{vib}}^0(\rho; s) + \hat{V}_{\text{bend}}^0(\beta; s) + \hat{V}_{\text{vb}}^0(s, \rho, \beta), & 0 \le \beta \le \beta^* \\ \hat{V}_{\text{tr}}^\pi(s) + \hat{V}_{\text{vib}}^\pi(\rho; s) + \hat{V}_{\text{bend}}^\pi(\beta; s) + \hat{V}_{\text{vb}}^\pi(s, \rho, \beta), & \beta^* \le \beta \le \pi \end{cases} \tag{7}$$

The first partitioning is used near the ABC pathway, while the second partitioning is used near the ACB pathway. The vibration-bend coupling terms V_{vb}^0 and V_{vb}^π are defined to make both partitionings exact. The angle β^* divides the angular range $(0, \pi)$ into two portions $(0, \beta^*)$ and (β^*, π) which lead as $s \to 0$ to the $M_{\alpha\gamma}$ or $M_{\alpha\beta}$ match surfaces.

The vibrational potential in equations (5) or (7) is computed in the collinear plane. However, it is also possible to define the vibrational basis relative to the angular average of the total interaction potential, $V_{\text{vib}}^0(\rho; s) = \langle P_0 | V(s, \rho, \gamma) | P_0 \rangle$.

3.3. Rotational Bases

3.3.1. Free-Rotor and Adiabatic Representations

We begin with the free-rotor Hamiltonian $\hat{H}_0 = \hat{T}_{st} + \hat{T}_{bend}$, whose eigenfunctions are products of normalized symmetric-top functions and normalized associated Legendre polynomials, $N^J_{MK}(\theta, \phi, \chi)\, P^K_j(\cos\gamma)$. This *symmetric-top–free-rotor diabatic basis* may be used to expand the rotational part of the scattering wave function. The close-coupled equations may be derived in this basis, and are discussed later in Section 4 [see equation (15)].

A local adiabatic rotational basis may also be used for expansion of the scattering wave function. [In the calculations of Harms, Elkowitz, and Wyatt,[12,27,28] the inertia coefficients A and I^* in equation (4) were evaluated on the reaction path.] Two alternatives routed for forming eigenstates of $\hat{H}_{rot} = \hat{T}_{rot} + \hat{V}_{bend}$ arise due to the order in which the dynamic coupling operator $(\hat{T}_{asym} + \hat{T}_{cor})$ and $\hat{V}_{bend}$ are diagonalized in the $\hat{H}^0$ basis. Along the first route, $(\hat{T}_{asym} + \hat{T}_{cor})$ is represented in the symmetric-top–free-rotor diabatic basis; this operator mixes basis elements of different K, but it does not mix subspaces of different j. Another way to view this diagonalization is based upon consideration of the orbital angular momentum operator $\mathbf{l} = \mathbf{J} - \mathbf{j}$. The representation of $\hat{l}^2$ in the $\hat{H}^0$ basis is not diagonal; in fact, the off-diagonal elements are identical [to within the factor A in equation (4) and Table 2] to those of $(\hat{T}_{asym} + \hat{T}_{cor})$. Thus diagonalization of $(\hat{T}_{asym} + \hat{T}_{cor})$ leads to eigenvectors which can be labeled by l. This *K-coupled diabatic basis* is given *exactly* by

$$\tilde{\Omega}^{JM}_{jl}(\theta, \phi, \chi, \gamma) = \sum_{K=-J}^{+J} N^J_{MK}(\theta, \phi, \chi)\, e^{-iK\pi/2} S_K \left(\frac{2l+1}{2J+1}\right)^{1/2} \times \langle lj0K | JK \rangle\, P^K_j(\cos\gamma) \tag{8}$$

where $S_K = (+1)$ for $K \leq 0$ or $(-1)^K$ for $K > 0$, and $\langle lj0K | JK \rangle$ is a Clebsch–Gordan coefficient.

In the final step along the first route, $\hat{V}_{bend}$ is diagonalized in the K-coupled diabatic basis; $\hat{V}_{bend}$ mixes states of differing j and l. The new states are

$$\Omega^{JM}_{jl} = \sum_{j'l'} d^J_{j'l',jl}(s)\, \tilde{\Omega}^{JM}_{j'l'} = \sum_K N^J_{MK} R^{JK}_{jl} \tag{9}$$

in which $R^{JK}_{jl}(\gamma; s)$ are hindered-internal-rotor wave functions. This basis of $\hat{H}_{rot}$ eigenvectors is the full rotational adiabatic basis. Indices j and l are no longer good quantum numbers except when $\hat{V}_{bend} \to 0$: $\Omega^{JM}_{jl} \to \tilde{\Omega}^{JM}_{jl}$ as $|s| \to \infty$.

A second path from the symmetric-top–free-rotor basis to the adiabatic representation will now be considered. We begin by diagonalizing $\hat{V}_{bend}$

in the $\hat{H}^0$ basis; $\hat{V}_{\text{bend}}$ mixes elements of differing j, but K is preserved as a good quantum number. The new eigenvectors are $N^J_{MK}G^K_j$, where G^K_j is a symmetric-top hindered-rotor function:

$$G^K_j(\gamma; s) = \sum_{j'} b^K_{j'j}(s)\, P^K_j(\cos\gamma) \tag{10}$$

This basis is adiabatic as far as $\hat{V}_{\text{bend}}$ is concerned, but it is diabatic with respect to $(\hat{T}_{\text{asym}} + \hat{T}_{\text{cor}})$. In the last stage of the alternative scheme, the dynamic coupling operator is diagonalized in the basis defined in equation (10); the resultant eigenvectors are again the full adiabatic states:

$$\Omega^{JM}_{jl} = \sum_K N^J_{MK} \sum_{j'} C^{JK}_{jlj'} G^K_{j'} \tag{11}$$

The dynamic coupling operator has matrix elements within the $N^J_{MK}G^K_j$ basis between states jK and $j'K'$; that is, both inter- and intra-j manifold coupling is present. However, as A separates from BC and $\hat{V}_{\text{bend}} \to 0$, only intramanifold coupling remains.

For the $H + H_2$ reaction, $V_{\text{bend}}(\gamma, s) = V_{\text{bend}}(\pi - \gamma; s)$, so the G^K_j have even or odd reflection symmetry about the $\gamma = \pi/2$ midplane. In forming hindered-rotor functions from associated Legendre polynomials, only even reflection symmetry polynomials ($j + K =$ even) are mixed to form even reflection symmetry G^K_j (or, $J + K =$ odd polynomials mix to form odd reflection symmetry G^K_j). Close-coupled equations for the translational wave functions obtained by expanding the scattering wave function in the adiabatic rotational basis will also be considered in Section 4 [see equation (14)].

3.3.2. *Asymmetric-Top Correlation Diagrams*

Adiabatic rotational states Ω^{JM}_{jl} have been computed for values of J from 0 to 20 for use in close-coupling studies of $H + H_2$ reactive collisions.[27] For each value of J, there are rotor states of even and odd parity (even or odd inversion symmetry with respect to the SF axes), which is determined by whether $j + l$ is even or odd, respectively. There is no mixing between states of different parity in either reactive or nonreactive collisions, since the Hamiltonian is invariant (in the absence of external fields) to inversion through the SF origin. The asymmetric-top rotational energy correlation diagram for positive-parity H_3 states with $J = 2$ is shown in Figure 4. The left side of the figure shows the level structure at moderate s, where the energies within each j manifold (with $|J - j| \leq l \leq J + j$) are slightly split, due to the centrifugal term $l(l + 1)\hbar^2/2\mu R^2$. As s increases toward the transition state ($s = 0$) each free-rotor state correlates with a specific bending level of the H_3 complex. If we follow Ω^{JM}_{jl} as $s \to 0$, the hindered-rotor densities $|R^{JK}_{jl}|^2$ and energy W^J_{jl} become (almost) identical

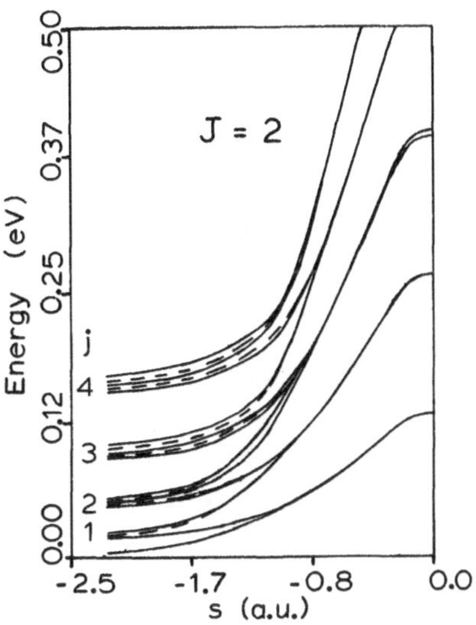

Figure 4. Asymmetric-top rotational energy correlation diagram[27] for $H + H_2$ at $J = 2$. Asymptotic $H + H_2$ states labeled by j and l ($l = |J - j|, \ldots, J + j$) on the left correlate with specific H_3 bending states at $s = 0$. Solid curves denote even nuclear purity ($j + l =$ even), while dashed curves denote odd purity ($j + l =$ odd).

to the density and energy from an asymptotically *different* top state (with the same overall parity). In Figure 4, the states $(jl, j'l') = (02, 11), (13, 20), (22, 24), \ldots$ become degenerate at $s = 0$. This pairwise degeneracy is a direct consequence of the reflection symmetry of the bending potential, $V_{\text{bend}}(\gamma) = V_{\text{bend}}(\pi - \gamma)$. In the NCC formulation, the pairwise degeneracy of the top states is used to build localized rotor functions for complexes ABC and ACB localized near the two reaction paths. This in turn greatly facilitates formation of the overlap matrix between the scattering wave function in the α tube with the β and γ tube scattering wave functions.[28]

3.3.3. Partitioned Rotors

A partitioned-rotor basis may be used to expand the angular part of the scattering wave function $\Psi_{\omega}^{JK}(s, \rho, \gamma)$ within the "bifurcation zone" ($s > s_B$) of an asymmetric reaction.[25] The free-rotor basis is partitioned into two subsets, one of which is complete (and localized) in the angular region $0 \leq \beta < \beta^*$, while the other is complete (and localized) for $\beta^* < \beta \leq \pi$. The partitioned free rotors are

$$S_j^{K0}(\beta) = \sum_{j'} a_{jj'}^{K0} P_{j'}^K(\beta), \qquad 0 \leq \beta < \beta^*$$
$$S_j^{K\pi}(\beta) = \sum_{j'} a_{jj'}^{K\pi} P_{j'}^K(\beta), \qquad \beta^* < \beta \leq \pi \tag{12}$$

where the superscripts $(0, \pi)$ again denote the localization region. The scattering wave function for $s > s_B$ is then a sum of components Ψ_ω^{JK0} and $\Psi_\omega^{JK\pi}$, where each component is expanded in the appropriate localized vibration–rotation basis:

$$\begin{aligned}\Psi_\omega^{JK} &= \Psi_\omega^{JK0} + \Psi_\omega^{JK\pi} \\ &= \sum_{vj} F_{vj}^{JK0}(s)\, H_v^0(\rho; s)\, S_j^{K0}(\beta) + \sum_{vj} F_{vj}^{JK\pi}(s)\, H_v^\pi(\rho; s)\, S_j^{K\pi}(\beta) \qquad (13)\end{aligned}$$

The set of partitioned rotors in equation (12) may be formed by diagonalizing the Heaviside step operator $\mathrm{H}(\beta^* - \beta)$ in the free-rotor basis.[25] Instead of using free-rotor partitioned sets within the two angular regions, it is also possible to define adiabatically deformed and localized rotor sets by diagonalizing a model bending potential (which has a high value near β^*) in the free-rotor basis.

4. Close-Coupling Equations

4.1. Wave Function Scaling and Basis Expansion

Once the internal basis sets have been selected, the major problem is to numerically integrate a system of close-coupled equations for the translational wave functions. For a scattering problem with a continuum coordinate R and one internal coordinate x, in which the local internal basis is $\{B_p(x; R), p = 1, 2, \ldots, N\}$, the scattering wave function at each R is expanded in the basis, $\Psi(R, x) = \sum T_p(R)\, B_p(x; R)$, where $\{T_p(R)\}$ are the translational wave functions. The close-coupled (CC) equations for the $T_p(R)$ are obtained by forcing Ψ to satisfy the Schrödinger equation at total energy E, followed by projection onto each basis function, one at a time: $\langle B_p | \hat{H} - E | \Psi \rangle_x = 0$, $p = 1, 2, \ldots, N$. The procedure is the same for the atom–diatom reaction, even though the algebra is much more tedious.

In the NCC formulation, the scattering wave function is expanded in a local hindered asymmetric-top–vibrator basis (ω denotes a set of channel indices, $\omega = vjl$)

$$\Psi_\omega^{JM}(\theta, \phi, \chi, s, \rho, \gamma) = \left(\frac{\eta}{gr^2}\right)^{1/2} \sum_{\omega'} T_{\omega\omega'}^J(s)\, \Omega_{j'l'}^{JM}(\theta, \phi, \chi, \gamma; s)\, H_{v'}(\rho; s) \qquad (14)$$

where the geometric scale factor[29] $(\eta/gr)^{1/2}$ is added to simplify the final equations for the translational wave functions $\{T_{vjlv'j'l'}^J(s)\}$. In regions I or II of Figure 1b, Schatz and Kuppermann[24] used a local-oscillator–free-rotor basis for the wave function expansion

$$\begin{aligned}\Psi_{vjK}^{JM}(\theta, \phi, \chi, R, r, \beta) = \frac{1}{Rr} \sum_{K'=-J}^{+J} D_{MK'}^J(\theta, \phi, \chi) \sum_{v'} \sum_{j'=|K|}^{\infty} g_{v'j'K'}^{JvjK}(R) \\ \times H_{v'j'}(r; R)\, Y_{j'}^{K'}(\beta, 0) \qquad (15)\end{aligned}$$

where $\{g^{Jvjk}_{v'j'K'}(R)\}$ are translational wave functions in BF coordinates. In the BF description, the channel index vjK may be used. Translational wave functions labeled by l in equation (14) are linear combinations (with Clebsch–Gordan weights) of the translational wave functions in equation (15). WLA[25] use an expansion equivalent to that in equation (15) for $s < s_B$ (at s_B conversion is made from the free-rotor basis Y^K_j to the two partitioned rotor bases). For $s > s_B$, the scattering wave function is partitioned into $\beta = 0$ and $\beta = \pi$ localized components, where the component localized in the $(0 \leq \beta \leq \beta^*)$ region is expanded (as usual, ω specifies the initial channel)

$$\Psi^{JM0}_{\omega}(\theta, \phi, \chi, s, \rho, \beta) = \left(\frac{\eta}{R^2 r^2}\right)^{1/2} \sum_{K=-J}^{+J} D^J_{MK}(\theta, \phi, \chi) \sum_v \sum_{j=|K|}^{\infty} g^{J\omega 0}_{jvK}(s) \times H^0_{vj}(\rho; s)\, S^{K0}_j(\beta) \tag{16}$$

In NCC, the close-coupled equations derived from equation (14) are

$$\frac{d^2 T^J_{\omega\omega'}}{ds^2} = \sum_{\omega''} \hat{V}^J_{\omega'\omega''} \frac{dT^J_{\omega\omega''}}{ds} + \sum_{\omega''} V^J_{\omega'\omega''} T^J_{\omega\omega''} \tag{17}$$

where the coupling matrices are

$$\hat{V}^J_{v'j'l'vjl} = -2\delta_{vv'} \sum_K \left\langle R^{JK}_{j'l'} \middle| \frac{d}{ds} \middle| R^{JK}_{jl} \right\rangle_{\gamma} - 2\delta_{jj'}\delta_{ll'} \left\langle H_{v'} \middle| \frac{d}{ds} \middle| H_v \right\rangle_{\rho} \tag{18}$$

$$\begin{aligned} V^J_{v'j'l'vjl} = {} & -\delta_{vv'} \sum_K \left\langle R^{JK}_{j'l'} \middle| \frac{d^2}{ds^2} \middle| R^{JK}_{jl} \right\rangle_{\gamma} - 2 \sum_K \left\langle R^{JK}_{jl'} \middle| \frac{d}{ds} \middle| R^{JK}_{jl} \right\rangle_{\gamma} \\ & \times \left\langle H_{v'} \middle| \frac{d}{ds} \middle| H_v \right\rangle_{\rho} - \delta_{jj'}\delta_{ll'} \left\{ \left\langle H_{v'} \middle| \frac{d^2}{ds^2} \middle| H_v \right\rangle_{\rho} \right. \\ & + \langle H_{v'} | \hat{V}_{\text{eff}} | H_v \rangle_{\rho} + \frac{2\mu}{\hbar^2} [E - (E_v + E_{v'})/2 - W^J_{jl} - V_{\text{tr}}(s)] \\ & \left. \times \langle H_{v'} | \eta^2 | H_v \rangle \right\} \end{aligned} \tag{19}$$

In equation (19) V_{eff} is an "effective potential" which involves metric coefficients and their derivatives. In deriving these equations, it was assumed that the local basis functions vary continuously with respect to s. However, if the adiabatic basis functions are calculated at one value of s within the sector from s to $s + h$, but do not vary with respect to s in this interval, then matrix elements in equation (18) involving first and second derivatives with respect to s of the local basis functions are zero. In the subsequent integration scheme, overlap matrices must be used at each sector boundary to adjust from one adiabatic basis to the new one in the next interval. It should also be noted that the total energy E appears in off-diagonal matrix elements in equation (19), in contrast to inelastic problems formulated in the usual $Rr\beta$ coordinates. The $\eta^{1/2}$ scale factor in equation (14) has the advantage that $\mathbf{V}^J$ in equation (19) is symmetric, if the adiabatic basis functions are independent of s within each sector.[29] In their $H + H_2$ studies, Elkowitz and Wyatt[12] used a different scale factor which did not lead to a symmetric coupling matrix.

Close-coupled equations may be similarly derived for the translational functions $\{g^{J\omega}_{jvK}(R)\}$ in equation (15). For the slightly more general case where a circular-arc reference curve is used in the polar region, with a wave function scale factor of $(\eta/R^2r^2)^{1/2}$ as in equation (14), and vibrational functions which are independent of s within each sector, we obtain the equations

$$\frac{d^2 g^{J\omega}_{vjK}}{dR^2} = \frac{2\mu}{\hbar^2}\sum_{v'j'}\left\langle vjK\left|\eta^2\left\{V_{vb} + V_{tr} + V_{bend} + V'_{eff}\right.\right.\right.$$
$$\left.\left.\left. + \frac{\hbar^2}{2\mu}\left[\frac{j(j+1)}{r^2} + \frac{J(J+1) + j(j+1) - 2K^2}{R^2}\right] - E\right\}\right|v'j'K\right\rangle$$
$$\times\, g^{J\omega}_{v'j'K} - \sum_{v'}\left\langle H_v\left|\frac{\eta^2}{R^2}\right|H_{v'}\right\rangle(\lambda^+_{JK}\lambda^+_{jK}g^{J\omega}_{v'jK+1} + \lambda^-_{JK}\lambda^-_{jK}g^{J\omega}_{v'jK-1}) \tag{20}$$

where $\lambda^{\pm}_{jK} = [j(j+1) - K(K\pm 1)]^{1/2}$, and where V'_{eff} is another effective potential. The first summation over $v'j'$ involves vibration–rotation coupling, but not K coupling, between translational functions labeled by JK. Kinematic coupling in the second sum mixes functions of the same j, with $K' = K \pm 1$. This type of coupling does not appear explicitly in equation (19) because it has already been incorporated in the generation of the hindered asymmetric-top functions Ω^{JM}_{jl} and energies W^J_{jl}.

4.2. *Propagation Equations*

The system of N close-coupled equations for the translational wave functions in equation (17) may be rewritten as a single matrix equation for a column vector containing the translational wave functions and their first derivatives[30]:

$$\frac{d}{ds}\begin{bmatrix}\mathbf{T}\\ \mathbf{T}'\end{bmatrix} = \begin{bmatrix}\mathbf{0} & \mathbf{1}\\ \mathbf{V} & \hat{\mathbf{V}}\end{bmatrix}\begin{bmatrix}\mathbf{T}\\ \mathbf{T}'\end{bmatrix} = \mathbf{A}\begin{bmatrix}\mathbf{T}\\ \mathbf{T}'\end{bmatrix} \tag{21}$$

where $\mathbf{A}$ is of dimension $2N \times 2N$, and $\mathbf{T}' = d\mathbf{T}/ds$. The coupling matrix $\hat{\mathbf{V}}$ contains matrix elements involving first derivatives of the adiabatic basis functions with respect to s. The local translational propagator over a small interval h is defined by the equation

$$\begin{bmatrix} \mathbf{T} \\ \mathbf{T}' \end{bmatrix}_{s+h} = \mathbf{P}(s+h, s) \begin{bmatrix} \mathbf{T} \\ \mathbf{T}' \end{bmatrix}_{s} \tag{22}$$

and the global propagator over a sequence of small intervals is assembled through multiplication of the local step propagators:

$$\begin{bmatrix} \mathbf{T} \\ \mathbf{T}' \end{bmatrix}_{s_m} = \prod_{i=1}^{m-1} \mathbf{P}(s_i + h, s_i) \begin{bmatrix} \mathbf{T} \\ \mathbf{T}' \end{bmatrix}_{s_1} \tag{23}$$

From equations (21) and (22), the local propagators satisfy the equations

$$\frac{d}{ds} \mathbf{P}(s_i + s, s_i) = \mathbf{A}(s)\, \mathbf{P}(s_i + s, s_i) \tag{24}$$

subject to the *initial condition* $\mathbf{P}(s_i, s_i) = \mathbf{1}$. The local and global propagators are energy dependent, but they do not depend upon the asymptotic *boundary conditions* (the S matrix) placed upon the scattering wave function. They can thus be calculated (at each value of E, J, and other conserved quantities) and stored for subsequent calculation of the S matrix.

If each arrangement tube is segmented into a series of boxes within which the local adiabatic basis is regarded as s independent, then $\hat{\mathbf{V}} = \mathbf{0}$ in equation (21), and matrix elements in equation (19) involving second derivatives of basis elements in $\mathbf{V}$ are also zero. At the boundaries between adjacent sectors, forcing continuity of the wave function and its normal derivative introduces an overlap matrix between adiabatic basis functions in adjacent sectors. The local propagator to advance from the right side of sector i to the right side of sector $i + 1$, including the sector-to-sector overlap matrix $\mathbf{O}(i + 1, i)$ is then

$$\begin{bmatrix} \mathbf{O}(i+1, i) & \mathbf{0} \\ \mathbf{0} & \mathbf{O}(i+1, i) \end{bmatrix} \mathbf{P}_i \tag{25}$$

where $\mathbf{P}_i$ is the propagator within sector i (right edge to left edge). The global propagator is again assembled through multiplication of the local propagators, as in equation (23).

The matching overlap matrices are slightly different from those in equation (25) at the Cartesian–polar boundary in Figure 1. The continuity conditions

$$\Psi_{\text{polar}} = \Psi_{\text{Cartesian}}$$

$$\frac{1}{\eta} \frac{\partial \Psi_{\text{polar}}}{\partial s} = \frac{\partial \Psi_{\text{Cartesian}}}{\partial s} \tag{26}$$

with the wave functions scaled as in equation (14) leads to the Cartesian–

polar overlap matrix

$$\begin{bmatrix} \mathbf{T} \\ \mathbf{T}' \end{bmatrix}_{\text{polar}} = \begin{bmatrix} \mathbf{C}_1 & \mathbf{0} \\ \mathbf{0} & \mathbf{C}_2 \end{bmatrix} \begin{bmatrix} \mathbf{T} \\ \mathbf{T}' \end{bmatrix}_{\text{Cartesian}} \tag{27}$$

where (if the vibrational and rotational basis is the same in the two sectors surrounding the match surface)

$$\begin{aligned} (\mathbf{C}_1)_{\omega\omega'} &= \delta_{jj'}\delta_{ll'}\langle H_v|\eta^{-1/2}|H_{v'}\rangle_\rho \\ (\mathbf{C}_2)_{\omega\omega'} &= \delta_{jj'}\delta_{ll'}\langle H_v|\eta^{+1/2}|H_{v'}\rangle_\rho \end{aligned} \tag{28}$$

The local propagators are obtained through numerical integration of equation (24), and this is the topic to which we now turn.

4.3. Integration Techniques

Three numerical methods have been used to integrate the close-coupled equations within each arrangement tube for three-dimensional reactions. Schatz and Kuppermann[10,14] used the Gordon[31] method to generate N linearly independent solutions (and their derivatives) $\mathbf{G}^+$ in advancing from an arbitrary linearly independent set at the start of region II in Figure 1b to the end of region IV. Another set of N linearly independent solutions $\mathbf{G}^-$ is similarly formed by advancing in the reverse direction from the end of region IV to the start of region II. Proper linear combinations of $\mathbf{G}^+$ and $\mathbf{G}^-$ in all arrangement tubes are then formed to simultaneously satisfy asymptotic boundary conditions and smoothness conditions across the matching surfaces.

The Magnus method[30] was used by Elkowitz and Wyatt[12] in studies of the three-dimensional $H + H_2$ reaction, following extensive applications of this method to collinear and coplanar reactions by Light and co-workers.[6-8,30] In the first Magnus approximation, the solution to equation (24) for the local propagator is the exponential matrix

$$\mathbf{P}(s + h, s) = e^{h\mathbf{A}(s+h/2)} \tag{29}$$

With a proper choice of coordinates, all diagonal elements of $\mathbf{A}$ vanish, $\mathrm{Tr}\,\mathbf{A} = 0$, and a unimodular local propagator $[\det(\mathbf{P}) = 1]$ is obtained, if the exponential is accurately evaluated. A unimodular global propagator is a necessary but not sufficient condition for a unitary S matrix. Several methods are available for evaluation of the exponential series in equation (29). For small h, the series expansion of the exponential may be used, but this is slow because $2N \times 2N$ matrices need be multiplied. For basis functions which do not vary within each sector, the wave function scaling in equation (14) leads to a real symmetric $N \times N$ coupling matrix $\mathbf{V}$. In this case, a faster way of evaluating $\exp[h\mathbf{A}]$ which only involves products of $N \times N$ matrices is possible.[15] In addition, $\mathbf{V}$ may be diagonalized[30] (which is equivalent to converting to a new local, energy-dependent internal basis). In the new local adiabatic basis (determined by the eigenvectors of $\mathbf{V}$, $\mathbf{P}$ is tridiagonal [with nonzero elements starting at positions, (1, 1), (1, $N + 1$), and ($N + 1$, 1)]. The propagation sequence then must include repeated basis transformations from sector-to-sector to keep changing into the new local adiabatic basis.

A significant recent development by Light and Walker[17] is the R-matrix propagation method. Across the propagation interval from a to b, the R matrix relates the wave functions to the derivatives at the two edges:

$$\begin{bmatrix} \mathbf{T}(a) \\ \mathbf{T}(b) \end{bmatrix} = \begin{bmatrix} \mathbf{R}_1 & \mathbf{R}_2 \\ \mathbf{R}_3 & \mathbf{R}_4 \end{bmatrix} \begin{bmatrix} \mathbf{T}'(a) \\ \mathbf{T}'(b) \end{bmatrix} \tag{30}$$

The global R matrix for each arrangement tube is generated recursively[32] from the local R matrix for step i (s to $s + h$), and the net R matrix from the initial step up to the beginning of step i (a to s). The local R matrix for step i (denoted $\mathbf{r}_i$) is easily formed from the local step propagator $\mathbf{P}_i$, which in turn is formed by first diagonalizing $\mathbf{V}$, using the eigenvalues to compute the "internal" propagator across the sector, and using the old and new eigenvectors to compute the step-to-step overlap matrix between adiabatic channels in adjacent sectors. The R matrix method has several important advantages: (i) exponential growth in closed channels (which sometimes plagues the straightforward propagator method by causing loss of linear independence in the solution vectors, thus necessitating repeated stabilizing transformations) is avoided; (ii) large steps may be taken since h is determined by $d\mathbf{V}/ds$ and not by $d\psi/ds$; (iii) if the global $\mathbf{R}$ is symmetric, then the open-channel S matrix will be unitary, independent of the completeness of the basis. Because of these attractive features, the R matrix method will probably be extensively used in future reactive scattering calculations.

4.4. *Numerical Details*

In the numerical integration of equation (17) with the Magnus method over the total energy range from 0.50 to 0.80 eV, Elkowitz and Wyatt[12] used step sizes of about $0.02a_0$ in the polar region, and about $0.20a_0$ in the Cartesian region. For each parity, as many as 50 vibration–rotation channels were included (divided evenly between even j and odd j rotor states), with as many as five vibrational channels (of which as many as four were energetically closed). From 2 to 12 asymmetric-top states were included in each vibrational manifold. (The asymmetric-top states were built by diagonalizing $\hat{H}_{\text{rot}}$ in a free-rotor basis containing from 20 to 30 free-rotor basis functions.) Full convergence was not obtained at some higher values of E and J with these basis sets.

In their extensive $H + H_2$ calculations, Schatz and Kuppermann[10,14] used enough vibration–rotation channels to produce no more than a 1% error in flux conservation and no more than a 10% error in symmetry (for nonnegligible transitions). For $J \geq 4$, this basis consisted of 18, 18, 12, and 12 free-rotor functions in $v = 0, 1, 2$, and 3 for even parity and 12, 12, 8, and 8 free-rotor functions in these vibrational levels for odd parity. Thus the total basis set size was 60 and 40 for even or odd parity, respectively. More vibrational and rotational channels were used for $J < 4$ (see Table II in Reference 14).

In their $H + H_2$ calculations, Stechel, Walker, and Light primarily used basis sets with three vibrational functions (contracted from a basis of 8 "primitive" local harmonic oscillator functions), with 7, 5, and 3 being the maximum value of the molecular rotational quantum numbers in the three manifolds.

5. *Matching Surface Continuity and Asymptotic Boundary Conditions*

After the $2N \times 2N$ propagator $\mathbf{P}_\lambda$ or R matrix $\mathbf{R}_\lambda$ has been computed in each arrangement channel ($\lambda = \alpha, \beta, \gamma$), the scattering wave function and its normal derivative are forced to be continuous across the three-arrangement channel matching surfaces, and the S matrix is then obtained by forcing the translational wave functions to obey asymptotic boundary conditions defined in the space-fixed frame. For incoming flux in internal channel $\omega = vjl$ of arrangement α, the open-channel translational functions asymptotically become

$$T^J_{\omega'\lambda,\omega\alpha}(R_\lambda) \sim k_{\omega_\alpha}^{-1/2}\left[e^{-i(k_{\omega\alpha}R_\alpha - l\pi/2)}\delta_{\omega\omega'}\delta_{\lambda\alpha} - \left(\frac{k_{\omega\alpha}}{k_{\omega'\lambda}}\right)^{1/2} S^J_{\omega'\lambda,\omega\alpha}e^{+i(k_{\omega'\lambda}R_\lambda - l'\pi/2)}\right] \tag{31}$$

where $S^J_{\omega'\lambda,\omega\alpha}$ is an element of the S matrix, and where $k_{\omega'\lambda}$ is the asymptotic translational wave number in channel $\omega'\lambda$. The outgoing flux in channel $\omega'\lambda$ relative to the incoming flux in channel $\omega\alpha$ is the transition probability

$$P^J_{\omega'\lambda,\omega\alpha} = \frac{(\text{out})_{\omega'\lambda}}{(\text{in})_{\omega\alpha}} = |S^J_{\omega'\lambda,\omega\alpha}|^2 \tag{32}$$

The arrangement channel matching technique is illustrated schematically in Figure 5, which shows the three match surfaces $M_{\alpha\gamma}$, $M_{\alpha\beta}$, and $M_{\beta\gamma}$ separating arrangement tubes α, β, and γ. The propagators $\mathbf{P}_\alpha$, $\mathbf{P}_\beta$, and $\mathbf{P}_\gamma$ advance the translational wave functions and derivatives from the asymptotic regions A_α, A_β, and A_γ to the match regions. If, for example,

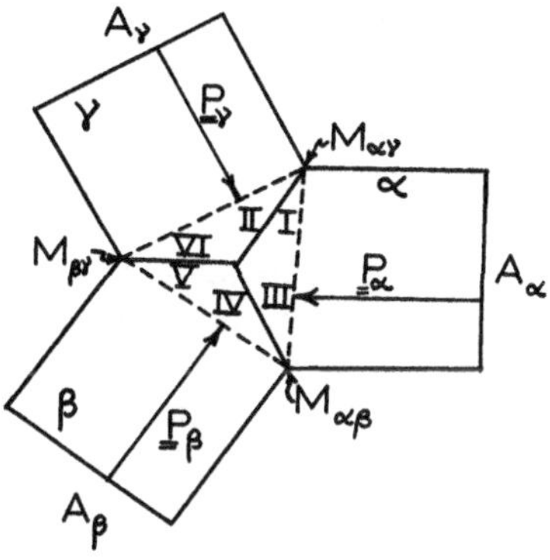

Figure 5. Schematic illustration of three arrangement tubes, with their asymptotic regions (A_λ) and tube propagators ($\mathbf{P}_\lambda$). The scattering wave functions and normal derivatives are required to be continuous across the three match surfaces ($M_{\alpha\beta}$, $M_{\alpha\gamma}$, $M_{\beta\gamma}$). (Illustration adapted from lecture notes by E. Stechel, University of Paris-Orsay, 1977.)

$\psi_{\alpha I}$ denotes the α-tube wave function projected onto $M_{\alpha\gamma}$, then continuity of the wave function and its normal derivative across this match surfaces is expressed by the equations

$$\begin{aligned} \psi_{\alpha \mathrm{I}} &= \psi_{\gamma \mathrm{II}} \\ \hat{\mathbf{n}} \cdot \nabla \psi_{\alpha \mathrm{I}} &= \hat{\mathbf{n}} \cdot \nabla \psi_{\gamma \mathrm{II}} \end{aligned} \tag{33}$$

Six equations of this type plus the asymptotic boundary conditions in equation (31) lead to an algebraic equation for the complete S matrix in the form[25]

$$S = \begin{pmatrix} S_{\alpha\alpha} & S_{\alpha\beta} & S_{\alpha\gamma} \\ S_{\beta\alpha} & S_{\beta\beta} & S_{\beta\gamma} \\ S_{\gamma\alpha} & S_{\gamma\beta} & S_{\gamma\gamma} \end{pmatrix} = \mathbf{K}^{+1/2}\mathbf{W}^{-1}\mathbf{W}^{*}\mathbf{K}^{-1/2} \tag{34}$$

where $\mathbf{K}$ is a diagonal matrix of channel wave numbers, and where $\mathbf{W}$ contains arrangement tube propagators or R matrices as well as arrangement channel match matrices.

Because of the high symmetry in the $H + H_2$ reaction, Elkowitz and Wyatt[12] were able to obtain the S matrix by first calculating (at each J and total parity) separate propagators for even j and odd j molecular rotor states for one arrangement tube (all three tubes are of course identical for this reaction), followed by explicit consideration of matching on only one of the three match surfaces. (Wolken and Karplus[11] and Schatz and Kuppermann[10,14] also took advantage of these symmetries in their calculations.) The α-tube internal wave function was expanded in hindered-rotor functions $R_{jl}^{JK\alpha}(\gamma; s)$, each of which becomes localized near region I ($\gamma = 0$) *and* region III ($\gamma = \pi$) in Figure 5. The portion of the wave function in region I, $\psi_{\alpha I}$, was expanded in localized oscillator (bending) functions on the $s = 0$ match surface (see Figure 3a), which in turn were reexpanded in product tube hindered-rotor functions $R_{jk}^{JK\gamma}(\gamma; s)$ at $s = 0$ on the γ tube. The overlap matrix between the delocalized hindered-rotor functions in the α-channel with those in the γ-channel can be written down analytically. This is made possible because: (a) the match surface in NCC occurs at *one* value of the propagation variable, $s = 0$, and (b), the hindered-rotor functions have simple localization properties with respect to the two reaction paths.

In their studies of the three-dimensional $H + H_2$ reaction, Schatz and Kuppermann[14] utilized the following method: scattering wave functions in sector IV of each arrangement tube (see Figure 1b) were reexpanded in terms of a complete set of basis functions $T_{vjK}^{\lambda\lambda'}(\zeta, \beta)$ in the two variables ζ and β (defined earlier in Section 2.4) on the matching surface $M_{\lambda\lambda'}$ (Sections 4.2 and 4.3 in Reference 14). (Note that on $M_{\alpha\gamma}$, for example, the angle η in Figure 1b is not constant, but is a function of β.) The vibration–bend basis functions were defined in the form

$$T_{vjK}^{\lambda\lambda'}(\zeta, \beta) = \phi_v(\zeta)\, D_{jK}^{\lambda\lambda'}(\beta) \tag{35}$$

where the angular functions, which form a complete set in the interval $0 \le \beta < \pi/2$, are, for example,

$$D_{jK}^{\lambda\lambda'}(\beta) = \begin{cases} 2^{1/2} P_j^K(\cos\beta), & j + K \text{ odd} \\ 0, & j + K \text{ even} \end{cases} \tag{36}$$

Incompleteness of the angular basis at $\beta = \pi/2$ is probably not significant since the high potential in close perpendicular approaches of H to H_2 prevents the scattering wave function from penetrating there.

6. Selected Results

6.1. Threshold Energies: $H + H_2$

In Figure 6 the low-energy behavior of the reaction probability, the cross length, and the cross section for the hydrogen exchange reaction in one, two, and three physical dimensions is shown. An interesting trend is observed: the threshold energy E_{th} (where P or σ is about 0.01) increases by close to 0.06 eV for each increase in dimension: $\sigma(3D) - \sigma(2D) \approx \sigma(2D) - P(1D) = 0.06$ eV. This trend directly correlates with the increasing bending zero-point energy at the transition state[13]: 0.00 eV in one dimension, 0.06 eV in two dimensions, and 0.12 eV in three dimensions. Buildup of the bending zero-point energy on the approach to $s = 0$ lessens the energy available for translation, thus increasing the total energy required to reach a specified value of the reactivity. A similar trend has been observed in one-dimensional and three-dimensional $F + H_2$ calculations: the threshold for the low-energy resonance $v = 0 \rightarrow v' = 2$ is about 0.08 eV higher for the three-dimensional calculation than for the one-dimensional model.[15] This difference (0.08 eV) is close to the bending zero-point energy of the linear FHH complex.

Figure 6. Reaction probability for the collinear $H + H_2$ ($v = 0$) reaction cross length (a_0) for the coplanar[10] $H + H_2$ ($v = 0$, $j = 0$) reaction, and cross section (a_0^2) for the three-dimensional[13] reaction. E is the total energy ($E_{trans} + 0.27$ eV). Note how the threshold energy increases by about 0.06 eV with each increase of physical dimension. The Porter–Karplus H_3 surface[41] was used in these calculations.

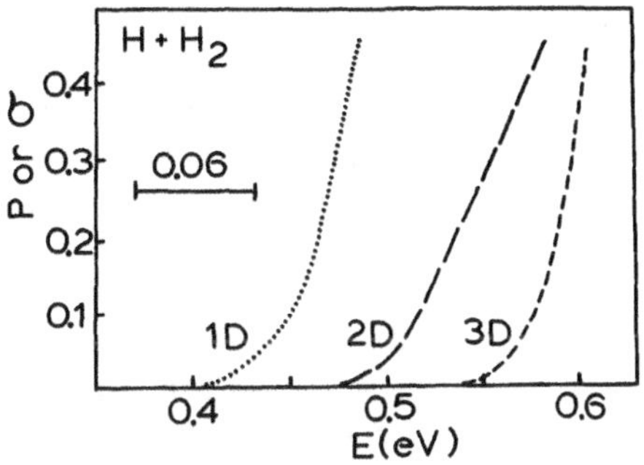

A number of other quantum results (including product rotational state distributions, product angular momentum polarization, differential cross sections, and thermal rate constants) for the three-dimensional $H + H_2$ reaction have been discussed by Schatz and Kuppermann.[14]

6.2. *Resonance Effects: $H + H_2$*

A very interesting feature in some reactive collisions, including most exoergic reactions, is the possibility of internal excitation resonances, the result of which is a rapid alteration in the reaction probability over a small energy range. Such Feshbach resonances arise, in a rough sense, when the system temporarily excites to a higher internal (e.g., vibration–rotation) state, thereby delaying the translational motion along the reaction coordinate. Resonances of this type are anticipated when, in a plot of the adiabatic channel energies vs. s, one or more channels which are closed asymptotically "open" temporarily over a range s. Near a resonance energy, these closed-channel wave functions grow and then decay rapidly in attempting to populate the temporarily open "well" in the correlation diagram.

Reactive resonances were first explicitly discussed by Levine and Wu[33] for collinear $H + H_2$ collisions. (Other resonance studies on this reaction are discussed by Truhlar and Wyatt.[2]) It is interesting to note that this potential surface, and others (e.g., $F + H_2$), broaden and then contract in the vibrational direction as the reaction zone is traversed. Temporary lowering of the local vibrational levels leads to "dynamic" wells which can support resonances. Hayes and Chapman[34,35] have recently predicted resonance energies in collinear $He + H_2^+$ collisions, at energies below the reaction threshold. Schatz and Kuppermann[16] have studied the low-energy resonance structure in collinear, coplanar, and three-dimensional $H + H_2$ collisions. The respective resonance (total) energy in $v = 0 \rightarrow v' = 1$ reactive transitions is 0.87, 0.92, and 0.97 eV, with resonance widths of 0.02, 0.04, and 0.03 eV. The coplanar and three-dimensional resonance energies are for $J = 0$, but these widths and energies are not very sensitive[16] to increases in J over the range 0–7 (these are the partial waves which contribute most strongly to the reactive cross section). Consequently, according to the calculations of Schatz and Kuppermann, this resonance is not "washed out" or extensively broadened in summing over J to obtain the total cross section. However, in the $v = 0 \rightarrow v' = 0$ channel, the resonance is washed out in the total cross section because the resonance amplitude vanishes by $J = 7$, while the nonresonant amplitude remains large up to $J \approx 17$. Thus an additional reason why the $v = 0 \rightarrow v' = 1$ resonance appears in the integral cross section is that its amplitude dominates the direct component at all J. Schatz and Kuppermann have also discussed the interference between resonant (compound state) and direct scattering processes in the vicinity of the $v = 0 \rightarrow v' = 1$ reactive resonance. The interference structure is shown as partial loops which do not enclose the origin in plots of S-matrix elements vs. energy.

In contrast, preliminary quantum results[15] on the three-dimensional $F + H_2$ reaction indicate that the relatively narrow low-energy $v = 0 \rightarrow v' = 2$ resonance in each partial wave is sensitive to J ($0 \leq J \leq 20$), so that considerable broadening may occur in the total cross section. Resonances are sensitive probes of local regions of the reaction zone, and their eventual experimental observation may provide very useful information about these regions.

6.3. *Quantized Whirlpools: $F + H_2$*

In Figure 7a is shown a perspective plot of the density associated with the quantum scattering wave function ($\rho = \psi^*\psi$) slightly above the energy of the $v = 0 \rightarrow v' = 2$ resonance in the collinear $F + H_2$ reaction. Extrema due to incoming-wave–elastically-scattered-wave interference is seen in the entrance valley (left side of figure), while the four-peaked ($v' = 2, 3$) product density is seen emerging into the product valley (right side of figure). A flux ($\mathbf{j}$) map over the interaction region of the potential surface is shown in Figure 7b. Flux circulation near three whirlpools is clearly shown. The center of each whirlpool is a nodal point of the scattering wave function. Hirschfelder and co-workers[36,37] have shown that the flux around whirlpools of this type is quantized in the sense that the hydrodynamic circulation integral takes on restricted values:

$$\oint \frac{\mathbf{j}}{\rho} \cdot d\mathbf{r} = \frac{M\hbar 2\pi}{\mu} \qquad (M = \pm 1, \pm 2, \ldots) \tag{37}$$

where the integration path is around any closed loop surrounding the nodal point at the "center" of the vortex. In the low-energy $F + H_2$ reaction, all whirlpools seem to be in their "ground state," $|M| = 1$. Flux passing

Figure 7. (a) Scattering probability density[40] for the collinear $F + H_2$ reaction at $E_{total} = 0.375$ eV, viewed from the F + H + H region in skewed coordinates. The $F + H_2$ reactant region is at the left and the FH + H product region emerges on the right. The maximum density (atomic units) is indicated in the figure. (b) Probability current vectors at 0.375 eV. The top center of (a) is being viewed from above, with the reactant region to the lower right. The probability current is $\mathbf{j} = \hbar/2\mu i\,[\psi^*\nabla\psi - \psi\nabla\psi^*]$. In (b), the vector function $\mathrm{Im}[\psi^*\nabla\psi]$ is plotted, which is the flux in units of $\hbar/\mu$.

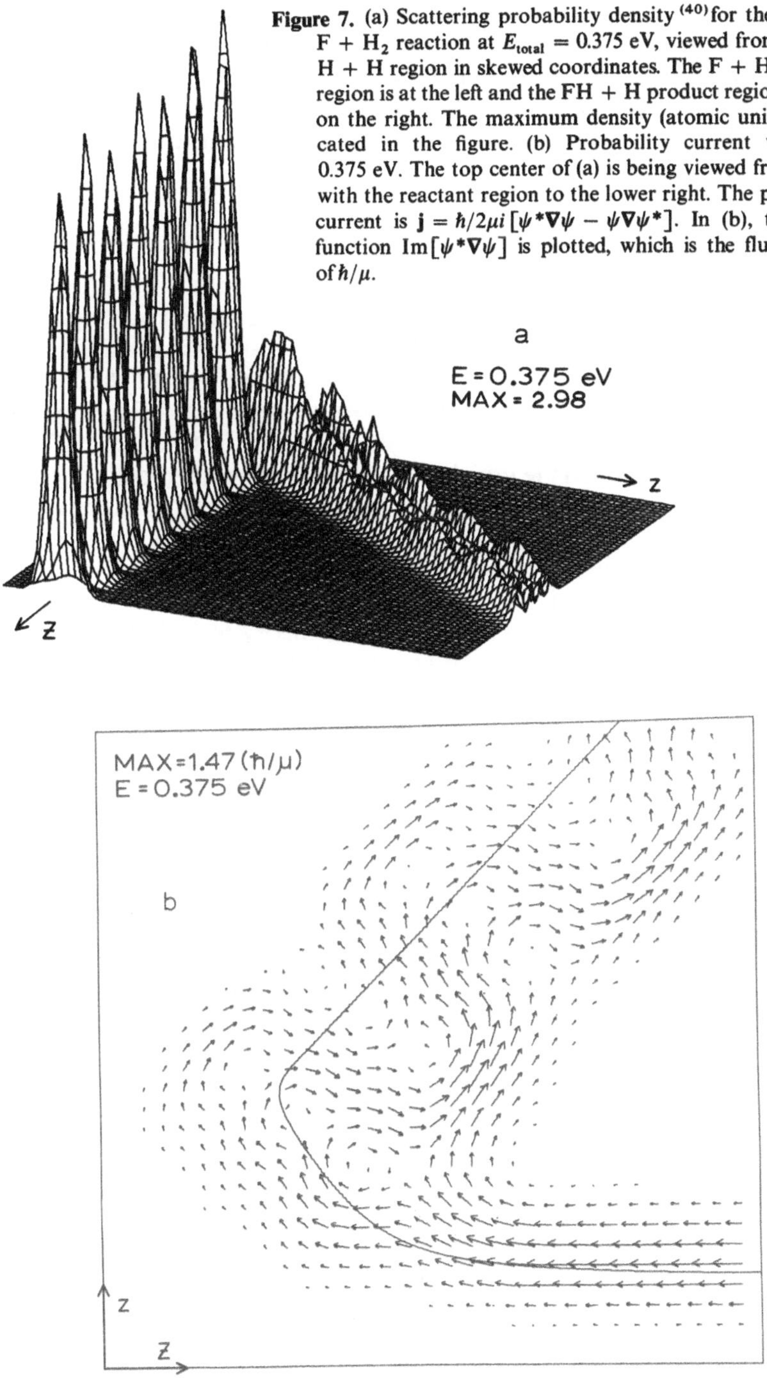

on to products must divert around these whirlpools, so that whirlpools act as "dynamic barriers" (as opposed to static potential barriers on the adiabatic potential surface) in opposing probability flow. Quantum whirlpools are probably common in three-dimensional reactive collisions although they have been studied so far only in the collinear $H + H_2$[38,39] and $F + H_2$ reactions.[40]

7. Concluding Remarks

The past several years has seen an enormous increase in our sophistication about *how to accurately calculate* dynamical features of chemical reactions from quantum mechanics. The first applications to the $H + H_2$ reaction have yielded new insights which will prove useful for approximate model calculations on other reactions, as discussed in Chapter 17. These studies (and others which are now being initiated) on state-to-state quantum reaction dynamics provide a detailed picture of energy interconversion and detailed quantum effects (threshold behavior, resonance structure, direct reaction–complex reaction competition, and transition state flux distributions) in atom–diatom collisions. The study of branching ratios in asymmetric reactions, the extension of the theory to include bent intermediate states (with the associated arrangement tube matching problem), and extensions to electronic nonadiabatic reactions, and to reactions in intense electromagnetic fields should provide exciting new results in the near future.

ACKNOWLEDGMENTS

This work was supported in part by research grants from the Robert A. Welch Foundation of Houston, Texas, and by the National Science Foundation.

For many constructive comments on the first draft of this chapter, I am very grateful to George C. Schatz, Aron Kuppermann, Robert B. Walker, and Ellen B. Stechel.

References

1. D.A. Micha, Quantum theory of reactive molecular collisions, *Adv. Chem. Phys.* **31**, 7–76 (1975).
2. D.G. Truhlar and R.E. Wyatt, History of H_3 kinetics, *Ann. Rev. Chem.* **27**, 1–43. (1976).
3. J.C. Light, Quantum theories of chemical kinetics, *Adv. Chem. Phys.* **19**, 1–31 (1971).
4. R.D. Levine, Molecular collisions and reactive scattering, in *MTP Int. Rev. of Science*, Vol. 1, W. Beyers Brown, editor, University Park Press, Baltimore, Maryland (1972).
5. T.F. George and J. Ross, Quantum dynamical theory of molecular collisions, *Ann. Rev. Phys. Chem.* **24**, 263–300 (1973).

6. R.P. Saxon and J.C. Light, Quantum calculations of planar reactive H + H_2. I. Theory, *J. Chem. Phys.* **56,** 3874–3884 (1972).
7. R.P. Saxon and J.C. Light, Quantum calculations of planar reactive H + H_2. II. Application, *J. Chem. Phys.* **56,** 3885–3897 (1972).
8. A. Altenberger-Siczek and J.C. Light, Quantum calculations of planar reactive H + H_2. III. Labeled nuclei and angular distributions, *J. Chem. Phys.* **61,** 4373–4383 (1974).
9. M. Baer, A coplanar quantum mechanical study of the exchange reaction HF + H, *J. Chem. Phys.* **65,** 493–495 (1976).
10. G.C. Schatz and A. Kuppermann, Quantum mechanical reactive scattering for planar atom plus diatom systems. II. Accurate cross sections for H + H_2, *J. Chem. Phys.* **65,** 4624–4641 (1976).
11. G. Wolken, Jr. and M. Karplus, Theoretical studies of H + H_2 reactive scattering, *J. Chem. Phys.* **60,** 361–367 (1974).
12. A.B. Elkowitz and R.E. Wyatt, Three-dimensional natural coordinate asymmetric top theory of reactions: Application to H + H_2, *J. Chem. Phys.* **63,** 702–721 (1975).
13. A.B. Elkowitz and R.E. Wyatt, Quantum mechanical reaction cross sections for the three-dimensional hydrogen exchange reaction, *J. Chem. Phys.* **62,** 2504–2506 (1975).
14. G.C. Schatz and A. Kuppermann, Quantum mechanical reactive scattering for three-dimensional atom plus diatom systems. II. Accurate cross sections for H + H_2, *J. Chem. Phys.* **65,** 4668–4682 (1976).
15. M.J. Redmon and R.E. Wyatt, Computational methods for reactive scattering, *Int. J. Quantum Chem. Symp. No.* **11,** 343–351 (1977).
16. G.C. Schatz and A. Kuppermann, Dynamical resonances in collinear, coplanar, and three-dimensional quantum mechanical reactive scattering, *Phys. Rev. Lett.* **35,** 1267–1270 (1975).
17. J.C. Light and R.B. Walker, An *R*-matrix approach to the solution of coupled equations for atom–molecule reactive scattering, *J. Chem. Phys.* **65,** 4272–4282 (1975).
18. F. Rebentrost and W.A. Lester, Nonadiabatic effects in the collision of $F(^2P)$ with $H_2(^1\Sigma_g^+)$. III. Scattering theory and coupled channel computation, *J. Chem. Phys.* **67,** 1302–1311 (1977).
19. D.L. Miller and R.E. Wyatt, Electronuclear basis for chemical reactions, *J. Chem. Phys.* **67,** 1302–1311 (1977).
20. P.L. DeVries and T.F. George, Quantum mechanical theory of structured atom-diatom collision system: $A + BC(^1\Sigma)$, *J. Chem. Phys.* **67,** 1293–1301 (1977).
21. Z.H. Top and M. Baer, Incorporation of electronically nonadiabatic effects into bimolecular reactive systems. II. The collinear ($H_2 + H^+$, $H_2^+ + H$) system, *Chem. Phys.* **25,** 1–18 (1977).
22. J.M. Bowman, S. Leasure, and A. Kuppermann, Large quantum effects in a model electronically nonadiabatic reaction: Ba + N_2O → BaO* + N_2, *Chem. Phys. Lett.* **43,** 374–376 (1976).
23. R.A. Marcus, Analytical mechanics of chemical reactions. III. Natural collision coordinates, *J. Chem. Phys.* **49,** 2610–2616 (1968).
24. G.C. Schatz and A. Kuppermann, Quantum mechanical reactive scattering for three-dimensional atom plus diatom systems. I. Theory, *J. Chem. Phys.* **65,** 4642–4667 (1977).
25. R.B. Walker, J.C. Light, and A. Altenberger-Siczek, Chemical reaction theory for asymmetric atom–molecule collisions, *J. Chem. Phys.* **64,** 1166–1181 (1976).
26. A. Kuppermann, A useful mapping of triatomic potential energy surfaces, *Chem. Phys. Lett.* **32,** 374–375 (1975).
27. S.H. Harms and R.E. Wyatt, Hindered asymmetric top wave functions for three-dimensional H + H_2, *J. Chem. Phys.* **62,** 3173–3180 (1975).
28. S.H. Harms, A.B. Elkowitz, and R.E. Wyatt, Asymmetric top states for chemical reactions: Angular representations, avoided crossings, and bifurcation, *Mol. Phys.* **31,** 177–188 (1976).

29. J.C. Light and R.B. Walker, Hermitian quantum equations for scattering in reaction coordinates, *J. Chem. Phys.* **65**, 1598–1599 (1976).
30. J.C. Light, Quantum calculations in chemically reactive systems, *Methods Comput. Phys.* **10**, 111–143 (1971).
31. R. Gordon, Quantum scattering using piecewise analytic solutions, *Methods Comput. Phys.* **10**, 81–110 (1971).
32. D.J. Zvijac and J.C. Light, *R*-matrix theory for collinear chemical reactions, *Chem. Phys.* **12**, 237–251 (1976).
33. R.D. Levine and S.F. Wu, Resonances in reactive collisions: Computational study of the $H + H_2$ reaction, *Chem. Phys. Lett.* **11**, 557–561 (1971).
34. F.M. Chapman and E.F. Hayes, Resonances in the collinear inelastic scattering of He by H_2^+ below the reaction threshold, *J. Chem. Phys.* **62**, 4400–4403 (1975).
35. F.M. Chapman and E.F. Hayes, Open and closed channel resonances in the collinear inelastic scattering of He by H_2^+, *J. Chem. Phys.* **65**, 1032–1034 (1976).
36. J.O. Hirschfelder, C.J. Goebel, and L.W. Bruch, Quantized vortices around wavefunction nodes. II, *J. Chem. Phys.* **61**, 5456–5459 (1974).
37. J.O. Hirschfelder and K.T. Tang, Quantum mechanical streamlines. IV. Collision of two spheres with square potential wells or barriers, *J. Chem. Phys.* **65**, 470–486 (1976).
38. E.A. McCullough, Jr. and R.E. Wyatt, Dynamics of the collinear $H + H_2$ reaction. I. Probability density and flux, *J. Chem. Phys.* **54**, 3578–3591 (1972).
39. A. Kuppermann, J.T. Adams, and D.G. Truhlar, Streamlines of probability current density and tunneling fractions for the collinear $H + H_2$ reaction, *ICPEAC Abstracts* **VIII**, 149–151 (1973).
40. R.E. Wyatt, Quantum mechanics of neutral atom–diatomic molecule reactions, in *State-to-State Chemistry*, P.R. Brooks and E.F. Hayes, editors, *ACS Symp. Series* **56**, 185–205 (1977).
41. R.N. Porter and M. Karplus, Potential energy surface for H_3, *J. Chem. Phys.* **40**, 1105–1115 (1964).

18
Direct-Mode Chemical Reactions II: Classical Theories

DONALD G. TRUHLAR and DAVID A. DIXON

1. Introduction

Other chapters of this book discuss accurate and approximate quantum-mechanical treatments of chemical reactions. Most of our understanding of chemical reactions, however, is based on classical models—both trajectory calculations (exact classical mechanics) and simpler classical models (approximate trajectories). Computational aspects of classical trajectory methods as applied to atom–molecule reactive collisions are discussed in Chapter 16 of this book. The present chapter emphasizes physical aspects of exact and approximate classical dynamical calculations for direct-mode chemical reactions. In particular it discusses the dependence of calculated reaction attributes on features of the potential energy hypersurfaces and the reliability of calculated reaction attributes. The reliability of the calculated results depends not only on the accuracy of the potential surface employed but also on the dynamical errors in the calculations. When numerically integrated exact classical trajectories are employed the dynamical approximation is simply the classical propagation of the motion. When simplified models are employed there are additional dynamical approximations whose validity must be examined carefully for the reaction or class of reactions under consideration. Statistical theories are sometimes applied to direct-mode reactions but are more generally applicable to complex-mode reactions, and are discussed in that context in Chapter 19. Thus the present chapter emphasizes the use of trajectory methods and nonstatistical approximate classical methods. Quasiclassical quantization of initial or final conditions of the trajectories,[1] semiclassical quantization

DONALD G. TRUHLAR and DAVID A. DIXON • Chemical Dynamics Laboratory, Department of Chemistry, University of Minnesota, Minneapolis, Minnesota 55455.

at both ends and inclusion of interference effects (as in classical S matrix theory),[2,3] and propagation of a whole wave packet centered at a classical trajectory[4] are ways to try to minimize the error caused by using classical mechanics. Of these techniques the quasiclassical method is the only one which is always straightforwardly applicable or at least which has already been well documented to be so. In keeping with the handbook character of this book, this is the trajectory technique that is emphasized in this chapter.

The classical treatment of chemical reactions is based solidly on the Born–Oppenheimer separation[5] of electronic and nuclear motions. According to this approximation, the potential energy for motion of the nuclei is the fixed-nuclei electronic energy as a function of nuclear positions plus internuclear electrostatic interactions. This is called the electronically adiabatic potential energy hypersurface or, for brevity, the potential surface. According to the Born–Oppenheimer approximation, the electronic state does not change during a collision, i.e., all collisions are electronically adiabatic. Even for nonadiabatic collisions, however, detailed theoretical progress is usually possible only when the collision can be visualized as occurring by intervals of propagation governed by single adiabatic potential surfaces and fairly localized regions of strong pairwise interaction of these surfaces.* The adiabatic electronic states are coupled by the internuclear motions,[10] and in the regions of strong nonadiabatic coupling a quantum-mechanical or semiclassical probability of change of electronic state must be considered; such a change is called a nonadiabatic transition, a diabatic passage, or a surface hopping.[11]

Thus the first question that must be answered in beginning to consider a chemical reaction is how many potential energy surfaces are involved in any important way. If more than one surface may be important one must understand their general natures and the nature of their nonadiabatic coupling. Only when these features of the problem have been at least qualitatively understood is it possible to intelligently decide on the further details of an appropriate classical model. In Section 2 of this chapter we discuss some approaches which can profitably be taken to obtain this initial level of qualitative understanding. In Sections 3 and 4 we discuss simple classical models which can be used to gain a first level of detailed or quantitative understanding of integral cross sections, angular differential cross sections, and energy disposal. In Sections 3 and 5 we discuss the use of one or more full potential energy surfaces and exact classical trajectories to try to model the reaction as quantitatively as possible with classical mechanics.

* Even so it is sometimes more convenient or even necessary to use diabatic (i.e., nonadiabatic) or quasiadiabatic potential surfaces to describe certain systems.[6-9] When we refer to potential surfaces or electronic states in this chapter, we mean the Born–Oppenheimer adiabatic ones.

2. Determination of Relevant Potential Surfaces

The number of surfaces that play an important role in the reaction dynamics must be characterized as this allows one to properly choose the dynamical model best suited to the problem. The number of surfaces that play a role in the entrance channel can be ascertained from the collision energy and from the electronic states of the reactants. For the simplest neutral atom–diatom collision, $H + H_2 \rightarrow H_2 + H$, the atom has a ^2S_u ground state, while the molecule has a $^1\Sigma_g^+$ ground state. The first excited state for the H atom is at 10.2 eV, while the lowest vertical triplet and singlet excitation energies of H_2 are 10.6 and 11.4 eV, respectively[(12)]; thus no excited states are accessible at thermal energies. This reaction can be expected to occur on a single potential surface in this energy regime, and any dynamical model that we choose need only involve the ground-state surface.

The first complication is the presence of electronic-state degeneracy in more than one of the reactants or a more than twofold degeneracy in either reactant. Then the degeneracy is almost always at least partially split by the atom–molecule interaction so that the system's motion is governed by more than one potential surface. The major exception is that all states of a system with an odd number of electrons are at least doubly degenerate in the absence of a magnetic field. This is known as Kramers' degeneracy.[(13)] Thus the degeneracy is not split in the $H + H_2$ system. When the degeneracy is split, but in the absence of nonadiabatic transitions, the probability of a collision being governed by any one surface is calculable from simple statistical factors. If nonadiabatic transitions occur these factors only determine the probability that a trajectory will start on a given surface. A simple example of this kind is the reaction $F + H_2$. Neglecting for a moment any effects due to spin–orbit coupling, a ground-state F atom is in a ^2P_u state. Since the H_2 ground state has symmetry $^1\Sigma_g^+$, the overall spin function is a doublet. The splitting of the spatial degeneracy depends on the direction of approach of reactants (and symmetry of the system) as follows: collinear ($C_{\infty v}$), $^2\Sigma^+$ and $^2\Pi$; perpendicular bisector (C_{2v}), $^2A_1$, $^2B_1$, and $^2B_2$; and any other (C_s), two $^2A'$ and one $^2A''$. At large separation these states are degenerate, but as the reactants approach they will split, and simple molecular orbital arguments can be employed to see how this occurs. Approach, for example, with $C_{\infty v}$ symmetry will involve the H_2 interacting most strongly with an F orbital containing a single electron or an F orbital containing two electrons (Figure 1a). The former approach leads to a $^2\Sigma^+$ surface with three electrons in orbitals that correlate with those of the ground-state products $H(^2S_u) + HF(^1\Sigma^+)$. The latter approach will initially be repulsive and correlates with excited states of the product. These are energetically inaccessible in thermal experiments so that only one surface is important for the reaction in the absence of

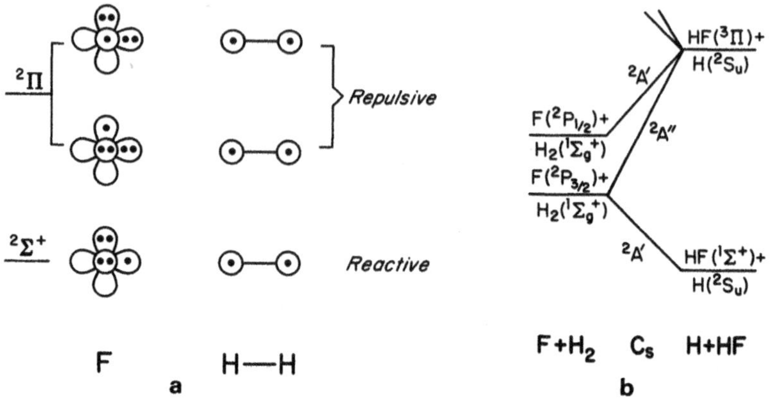

Figure 1. Orbital and state correlations for the reaction $F + H_2 \rightarrow FH + H$. (a) Qualitative orbitals for the $C_{\infty v}$ (collinear) path. For F only the $2p$ orbitals are shown, two in the plane and one out of the plane, and for H the $1s$ orbitals are shown. In the Π state, an orbital containing two electrons approaches the closed shell H_2, yielding a repulsive interaction. The reactive Σ state involves the interaction with H_2 of an orbital with only one electron. (b) State correlation diagram in C_s (planar) symmetry. Note that the spin–orbit coupling splits the ^2P_u state of F into $^2P_{3/2}$ and $^2P_{1/2}$. The product Π state correlates with both reactant spin–orbit states. In correlation diagrams the vertical direction corresponds to energy, and the horizontal direction corresponds to a generalized reaction coordinate.

nonadiabatic transitions and spin–orbit coupling.(14) The spin–orbit splitting of the ^2P_u state of F into two states separated by 0.050 eV makes things more complicated.(14,15) The correlation of the spin–orbit states with the triatomic molecular states is shown in Figure 1b. In order to interpret results for a system with a thermal population of F atoms both states must be included, and this should be treated as a two- or three-surface problem (see Section 3).(16,17) In any event the proper statistical factor for each possible electronic state of the atom–molecule system is the first theoretical datum needed to treat the collisions.

In a similar fashion electronic states for the products must be examined and their energies must be compared with the quantity $E_{\text{collision}} + E_{\text{exo}}$, where $E_{\text{collision}}$ is the sum of the relative translational energy E_{rel} and the internal excitation energy of the reactants, and E_{exo} is the ground-state-to-ground-state exoergicity of the reaction. For very exothermic reactions or for high-energy collisions, it is possible to produce electronically excited atomic or molecular states in the products; this may lead to chemiluminescence. As an example, consider the reaction $H + I_2 \rightarrow HI + I$ for which E_{exo} is 1.6 eV. This energy is larger than the $^2P_{3/2}$–$^2P_{1/2}$ spin–orbit splitting of 0.94 eV for the ^2P_u ground state of I, and the $^2P_{1/2}$ excited state could be populated by the reaction. For a system where experimental results are not known, the excited surface leading to this product should be included in the model. However, one often has recourse to experimental results which provide information about whether a given state must be included. For many reactions producing atomic halogen products the excited atomic state of the halogen has not been observed; however, in the F + HBr reaction it has been observed,[18] and it would have to be included.

When spectroscopic or thermochemical data about the ground or excited states of the intermediate are available, even more information about which surfaces should be considered can be gained from electronic-state correlation diagrams.[19-22] Consider the reaction $O + H_2 \rightarrow OH + H$ which could conceivably pass through a stable intermediate, $H_2O(^1A)$.[23] We again neglect spin–orbit coupling. The ground states of the reactants are $O(^3P_g) + H_2(^1\Sigma_g^+)$, which correlate in C_{2v} symmetry with the first excited state of $H_2O(^3B_1)$, which is unbound and with two higher energy states (see Figure 2). Although the $^3B_1$ surface correlates with products $OH(^2\Pi) + H(^2S_u)$, the deep well due to $H_2O(^1A_1)$ is not available and the reaction should proceed in a direct mode. A similar correlation of ground-state reactants and products with the lowest-energy triplet state of H_2O also holds for a $C_{\infty v}$ approach. The first excited state of the reactants corresponds to $O(^1D) + H_2(^1\Sigma_g^+)$ and 1/5 of the collisions involving this pair of reactant states would begin on the ground-state $(^1A_1)$ surface of H_2O. For C_{2v} geometries this surface correlates with excited-state products $OH(^2\Sigma^+) + H(^2S_u)$. However, an excited state of $H_2O(^1A_1)$ does correlate with ground-state products in C_{2v} symmetry, and in C_s symmetry these

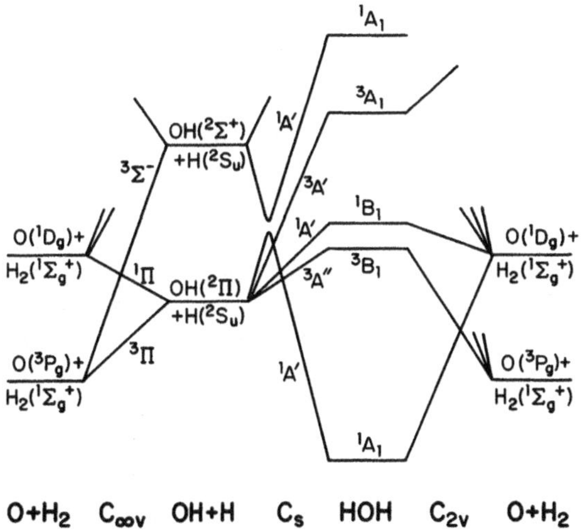

Figure 2. State correlation diagram for the reaction $O + H_2 \rightarrow OH + H$. Reactant states are shown correlating to product states via a $C_{\infty v}$ (collinear) path or a combination C_{2v} and C_s path through the water intermediate of C_{2v} symmetry. Note the avoided crossing in the C_s portion of the path.

two $^1A_1$ states become $^1A'$ and exhibit an avoided crossing in the exit channel. Another excited-state reactant channel ($^1B_1$) does correlate directly to ground-state products, and thus in the absence of any other electronic information both states should be considered in a dynamical study of ^1D oxygen atoms. Further complications arise in the reaction $O + Br_2 \rightarrow OBr + Br$, in which ground state $O(^3P_g)$ cannot adiabatically gain access to the well of a stable intermediate $OBr_2(^1A_1)$, just as in the O + H reaction. However, at low collision energies, the reaction shows the behavior exhibited by a long-lived complex. Any dynamical model proposed to explain this must either (i) involve spin–orbit coupling or nonadiabatic transitions and at least two surfaces, or (ii) involve reaction through a bound triplet intermediate. Further progress in understanding the reaction can be obtained only from detailed experimental results or electronic structure calculations.[24-26]

In many cases the information obtained from state correlation diagrams can be augmented by considering orbital correlations[22,27-29] for the same reaction. Mahan has applied both methods to the reaction $C^+(^2P_u) + H_2(^1\Sigma_g^+) \rightarrow CH^+ + H$.[29] The question to be answered is: Are the dynamics governed by the two states ($^2A_1$ and $^2B_1$) of CH_2^+ which are bound with respect to reactants? As for $F + H_2$, the ground-state reactants interact in C_s symmetry according to two $^2A'$ surfaces and one $^2A''$ surface. For C_{2v} geometries these become $^2A_1$, $^2B_2$, and $^2B_1$, respectively. The orbital correlations for the $^2A_1$ and $^2B_2$ surfaces show the behavior expected for a forbidden reaction, viz., occupied orbitals of the reactants correlate with excited orbitals of CH_2^+. For C_{2v} approaches, these surfaces should be repulsive. For the $^2B_2$ surface, however, the ground-state orbitals correlate with the ground-state orbitals of the $^2B_2$ state of CH_2^+, which has two bonding electrons (the same number as the reactants), no antibonding electrons, and all the rest nonbonding electrons. Thus for C_{2v} approaches, this surface might be expected to be about flat. For less symmetric approaches, however, the $^2A_1$ and $^2B_2$ states correlate with two $^2A'$ states. Since these states have the same symmetry they exhibit an avoided crossing. As a result the lowest-energy $^2A'$ state might be expected to be flat at large distances and bound at small distances. This provides a possible low-energy path for the system to reach the deep potential well of ground-state CH_2^+. These arguments based on correlation diagrams are supported by *ab initio* electronic structure calculations[30] and experiments.[31,32]

Another example of the use of orbital correlation diagrams is the reaction $H + Cl_2$ for which sideways insertion is predicted to be a high-energy process.[33] Application of orbital correlations to collinear approaches in atom–diatom collisions is generally more ambiguous; the principle of conservation of orbital symmetry[27] cannot be employed as no symmetry elements bisect bonds that are being made and bonds that are being broken. Application of the orbital phase continuity principle[34] to the reaction $H + H_2 \rightarrow H_2 + H$ does suggest that the reaction is allowed; however, no prediction of the barrier height can be made from such simple orbital correlations. Orbital correlation diagrams are most useful in atom–molecule reactions for eliminating geometries from consideration at low collision energies, but they are only qualitative and should be used with discretion.

Although orbital and state correlations do not predict barrier heights some characteristics of the potential surface can be deduced on the basis of other simple generalizations. In general, exoergic reactions tend to have their barriers, if any, located in the entrance channel, while the barriers for endoergic reactions tend to be located in the exit channel.[35] Other useful generalizations are available,[36] e.g., more exoergic reactions tend to have smaller intrinsic barriers. Simple estimates of the thermodynamics of possible intermediates and transition states can be gained from spectroscopic observations of the intermediate, as in ion–molecule reactions, or by using the group additivity method[37] together with general structural considerations. Other features of the surface can also be ascertained from general electronegativity considerations. For example, in the lowest-energy configurations of trihalogens the least electronegative atom is in the center[38] and, thus in the reaction of Cl + IBr, ICl may be expected to be formed preferentially over the more exoergic product BrCl. Experimentally this is the result.[39] Such electronegativity arguments have been used in discussing the reactions of oxygen[24-26] and hydrogen[40] atoms and CH_3 radicals[41] with the interhalogens.

Further understanding of the properties of the potential energy surface can come from simple molecular orbital theories. The bond-breaking and bond-making portions of many atom–molecule reactions are dominated by covalent forces, as in the reactions $H + X_2$ and $X + H_2$ where X is a halogen. In other cases, e.g., $M + X_2$ and M + RX where M is an electropositive metal and R is an alkyl group, the Coulombic forces are dominant. The latter group of reactions can be qualitatively described by an electron jump model,[42] as discussed in Section 4 below. Reactions dominated by covalent forces can be usefully described in terms of molecular orbital theory. A qualitative ordering of the molecular orbitals in a transition state or intermediate for a triatomic as a function of bond angle can be obtained using Walsh's rules.[43-45] The next level of refinement requires computer calculations at the minimum-basis-set level. In order to obtain correct qualitative descriptions of the molecular orbitals and charge densities at geometries along the reaction coordinate, one needs to perform an open-shell self-consistent-field calculation.[46] If dissociation is to be studied, a generalized valence-bond or multiconfiguration self-consistent-field calculation may be required.[47] An example of an approximate minimum-basis-set method which should be sufficiently accurate for qualitative descriptions of the molecular orbitals and is also computationally efficient is the partial retention of diatomic differential overlap method (PRDDO).[48] A rapid computer program which employs this method and which is set up for proper open-shell treatment of either one- or two-configuration self-consistent-field calculations is available.[49] Although these calculations are not likely to provide quantitative information about the reaction,* they may provide a qualitative description about the electronic structure at points on the surface. The simplest example of a reaction dominated by covalent forces is the reaction $H + H_2 \rightarrow H_2 + H$, for which a minimum-basis-set linear-combination-of-atomic-orbitals description of the transition state molecular orbitals neglecting overlap is given by

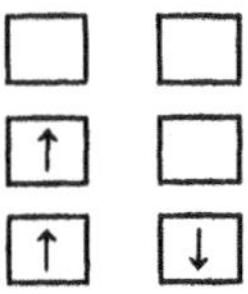

$$\psi_3 = (6)^{-1/2}(\phi_A - 2\phi_B + \phi_C)$$
$$\psi_2 = (2)^{-1/2}(\phi_A - \phi_C)$$
$$\psi_1 = (6)^{-1/2}(\phi_A + 2\phi_B + \phi_C)$$

where the boxes show the occupancy of the molecular orbitals ψ_i, and ϕ_A, ϕ_B, and ϕ_C are atomic orbitals on centers A, B, and C, respectively. These molecular orbitals are analogous to those of the allyl radical. At a very simple level, the above orbital description is approximately applicable to all colinear reactions involving doublet-state atoms with large ionization potentials and closed-shell diatomic molecules. The density of the highest occupied molecular orbital will vary with the electronegativities of the species, but detailed restricted Hartree–Fock SCF calculations on several systems show that as the reaction proceeds the density in this orbital moves smoothly from one end atom to the other without any significant density on the central atom, i.e., with the node always very close to the center atom.[51] This shows how cautious one must be of using simpler molecular orbital theories which might predict that ψ_2 looks like an antibonding orbital of BC. Such possible correlations of the reactive system's molecular orbitals with those for stable species with known geometries or with photodissociative excited states can be quite suggestive in trying to understand the shape of the potential surface and how it affects the reaction dynamics.[40] In turn they might suggest what kind of a classical model is most appropriate for treating the dynamics.

Molecular orbital theory and symmetry arguments can also be used to characterize the nature of the avoided crossings and strong interactions of potential surfaces.[52-54]

Assuming that the set of surfaces necessary to describe the reaction dynamics has been characterized, the appropriate method for treating the dynamics can be chosen. The first choice to be made is whether the reaction

* For example, for $H + H_2$, even with an essentially converged one-electron basis of Slater-type orbitals or Gaussians, a one-configuration restricted Hartree–Fock SCF calculation yields a barrier height of 1.0 eV and an eight-configuration SCF calculation yields a barrier height of 0.72 eV, as opposed to the accurate value of 0.42 eV.[50] But the single-configuration calculation gives the position and force constants of the transition state more accurately than it gives the barrier height.

proceeds via a direct mechanism or a complex. The presence of deep wells on the potential surface may lead to long-lived-complex formation. However, the energetics of the reaction, e.g., a high initial translational energy, and dynamical effects may prevent a long-lived complex from occurring, and even reactions governed by potential surfaces with wells may proceed by a direct mechanism. Reactions that do proceed by complexes usually exhibit forward–backward symmetry in the center-of-mass angular scattering distribution, and often show a statistical partitioning of the product energy.[55] Such reactions are treated in Chapter 19.

Reactions involving excited-state reactants or highly degenerate reactants can become very complicated. For example, the reaction $O(^3P_g) + H_2^+(^2\Sigma_g^+) \rightarrow OH^+ + H$ may be considered to involve excited-state reactants since $O^+(^4S_u) + H_2(^1\Sigma_g^+)$ lies about 2 eV lower in energy. There are several adiabatic pathways between the former reactants and $OH^+ + H$ products. But the experimental low-energy cross section appears to be too large to be explained in terms of electronically adiabatic collisions.[56] Another example is provided by the reactions of La and Sc atoms with O_2. Several low-lying states of the atoms may be excited even under thermal conditions, the states are appreciably split by spin–orbit coupling, and the reactions are so exothermic that LaO can be formed in $^2\Delta$, $^2\Sigma^+$, and two $^2\Pi$ states and ScO can be formed in $^2\Sigma^+$ and $^2\Pi$ states. These molecular electronic states are also appreciably split by spin–orbit coupling. Fifteen doubly degenerate MO_2 states correlate with the $^2D_{5/2}$ and $^2D_{3/2}$ sublevels of M alone. And even more states must be considered to analyze the collisions. Again, after more detailed considerations, it seems to be possible to conclude that nonadiabatic processes make significant contributions to the observed product populations.[57] The treatment of such multisurface reactions is discussed in more detail in Section 3, and the treatment of reactions believed to occur in a single surface is considered again in Sections 4 and 5.

3. *Models for Multisurface Reactions*

As discussed at the end of the previous section, in many cases one must consider several potential surfaces and nonadiabatic transitions between them. For example, relative rate constants for forming product MO electronic states in reactions of Group IIIB metal atoms M with ground-state O_2 seem to agree with prior statistical distributions with no dynamical constraints.[57] The product vibrational distributions in both the ground and the excited electronic states for these reactions seem to be in at least approximate accord with statistical phase-space theory.[57,58] Another example is the reaction $X + M_2 \rightarrow MX + M$ where M is an alkali atom and X is a halogen. In these cases a statistical partitioning of product atoms between the ground and excited electronic states has been claimed to have been observed.[59] Both statistical models[60] (see Chapter 19) and information theory models[61] (see Chapter 22) have been employed in a treatment of the populations of the excited states relative to the ground state for these reactions. The most significant feature of these reactions may be the large exoergicity combined with the low electronic excitation energies of the product atoms. If one can generalize from the available results, it seems to be a common case, rather than a rare exception, that once the system gets into one or more of a fairly dense set of excited electronic surfaces, the electronic energy begins to be partitioned more statistically than adiabatically. Further, one does not have to go very high in excitation energy for many systems before the electronic states become very dense. In such cases one may be able to treat the reaction in terms of two sets of surfaces with strong nonadiabatic coupling within the higher-energy set and possibly weaker nonadiabatic coupling of this set to the lower-energy set. Within the higher-energy set one can try to analyze the reaction by phase-space theory or information theory with statistical thermodynamic prior expectations.

In many or most cases, of course, the final electronic state distributions are definitely nonstatistical.[11,62,63] One might still try to apply statistical methods to the rotational–vibrational distributions in specified product electronic states.[57,58,64] This corresponds to a modified phase-space theory which is statistical except for a weighting factor W_a which depends only on electronic quantum numbers. Another variation would be to let the nonstatistical weight W_a also depend on rotational–vibrational state and final relative velocity according to some model.[65] Similarly, one can employ information theory to analyze deviations from statistical or model predictions.[57,58,66] If the number of interacting surfaces can be reduced to a more manageable number, then it becomes possible to make a more detailed treatment of the dynamics by time-dependent semiclassical mechanics. For surfaces that couple strongly or have narrowly avoided crossings, some nonadiabatic behavior can be expected. Such behavior occurs for low-energy collisions with high probability only if the adiabatic surfaces approach each other closely (in general, in regions of avoided crossings). Although diabatic states have been very useful in describing atom–atom collisions involving curve crossings, this type of surface is not as useful, in general, in atom–molecule collisions except when obtained by a formal transformation of the adiabatic surfaces. For atom–molecule systems with $(3N - 6)$ internal nuclear degrees of freedom the avoided crossings, if any,[54] occur at seams in the multidimensional internal-coordinate space instead of at a point as in atom–atom collisions. For many systems, the seam is expected to be well-localized, and diabatic behavior thus occurs only in these local regions, while adiabatic behavior may be expected elsewhere. Thus a reasonable basis for classical models is to use diabatic representations only in the localized seam regions and adiabatic surfaces for the remaining regions of the hypersurface.

The method that has been most widely applied is the surface-hopping trajectory model.[11,67,68] In this method, the electrons are treated quantum mechanically, while the nuclear motion is treated by classical mechanics. To explain the method we start by considering another one, the so-called classical-path semiclassical method. In this method one assumes that the internuclear motion $\mathbf{R}(t)$ as a function of time t is initially governed by the potential surface corresponding to the initial electronic state i. The electronic motion can then be calculated from the time-dependent electronic Schrödinger equation

$$H_{\text{el}}(\mathbf{r}, \mathbf{R}(t))\,\Phi(\mathbf{r}, t) = i\hbar\partial\Phi(\mathbf{r}, t)/\partial t$$

where $\mathbf{r}$ denotes all the electronic degrees of freedom. If one expands the electronic wave function $\Phi(\mathbf{r}, t)$ in terms of Born–Oppenheimer adiabatic states $\Phi_j[\mathbf{r}, \mathbf{R}(t)]$ as

$$\Phi(\mathbf{r}, t) = \sum_j a_j(t)\,\phi_j[\mathbf{r}, \mathbf{R}(t)]\exp\{-(i/h)\int^t E_j[\mathbf{R}(t)]\,dt\}$$

where $E_j(\mathbf{R})$ are the adiabatic potential surfaces, one obtains a set of coupled differential equations for the $a_j(t)$, which are the probability amplitudes for the adiabatic states:

$$\frac{da_j}{dt} = -\sum_{i \neq j} a_i \mathbf{V}\cdot\left\langle \phi_j \frac{\partial}{\partial \mathbf{R}} \phi_i \right\rangle \exp\left[-(i/\hbar)\int^t (E_i - E_j)\,dt\right]$$

The coupling term between surfaces is simply the nuclear velocity $\mathbf{V}$ multiplied by the nuclear derivative coupling matrix element evaluated in the basis of adiabatic electronic wave functions. For the simplest case, only two surfaces are treated and there are only two coupled equations for the probability amplitudes. One integrates these numerically simultaneously with the coupled classical equations for the trajectory. The initial condition is $a_j = \delta_{ji}$, where δ_{ji} is the Kronecker delta. As soon as any other a_j becomes nonzero, the method becomes inconsistent since the probability amplitude for more than one electronic state is nonzero and there is no unique choice for one potential surface to use to govern the internuclear motion.* To circumvent this one makes an extra classical approximation. One assumes that the system stays in the initial adiabatic electronic state until a seam $S(\mathbf{R})$ is reached. Then one computes a semiclassical probability of an electronic-state change and splits the trajectory into two branches, each propagating on a single adiabatic potential surface. After leaving the seam each branch of the initial trajectory is continued as a new trajectory until a seam is again reached where branching can again occur. This sequence is continued until all branches of the trajectory are complete. Each branch

* In principle one should make the potential which governs the internuclear motion self-consistent with the set of a_j,[69] but there is no practical scheme for this which is generally applicable and well tested.

receives a probability weighting factor based on the probabilities for continuing on each surface at each seam. This method is called trajectory surface hopping. Since probabilities rather than amplitudes are employed in weighting the branches, the motion of the nuclei is essentially being treated classically. If the number of atoms or number of potential surfaces increases, it may become too difficult to define the seam in advance and one may have to determine the surfaces of avoided intersection during the course of the trajectory calculations.[70]

Two instructive applications of the trajectory-surface-hopping method have been to the reactions[67] $H^+ + D_2 \rightarrow HD + D^+$, $HD^+ + D$, or $H + D_2^+$ and isotopic analogs of these reactions,[71] and to the reactions $Ar^+ + H_2 \rightarrow ArH^+ + H$ or $Ar + H_2^+$.[72] Several aspects of the applications to these systems should be noted. First, in both systems Preston and coworkers computed the probability of a surface hop using a pair of coupled time-dependent semiclassical equations for the probability that the system is in a given electronic state. The results of these numerical calculations were used to parametrize the Landau–Zener expression (see Chapter 13) for the probability of a surface hop at a seam. For the $H^+ + D_2$ reaction, it was found that the surface-hopping probability is essentially the same using either the Landau–Zener expression with parameters fit to the results of the classical-path semiclassical calculation, or the Landau–Zener expression with parameters taken from the surface in the usual way. It should not be assumed in general that a Landau–Zener approximation that is not fit to semiclassical calculations will work well. Second, the trajectory-surface-hopping calculations are in excellent agreement with experiment for absolute reaction cross sections, product translational-energy distributions, and product angle-velocity contour maps. Third, the sensitivity of the results to the surface-hopping probability $P(\mathbf{R}, \dot{\mathbf{R}})$ was studied for $H^+ + H_2$. The Landau–Zener form of $P(\mathbf{R}, \dot{\mathbf{R}})$ that was used may be written $\exp[-C(\mathbf{R})/V_\perp]$, where $V_\perp$ is the velocity perpendicular to the seam and $C(\mathbf{R})$ is the function which was parametrized to the classical-path semiclassical calculations. If $C(\mathbf{R})$ is changed by a factor of 5, the cross sections for various product arrangements are changed by no more than 20%. To explain similar results, Tully and Preston argued that for atom–molecule collisions electronic transition probabilities will usually be determined more by the relative number of trajectories which enter regions of strong diabatic behavior than by the exact values of the surface-hopping probabilities at particular seam crossings. If this is true in general, it provides a significant simplification of the treatment of electronically nonadiabatic reactions. A fourth qualitative point emerges from these studies. For both systems the relevant seam turns out to be perpendicular to a vibrational degree of freedom of the reactants. This contrasts sharply with the many simple models of curve crossing that emphasize the radial coordinate of relative separation of the reactants.

George and co-workers have developed a decoupling approximation to the classical S matrix theory of surface hopping.[73] The classical S matrix theory involves complex-valued trajectories, root searches, analytic continuation of potential surfaces, and interference effects, and thus has severe computational drawbacks to its general applicability. In contrast, the decoupling scheme bears a strong similarity to the Tully–Preston trajectory-surface-hopping approach for those cases where both can be applied. The decoupling scheme is more general since it can be applied to cases where there is no avoided crossing, or to study effects of classically energetically inaccessible excited electronic states.[73,74] Of course, cases without an avoided crossing can also be treated by the classical-path semiclassical method already mentioned, which involves coupled classical equations of motion for nuclear coordinates and quantum-mechanical equations of motion for the electronic coordinates. Both the classical-path semiclassical method and the decoupling method have been applied to the nonadiabatic reaction $F(^2P_{1/2}) + H_2 \rightarrow FH + H$ mentioned in Section 2 of this chapter.[16,17,73]

Komornicki *et al.*[73] have tested the decoupling procedure against quantum-mechanical calculations for nonreactive systems and found good accuracy except for a case involving strong electronic-to-vibrational energy transfer with a small energy-level defect. Similar effects could also occur in reactive systems. The only test of the trajectory-surface-hopping method against accurate quantum results for a reactive system is provided by the study of Bowman *et al.*[75] of the collinear reaction $Ba + N_2O \rightarrow BaO(X\,^1\Sigma) + N_2$ or $BaO(a\,^3\Pi) + N_2$, where N_2 was treated as a point mass. In this study the usual Landau–Zener approximation was used for the surface-hopping probability. Two surfaces are involved and large quantum effects are observed. Since the seam in the exit channel in this case is displaced from the equilibrium position of BaO, vibrational energy in the product BaO is necessary to cause the transition between surfaces. The classical model shows lower electronically diabatic reaction probabilities than the quantum-mechanical model does. It was suggested that this deficiency may be remedied by allowing hops in a band about the seam rather than just at the seam. As in any classical model of the nuclear motion care must be taken when important dynamic effects are associated with regions near classical turning points. All of the problems with trajectories governed by single potential surfaces may also affect trajectory-surface-hopping calculations. These problems are discussed in Section 5 of this chapter.

It is of course not necessary to employ exact trajectories with the surface-hopping approach. Exact propagation of trajectories between regions of localized coupling could be replaced by model calculations, as discussed next for single-surface interactions.

4. Simple Models for Single-Surface Reactions

4.1. Entrance-Channel Models

Before discussing approximate models used in describing distributions of product velocity and scattering angle, we first describe some simple models based on the behavior of the system in the entrance channel. This includes some simple models for the total integral reaction cross section.

One simple model, the harpooning model,[42,76,77] accounts reasonably well for entrance-channel behavior in exoergic reactions M + XY → MX + Y where M is an alkali atom and XY is a covalently bonded molecule. In these reactions, the atom and diatomic approach on a covalent surface, but the products MX + Y are on an ionic surface. At the distance where the ionic and covalent curves cross, the electron is assumed to jump from the atom to the molecule in a process denoted as the harpooning mechanism. If the jump occurs at large distances and the ionization potential (IP) of M exceeds the electron affinity (EA) of XY, the distance R between M and the center of mass of XY at the crossing point is given by

$$R_c = \frac{e^2}{\mathrm{IP_M} - \mathrm{EA_{XY}}} = \frac{14.4\,\text{Å eV}}{\mathrm{IP_M} - \mathrm{EA_{XY}}}$$

where e is the electronic charge. For these systems the IP is low and for halogen targets the EA is large so that R_c is large. Since the atom and diatomic are assumed to interact negligibly until R_c is reached, the cross section for strongly coupled collisions is given by $Q_{sc} = \pi R_c^2$. If the electron jump occurs at small R_c where Coulombic forces are not dominant then valence forces will play a role and different behavior can be expected. The electron jump model has recently been used to try to explain selective formation of excited states in the reactions of Group III B metals with various oxidants.[78,79] Another use of the electron jump model including multiple electron jumps has been made for reactions of alkali dimers with halogens and alkaline earths with halogens.[80]

Several simple models are available for estimating the reaction cross section.[81-86] Consider a general attractive potential $V(R)$ governing the collision of two spheres of radii r_a and r_b. The cross section for contact of the two spheres is given at relative translational energy E_{rel} by

$$Q_{hs}(E_{\mathrm{rel}}) = \pi(r_a + r_b)^2\left[1 - \frac{V(r_a + r_b)}{E_{\mathrm{rel}}}\right]$$

Another simple result is obtained by assuming $V(R) = 0$ when $R > (r_a + r_b)$, and that reaction occurs only when the component of relative kinetic energy along the line of centers at contact is greater than a threshold value E_0. This leads to the following expression for the reaction cross section:

$$Q(E_{\rm rel}) = \begin{cases} 0, & E_{\rm rel} < E_0 \\ \pi(r_a + r_b)^2[1 - (E_0/E_{\rm rel})], & E_{\rm rel} > E_0 \end{cases}$$

For reactions of ions with nonpolar neutral species, the cross section for close collisions may often be approximated by the Langevin model.[81,83,85] If we assume a spherical neutral, the collision is governed at large distances by the potential $V(R) = q^2\alpha/(2R^4)$, where α is the polarizability of the neutral and q is the charge of the ion. At a certain impact parameter b_0 given by $(2q^2\alpha/E_{\rm rel})^{1/4}$, a closed orbit with an orbiting distance equal to $R_c = b_0/2^{1/2}$ will occur. For $b < b_0$, the molecule will spiral in and reaction may occur, while for $b > b_0$ the particles never approach any closer than R_c. If a constant fraction P_0 (independent of initial energy and angular momentum over the range of interest) of close collisions leads to reaction, then, assuming no activation energy, the reaction cross section is given by

$$Q(E_{\rm rel}) = P_0 \pi q \left(\frac{2\alpha}{E_{\rm rel}} \right)^{1/2}$$

Thus the reaction cross section decreases with increasing translational energy. Although the Langevin theory gives a qualitative description of the upper limit for the reaction cross section, it should be used with caution. It should only be applied at thermal energies since at higher energies the asymptotic form of the potential is not valid at the calculated orbiting distance. The Langevin model does not apply to processes involving electron transfer at distances beyond the centrifugal barrier. Care must be taken to use a proper statistical weighting for the cross section according to the number of asymptotic electronic states that lead to reaction (see Section 2). A number of modified versions of this theory have been used. The interaction of an ion with a molecule with a permanent dipole moment $\mathbf{D}$ leads to the long-range potential $V(\mathbf{r}) = -\alpha e^2/2R^4 - q\mathbf{D}\cdot\mathbf{R}/R^3$, which is orientation dependent. Various methods have been proposed[83,87,88] to treat such cases but none is completely satisfactory because the dynamic effects of the dipole motion are hard to include properly. A more clear-cut case involves the interaction of molecular ions with open-shell atoms which have quadrupole moments.[89] This gives a term in the potential proportional to R^{-3}. At very low kinetic energies (< 0.10 eV) the cross section may be smaller than the Langevin cross section because some collisions are being governed by repulsive interactions.

The effect of internal excitation of the reactants may be included in entrance-channel models by including one or more of the following effects: (i) The internal excitation energy is available, at least in part, in overcoming the barrier to reaction, thereby increasing the reaction probability at low translational energy or high impact parameter or both. (ii) Vibrationally excited reactants have extended bond lengths and hence are larger and more reactive. (iii) The increased velocity distributions of the atoms of an excited molecular reactant may alter the velocity of these atoms relative to the attacking atom at crucial stages of some collisions. But no quantitative statements can be made based on such simple considerations. Some workers have been led to serious errors by assuming all internal excitation energy is available for overcoming barriers.

4.2. Channel-to-Channel Models and Angular Momentum Considerations

In this section we consider the additional information which can be easily gained by considering not only the asymptotic or dominant interaction of the reactants and their energy and angular momentum, but also the same quantities for the products.

A number of simple generalizations about reaction product distributions can be made by investigating the conservation of angular momentum.[77] For an atom–molecule collision, the total angular momentum $\boldsymbol{\mathcal{J}}$ is given by $\boldsymbol{\mathcal{J}} = \mathbf{J} + \mathbf{L} = \mathbf{J}' + \mathbf{L}'$, where $\mathbf{J}$ is the rotational angular momentum, $\mathbf{L}$ is the orbital angular momentum of relative translational motion, and unprimed quantities refer to reactants and primed quantities refer to products. The magnitude of $\mathbf{L}$ is given by $\mu_R V_R b$, where μ_R is the reduced mass for relative motion, V_R the relative speed, and b the impact parameter. For linear products, rotational energies E'_{rot} are given by $J'(J' + 1)/2I'$, where I' is the product's moment of inertia. For the reaction $\mathcal{H} + \mathcal{H}\mathcal{L} \to \mathcal{H}\mathcal{H} + \mathcal{L}$, where $\mathcal{H}$ is a heavy particle and $\mathcal{L}$ a light one, under the usual conditions one finds $L >> J$. Since the products have a small μ'_R, L' will thus be small, and conservation of angular momentum gives $\boldsymbol{\mathcal{J}} \cong \mathbf{L} \cong \mathbf{J}'$. For this type of reaction $\mathbf{J}'$ is predicted to be strongly polarized with $\mathbf{J}'$ approximately perpendicular to $\mathbf{V}_R$. For the reverse case $\mathcal{L} + \mathcal{H}\mathcal{H} \to \mathcal{L}\mathcal{H} + \mathcal{H}$, one must consider the rotational population of $\mathcal{H}\mathcal{H}$. If $\mathcal{H}\mathcal{H}$ is supplied by a nozzle beam so that the rotations of $\mathcal{H}\mathcal{H}$ are relaxed, then J will be small and the magnitudes of $\mathbf{J}$ and $\mathbf{L}$ could be comparable. The product angular momentum will still be dominated by $\mathbf{L}'$ giving $\boldsymbol{\mathcal{J}} \cong \mathbf{L} + \mathbf{J} \cong \mathbf{L}'$. For beam studies of reactions of the type $M + X_2 \to MX + X$, where M is an alkali and X a halogen, $\boldsymbol{\mathcal{J}} \cong \mathbf{L}$ since J is low and the reaction cross section is large. However, in the products a large value for J' need not imply a large value for E'_{rot}. For example, if $J' = 150\hbar$, then $E'_{\text{rot}} = 1.5$ kcal/mol for CsI. Since $\mathbf{L}'$ and $\mathbf{J}'$ are vectors they may be added in a variety of ways with the limits corresponding to $\mathcal{J} = J' + L'$ (parallel) and $\mathcal{J} = J' - L'$ (antiparallel). Thus if the magnitude or direction of $\mathbf{L}'$ is not known, the value of $\mathbf{J}'$ estimated from parallel addition may be wrong.

By an extension of this kind of reasoning one can predict the qualitative shape of the opacity function, i.e., the probability $P(b)$ of the reaction A + BC → AB + C as a function of impact parameter b.[90,91] If the initial reduced mass $\mu_{A,BC}$ is greater than the final reduced mass $\mu_{AB,C}$ and/or the reaction is quite endoergic, then ability to accommodate orbital angular momentum **L**′ in the exit channel may be the limiting constraint. The value of $P(b)$ computed by a statistical theory may have roughly the behavior $P(b) \cong P_0[1 - (b/b_{max})]$, where P_0 and b_{max} are decreasing functions of initial relative translation energy. In such a case the reaction cross section may be decreased compared to what is expected solely on the basis of an entrance-channel model. However, if $\mu_{A,BC} < \mu_{AB,C}$ or a large exoergicity is present, or the reactants have a large internal excitation energy, then $P(b)$ may tend to be more roughly constant out to some impact parameter b_{max}, where in this case b_{max} is dependent mainly on entrance-channel factors. Intermediate cases may also arise but the two cases above are useful limits. They have been called type I and type II, respectively.[91]

The simplest model which includes exit-channel effects is the spectator-stripping model.[92,93] It has been applied to a number of reactions that exhibit forward scattering, while in the form of the elastic spectator model[77] it has also been applied to rebound reactions. For A + BC → AB + C, the spectator-stripping model assumes that B is abstracted so suddenly by A that C continues with no momentum transfer. The elastic spectator model assumes that C undergoes elastic scattering by the AB molecule (rather than continuing along the initial relative velocity vector) in order to give rebound behavior. The internal energy of the products in the spectator-stripping model is calculated from the conservation of momentum and conservation of energy. The internal energy E'_{int} is given by

$$E'_{int} = \frac{M_B(M_{ABC})}{M_{AB}M_{BC}} E_{rel} + \Delta D$$

where ΔD equals the bond-strength difference and is the absolute magnitude of the exoergicity or minus the endoergicity. At a certain high translational energy, E'_{int} is predicted to equal D_{AB} and no bound product is expected to be produced. This energy is given by

$$E_{rel}(\text{upper}) = \frac{M_{AB}M_{BC}}{M_B(M_{ABC})} D_{AB}$$

This upper limit has been discussed elsewhere,[94,95] and, for example, does correlate with the behavior observed in the reaction $U^+ + N_2 \rightarrow UN^+ + N$.

Muckerman[95] has studied the F + HD reaction using classical trajectories at very high relative collision energies (up to 68 eV) and has calculated the reaction cross section $Q(E_{rel})$. He found that $Q(E_{rel})$ is still reasonably large at five times E_{rel}(upper), as given by the spectator-stripping model. The spectator-stripping model was able to account for the behavior of the product translational energy reasonably well up to near the limit E_{rel}(upper). Beyond the limit imposed by spectator stripping, another model was found to qualitatively explain the behavior. The F impinges on HD essentially collinearly (but with nonzero impact parameter) and strikes the molecule with such force that the near atom in the molecule is ejected and the far atom then combines with the F to form the product. This mechanism also explains why the HF product becomes dominant at high collision energies in contrast to the results at low collision energies for the ground rotation–vibration state, where DF is the dominant product (see Section 5.2). This model is opposite in spirit to the spectator stripping model but is more akin to the sequential impulse model described next.

The sequential impulse model[(96,97)] has received considerable attention and has been applied to a number of reactions. The reaction A + BC → AB + C is treated in terms of hard collisions[(96)] where atom A collides with B impulsively in an elastic potential, and B then collides with C in a similar fashion. The product AB is formed only if the relative kinetic energy of AB is less than the dissociation energy of AB. In comparison with classical trajectories,[(95)] a simple form of the model[(97)] does not correctly predict the energy dependence or high-energy cutoff in the cross section for the F + HD reaction. Some refinements of the model have been presented.[(22,98-104)] For example, for the reaction $Ar^+ + D_2$, attractive ion-induced-dipole forces were included in the entrance channel by an approximate trajectory.[(99)]

In principle, the sequential impulse model could be extended to apply even to collisions involving snarled trajectories. However, the advantages of a simple theory with a few well-defined parameters could be lost as one introduced more and more of the attractive forces responsible for successive impacts. Consider, for example, the "migration" mechanism for an A + BC reaction. This consists of an abortive reaction of A with B followed by eventual reaction with C such that the fragment eventually ejected is at first attracted to the attacking atom, and only later repelled by it. Kuntz *et al.*[(105)] have argued that even this simple complication of the trajectory is hard to treat analytically in a realistic way. Most such effects are most easily studied by actually solving the classical equations of motions for assumed potential energy surfaces for representative sets of collisions, i.e., by the Monte Carlo trajectory method discussed in Section 5.

Another very simple model of reactions is based on what has been called the optical potential model of elastic scattering.[(42,77,106-109)] It can be considered as the classical analog of the distorted-wave theory, which is treated in Chapter 15. Consider the reaction of A with BC. Scattering parameters in the entrance channel are the initial relative translational energy E_{rel} and the impact parameter b. Some fraction $P(b, E_{rel})$ of collisions leads to reaction, and the products AB + C separate along a trajectory with exit impact parameter b' and asymptotic relative translational energy E'_{rel}. The differential cross sections for elastic and reactive scattering are given by

$$I_{elastic}(\theta) = [1 - P(b, E_{rel})]\, d(\pi b^2)/d\Omega$$

and

$$I_{reactive}(\theta') = P(b, E_{rel})\, d(\pi b^2)/d\Omega'$$

respectively, where $d\Omega = 2\pi \sin\theta\, d\theta$ and $d\Omega' = 2\pi \sin\theta'\, d\theta'$, θ is the center-of-mass scattering angle for elastic scattering, and θ' is the center-of-mass reactive scattering angle; inelastic scattering is neglected. The scattering angle θ' is given approximately by the sum of two half-collisions:

$$\theta' = \tfrac{1}{2}\theta + \tfrac{1}{2}\theta_p$$

where θ_p is the scattering angle for a time-reversed elastic trajectory of the products. Elastic scattering for small b is usually dominated by the repulsive wall; and so for rebound reactions for which $P(b, E_{rel})$ is zero except at small b, we assume that reaction is also governed by repulsive forces. If we assume hard-sphere collisions, this leads to the following expression for θ:

$$\theta = \begin{cases} 2\arccos(b/\sigma), & b \le \sigma \\ 0, & b > \sigma \end{cases}$$

where σ is the hard-sphere collision diameter for A + BC. The exit impact parameter is also assumed to be small, so

$$\theta_p = 2\arccos(b'/\sigma')$$

where σ' is the hard-sphere collision diameter for AB + C. For linear reactive encounters $b' = b'(b)$, but for bent collision complexes azimuthal averaging about the axis from A to the center of mass of BC gives a range of b' for a single initial b. However, averaging over this range gives about the same results as found for linear configurations. If reaction can occur only for $b \le \sigma/2^{1/2}$ and $b' < \sigma'/2^{1/2}$, then scattering is confined entirely to the backward hemisphere, i.e., the AB final velocity vector deviates by more than 90° from the initial velocity of A in the center-of-mass system. This is called rebound behavior. For $\sigma' = \sigma$ and $P(b, E_{rel})$ roughly constant for $b \le \sigma$, this hard-sphere model gives approximately isotropic scattering. For large reaction cross sections or scattering primarily into the forward hemisphere, as in stripping, $P(b, E_{rel})$ must be appreciable at large b. In this case θ must be estimated on the basis of long-range forces. Even for small b', detailed considerations predict a forward-peaking contribution from scattering at large b that is superimposed on the flatter distribution due to scattering at small values of b.

In this section we have discussed some attempts to use simple considerations based on conservation laws, mass ratios, and dominant interactions in the entrance and exit channel to predict the gross features of chemical reactions. To go beyond these simple models one must introduce three-body aspects of reactions more explicitly. For example, one can use the effective two-body interactions in the entrance and exit channels to define those entrance-channel states (or regions of phase space) and exit-channel states (or regions of phase space) which are strongly coupled, and then use a statistical approximation to compute state-to-state transition probabilities and cross sections. This yields statistical phase-space theory.[60,62-64,90,91,110-112] This theory is a logical extension of the considerations of this section in the sense that energy and angular-momentum conservation and dominant channel effects are included. However, the statistical assumption is not generally valid for the three-body dynamics for direct-mode reactions. Statistical phase-space theory is discussed further in Chapter 19 in connection with complex-mode reactions. There are other approaches, also logical extensions of this section, which are more appropriate for direct-mode reactions. First one may use a statistical phase-space calculation or a set of state-to-state reaction probabilities computed on a statistical thermodynamic model as a prior expectation[113] against which to measure information-theoretic surprisals and entropy deficiencies. One can then attempt to interpret the deviations of experimental results from prior expectations in terms of dynamic constraints on the reaction. The information-theoretic approach is discussed more fully in Chapter 22. Alternatively, one may attempt to incorporate dynamic effects by a nonstatistical weighting of the channel states or channel phase-space regions which are strongly coupled through the three-body interactions. This leads to nonstatistical phase-space theory, which has also been called the intermediate-coupling probability-matrix approach.[65] In this approach one can still account for all requirements of angular momentum and energy conservation and for dominant two-body channel effects, but also introduce some simple dynamic factors or biases affecting the transitions between reactant and product configurations. A third approach is to introduce even more specific dynamic models of the three-body classical mechanics. Such approaches are discussed in the following section, 3.3. The ultimate refinement of such simple dynamical classical models is the calculation of exact classical trajectories for full assumed potential energy surfaces as discussed in Chapter 16 and in Sections 3 and 5 of this chapter.

4.3. More Detailed Dynamic Models of Product-State Distributions

We have already discussed the simple way in which the spectator-stripping model predicts the product internal energy. Several of the other simple models totally ignore this question. To get a more realistic model of product vibrational energy, one must consider the nature of the potential surface and the characteristics of its energy release in more detail. One attempt to do this for the class of $A + BC \rightarrow AB + C$ reactions which are governed by covalent forces and for which collinear reaction is favored is the FOTO (forced oscillation of a tightening oscillator) model.[114] This model requires only a single parameter, the bond order n_{10}^{*} of the product at some formation time. Repulsion between B and C is assumed to occur as the A–B bond is being formed, and this causes oscillation of the A–B bond. Initially the vibrational energy of the A–B bond is E_{vib}^{*}. This is increased to E'_{vib}, the vibration energy of the products, by forced oscillation for τ sec, where τ is determined by the time required to increase the A–B bond order from n_{10}^{*} to unity. The model actually involves solving a simple set of integral equations for a tightening harmonic oscillator for the time interval in which the repulsive force is present ($0 < t \leq \tau$). The parameter n_{10}^{*} is chosen to give the correct value of E'_{vib} and thus it is an empirical parameter in the present form of the theory. There are a number of noteworthy assumptions of the theory in the FOTO model. The potential surface underlying the model is provided by a BEBO-type prescription† for the relationships between bond order, bond energy, and bond length, and Badger's rule is used to relate the force constant to bond length. The BEBO model does not always predict a qualitatively correct surface, and even when it does the surface features are quantitatively sensitive to the exact values of the parameters. Another assumption is that the dependence of A–B bond order on time is fixed by postulating that the A–B bond strength increases linearly with time in the interval from 0 to τ. Another important limitation is the restriction to collinear events. The collinear model does not work well for the mass combination ($\mathscr{H} + \mathscr{L}\mathscr{H}$). Finally we note, as discussed especially in Section 5.3, that overemphasizing a single feature of the potential surface generally leads to an oversimplication of the actual dynamics. This model and all the others can, of course, be refined. The list of limitations just given is not meant to single out one model for criticism, but rather to illustrate the difficulties in trying to obtain a single simple model with wide generality. The FOTO model is in substantial qualitative accord with the dynamical features of three-dimensional trajectories on semiempirical LEPS potential surfaces‡ for $H + X_2$ (X = halogen) reactions, and so it does provide an insightful "interpretation" of those trajectories.

A simple model that is similar in spirit to FOTO was developed by Parrish and Herm[116] to explain internal excitation in product alkyl groups R produced in the reaction $Cs + IR \rightarrow CsI + R$. The reaction is treated as a four-particle collinear system with harmonic forces, and the vibrational motion is treated using a normal-coordinate analysis. A repulsion parametrized to experimental results occurs between R and I as in the DIPR model described next, and this is the driving force for the oscillators.

† The bond-energy–bond-order method is discussed in detail in Johnston's text.[115]

‡ The LEPS model for atom–diatom potential surfaces is explained elsewhere.[36]

This model provides reasonable agreement with experiment if a highly repulsive surface is assumed.

The final simple model which we shall discuss is the DIPR (direct interaction with product repulsion) model and a version of it called the DIPR-DIP (distributed as in photodissociation) model. This model is distinguished by the fact that it can be used to predict not just one or a few reaction attributes but a whole range of them: angular distributions, recoil velocity distributions, product rotational energies, and even final rotational orientation. The basic physics that underlie the model are that product distributions in reactions of the kind $A + BC \rightarrow AB + C$ are mainly determined by a monotonically decreasing B–C repulsive force which acts along the axis of the breaking bond. The DIPR model[117-120] was originally proposed to provide a simple description of some feature of three-dimensional classical trajectories for reactions like $M + XC \rightarrow MX + C$ (M = alkali, XC = halogen or halide) in which the entrance channel is governed by an electron jump. Later it was recast in an impulsive, pseudo-hard-sphere framework.[121] Allowance for a distribution of repulsive energy release leads to the DIPR-DIP model,[122-124] named after one possible way of estimating this distribution. We will call the model DIPR except when we wish to emphasize the distribution of repulsive energy release. An important advantage of the model is that other theoretical or empirical methods can also be used to estimate this distribution. Thus the utility of the model should not be confused with the validity of any one method of estimating its parameters. When experimental data are available, all the parameters may be taken as empirical.

Now we discuss the DIPR-DIP model in more detail.[122-124] The dynamical assumptions of the model are as follows. (i) BC is initially in the ground rotation–vibration state. (ii) No interaction between A and BC occurs until the system reaches a prespecified value R_c of the distance R between A and BC (usually R_c is the electron jump distance). (iii) A repulsive force between B and C is turned on at $R = R_c$ to force rapid decomposition of BC, and the product AB is formed as B and C separate. (iv) No interaction between A and C is included. In this model, the angular distribution and most probable final relative translational energy are determined by the dependence of the reaction probability on the initial orientation of BC and by the distribution of values of the total repulsive energy release $\mathscr{R}$. In order to calculate various product rotational distributions, the repulsive force as a function of separation coordinate must also be known.

The appropriate coordinate systems are shown in Figure 3, while formulas for the center-of-mass scattering angle θ of AB with respect to the initial direction of A and for the final relative energy E'_{rel} are given in Table 1. The differential cross section per steradian is given by

$$I(\theta) \propto \frac{1}{\sin\theta} \int_0^{\mathscr{R}_{\max}} \int_0^{\pi} P(\xi)\, P(\mathscr{R})\, \delta[\theta - \theta(\xi, \mathscr{R})]\, d\xi\, d\mathscr{R}$$

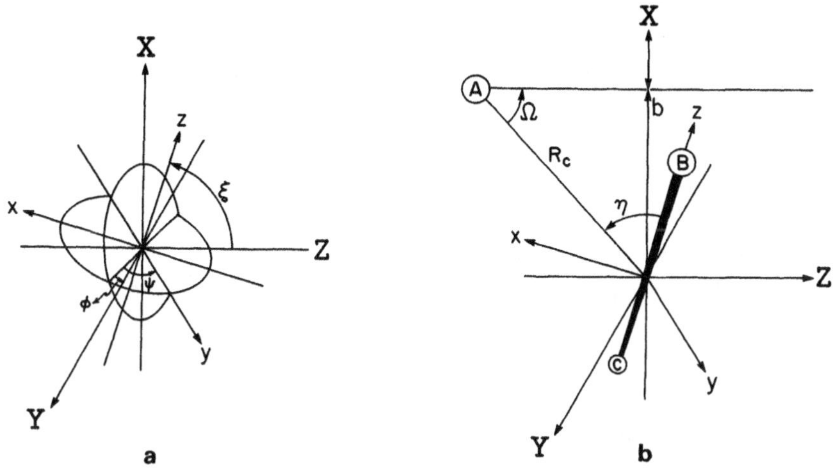

Figure 3. Coordinate systems for the DIPR model for the reaction A + BC → AB + C. The space-fixed axes are represented by X, Y, Z while the molecule-fixed axes are given by x, y, z. (a) Eulerian angle diagram showing relationship of the two coordinate systems. The XY and xy planes are shown by ellipses. (b) Geometric variables in the DIPR model. R_c is the distance between A and the center of mass of BC, and b is the impact parameter. In part (b) only the positive halves of the x, y, and z axes are shown.

where $P(\xi)$ is the differential cross section per unit Eulerian angle ξ and $P(\mathscr{R})$ is the probability function for total repulsive energy release. Various forms for $P(\xi)$ corresponding to one end reactive, both ends reactive, broadside, and isotropic reactivity patterns are given in Table 2. The probability function for the final translational energy distribution at a given scattering angle θ is given by

$$F(E'_{\text{rel}}, \theta) \propto \frac{1}{U_{\text{AB}}[\xi(\theta, \mathscr{R}), \mathscr{R}]} \int_0^{\mathscr{R}_{\max}} \frac{P[\xi(\theta, \mathscr{R})]\, P(\mathscr{R})\, \delta\{U_{\text{AB}} - U_{\text{AB}}[\xi(\theta, \mathscr{R}), \mathscr{R}]\}\, d\mathscr{R}}{|(d\theta/d\xi)|_{\xi = \xi(\theta, \mathscr{R})}}$$

where the expression for $U_{\text{AB}}(\xi, \mathscr{R})$ is given in Table 1. Note that U_{AB} is the final velocity of AB in the center-of-mass coordinate system.

Information about the product rotational distributions can be obtained by modeling the release of the repulsive force as a function of the separation coordinate. This involves an empirical parameter $\Delta\tau$. It is assumed that the equations of motion of atom C for $t > 0$ in a coordinate system fixed to the BC subsystem with the z axis chosen as the B–C axis are $d^2x/dt^2 = d^2y/dt^2 = 0$ and $d^2z/dt^2 = F_0 e^{-t/\Delta t}$. The components J'_X, J'_Y, and J'_Z of product rotational angular momentum $\mathbf{J}'$ in the space-fixed coordinate system of Figure 3 are given in Table 3, while the final rotational energy is given by

$$E'_{\text{rot}} = |J'|^2/2I_{\text{AB}}$$

Table 1. Scattering Angle and Recoil Energy for the DIPR-DIP Model[a]

Scattering angle:

$$\theta(\xi, \mathscr{R}) = \arctan\left[\sin\xi/(p + \cos\xi)\right]$$

$$p = (E_{\text{rel}}M_{\text{A}}M_{\text{C}}/\mathscr{R}M_{\text{B}}M_{\text{ABC}})^{1/2}$$

Product relative translational energy:

$$E'_{\text{rel}}(\xi, \mathscr{R}) = (M_{\text{C}}M_{\text{ABC}}/M_{\text{AB}}) \times \left[M_{\text{B}}\mathscr{R}/M_{\text{C}}M_{\text{BC}} + M_{\text{A}}E_{\text{rel}}/M_{\text{ABC}}M_{\text{BC}} + (2\cos\xi/M_{\text{BC}})(E_{\text{rel}}\mathscr{R}M_{\text{A}}M_{\text{B}}/M_{\text{C}}M_{\text{ABC}})^{1/2}\right]$$

Final velocity of AB in the center of mass:

$$U_{\text{AB}}(\xi, \mathscr{R}) = (M_{\text{C}}/M_{\text{AB}}) \times \left[2M_{\text{B}}\mathscr{R}/M_{\text{BC}}M_{\text{C}} + 2E_{\text{rel}}\, M_{\text{A}}/M_{\text{ABC}}M_{\text{BC}} + (4\cos\xi/M_{\text{BC}})(E_{\text{rel}}\mathscr{R}M_{\text{A}}M_{\text{B}}/M_{\text{C}}M_{\text{ABC}})^{1/2}\right]^{1/2}$$

[a] $M_{ijk} = M_i + M_j + M_k$.

Table 2. Probability Distributions Associated with Reactivity Patterns

Reactivity pattern	$P(\eta)$	$P(\xi)$[a]
B end reactive	$\cos^2(\eta/2)$	$R_c^2[1 - (2/3)\cos\xi]\sin\xi$
Both ends reactive	$\cos^2(\eta)$	$R_c^2(1 + \cos^2\xi)\sin\xi$
Broadside	$\sin^2(\eta)$	$R_c^2(2 + \sin^2\xi)\sin\xi$
Isotropic	1	$R_c^2\sin\xi$

[a] Unnormalized.

Table 3. Formulas for Product Rotational Angular Momenta Components J'_i in the DIPR-DIP Model

$$J'_X = (\sin\xi\sin\phi M_{\text{A}}M_{\text{C}}/M_{\text{AB}})\{-(M_{\text{B}}r_{\text{BC}}/M_{\text{BC}})(2E_{\text{rel}}M_{\text{ABC}}/M_{\text{A}}M_{\text{BC}})^{1/2} + (1/M_{\text{C}})(2\mathscr{R}M_{\text{B}}M_{\text{C}}/M_{\text{BC}})^{1/2}[(2E_{\text{rel}}M_{\text{ABC}}/M_{\text{A}}M_{\text{BC}})^{1/2}\Delta\tau - b]\}$$

$$J'_Y = (M_{\text{A}}M_{\text{C}}/M_{\text{AB}})\big((M_{\text{A}}b/M_{\text{ABC}})(2E_{\text{rel}}M_{\text{ABC}}/M_{\text{A}}M_{\text{BC}})^{1/2}(1 - M_{\text{AB}}M_{\text{BC}}/M_{\text{A}}M_{\text{C}}) - \sin\xi\cos\phi\{(M_{\text{B}}r_{\text{BC}}/M_{\text{BC}})(2E_{\text{rel}}M_{\text{ABC}}/M_{\text{A}}M_{\text{BC}})^{1/2} + (1/M_{\text{C}})(2\mathscr{R}M_{\text{B}}M_{\text{C}}/M_{\text{BC}})^{1/2}[(2E_{\text{rel}}M_{\text{ABC}}/M_{\text{A}}M_{\text{BC}})^{1/2}\Delta\tau - b]\} + (\cos\xi b/M_{\text{C}})(2\mathscr{R}M_{\text{B}}M_{\text{C}}/M_{\text{BC}})^{1/2}\big)$$

$$J'_Z = (\sin\xi\sin\phi M_{\text{A}}b/M_{\text{AB}})(2\mathscr{R}M_{\text{B}}M_{\text{C}}/M_{\text{BC}})^{1/2}$$

where I_{AB} is the moment of inertia of the product AB. The cosine of the polarization angle of the final rotational angular momentum with respect to the initial relative velocity is given by

$$\cos \chi = J'_Z/|J'|$$

The distribution function for the product rotational energy E'_{rot} at a given value of θ is given by[124]

$$F(E'_{rot}, \theta) \propto \int_0^{b_{max}} \int_0^{\mathscr{R}_{max}} \int_0^{2\pi} P\{\eta[\xi(\theta, \mathscr{R}), \phi, b]\} P(\mathscr{R}) P(b) \sin \xi(\theta, \mathscr{R}) \times \delta\{E'_{rot} - E'_{rot}[\xi(\theta, \mathscr{R}), \mathscr{R}, \phi, b]\} \, d\phi \, d\mathscr{R} \, db$$

where $P(b)$ is given by $P(b) = 2\pi b$ for $b < b_{max}$ and $P(b) = 0$ for $b > b_{max}$ and $P(\eta)$ is the reaction probability for a collision with a given value of η at the onset of reaction. $P(\eta)$ is given in Table 2 for the set of reactivity patterns which give the set of $P(\xi)$ mentioned earlier. The value of η is given by

$$\eta(\xi, \phi, b) = \arccos\left[-(\cos \xi)\frac{(R_c^2 - b^2)^{1/2}}{R_c} + (\sin \xi)\frac{b}{R_c}\cos \phi\right]$$

The probability function for $\cos \chi$ is given by[124]

$$F(\cos \chi) \propto \int_0^{b_{max}} \int_0^{\mathscr{R}_{max}} \int_0^{2\pi} \int_0^{\pi} P[\eta(\xi, \phi, b)] \sin \xi P(\mathscr{R}) P(b) \times \delta[\cos \chi - \cos \chi(\xi, \mathscr{R}, \phi, b)] \, d\xi \, d\phi \, d\mathscr{R} \, db$$

The above four-dimensional integral can be evaluated as a sum over ξ plus a numerical integration over the remaining three-dimensional integral. Product vibrational energy distributions can be obtained by energy conservation.

The distribution function $P(\mathscr{R})$ can be obtained from experiment by velocity analysis or from simple theoretical models. Herschbach and co-workers[122-125] have suggested calculating $P(\mathscr{R})$ in a quasidiatomic "reflection" approximation that has been used to explain photodissociation and electron attachment. In their model $P(\mathscr{R})$ is approximated as a Gaussian function centered at $\mathscr{R}_{mp}$, which for systems involving an electron jump is given by

$$\mathscr{R}_{mp} = EA_B - D_{BC} - EA_{BC}$$

where EA_B is the electron affinity of B and D_{BC} and EA_{BC} are the bond strength and vertical electron affinity of BC. The other input parameter is the value of $\Delta\tau$ which can be considered as the time necessary to dissociate BC^-. If BC^- is unbound then $\Delta\tau$ is positive, and it can be considered as a "speed-up" time due to a repulsive force, but a negative $\Delta\tau$ implies that attractive forces are present and it can be considered as a delay time.

In the electron jump model the value of $\Delta\tau$ depends on the potential curve for BC^-.

It is possible to incorporate into DIPR more entrance-channel effects such as initial rotational energy or forces occurring before R_c.[123] Entrance-channel attraction or repulsion can be incorporated by varying the velocity of C at the initiation of energy release.[123] Recrossing effects[126] could be incorporated in $P(\eta)$. However, forces affecting the repulsion between B and C, nonspherical A–BC forces in the entrance channel, long-range attraction between AB and C in the exit channel, and the migration mechanism cannot be included.

One interesting aspect of the DIPR model is its prediction of a correlation between E'_{rel} and θ. For example, at $\theta = 0°$, the final speed of C in the center-of-mass coordinate system is the sum of its speed at the onset of reaction and the speed increment due to release of $\mathscr{R}$; but at $\theta = 180°$ it is the difference.

The DIPR-DIP model has been compared to experimental results for the $M + CH_3I$ reaction, where M is an alkali atom. Good agreement was found for angular distributions, product velocities, and rotational-energy distributions.[120,122-124] For the rotational-energy distributions in $K + Br_2$ and $Cs + CCl_4$ the agreement is less satisfactory,[124] even though the angular distribution agrees well with trajectory studies for $K + Br_2$.[122,127] The model's prediction of $\cos^2\chi$ for the $Cs + CH_3I$ reaction is inaccurate.[128] However, neglect of initial rotation in CH_3I may be important in these vector–vector correlations. Further modifications of the model have also been made to treat internal excitation in the alkyl group R in the reaction $M + RI \rightarrow MI + R$.[122] The original DIPR model also gave good agreement with trajectory results for the reaction M + XR, where R was a point particle.[117] Postulating an analogy between the product velocity distributions in the reaction $H + X_2 \rightarrow HX + X$ and photodissociation of X_2, the DIPR model was also applied to this reaction.[39,125] Although such models are apparently quite useful in examining the above reactions, further testing of the parameters and more applications would be useful. It would also be interesting to employ this type of model for multisurface reactions.

It is interesting to consider two limits of the DIPR model which may be classified according to the value of the quantity p in Table 1. This quantity may be understood by writing the final speed of C in the center-of-mass coordinate system as

$$V'_C = (V^2_{BC} + V^2_{\mathscr{R},C} + 2V_{BC}V_{\mathscr{R},C}\cos\xi)^{1/2}$$

where V_{BC} is the initial speed of BC in the center-of-mass coordinate system, and the increment $V_{\mathscr{R},C}$ is due to the repulsive energy release; then $p = V_{BC}/V_{\mathscr{R},C}$. For $p \gg 1$, the DIPR model reduces to the stripping limit. For $p \ll 1$, the repulsive energy release is the dominant factor and one should consider the validity of the impulsive approximation used in treating this energy release.[125] The impulsive approximation will be most valid if $\Delta\tau$ is much less than a vibrational period τ_{AB} of AB. If $\Delta\tau$ is comparable to τ_{AB} then there will be appreciable interaction between AB and C in the exit channel; for this case dynamical effects such as included in the FOTO model may be important.

The models discussed in this section are useful in that they provide simple pictorial interpretations of some of the observed features of experimental results and trajectory studies of atom–molecule reactions. The DIPR-DIP model was discussed in great detail because it shows how one can attempt to put electronic structure concepts into a simple model and extract the whole range of observable reaction attributes. In favorable cases one might also predict these various parameters in an *a priori* fashion. The inherent simplicity of these models is also a flaw, however. They are of course not generally trustworthy for *a priori* quantitative predictive purposes. Further, by trying to emphasize a single dominant feature of the reaction, they all oversimplify one or more aspects of reactive collisions which may in particular cases be crucial. Thus while they often provide a useful low-order interpretation, they do not provide a final or complete one.

One limitation of all the above models is their assumption of ultradirect collisions. Not only is the reaction assumed to be direct but the models involve a monotonically increasing reaction coordinate. In contrast, trajectory studies have often shown that even direct-mode reactions may be more complicated. Another serious limitation of the above models is they have been developed primarily for reactions in which the internal excitation energy of the reactants does not play an important role. Extensions are of course possible, but so far the role of reactant internal energy has been treated theoretically primarily by the use of numerical trajectories, statistical and nonstatistical phase-space theories, and information-theoretic methods.

5. Numerical Trajectories

In cases where one potential energy surface or only a few surfaces with fairly-well-localized interactions are involved, and some method for generating the full surface or surfaces is available, it is possible to actually calculate exact classical trajectories. Such trajectories are discussed in the context of single-surface reactions in this section. There are numerous reviews which emphasize the results of trajectory studies of chemical reactions.[68,129-145] We will not attempt to summarize everything of importance in those reviews or the studies reviewed. Instead, in the rest of this section we attempt to emphasize the reliability or lack thereof of the computed results based on considerations of the sensitivity to potential surface and the accuracy of the classical dynamics. Information about the latter can be obtained from three sources: (a) known inadequacies of classical mechanics, (b) comparison to more accurate results (in three dimensions), either experimental results or theoretical quantum-mechanical ones, and (c) by comparison of collinear trajectory calculations to accurate collinear quantum-mechanical results. The comparisons of trajectory calculations to more accurate quantum-mechanical calculations is particularly instructive because both dynamical methods can be applied assuming the same potential energy surface. This provides a direct measure of the error in the classical dynamics. Comparisons to experiment are more suspect because the differences are due not only to classical dynamics but also to the generally unknown errors in the potential energy surface. It is also difficult to know exactly what should be included in the averaging process if the initial state populations are not known exactly. This is especially important in experiments involving partially relaxed beams. For highly averaged results on systems with no light atoms for which excited electronic states play no role, the differences from experiment are probably due mainly to the potential surface.

There are some general criteria which can be applied to help determine whether classical or quasiclassical methods are appropriate in a given case. In Chapter 16 are listed five criteria, which are not mutually exclusive. First, the de Broglie wavelength for all degrees of freedom should be small compared to the distances over which the potential energy changes appreciably. Second, the methods work best for reaction attributes which result from averaging over most of the details of the behavior. For example, the predicted average vibrational energy of the products is more likely to be accurate than is the complete distribution of product vibrational energies. Third and fourth, the methods fail for classically forbidden regions and in threshold regions, in the classical S matrix senses of the terms. In this context, threshold regions are regions of parameter space which have at least one boundary at which the desired outcome does not occur in the trajectory calculation. By parameter space we mean initial conditions of the trajectories, masses, or parameters of the potential energy hypersurface. For example, it may be that below a certain energy no trajectories lead to products with vibrational energy larger than the quantized energy of the first excited vibrational state, but above this energy some trajectories do. Then products in this vibrational level are said to be classically forbidden or classically inaccessible below this energy, and nearby energies are in a threshold region for this level of the products even though the reaction threshold may be much lower in energy. Fifth, the methods should not be applied to predict detailed information about resonances or quantum interference effects. All of these criteria are somewhat ambiguous to apply, but perhaps the examples which comprise most of this section will give a better "feeling" for the general applicability of the quasiclassical trajectory technique.

In this section we consider not only trajectory studies but also some of the simple explanations which have been advanced for their understanding. Trajectory studies have been valuable in indicating several important features of reactive collisions which must be present in any qualitatively correct treatment. The simplest features, like the models of Section 4, may be understood by focusing on the ultradirect trajectories. But the trajectory studies have also shown the importance of secondary encounters in many cases. Thus even though the reaction may be direct, it need not be as direct as simple models involving smooth monotonic approach and retreat. Secondary encounters have been further classified as migration, clutching, and clouting.[136,146] Multiple secondary encounters (snarled trajectories) are often observed, even for reactions which are basically direct-mode reactions and which occur on potential surfaces without deep wells. Some collisions involving secondary encounters have a close relation to quantum-mechanical resonances involving virtual excitation of vibrational modes of the complex[147-149]; these are usually called Feshbach resonances. But the exact relationship is not clear except in very favorable cases.[150]

Trajectory studies are also helpful in elucidating the dynamics on surfaces with potential wells.[151-156] Even in the presence of wells, the long-range forces and the centrifugal barriers and chemical barriers, if any, in the entrance and exit channels are important in determining both the tendency of complex formation and the kind of dynamics observed. However, if the potential well leads primarily to long-lived complexes instead of direct reactions, many features of the angular and energy distribution can be predicted on the basis of simpler considerations, and it may be unnecessary to compute trajectories to understand this.[157] This subject is further discussed in Chapter 19.

In this section we first consider distributions of scattering angle of reaction products; then we consider energy utilization and disposal. The examples and references are not intended to be comprehensive. In fact, throughout this chapter, whenever possible, we deliberately emphasize a few particularly clear examples and some more recent references from which earlier work can be traced. Hopefully these illustrate the kinds of uses to which classical models can profitably be put without providing an exhaustive list.

5.1. Angular Distributions

The first qualitative results of molecular beam chemistry to whose understanding the trajectory technique was addressed was the question of backward vs. forward scattering.[42] Although trajectory calculations have been successful in calculating whether the reactive scattering is predominantly forward or backward in a number of cases, one should not be too optimistic about predicting even this qualitative trend in cases where the experimental result is not already known. Until quite recently, most reactions studied in molecular beams had large cross sections under thermal conditions, and hence may be assumed to have little or no potential energy barrier to reaction. Angular distributions for such reactions may be quite sensitive to the long-range dispersion and induction interactions and also to the long-range tail of the valence interaction. The latter is very hard to estimate quantitatively, even with state-of-the-art *ab initio* methods. As a consequence the potential surfaces used for trajectory studies are often quite inaccurate at large and intermediate distances and this may be reflected in unreliable angular distributions.[152,158] Another problem occurs for potential surfaces with spurious wells. These may lead to too many long-lived trajectories with a consequent broadening of the angular distribution, i.e., extra backward scattering for reactions where the direct mechanism leads to predominantly forward scattering, and vice-versa, or can lead to excess long-range attractive character if the wells are located in the entrance or exit channel.

In many cases the general character of the angular distribution of a direct-mode reaction may be well correlated by a simple dynamical model such as spectator stripping, the DIPR-DIP approximation, or a rebound model. This, however, should not be interpreted as if the angular distribution is not highly dependent on the range of the forces. On the contrary, the success of such a simple model for a given reaction tells us something about the range of the forces for that reaction.

The mass case $\mathscr{L} + \mathscr{H}\mathscr{H}$ is probably the most favorable one for trying to understand the relationship of the differential cross section to the short-range part of the potential surface. Many workers have emphasized that for this case conservation of linear momentum could cause the angular distribution to reflect the geometry at the onset of product separation. For this reason, Blais and one of the coauthors undertook a detailed study of the effect for this mass combination (in particular for $H + Br_2 \rightarrow HBr + Br$) on the scattering-angular distributions of variations of the short-range and intermediate-range potential surface characteristics, including the bond angle at the col and the narrowness of the bond-angular width of the pass around the col.[159] A good correlation between the bond angle at the onset of reaction and the scattering angle for individual collisions was found. Some of the results for mean molecular (HBr) center-of-mass scattering angle $\bar{\theta}$ as a function of surface type under molecular beam collision conditions are shown in Table 4. This correlation of $\bar{\theta}$ with surface is not strong enough to suggest that one should have confidence in the assignments of probable transition state bond angles that have been made for $\mathscr{L} + \mathscr{H}\mathscr{H}$ reactions as interpretations of observed differential cross sections. This illustrates the general rule that inversion of experimental reactive differential cross sections for some unique feature of the short-range potential surface is very difficult.

A key concept which has been elucidated by trajectory studies is

Table 4. Correlation of Mean Molecular Scattering Angle $\bar{\theta}$ with Surface Type for a Trajectory Study on $H + Br_2 \rightarrow HBr + Br$[a]

	Reactive entrance valley		
Surface	Bond angular width	Most favorable bond angle[b]	$\bar{\theta}$
1	Wide	all	94
2	Wide	160°	110
3	Wide	140°	104
4	Wide	120°	100
5	Wide	100°	98
6	Medium	140°	116
7	Medium	120°	106
8	Medium	100°	95
9–13	Narrow	140°	110–117

[a] Reference 159.
[b] Rounded.

the trend of increasingly forward scattering of the product which contains the new bond as the incident kinetic energy is increased.(95,117,156,160-164) This is in accord with the predictions of simple models, for example, the DIPR model. Trajectory studies have also shown that angular distributions which are backwards peaked at low energies often becomes increasingly more forward and broader as reactant vibrational energy is increased.(156,160-162,164,165)

As discussed above, the dependence of the differential cross section on the potential surface is still very poorly understood in general. And potential energy surfaces are not known well enough that the classical error in trajectory treatments of differential cross sections can be studied by comparing trajectory calculations to experiment. One measure of the classical error in reactive differential cross sections is obtained by comparing the quasiclassical trajectory calculations(166) and accurate quantum-mechanical calculations(167) of the differential cross section for the $H + H_2$ reaction where both sets of calculations employed the same assumed potential energy surface. Such a comparison has been made by Schatz and Kuppermann(167) for a relative translational energy of 10 kcal/mol. The potential surface employed had a symmetrical col with classical barrier height 9.1 kcal/mol, but because some of the initial zero point energy is available for overcoming the barrier, the collision conditions are actually about 4 kcal/mol above the phenomenological threshold. The agreement of the exact quantum and the quasiclassical reactive differential cross sections is excellent.

5.2. Rotational Energy

The effect of initial rotational energy on reaction dynamics has not been studied as systematically as the effect of initial vibrational energy. This is due not only to the usually small size of the rotational quanta and the lack of extensive experimental data, but also to the fact that the usefulness of rotational energy in promoting reaction is not easily related to specific features of the potential energy surface.

The reaction of F with various isotopes of the hydrogen molecule (H_2, D_2, HD) is a good test case since different values of the rotational quantum number J correspond to relatively large energy differences which could appreciably affect the reaction. At an initial relative kinetic energy of 2 kcal/mol, Muckerman[168] found for $J = 0$ to $J = 4$ that the reaction cross section Q decreases monotonically with J for F + D_2 and F + HD, while for F + H_2 it increases from $J = 0$ to $J = 1$ and then decreases with increasing J. Schreiber[162] and Jaffe and Anderson[169] observed the same trends in $Q(J)$ in a study of the F + H_2 reaction. The increase in $Q(J)$ in going from $J = 0$ to $J = 1$ has been attributed[162,168] to a favorable probability for collinear reaction. Since the surfaces employed favor collinear reactions, the cross section would be increased. The decrease of $Q(J)$ at higher J has been interpreted similarly. In a study of the F + D_2 reaction, Blais and one of the authors[161] found a decrease in Q in going from $J = 0$ to $J = 1$ and an increase in Q from $J = 1$ to $J = 5$. They noted that their surface contains a long-range attractive well of depth 4 kcal/mol in the entrance channel and suggested that this could be among the reasons for the differences in $Q(J)$. Schreiber[162] examined this effect by artificially placing an entrance-channel well in one of his surfaces and found $Q(J)$ varying, as found by Blais and one of the authors.[161] Thus long-range attractive forces play a role in determining the effect of reagent rotation on the cross section.

The F + H_2 reactions also provide information about the effect of the initial J state on the center-of-mass angular distribution $I(\theta)$ and on product ratios. Blais and one of the authors[161] found a slight forward shift in $I(\theta)$ with increasing J, while Jaffe and Anderson[169] found a slight backwards shift in $I(\theta)$ with increasing J. Whether this is due to variations in isotopes or variations in surfaces has not been determined. The reaction of F with HD can yield two products, and the variation of the two cross sections with J is quite interesting. For $J = 0$ the DF product dominates, while HF dominates for $J = 1$, and at higher J ($J = 3, 4$), HF is formed almost exclusively. This phenomenon was first noted by Muckerman[170] and was confirmed by Schreiber.[162] The decrease in DF formation has been attributed[162,170] to a shielding effect. Since D is closer to the center of mass of HD than is H, the H cuts out a larger path, and as J is increased, collision of F with H is more likely than collision of F with D.

Further studies of the effect of initial rotation on reaction probability have been presented for F + HCl and H + Cl_2.[171] For both reactions thermal averages of J were employed, and for the latter reaction little variation in reaction probability with reactant rotation was noted. Although the effects were small, it was concluded for the former reaction that the rate constant $k(J)$ decreases with increasing J if the value of J is small, while $k(J)$ increases with increasing J for high values of J.

Another interesting question concerns the effect of orientation of the rotational angular momentum **J**. One aspect of this was studied by Hijazi and Polanyi.[172] For the case $A + BC \rightarrow AB + C$ with three equal masses and a repulsive potential surface (see Section 5.3 for an explanation of attractive and repulsive surfaces), it was found that the parallel or antiparallel orientation of **J** and **L**, where the latter is the initial orbital angular momentum of the relative motion, is the best orientation for reaction; thus coplanar collisions are the most reactive. On an attractive surface there was less dependence of reaction probability on relative orientation of **J** and **L**.

The effect of excess initial rotational energy on product energy distributions has not been extensively studied. Consider one study[164] of F + H_2 as an example. For a thermal distribution of relative translational energies corresponding to a temperature of 300°K it was found that increasing the reactant rotational energy by 3.4 kcal/mol led to increases in the initial relative translational energy for reacting systems, the mean product translational energy, and the mean product rotational energy by 1.7, 1.8, and 4.2 kcal/mol, respectively, but that product vibrational energy decreased, and mean molecular scattering angle was approximately invariant. But there are not enough systematic studies to be able to generalize.

A comparison of quantum-mechanical scattering in three dimensions with quasiclassical trajectories[166] has been made for H + H_2 on a potential surface with no well in the entrance or exit channel.[167] Both methods find $Q(J)$ decreasing in going from $J = 0$ to $J = 2$ at a given relative translational energy E_{rel} in the energy range 0.50–0.80 eV. Excellent agreement for the variation of $Q(J, E_{rel})$ with increasing E_{rel} is found for both $J = 0$ and $J = 1$. However, a significant discrepancy is found for $J = 2$ in the threshold region $E_{rel} \leq 0.42$ eV, where the quasiclassical cross sections are too small. Of course, it is in the region of small cross sections that the quasiclassical results are least accurate with the poorest statistics. It should also be noted that tunneling effects should be considered for this reaction at low energies. These results provide a warning that quasiclassical predictions of $Q(J, E_{rel})$ are not always quantitatively reliable.

Product rotational distributions are often dominated by angular-momentum constraints, especially for the mass combination $\mathscr{L} + \mathscr{H}\mathscr{H} \rightarrow \mathscr{L}\mathscr{H} + \mathscr{H}$ where $\mathscr{L}$ is a light atom and $\mathscr{H}$ is a heavy atom or group. Other mass combinations are sometimes more favorable for gaining information about the dynamical effects governing rotational distributions. The trajectory studies of $F + H_2, D_2$ have provided some information about the correlation of final rotational quantum number J' with final vibrational quantum number v' and with initial J and E_{rel}, but it is not well understood.[161,162,164,168,169,171] For $H + Cl_2$ and $F + HCl$ both experiment and theory[171] agree that initial translational energy enhances product translational energy more than enhancing product rotational energy. Even though such trends have not been studied as systematically as one would like, wells in the exit channel and attractive forces between separating products may play roles in final rotational populations similar to the role played by attractive wells in the entrance channel in determining the effect of initial rotation in $F + D_2$.

It is now possible to measure experimentally the alignment of the rotational angular momentum J' of a dipolar product using molecular beam electric deflection techniques.[128,173] The interpretation of these results has involved statistical methods[174,175] and the DIPR-DIP model.[128] Three sets of trajectory studies are presently available; one discusses the $H + Br_2$ reaction,[176] one discusses the mass combinations $\mathscr{H} + \mathscr{H}\mathscr{L}$, $\mathscr{L} + \mathscr{H}\mathscr{H}$, and especially $\mathscr{L} + \mathscr{L}\mathscr{L}$ on attractive and repulsive surfaces,[172] and one discusses the $F + H_2$ reaction.[164] The study of the $H + Br_2$ reaction[176] showed that the results were not consistent with statistical models, and also showed, for example, that the square $(J'_{z'})^2$ of the projection of $\mathbf{J}'$ on $\mathbf{V}'_{rel}$ was strongly dependent on center-of-mass scattering angle. For the $\mathscr{L}^A + \mathscr{L}^B\mathscr{L}^C$ reaction on a repulsive surface,[172] it was found that $\mathbf{J}'$ is predominantly antiparallel to $\mathbf{L}'$; this was attributed to the release of the repulsion between $\mathscr{L}^B$ and $\mathscr{L}^C$ in the exit channel. A similar result was found for $F + H_2 \rightarrow FH + H$ and was again attributed to repulsive energy release.[164] Although these studies are suggestive, further trajectory work is required to explore the effects of variations of the potential surface and reagent energy on vector–vector correlations, and more comparison with simpler models needs to be done.

5.3. Vibrational Energy

Vibrational energy release and utilization probably correlates with the short-range part of the potential surface better than any other reaction attribute. As early as the 1930s it was appreciated by some workers that energy released as the reactants approach was more likely to be channeled into product vibration than energy released as the products recede. For example, an attempt to interpret the chemiluminescence in the diffusion flame reactions of sodium dimers with halogen atoms led Goodeve[177] to postulate that a chlorine atom, on approaching Na_2,

> . . . attracts the nearer Na atom and repels the other, but the repulsive force is very weak compared with the attractive force. The result is that the first Na atom is drawn violently to

the Cl atom, leaving the other atom at a standstill. Most of the heat of reaction appears as vibrational energy in the NaCl molecule.... The same will be true to a greater or lesser extent in all highly exothermic reactions....

This prediction has been generally confirmed and the explanation is qualitatively correct. Our present level of understanding allows for greater nuances and a more detailed discussion. Goodeve's argument was reformulated by Evans and Polanyi,[178] who, interpreting the same experiment, proposed an early downhill potential surface on which the representative Na atom entering the valley from transition state to products is subject to no repulsion forces along the axis of the valley. In this case they proposed that virtually all the exothermicity would be converted to vibrational energy of the nascent NaX. This condition—the weakness or absence of repulsion of the newly formed products—has been widely applied. It leads to a division of potential energy release into early downhill or attractive ($\mathscr{A}$) and late downhill or repulsive ($\mathscr{R}$) with the former being preferentially channeled into product vibration and the latter into product rotation. This $\mathscr{A}-\mathscr{R}$ correlation was resurrected in the 1950s[179] and is often still invoked in this simple form today. However, J. Polanyi and co-workers have shown that this simple $\mathscr{A}-\mathscr{R}$ correlation is inadequate. In particular, the classic work by Kuntz, Nemeth, Polanyi, Rosner, and Young[180] showed by trajectory calculations involving eight different extended LEPS potential energy surfaces that even late downhill potential energy surfaces, e.g., their surface 2, often lead to high vibrational excitation of the products. They showed, however, that the results of all their trajectory calculations could be explained by dividing the energy release into attractive ($\mathscr{A}$), mixed ($\mathscr{M}$), and repulsive ($\mathscr{R}$). They defined mixed energy release as "that portion of the energy of reaction which is released while the products are separating but the reagents are still attracting one another." Thus mixed energy release is associated with simultaneous (as opposed to consecutive) bond making and bond breaking. Kuntz *et al.* explained the cases of late downhill surfaces leading to high vibrational excitation of the products in terms of mixed energy release. They found that both $\mathscr{A}$ and $\mathscr{M}$ energy release tended to be channeled into product vibration. Mixed energy release is favored by small values of the skewing angle

$$\beta = \arctan\left(\frac{M_B}{M_A} + \frac{M_B}{M_C} + \frac{M_B^2}{M_A M_C}\right)^{1/2}$$

(where M_D is the mass of D) and small values of the exit-valley lengthening factor

$$F = \left(\frac{M_{AB} M_C}{M_A M_{BC}}\right)^{1/2}$$

and unfavored by large β and large F.* Thus, in particular, mixed energy release is unfavored for the mass combination $\mathscr{L} + \mathscr{H}\mathscr{H}$ (called the light-atom anomaly) where $\mathscr{L}$ is a light atom and $\mathscr{H}$ is a heavy one. For this mass combination then the $\mathscr{A}$–$\mathscr{R}$ correlation (without $\mathscr{M}$) is sufficient. In addition to elucidating these important dynamic correlations Kuntz *et al.* showed how the amount of $\mathscr{A}$, $\mathscr{M}$, and $\mathscr{R}$ energy release could be predicted from the potential energy surface itself or from representative collinear trajectories. This classic paper is well worth detailed study by everyone interested in product vibrational energies. Unfortunately the conclusions are often misstated by experimentalists who, ignoring the role of mixed energy release, cite the paper and interpret any evidence of high product vibrational excitation, even for a case with small F, as evidence for an early downhill potential surface. Although the paper by Kuntz *et al.* involved coplanar trajectories the results are supported by later three-dimensional trajectories of the Polanyi group, which they have reviewed elsewhere,[134,136,137,144,164] and of others.[161,170] We will discuss below, however, the fact that the correlations they observed do not always hold when one examines a broader range of potential energy surface topologies.

Another point of view on the question of product vibrational excitation is provided by the use of natural collision coordinates.[181] In these coordinates the reactant vibrational degree of freedom transforms during the reaction into the product vibrational degree of freedom and is represented by a natural collision coordinate t which is measured transverse to the reaction coordinate s. If the motion associated with the transverse vibrational coordinate were adiabatic with respect to the reaction coordinate, then the initially unexcited vibrational degree of freedom would remain unexcited. In this respect, product vibrational excitation may be viewed as a manifestation of vibrational nonadiabaticity.[182-184] This argument may be quantified by considering the Hamiltonian for a collinear collision in natural collision coordinates[182,185-188]:

$$H = T_s + V(t = 0) + H_{\text{vib}}$$

where T_s is the kinetic energy associated with motion along the reaction coordinate, $V(t = 0)$ is the potential energy on the minimum-energy reaction path, and H_{vib} is the kinetic and potential energy associated with transverse motion. T_s may be expanded in a Taylor series

$$\begin{aligned} T_s &= T_s(t = 0) + \left.\frac{\partial T}{\partial t}\right|_{t=0} t + \cdots \\ &= T_s(t = 0) + 2\kappa(s)\, T_s(t = 0)\, t + \cdots \end{aligned}$$

* The skew angle and factor F are the ones that occur in the scaled and skewed coordinate system for which the kinetic energy has no cross terms and has the same reduced mass in every direction.

where $\kappa(s)$ is the curvature of the reaction path. The transverse motion would be adiabatic if H were separable in s and t. There are two kinds of coupling: one, in H_{vib}, results from the fact that the vibrational force constant is a function of s, and the other results from terms such as $2\kappa(s)\,T_s(t=0)\,t$. The latter term has received the most attention. It is a centrifugal force in the (s, t) space and is equivalent to local displacement of the minimum of the vibrational potential. It predicts that vibrational excitation should correlate with curvature of the reaction path and explains[182,186,187,189] those dynamical studies[189-191] on model potential surfaces which show that it does. Closer examination of the centrifugal coupling term shows that nonadiabaticity is expected to be greatest when the curvature and the local kinetic energy along the reaction coordinate are simultaneously high. This is analogous to the conditions which favor high transverse excitation of a bobsled, and vibrational nonadiabaticity has sometimes been called the bobsled effect. From another point of view the centrifugal contribution leads to bigger corrections for lower internal energies (at a fixed total energy) and thus to vibrationally nonadiabatic effective potential curves which approach closely and interact strongly, causing state changes. For the $H + H_2$ reaction $\kappa(s)$ is nonnegligible only within about $0.8a_0$ of the saddle point $(s = 0)$. But for this system $T_s(s = 0.8a_0) \cong T_s(s = 0) +$ 1.0 kcal/mol, which is small, so we conclude that the reaction is vibrationally adiabatic. This appears to be a good approximation in the classical threshold region.[142] The next paragraph discusses a more interesting example, in which the bond energy changes.

Duff and one of the authors[189] have attempted to compare the $\mathscr{A}-\mathscr{M}-\mathscr{R}$ and curvature models by running collinear trajectories for eight mass combinations on a pair of endoergic surfaces which had essentially identical asymptotic diatomic properties and saddle point properties. According to the $\mathscr{A}-\mathscr{R}$ partition of energy release, both surfaces are 100% repulsive. Thus they put 10% or less of the available energy in product vibration for the $\mathscr{L} + \mathscr{H}\mathscr{H}$ mass combination. Further the percentage of energy in product vibration is less than 30% for both surfaces for the $\mathscr{L} + \mathscr{H}\mathscr{L}$ mass combination. However, it is as large as 92% for the other mass combinations. This is explained qualitatively by the $\mathscr{A}-\mathscr{M}-\mathscr{R}$ correlation. One surface, that of Raff *et al.*, has about 55% attractive and mixed energy release, where the sum of these quantities is defined as all energy released as the system progresses along the minimum-energy path until the distance between the two newly bonded atoms has decreased to the outer-classical-turning-point distance of the final diatomic molecule. The other surface, a rotated Morse curve (RMC), has about 97% $\mathscr{A} + \mathscr{M}$ energy release. Further study shows, however, that mixed energy release is not always preferentially channeled into product vibration. The biggest differences between the surfaces are for the $\mathscr{L} + \mathscr{L}\mathscr{L}$, $\mathscr{L} + \mathscr{L}\mathscr{H}$, and $\mathscr{H} + \mathscr{H}\mathscr{H}$ mass combinations. In these cases the Raff *et al.* surface puts an average of 57% of the available energy into vibration but the RMC surface puts only 24% into vibration. The difference is opposite in direction to that expected from the $\mathscr{A}-\mathscr{M}-\mathscr{R}$ correlation of reaction attributes with surface features. But a better correlation of product vibrational energy for these two surfaces and three others was obtained by considering the maximum of $-\kappa(s)\, T_s(t = 0)$, where the negative sign is chosen because it corresponds to the reaction path turning toward products. The Raff *et al.* surface maximizes this quantity at $s = 0.5$ Å (i.e., 0.5 Å past the saddle point) where 15 out of 36 kcal/mol have already been released. But the RMC surface maximizes it at $s = 0.2$ Å, where only 5 kcal/mol have been released. In other words, the results are explained by the fact that the position of maximum curvature is more uphill on the RMC surface, and when the turns come too soon the "bobsled run" does not lead to significant transverse oscillation. Three more general conclusions are also illustrated by this work: (i) Surface shapes yielding very similar results for some mass combinations may yield very different results for others. (ii) Generalizations made on the basis of only extended LEPS surfaces or other too restricted ranges of surface topologies may break down when applied to other surface topologies. In a given family of model surfaces, it is possible that when one noticeable feature of the surface is changed, many other ones also change in a correlated way. Thus even if some other features are responsible for a dynamic effect, the dynamic results may also correlate well with a more noticeable feature. (iii) No *one* of the following list of surface features (all of which have been singled out at one time or another) can explain all the results for product vibrational energies: skew angle β; reagent, product, and saddle point properties, including saddle point position; positions of certain special energy contours, e.g., as used to compute $\% \mathscr{A} + \mathscr{M}$ from the surface; attractive and mixed energy release in representative trajectories; curvature of reaction path; or even the two properties must successful in explaining the above example,[189] viz., how uphill is the position of maximum curvature of the reaction path and the curvature of the repulsive energy contours when all three atoms are close, i.e., the inner repulsive wall.

Sathyamurthy *et al.*[192] have studied trajectories for two different potential surfaces for the endoergic reaction $He + H_2^+ \rightarrow HeH^+ + H$. They found again that important dynamical differences in the role of vibrational energy can be attributed to differences in the inner repulsive wall. It should be emphasized that there is little hard data about what real potential surfaces look like in this crucial corner-turning region. And one should be especially cautious about believing the "predictions" of semiempirical potential surfaces about this region.

Of course the final-state vibrational distribution is also influenced by all the complications mentioned earlier in this chapter, for example, multiple impacts in a given collision. We have tried to focus on the simplest cases. Reaction of atoms with polyatomic molecules also involve more complicating features not discussed here.[135,193-196]

Utilization of vibrational energy in endoergic reactions and vibrational energy disposal in exoergic reactions are of course just two different ways of looking at the same process. Another aspect of the problem, which is somewhat different, concerns the disposal of excess energy over and above that required to cause a reaction. These questions have also been explored by trajectory studies.(155-156,159-164,168-171,193,194,197-199) A full set of references would be out of place here. To explain some of their results obtained using extended LEPS potential surfaces, Polanyi's group has developed the concepts of "induced" attractive and repulsive energy release.(146,198)

An interesting example of how much still needs to be learned is provided by the reaction Ba + FH → BaF + H, which is experimentally observed to release only 12% of the available energy into product vibration.(200,201) Yet, from the trajectory results available, the mass combination $\mathscr{H} + \mathscr{H}\mathscr{L}$ seems to be especially favorable for mixed energy release into product vibration, even for surfaces which are repulsive in the $\mathscr{A}$–$\mathscr{R}$ classification. It seems that a greater variety of possible surface topographies needs to be systematically explored by the trajectory method. For example, it has been suggested that the unexpected results for Ba + FH might be due to an exceptionally narrow entrance valley combined with a highly repulsive surface so that both mixed and attractive energy release are minimized even for this mass combination.(202)

As mentioned already, the accuracy of the trajectory method can be assessed by comparing trajectory calculations to accurate quantum-mechanical calculations. Some of the best tests are for inelastic nonreactive collisions. These tests show that the quasiclassical trajectory method is often accurate when the transition probabilities are large enough. There are also some tests for vibrational state distributions in reactive collisions.(142,203-210) We will consider first the exothermic reaction H + $Cl_2(v = 0, 1, 2)$ → HCl + Cl. This was studied on a LEPS potential surface with a 2.5 kcal/mol classical barrier height and 37% attractive energy release according to the $\mathscr{A}$–$\mathscr{R}$ scheme.(206,207) The quasiclassical trajectory and quantum-mechanical calculations of the first moment of the product vibrational energy are given in Table 5. The product-state distributions for thermal reactants are given in Table 6. The results show that for ground-state reactants the quasiclassical final-state distribution is only qualitatively correct at 300°K but is quantitatively accurate at 1000°K. However, the first moment is quite accurate even at 300°K. For excited-state reactants the errors in the first moments are a little larger. Tables 5 and 6 illustrate the fact that trajectories are more accurate for highly averaged results like the mean vibrational energy of the products than they are for the detailed results such as the full vibrational distribution. The better accuracy at higher temperature is a result both of averaging over a broader distribution of initial translational energies and of obtaining relatively less contributions from threshold regions. The individual state-to-state reaction probabilities

Table 5. Percentage of Available Energy That Appears as Product Vibrational Energy for the Thermal Collinear Reaction $H + Cl_2(v) \rightarrow HCl + Cl$[a]

Method:	Quantum-mechanical	Quasiclassical trajectory	Classical *S* matrix[b]
	$T = 300°K$		
$v = 0$	58[c]	58	58
1	64	65	67
2	66	73	63
	$T = 1000°K$		
$v = 0$	57	58	58
1	64	63	65
2	65	68	64

[a] Reference 206.
[b] Without analytic continuation.
[c] This quantity is $(\langle E'_{vib}\rangle/\langle E_{tot}\rangle) \times 100\%$ where $\langle E'_{vib}\rangle$ is the average final vibrational energy and $\langle E_{tot}\rangle$ is the average total energy. $\langle E'_{vib}\rangle$ is also called the first moment of the final vibrational energy.

Table 6. Thermal Rate Coefficients (in cm molecule^{-1} sec^{-1}) for Production of Specific Vibrational States in the Thermal Collinear Reaction $H + Cl_2 \rightarrow HCl + Cl(v')$[a]

Method:	Quantum-mechanical	Quasiclassical trajectory	Classical *S* matrix[b]
	$T = 300°K$		
$v' = 2$	74	31	2
3	575	459	371
4	705	650	444
5	84	56	38
6	2	4	1
	$T = 1000°K$		
$v' = 2$	2,080	2,260	675
3	9,460	8,820	8,620
4	11,300	11,800	10,200
5	6,590	6,190	4,650
6	1,980	2,180	1,520

[a] Reference 206.
[b] Without analytic continuation.

at fixed energy are much less accurate than the averaged results in the two tables.[206] Attempts to improve on the quasiclassical quantization in threshold regions by using reverse trajectories quantized in the HCl + Cl channel or by using classical *S* matrix theory without analytic continuation do not lead to significant overall improvement in this case. The reaction probability in the threshold region can, in favorable cases, probably be calculated much more accurately using complex time paths, analytically continued classical mechanics, and classical *S* matrix theory.[2,3] However, the computations for real three-dimensional systems are generally impractically difficult to carry out.

Another good example is the endoergic reaction $I + H_2 \rightarrow IH + H$. A generalization of the results of many trajectory studies of this and other reactions is that vibrational excitation energy is more effective than translational excitation energy in causing endothermic reactions, so much so in fact that vibrationally excited species may dominate thermal rate coefficients.[189,197,198,210-213] Examples are shown in Table 7. Further, Table 7 shows that quantum-mechanical rate constants for $I + H_2$ confirm this domination by excited states for the surface of Raff *et al.* But the prediction of which state does dominate the rate constant depends on the dynamical method used and on the surface.

The last example is unfortunately quite typical of many cases where we wish to use theory to understand the dynamics of atom–molecule

Table 7. Fractional Contribution of Each Initial Vibrational State v to the Thermal Reaction Rate for the Collinear Reaction $I + H_2(v) \rightarrow IH + H$[a]

	Potential surface of Raff *et al.*		Potential surface from rotated Morse curve		
	Quantum-mechanical	Quasiclassical trajectory	Quantum-mechanical	Quasiclassical trajectory	Quasiclassical trajectory-reverse
			$T = 300°$		
$v = 0$	0.00	0.00	0.31	0.57	0.33
1	0.29	0.78	0.52	0.41	0.63
2	0.71	0.22	0.16	0.02	0.04
3	0.00	0.00	0.00	0.00	0.00
			$T = 1000°K$		
$v = 0$	0.00	0.00	0.39	0.46	0.32
1	0.29	0.55	0.45	0.37	0.62
2	0.69	0.42	0.14	0.14	0.03
3	0.02	0.02	0.01	0.03	0.02

[a] Reference 210.

reactions. There are important uncertainties in the potential surface and often there is the likely possiblity of quantitative errors in the dynamical treatment. Thus the methods discussed in this chapter do not at this time provide *a priori* predictions of detailed dynamical behavior. The possibility for *a priori* prediction will depend greatly on the ability of chemists to develop methods for constructing potential surfaces of chemical accuracy. However, lest this appear too pessimistic, we note that much of our qualitative understanding of the dynamics of atom–molecule reactions has come from classical trajectory studies or from simple models. Without these studies, experimental results such as angle-velocity distributions and product vibrational distributions would simply be a collection of empirical data. Classical models of the dynamics have been very important in interpreting experimental results and in providing details about the results of a single collision between an atom and a molecule. Even molecular beam studies average over some of this information, and without trajectory studies the details of the dynamics could not be fully examined. As we have discussed, these methods are progressing beyond single-surface studies into the complicated realm of multisurface reactions and electronic nonadiabaticity. Here, too, trajectory studies are becoming significant in elucidating the details of the dynamics. In the future, trajectory studies will help to guide the calculation of potential surfaces by showing which surface features most strongly influence the observable dynamics. Already our qualitative understanding of single-surface atom–molecule reactions is growing rapidly and the classical methods discussed in this chapter are another tool in the chemical dynamicist's tool box for studying reactions.

ACKNOWLEDGMENTS

The authors thank Professor Dudley R. Herschbach, Dr. James T. Muckerman, and Professor John C. Polanyi for supplying unpublished theses (References 23, 122–124, and 162) discussed here. The authors are grateful to Professor W. Ronald Gentry for a critical reading of the original manuscript. David A. Dixon acknowledges an Alfred P. Sloan Research Fellowship. This work was supported in part by the National Science Foundation by a research grant to Donald G. Truhlar.

References

1. D.G. Truhlar, Quasiclassical predictions of final vibrational state distributions in reactive and nonreactive collisions, *Int. J. Quantum Chem. Symp.* **10,** 239–250 (1976).
2. W.H. Miller, Semiclassical methods in reactive and non-reactive collisions, in *The Physics of Electronic and Atomic Collisions, VIII ICPEAC, Invited Lectures and Progress Reports,* B.C. Cobić and M.V. Kurepa, editors, Institute of Physics, Beograd (1973), pp. 503–528.

3. W.H. Miller, Classical *S*-matrix in molecular collisions, *Adv. Chem. Phys.* **30,** 77–136 (1975).
4. E.J. Heller, Time-dependent approach to semiclassical dynamics, *J. Chem. Phys.* **62,** 1544–1555 (1975).
5. C.J. Ballhaüsen and A.E. Hansen, Electronic spectra, *Ann. Rev. Phys. Chem.* **23,** 15–38 (1972).
6. T.F. O'Malley, Diabatic states of molecules—quasistationary electronic states, *Adv. At. Mol. Phys.* **7,** 223–249 (1971).
7. R.W. Numrich and D.G. Truhlar, Mixing of ionic and covalent configurations for NaH, KH, and MgH^+, *J. Phys. Chem.* **79,** 2745–2766 (1975).
8. G.C. Schatz and J. Ross, Franck–Condon factors in studies of dynamics of chemical reactions. I. General theory, application to collinear atom–diatom reactions, *J. Chem. Phys.* **66,** 1021–1036 (1977).
9. Z.H. Top and M. Baer, Incorporation of electronically nonadiabatic effects into bimolecular reactive systems. I. Theory, *J. Chem. Phys.* **66,** 1363–1371 (1977).
10. J.C. Tully, Diatomics-in-molecules potential surfaces. II. Nonadiabatic and spin–orbit interactions, *J. Chem. Phys.* **59,** 5122–5134 (1973).
11. J.C. Tully, Nonadiabatic processes in molecular collisions, in *Modern Theoretical Chemistry,* Vol. 2, *Dynamics of Molecular Collisions, Part B,* W.H. Miller, editor, Plenum Press, New York (1976), pp. 217–267.
12. G. Herzberg, *Spectra of Diatomic Molecules,* Van Nostrand, Princeton, New Jersey (1950), pp. 530–532.
13. D.G. Truhlar, C.A. Mead, and A. Brandt, Time-reversal invariance, representations for scattering wavefunctions, symmetry of the scattering matrix, and differential cross-sections, *Adv. Chem. Phys.* **33,** 295–344 (1975).
14. D.G. Truhlar, Multiple potential energy surfaces for reactions of species in degenerate electron states, *J. Chem. Phys.* **56,** 3189–3190 (1972); Erratum: **61,** 440 (1974).
15. J.T. Muckerman and M.D. Newton, Comment on "Multiple potential energy surfaces for reactions of species in degenerate electronic states" by D.G. Truhlar, *J. Chem. Phys.* **56,** 3191–3192 (1972).
16. J.C. Tully, Collisions of $F(^2P_{1/2})$ with H_2, *J. Chem. Phys.* **60,** 3042–3050 (1974).
17. A. Komornicki, T.F. George, and K. Morokuma, How much do multiple electronic surface influence chemical reactivity? $F + H_2$: A case study, *J. Chem. Phys.* **65,** 4312–4314 (1976).
18. J.P. Sung and D.W. Setser, Electronic and vibrational energy disposal in the reactions of fluorine atoms with hydrogen bromide and hydrogen iodide, *Chem. Phys. Lett.* **48,** 413–419 (1977).
19. R.J. Donavon and D. Husain, Recent advances in the chemistry of electronically excited atoms, *Chem. Rev.* **70,** 489–516 (1970).
20. D. Husain, The reactivity of electronically excited species, *Ber. Bunsenges. Phys. Chem.* **81,** 168–177 (1977).
21. J.J. Kaufman, Potential-energy surface considerations for excited state reactions, *Adv. Chem. Phys.* **28,** 113–169 (1975).
22. B.H. Mahan, Ion-molecule collision phenomena, in *International Review of Science: Physical Chemistry, Series Two,* Vol. 9, *Chemical Kinetics,* D.R. Herschbach, editor, Butterworths, London (1976), pp. 25–65.
23. P.A. Whitlock, Theoretical investigations of the energetics and dynamics of the reactions $O(^3P, ^1D) + H_2$ and $C(^1D) + H_2$, Ph.D. Thesis, Research Institute for Engineering Sciences, Wayne State University, Detroit, Michigan (1977).
24. D.D. Parrish and D.R. Herschbach, Molecular beam chemistry: Persistent collision complexes in reactions of oxygen atoms with bromine molecules, *J. Am. Chem. Soc.* **95,** 6133–6134 (1973).
25. D.A. Dixon, D.D. Parrish, and D.R. Herschbach, Possibility of singlet-triplet transi-

tions in oxygen exchange reactions, *Faraday Discuss. Chem. Soc.* **55**, 385–387 (1973).
26. D.St.A.G. Radlein, J.C. Whitehead, and R. Grice, Reactive scattering of oxygen atoms: $O + I_2, ICl, Br_2$, *Mol. Phys.* **29**, 1813–1828 (1975).
27. R.B. Woodward and R. Hoffmann, *The Conservation of Orbital Symmetry*, Academic Press, New York (1970).
28. B.H. Mahan, Molecular orbital correlations and ion–molecule reaction dynamics, *J. Chem. Phys.* **55**, 1436–1446 (1971).
29. B.H. Mahan, Electronic structure and chemical dynamics, *Acc. Chem. Res.* **8**, 55–61 (1975).
30. D.H. Liskow, C.F. Bender, and H.F. Schaeffer III. Potential energy surfaces related to the ion–molecule reaction $C^+ + H_2$, *J. Chem. Phys.* **61**, 2507–2513 (1974).
31. C.R. Iden, R. Liardon, and W.S. Koski, Complex formation in the reaction $C^+(D_2, D) CD^+$, *J. Chem. Phys.* **54**, 2757–2758 (1971).
32. B.H. Mahan and T.M. Sloane, Dynamics of the C^+–H_2 reaction *J. Chem. Phys.* **59**, 5661–5676 (1973).
33. C. Maltz, Barrier to complex formation in the hydrogen atom–chlorine molecule reaction, *Chem. Phys. Lett.* **9**, 251–253 (1971).
34. W.A. Goddard III, Selection rules for chemical reactions using the orbital phase continuity principle, *J. Am. Chem. Soc.* **94**, 793–807 (1972).
35. G.S. Hammond, A correlation of reaction rates, *J. Am. Chem. Soc.* **77**, 334–338 (1955).
36. C.A. Parr and D.G. Truhlar, Potential energy surfaces for atom transfer reactions involving hydrogens and halogens, *J. Phys. Chem.* **75**, 1844–1860 (1971).
37. S.W. Benson, *Thermochemical Kinetics*, 2nd edition, Wiley Interscience, New York (1976).
38. A.I. Popov, Interhalogen compounds and polyhalide anions, in *MTP International Review of Science: Inorganic Chemistry, Series One*, Vol. 3, *Main Group Elements, Group VIII, and Noble Gases*, V. Gutmann, editor, Butterworths, London (1971), pp. 53–84.
39. Y.T. Lee, P.R. LeBreton, J.D. McDonald, and D.R. Herschbach, Molecular beam kinetics: Evidence for preferred geometry in interhalogen exchange reactions, *J. Chem. Phys.* **51**, 455–456 (1969).
40. J.D. McDonald, P.R. LeBreton, Y.T. Lee, and D.R. Herschbach, Molecular beam kinetics: Reactions of deuterium atoms with halogen molecules, *J. Chem. Phys.* **56**, 769–788 (1972).
41. C.F. Carter, M.R. Levy, and R. Grice, Reactive scattering of methyl radicals: $CH_3 + ICl, IBr, I_2$, *Discuss. Faraday Soc.* **53**, 357–368 (1973).
42. D.R. Herschbach, Reactive scattering in molecular beams, *Adv. Chem. Phys.* **10**, 319–393 (1966).
43. A.D. Walsh, The electronic orbitals, shapes, and spectra of polyatomic molecules, *J. Chem. Soc.* **1953**, 2260–2288 (1953).
44. R.J. Buenker and S.D. Peyerimhoff, Molecular geometry and the Mulliken–Walsh molecular orbital model. An *ab initio* study, *Chem. Rev.* **1974**, 127–188 (1974).
45. A.D. Walsh, Some notes on the electronic spectra of small polyatomic molecules, in *International Review of Science: Physical Chemistry, Series Two*, Vol. 3, *Spectroscopy*, D.A. Ramsay, editor, Butterworths, London (1976), pp. 301–316.
46. G.G. Balint-Kurti, Potential energy surfaces for chemical reaction, *Adv. Chem. Phys.* **30**, 137–184 (1975).
47. W.A. Goddard III and R.C. Ladner, A generalized orbital description of the reactions of small molecules, *J. Am. Chem. Soc.* **93**, 6750–6756 (1971).
48. T.A. Halgren and W.N. Lipscomb, Self-consistent-field wavefunctions for complex molecules: The approximation of partial retention of diatomic differential overlap, *J. Chem. Phys.* **58**, 1569–1591 (1973).

49. T.A. Halgren and D.A. Kleier, program to be submitted to Quantum Chemistry Program Exchange, Indiana University, Bloomington, Indiana.
50. P. Siegbahn and B. Liu, An accurate three-dimensional potential energy surface for H_3, *J. Chem. Phys.* **68**, 2457–2465 (1978).
51. D.A. Dixon, unpublished.
52. T. Carrington and J.C. Polanyi, Chemiluminescent reactions, in *MTP International Review of Science: Physical Chemistry, Series One*, Vol. 9, *Chemical Kinetics*, J.C. Polanyi, editor, Butterworths, London (1972), pp. 135–171.
53. L. Salem, C. Leforestier, G. Segal and R. Wetmore, On avoided surface crossings, *J. Am. Chem. Soc.* **97**, 479–487 (1975).
54. C.A. Mead, The "non-crossing" rule for electronic potential energy surfaces: The role of time-reversal invariance, *J. Chem. Phys.* (in press), and references therein.
55. R. Wolfgang, Energy and chemical reactions. II. Intermediate complexes *vs.* direct mechanisms, *Acc. Chem. Res.* **3**, 48–54 (1970).
56. D.J. McClure, C.H. Douglass, and W.R. Gentry, The dynamics of the reaction $D_2^+ + O(^3P) \rightarrow OD^+ + D$, and the influence of the atomic quadrupole moment on the cross section at very low kinetic energies, *J. Chem. Phys.* **67**, 2362–2370 (1977).
57. D.M. Manos and J.M. Parson, Crossed molecular beam study of chemiluminescent reactions of group IIIB atoms with O_2, *J. Chem. Phys.* **63**, 3575–3585 (1975).
58. K. Liu and J.M. Parson, Laser fluorescence detection of nascent product state distributions of Sc and Y with O_2, NO, and SO_2, *J. Chem. Phys.* **67**, 1814–1828 (1977).
59. W.S. Struve, J.R. Krenos, D.L. McFadden, and D.R. Herschbach, Molecular beam kinetics: Angular distributions and chemiluminescence in reactions of alkali dimers with halogen atoms and molecules, *J. Chem. Phys.* **62**, 404–419 (1975).
60. J.R. Krenos and J.C. Tully, Statistical partitioning of electronic energy: Reactions of alkali dimers with halogen atoms, *J. Chem. Phys.* **62**, 420–424 (1975).
61. M.B. Faist and R.D. Levine, On the product electronic state distribution in reactions of alkali dimers with halogen atoms, *Chem. Phys. Lett.* **47**, 5–10 (1977).
62. D.G. Truhlar and A. Kuppermann, Application of the statistical phase space theory to reactions of atomic hydrogen with deuterium halides, *J. Phys. Chem.* **73**, 1722–1734 (1969).
63. D.C. Fullerton and T.F. Moran, Application of the statistical phase-space theory to the reactions of rare-gas ions with nitrogen molecules, *J. Chem. Phys.* **54**, 5221–5230 (1971).
64. G.H. Saban and T.F. Moran, Vibrational distribution of N_2^+ $B^2\Sigma$ products from Ne^+–N_2 collisions, *J. Chem. Phys.* **47**, 895–898 (1972).
65. D.G. Truhlar, Intermediate-coupling probability matrix approach to chemical reactions: Dependence of the reaction cross section for $K + HCl \rightarrow KCl + H$ on initial translational and vibrational energy, *J. Am. Chem. Soc.* **97**, 6310–6317 (1975).
66. R.D. Levine, The formation of electronically excited products in chemical reactions: An information theoretic analysis, in *Summer Colloquium on Electronic Transition Lasers, 2nd, Woods Hole, Massachusetts, 1975*, J.I. Steinfeld, editor, M.I.T. Press, Cambridge, Massachusetts (1976), pp. 251–268.
67. J.C. Tully and R.K. Preston, Trajectory surface hopping approach to non-adiabatic molecular collisions: The reaction of H^+ with D_2, *J. Chem. Phys.* **55**, 562–572 (1971).
68. J.C. Tully, Trajectories in ion-molecule reactions, *Ber. Bunsenges. Phys. Chem.* **77**, 557–565 (1973).
69. P. Pechukas, Time-dependent semiclassical scattering theory. II. Atomic collisions, *Phys. Rev.* **181**, 174–185 (1969).
70. J.R. Stine and J.T. Muckerman, On the multidimensional surface intersection problem and trajectory "surface hopping," *J. Chem. Phys.* **65**, 3975–3984 (1976).
71. J.R. Krenos, R.K. Preston, R. Wolfgang, and J.C. Tully, Molecular beam and trajectory studies of reactions of H^+ with H_2, *J. Chem. Phys.* **60**, 1634–1659 (1974).

72. S. Chapman and R.K. Preston, Nonadiabatic molecular collisions: Charge exchange and chemical reaction in the $Ar^+ + H_2$ system, *J. Chem. Phys.* **60**, 650–659 (1974).
73. A. Komornicki, T.F. George, and K. Morokuma, Decoupling scheme for a semiclassical treatment of electronic transitions in atom–diatom collisions: Real-valued trajectories and local analytic continuation, *J. Chem. Phys.* **65**, 48–54 (1976).
74. J.R. Laing, J.-M. Yuan, I.H. Zimmerman, P.L. DeVries, and T.F. George, Semiclassical theory of electronic transitions in molecular collisions: Combined effects of tunneling and energetically inaccessible electronic states, *J. Chem. Phys.* **66**, 2801–2805 (1977).
75. J.M. Bowman, S.C. Leasure, and A. Kuppermann, Large quantum effects in a model electronically nonadiabatic reaction: $Ba + N_2O \rightarrow BaO + N_2$, *Chem. Phys. Lett.* **43**, 374–376 (1976).
76. J.L. Magee, The mechanism of reactions involving excited electronic states: The gaseous reactions of the alkali metals and halogens, *J. Chem. Phys.* **8**, 687–698 (1940).
77. D.R. Herschbach, Molecular beam studies of internal excitation of reaction products, *Appl. Opt. Suppl.* **2**, 128–144 (1965).
78. J.L. Gole, Development of visible chemical lasers from reactions yielding visible chemiluminescence, in *Summer Colloquium on Electronic Transition Lasers II, Third, Snowmass, Colorado, 1976,* L.E. Wilson, S.N. Suchard, and J.I. Steinfeld, editors, M.I.T. Press, Cambridge, Massachusetts (1977), pp. 136–166.
79. J.L. Gole, On the formation of the group IIA dihalides—correlation of reaction dynamics and molecular electronic structure, *Chem. Phys.*, submitted for publication.
80. R. Grice, Reactive scattering, *Adv. Chem. Phys.* **30**, 247–312 (1975).
81. G. Gioumousis and D.P. Stevenson, Reactions of gaseous molecule ions with gaseous molecules. V. Theory, *J. Chem. Phys.* **29**, 294–299 (1958).
82. E.F. Greene and A. Kuppermann, Chemical reaction cross sections and rate constants, *J. Chem. Ed.* **45**, 361–369 (1968).
83. E.W. McDaniel, V. Cermak, A. Dalgarno, E.E. Ferguson, and L. Friedman, *Ion–Molecule Reactions,* Wiley-Interscience, New York (1970), pp. 198–213.
84. R.D. Levine and R.B. Bernstein, Post-threshold energy dependence of the cross section for endoergic processes: Vibrational excitation and reactive scattering, *J. Chem. Phys.* **56**, 2281–2287 (1972).
85. M. Henchman, Rate constants and cross sections, in *Ion–Molecule Reactions,* Vol. 1, J.L. Franklin, editor, Plenum Press, New York (1972), pp. 101–259.
86. R.M. Harris and D.R. Herschbach, *Faraday Discuss. Chem. Soc.* **55**, 121–123 (1973).
87. T. Su and M.T. Bowers, Theory of ion-polar molecule collisions: Comparison with experimental charge transfer reactions of rare gas ions to geometric isomers of difluorobenzene and dichloroethylene, *J. Chem. Phys.* **58**, 3027–3037 (1973).
88. M.T. Bowers and T. Su, Theory of ion polar molecule collisions, in *Interaction Between Ions and Molecules,* P. Ausloos, editor, Plenum Press, New York (1975), pp. 163–183.
89. W.R. Gentry and C.F. Giese, Long-range interaction of ions with atoms having partially filled *p* subshells, *J. Chem. Phys.* **67**, 2355–2361 (1977).
90. D.G. Truhlar, Statistical phase-space theory of the reaction $C^+ + D_2$ including threshold behavior, *J. Chem. Phys.* **51**, 4617–4623 (1969).
91. D.G. Truhlar, Enhancement of the reaction cross section of $He + H_2^+ \rightarrow HeH^+ + H$ by vibrational excitation of H_2^+ according to the statistical phase-space theory, *J. Chem. Phys.* **56**, 1481–1486 (1972); Erratum: A.F. Wagner and D.G. Truhlar, *ibid.* **57**, 4063–4064 (1972).
92. A. Henglein, Stripping effects in ion-molecule reactions, in *Advances in Chemistry Series,* Vol. 58, P.J. Ausloos, editor, American Chemical Society, Washington, D.C. (1966), pp. 63–79.
93. R.E. Minturn, S. Datz, and R.L. Becker, Alkali-atom-halogen-molecule reactions in

molecular beams: The spectator stripping model, *J. Chem. Phys.* **44**, 1149–1159 (1966).

94. P.B. Armentrout, R.V. Hodges, and J.L. Beauchamp, Endothermic reaction of uranium ions with N_2, D_2 and CD_4, *J. Chem. Phys.* **66**, 4683–4688 (1977).
95. J.T. Muckerman, Classical dynamics of the reaction of fluorine atoms with hydrogen molecules. III. The hot-atom reactions of ^{18}F with HD, *J. Chem. Phys.* **57**, 3388–3396 (1972).
96. D.R. Bates, C.J. Cook, and F.J. Smith, Classical theory of ion–molecule rearrangement collisions at high impact energies, *Proc. Phys. Soc. London* **83**, 49–57 (1964).
97. J.C. Light and J. Horrocks, Molecular rearrangement collisions at high impact energies, *Proc. Phys. Soc. London* **84**, 527–530 (1964).
98. R.J. Suplinskas, Kinematic model for atom–diatom reactions, *J. Chem. Phys.* **49**, 5046–5053 (1968).
99. T.F. George and R.J. Suplinskas, Kinematic model for reaction. II. Ion–molecule reactions involving H_2 and D_2, *J. Chem. Phys.* **51**, 3666–3670 (1969).
100. T.F. George and R.J. Suplinskas, Kinematic model for reaction. III. Detailed dynamics of the reaction of Ar^+ with D_2, *J. Chem. Phys.* **54**, 1037–1045 (1971).
101. T.F. George and R.J. Suplinskas, Kinematic model for reaction. IV. Orientation and isotope effect in the $Ar^+ + HD$ reaction, *J. Chem. Phys.* **56**, 1046–1049 (1971).
102. K.T. Gillen, B.H. Mahan, and J.S. Winn, Dynamics of the $O^+ + H_2$ reaction. II. Reactive and nonreactive scattering of $O^+(^4S_{3/2})$ at relative energies above 13 eV, *J. Chem. Phys.* **59**, 6380–6396 (1973).
103. B.H. Mahan, An analysis of direct ion–molecule reactions, in *Interaction Between Ions and Molecules,* P. Ausloos, editor, Plenum Press, New York (1975), pp. 75–93.
104. B.H. Mahan, W.E.W. Ruska, and J.S. Winn, Sequential model of direct reactions, *J. Chem. Phys.* **65**, 3888–3896 (1976).
105. P.J. Kuntz, E.M. Nemeth and J.C. Polanyi, Distribution of reaction products. IV. Reactions forming an ionic bond, M + XC(2D), *J. Chem. Phys.* **50**, 4607–4622 (1969).
106. M.T. Marron, Simple collision theory of reactive hard spheres, *J. Chem. Phys.* **52**, 4060–4061 (1970).
107. R. Grice and D.R. Hardin, Model for rebound reactions: M + RI, *Mol. Phys.* **21**, 805–816 (1971).
108. A. Gonzalez Ureña and F.J. Aolz, Simple cross-section model for elementary reactions, *Chem. Phys. Lett.* **51**, 281–286 (1977).
109. J.L. Kinsey, G.H. Kwei, and D.R. Herschbach, Molecular beam kinetics: Optical model for reactive scattering of alkali atoms and alkyl iodides, *J. Chem. Phys.* **64**, 1914–1924 (1976).
110. J. Lin and J.C. Light, Phase-space theory of chemical kinetics. III. Reactions with activation energy, *J. Chem. Phys.* **45**, 2545–2559 (1966).
111. J.C. Light, Statistical theory of bimolecular exchange reactions, *Discuss. Faraday Soc.* **44**, 14–29, 82 (1968).
112. R.D. Levine, *Discuss. Faraday Soc.* **44**, 81–82 (1968).
113. A. Ben-Shaul, R.D. Levine, and R.B. Bernstein, Prior-expectation distribution functions for energy disposal and energy consumption in reactive molecular collisions, *J. Chem. Phys.* **61**, 4937–4938 (1974).
114. M.D. Pattengill and J.C. Polanyi, A simple model of chemical reaction: FOTO (forced oscillation of a tightening oscillator), *Chem. Phys.* **3**, 1–18 (1974).
115. H.S. Johnston, *Gas Phase Reaction Rate Theory,* Ronald Press, New York (1966).
116. D.D. Parrish and R.R. Herm, Harmonic forces linear model for reactions of Cs atoms with alkyl iodides, *J. Chem. Phys.* **53**, 2431–2435 (1970).
117. P.J. Kuntz, M.H. Mok, and J.C. Polanyi, Distribution of reaction products. V. Reactions forming an ionic bond, M + XC(3D), *J. Chem. Phys.* **50**, 4623–4652 (1969).
118. P.J. Kuntz, Ion–molecule reaction dynamics: A comparison of two direct interaction models, *Chem. Phys. Lett.* **4**, 129–131 (1969).

119. P.J. Kuntz, Analytical properties of a direct interaction model for gas-phase chemical reaction A + BC → AB + C, *Trans. Faraday Soc.* **66**, 2980–2996 (1970).
120. P.J. Kuntz, The K + ICH_3 → KI + CH_3 reaction: Interpretation of the product angular and energy distributions in terms of a direct interaction model, *Mol. Phys.* **23**, 1035–1050 (1972).
121. M.T. Marron, Derivation of the DIPR model for reactive scattering, *J. Chem. Phys.* **58**, 153–157 (1973).
122. R. Harris, Optical models in molecular scattering, Ph.D. Thesis, Harvard University, Cambridge, Massachusetts (1970).
123. N.D. Weinstein, Angular momentum in chemical reactions, Ph.D. Thesis, Harvard University, Cambridge, Massachusetts (1972).
124. D.S.Y. Hsu, Product rotational polarization in reactive scattering, Ph.D. Thesis, Harvard University, Cambridge, Massachusetts (1974).
125. D.R. Herschbach, Reactive scattering, *Faraday Discuss. Chem. Soc.* **55**, 233–251 (1973).
126. R.A. LaBudde, P.J. Kuntz, R.B. Bernstein, and R.D. Levine, Classical trajectory study of the K + CH_3I reaction, *J. Chem. Phys.* **59**, 6286–6298 (1973).
127. M. Godfrey and M. Karplus, Theoretical investigation of reactive collisions in molecular beams: K + Br_2, *J. Chem. Phys.* **49**, 3602–3609 (1968).
128. D.S.Y. Hsu, G.M. McClelland, and D.R. Herschbach, Molecular beam kinetics: Angle–angular momentum correlation in reactive scattering, *J. Chem. Phys.* **61**, 4927–4928 (1974).
129. M. Karplus, Structural implications of reaction kinetics, in *Structural Chemistry and Molecular Biology,* A. Rich and N. Davidson, editors, W.H. Freeman, San Francisco (1968), pp. 837–847.
130. H. Heydtmann, Monte-Carlo-Rechnungen in der chemischen Kinetik, in *Chemische Elementarprozesse,* H. Hartmann, editor, Springer-Verlag, Heidelberg and New York (1968), pp. 143–156.
131. D.L. Bunker, Trajectory studies, in *Molecular Beams and Reaction Kinetics,* C. Schlier, editor, Academic Press, New York (1970), pp. 355–371.
132. M. Karplus, Special results of trajectory methods, in *Molecular Beams and Reaction Kinetics,* C. Schlier, editor, Academic Press, New York (1970), pp. 372–391.
133. D.L. Bunker, Classical trajectory methods, *Methods Comput. Phys.* **10,** 287–325 (1971).
134. P.J. Kuntz, The computer simulation of reactive collisions, in *The Physics of Electronic and Atomic Collisions, VII ICPEAC, Invited Papers and Progress Reports,* T.R. Govers and F.J. deHeer, editors, North-Holland Publishing Co., Amsterdam (1972), pp. 427–442.
135. J.C. Keck, Monte Carlo trajectory calculations of atomic and molecular excitation in thermal systems, *Adv. Atom. Molec. Phys.* **8,** 39–69 (1972).
136. J.C. Polanyi, Some concepts in reaction dynamics, *Acc. Chem. Res.* **5**, 161–168 (1972).
137. J.C. Polanyi and J.L. Schreiber, The dynamics of bimolecular collisions, in *Physical Chemistry—An Advanced Treatise,* Vol. VIA, *Kinetics of Gas Reactions,* H. Eyring, W. Jost, and D. Henderson, editors, Academic Press, New York (1974), pp. 383–487.
138. R.N. Porter, Molecular trajectory calculations, *Ann. Rev. Phys. Chem.* **25**, 317–356 (1974).
139. D.L. Bunker, Simple kinetic models from Arrhenius to the computer, *Acc. Chem. Res.* **5,** 195–201 (1974).
140. I.W.M. Smith, The production of excited species in simple chemical reactions, *Adv. Chem. Phys* **28**, 1–112 (1975).
141. P.J. Kuntz, The classical trajectory method, in *Interaction Between Ions and Molecules,* P. Ausloos, editor, Plenum Press, New York (1975), pp. 123–137.

142. D.G. Truhlar and R.E. Wyatt, History of H_3 kinetics, *Ann. Rev. Phys. Chem.* **27**, 1–43 (1976).
143. R.N. Porter and L.M. Raff, Classical trajectory methods in molecular collisions, in *Modern Theoretical Chemistry*, Vol. 2, *Dynamics of Molecular Collisions, Part B*, W.H. Miller, editor, Plenum Press, New York (1976), pp. 1–52.
144. P.J. Kuntz, Features of potential energy surfaces and their effect on collisions, in *Modern Theoretical Chemistry*, Vol. 2, *Dynamics of Molecular Collisions, Part B*, W.H. Miller, editor, Plenum Press, New York (1976), pp. 53–120.
145. R.L. Wilkins, Classical dynamics of bimolecular reactions, in *Handbook of Chemical Lasers*, R.W.F. Gross and J.F. Bott, editors, Wiley-Interscience, New York (1976), pp. 551–578.
146. C.A. Parr, J.C. Polanyi, and W.H. Wong, Distribution of reaction products (theory). VIII. Cl + HI, Cl + DI, *J. Chem. Phys.* **58**, 5–20 (1973).
147. R.D. Levine and S.-f. Wu, Resonances in reactive collisions: Computational study of the H + H_2 reaction, *Chem. Phys. Lett.* **11**, 557–561 (1971).
148. J.W. Duff and D.G. Truhlar, Classical S matrix: Numerical applications to classically allowed chemical reactions, *Chem. Phys.* **4**, 1–23 (1974).
149. J.R. Stine and R.A. Marcus, Semiclassical S matrix theory for a compound state resonance in the reactive collinear H + H_2 collision, *Chem. Phys. Lett.* **29**, 575–579 (1974).
150. J.W. Duff and D.G. Truhlar, Classical S matrix: Application of the Bessel uniform approximation to a chemical reaction, *Chem. Phys. Lett.* **40**, 251–256 (1976).
151. I.C. Csizmadia, J.C. Polanyi, A.C. Roach, and W.H. Wong, Distribution of reaction products (theory). VII. $D^+ + H_2 \rightarrow DH + H^+$ using an *ab initio* potential energy surface, *Can. J. Chem.* **47**, 4097–4099 (1969).
152. T.B. Borne and D.L. Bunker, Trajectory studies of halogen atom–molecule exchange reactions, *J. Chem. Phys.* **55**, 4861–4866 (1971).
153. G.H. Kwei, B.P. Boffardi, and S.F. Sun, Classical trajectory studies with long-lived collision complexes. I. Reactions of K with NaCl molecules, *J. Chem. Phys.* **58**, 1722–1734 (1973).
154. A. Gelb and J.S. Alper, Energy transfer in reactive and nonreactive collisions of Na with Na_2, *Chem. Phys.* **14**, 365–373 (1976).
155. A. Gelb and J.S. Alper, Reaction probability and energy transfer in collisions of sodium atoms with sodium dimers, *Chem. Phys.* **19**, 387–395 (1977).
156. J.C. Whitehead, Classical trajectory studies of alkali atom–alkali dimer exchange reactions: Na + Li_2 and Li + Na_2, *Mol. Phys.* **31**, 549–569 (1976).
157. W.B. Miller, S.A. Safron, and D.R. Herschbach, Exchange reactions of alkali atoms with alkali halides: A collision complex mechanism, *Discuss. Faraday Soc.* **44**, 108–122 (1968).
158. N.C. Blais and D.L. Bunker, Monte Carlo calculations. III. A general study of bimolecular exchange reactions, *J. Chem. Phys.* **39**, 315–323 (1963).
159. N.C. Blais and D.G. Truhlar, Monte Carlo trajectories: The reaction $H + Br_2 \rightarrow HBr + Br$, *J. Chem. Phys.* **61**, 4186–4203 (1974); Erratum: **65**, 3803–3804 (1976).
160. J.C. Polanyi and W.H. Wong, Location of energy barriers. I. Effect on the dynamics of reactions A + BC, *J. Chem. Phys.* **51**, 1439–1450 (1969).
161. N.C. Blais and D.G. Truhlar, Monte Carlo trajectories: Dynamics of the reaction F + D_2 on a semiempirical valence-bond potential energy surface, *J. Chem. Phys.* **58**, 1090–1108 (1973).
162. J.L. Schreiber, Classical trajectory studies of chemical reactions, Ph.D. Thesis, University of Toronto (1973).
163. D.J. Malcome-Lawes, Dynamics of hydrogen isotopic exchange reactions at high energies, *J. Chem. Soc. Faraday Trans. II* **71**, 1183–1199 (1975).
164. J.C. Polanyi and J.L. Schreiber, The reaction $F + H_2 \rightarrow HF + H$: A case study in reaction dynamics, *Faraday Discuss. Chem. Soc.* **62**, 267–290 (1977).

165. F. Schneider, U. Havemann, L. Zülicke, V. Pacák, K. Birkenshaw, and Z. Herman, Dynamics of the reaction H_2^+ (He, H)HeH^+: Comparison of beam experiments with quasiclassical trajectory studies, *Chem. Phys. Lett.* **37**, 323–328 (1976).
166. M. Karplus, R.N. Porter, and R.D. Sharma, Exchange reactions with activation energy. I. Simple barrier potential for (H, H_2), *J. Chem. Phys.* **43**, 3259–3287 (1965).
167. G.C. Schatz and A. Kuppermann, Quantum mechanical reactive scattering for three-dimensional atom plus diatom systems. II. Accurate cross sections for H + H_2, *J. Chem. Phys.* **65**, 4668–4692 (1976).
168. J.T. Muckerman, Monte Carlo calculations of energy partitioning and isotope effects in reactions of fluorine atoms with H_2, HD, and D_2, *J. Chem. Phys.* **54**, 1155–1164 (1971).
169. R.L. Jaffe and J.B. Anderson, Classical trajectory analysis of the reaction F + H_2 → HF + H, *J. Chem. Phys.* **54**, 2224–2236 (1971); Erratum: **56**, 682 (1972).
170. J.T. Muckerman, Classical dynamics of the reaction of fluorine atoms with hydrogen molecules. II. Dependence on the potential energy surface, *J. Chem. Phys.* **56**, 2997–3006 (1972).
171. A.M.G. Ding, L.J. Kirsch, D.S. Perry, J.C. Polanyi, and J.L. Schreiber, Effect of changing reagent energy on reaction probability and product energy distribution, *Faraday Discuss. Chem. Soc.* **55**, 252–276 (1973).
172. N.H. Hijazi and J.C. Polanyi, Magnitude and orientation of rotation in exchange reactions A + BC → AB + C. II, *Chem. Phys.* **11**, 1–16 (1975).
173. D.S.Y. Hsu, N.D. Weinstein, and D.R. Herschbach, Rotational polarization of reaction products: Analyses of electric deflection profiles, *Mol. Phys.* **29**, 257–278 (1975).
174. D.A. Case and D.R. Herschbach, Statistical theory of angular momentum polarization in chemical reactions, *Mol. Phys.* **30**, 1537–1564 (1975).
175. D.A. Case and D.R. Herschbach, Statistical theory of angular distributions and rotational orientation in chemical reactions, *J. Chem. Phys.* **64**, 4212–4222 (1976).
176. N.C. Blais and D.G. Truhlar, Monte Carlo trajectories: Alignment of HBr rotational angular momentum as a function of scattering angle for the reaction H + Br_2 → HBr + Br, *J. Chem. Phys.* **67**, 1540–1546 (1977).
177. C.F. Goodeve, Three dimensional models of the potential energy of triatomic systems, *Trans. Faraday Soc.* **30**, 60–68 (1934).
178. M.C. Evans and M. Polanyi, Notes on the luminescence of sodium vapor in highly dilute flames, *Trans. Faraday Soc.* **35**, 178–185 (1939).
179. J.C. Polanyi, Energy distribution among reagents and products of atomic reactions, *J. Chem. Phys.* **31**, 1338–1351 (1959).
180. P.J. Kuntz, E.M. Nemeth, J.C. Polanyi, S.D. Rosner, and C.E. Young, Energy distribution among products of exothermic reactions. II. Repulsive, mixed, and attractive energy release, *J. Chem. Phys.* **44**, 1168–1184 (1966).
181. R.A. Marcus, Analytical mechanics and almost vibrationally-adiabatic chemical reactions, *Discuss. Faraday Soc.* **44**, 7–13 (1968).
182. R.A. Marcus, On the analytical mechanics of chemical reactions: Classical mechanics of linear collisions, *J. Chem. Phys.* **45**, 4500–4504 (1966).
183. R.D. Levine, Radiationless transitions and population inversions: Two examples of internal conversions, *Chem. Phys. Lett.* **10**, 510–515 (1971).
184. G.L. Hofacker and R.D. Levine, Diabatic transition-state theory and the concept of temperature, *Chem. Phys. Lett.* **15**, 165–170 (1972).
185. G.L. Hofacker and N. Rösch, Semiclassical evaluation of the translational–vibrational coupling from the potential energy surface of an atom–diatom collision, *Ber. Bunsenges. Physik. Chem.* **77**, 661–664 (1973).
186. G.L. Hofacker and K.W. Michel, Prediction of vibrational inversion among products of exothermic radical reactions by a classical model, *Ber. Bunsenges. Physik. Chem.* **78**, 174–175 (1974).

187. A. Ben-Shaul and G.L. Hofacker, Statistical and dynamical models of population inversion, in *Handbook of Chemical Lasers,* R.W.F. Gross and J.F. Bott, editors, Wiley-Interscience, New York (1976), pp. 579–617.
188. M.V. Basilevsky, Perturbation theory for vibrational transitions induced by bimolecular reactions, *Mol. Phys.* **26,** 765–775 (1973).
189. J.W. Duff and D.G. Truhlar, Effect of curvature of the reaction path on dynamic effects in endothermic reactions and product energy in exothermic reactions, *J. Chem. Phys.* **62,** 2477–2491 (1975).
190. F.T. Wall and R.N. Porter, Sensitivity of exchange-reaction probabilities to the potential-energy surface, *J. Chem. Phys.* **39,** 3112–3117 (1963).
191. C.C. Rankin and J.C. Light, Quantum solution of collinear reactive systems: $H + Cl_2 \rightarrow HCl + Cl$, *J. Chem. Phys.* **51,** 1701–1719 (1969).
192. N. Sathyamurthy, J.W. Duff, C.W. Stroud, and L.M. Raff, On the origin of the dynamical differences on the diatomics-in-molecules and splinefitted *ab initio* surfaces for the $He + H_2^+$ reaction, *J. Chem. Phys.* **67,** 3563–3569 (1977).
193. T. Valencich and D.L. Bunker, Trajectory studies of hot atom reactions. II. An unrestricted potential for CH_5, *J. Chem. Phys.* **61,** 21–29 (1974).
194. D.L. Bunker and E.A. Goring-Simpson, Alkali–methyl iodide reactions, *Faraday Discuss. Chem. Soc.* **55,** 93–99 (1973).
195. L.M. Raff, Theoretical investigations of the reactions dynamics of polyatomic systems: Chemistry of the hot atom ($T^* + CH_4$) and ($T^* + CD_4$) systems, *J. Chem. Phys.* **60,** 2220–2244 (1974).
196. J.D. McDonald and R.A. Marcus, Classical trajectory study of internal energy distributions in unimolecular processes, *J. Chem. Phys.* **65,** 2180–2192 (1976).
197. R.L. Wilkins, Monte Carlo calculations of reaction rates and energy distribution among reaction products. II. $H + HF(v) \rightarrow H_2(v') + F$ and $HF(v) \rightarrow HF(v') + H$, *J. Chem. Phys.* **58,** 3038–3046 (1973).
198. D.S. Perry, J.C. Polanyi, and C.W. Wilson, Jr., Location of energy barriers. VI. The dynamics of endothermic reactions $A + BC$, *Chem. Phys.* **3,** 317–331 (1974).
199. D.S. Perry and J.C. Polanyi, Energy distribution among reaction products. IX. $F + H_2$, HD, and D_2, *Chem. Phys.* **12,** 419–431 (1976).
200. H.W. Cruse, P.J. Dagdigian, and R.N. Zare, Crossed beam reactions of barium with hydrogen halides: Measurement of internal state distribution by laser-induced fluorescence, *Faraday Discuss. Chem. Soc.* **55,** 277–292 (1973).
201. J.G. Pruett and R.N. Zare, State-to-state reaction rates: $Ba + HF(v = 0, 1) \rightarrow BaF$ ($v = 0$–12) + H, *J. Chem. Phys.* **64,** 1774–1783 (1976).
202. J.C. Polanyi, Molecular beam scattering, *Faraday Discuss. Chem. Soc.* **55,** 389–409 (1973).
203. M. Baer, U. Halavee, and A. Persky, The collinear $Cl + XY$ system ($X, Y = H, D$): A comparison between quantum mechanical, classical, and transition state theory results, *J. Chem. Phys.* **61,** 5122–5131 (1974).
204. P.A. Whitlock and J.T. Muckerman, Comparison of quasiclassical trajectory and classical S matrix treatments of collinear collisions of F and D_2, *J. Chem. Phys.* **61,** 4618–4629 (1974).
205. G.C. Schatz, J.M. Bowman, and A. Kuppermann, Exact quantum, quasiclassical, and semiclassical reaction probabilities for the collinear $F + H_2 \rightarrow FH + H$ and $F + D_2 \rightarrow FD + D$ reactions, *J. Chem. Phys.* **63,** 674–684, 685–696 (1975).
206. D.G. Truhlar, J.A. Merrick, and J.W. Duff, Comparison of trajectory calculations, transition state theory, quantum mechanical reaction probabilities, and rate constants for the collinear reaction $H + Cl_2 \rightarrow HCl + Cl$, *J. Am. Chem. Soc.* **97,** 6771–6783 (1976); J.C. Gray, D.G. Truhlar, and M. Baer, Test of trajectory calculations against quantum mechanical state-to-state and thermal collinear reaction rates for $H + Cl_2 \rightarrow HCl + Cl$, *J. Phys. Chem.*, in press.

207. H. Essén, G.D. Billing, and M. Baer, Comparison of quantum mechanical and quasi-classical calculations of collinear reaction rate constants for the $H + Cl_2$ and $D + Cl_2$ systems, *Chem. Phys.* **17**, 443–449 (1976).
208. G.E. Kellerhalls, N. Sathyamurthy, and L.M. Raff, Comparison of quantum mechanical and quasiclassical scattering as a function of surface topology, *J. Chem. Phys.* **64**, 818–825 (1975).
209. N. Sathyamurthy, R. Rangarajan, and L.M. Raff, Reactive scattering calculations on a splinefitted *ab initio* surface: The $He + H_2^+ (v = 0, 1, 2) \rightarrow HeH^+ + H$ reaction, *J. Chem. Phys.* **64**, 4606–4611 (1976).
210. J.C. Gray. D.G. Truhlar, L. Clemens, J.W. Duff, F.M. Chapman, Jr., G. Morrell, and E.F. Hayes, *J. Chem. Phys.* **69**, 240–252 (1978).
211. J.B. Anderson, Energy requirements for chemical reaction: $H + HF \rightarrow H_2 + F$, *J. Chem. Phys.* **52**, 3849–3850 (1970).
212. L.M. Raff, L.B. Sims, D.L. Thompson, and R.N. Porter, Dynamic effects in exchange reactions: dependence of thermal rate constants upon vibrational excitation, *J. Chem. Phys.* **53**, 1606–1607 (1970).
213. R.N. Porter, L.B. Sims, D.L. Thompson, and L.M. Raff, Classical dynamical investigations of reaction mechanism in three-body hydrogen–halogen systems, *J. Chem. Phys.* **58**, 2855–2869 (1973).

19

Complex-Mode Chemical Reactions: Statistical Theories of Bimolecular Reactions

JOHN C. LIGHT

1. Introduction

Statistical theories of chemical reactions have had an exceptionally long tenure as one of the means of theoretical interpretation of experimental results. In fact it is only in the past decade that, with the availability and scientific understanding of computers, theoreticians have had the temerity to attempt to go beyond either statistical or elementary model theories for reactions of simple systems. However, even in the age of Cray I computers statistical theories of chemical reactions are applied often, refined periodically, and occasionally relied upon. The reason for this is simple. In spite of occasionally somewhat tedious algebra, statistical theories provide an easily understood, if extreme, model against which to compare reality, i.e., experiment. Statistical theories can relatively easily be tailored to conform to the experimental information at hand and do provide a quantitative measure of one relatively intuitive possible outcome.

Statistical theories can be distinguished from "exact" dynamical theories by the explicit admission that probabilistic assumptions are made in their development. Even more, statistical theories proudly replace gigantic chunks of single-particle dynamics with relatively elementary hypotheses relying on averaged initial conditions and imprecise measure-

JOHN C. LIGHT • The James Franck Institute and The Department of Chemistry, The University of Chicago, Chicago, Illinois 60637.

ments of final conditions to remain a diagnostic and sometimes predictive tool in the kineticists' arsenal. The use of statistical theories bypasses the extraordinarily difficult problem of detailed classical or quantum dynamics at the price of reliability—statistical theories only provide model descriptions (which may be accurate) which can be checked against measurements. Eventually, perhaps, a body of experimental knowledge can be accumulated which will permit one to decide, beforehand, with reasonable assurance, that a particular statistical theory will apply.

No "pure" statistical theory of dynamics exists—all involve constraints or assumptions beyond those imposed by time-reversal invariance and the conservation of energy and angular momentum imposed by classical and quantum mechanics. It is the main purpose of this chapter to define the domains of applicability of these theories, reference their development, and give references to the relevant techniques. With apologies to the readers and the editor, and to many scientists who have contributed, but are not referenced, we begin by examining the types of experimental information available and the constraints that this places on theory. We then define some of the theoretical approaches to the problem, and, in Section 3, describe some of the basic working procedures for applying statistical theories.

2. *Averaging and Constraints*

Although all exact approaches to reaction dynamics produce information equivalent to the scattering matrix from which the most detailed observables can be constructed, statistical theories can be constructed to yield more or less detailed information using more or less detailed initial distributions. Thus transition state theory and the RRKM theory in general are used with thermal initial distributions and yield directly thermal rate constants. The more recent statistical or phase-space theories may start with state-selected reactants at fixed translational energy and yield differential cross sections for products in specified states.

Since transition state theory has been in the literature for many years, we shall not discuss it here. Some of the more complete discussions are in References 1–3, and a good review of the assumptions and validity of the underlying hypotheses is given by Pechukas.[4] There are a few recent developments, e.g., the careful application to ion–molecule reactions by Bates,[5] and the discussion of the separability approximation by Miller.[6] Transition state theory, because it focuses on the properties of the complex in thermal equilibrium, cannot easily be used to predict state-to-state cross sections.

The RRKM theory of thermal unimolecular decomposition[7,8] has also been reviewed[3,9,10] and applied extensively. As in transition state theory (from which it is, essentially, derived) RRKM theory requires the use of the properties of the energized complex (excited molecule) and a definition of the transition state, that hypersurface in phase space which, once crossed in the product direction, is never recrossed. Since the RRKM theory has been cast in the form of averages over rates from microcanonical (constant E) distributions, it is much closer to the statistical theories discussed below than the transition state theory. The primary differences lie in the common (but not necessary) neglect of angular momentum constraints and the use of various transition states in the RRKM theory. Since RRKM theory is applied to unimolecular decomposition, the angular momentum of the energized molecule will depend on the mode of excitation (photoexcitation, collisional excitation, or chemical activation, for example) but will often be small enough so that angular momentum constraints are not important.

3. Statistical Theories

In more recent work, Marcus[11-13] has developed a microcanonical version of transition state theory and generalized the RRKM theory to yield state-to-state cross sections. In this form the theories become quite comparable to the statistical (or phase-space) theory of reactions proposed by Nikitin[14] and Pechukas and Light.[15,16]

In this section we restrict the discussion of statistical theories to those based essentially on the definition used by Marcus[13] and Levine,[17] among others, that the squares of the S-matrix elements are approximated by an expression of the form [see Reference 13, equation (4.1)]

$$|S^{JE}_{jnl\alpha \leftarrow j_0 n_0 l_0 \alpha_0}|^2 = (1 - \omega^{\alpha J}_{jnlE})\,\delta_{jj_0}\delta_{nn_0}\delta_{ll_0}\delta_{\alpha\alpha_0} + \omega^{\alpha J}_{jnlE}\,\omega^{\alpha_0 J}_{j_0 n_0 l_0 E} \Big/ \sum_{\alpha} N^{\ddagger}_{\alpha J} \quad (1)$$

where, for fixed energy E and total angular momentum J, a reactant (final) state with quantum numbers $\alpha_0 j_0 n_0 l_0 JE$ ($\alpha jnlJE$) has probability $\omega^{\alpha_0 J}_{j_0 n_0 l_0 E}$ of forming a transient complex, and probability $\omega^{\alpha J}_{jnlE}/\sum_\alpha N^{\ddagger}_{\alpha J}$ of decomposing to the specified final state; α is a chemical channel index, and $N^{\ddagger}_{\alpha J}$ is the total number of quantum states in the exit channel at that "boundary" or transition state of the complex defined by

$$N^{\ddagger}_{\alpha J}(E) = \sum_{jnl} \omega^{\alpha J}_{jnlE} \quad (2)$$

This definition satisfies the two general requirements for a "correct" statistical theory—it satisfies miscroscopic reversibility, and a statistical assumption has been made in decoupling the initial and final states, i.e., working with probabilities of complex formation and decay. This expression is considerably more general than that given earlier[15] in that the ω's are not restricted to being zero or one (although this is a common choice), and the evaluation is made at a transition state which may not be the asymptotic (or "loose" transition state) molecular states.

There are a number of statistical treatments in the literature, with the quantum treatments essentially equivalent to some prescription for evaluating the ω's above. We shall attempt to list some of those here, giving the assumptions (but not the equations), and refer the reader to the relevant literature for applications. The current quantum statistical theories can be categorized by the transition state as "loose" or "tight."

3.1. "Loose" Transition States

In the statistical theory of Light and co-workers[15,16] the transition state was taken at the top of the barrier to radial motion formed from the effective orbital angular momentum potential and the long-range attractive potentials. The states were characterized by their *asymptotic* properties, and the ω's were set to 1 or 0 depending on whether or not the translational energy exceeded the barrier height:

$$E_{\text{Tr}} = E - E_{\text{int}}(nj\alpha) \geq \text{Barrier maximum}\,(E, l)$$

No orientation effects and no partial conservation properties were assumed. The working equations for three-atom complexes are given in References 15, 16, and 18. This model was extended to a four-atom complex by Dagdigian *et al.*[19] (see their excellent appendix). Chesnavich and Bowers[20] have extended a classical form of this theory to symmetrical polyatomic systems.

The "loose" transition state statistical theory described in the references given above is the easiest to apply and requires the fewest detailed assumptions about the nature of the collision complex. The state-to-state cross sections are given by a simple sum[18]

$$\sigma(\alpha, n, j, l; \alpha_0, n_0, j_0, l_0, E) = \frac{\pi\hbar^2}{2\mu_0 E_0 (2j_0 + 1)} \sum_J (2J + 1) P^J(\alpha n j l; \alpha_0 n_0 j_0 l_0 E) \tag{3}$$

where the sum on J is restricted by

$$|j_0 - l_0| \leq J \leq |j_0 + l_0|$$

E_0 is the initial translation energy, and

$$P = 1/N(E,J)$$

if both initial and final states are accessible from the complex, and zero otherwise. $N(E,J)$ is the total number of states accessible from the complex of total energy E, total angular momentum J.

This theory has been most successfully applied to low-energy ion–molecule reactions, many of which (but by no means all) appear to proceed through a true complex. This theory will never predict inverted product distributions. At energies above the dissociation limit the above theory breaks down since the dissociative continuum cannot be treated directly.

Rebick and Levine[21] developed a modified statistical theory in which dissociation can be treated.

3.2. *"Tight" Transition States*

More flexibility can be introduced into statistical theories by introducing some properties of the complex into the definition of the transition state. Equation (1) still applies, but the definition of the probabilities and the properties of the states (e.g., their energies) are changed from their asymptotic values. The quantum states defined at the transition state surface are then connected adiabatically to the asymptotic states for the analysis of the final distributions. This approach has been explored by Marcus and Worry,[12,13] by Safron *et al.*,[22,23] by Wagner and Parks,[24] and others. The most general theoretical description appears to be that of Marcus,[13] where the theory is developed in considerable detail and equations for both tight and loose transition states are given. Dagdigian *et al.*[19] give a useful compilation of the results for three- and four-atom complexes based on Safron's model.

The advantage of using a statistical theory containing a dependence on the properties of the complex is, of course, that it should be able to model the true system more accurately. The disadvantage is that one has little information *a priori* on whether the assumed transition state is correct. This type of theory, however, has been more successful at modeling certain types of reactions[13,23] than the loose transition state theory.

Finally, Miller[25] has recently proposed an interesting connection between the statistical and direct reaction mechanisms which becomes the normal transition state theory in one limit. Although this has not yet been applied to realistic systems, it is an intriguing concept, and includes a variational criterion for choosing the transition state.

Another use, of course, for statistical theories is to serve as *prior* distributions in the information theoretic description of chemical reactions. Since this approach is discussed elsewhere in the book, we only mention here that Pollak[26] has recently analyzed several possible statistical distributions and their adequacy as prior distributions. Apparently the class of statistical theories described above is the most appropriate.

It is hoped that this brief description will allow the interested practitioner to find the relevant material in the literature with a minimum of searching.

ACKNOWLEDGMENT

This work was supported in part by the National Science Foundation.

References

1. S. Glasstone, K. Laidler, and H. Eyring, *The Theory of Rate Processes*, McGraw-Hill, New York.
2. K. Laidler, *Chemical Kinetics*, McGraw-Hill, New York (1965).
3. D.L. Bunker, *Theory of Elementary Gas Reactions*, Pergamon Press, Oxford (1966).
4. P. Pechukas, in *Dynamics of Molecular Collisions, Part B* (Volume 2 of *Modern Theoretical Chemistry*), Wm. Miller, editor, Plenum Press, New York (1976).
5. D.R. Bates, Transition state theory for ion–molecule reactions, *Proc. R. Soc. London A* **360,** 1 (1978).
6. W. H. Miller, Quantum mechanical transition state theory and a new semiclassical model for reaction rate constants, *J. Chem. Phys*, **61,** 1823 (1974); Semiclassical limit of quantum mechanical transition state theory for nonseparable systems, *J. Chem. Phys.* **62,** 1899 (1975).
7. R.A. Marcus, Dissociation and isomerization of vibrationally excited species, *J. Chem. Phys.* **20,** 359 (1952).
8. G,M. Wieder and R.A. Marcus, Dissociation and isomerization of vibrationally excited species. II. Unimolecular reaction rate theory and its application, *J. Chem. Phys.* **37,** 1835 (1962).
9. E.E. Nikitin, *Theory of Thermally Induced Gas Phase Reactions*, Indiana University Press, Bloomington (1966),
10. W.L. Hase, Dynamics of unimolecular reactions, in *Dynamics of Molecular Collisions, Part B* (Volume 2 of *Modern Theoretical Chemistry*), Wm. Miller, editor, Plenum Press, New York (1976).
11. R.A. Marcus, Chemical reaction cross sections, quasiequilibrium, and generalized activated complexes, *J. Chem. Phys.* **45,** 2138 (1966),
12. R.A. Marcus, On the theory of energy distributions of products of molecular beam reactions involving transient complexes, *J. Chem. Phys.* **62,** 1372 (1975).
13. G. Worry and R.A. Marcus, On the theory of translational energy distributions of product molecules of molecular beam reactions involving transient complexes. II, *J. Chem. Phys.* **67,** 1636 (1977).
14. E.E. Nikitin, Relaxation in a double potential well, *Theor. Expt. Chem. (USSR)* **1,** 5 (1965).
15. P. Pechukas and J. Light, On detailed balancing and statistical theories of chemical kinetics, *J. Chem. Phys.* **42,** 3281 (1965).
16. J.C. Light, Statistical theories of bimolecular exchange reactions, *Discuss. Faraday Soc.* **44,** 14 (1967).
17. R.D. Levine, *Quantum Mechanics of Molecular Rate Processes*, Oxford University Press, Oxford (1968); see also *Discuss. Faraday Soc.* **44,** 81 (1967).
18. P. Pechukas, J. Light, and C. Rankin, Statistical theory of chemical kinetics: Application to neutral-atom–molecule reactions, *J. Chem. Phys.* **44,** 794 (1966).
19. P. Dagdigian, H. Cruse, A. Schultz, and R. Zare, Product state analysis of BaO from the reactants Ba + CO_2 and Ba + O_2, *J. Chem. Phys.* **61,** 4450 (1974).
20. W. Chesnavich and M. Bowers, Statistical phase space theory of polyatomic systems: Rigorous energy and angular momentum conservation in reactions involving symmetric polyatomic species, *J. Chem. Phys.* **66,** 2306 (1977).
21. C. Rebick and R. Levine, Collision induced dissociation: A statistical theory, *J. Chem. Phys.* **58,** 3942 (1973).
22. S. Safron, Ph.D. Thesis, Harvard University, Cambridge, Massachusetts, 1969 (unpublished).
23. W.B. Miller, S.A. Safron, and D.R. Herschbach, Exchange reactions of alkali atoms with alkali halides: A collision complex mechanism, *Discuss. Faraday Soc.* **44,** 108 (1967).

24. A. Wagner and E. Parks, A classical statistical theory for chemical reactions, *J. Chem. Phys.* **65,** 4343 (1976).
25. W.H. Miller, Unified statistical model for "complex" and "direct" reaction mechanisms, *J. Chem. Phys.* **65,** 2216 (1976).
26. E. Pollak, Prior statistical distributions for the collision of an atom with a diatom, *J. Chem. Phys.* **68,** 547 (1978).

20

Collision-Induced Dissociation I: Quantal Treatment

DENNIS J. DIESTLER

1. Introduction

In this chapter we are concerned with the elementary bimolecular-collision process

$$A + BC \rightarrow A + B + C \tag{1}$$

i.e., collision-induced dissociation (CID), which plays a basic role in the kinetics of high-temperature gases. It is of special importance in connection with unimolecular decay, relaxation in shock waves and chemical lasers, and three-body recombination, which is the reverse of CID. In recent years molecular beam data[1-6] on chemical systems exhibiting CID have become available. Consequently, the calculation of cross sections for CID has assumed increasing importance. Even so, the state of the art of computing CID cross sections for systems of chemical interest remains relatively undeveloped compared to that of treating inelastic (Chapters 4–13) and/or reactive (Chapters 14–19) scattering. A number of calculations of CID cross sections have, in fact, appeared. These have been based on a variety of methods: statistical (phase-space),[7,8] optical-model,[9] classical-trajectory (quasi-classical),[10-13] semiclassical,[14-19] and quantal.[20-29] As the statistical, optical-model, and semiclassical methods are treated extensively elsewhere in this volume (see Chapters 12, 16, and 19), we shall focus here on quantal techniques. The following chapter deals with quasiclassical treatments of CID.

DENNIS J. DIESTLER • Department of Chemistry, Purdue University, West Lafayette, Indiana 47907.

2. Quantal Description of the CID Process

For simplicity and clarity, we shall restrict our consideration initially to the prototypical model of the theoretical dynamicist, namely, the collinear (i.e., one-dimensional) collision of an atom A with a diatomic molecule BC (see Figure 1). The generalization of the formal results to two- and three-dimensional collisions is straightforward. (Of course, the effort required to compute cross sections for the higher-dimensional systems is, in general, orders of magnitude greater than for the one-dimensional analogs.)

Suppose A impinges on BC initially in the bound (vibrational) level v. If the initial translational energy [of A relative to the center of mass (c.m.) of BC] is sufficiently great, the diatom can break up according to

$$A + BC(v) \rightarrow A + B + C \tag{2}$$

with all three "product" atoms emerging *free* from the encounter. The total probability for process (2) may be expressed formally as

$$P(\alpha v \rightarrow 0) = (2\pi)^2\, \sigma_\alpha q_\alpha^{-1} \int dq'_\alpha \int dp'_\alpha\, \delta(E'_0 - E_\alpha)\, |\langle \Phi_0 | \mathbf{T}_{0\alpha} | \Phi_{\alpha v} \rangle|^2 \tag{3}$$

In equation (3) the subscripts α and 0 denote the initial (A + BC) and final (A + B + C) particle arrangements, respectively. The momenta

$$p_\alpha = (m_B p_C - m_C p_B)\,/\,M_{BC}$$

$$q_\alpha = [m_A(p_B + p_C) - M_{BC}\, p_A]\,/M$$

are, respectively, conjugate to the coordinates

$$\begin{aligned} R_\alpha &= r_C - r_B \\ r_\alpha &= (m_B r_B + m_C r_C)/M_{BC} - r_A \end{aligned} \tag{4}$$

where m_X, r_X, and p_X refer to the mass, coordinate, and conjugate momentum

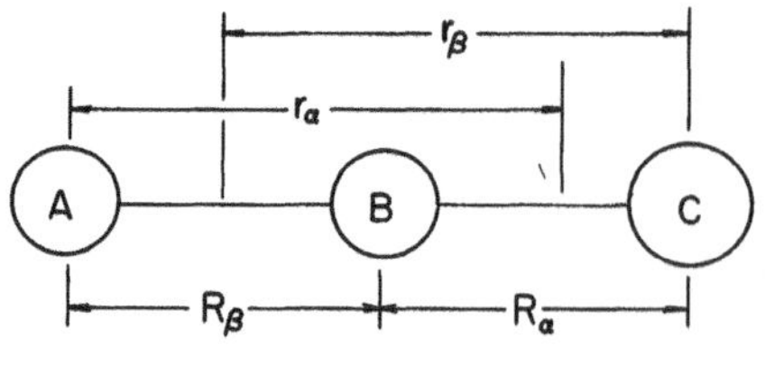

Figure 1. Schematic diagram of a collinear encounter involving three atoms. The distances of A and C from the centers of mass of BC and AB, respectively, are denoted by r_α and r_β.

of atom X (=A, B, C) and

$$M = m_A + m_B + m_C$$

$$M_{XY} = m_X + m_Y$$

The transition operator is defined by

$$\mathbf{T}_{0\alpha} = \mathbf{V}(1 + \mathbf{G}\bar{\mathbf{V}}_\alpha)$$

where the resolvent $\mathbf{G}$ is

$$\mathbf{G} = \lim_{\eta \to 0^+} [E + i\eta - \mathbf{H}]^{-1}$$

$\mathbf{H}$ is the total Hamiltonian and $\mathbf{V}$ is the total potential energy, which can be expressed as

$$\mathbf{V} = \mathbf{V}_\alpha + \bar{\mathbf{V}}_\alpha$$

$\mathbf{V}_\alpha$ being the potential binding (below the CID threshold) the atoms of BC and $\bar{\mathbf{V}}_\alpha$ the interaction (or channel) potential.

The initial state of the system is

$$|\Phi_{\alpha v}\rangle = |v\rangle \otimes |q_\alpha\rangle \tag{5}$$

where $|v\rangle$ is the bound state of BC and $|q_\alpha\rangle$ is a plane-wave state describing the free translation of A relative to the c.m. of BC. The final state

$$|\Phi'_0\rangle = |p'_\alpha; q'_\alpha\rangle = |p'_\alpha\rangle \otimes |q'_\alpha\rangle$$

is a product of plane-wave states governing the free translation of B relative to C ($|p'_\alpha\rangle$) and of A relative to the c.m. of BC ($|q'_\alpha\rangle$). The initial and final channel energies satisfy the relation

$$E = E_\alpha = \tilde{E}_v + q_\alpha^2/2\sigma_\alpha = E'_0 = p_\alpha'^2/2\mu_\alpha + q_\alpha'^2/2\sigma_\alpha \tag{6}$$

where $\tilde{E}_v$ is the initial vibrational energy of BC and

$$\sigma_\alpha = m_A M_{BC}/M$$

$$\mu_\alpha = m_B m_C/M_{BC}$$

From equation (3) one can derive the differential probability

$$\frac{dP(\alpha v_\alpha \to 0)}{dp'_\alpha} = (2\pi)^2\,\sigma_\alpha^2 q_\alpha^{-1}\left[2\sigma_\alpha\left(\frac{q_\alpha^2}{2\sigma_\alpha} + \tilde{E}_v - \frac{p_\alpha'^2}{2\mu_\alpha}\right)\right]^{-1/2}$$

$$\times\,\theta\left(\frac{q_\alpha^2}{2\sigma_\alpha} + \tilde{E}_v - \frac{p_\alpha'^2}{2\mu_\alpha}\right)|\langle\Phi'_0|\mathbf{T}_{0\alpha}|\Phi_{\alpha v}\rangle|^2_{E'_0 = E_\alpha} \tag{7}$$

where θ is the Heaviside unit step function. The three-dimensional analogs of equations (3) and (7) are the total CID cross section

$$\sigma(\alpha n \to 0) = (2\pi)^4 \sigma_\alpha q_\alpha^{-1} \int d\mathbf{q}_\alpha \int d\mathbf{p}_\alpha \delta(E_0' - E_\alpha) |\langle \Phi_0' | \mathbf{T}_{0\alpha} | \Phi_{\alpha n} \rangle|^2 \quad (3')$$

and the fivefold differential cross section

$$\frac{d\sigma(\alpha n \to 0)}{p_\alpha'^2 dp_\alpha' d\Omega_{\mathbf{p}_\alpha'} d\Omega_{\mathbf{q}_\alpha'}} = (2\pi)^4 \sigma_\alpha^2 q_\alpha^{-1} [2\sigma_\alpha(q_\alpha^2/2\sigma_\alpha + \tilde{E}_n - p_\alpha'^2/2\mu_\alpha)]^{1/2}$$

$$\times\, \theta\left(\frac{q_\alpha^2}{2\sigma_\alpha} + \tilde{E}_n - \frac{p_\alpha'^2}{2\mu_\alpha}\right) |\langle \Phi_0' | \mathbf{T}_{0\alpha} | \Phi_{\alpha n} \rangle|^2_{E^{0'} = E_\alpha} \quad (7')$$

In equations (3′) and (7′) the notation is analogous to that developed previously for the collinear model, with the exception that n now stands for the *set* of quantum numbers $\{vJM\}$ required to specify the initial vibrational state of BC. Also, in equation (7′) $\Omega_{\mathbf{x}}$ is the solid angle in which the vector $\mathbf{x}$ lies.

In determining the quantal CID probability (or cross section) our task reduces essentially to computing the transition-matrix element

$$T_{0';\alpha v} = \langle \Phi_0' | \mathbf{T}_{0\alpha} | \Phi_{\alpha v} \rangle \quad (8)$$

which can be expressed alternatively as

$$T_{0';\alpha v} = \langle \Phi_0' | \mathbf{V} | \Psi_{\alpha v}(+) \rangle$$

where $|\Psi_{\alpha v}(+)\rangle$ is the full stationary scattering vector satisfying the Schrödinger equation

$$\mathbf{H}\,|\Psi_{\alpha v}(+)\rangle = E|\Psi_{\alpha v}(+)\rangle \quad (9)$$

along with "outgoing" boundary conditions.* Furthermore, it can be shown[29] that

$$T_{0':\alpha v} = \langle q_\alpha'; k_\alpha'(-) | \mathbf{V}_\alpha | \Psi_{\alpha v}(+) \rangle \quad (10)$$

where

$$|q_\alpha'; k_\alpha'(-)\rangle = |q_\alpha'\rangle \otimes |k_\alpha'(-)\rangle$$

describes the final state as a product of the full two-body continuum state $|k_\alpha'(-)\rangle$ (labeled by the asymptotic wave number k_α') and the corresponding plane-wave state. The energy-conservation relation (6) now becomes

$$E = \tilde{E}_v + q_\alpha^2/2\mu_\alpha$$
$$= \hbar^2 k_\alpha'^2/2\mu_\alpha + q_\alpha'^2/2\sigma_\alpha$$

* The (±) sign associated with the scattering vector indicates whether it satisfies outgoing (+) or incoming (−) boundary conditions.

3. *Quantal Approximation Methods*

A variety of methods of approximating $T_{0';\alpha v}$ are available. We shall review a selection of these briefly before going on to a more detailed consideration of highly accurate numerical methods based on the close-coupling expansion.

3.1. *Born Approximation (BA)*[30]

Perhaps the simplest approximation is that of Born, in which one replaces $\mathbf{T}_{0\alpha}$ by $\mathbf{V}$ in equation (8), to get

$$T_{0';\alpha n} \simeq T^{\rm BA}_{0';\alpha n} = \langle \Phi'_0 | \bar{\mathbf{V}} | \Phi_{\alpha n} \rangle \tag{11}$$

In applying the BA our problem thus reduces to the evaluation of the integral

$$\langle \Phi'_0 | \bar{\mathbf{V}} | \Phi_{\alpha n} \rangle = (2\pi\hbar)^{-9/2} \int d\mathbf{r}_\alpha \int d\mathbf{R}_\alpha \exp(-i\mathbf{q}'_\alpha \cdot \mathbf{r}_\alpha/\hbar) \exp(-i\mathbf{p}'_\alpha \cdot \mathbf{R}_\alpha/\hbar)$$
$$\times \bar{V}(\mathbf{r}_\alpha, \mathbf{R}_\alpha) \exp(i\mathbf{q}_\alpha \cdot \mathbf{r}_\alpha/\hbar)\, \phi_n(\mathbf{R}_\alpha) \tag{12}$$

where $\phi_n(\mathbf{R}_\alpha)$ is the initial bound-state wave function of BC.

Bauer[20,21] has treated CID for the system

$$H + H_2 \rightarrow 3H$$

by the BA, assuming a factorable form for the interaction, i.e., $\bar{V}(\mathbf{r}_\alpha, \mathbf{R}_\alpha) = U(\mathbf{r}_\alpha)\, W(\mathbf{R}_\alpha)$. In this case the six-dimensional integral (12) simplifies to a product of two three-dimensional ones.

3.2. *Distorted-Wave Born Approximation (DWBA)*[30]

Let us suppose that the interaction potential can be written as

$$\bar{\mathbf{V}}_\alpha(\mathbf{r}_\alpha, \mathbf{R}_\alpha) = \bar{\mathbf{V}}^0_\alpha(\mathbf{r}_\alpha) + \bar{\mathbf{V}}^1_\alpha(\mathbf{r}_\alpha, \mathbf{R}_\alpha) \tag{13}$$

where $\bar{\mathbf{V}}^0_\alpha$ is a function of only $\mathbf{r}_\alpha$. Then the DWBA is simply

$$T_{0';\alpha n} \simeq T^{\rm DWBA}_{0';\alpha n} = \langle \Psi^{(0)}_{0'}(-) | \bar{\mathbf{V}}^1_\alpha | \Psi^{(0)}_{\alpha n}(+) \rangle$$

where

$$|\Psi^{(0)}_{\alpha n}(+)\rangle = |\mathbf{q}_\alpha(+)\rangle \otimes |n\rangle$$

and

$$|\Psi^{(0)}_{0'}(-)\rangle = |\mathbf{q}'_\alpha(-)\rangle \otimes |\mathbf{k}'_\alpha(-)\rangle$$

The states $|\mathbf{q}_\alpha(\pm)\rangle$, which satisfy the Schrödinger equation

$$(\mathbf{q}_\alpha^2/2\sigma_\alpha + \bar{\mathbf{V}}_\alpha^0)\,|\mathbf{q}'_\alpha(\pm)\rangle = q_\alpha'^2/2\sigma_\alpha|\mathbf{q}'_\alpha(\pm)\rangle \qquad (14)$$

describe the translation of A in the field of the distortion potential $\bar{V}_\alpha^0$. We note that the only difference between the BA and the DWBA is that the free translational states $|\mathbf{q}_\alpha\rangle$ used in the BA are replaced by the "distorted" states $|\mathbf{q}_\alpha(\pm)\rangle$ in the DWBA. Clearly, if $\bar{V}_\alpha^1 = 0$, then

$$T_{0';\alpha n}^{\mathrm{DWBA}} = T_{0';\alpha n}^{\mathrm{BA}}$$

Lin[22] has applied the DWBA to a collinear model for the CID processes

$$X + H_2 \rightarrow X + 2H \qquad X = \mathrm{He, Ar}$$

taking V_α to be a Morse potential and $\bar{V}_\alpha$ a repulsive exponential. The results seem to be quite sensitive to the choice of the distortion potential ($\bar{V}_\alpha^0$), which is, of course, arbitrary.

3.3. Variational Methods

Bryce and co-workers[23,24] have employed the Kohn variational principle to study CID in a collinear model with short-range forces. In this method, one approximates the dissociation amplitude by

$$T_{0';\alpha n} \simeq (T_{0';\alpha n})_{\mathrm{tr}} - \langle \Psi_{0,\mathrm{tr}}(-)|(E - \mathbf{H})|\Psi_{\alpha n,\mathrm{tr}}(+)\rangle \qquad (15)$$

where the subscript tr denotes a "trial" quantity. The trial scattering wave functions, i.e.,

$$\langle \mathbf{r}_\alpha, \mathbf{R}_\alpha|\Psi_{\mathrm{tr}}(\pm)\rangle \equiv \Psi_{\mathrm{tr}}^{(\pm)}(\mathbf{r}_\alpha, \mathbf{R}_\alpha)$$

must satisfy the same asymptotic conditions as the true wave functions and, of course, the matrix element of $(E - \mathbf{H})$ indicated in equation (15) must exist.

3.4. Impulse Approximation (IA)

In developing the IA, we shall deal with the one-dimensional case. The scattering vector is approximated by

$$|\Psi_{\alpha v}(+)\rangle \simeq \int dq_\beta \int dp_\beta \langle q_\beta; p_\beta|\Phi_{\alpha v}\rangle\, |\psi(q_\beta, p_\beta)(+)\rangle \qquad (16)$$

where $|\psi(q_\beta, p_\beta)(+)\rangle$ describes the scattering of atom A off the struck atom B, atom C participating only as a "spectator" moving freely with relative momentum q_β. The momenta q_β and p_β are conjugate to r_β and

R_β, respectively, where in analogy to equation (4) (see Figure 1)

$$R_\beta = r_B - r_A$$

$$r_\beta = r_C - (m_A r_A + m_B r_B)/M_{AB}$$

Note that the states $|\psi(q_\beta, p_\beta)(+)\rangle$ are weighted with the probability amplitude $\langle q_\beta; p_\beta|\Phi_{\alpha v}\rangle$ that the free plane-wave states

$$|q_\beta; p_\beta\rangle = |q_\beta\rangle \otimes |p_\beta\rangle$$

are contained in the initial state $|\Phi_{\alpha v}\rangle$ [see equation (5)]. If we assume that the interaction potential $\bar{V}_\alpha$ depends only on R_β, i.e., atom A sees only atom B, we have from equations (8) and (16)

$$T_{0';\alpha v} \simeq T^{\text{IA}}_{0';\alpha v} = M_{BC} m_C^{-1} \int dp_\beta \langle k'_\alpha|\tilde{p}_\alpha\rangle \tau^+(\bar{p}_\beta, p_\beta)\langle \bar{p}_\alpha|v\rangle$$

where the two-body T-matrix element

$$\tau^+(\bar{p}_\beta, p_\beta) \equiv \langle \bar{p}_\beta|\bar{\mathbf{V}}_\alpha|\phi(p_\beta)(+)\rangle$$

describes scattering of A off B, $|\phi(p_\beta)(+)\rangle$ being the full (outgoing) two-body scattering vector, and

$$\tilde{p}_\alpha = \tilde{q}_\beta - m_A q_\alpha/M_{AB}$$

$$\bar{p}_\beta = m_B M q'_\alpha/M_{AB}M_{BC} - m_C\tilde{p}_\alpha/M_{BC}$$

$$\bar{p}_\alpha = m_B M \tilde{q}_\beta/M_{AB}M_{BC} - m_A p_\beta/M_{AB}$$

$$\tilde{q}_\beta = M_{BC}(q_\alpha - p_\alpha)/m_C$$

Eckelt and Korsch[25] have applied the IA to a collinear model, taking V_α to be a Morse potential and $\bar{V}_\alpha$ to be impulsive.

4. *Close-Coupling Methods*

So far we have been concerned with approximate techniques, the validity of which is difficult to ascertain. The purpose of this section is to develop a computational procedure which can be systematically improved to give results (i.e., CID probabilities or cross sections) as accurate as desired. Our approach is to solve the Schrödinger equation (9) by a simple extension of the standard close-coupling method of handling inelastic scattering below CID threshold (see Chapters 6–12). Again, to simplify the development, we shall consider the collinear case. We now expand the total scattering wave function as

$$\langle r_\alpha, R_\alpha|\Psi_{\alpha v}(+)\rangle = \Psi^{(+)}_{\alpha i}(r_\alpha, R_\alpha)$$

$$= \sum_{v'=1}^{N_b} f_{v'}(r_\alpha)\,\phi_{v'}(R_\alpha) + \int dk_\alpha f(k_\alpha; r_\alpha)\,\phi(k_\alpha; R_\alpha) \qquad (17)$$

where $\phi_{v'}(R_\alpha)$ and $\phi(k; R_\alpha)$ are, respectively, bound- and continuum-state eigenfunctions of the Hamiltonian of the isolated diatomic

$$\mathbf{H}_\alpha^0 = \mathbf{p}_\alpha^2/2\mu_\alpha + \mathbf{V}_\alpha$$

The functions $f_{v'}(r_\alpha)$ and $f(k_\alpha; r_\alpha)$ are to be determined. It is important to realize that in writing equation (17) we are excluding the possibility of chemical reactions, i.e., the potential energy is assumed to allow bound states only of BC and not of AB or AC. Our eschewing the possibility of rearrangement scattering is not simply a matter of convenience. There are apparently subtle problems of principle involved in handling both CID and reactive scattering simultaneously.

Now, substituting expression (17) for $\Psi_{\alpha v}^{(+)}$ into the Schrödinger equation (9) and projecting both members of the resulting equation successively onto the bound and continuum states of BC, we obtain the close-coupled equations, which are conveniently written in matrix form as

$$[\mathbf{1}\, d^2/dr_\alpha^2 + \bar{\mathbf{E}} + \bar{\mathbf{V}}(r_\alpha)] \cdot \mathbf{f}(r_\alpha) = \mathbf{0} \tag{18}$$

where

$$\begin{aligned}
(\mathbf{1})_{vv'} &= \delta_{vv'} \qquad v, v' \le N_b \\
(\mathbf{1})_{k:k'} &= \delta(k - k') \qquad k, k' \ge 0 \\
(\bar{\mathbf{E}})_{vv'} &= 2\mu\hbar^{-2}(E - \tilde{E}_v)\,\delta_{vv'} \qquad v, v' \le N_b \\
(\bar{\mathbf{E}})_{k;k'} &= 2\mu\hbar^{-2}[E - \tilde{E}(k)]\,\delta(k - k') \qquad k, k' > 0 \\
[\bar{\mathbf{V}}(r_\alpha)]_{vv'} &= 2\mu\hbar^{-2}\langle v|\bar{\mathbf{V}}_\alpha|v'\rangle \qquad v, v' \le N_b \\
[\bar{\mathbf{V}}(r_\alpha)]_{k;v'} &= 2\mu\hbar^{-2}\langle k|\bar{\mathbf{V}}_\alpha|v'\rangle \qquad v' \le N_b \quad k \ge 0 \\
[\bar{\mathbf{V}}(r_\alpha)]_{v;k'} &= 2\mu\hbar^{-2}\langle v|\bar{\mathbf{V}}_\alpha|k'\rangle \qquad v \le N_b \quad k' \ge 0 \\
[\bar{\mathbf{V}}(r_\alpha)]_{k;k'} &= 2\mu\hbar^{-2}\langle k|\bar{\mathbf{V}}_\alpha|k'\rangle \qquad k, k' \ge 0
\end{aligned}$$

The column vector $\mathbf{f}$ comprises the sought functions $f_v(r_\alpha)$ and $f(k; r_\alpha)$. In equations (18) and (19) we have taken the two-body continuum eigenvectors to be normalized according to

$$\langle k|k'\rangle = \int dR_\alpha\, \phi^*(k, R_\alpha)\, \phi(k', R_\alpha) = \delta(k - k')$$

and the zero of energy so that the continuum starts at $\tilde{E}(k) = \hbar^2 k^2/2\mu_\alpha = 0$. Also, to simplify the notation the subscript α on k has been suppressed.

From the asymptotic form of **f**,

$$\lim_{r_\alpha \to \infty} f_{v'}(r_\alpha) = (2\pi\hbar)^{-1/2}\left[\delta_{vv'}\exp(-iq_{\alpha v}r_\alpha/\hbar) - 2\pi i\mu_\alpha q_{\alpha v'}^{-1}\langle q_{\alpha v'};v'|\bar{\mathbf{V}}_\alpha|\Psi_{\alpha v}(+)\rangle \times \exp(iq_{\alpha v'}r_\alpha/\hbar)\right] \qquad v, v' \le N_b$$

$$\lim_{r_\alpha \to \infty} f(k;r_\alpha) = -2\pi i\mu_\alpha[q_\alpha(k)]^{-1}\langle q_\alpha(k);k|\bar{\mathbf{V}}_\alpha|\Psi_{\alpha v}(+)\rangle(2\pi\hbar)^{-1/2} \times \exp[iq_\alpha(k)r_\alpha/\hbar] \qquad k \ge 0 \tag{20}$$

the required T-matrix elements [see equation (10)] can in principle be found.* However, in order to determine **f**, we must solve the infinite, continuous set of close-coupled equations (18). In practice, this can be accomplished by "discretizing" the continuous portion of equations (18).

One way that discretization of the close-coupled equations can be achieved is simply by transforming from the continuous (k) representation to a discrete (n) one by expanding $f(k;r_\alpha)$:

$$f(k;r_\alpha) = \sum_n L_n(k)\,g_n(r_\alpha) \tag{21}$$

where L_n constitute a complete orthonormal set of functions (to be chosen). Combining equations (17) and (21). we can recast the stationary collision wave function in the form

$$\Psi_{\alpha v}^{(+)} = \sum_{v'=1}^{N_b} f_{v'}(r_\alpha)\,\phi_{v'}(R_\alpha) + \sum_{n>N_b} g_n(r_\alpha)\,\chi_n(R_\alpha)$$

in terms of *pseudocontinuum* states χ_n, which are related to the L_n by a Fourier-type transform

$$\chi_n(R_\alpha) = \int dk\, L_{n-N_b}(k)\,\phi(k;R_\alpha) \qquad n = N_b + 1, \ldots \tag{22}$$

We can choose the set $\{L_n\}$ such that $\mathbf{H}_\alpha^0$ is *diagonal* in the basis $\{\chi_n\}$. Then in the new, totally discrete representation, the close-coupled equations become

$$[\mathbf{1}d^2/dr_\alpha^2 + \mathbf{E}' - \bar{\mathbf{V}}'(r_\alpha)]\cdot\mathbf{g}(r_\alpha) = \mathbf{0} \tag{23}$$

where $\mathbf{1}$, $\mathbf{E}'$, $\bar{\mathbf{V}}'$, and $\mathbf{g}$ are the totally discrete analogs of $\mathbf{1}$, $\mathbf{E}$, $\bar{\mathbf{V}}$, and $\mathbf{f}$, respectively. In particular, we have

$$(\mathbf{E}')_{nm} = 2\mu\hbar^{-2}(E - \tilde{E}_n)\,\delta_{nm}$$

* In general, the full two-body continuum state vector should be written as $|k(\pm)\rangle$, the signs indicating the boundary conditions. In the collinear models usually considered, $\lim_{R_\alpha \to -\infty} V_\alpha(R_\alpha) = \infty$, which implies that $|k(+)\rangle = |k(-)\rangle$. Hence for notational simplicity the signs are dropped in the text.

where $\tilde{E}_n$ are eigenvalues of the true bound states for $n \le N_b$ and of the pseudocontinuum states for $n > N_b$. The elements of $\mathbf{g}$ are

$$(\mathbf{g})_n = \begin{cases} f_n & n \le N_b \\ g_n & n > N_b \end{cases}$$

In general, $\bar{\mathbf{V}}(r_\alpha)$ given by equation (19) (or the analogous discrete element) tends to diverge no matter how large r_α becomes, even for "reasonable" interaction potentials. This is true since for any fixed value of r_α, $\bar{V}$ increases without bound as R_α increases [and hence A approaches, and maybe penetrates B (see Figure 1)], while the two-body continuum wave function $\phi(k; R_\alpha)$ merely oscillates, remaining of order unity. In other words, the true continuum states constitute a poor basis in which to expand $\Psi'^{(+)}_{\alpha v}$, since they "permit" the system to assume configurations of infinitely high energy. Hence one would like to choose the set $\{L_n\}$ (or equivalently $\{\chi_n\}$) so that the interaction-potential matrix exists. This requires that the $\{\chi_n\}$ be localized so as to prevent the system's visiting truly inaccessible regions.

An alternative, and more direct, way to discretize equations (18) is simply to diagonalize $\mathbf{H}^0_\alpha$ in an arbitrary, discrete (square-integrable) basis. The lowest N_b eigenvalues

$$\langle \chi'_i | \mathbf{H}^0_\alpha | \chi'_j \rangle = \lambda_i \delta_{ij} \qquad i, j \le N$$

are upper bounds (by the variational theorem) on the true bound-state energies E_v of the isolated diatomic; the corresponding eigenvectors $|\chi'_j\rangle$, of course, approximate the true bound-state eigenvectors. The remaining set of $(N - N_b)$ eigenstates are (localized) pseudocontinuum states and are precisely analogous to the χ_n defined by equation (22). In the basis $\{\chi'_j\}$ the close-coupled equations assume a discrete form identical to equations (23).

Regardless of whether we use the "transformation" or "diagonalization" methods, we obtain a discrete set of close-coupled equations (23). These can be solved using standard procedures (see Chapter 6) to yield the "nonphysical" T-matrix elements $\langle q_{\alpha i}; \chi_j | \bar{\mathbf{V}}_\alpha | \Psi_{\alpha v}(+) \rangle$. The physically relevant elements are, in turn, calculated from these by

$$\langle q_{\alpha v'}; v' | \bar{\mathbf{V}}_\alpha | \Psi_{\alpha v}(+) \rangle = \sum_n L_{nv'} \langle q_{\alpha v'}; \chi_n | \bar{\mathbf{V}}_\alpha | \Psi_{\alpha v}(+) \rangle$$

$$\langle q_\alpha(k'); k' | \bar{\mathbf{V}}_\alpha | \Psi_{\alpha v}(+) \rangle = \sum_n L_n(k) \langle q_\alpha(k'); \chi_n | \bar{\mathbf{V}}_\alpha | \Psi_{\alpha v}(+) \rangle$$

where

$$L_{nv'} \equiv \langle v' | \chi_n \rangle \qquad v' \le N_b$$

$$L_n(k) \equiv \langle k | \chi_n \rangle$$

$$= \int dR_\alpha \, \phi^*(k; R_\alpha) \, \chi_n(R_\alpha)$$

The T-matrix elements $\langle q_{\alpha v'}; \chi_n|\bar{\mathbf{V}}_\alpha|\Psi_{\alpha v}(+)\rangle$ and $\langle q_\alpha(k'); \chi_n|\bar{\mathbf{V}}_\alpha|\Psi_{\alpha v}(+)\rangle$ are, in fact, off the energy shell, whereas the numerically computed pseudo-elements $\langle q_{\alpha n}; \chi_n|\bar{\mathbf{V}}_\alpha|\Psi_{\alpha v}(+)\rangle$ are on-shell. We approximate the physically relevant elements by

$$\langle q_{\alpha v'}; v'|\bar{\mathbf{V}}_\alpha|\Psi_{\alpha v}(+)\rangle = \sum_n L_{nv'} \langle q_{\alpha n}; \chi_n|\bar{\mathbf{V}}_\alpha|\Psi_{\alpha v}(+)\rangle \tag{24a}$$

$$\langle q_\alpha(k'); k'|\bar{\mathbf{V}}_\alpha|\Psi_{\alpha v}(+)\rangle = \sum_n L_n(k') \langle q_{\alpha n}; \chi_n|\bar{\mathbf{V}}_\alpha|\Psi_{\alpha v}(+)\rangle \tag{24b}$$

As the size of the basis set $\{\chi_n\}$ increases, each $|\chi_n\rangle$ becomes a better approximation to the true two-body state, i.e.,

$$L_{nv'} \to \delta_{nv'} \qquad n, v' \leq N_b$$

$$L_n(k) \to \delta(k - k_n) \qquad k > 0$$

and hence equations (24) become a better approximation.

Finally, using equations (3), (10), and (24b), we can compute the differential CID probability. If we are interested only in the total CID probability (P_{vD}), we may compute that most conveniently from the relation

$$P_{vD} = 1 - \sum_{v'} P_{vv'}$$

where the inelastic (and elastic) transition probabilities are explicitly

$$P_{vv'} = (2\pi)^2 \mu_\alpha^2 |q_{\alpha v}|^{-1} |q_{\alpha v'}|^{-1} |\langle q_{\alpha v'}; v'|\bar{\mathbf{V}}_\alpha|\Psi_{\alpha v}(+)\rangle|^2$$

We have applied[27] the "transformation" method of discretizing equation (12) to the following collinear model: V_α is a semi-infinite square well; $\bar{V}_\alpha$ is a repulsive exponential. We have also used[29] the "diagonalization" method to treat a one-dimensional model in which V_α is a Morse potential and $\bar{V}_\alpha$ a repulsive exponential. In both models, the results appear to converge to 1 or 2%.

5. *Concluding Remarks*

So far, the close-coupling method has been applied only to very special collinear models.[27,29] A severe difficulty which plagues close-coupling calculations on systems of practical interest is the excessively large number of channels that must be dealt with explicitly. This is especially problematic in the case of CID, in which all bound states must be accounted for. Some practicable method of handling the large number of channels must be developed (see, for example, Chapters 8–10).

Another very important problem, which we have studiously avoided here, is the close-coupling treatment of systems in which reactive and dissociative collisions can occur simultaneously. For the majority of "chemically interesting" systems the potential-energy surface is such as to allow bound states of several different combinations (arrangements) of the constituent atoms. For example, in the prototypical collinear model (see Figure 1) two arrangements are generally possible: reactants A + BC and products AB + C. It would seem that the continuum contribution, i.e.,

$$\int dk_\alpha f(k_\alpha, r_\alpha, \phi(k_\alpha; R_\alpha)),$$

in arrangement A + BC should implicitly describe both reaction

$$A + BC(v) \rightarrow AB(v') + C$$

and dissociation (1). Whether the extraction of reaction probabilities from this term is practicable remains to be investigated.

References

1. W.A. Chupka, J. Berkowitz, and M.E. Russell, A study of some reactions of H_2^+ in selected vibrational states, in *Abstracts of the VI International Conference on the Physics of Electronic and Atomic Collisions*, I. Amdur, editor, MIT Press, Cambridge, Mass. (1969), pp. 71–72.
2. F.P. Tully, Y.T. Lee, and R.S. Berry, Crossed molecular beam study of collision-induced dissociation of alkali halides, *Chem. Phys. Lett.* **9**, 80–84 (1971).
3. E.K. Parks, N.J. Hansen, and S. Wexler, Collision-induced ion pair formation of thallium halides: Absolute cross sections, *J. Chem. Phys.* **58**, 5489–5501 (1973).
4. E.K. Parks, A. Wagner, and S. Wexler, Collision-induced ion pair formation of thallium halides: Threshold behavior, *J. Chem. Phys.* **58**, 5502–5512 (1973).
5. J. Schöttler and J.P. Toennies, Dissociation of hydrogen molecules in collision with light alkali ions. I. $Na^+ - D_2$, *Chem. Phys.* **2**, 137–157 (1973).
6. J. Schöttler and J.P. Toennies, Dissociation of hydrogen molecules in collisions with light alkali ions. II. $Li^+ - H_2$, *Chem. Phys.* **4**, 24–35 (1974).
7. T.F. Moran and D.C. Fullerton, Statistical phase-space model of the collision-induced dissociation of excited O_2^+ and NO^+ ions, *J. Chem. Phys.* **54**, 5231–5236 (1971).
8. C. Rebick and R.D. Levine, Collision-induced dissociation: A statistical theory, *J. Chem. Phys.* **58**, 3942–3952 (1973).
9. R.D. Levine and R.B. Bernstein, Collision-induced dissociation: A simplistic optical-model analysis, *Chem. Phys. Lett.* **11**, 552–556 (1971).
10. S.W. Benson and G.C. Berend, Vibrational energy exchange of highly excited anharmonic oscillators, *J. Chem. Phys.* **40**, 1289–1298 (1964).
11. H. Fan, Calculation of collisional dissociation of alkali halide molecules by one-dimensional models, *J. Chem. Phys.* **55**, 4628–4632 (1971).
12. W.H. Wong and G. Burns, The dynamics of dissociation of diatomic molecules and mass effect, *J. Chem. Phys.* **62**, 1712–1713 (1975).
13. R.E. Howard, R.E. Roberts, and M.J. DelleDonne, Three-body effects in the exchange and dissociation encounters for $Ar + Ar_2$, *J. Chem. Phys.* **65**, 3067–3074 (1976).
14. L.L. Johnson and R.E. Roberts, A semiclassical treatment of collision induced dissociation, *Chem. Phys. Lett.* **7**, 480–482 (1970).
15. R.D. Levine, On the theory of collision-induced dissociation, *Chem. Phys. Lett.* **11**, 109–112 (1971).
16. W.G. Valence and J. Lin, Bound state to continuum transition probabilities, *Physica (Utrecht)* **52**, 620 (1971).

17. C.S. Lin, Semi-classical study of molecular dissociation: H_2 + He, *Chem. Phys.* **3**, 125–130 (1974).
18. K.R. Dastidar (Mullick) and A.K. Barua, Dissociation of H_2^+ ion by collision-induced vibrational excitation, *J. Chem. Phys.* **62**, 4373–4379 (1975).
19. I. Rusinek and R.E. Roberts, Semiclassical calculation for collision induced dissociation, *J. Chem. Phys.* **65**, 872–880 (1976).
20. E. Bauer, Dissociation of hydrogen molecules by vibrational excitation and three-body recombination coefficient, *Phys. Rev.* **84**, 315–321 (1951).
21. E. Bauer, Dissociation of hydrogen molecules by vibrational excitation and three-body recombination coefficient. II, *Phys. Rev.* **85**, 277–280 (1952).
22. J. Lin, Bound state to continuum collisional probabilities. II. Morse oscillators, *Physica (Utrecht)* **62**, 369–378 (1972).
23. W.M. Bryce and F. Mandl, A Kohn variational principle for three-body break-up processes, *J. Phys. B* **5**, 912–920 (1972).
24. C.F. Bottcher, W.M. Bryce, and F. Mandl, Variational solution of a model breakup process, *J. Phys. B.* **7**, 769–781 (1974).
25. P. Eckelt and H.-J. Korsch, Collision-induced dissociation: Differential energy transfer probabilities, *Chem. Phys. Lett.* **18**, 584–588 (1973).
26. L.W. Ford, D.J. Diestler, and A.F. Wagner, A model study of collision-induced dissociation of a diatomic molecule by an atom, *J. Chem. Phys.* **63**, 2019–2034 (1975).
27. E.-W. Knapp, D.J. Diestler, and Y.-W. Lin, Close-coupling calculation of quantum-mechanical probabilities for collision-induced dissociation, *Chem. Phys. Lett.* **49**, 379–383 (1977).
28. G. Wolken, Jr., Quantum mechanical studies of collision induced dissociation: Discretization theory, *J. Chem. Phys.* **63**, 528–533 (1975).
29. E.-W. Knapp and D.J. Diestler, Quantum-mechanical treatment of collision-induced dissociation, *J. Chem. Phys.* **67**, 4969 (1977).
30. J.R. Taylor, *Scattering Theory*, John Wiley and Sons, New York (1972).

21
Collision-Induced Dissociation II: Trajectories and Models

P. J. KUNTZ

1. Classical Trajectory Calculations

1.1. General Considerations

1.1.1. Introduction

In this chapter we treat collision-induced dissociation (CID) within the framework of classical mechanics. Such an approach is of great practical use not only in the calculation of CID cross sections for their own sake, but also in the fields of hot-atom chemistry, high-energy molecular beam reactions, electron recombination and detachment, and charge transfer, where CID, although not the main process of interest, is present alongside other processes and must be properly taken into account. Indeed, CID cannot be treated independently of other product channels, since dissociation competes with these as soon as the collision energy increases beyond the dissociation threshold. A decent CID calculation must treat this competition adequately.

The most useful classical treatment of collision processes is by the classical trajectory method, which is treated in Chapters 10, 12, and 16 of this book. Most of the trajectory calculations in the literature have been designed primarily for the estimation of reaction cross sections below the dissociation threshold energy; however, several studies at high energies have had to treat the accompanying CID process,[1-22] and a few trajectory papers

P. J. KUNTZ • Hahn-Meitner-Institut für Kernforschung, Glienicker Str. 100, 1000 Berlin 39, West Germany.

have been chiefly concerned with CID itself.[23-27] The essence of the method is to follow the trajectory of the system through phase space starting off in a region corresponding to reactant molecules approaching each other but still sufficiently far apart that there is negligible interaction between them (i.e., the reactant asymptotic region). The trajectory is followed by numerical integration of the classical equations of motion until it is in one of several possible product asymptotic regions, corresponding to molecules separating from each other and having negligible interaction between them. In the asymptotic regions, the variables defining a trajectory fall into two groups: the action variables, which are constants of the motion, and the angle variables, which are characterized by distributions. The result of computing a large number of trajectories all having the same reactant action variables is effectively to transform the reactant *angle* variable distribution (which is known) into a distribution over product *action* variables (which is desired). In this way, cross sections for production of molecules in specific ranges can be obtained. Background material on this subject is given in recent reviews,[28,29] and practical instructions are in Chapter 16. A nice formulation in terms of transformation of distributions is given in a paper by Wagner and Parks.[30]

Application to the CID process requires a careful determination of the dissociation channel, which is discussed in Section 1.2. This aspect of trajectory calculations has been variously treated in the literature, probably because CID has not been the main process of interest. An added complication in the classical method is that a product molecule may have an internal energy greater than its dissociation energy but may still be bound by a maximum in the effective potential which is due to the rotational angular momentum of the product molecule. Such a molecule is referred to as quasibound. Many papers in the literature have taken quasibound molecules to contribute to the dissociation channel, arguing that, in real life, quantum-mechanical tunneling would eventually cause the breakup of such a molecule; other papers have assumed these molecules to be bound. It is my opinion that the former procedure is usually the safest; however, one should do a rough calculation of the lifetime of the quasibound molecule. If this lifetime is longer than the mean time between collisions, then it may be necessary to study the fate of quasibound molecules by further trajectory calculations.

1.1.2. Choosing a Potential Energy Surface

The choice of a potential energy (PE) surface is the hardest part of any collision-theory calculation. If one is fortunate enough to have an *ab initio* surface at hand, then it is only necessary to find an adequate representation of it which allows numerical integration of the equations of motion and one can proceed with the dynamics. If, however, some sort of model representation is used, then it may be wise to bear in mind some PE surface effects, since no amount of effort or sophistication on the dynamics part of the calculation can overcome restrictions or deficiencies introduced by the potential function.

We consider first of all a CID process which occurs by way of a single PE surface. In general, it is very important that the PE surface be able to describe other product channels, particularly chemical reaction, since these compete with CID. Consider for simplicity the three-body system A + BC → A + B + C. If the potential energy function were chosen to be a sum of two-body interactions ($U = U_{AB} + U_{BC} + U_{CA}$), then there would be no provision for a "tearing apart" of the BC molecule by a mechanism which weakens the BC bond as A approaches. The only alternative is for A to "smash" BC apart by the transfer of as much relative kinetic energy to BC internal energy as is possible in a single collision. Several trajectory calculations have demonstrated that many molecules prefer to be "torn apart" by incipient chemical reaction rather than "smashed" apart.[2,21-23] A noteworthy exception is the case where BC has internal energy almost equal to the dissociation energy (or greater if BC is quasibound), and the AB and AC interactions are mainly repulsive. In such circumstances, the pairwise additive potential may be quite adequate and has often been used in studies of three-body recombination.[4-9] At low energies, however, three-body effects are important.[20]

A second consideration is the long-range behavior of the potential energy of the bound diatomic pairs. The effective potential of a diatomic molecule is a sum of the potential function itself, U, plus a repulsive term proportional to the square of the product rotational angular momentum J:

$$U_{\text{eff}}(R) = U(R) + J^2/(2\mu R^2) \quad (1)$$

If this function has a local maximum, it is possible for the diatomic to be quasibound. A maximum is only possible if as $R \to \infty$, $U(R) \sim C/R^N$, $N \geq 2$.[31] Hence the long-range behavior of $U(R)$ will determine the importance of quasibound molecules to the CID cross section.

There often arise cases in which more than one potential energy surface is needed to account for CID. An example is the collision-induced predissociation (CIP) of I_2 which results from collisions of an excited iodine molecule with a rare-gas atom.[27] The rare-gas atom allows a transition from the BO_u^+ state of I_2 to a repulsive state of O_g^+ symmetry through which dissociation occurs:

$$I_2(BO_u^+) + M \to I_2(O_g^+) + M \to I + I + M$$

This type of process can only be properly described by a trajectory surface-hopping calculation[32]: as M approaches the vibrating I_2 molecule, the coupling between the two I_2 states grows, resulting in a region of avoided crossing (a "crossing seam") of the two MI_2 electronic states involved. This crossing seam runs parallel to the entrance channel and extends into the reactant asymptotic region. Hence I_2 molecules having sufficient vibrational energy (either before collision with M or as a result of colliding with M) to access the crossing seam (twice in every vibration) stand a very good chance of dissociating (i.e., remaining on the lower MI_2 surface during an extension of the I_2 bond) as long as atom M is in the vicinity. In this type of CID, atom M plays the role of a catalyst: it opens a pathway for the process but does not have to actively promote dissociation through collisional excitation of I_2, although that is not excluded.

Collision-induced predissociation should always be looked for if the molecule being dissociated is in an electronically excited state. It often occurs accompanied by charge-transfer or ion-pair formation.[33,34] An example is the dissociation of He_2^+ molecules by collision with Ne atoms:

$$Ne + He_2^+ \rightarrow Ne^+ + He + He$$

In this system, the surfaces correlating with $Ne^+ + He_2$ and $Ne + He_2^+$ cross each other in the entrance channel. As Ne approaches He_2^+ a splitting of surfaces develops just as in the $M + I_2$ case. The He_2^+, under the influence of the Ne atom, passes smoothly onto the He_2 (repulsive) curve during one of its outward vibrations, and dissociation occurs. Failure to recognize the possibility of CIP in general invalidates any CID calculation if a CIP pathway exists, for CIP cross sections can be very large. Also, the

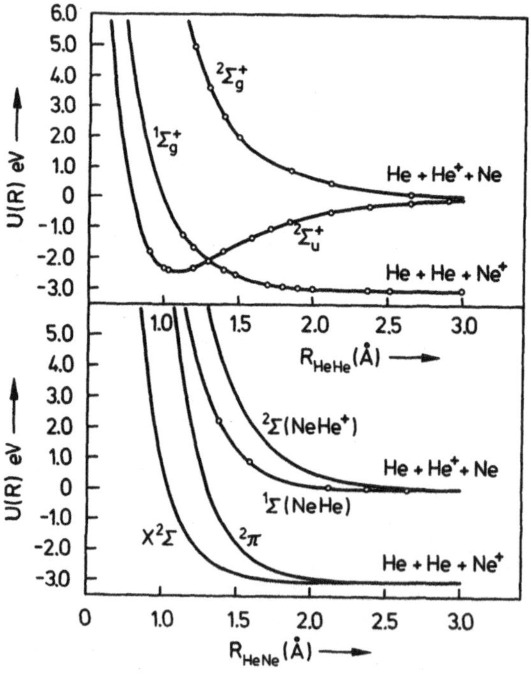

Figure 1. Asymptotic regions of the PE surfaces for the $NeHe_2^+$ system. These cuts through the entrance channel (upper panel) and exit channel (lower panel) were constructed solely from diatomic curves and ionization potentials (after Kuntz and Whitton[34]).

dynamics of CIP are quite different from those of CID: the dependence of the dissociation cross section on collision energy and on the internal energy of the molecule is different in the two cases. For CID on a single PE surface, the energy for dissociation is provided by the motion of the atoms involved (either internal energy of the molecule or relative collision energy), whereas in CIP the energy for dissociation comes from potential energy.

Obviously, deciding whether a multisurface calculation is necessary is a rather important step in CID work. A recommendation which I strongly urge is to write down all of the states of the atoms which could possibly be produced as dissociation products and then to draw a graph of cuts of all the *three-body* PE surfaces (i.e., taking the system $A + BC \rightarrow AB + C$ as an example) through the two-body asymptotic regions. This procedure requires a knowledge only of the two-body interactions and atomic excitation energies or ionization potentials. For example, for the system $[Ne + He_2]^+$ one needs the ionization potentials of Ne and He, and the diatomic curves for the $X\,^2\Sigma_u^+$ and $^2\Sigma_g^+$ states of He_2^+ as well as the $X\,^1\Sigma_g^+$ state of He_2. For the channel $[NeHe + He]^+$ one needs in addition the $X\,^1\Sigma^+$ state of NeHe, and the $^2\Sigma^+$ and $^2\Pi$ states of $NeHe^+$. Cuts through both channels for this system are shown in Figure 1, from which it is obvious that a crossing seam will develop in the entrance channel. This suggestion is not a foolproof way of detecting crossings but it is simple to implement and can help avoid blunders at the earliest stage of a calculation.

1.2. Identification of the Product Channel for CID

1.2.1. General Strategy

Two problems common to all trajectory calculations are determining when to stop the integration and assigning the correct product channel. This requires particular care in CID calculations, since bound (or quasi-bound) diatomic molecule fragments with large internal energy can have enormous vibration amplitudes, so that the atom–molecule distance at which a trajectory is terminated must be large enough to preclude atom–diatomic collisions during a subsequent molecular vibration. I am not aware of a procedure in the literature which addresses this problem systematically and in all generality. The methods employed for the end tests appear to vary somewhat from paper to paper. Nevertheless, the principal requirements are outlined by Porter and Raff.[28] In the following analysis I have tried to incorporate their ideas and in addition have included some tests found in a paper by Malcolme-Lawes.[14]

The end tests separate into two main parts: tests to see whether the trajectory has left the "strong-coupling region" (defined below), and, if so, a set of tests enabling the assignment of a product channel. A trajectory which fails one of these tests must be integrated longer and the tests applied again until they are passed. The first group of tests requires very little computation, whereas the second is more time consuming. One should therefore choose the parameters of the first group such that the trajectory can be terminated very shortly after the first set of tests is passed. On the other hand, the first group of tests should not be so stringent as to require excessive integration of the trajectory. The second group of tests is arranged in a hierarchy designed to eliminate the possibility of bound and quasi-bound channels before assigning the trajectory to CID. The scheme outlined here is meant to be used within the context of any standard or surface-hopping trajectory calculation, the emphasis being on the special problems arising in CID.

The notation used throughout the remainder of this chapter refers specifically to the atom–molecule three-body system A + BC → AB + C, AC + B, A + BC, or A + B + C. The atoms A, B, C are numbered 1, 2, 3, respectively, as are the molecules BC, CA, and AB. The interatomic distances R_{BC}, R_{CA}, R_{AB} are labeled R_1, R_2, R_3, respectively. The Cartesian velocities of A, B, C (in the ABC center-of-mass coordinate system) are $\mathbf{V}_1$, $\mathbf{V}_2$ and $\mathbf{V}_3$, respectively. The masses of A, B, C are m_1, m_2, and m_3, and $M = m_1 + m_2 + m_3$.

1.2.2. The Strong-Coupling Region (SCR)

The strong-coupling region is that region of *phase space* where the forces are able to bring about a rearrangement or a change of potential surface (nonadiabatic transition). Outside of the SCR the system is separable into an atom plus diatomic or into three separate atoms moving on a single potential surface; there may still be interaction between the separating fragments, but not enough to prevent the trajectory from reaching one of the asymptotic regions. This occurs, for example, when ions separate from each other with sufficient kinetic energy to overcome their long-range attraction.

The test for being in the SCR is somewhat arbitrary. It can be very crude or rather sophisticated. To save computer time it is recommended to break up the test into a weak test followed by a more rigorous test. The weak test should involve very little computation and should include regions of phase space where the trajectory will almost certainly be in the strong-coupling region. For a three-body system a very simple procedure is to find the middle distance R_{mid} of the three R_1, R_2, R_3 and compare it with a test value $\bar{R}_{mid}$. If $R_{mid} \leq \bar{R}_{mid}$ then the system is still in the SCR. Clearly, $R_{mid} > \bar{R}_{mid}$ as the trajectory moves toward any of the asymptotic regions.

The stronger test should be a compromise between a complicated but fairly exact description of the boundary of the SCR and a simpler description of a boundary which includes all of the SCR but perhaps some of phase space outside the SCR as well. A complicated boundary enables quicker termination of the trajectory, but it would in general require repeated calculations of channel energies, which can be expensive if done too often. A fairly good strong test based only on the molecular shape begins by associating with each diatomic pair i a two-body strong-interaction radius, $\bar{R}_i$. From the internuclear distances R_i, form the quantities $R_i' = |R_j - R_k|$, where $i = (i' + 1)(\text{mod}\,3) + 1, j = (i' + 2)(\text{mod}\,3) + 1, k = (i' + 3)(\text{mod}\,3) + 1$, i' going from 0 to 2 (e.g., when $i' = 1$, $i = 3$, $j = 1$, $k = 2$). Then, if any two of the quantities $R_i (i = 1, 2, 3) \leq \bar{R}_i$, then one is still in the SCR. This test is similar to one proposed by Wagner and Parks,[30] who require any two of $R_i \leq \bar{R}_i$; however, the present test is stronger, since the triangle conditions imply that $R_i \geq R_i'$ so that if $R_i < \bar{R}_i$, then $R_i' \leq R_i < \bar{R}_i$ as well.

The above definition of the SCR is fairly realistic for systems where the vibrational amplitudes of the product molecules are not very large: however, it is still possible to have a trajectory which leaves the SCR (so defined) and then reenter it a short time later due to the vibrational motion of a highly excited molecule. This is particularly true of lopsided molecules, and is just the sort of situation which is likely to occur quite often in CID trajectory calculations. To avoid it, the following definition of the SCR is recommended: The numbers $\bar{R}_i$ are chosen to be the largest distances at which the two-body forces are significant; they can be obtained from the diatomic potential functions. The three bound channels (A + BC, B + AC, C + AB) are then investigated in turn ($i = 1, 2, 3$). If $R_i > \bar{R}_i$, then the next value of i is investigated; if $R_i > \bar{R}_i$ for all i, then the trajectory is outside the SCR. If $R_i < \bar{R}_i$ for some i, then the distance of atom i to the center of mass of molecule i, R_i^M, is calculated from the interatomic distances and the masses m_j and m_k:

$$(R_i^M)^2 = (1 - \alpha_k)(R_k^2 - \alpha_k R_i^2) + \alpha_k R_j^2 \tag{2}$$

where $\alpha_k = m_k/(m_j + m_k)$. The molecule i requires space about its center of mass in which to vibrate and rotate, and this will be determined by the number $\bar{R}_i$. The closest distances which atoms j and k can approach atom i owing to the motion of j relative to k are

$$\rho_{ij} = R_i^M - \alpha_k \bar{R}_i \tag{3}$$

$$\rho_{ik} = R_i^M - \alpha_j \bar{R}_i \tag{4}$$

where $\alpha_j = m_j/(m_j + m_k)$. In order for the system to be outside the SCR, two conditions must therefore be met:

$$\begin{aligned} \rho_{ij} &> \bar{R}_k \\ \rho_{ik} &> \bar{R}_j \end{aligned} \tag{5}$$

The tests described here and outlined in the flow diagram in Figure 2 must be passed for all channels for which $R_i < \bar{R}_i$ before the system is deemed to be outside the SCR. They are very easy to implement and are particularly useful in CID calculations.

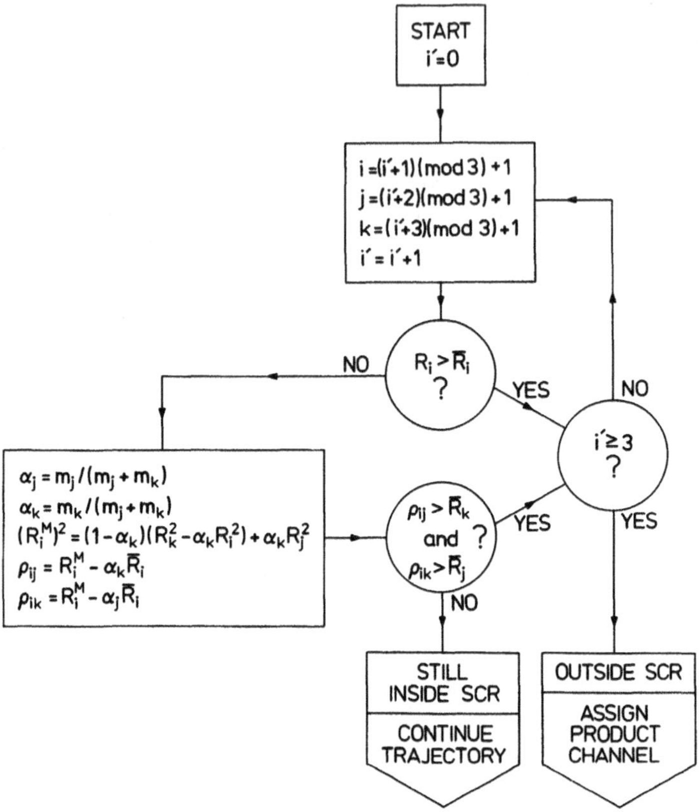

Figure 2. Flow chart for determining when a trajectory is outside the strong-coupling region (SCR).

An illustrative example of a system where the SCR must be carefully defined is given by KI_2 (label K by 1, I by 2, I by 3). Surface-hopping trajectories have been done on this system employing two potential energy surfaces which exhibit an avoided-crossing seam.[26] The lowest collinear adiabatic surface is shown in Figure 3. The dotted line shows the approximate position of the crossing seam. The reactant channel corresponds to $K + I_2$, the bound product channel to $K^+I^- + I$, and the dissociation channel to $K + I + I$. The corresponding channels on the upper surface (not shown) are $K^+ + I_2^-$, $KI + I$ and $K^+ + I^- + I$. Cuts of the two surfaces in the entrance and exit channels are shown in Figure 4. Because of the protuberance of the crossing seam into the entrance channel, the SCR must extend to KI distances of at least 18 Å. Even at $R_2 = 18$ Å the interaction between K^+ and I^- is 0.8 eV, so that the numbers $\bar{R}_2$, $\bar{R}_3$ must even be greater if secondary collisions are to be provided for; however, as a compromise an SCR defined by $\bar{R}_2 = \bar{R}_3 = 20$ Å would probably suffice. The maximum I_2 distance for significant I_2 (or I_2^-) inter-

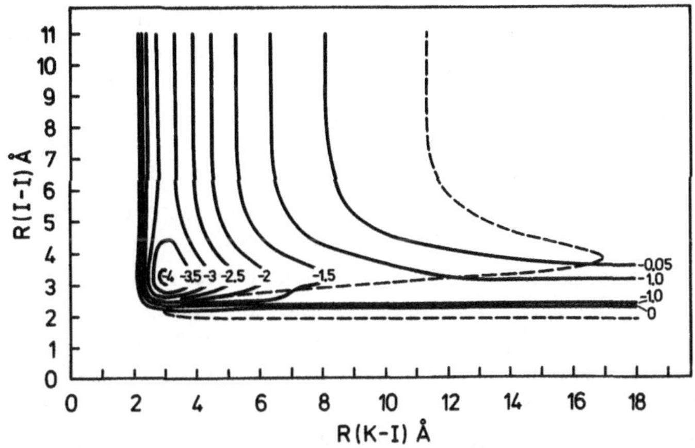

Figure 3. Lowest adiabatic potential surface for $K + I_2$ in an unsymmetrical collinear geometry. The contours are in eV. A trajectory leading from the lower-right-hand corner to the upper-left-hand corner would describe the process $K + I_2 \rightarrow KI + I$. The upper-right-hand corner corresponds to $K + I + I$.

action is seen from Figure 4 to be about 7 Å, so that $\bar{R}_1 = 7$ Å would do. For the rough test, $\bar{R}_{mid} = 10$ Å would be satisfactory.

Note that the SCR is defined by those interaction potentials having the longest range and that one SCR applies to all surfaces. It is obviously possible to use parameters $\bar{R}_i$ which depend on energies, but this is rather

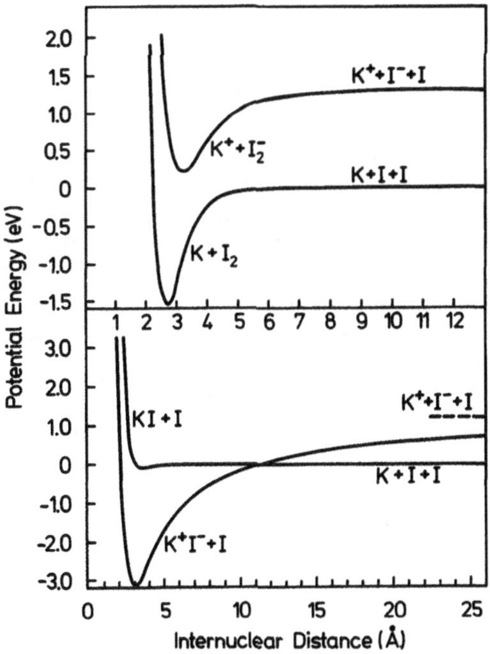

Figure 4. Asymptotic regions of the PE surfaces for KI_2. The entrance channel cuts (upper panel) show the I_2 and I_2^- curves; the exit channel cuts (lower panel) the two Σ states of KI.

complicated, and in any case cannot be outlined here in all generality. Most systems are not as pathological as KI_2 and have a SCR occupying a smaller volume of phase space.

1.2.3. Quasibound Molecules

As pointed out, classical mechanical trajectory calculations often produce molecules which are bound only by virtue of a "rotational barrier" in the effective potential [equation (1)]. These molecules, if left alone, would eventually decay via tunneling and can therefore be included in the CID cross section. The test for quasibound molecules involves finding the maximum $\hat{U}_{\text{eff}}$ in the effective potential and its position $\hat{r}_{\text{eff}}$, and then comparing $\hat{U}_{\text{eff}}$ with the internal (vibrational–rotational) energy of the molecule, E_i^{in}. The molecule is quasibound if three conditions are fulfilled: (i) E_i^{in} is greater than the dissociation energy of the molecules, D_i; (ii) $E_i^{\text{in}} < \hat{U}_{\text{eff}}$; and (iii) $R_i < \hat{r}_{\text{eff}}$. It is assumed that the trajectory is in the asymptotic region of some product channel i. $U_s(R_i)$ is a two-body potential curve corresponding to channel i and potential surface s (i.e., the relative motion of atoms j and k is completely determined by the potential U_s). We assume the potential is zero at its minimum, which occurs at a distance r_s and lies at an energy D_s below the dissociation limit. Let the velocity of atom j relative to k be $\mathbf{v}_i = \mathbf{V}_j - \mathbf{V}_k$. The internal energy of molecule i is then defined as

$$E_i^{\text{in}} = \tfrac{1}{2}\mu_i \mathbf{v}_i \cdot \mathbf{v}_i + U_s(R_i) \tag{6}$$

where $\mu_i = m_j m_k/(m_j + m_k)$. $\hat{r}_{\text{eff}}$ and $\hat{U}_{\text{eff}}$ are found from the maximum of the function

$$U_{\text{eff}} = U_s(R_i) + J^2/(2\mu_i R_i^2) \tag{7}$$

Typical graphs of $U_{\text{eff}}(R_i)$ are shown in Figure 5 for an I_2 potential. To find $\hat{r}_{\text{eff}}$ and $\hat{U}_{\text{eff}}$ numerically, it is suggested to start at $R_i = r_s$, incrementing R_i positively until two points are found for which (i) $U_{\text{eff}} > D_s$, (ii) $d^2U_{\text{eff}}/dR_i^2 < 0$, and (iii) which bracket the maximum (dU_{eff}/dR_i will differ in sign for the two points). Any standard method for finding the maximum between these points can then be used.

The approximate lifetime of a quasibound molecule can be estimated by calculating the probability of tunneling through the "rotational barrier," θ, and the period of vibration of the molecule i, τ.[31] Every τ sec the system has the probability θ of dissociating into atoms, so that the lifetime is $\bar{\tau} = \tau/\theta$. Let the zeros of the function $(E_i^{\text{in}} - U_{\text{eff}})$ be $r_1 \leq r_2 \leq r_3$. Then

$$\theta = \exp(-2G)$$

where

$$G = (1/\hbar) \int_{r_2}^{r_3} [2\mu_i(U_{\text{eff}} - E_i^{\text{in}})]^{1/2}\, dR \tag{8}$$

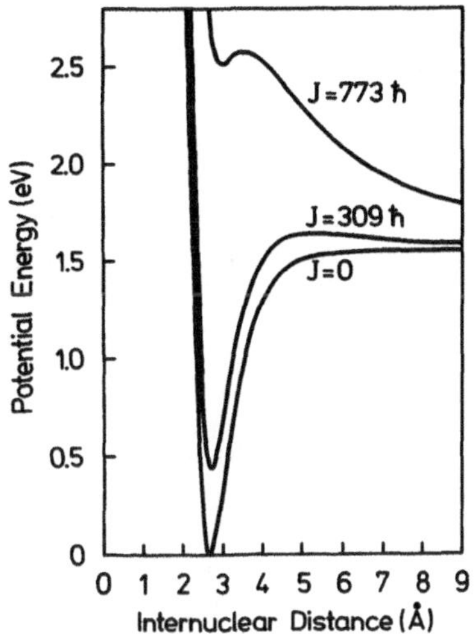

Figure 5. Typical effective potentials for one $U_s(R)$ for $J = 0, 309\hbar, 773\hbar$ [see equation (7)].

This formula is valid for $G > 1$. The vibrational period is

$$\tau = 2\int_{r_1}^{r_2} [(2/\mu_i)(E_i^{\text{in}} - U_{\text{eff}})]^{-1/2}\,dR \tag{9}$$

The two integrals must be evaluated numerically. The first presents no problems, but the second has an integrand which approaches infinity at the integration limits. To obviate this difficulty a change of variables is made:

$$R(x) = \tfrac{1}{2}(r_1 + r_2) + \tfrac{1}{2}(r_2 - r_1)\sin x \tag{10}$$

The integral then becomes

$$\tau = 2\int_{-\pi/2}^{\pi/2} I(x)\,dx \tag{11}$$

where

$$I(x) = \tfrac{1}{2}(r_2 - r_1)\{(2/\mu_i)[E_i^{\text{in}} - U_{\text{eff}}(R(x))]\}^{-1/2}\cos x \tag{12}$$

The value of $I(x)$ at the turning points ($x = \pm\pi/2$) is found by L'Hospital's rule to be

$$I\left(\pm\frac{\pi}{2}\right) = \left[\frac{(r_2 - r_1)/\mu_i}{2U'_{\text{eff}}(x = \pm\pi/2)}\right]^{1/2} \tag{13}$$

where $U'_{\text{eff}} = (dU_{\text{eff}}/dR)$.

1.2.4. Identification of CID Channel

We are now in a position to discuss the stopping criteria for the trajectory and the channel identification. It is assumed that the integration of the trajectory is interrupted at regular intervals for the purpose of applying the stopping criteria. The procedure followed here is to apply a set of increasingly stringent tests for the trajectory which ultimately will allow the assignment of the trajectory to a specific asymptotic product channel. A flow chart outlining the tests is given in Figure 6.

The first test is assumed to be capable of quick application and asks the question "Is the trajectory definitely in the SCR?" If yes, integration must continue; if no, then the stronger test is applied: "Is the trajectory out of the SCR?" If yes, then the process of assigning a channel begins.

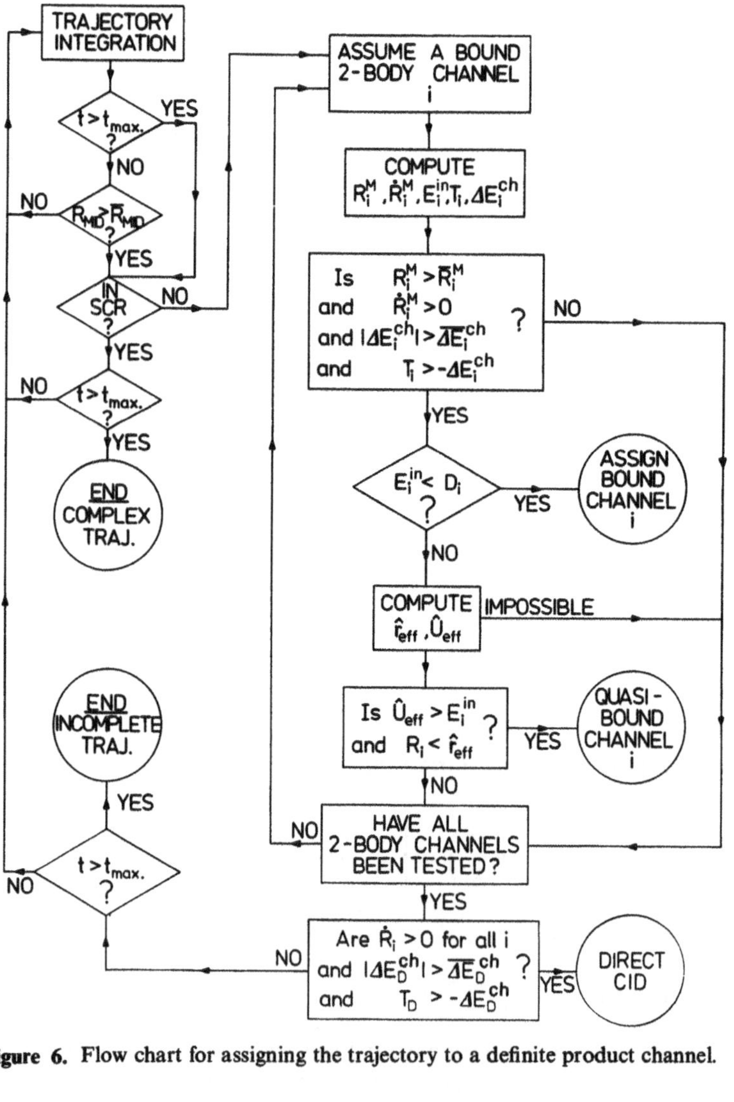

Figure 6. Flow chart for assigning the trajectory to a definite product channel.

To assign the CID channel to a trajectory correctly, it is necessary to check first for the existence of a bound channel. This is done for each molecule i taken in order of increasing molecular distances R_i. In order for a channel to qualify as bound or quasibound, three tests must be passed: First, the distance of atom i from the center of mass of molecule i must be increasing, $\dot{R}_i^M > 0$; i.e., the product fragments must be separating. Second, the channel energy $E_i^{\rm ch}$ must lie within a distance $\overline{\Delta E_i^{\rm ch}}$ of the value expected in the asymptotic region. The channel energy is $E_i^{\rm ch} = E_i^{\rm in} + T_i$, where T_i is the relative kinetic energy of the product fragments and is defined in terms of the velocity of atom i:

$$T_i = \tfrac{1}{2}[m_i(m_j + m_k)/M]\, \mathbf{V}_i^R \cdot \mathbf{V}_i^R \tag{14}$$

where

$$\mathbf{V}_i^R = \mathbf{V}_i M/(m_j + m_k) \tag{15}$$

The error in $E_i^{\rm ch}$ incurred by stopping the trajectory at a finite distance between product fragments is

$$\Delta E_i^{\rm ch} = E_i^{\rm ch} - \lim_{R_j, R_k \to \infty} E_i^{\rm ch} \tag{16}$$

where the limiting value can be obtained from the potential energy surface in the asymptotic region together with the amount of energy $(E_1^{\rm in} + T_1)$ available in the entrance channel, assuming total energy is conserved. Third, the final translational energy T_i must exceed $-\overline{\Delta E_i^{\rm ch}}$, i.e., the trajectory must have enough energy to reach the product asymptotic region.

If the previous tests are passed, then the trajectory is in the bound channel i if in addition $E_i^{\rm in} < D_i$, the dissociation energy of the molecule i. If $E_i^{\rm in} > D_i$, then it is necessary to calculate the position and height of the rotational barrier ($\hat{r}_{\rm eff}$ and $\hat{U}_{\rm eff}$) and then apply the tests described in Section 1.2.3 to determine whether the molecule is quasibound. If a maximum cannot be found, or if the tests are not passed, then it is possible that direct dissociation has occurred; however, the remaining bound channels need to be investigated before it can be concluded unambiguously that direct CID has occurred. If all bound and quasibound channels are unacceptable, two further conditions are imposed: (i) all internuclear distances must be increasing, and (ii) the channel energy error for dissociation must be small enough. The channel energy in this case is simply the total kinetic energy

$$E_D^{\rm ch} = \tfrac{1}{2}\sum_i m_i \mathbf{V}_i \cdot \mathbf{V}_i \tag{17}$$

The channel assignment procedure can be speeded up by assigning a flag to all bound diatomic potentials. Immediately the tentative channel i is assigned, the relevant diatomic curve should be checked; if it is not bound, then the tests for channel i can be bypassed. Note that in the case of a multisurface calculation, the relevant potential curve for the diatomic molecule is indexed by the potential surface as well as by the channel index i. The proper surface is of course determined by the trajectory itself.

The channel energy parameter $\overline{\Delta E_i^{ch}}$ determines how close the trajectory must approach the asymptotic region before a channel assignment is possible. In the KI_2 example the ionic CID channel ($K^+ + I^- + I$) still has a K^+–I^- interaction energy of ≈ 0.8 eV for $R_{KI} = 18$ Å, i.e., if the trajectory were to be stopped just outside the boundary of the SCR, the $\overline{\Delta E_i^{ch}}$ would have to be at least 0.8 eV. Such a value would allow a good determination of the CID cross section, but not of the angular scattering of the ions. For that, the trajectory would have to be integrated much longer, so that $\overline{\Delta E_i^{ch}}$ would have to be set to a very small value.

The procedure outlined here should enable one to treat the CID channel fairly rigorously; the tests given here include in a systematic fashion most of those found in the literature on trajectory calculations. The actual CID cross sections and other properties (e.g., angular scattering) can be computed by the normal trajectory methods, which are explained in Chapters 16 and 18.

2. Classical Models for CID

2.1. Overview

2.1.1. Availability

It is very difficult to recommend the use of particular models for CID in general. This is not because of a lack of models available. Although there are few models which treat CID directly, there are a large number of models designed to treat chemical reaction, and most of these imply something about CID as well. The problem lies in the fact that none of the particular models have been well tested even for application to chemical reaction, much less to CID, and many are designed for application to specific systems and are not easily generalized.

2.1.2. Dynamical Models

There are many dynamical models of chemical reactions which could be extended to CID. They are characterized by being oversimplified, arbitrary, and as far as CID is concerned, incomplete: oversimplified because they are usually designed for quick computation, arbitrary because they often do not refer to the potential energy function or to well-defined reductions in the equations of motion, and incomplete because they are often chosen to clarify a particular observation on a chemical reaction and, therefore, concentrate on a single dynamical mechanism to the exclusion of all others.

The incompleteness of the models is the most damaging characteristic as far as their application to CID is concerned. One can see this by referring to one of the simplest and most widely used mechanisms: the spectator stripping model.[35] This assumes that in the process $A + BC \rightarrow AB + C$, no change in velocity of atom C occurs. Application of the conservation of linear momentum immediately implies that the product translational energy is $T_i = T_1 m_A m_C/[(m_A + m_B)(m_B + m_C)]$ where T_1 is the relative kinetic energy of the reactants. The mass factor is less than 1, so that as T_1 increases, more and more of the available energy becomes internal energy of AB. Clearly, there exists some value of T_1 above which only CID will occur. Several models related to the stripping model[36] make similar predictions about CID and are capable of delivering a CID cross section over a range of collision energies E. They all do this, however, by assigning to the CID channel all trajectories which cannot produce molecules AB by the mechanism in question. This has the effect of seriously overestimating the CID cross section, for there may be other mechanisms capable of producing bound AB molecules. Therefore, the simple dynamical models cannot be recommended for CID calculations in general.

2.1.3. Hard-Sphere Models

These models are not as arbitrary or as simplified as other dynamical models and they become representative of real trajectories in the limit of high collision energies. The basic idea is to assume a hard-sphere interaction between each pair of atoms, the size of the atom being specified by a fixed atomic radius. The collision of A with BC is then treated as a sum of successive two-body encounters, and after the collision the relative energies of each pair are calculated; if each pair-energy is greater than the corresponding dissociation energy, then dissociation has occurred.[37-44]

The simplest treatment has been developed by Mahan and co-workers[38,39] and is based on an idea by Bates, Smith, and Cook.[37] A particular sequence of encounters is assumed to predominate: A hits B, B hits C, AB is tested for being bound or not. This straightforward version admits of an analytical expression for the differential cross section for formation of AB; however, the neglect of other hard-sphere encounter sequences invalidates its possible application to CID calculations.

A more complicated treatment has been applied by Suplinskas.[42] He allows for a number of different hard-sphere collision sequences but has to obtain his results by numerically averaging over many "hard-sphere trajectories." Even this model, however, can only treat a small number of the possible sequences. It does not handle repeated collisions of one sphere with the other two (chattering), an effect which reduces CID cross sections.[44] Moreover, for applications to ion–molecule reactions, the model is only used to handle the rearrangement part of the collision, the approach of the reactants and separation of products occurring under the influence of long-range forces.[43] This part of the calculation requires numerical integration of the equations of motion, i.e., the model has become almost as complicated as a full trajectory calculation.

2.1.4. Recommendations

The problems alluded to above are summarized in the frequently occurring dilemma that a model which contains sufficient dynamical flexibility to be realistic is often too complicated to apply readily. There does not seem to be an easy way out of this dilemma for CID. The goals, however, are clear enough: (i) Since CID cross sections are overestimated by models which inadvertently ignore mechanisms leading to bound channels, we seek a model which investigates all possible mechanisms consistent with classical dynamics (a trajectory calculation is such a model). (ii) The desired method ought to be significantly faster than the classical trajectory method; therefore, stepwise numerical integration of dynamical equations must be avoided. (iii) The time development of collisions must proceed in a physically reasonable fashion; if possible the motion should be governed by approximations which can be related to a potential energy surface.

These goals might be achieved by a number of dynamical models. I offer here one model which is capable of treating CID consistently with the above goals, namely, the square-well trajectory model.[45] This is exactly the same as a trajectory calculation except that the potential function is reduced to a set of adjoining square wells; this enormously simplifies the calculation of the trajectory but still allows for collisions of atoms with vibrating–rotating molecules. It is also possible to have CID occurring directly or via quasibound molecules. This model is described ln detail in Section 2.2.

2.2. The Square-Well Trajectory Model

2.2.1. The Potential Energy Function

In this model the entire coordinate space is divided into regions within each of which the potential energy is constant. The potential surface is thus a piecewise continuous function. The model has the advantage of being related to a realistic potential function and this relationship can be improved by introducing a larger number of square-well regions.

The simplest potential representation is given by Rebick, Levine, and Bernstein,[45] who divide configuration space into the regions A + B + C (dissociation), A + BC, B + AC, C + AB (atom–diatomic channels), ABC, BCA, CAB (two simultaneous pair interactions), $\widehat{ABC}$ (three simul-

Table 1. Potential Energy Function for the Square-Well Model

$U(R_1,R_2,R_3)$	R_1	R_2	R_3	Configuration
∞	$< \hat{R}_1$	—	—	Hard core
∞	—	$< \check{R}_2$	—	Hard core
∞	—	—	$< \check{R}_3$	Hard core
U_1	$\check{R}_1 < R_1 < \hat{R}_1$	$> \hat{R}_2$	$> \hat{R}_3$	A + BC
U_2	$> \hat{R}_1$	$\check{R}_2 < R_2 < \hat{R}_2$	$> \hat{R}_3$	B + CA
U_3	$> \hat{R}_1$	$> \hat{R}_2$	$\check{R}_3 < R_3 < \hat{R}_3$	C + AB
0	$> \hat{R}_1$	$> \hat{R}_2$	$> \hat{R}_3$	A + B + C
U_{12}	$\check{R}_1 < R_1 < \hat{R}_1$	$\check{R}_2 < R_2 < \hat{R}_3$	$> \hat{R}_3$	BCA
U_{23}	$> \hat{R}_1$	$\check{R}_2 < R_2 < \hat{R}_2$	$\check{R}_3 < R_3 < \hat{R}_3$	CAB
U_{31}	$\check{R}_1 < \check{R}_1 < \hat{R}_1$	$> \hat{R}_2$	$\check{R}_3 < R_3 < \hat{R}_3$	ABC
U_{123}	$\check{R}_1 < R_1 < \hat{R}_1$	$\check{R}_2 < R_2 < \hat{R}_2$	$\check{R}_3 < R_3 < \hat{R}_3$	$\overline{\text{ABC}}$

taneous pair interactions), and a hard-core region. With labels 1, 2, 3 for A, B, C, the distances B–C, C–A, and A–B are R_1, R_2, and R_3, respectively. Each atom consists of an impenetrable core and an outer interactive sphere of radii $\check{r}_i$ and $\hat{r}_i$, respectively. For each pair of atoms jk, two-body distance parameters can be defined as $\check{R}_i = \check{r}_j + \check{r}_k$ and $\hat{R}_i = \hat{r}_j + \hat{r}_k$. The potential function $U(R_1, R_2, R_3)$ can then assume one of the following values: 0, ∞, U_1, U_2, U_3, U_{12}, U_{23}, U_{31}, U_{123}, as shown in Table 1.

2.2.2. *Quasibound Molecules*

The internal energy of molecule i is E_i^{in} and is given by equation (6) taking $U_s(R_i) = 0$ (since U_s is measured from the bottom of the potential well). The effective potential is discontinuous at the distance $R_i = \hat{R}_i$ and hence $r_{\text{eff}} = \hat{R}_i$ and $U_{\text{eff}} = U_s + J^2/(2\mu_i \hat{R}_i^2)$. With these definitions, the conditions for molecule i being quasibound are the same as described in Section 1.

2.2.3. *Method of Calculation*[46]

Within the various regions of configuration space used to define the potential function, the atoms move in straight lines with constant velocities. When any one of the distances R_i reaches its boundary points $\check{R}_i$ or $\hat{R}_i$, the trajectory receives a discontinuous change in the component of the relative velocity $\mathbf{v}_i = \mathbf{V}_j - \mathbf{V}_k$ parallel to the line of centers connecting j and k:

$$\mathbf{v}_{i\parallel} = (\mathbf{v}_i \cdot \mathbf{R}_i)\mathbf{R}_i/R_i^2 \equiv B_i \mathbf{R}_i/R_i^2 \tag{18}$$

The changes in $\mathbf{V}_j$ and $\mathbf{V}_k$ ($\Delta\mathbf{V}_j$ and $\Delta\mathbf{V}_k$) at the boundaries $\check{R}_i$ and $\hat{R}_i$ are made in such a way as to conserve linear momentum, angular momentum, and energy, which means that

$$\Delta\mathbf{V}_k = -m_j\,\Delta\mathbf{V}_j/m_k \tag{19}$$

The internal kinetic energy of molecule i must increase or decrease by an amount equal to the change in potential energy ΔU. This means that the following equation for the change in relative velocity is obeyed:

$$\Delta v_{i\parallel}^2 + 2B_i\,\Delta v_{i\parallel}/R_i + 2\Delta U/\mu_i = 0 \tag{20}$$

where $\mu_i = m_j m_k/(m_j + m_k)$ and $\Delta\mathbf{v}_{i\parallel} = \Delta v_{i\parallel}\mathbf{R}_i/R_i$. The change in $\mathbf{V}_j$ is then given by

$$\Delta\mathbf{V}_j = m_k\,\Delta v_{i\parallel}\,\mathbf{R}_i/[R_i(m_j + m_k)] \tag{21}$$

If $\Delta U > 0$, then $\Delta v_{i\parallel}$ must have the opposite sign to B_i, while if $\Delta U < 0$, $\Delta v_{i\parallel}$ must have the same sign:

$$\Delta v_{i\parallel} = -\operatorname{sgn}(B_i)\operatorname{sgn}(\Delta U)\,|\Delta v_{i\parallel}| \tag{22}$$

Insertion into equation (20) gives

$$|\Delta v_{i\parallel}| = \frac{|B_i|}{R_i}\operatorname{sgn}(\Delta U)\left[1 - \left(1 - \frac{2\Delta U R_i^2}{\mu_i B_i^2}\right)^{1/2}\right] \tag{23}$$

The procedure for changing the velocities $\mathbf{V}_j$ and $\mathbf{V}_k$ when a boundary is traversed is therefore as follows: Calculate the potential energy change ΔU, the quantities B_i and μ_i, and set the value of R_i in the above equations to its value at the boundary in question (i.e., $R_i = \check{R}_i$ or $\hat{R}_i$).

If $1 > 2\Delta U R_i^2/\mu_i B_i^2$, then from equations (20)–(23)

$$\Delta\mathbf{V}_j = -\left(\frac{m_k}{m_j + m_k}\right)\mathbf{R}_i B_i\left[1 - \left(1 - \frac{2\Delta U R_i^2}{\mu_i B_i^2}\right)^{1/2}\right]\Big/R_i^2 \tag{24}$$

$\Delta\mathbf{V}_k$ is then obtained from (19). If $1 < 2\Delta U R_i^2/\mu_i B_i^2$ (which can only happen if $\Delta U > 0$), it means that there is not enough energy to cross the boundary. In this case, the velocity change is $\Delta\mathbf{v}_{i\parallel} = -2\mathbf{v}_{i\parallel}$. This leads at once to

$$\Delta\mathbf{V}_j = -2m_k B_i \mathbf{R}_i/[(m_j + m_k)R_i^2] \tag{25}$$

The procedure for following a trajectory is to start off in the asymptotic region A + BC at time t_0 and then to apply the two-body collision tests. These consist of calculating the times at which the distances R_i will reach the boundaries $\check{R}_i$ and $\hat{R}_i$; let these times be $\check{t}_i$ and $\hat{t}_i$, $i = 1, 2, 3$. The next two-body collision will occur for the pair i corresponding to the shortest of these times. The times are calculated from

$$\check{t}_i = [-B_i \pm (B_i^2 - v_i^2\check{C}_i)^{1/2}]/v_i^2 \tag{26}$$

where $\check{C}_i = R_i^2 - \check{R}_i^2$ with a similar expression for $\hat{t}_i$. The sign in the formula above is chosen so that $\check{t}_i$ (and $\hat{t}_i$) > 0. If $(B_i^2 - v_i^2\check{C}_i) < 0$, then the distance R_i will not reach the value $\check{R}_i$ on this segment of the straight-line trajectory. The values of the other distances at the new time $t = t_0 + \check{t}_i$ are

$$\mathbf{R}_j(t) = \mathbf{R}_j(t_0) + \mathbf{v}_j(t - t_0) \tag{27}$$

This equation is used to calculate the new molecular configuration at time t.

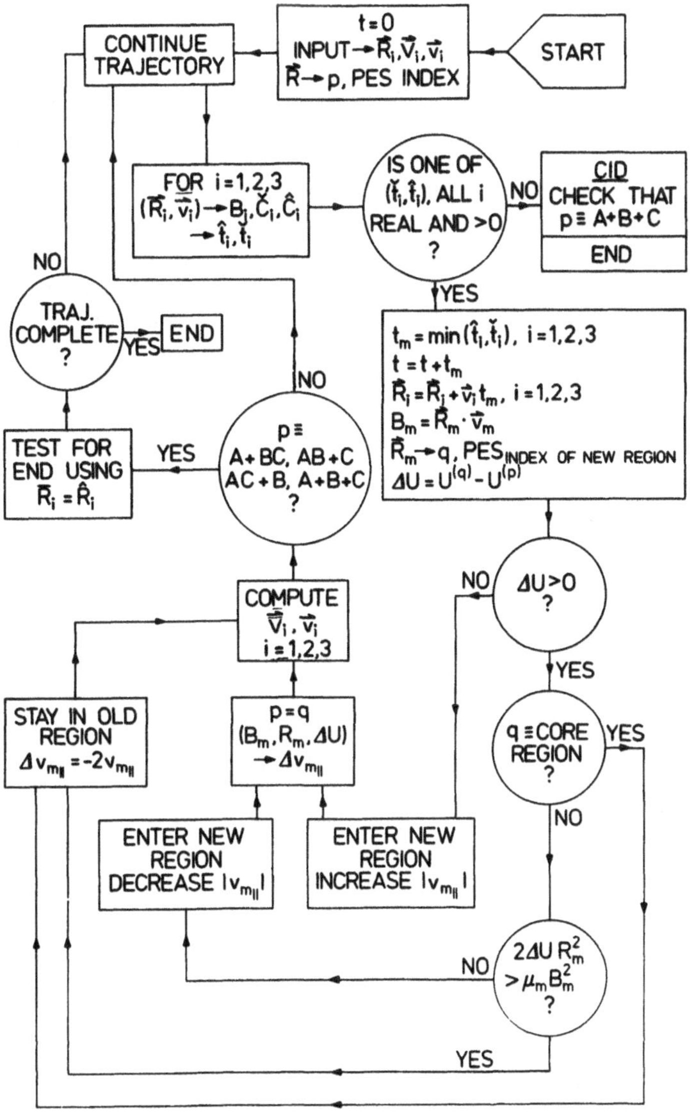

Figure 7. Flow chart for the square-well trajectory model. The symbol PES stands for potential energy surface; p and q are channel labels.

The computation procedure is then (i) from the values of B_i, $\mathbf{R}_i$, and $\mathbf{v}_i$ at t_0 find the new values of C_i and the time $t > t_0$ at which the next two-body event occurs; (ii) move the system to the new time t, recalculating B_i, $\mathbf{R}_i$; (iii) decide what kind of event occurs at time t and calculate the new velocities $\mathbf{v}_i$ and projections B_i; (iv) test for completion of trajectory. If the trajectory is not complete, go back to (i). A flow chart outlining the square-well trajectory model is given in Figure 7.

2.3. Statistical Approaches

There are no well-tested statistical methods for calculating CID cross sections. Nevertheless there have been some attempts in this direction, and recently a statistical model has been proposed which should handle dissociation.[30] For this reason, a brief mention will be made of this topic.

The basic idea of a statistical model within a classical mechanical work is to replace the dynamical determinism associated with classical trajectories by an assumption based on statistical considerations about the outcome of trajectories. In order to do this, a strong-coupling region (SCR) is defined such that (i) only trajectories which enter the SCR are considered, and (ii) the products which leave the SCR are assumed to be distributed over all trajectories which have the same energy and angular momentum as the original trajectory entering the SCR. The distribution function used is most often taken to be the uniform distribution.

The effectiveness of a statistical model is therefore dependent on how flexibly the SCR can be chosen. It is desirable to have an SCR which takes cognizance of the size and shape of molecules. Wagner and Parks have recently proposed a classical model with these properties, and they present explicit but lengthy expressions for cross sections for reaction and dissociation.[30] Their model is essentially a reformulation of the classical trajectory method and uses a statistical expression in place of the classical (deterministic) probability $P(\gamma|\gamma')$ of going from a point γ in the reactant region of trajectory phase space to a point γ' in the product region:

$$P(\gamma\,|\,\gamma') \propto P_{\mathrm{SCR}}(\gamma)\, P_{\mathrm{SCR}}(\gamma')$$

where $P_{\mathrm{SCR}}(\gamma) = 1$ if a trajectory defined by the variables $\boldsymbol{\gamma}$ passes through the SCR, and $P_{\mathrm{SCR}}(\gamma) = 0$ otherwise. Their formulation is very attractive, for it shows that time-reversal symmetry is obeyed and it allows great freedom in the choice of the SCR. For this reason the method is to be recommended over other methods. It would be out of place here to give more details of this method for two reasons: (i) most of their formulation would involve another presentation of the material in Chapters 16 and 18, and (ii) at the time of writing, there have been no applications of the model to physically realistic choices of the SCR, i.e., the method is not well tested.

The Wagner and Parks formulation offers a vantage point for the analysis of other CID statistical models. The phase-space theory of Light[47] cannot cope with CID, but an adaptation of this theory by Rebick and Levine[48] can. This (quantum-mechanical) formulation involves using the concept of the generalized angular momentum[49] in order to be able to index the CID channel by discrete quantum numbers, thereby enabling the treatment of direct dissociation. Unfortunately, application of the model's assumptions in a classical context is shown to put inconsistent demands on the SCR; furthermore, the SCR that can be used is very restrictive.

Finally, Levine and Bernstein[50] have used the generalized angular momentum, $\Lambda = \hbar\lambda$, to express CID cross sections in terms of the collision-induced recombination (CIR) probability of the reverse collision:

$$\sigma_{\text{CID}} = \frac{\pi}{k^2} \sum_{\lambda} g(\lambda)\, P_{\lambda}(\text{CIR}) \tag{28}$$

$$k^2 = m_{\text{A}}(m_{\text{B}} + m_{\text{C}})E \,/\, \hbar^2(m_{\text{A}} + m_{\text{B}} + m_{\text{C}}) \tag{29}$$

$$g(\lambda) = (\lambda + 1)(\lambda + 2)^2 (\lambda + 3)/12 \approx \lambda^4/12 \tag{30}$$

Here, $P_{\lambda}(\text{CIR})$ is the probability of going from A + B + C characterized by angular momentum $\Lambda = \hbar\lambda$ to all possible products A + BC with orbital angular momentum $L \approx \hbar l$. Replacing the sum over λ by an integral and assuming that only the probability at the maximum λ (for which the centrifugal barrier can be mounted), λ_m, is important, one has

$$\sigma_{\text{CID}} \approx (\pi/k^2)\lambda_m^5 P_{\lambda_m}/60 \tag{31}$$

The maximum λ_m can be related to the maximum impact parameter R_0 by

$$\lambda_m^2 \approx 2\mu R_0^2(E_{\text{tot}} - E_{\text{o}})$$

where E_{tot} is the total energy of A + BC, E_{o} the observed threshold for dissociation, and μ is a three-body reduced mass. Hence

$$\sigma_{\text{CID}} \propto (E_{\text{tot}} - E_{\text{o}})^{5/2}/E \tag{32}$$

where E is the relative collision energy for A + BC. Parametrizations of the type

$$\sigma_{\text{CID}} = Af(E_{\text{vib}})(E_{\text{tot}} - E_{\text{o}})^n/E \tag{33}$$

where $f(E_{\text{vib}})$ contains the vibrational dependence of σ_{CID}, have been found useful in describing the post-threshold behavior of experimental CID cross sections.[51,52]

3. *Summary*

In the context of classical mechanics, the trajectory method provides a very general and physically realistic method of treating CID. The flow chart given in Section 1 is designed to cope with the special problems which arise when the product molecules have very large amounts of internal energy, as is expected in calculations of CID.

Models for CID must be used with caution since they often treat the dynamics incompletely or in a biased way and they often severely restrict the shape of the strong-coupled region (i.e., the type of potential function). It is recommended here to retain flexibility of the potential energy function and to simplify the dynamical equations of motion. The square-well model presented here provides one way of doing this.

ACKNOWLEDGMENTS

I would like to thank Dr. John Kendrick, Mrs. G. Snoei, Mrs. H. Gadewoltz, and Mrs. K. Gfrörer for their assistance in the preparation of the manuscript.

References

1. R.T.V. Kung and J.B. Anderson, Phase-space theory of atomic dissociation and recombination reactions, *J. Chem. Phys.* **60,** 3731–3743 (1974).
2. R.A. LaBudde, P.J. Kuntz, R.B. Bernstein, and R.D. Levine, Classical trajectory study of the K + CH_3I reaction, *J. Chem. Phys.* **59,** 6286–6298 (1973).
3. T. Valencich and D.L. Bunker, Trajectory studies of hot atom reactions. II. An unrestricted potential for CH_5, *J. Chem. Phys.* **61,** 21–29 (1974).
4. A.G. Clarke and G. Burns, Trajectory studies of atomic recombination reactions, *J. Chem. Phys.* **55,** 4717–4730 (1971).
5. A. Gelb, R. Kapral, and G. Burns, Nonequilibrium effects in atomic recombination reactions, *J. Chem. Phys.* **56,** 4631–4635 (1972).
6. A.G. Clarke and G. Burns, Trajectory studies of atomic recombination. II, *J. Chem. Phys.* **56,** 4636–4645 (1972).
7. A.G. Clarke and G. Burns, Trajectory studies of atomic recombination reactions. III, *J. Chem. Phys.* **58,** 1908–1913 (1973).
8. W.H. Wong and G. Burns, Trajectory studies of atomic recombination of I atoms. IV, *J. Chem. Phys.* **58,** 4495–4467 (1973).
9. W.H. Wong and G. Burns, Trajectory studies of atomic recombination reactions. V, *J. Chem. Phys.* **59,** 2974–2976 (1973).
10. A. Jones and J.L.J. Rosenfeld, Monte Carlo simulation of H-atom recombination, in *Abstracts of VII International Conference on the Physics of Electronic and Atomic Collisions,* North Holland, Amsterdam (1971), p. 314.
11. M. Karplus, R.N. Porter, and R.D. Sharma, Energy dependence of cross sections for T + H_2, T + D_2 collisions, *J. Chem. Phys.* **45,** 3871–3873 (1966).
12. G.R. North and J.J. Leventhal, Classical superposition phenomena in H_2^+ (v = 0)–He reactive collisions, *Chem. Phys. Lett.* **23,** 600–602 (1973).
13. G.R. North, H.H. Harris, J.J. Leventhal, and P.B. James, Model for H_2^+ (v = 0)–He collisions above 2 eV, *J. Chem. Phys.* **61,** 5060–5065 (1974).
14. D.J. Malcolme-Lawes, Hydrogen isotopic exchange reactions at high energies, *J. Chem. Soc. Faraday Trans.* **2** **71,** 1183–1199 (1975).

15. J.T. Muckerman, Classical dynamics of hot atom reactions of F with HD, *J. Chem. Phys.* **57**, 3388–3396 (1972).
16. P.A. Whitlock, J.T. Muckerman, and R.E. Roberts, Classical mechanics of recombination $H + H + M \rightarrow H_2 + M$, *J. Chem. Phys.* **60**, 3658–3673 (1974).
17. R.K. Preston and J.S. Cohen, Chemi-ionization in atom–diatomic collisions, *J. Chem. Phys.* **65**, 1589–1590 (1976).
18. J.R. Krenos, R.K. Preston, R. Wolfgang, and J.C. Tully, Molecular beam and trajectory studies of $H^+ + H_2$, *J. Chem. Phys.* **60**, 1634–1659 (1974).
19. R.K. Preston and R.J. Cross, Jr., Charge exchange and chemical reaction: $D_2^+ + H$, *J. Chem. Phys.* **59**, 3616–3622 (1973).
20. R.E. Howard, R.E. Roberts, and M.J. Delle Donne, 3-body effects in exchange and dissociation encounters for $Ar + Ar_2$, *J. Chem. Phys.* **65**, 3067–3074 (1976).
21. N.C. Blais and D.G. Truhlar, Trajectory study of $Ar + H_2$ collisions. I, *J. Chem. Phys.* **65**, 5335–5356 (1970).
22. N.C. Blais and D.G. Truhlar, Monte Carlo trajectory study of $Ar + H_2$ collisions. II, *J. Chem. Phys.* **66**, 772–778 (1977).
23. N.J. Brown and R.J. Munn, Molecular dynamics: The dissociation of H_2 by He, *J. Chem. Phys.* **56**, 1983–1987 (1972).
24. A. Gelb, R. Kapral, and G. Burns, Dissociation of vibrationally–rotationally excited $I_2(B^3\Pi O_u^+)$, *J. Chem. Phys.* **59**, 2980–2985 (1973).
25. W.H. Wong and G. Burns, Dynamics of dissociation of diatomic molecules and mass effect, *J. Chem. Phys.* **62**, 1712–1713 (1975).
26. C. Evers, Trajectory surface hopping study of $M + I_2$ collisions ($M = Na, K, Cs$), *J. Chem. Phys.* **21**, 355–371 (1977).
27. B. Garetz, M. Rubinson, and J.I. Steinfeld, Classical trajectory surface hopping applied to CID, *Chem. Phys. Lett.* **28**, 120–124 (1974).
28. R.N. Porter and L.M. Raff, Classical trajectory methods in molecular collisions, in *Dynamics of Molecular Collisions, Part B*, W. H. Miller, editor, Plenum Press, New York (1976), Chap. 1.
29. R.N. Porter, Molecular trajectory calculations, *Ann. Rev. Phys. Chem.* **25**, 317–355 (1974).
30. A.F. Wagner and E.K. Parks, A classical statistical theory for chemical reactions, *J. Chem. Phys.* **64**, 4343–4361 (1976).
31. D.E. Stogryn and J.O. Hirschfelder, Contribution of bound, metastable, and free molecules to the second virial coefficient, *J. Chem. Phys.* **31**, 1534–1545 (1959).
32. J.C. Tully, Nonadiabatic processes in molecular collisions, in *Dynamics of Molecular Collisions, Part B*, W.H. Miller, editor, Plenum Press, New York (1976), Chap. 5.
33. E.K. Parks, N.J. Hansen, and S. Wexler, Collision-induced ion pair formation of thallium halides, *J. Chem. Phys.* **58**, 5489–5501 (1973).
34. P.J. Kuntz and W.N. Whitton, Interpretation of CID charge-transfer processes in rare-gas molecule-ions, *Chem. Phys.* **16**, 301–310 (1976).
35. A. Henglein, Stripping effects in ion–molecule reactions, in *Ion–Molecule Reactions in the Gas Phase*, Vol. 58 of *Advances in Chemistry Series*, R.F. Gould, editor, American Chemical Society, Washington D.C. (1966), Chap. 5.
36. P.J. Kuntz, A direct interaction model for gas-phase chemical reactions, *Trans. Faraday Soc.* **66**, 2980–2996 (1970).
37. D.R. Bates, C.J. Cook, and F.J. Smith, Classical theory of ion–molecule rearrangement at high energies, *Proc. Phys. Soc.* **83**, 49–57 (1964).
38. B.H. Mahan, An analysis of direct ion–molecule reactions, in *Interactions between Ions and Molecules*, Pierre Ausloos, editor, Plenum Press, New York (1975), pp. 75–93.
39. B.H. Mahan, W.E.W. Ruska, and J.S. Winn, Sequential impulse model of direct reactions, *J. Chem. Phys.* **65**, 3888–3896 (1976).
40. D.J. Malcolme-Lawes, Computer simulation of reactions of hot hydrogen atoms, *J. Chem. Phys.* **57**, 5522–5530 (1972).

41. D.J. Malcolme-Lawes, High energy reaction kinetics using a hard-sphere model, *J. Chem. Soc. Faraday Trans. 2* **68,** 1613–1622 (1972).
42. R.J. Suplinskas, Kinematic model for atom–diatom reactions, *J. Chem. Phys.* **49,** 5046–5053 (1968).
43. T.F. George and R.J. Suplinskas, Kinematic model for reaction. III. $Ar^+ + D_2$, *J. Chem. Phys.* **54,** 1037–1045 (1971).
44. G.M. Kendall, Chattering in hard sphere reactions, *J. Chem. Phys.* **58,** 3523–3524 (1973).
45. C. Rebick, R.D. Levine, and R.B. Bernstein, Energy requirements and energy disposal: Reaction probability matrices and a computational study of a model system, *J. Chem. Phys.* **60,** 4977–4989 (1974).
46. B.J. Alder and T.E. Wainwright, Studies in molecular dynamics. I. General method, *J. Chem. Phys.* **31,** 459–466 (1959).
47. J.C. Light, Phase-space theory of chemical kinetics, *J. Chem. Phys.* **40,** 3221–3229 (1964).
48. C. Rebick and R.D. Levine, Collision induced dissociation: A statistical theory, *J. Chem. Phys.* **58,** 3942–3952 (1973).
49. F.T. Smith, Generalized angular momentum in many-body collisions, *Phys. Rev.* **120,** 1058–1069 (1960).
50. R.D. Levine and R.B. Bernstein, Collision-induced dissociation: A simplistic optical model analysis, *Chem. Phys. Lett.* **11,** 552–556 (1971).
51. W.B. Maier II, Dissociative ionization of N_2 and N_2O by rare gas ion impact, *J. Chem. Phys.* **41,** 2174–2181 (1964).
52. E.K. Parks, A. Wagner, and S. Wexler, Collision-induced ion pair formation of thallium halides: Threshold behavior, *J. Chem. Phys.* **58,** 5502–5513 (1976).

22

Information-Theoretic Approach: Application to Molecular Collisions

R. D. LEVINE and J. L. KINSEY

1. Introduction

What we now know as information theory[1,2] was introduced by Shannon[3] and Wiener[4] in the late 1940s. Like thermodynamics, the need for the theory was motivated by very practical considerations. Just as the question "Are there natural limits to the conversion of heat into mechanical energy?" prompted Carnot's formulation of the second law of thermodynamics, the question "Are there theoretical limitations to the number of messages that can be reliably transmitted via a communication channel?" led Shannon to the second theorem of information theory. It is not our purpose here to review the rest of the story but to present a utilitarian summary of the current practice of information theoretic methods in the study of molecular collisions. The area has been previously reviewed[5-9] and a "handbook" of results at a given total energy is available.[10] In this chapter we focus attention on the theory from the experimentalist's point of view, although the "experimental" data to be treated may well include computational results, e.g., from classical trajectory calculations. We intend to describe the current usage and, where appropriate, to make specific recommendations. This chapter is thus a handbook and does not always provide detailed derivations, nor does it provide a literature survey (which can be

R. D. LEVINE and J. L. KINSEY • Department of Chemistry, Massachusetts Institute of Technology, Cambridge, Massachusetts 02139. Permanent address of R. D. Levine: Department of Physical Chemistry, The Hebrew University, Jerusalem, Israel.

found, e.g., in Reference 9). We apologize to those of our colleagues whose work is not explicitly cited.

The primary objects of our considerations will be the rates of various kinds of collision processes, or more precisely, the specific rate coefficients $k(\mathbf{a})$ for a process of the $\mathbf{a}$th kind. The label $\mathbf{a}$ might, for example, designate collisions which result in population of the $\mathbf{a}$th vibrational level of one of the species involved, starting, say, from thermal reagents. Typically, one can observe more than one outcome and so one need consider the relative frequency $P(\mathbf{a})$ of the $\mathbf{a}$th outcome from among the set of all possible outcomes,

$$P(\mathbf{a}) = k(\mathbf{a}) \Big/ \sum_{\mathbf{a}} k(\mathbf{a}) \tag{1}$$

The first part of this review considers the ways in which information theory can be useful in discussing these frequencies (they are, after all, a message transmitted by nature). In the second part, we shall explicitly recognize that the relative frequency of the different outcomes may well depend on the initial conditions and extend the discussion to allow for an examination of this dependence. The stress of the chapter is thus on the use of information theory in the analysis of experimental results. Less attention will be given to the complementary task of inducing further details given the data at hand. We shall, however, consider the principles (if not all the technical details) involved in generating inductive distributions.

1.1. Overview

Information is provided by a choice among several possible outcomes.[1,2] A quantitative measure for the amount of information need take into account the possibility that some of the outcomes may be intrinsically more probable than others. Different values of the outcome index $\mathbf{a}$ [cf. equation (1)] may well correspond to different degrees of specificity, and means of keeping track of this possibility must be introduced. This is quantitatively expressed by the "prior" distribution, $P^0(\mathbf{a})$ which obtains in a completely nonspecific (and hence noninformative) process. We shall have more to say about the prior distribution but for now we can simply think of it as the most chaotic distribution of outcomes. (Explicit results for prior distributions are compiled in Appendix A.)

1.1.1. Surprisal Analysis

The quantitative amount of information provided by a measurement of $P(\mathbf{a})$, the probability of the $\mathbf{a}$th outcome is defined, in relation to the prior case, by[1,2]

$$-I(\mathbf{a}) = \ln[P(\mathbf{a})/P^0(\mathbf{a})] \tag{2}$$

Given a set of experimental data, the first step in the analysis is the computation of the "surprisal" $I(\mathbf{a})$ for the different outcomes $\mathbf{a}$.

Just as complicated looking spectra can be systematized with a judicious use of a handful of spectroscopic constants, a detailed analysis of rate processes can boil down to a small number of "surprisal parameters." A remarkable number of systems exhibit linearity of the surprisal with respect to a physical observable $A(\mathbf{a})$ associated with $\mathbf{a}$,[5-10]

$$I(\mathbf{a}) = \lambda_0 + \lambda A(\mathbf{a}) \tag{3}$$

Only the parameter λ in (3) is independent, and need be determined from the data. The magnitude of λ_0 is completely specified by the magnitude of λ and the condition that the distribution $P(\mathbf{a})$ be normalized. Explicitly, from (3) and (2)

$$\exp(\lambda_0) = \sum_{\mathbf{a}} P^0(\mathbf{a}) \exp[-\lambda A(\mathbf{a})] \tag{4}$$

Even when more than one term is needed in (3) (and this will often be the case if $\mathbf{a}$ is a composite index, e.g., the joint final vibrational–rotational distribution) the number of terms required for accurate reproduction of the data (cf. Appendix B) is far smaller than the range of the index $\mathbf{a}$.

1.1.2. *Information Content*

Just as the spectroscopic constants need ultimately to be understood in more fundamental terms (i.e., an energy level diagram derived from a Hamiltonian), so the surprisal parameters need to be derived from a dynamical approach. The first step is the identification of those properties $A(\mathbf{a})$ that appear in the surprisal, equation (3). Consider therefore the amount of information obtained when the entire set of frequencies $P(\mathbf{a})$ $[\sum_{\mathbf{a}} P(\mathbf{a}) = 1]$ is determined. This is just the average, over all outcomes, of the information per outcome, equation (2). Hence

$$I[\mathbf{a}] = \sum_{\mathbf{a}} P(\mathbf{a}) \ln[P(\mathbf{a})/P^0(\mathbf{a})] \tag{5}$$

The quantity $I[\mathbf{a}]$ can be computed irrespective of the functional form of the surprisal. When multiplied by the gas constant R [$R = 1.987$ eu (entropy units)] the result is known as the "entropy deficiency".[10] While the surprisal offers a measure of the deviance of a particular outcome from the prior (or most chaotic) distribution, the entropy deficiency provides an overall or "integral" measure of specificity.

1.1.3. *Constraints: Informative Observables*

Suppose the results of surprisal analysis admit of the representation

$$-\ln[P(\mathbf{a})/P^0(\mathbf{a})] = \lambda_0 + \sum_{r=1}^{n} \lambda_r A_r(\mathbf{a}) \tag{6}$$

where the A_r's are n distinct properties. Substituting (6) in (5),

$$I[\mathbf{a}] = -\lambda_0 - \sum_r \lambda_r \langle A_r \rangle \tag{7}$$

where $\langle A_r \rangle$,

$$\langle A_r \rangle = \sum_{\mathbf{a}} P(\mathbf{a})\, A_r(\mathbf{a}) \tag{8}$$

is the average value of $A_r(\mathbf{a})$ over the experimental distribution. In information-theoretic terms the magnitude of λ_r can be formally defined, using (7), by

$$\lambda_r = -\partial I[\mathbf{a}]/\partial \langle A_r \rangle \tag{9}$$

In order that λ_r be nonzero, i.e., in order for A_r to appear in the surprisal analysis, it is necessary that the magnitude of its average value affect the amount of information provided by the experiment [i.e., the right-hand side in (9) need be nonzero]. Such a property is known[8] as an "informative observable" or as a "constraint."

This is as far as one can go with information-theoretic analysis. The surprisal and the entropy deficiency provide a local and an integral measure of specificity. Surprisal analysis identifies those observables that account for the deviance from the prior distribution. If these observables all had their average value correctly given by the prior distribution, there would be no deviance. The actual and prior distributions would agree. The major empirical finding of surprisal analysis is that there are typically only very few informative observables and the specificity of the process can be summarized by a few surprisal parameters [the λ_r in equation (6)] or, equivalently, by the magnitudes of the average values of a few informative observables [the $\langle A_r \rangle$ in equation (7)]. It is important to note that either set provides a complete characterization. If one knows the λ_r's, one can compute the $\langle A_r \rangle$ [by using equation (6) in (8)], and if one knows the $\langle A_r \rangle$'s, one can compute the λ_r (an optimal numerical procedure is provided in Appendix C). Figure 1 shows a set of products' vibrational state distributions computed for a given magnitude of $\langle E_v \rangle$, the mean energy in the final vibrational excitation. Also shown are λ_v and $I[v]$ computed in terms of $\langle E_v \rangle$.

1.1.4. Inductive and Deductive Reasoning

The other respect in which information theory can be useful is as a predictive tool. These applications carry the theory beyond the enumeration of possible messages or the analysis of actual messages. To clarify some past misunderstandings, one should note that predictions can be made in either one of two modes. The first is via a process of inductive

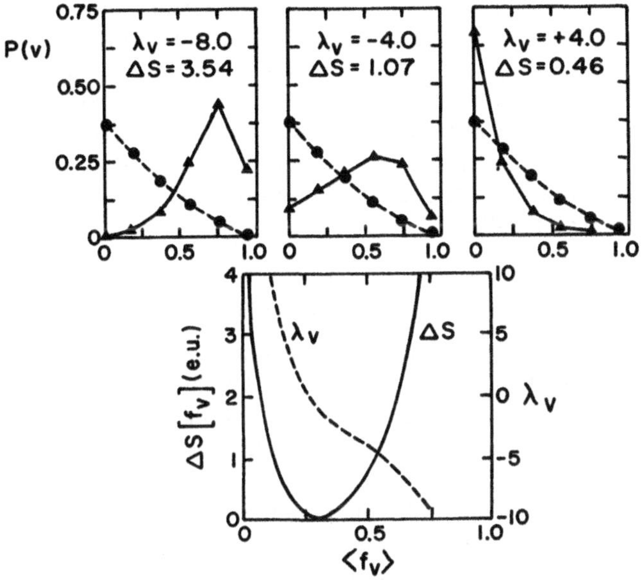

Figure 1. Top row: products' vibrational state distributions, $P(v) = P^0(v)\exp(-\lambda_0 - \lambda_v f_v)$ (triangles) induced for a given magnitude of mean final vibrational energy, $\langle E_v \rangle$, vs. $f_v = E_v/E$. Solid dots indicate $P^0(v)$. Shown also are the resulting magnitudes of λ_v and of $\Delta S = RI[v]$. Bottom: the general dependence of λ_v (broken line) and of ΔS on $\langle f_v \rangle$. When $\langle f_v \rangle = 1/3$ we induce $P(v) = P^0(v)$. Any deviance of $\langle f_v \rangle$ from 1/3 results in a deviance of $P(v)$ from $P^0(v)$. ΔS, the integral measure of deviance, is never negative, but λ_v is positive only if $\langle f_v \rangle < 1/3$ and is negative otherwise.

reasoning.[11-14] Given the data (or the model) at hand, what can one induce regarding yet unobserved aspects of the collision? It is to be emphasized that such a process is one of induction. It offers the most conservative (i.e., most probable or most likely) estimate that can be made using the data at hand. As such it differes from the more familiar deductive reasoning that one commonly employs in the physical sciences. It is therefore important to note that predictions of the surprisal can also be made in an alternative or deductive mode[15,16] and that one can demonstrate the convergence of the inductive procedure to the results obtained by deduction. In plain English, when the data at hand include all the informative observables, inductive reasoning, as discussed below, will yield the correct description, identical to the one provided by a dynamical theory.

The question for deductive reasoning is whether one can start with the initial, precollision state of the system and, given a Hamiltonian, directly identify the informative observables and predict their magnitude after the collision. The answer appears to be yes.[15,16] One can rewrite collision theory so as to limit attention to the evolution of the surprisal and hence of the informative observables. When these are employed in the maximal entropy procedure,[11-14] as described below, the result is identical to that of an exact dynamical computation. Often, of course, one is only interested in an approximate dynamical computation (e.g., the close-coupling procedure), and similar approximations are available for the surprisal.[15,16] The approach is, however, in its early infancy and its main import here is in providing the demonstration that the process of inductive reasoning (i.e., the maximal entropy procedure) will converge to an exact dynamical result.

1.1.5. Surprisal Synthesis via Inductive Reasoning

Inductive reasoning uses any pieces of available data as constraints. These may be incomplete measurements. Say the final-state distribution was separately measured for two reaction paths: what can one induce about the branching ratio[6,10,17-23]? Or, the overall bulk rate of energy transfer is known: what can one induce about the individual state-to-state rate constants[24,25]? In addition, the incomplete characterization can come from dynamical models. For example, a variety of dynamical models suggest that mean momentum for the relative motion of the receding products, $\langle k' \rangle = \int \alpha k' P(k')\, dk'$, increases linearly with the momentum of the colliding reagents:

$$\langle k' \rangle = \alpha k + \rho \tag{10}$$

ρ is a measure of the "repulsive" energy release[26] and $\rho = 0$ in the "stripping" limit. Equation (10) is thus a condition on the distribution, $P(k')$ of the final momentum, and hence can be invoked in order to induce $P(k')$ (or the reactive/nonreactive branching ratio[27]). The technical point here is that there is no such thing as "too much of a good thing"; any piece of data can be used. At worst, it will be noninformative, in which case the associated Lagrange parameter will turn out to be zero. No harm, however, can be caused by superfluous constraints.

The method consists in seeking that distribution of minimal information content that is consistent with the given constraints. The practical details are provided in Appendix C. Given constraints $\langle A_1 \rangle$ to $\langle A_n \rangle$ the resulting distribution is of the functional form (6) with the magnitude of the $n + 1$ parameters chosen such that the distribution is normalized and the magnitude of the n expectation values computed using $P(\mathbf{a})$ in the form (6) agrees with their given values. The resulting distribution has a higher entropy (i.e., is more chaotic or more probable) than any other distribution which is also consistent with the specified magnitudes of the constraints.[6,8,10-14] Even when the magnitude of the average values is not given, the mere identification of the observables that serve as constraints suffices to determine the functional form of the distribution.

1.2. Example: Rotational Energy Transfer

Rotational excitation provides a good illustration of the procedure of surprisal analysis and some of the inferences that may follow from such an analysis. The prior probability to populate the final rotational state j' at a given total energy E is proportional to the degeneracy $2j' + 1$ of the final rotational state and to the density of final translational states at the final translational energy E'_T:

$$P^0(j') = A(E)(2j' + 1)(E'_T)^{1/2} \tag{11}$$

Here $A(E)$ is a normalization constant $[\sum P^0(j') = 1]$ which is independent of j'; E is the total energy, and for a rotor in the ground vibrational state $E'_T = E - E_{j'}$, where $E_{j'}$ is the final rotational energy. To determine the j' dependence of the surprisal, one can just as easily use the rate constants themselves,

$$I(j', j; E) = -\ln[k(j \to j'; E)/k^0(j \to j'; E)] \tag{12}$$

If we take for the prior rate constants

$$k^0(j \to j'; E) = R^0(2j' + 1)(E'_T)^{1/2} \tag{13}$$

with some j'-independent constant R^0, then equation (11) clearly holds.

Figure 2 shows data taken from a quantum-mechanical close-coupling computation[28] of low-energy He–HF collisions. The results provide an example of a linear surprisal plot where the abscissa is the absolute value of the reduced variable Δg_R:

$$\Delta g_R(j, j') = (E_j - E_{j'})/E$$

the fraction of the total energy transferred in the collision. The sort of

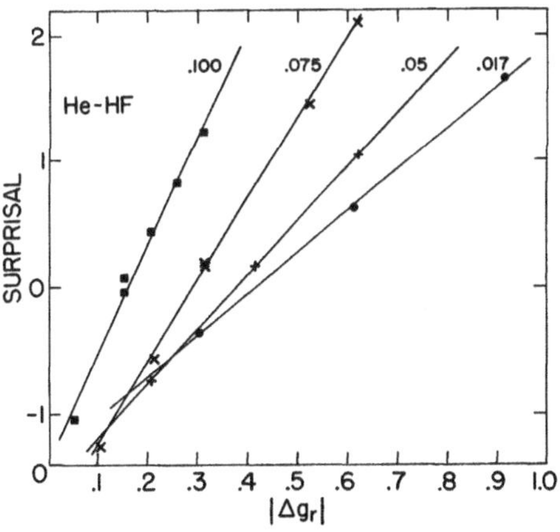

Figure 2. Surprisal plots for rotational state distribution in He + HF(j) → He + HF(j') collisions. Shown is $-\ln[P(j \to j'; E)/P^0(j \to j'; E)]$ vs. $|\Delta g_R| = |E_j - E_{j'}|/E$ (from the computations of L. A. Collins and N. F. Lane[28]) at total energies of 0.017, 0.05, 0.075, and 0.100 eV. The straight lines are least-squares best fits.

behavior exhibited in Figure 2 has been found to hold for rotational energy transfer in a large number of systems.[29]

As indicated by the discussion in Section 1.1, such an observation implies the presence of a dynamic constraint, in this case the magnitude of the rotational energy transfer. This, in turn, suggests that a model that accurately determines the average rotational energy transfer should be able to account for all the individual $j \to j'$ rates. A very simple model[30,31] can be based on the sum

$$\Sigma(j) = \sum_{j'} (E_{j'} - E_j)\, k(j \to j'; E) \tag{14}$$

As j goes from the lowest ($j = 0$) to the highest value allowed at a given energy, $\Sigma(j)$ must necessarily change its sign (from positive to negative). The particular point where the sum vanishes is where the net rate of energy transfer is zero. Hence the point is uniquely determined by detailed balance irrespective of the dynamics.[25,30] What is left for the model is to specify the j dependence of $\Sigma(j)$. Using a linear approximation is found sufficient to provide a satisfactory account.[30,31]

Figure 3 dispels the impression that surprisal plots must always be linear. Shown is an analysis of experimental results[32] obtained by laser-induced fluorescence studies of

$$\text{LiH}(j = 1) + \text{Ar} \to \text{LiH}(j') + \text{Ar}$$

collisions. The curvature in the surprisal plot has also been observed in a number of computational studies (e.g., References 33 and 34), in theoretical investigations[20,35] and in other experimental studies, e.g.,

$$\text{Na}_2^*(A\,^1\Sigma_u^+, j = 38) + \text{Xe} \to \text{Na}_2^*(A\,^1\Sigma_u^+, j') + \text{Xe}$$

collisions.[36] The origin of the second dynamic constraint required to account for the curvature is the subject of current studies.[9,35,37] In Section

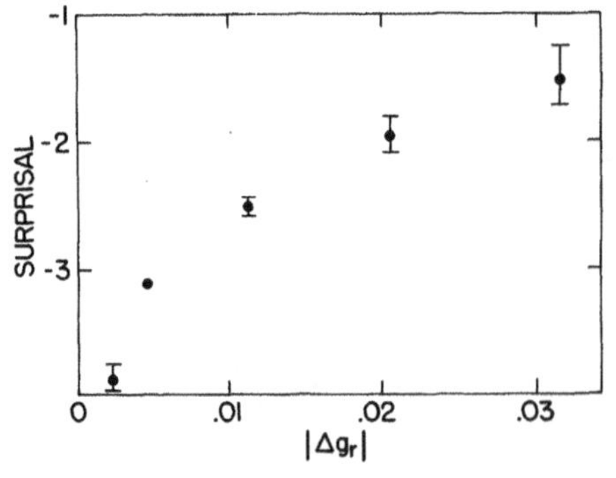

Figure 3. Surprisal plot for the (unnormalized) final rotational state distribution in LiH ($j=1$) + Ar → LiH (j') + Ar collisions. Shown is $-\ln[\sigma(j\to j';E)/(2j'+1)(E-E_{j'})^{1/2}]$ vs. $|\Delta g_R| = |E_j - E_{j'}|/E$ at a nominal total energy E of 0.80 eV (adapted from B. E. Wilcomb and P. J. Dagdigian[32]). Since neither the actual nor the prior distribution have been normalized, the intercept at $|\Delta g_R| = 0$ is not meaningful. The slope of the plot is, however, independent of the normalization. In Section 2 we shall reiterate this point: it is not necessary to observe all final states in order to make a surprisal plot.

2.5, where state-to-state processes are considered, it is suggested that a single dynamic constraint plus the constraint of detailed balance suffice to account for the curvature.[20] The primary observation is, however, that the deviance from the prior behavior is essentially of an "exponential gap" functionality, i.e., $\propto \exp(-\theta_R|\Delta g_R|)$,[34] and that the actual $E_{j'}$ distribution $P(E_{j'}) \propto (2j' + 1)P(j')$ will be nearly symmetric about E_j (particularly so at higher energies $E \gg E_{j'}$, where E'_T is only slowly varying with $E_{j'}$).

1.3. The Prior Distribution

Typical experiments are carried out with only a partial resolution of the accessible final states,[38] and the same holds for many computational studies (where the final m_j quantum numbers or the directions of the relative final momentum are often summed over). The frequencies one is concerned with are thus the relative rates into different groups of final (quantum) states. It is intuitively reasonable to assume that a group Γ consisting of very many final states is more likely to be populated during the collision. The prior distribution serves to eliminate this experimentally built-in bias. It populates every possible final group with a relative rate which is directly proportional to the number of accessible final quantum states which belong to the group. It is the deviation from the prior distribution which is indicative of dynamical constraints during the collision.

1.3.1. Counting Final States

The physically motivated, qualitative argument for the introduction of a prior distribution has an information theoretic, quantitative counterpart. Given a theoretical distribution $Q(\mathbf{f})$ over the final quantum states, one can compute its entropy[1-3,5-14] as

$$S = -R \sum_{\mathbf{f}=1}^{N} Q(\mathbf{f}) \ln Q(\mathbf{f}) \tag{15}$$

To simplify the discussion we shall, for the time being, treat the index $\mathbf{f}$ as a discrete index and take the finite number of accessible final quantum states to be N. How is one to proceed to compute S if all that is given are the (lower-resolution) experimental frequencies $P(\Gamma)$, where each Γ is a (nontrivial) set of final quantum states? We need the $P(\mathbf{f})$'s, yet these were not provided by the experiment. All we know about these unknown $P(\mathbf{f})$'s is that for every Γ

$$P(\Gamma) = \sum_{f \in \Gamma} P(\mathbf{f}) \tag{16}$$

Here the $P(\Gamma)$'s are frequencies determined by the experiment, and the notation $\mathbf{f} \in \Gamma$ means summation over those final states in the group Γ.

It is easy to demonstrate that among all the distributions of final quantum states which are consistent with the observed frequencies of the final groups, the unique distribution of maximal entropy has all the states **f** within any group Γ equally probable. Hence the assumption

$$P(\mathbf{f}) = P(\Gamma)/g_\Gamma \tag{17}$$

for all $\mathbf{f} \in \Gamma$ is the least biased one which can be made where g_Γ,

$$g_\Gamma = \sum_{\mathbf{f} \in \Gamma} 1 \tag{18}$$

is the number of quantum states in the group Γ. Any other assignment would correspond to a less than maximal entropy for the distribution of quantum states within the group Γ and hence imply that we do have some information on that distribution which, by assumption, we do not.

An explicit example of the counting of final states was already provided in the previous section. Other detailed examples are scattered throughout the chapter and are collected together in Appendix A.

1.3.2. Counting Translational States

At this point a caveat is in order. For all practical purposes, the translational states of real systems are continuous rather than discrete. Hence the use of discrete indices to count "states" involves some kind of envisioned quantization of this continuum. One commonly employed device for accomplishing this is to imagine the system enclosed in a box with totally reflecting walls and with dimensions much larger than molecular sizes. Translational "states" are then defined as acceptable solutions of the free-molecule Schrödinger equation inside the box, with normalization corresponding to a volume h^3 in phase space (h per translational degree of freedom). In the limit of a very large box, these discrete translational states converge to a continuum with a *density* of states (number per unit volume and per translational energy interval) given by[5-10,38]

$$\begin{aligned} \rho(E_T) &= h^{-3} 4\pi p^2 (dp/dE_T) \\ &= (\pi^{-1} k^2/\mathbf{v})/h \\ &= A_T E_T^{1/2} \end{aligned} \tag{19}$$

Here $E_T = p^2/2\mu = \hbar^2 k^2/2\mu$ is the translational energy, $\mathbf{v} = p/\mu$ is the velocity, and $A_T \propto \mu^{3/2}$ (or $A_T = \mu^{3/2}/2^{1/2}\pi^2\hbar^3$). This "degeneracy" of translational $\hbar^3$ states is the basis for the prior distribution in what follows and in the previous studies[5-10] using the information-theoretic point of view. It must be pointed out, however, that workers reasonably may and do differ in the precise manner translational states are to be counted, the point being that in this chapter we are concerned not with any arbitrary distribution over quantum states but specifically with the nascent final-state distribution after a binary collision. One may therefore argue that this information should be incorporated into the way translational states are counted. When this is done, the resulting distribution is typically referred to as a "statistical theory."[39-43] It is the "most chaotic" distribution possible subject to quite general dynamical limitations imposed on physical grounds (an example is discussed later). The problem is that there may well be legitimate differences among the precise limitations one may wish to impose. In particular, these may well depend on the dynamics of the particular system.[43,44] We therefore *recommend* that the prior distribution be defined, using the procedure of counting all the final quantum in the manner just outlined, i.e., including the translational degeneracy. We do this notwithstanding the fact that for a collision process, where $P(\mathbf{f})$ is the relative rate into the final state $\mathbf{f}$, there are alternative defensible *theories*. Indeed, this is precisely the point. The prior distribution is not offered as a theory but as a reference.[45] It is the most chaotic distribution of quantum states. If for any reason the actual distribution is deviant from the prior one, let the analysis show it. If the origin of the deviance is physically clear, it can always be introduced as a constraint and the maximal entropy distribution would lead to an inducted distribution in accord with the observation. Such a distribution, being of maximal entropy *subject to constraints* is no longer a prior distribution.

Throughout the remainder of this chapter we shall reserve the designation "state" for a description in which both the internal and translational degrees of freedom are regarded as discrete and fully specified, with indices $\mathbf{i}$ and $\mathbf{f}$, respectively, denoting possible reactant or product quantum states in a reaction.

1.3.3. Entropy and Entropy Deficiency

The entropy of $P(\Gamma)$, the experimental distribution, can now be computed, using (15) and (17), as

$$\begin{aligned} S(\Gamma) &= -R\sum_{\mathbf{f}} P(\mathbf{f})\ln P(\mathbf{f}) \\ &= -R\sum_{\Gamma}\sum_{\mathbf{f}\in\Gamma} [P(\Gamma)/g_\Gamma]\ln[P(\Gamma)/g_\Gamma] \qquad (20) \\ &= -R\sum_{\Gamma} P(\Gamma)\ln[P(\Gamma)/g_\Gamma] \end{aligned}$$

The replacement of $P(\mathbf{f})$ by $P(\Gamma)/g_\Gamma$ and the summation over $\mathbf{f}$, $\mathbf{f}\in\Gamma$, required the explicit incorporation of the assumption of no variation of $P(\mathbf{f})$ within the group Γ. It is convenient to rephrase expression (20) by the explicit introduction of the prior distribution $P^0(\Gamma)$ defined by

$$P^0(\Gamma) = g_\Gamma \Big/ \sum_\Gamma g_\Gamma \equiv g_\Gamma/N \tag{21}$$

or, cf. (17),

$$P^0(\mathbf{f}) = 1 \Big/ \sum_\Gamma g_\Gamma = 1/N \tag{22}$$

The prior distribution is the one of globally maximal entropy, where all final states are equally probable. With the definition (21)

$$\begin{aligned} S(\Gamma) &\equiv -R\sum_\Gamma P(\Gamma)\ln[P(\Gamma)/NP^0(\Gamma)] \\ &= R\ln N - R\sum_\Gamma P(\Gamma)\ln[P(\Gamma)/P^0(\Gamma)] \end{aligned} \tag{23}$$

We see that

$$\Delta S[\Gamma] = R\sum_\Gamma P(\Gamma)\ln[P(\Gamma)/P^0(\Gamma)] \tag{24}$$

is the (nonnegative) difference between the maximal ($R\ln N$) and actual ($S[\Gamma]$) magnitudes of the entropy.

In information-theoretic terminology, $\Delta S[\Gamma]$ is the information (in entropy units) provided when the distribution $\{P(\Gamma)\}$ has been measured. The origin of the terminology stems from the following considerations. In the absence of any information, there are N possible outcomes to the experiment, all equally probable. If only one particular outcome is observed to occur, we obtain $R\ln N$ units of information.[46] The actual experiment has resulted in a distribution over the possible outcomes. The entropy $S[\Gamma]$ is the amount of information still missing when all we know is just the distribution. The difference $\Delta S[\Gamma]$,

$$\Delta S[\Gamma] = R\ln N - S[\Gamma] \tag{25}$$

is therefore the amount of information provided by the knowledge of the distribution. The inequality

$$\ln[P(\Gamma)/P^0(\Gamma)] \geq 1 - P^0(\Gamma)/P(\Gamma) \tag{26}$$

with equality holding if (and only if) $P(\Gamma) = P^0(\Gamma)$ verifies that $\Delta S[\Gamma]$ is nonnegative and equals zero only when the measured frequencies equal the prior distribution. Any deviance is reflected in a finite positive value of $\Delta S[\Gamma]$ (cf. Figure 1).

1.3.4. Theoretical Distributions

Even without any prior dynamical information, we know that the fate of any system will be limited by restrictions imposed by conservation laws. The prior distributions introduced above represent the least biased assumption consistent with incorporation of a single conservation law, that for the total energy. Since there are undoubtedly other conservation principles operative for any given system, this prior is likely to be too permissive in that there may well be parts of phase space which conserve energy but cannot be reached by the collision process under investigation, because doing so would violate one or more of these additional conservation laws. Two general types of such constants of the motion are the conservation of the total angular momentum on the one hand and the constants of motion typical of classical systems which, in classical mechanics, exclude certain regions of the final phase space.(47) If these extra constants of the motion could be included, they would yield a more sophisticated prior distribution. Except for the total angular momentum, none of these other conservation laws has a sweeping generality. Furthermore, none has the same ease of incorporation as the conservation of energy. In particular, imposing the conservation of total angular momentum inevitably requires a theoretical decision regarding the "cutoff."(39) Hence we choose to regard any refinement of the prior distribution (subject to conservation of energy) as a theoretical distribution rather than as a "superior" prior.

We shall employ the notation $Q(\Gamma)$ to represent a theoretical (as opposed to an observed) distribution.

It is inevitably tempting to regard any predicted distribution in much the same light as a prior expectation. Information theory recognizes the legitimacy of this point of view(8,12,14,48) and introduces the quantity H,

$$H = \sum_{\Gamma} P(\Gamma) \ln[P(\Gamma)/Q(\Gamma)] \tag{27}$$

as the measure of the information provided by the observed frequencies $P(\Gamma)$ when the theoretical probabilities $Q(\Gamma)$ are known beforehand. H is nonnegative and vanishes if and only if theory and experiment are in complete accord $[P(\Gamma) \equiv Q(\Gamma)$ for all $\Gamma]$. It is further discussed in Appendix B. It must, however, be explicitly recognized that H is not the entropy deficiency [unless, of course, $Q(\Gamma) \equiv P^0(\Gamma)$]. The nomenclature of the "persuasion of the theory to the data" is hereby proposed for H to avoid any possible source of confusion between $\Delta S[\Gamma]$, the entropy deficiency which is the integral measure of the specificity of the distribution of outcomes, and the persuasion H, a quantity of interest only to the originator of the theory leading to the distribution $Q(\Gamma)$. An intermediate position is occupied by those theories obtained by maximizing the entropy subject to constraints (Appendix C). There H appears as the difference between the entropy deficiency computed for the experimental data and the entropy deficiency computed for the theoretical distribution. For such theories only, one can assign a thermodynamic significance to the persuasion.[8,49]

1.3.5. Transition State Theory

A particularly useful theory which has been extensively discussed in the literature[39-44,50,51] is the "transition state" theory. In its simplest form it seeks to add just one physically motivated constraint: that molecules have a finite size or, explicitly, that around the center of mass of the colliding molecules one can imagine a large "sphere of no return." Whenever the relative separation of the two fragments exceeds the radius of the sphere, they will continue to form products. If the radius of the sphere is taken to be energy independent, (as in "tight" transition state theory) the predicted products' state distribution will be quantitatively different from the prior distribution (even though it is qualitatively similar). When the radius of the sphere is determined in terms of the centrifugal barrier (as in "loose" transition state or as in "phase-space" theory[39,41]) and is thus energy dependent (decreasing as E_T increases), the results differ even less from the prior distribution. In somewhat more quantitative terms, let $b(\alpha)$ be the maximal impact parameter leading to products of type α, and $\mathbf{v}_\alpha$ the products' relative velocity. Then $P^0(\alpha) \propto \mathbf{v}_\alpha b^2(\alpha)$, while for transition state theory $P(\alpha) \propto \mathbf{v}_\alpha^2 b^2(\alpha)$ [in phase-space theory[39] $b(\alpha)$ is typically a decreasing function of $\mathbf{v}_\alpha$]. The same factor of $\mathbf{v}_\alpha$ is present in discussing energy consumption. At a given total energy, all initial states react with the same prior rate (cf. Section 2.3.1), while in the phase-space theory the rate is proportional to $\mathbf{v}_\alpha b^2(\alpha)$ [where again $b(\alpha)$ is a decreasing function of $\mathbf{v}_\alpha$, so that $\mathbf{v}_\alpha b^2(\alpha)$ is only weakly $\mathbf{v}_\alpha$ dependent[39]].

1.4. Example: Products' Translational Energy Distribution

At a sharply defined total energy, the final kinetic energy E_T is limited to such discrete values as are allowed by energy conservation $E = E_T + E_I$. Here E_I is the internal (quantized) energy of the products. Given the finite resolution of both the initial velocity selection and the final velocity analysis, the measured translational energy distribution can be approximated as a continuous distribution. This will be a smooth, structureless distribution provided that the number of products' internal states per energy interval is sufficiently high and sufficiently slowly varying with energy. Except for low-energy collisions involving diatomic hydrides or when E_T is quite near E, the required condition does obtain. The results of a detailed molecular beam experiment (not using internal-state detection, e.g., Reference 52) are thus typically represented as a flux contour map.[51,53,54] The two most common forms are a "Cartesian" plot showing contours of $d^3\sigma/d\mathbf{k}$ and a polar plot showing contours of $d^3\sigma/d^2\omega dE_T$:

$$d^3\sigma/d^2\omega dE_T = (d^3\sigma/d\mathbf{k})(d\mathbf{k}/d^2\omega dE_T)$$
$$= (d^3\sigma/d\mathbf{k})[\rho_T(E_T)/4\pi] \tag{28}$$

Here, as usual [cf. equation (19)] $\rho_T(E_T)$,

$$\rho_T(E_T)\, dE_T(d^2\omega/4\pi) = d\mathbf{k} = k^2 dk d^2\omega \tag{29}$$

with $E_T = \hbar^2k^2/2\mu$ the density of final translational states, $d\mathbf{k}$ a volume element in the final momentum space, and $d^2\omega$ an element of solid angle in the direction of **k**. Multiplying (28) by the initial velocity, one obtains the rate coefficient for formation of products in the specified range.

1.4.1. Counting Final States

In terms of the scattering-theoretic differential cross section introduced[51,53] as the flux into a given final state n with a final momentum in the range $\mathbf{k}, \mathbf{k} + d\mathbf{k}$,

$$(d^3\sigma/d\mathbf{k})\,\Delta E_T = \sum_{n\in\Gamma} \sigma(\rightarrow n, \mathbf{k}) \tag{30}$$

The summation on the right is restricted to final states in the group Γ defined such that the internal energy of states within the group is within ΔE_T of $E - E_T$, $E - (E_T + \Delta E_T) < E_I < E - E_T$. To compare $d^3\sigma/d\mathbf{k}$ to an "average" cross section in the stated interval, one can rewrite (30) as

$$(d^3\sigma/d\mathbf{k})\,\Delta E_T = \left[\sum_{n\in\Gamma} \sigma(\rightarrow n, \mathbf{k}) \Big/ \sum_{n\in\Gamma} 1\right] \cdot \sum_{n\in\Gamma} 1$$
$$= \left[\sum_{n\in\Gamma} \sigma(\rightarrow n, \mathbf{k}) \Big/ \sum_{n\in\Gamma} 1\right] \rho_I(E - E_T)\,\Delta E_T \tag{31}$$

Here $\rho_I(E_I)$ is the density of internal states

$$\rho_I(E_I)\,\Delta E_T = \sum_{n\in\Gamma} 1 \tag{32}$$

or

$$\sum_n \rightarrow \int dE_I \rho_I(E_I) \tag{33}$$

The second line of (31) defines the continuum approximation where the internal energy E_I is taken as a continuous variable, so that the limit $\Delta E_T \rightarrow 0$ is well defined for both (31) and (32). This is, of course, the case for a classical trajectory computational experiment and serves as a realistic approximation otherwise.

1.4.2. The Prior Distribution

The prior distribution is defined by the assumption of equal flux into any final quantum state. Introducing R^0,

$$R^0 = \sum_n k^0(\rightarrow n, \mathbf{k}) \Big/ \sum_n 1 \tag{34}$$

the prior rate coefficient into a final state

$$d^3k^0/d\mathbf{k} = R^0\rho_I(E - E_T) \tag{35}$$

From (35), the total prior rate coefficient (integrated over all final states) is

$$\begin{aligned} k^0(E) &= \int d\mathbf{k} R^0 \rho_I(E - E_T) \\ &= R^0 \int (d^2\omega/4\pi) \int dE_T \rho_T(E_T)\,\rho_I(E - E_T) \\ &= R^0\rho(E) \end{aligned} \tag{36}$$

where $\rho(E)$ is the total density of states at the energy E

$$\rho(E) = \int \rho_T(E_T)\,\rho_I(E - E_T)\,dE_T \tag{37}$$

The polar prior differential cross section is given by

$$d^3\sigma^0/d^2\omega dE_T = (R^0/\mathbf{u})\,\rho_T(E_T)\,\rho_I(E - E_T)/4\pi \tag{38}$$

(where $\mathbf{u}$ is the *initial* relative velocity) and is shown as a contour plot in Figure 4. As long as only the final-state distribution is under consideration,

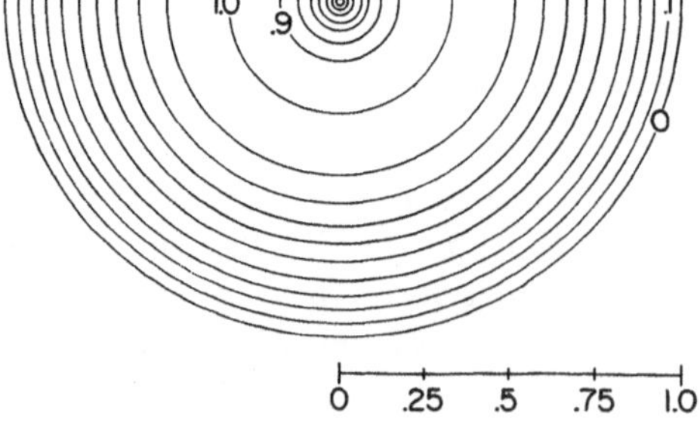

Figure 4. Products flux [energy–angle, $P^0(E_T, \omega)$] relative contour map, for an atom–diatom collision, in the prior limit [cf. equation (38)]. In the RRHO limit $P^0(E_T, \omega) \propto f_T^{1/2}(1 - f_T)$ [cf. Appendix A, equation (A.8)]. The most probable contour, at $f_T = 1/3$, has been assigned the value 1. The velocity-angle contour map [$P^0(v, \omega) \propto f_T(1 - f_T)$] will also correspond to concentric circles, with the most probable contour at $f_T = 1/2$.

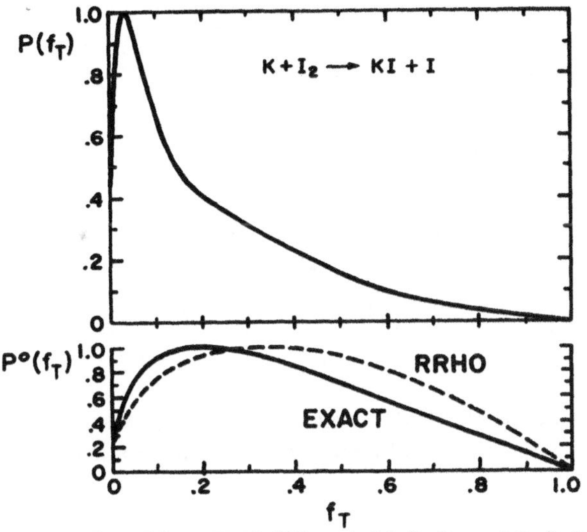

Figure 5. Observed [adapted from K. T. Gillen, A. M. Rulis, and R. B. Bernstein, *J. Chem. Phys.* **54**, 2831 (1971)] and prior (bottom panel) final translational energy distribution for the $K + I_2 \rightarrow KI + I$ reaction. (Plot adapted from U. Dinur, R. Kosloff, R. D. Levine, and M. J. Berry[19].) The ionic KI molecule is only poorly approximated by the RRHO level scheme (cf. Appendix A). Hence an exact counting of the KI internal levels (solid curve in bottom panel) leads to a somewhat different prior distribution than that determined in the RRHO limit (dashed line). Note, however, that the total energy (~45 kcal/mol) is high.

one can use either the cross section or the rate coefficient. Once, however, the role of the initial state is also being probed, the rate coefficient emerges as the more "natural" quantity, as will become evident in the second part of this chapter. In anticipation of these developments, we have introduced the prior rate constant at this earlier junction.

Less detailed (so called "marginal") distributions are obtained by summing (i.e., integrating) over some of the variables. For example

$$\begin{aligned} P^0(E_T) &= \int d^2\omega (d^3k^0/dE_T d^2\omega)/k^0(E) \\ &= (dk^0/dE_T)/k^0(E) \\ &= \rho_T(E_T)\,\rho_I(E - E_T)/\rho(E) \end{aligned} \tag{39}$$

shown vs. E_T in Figure 5 and

$$\begin{aligned} P^0(\omega) &= \int dE_T (d^3k^0/dE_T d^2\omega)/k^0(E) \\ &= (d^2k^0/d^2\omega)/k^0(E) \\ &= 1/4\pi \end{aligned} \tag{40}$$

The introduction of a continuous probability distribution requires proper definition and means of practical computation of the entropy deficiency. A detailed discussion with worked out examples is available.[55]

1.4.3. Translational Surprisal

Measured translational energy distribution usually, but not always, deviates from the prior distribution, e.g., Figure 5.[19] In a great many cases, the deviance could be well represented by[26,27,56]

$$-\ln[P(E_T)/P^0(E_T)] = \gamma_0 + \tilde{\gamma}_T \kappa^{-2}(k-\kappa)^2$$
$$= \gamma_0 + \tilde{\gamma}_T \varepsilon^{-1}(E_T^{1/2} - \varepsilon^{1/2})^2 \qquad (41)$$

where κ is a parameter with the dimensions of k; γ_0 and $\tilde{\gamma}_T$ are dimensionless parameters. ε often varies linearly with initial kinetic energy (Figure 6).

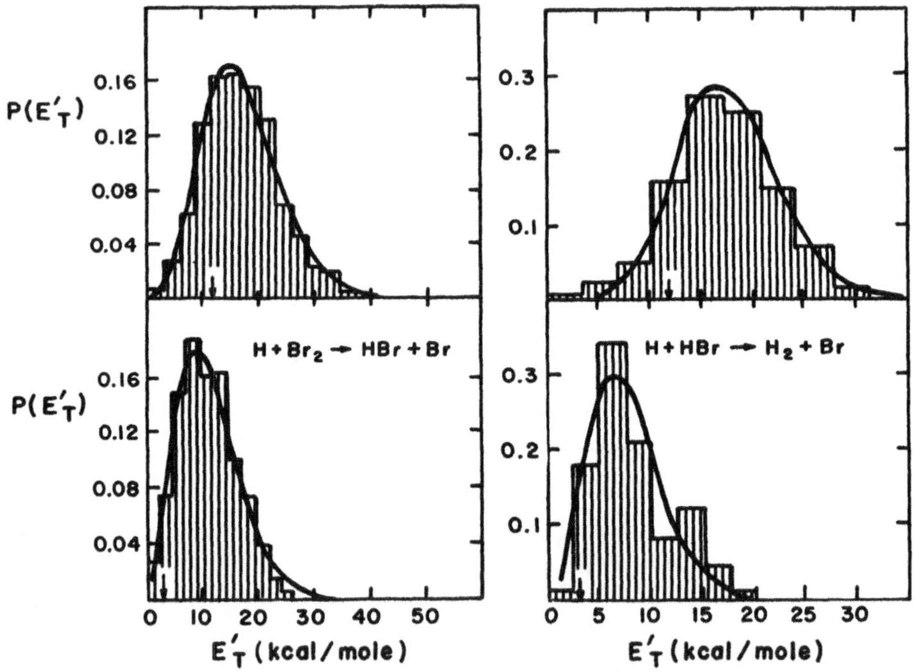

Figure 6. Histogramic representation of $P(E_T')$ determined via classical trajectory computations [J. M. White, *J. Chem. Phys.* **65**, 3674 (1976)] and the (continuous) fit provided by equation (40) (plot adapted from E. Pollak and R. D. Levine[56]). The initial translational energy is indicated by an arrow, and the most probable final translational energy increases as E_T increases. (This is well simulated by a linear increase of κ with k.) For H + Br_2, $\Delta S[E_T'] = 1.26$ and 0.95 eu and the "goodness-of-fit" (cf. Appendix B) $H = 0.05$ and 0.03 eu. For H + HBr, $\Delta S[E_T'] = 0.63$ and 0.58 eu and $H = 0.12$ and 0.14 eu, respectively.

Another functional form[57] which has been suggested to give better fits for "backwards" scattering is provided by

$$-\ln[P(E_T)\,E_T^{1/2}] = \gamma_0 + \tilde{\gamma}_T \kappa'^{-2}(k - \kappa')^2 \tag{42}$$

No quantitative estimates have been provided for the relative quality of fit (Appendix B) for the two functional forms. However, since the fit provided by the form (41) is already within the probable limits of error of both the best experimental and computational data[56] we recommend its general use for the time being in the analysis of translational energy distributions. This advice may require revision, however, if the form (42) (which is not cast in "surprisal" language) does prove to have a statistically significant superiority.

The functional form

$$-\ln[P(E_T)/P^0(E_T)] = \lambda_0 + \lambda_T E_T/\langle E\rangle \tag{43}$$

where $\langle E\rangle = -\Delta E_0 + \langle E_{\text{reagents}}\rangle$, is often sufficient. The reason is that the appearance of terms linear in k in (41) stems from the finite value of the parameter κ. The low initial energy dependence of κ is primarily a measure of the impulse imparted during the collision, i.e., $\hbar^2\kappa^2/2\mu$ is, at low initial energies, the repulsive energy release. As $\kappa \to 0$, the form (41) reduces to (43), with λ_T being the finite limit of $\tilde{\gamma}_T\kappa^{-2}$ as $\kappa \to 0$.

Not every system will have the high density of products' internal states required to give the sort of smooth, structureless translational energy distributions treated in the foregoing discussion. When internal states of the products are coarsely spaced compared to the spread in initial energy and the translational energy resolution, it is possible for the distribution in translational energy to exhibit definite structure arising from population of different individual internal states or groups of states. Although such behavior is uncommon, the appearance of individual "peaks" owing to formation of different vibrational states of DF from $F + D_2$ has been observed in the velocity-angle distributions of the DF product.[58] A computed translational-energy distribution for the same reaction, based on the infrared chemiluminescence data, shows the same structure.[51,58]

1.5. *Guideposts for Surprisal Analysis*

This section is a flow chart of surprisal analysis for products' final states with the more venturesome forays indicated as options.

1.5.1. *Determine the Frequencies*

Specify the outcomes under consideration. If not already available, obtain the frequencies $P(\Gamma)$ from the rate data by normalization,

$$P(\Gamma) = k(\Gamma) \Big/ \sum_{\Gamma} k(\Gamma) \tag{44}$$

In equation (44), the label Γ refers to a specific final group of states, since we have not yet extended our considerations to include initial states. For the simple case of data on energy consumption from various initial states most of these guidelines can be taken over essentially intact. The only required difference, which originates in a subtlety of normalization convention to be discussed later, is to eliminate this first step, i.e., do not bother with the frequencies; where the guidelines say "form $-\ln(P/P^0)$," just use $-\ln(k/k^0)$ instead. The two will differ only by a state-independent normalization.

1.5.2. Determine the Conditions

(a) *Is the process at a single value of the total energy*? This is usually true for computational data. It is also often a reasonably good start for highly exoergic reactions where the spread in initial energy is only a small fraction of the total energy available for the products. However, this approximation will be in error for states near the nominal value of the energy employed.

(b) *If not a single total energy process*, decide on the actual conditions, e.g., a thermal distribution for all reagents' degrees of freedom, etc.

1.5.3. Find the Prior

For the reasons discussed in Section 1.3, we recommend that in the interest of consistency a prior distribution based on the density of final states (the most entropic distribution subject to conservation of total energy) be employed.

(a) *Given total energy or given temperature*: For processes at a given total energy and processes where the reagents have a thermal distribution except for selected degrees of freedom, use Appendix A or some other source[10,25,38,59,60] to determine the prior, for each data point.

(b) *For other experimental conditions*: Determine the appropriate prior by averaging the prior rate (at a given total energy) over the experimental distribution of reagent states. Several examples are worked out in the second part, and the general case is stated in Section 2.3.1.

1.5.4. Determine the Surprisal

Form $-\ln[P(\Gamma)/P^0(\Gamma)]$ for each data point. To within the accuracy of $P(\Gamma)$ is the surprisal different from zero?

(a) *No*? You have an additional example of completely chaotic final-state distribution. There are other examples (9) for all possible degrees of freedom but, on the whole, such processes are in a minority. Hence check your conclusion by computing the entropy deficiency. It will not be zero due to the experimental uncertainties in the probabilities. In general [cf. equation (B.3)] the error $\delta\Delta S$ in ΔS due to experimental uncertainty $\delta P(\Gamma)$ in the probabilities is

$$\delta\Delta S = (1/2)\, R \sum_{\Gamma} P(\Gamma)\, [\delta P(\Gamma)/P(\Gamma)]^2$$

$\delta\Delta S$ (in eu) is seen to equal the (average) square fractional error in the probabilities. Hence when the probabilities $P(\Gamma)$ are known to, say, 10%, ΔS could be regarded as zero if it is smaller than 10^{-2} eu.

(b) *The results do deviate from the prior distribution*: The process does show specificity of energy disposal. Compute the entropy deficiency as an integral measure of this specificity and proceed to characterize the deviance via surprisal analysis. (Omit the entropy computation for the simple analysis of energy *consumption* or wherever you do not have sufficient data to determine a normalized distribution.)

1.5.5. *Surprisal Analysis*

(a) *Decide on a relevant variable* $A(\Gamma)$, e.g., $|\Delta g_R(\Gamma)|$ in the example of rotational energy transfer. Plot $-\ln[P(\Gamma)/P^0(\Gamma)]$ vs. $A(\Gamma)$. Do try several possible choices for $A(\Gamma)$.

An important guideline, backed by theory[16] is that processes of the same family typically have the same constraints. If a similar process were previously analyzed (see, e.g., Reference 9 for the latest reference compilation) you might try the same variables. In particular, repulsive release reactions will typically require more than one constraint (cf. Section 2.5). Also, if reagent state selection has been carried out, do at least glance through Section 2 (Section 2.5 in particular).

(b) *Examine the resulting plot.* Is it smooth? If so, go to (c) below. If not, consider whether another variable might not do better. What you are looking for is a variable that accounts for the systematics of the deviance of the products' state distribution from the most uniform population of their phase space. If no such clear-cut variable emerges, congratulate yourself on having chosen a fascinating problem to address. Your surprisal is not linear. As one option go to (e). A second option, depending on your temperament and on previous results for similar systems, is to reexamine the data.

(c) *Is the smooth surprisal plot almost linear*? If not, go to (d). If it is, this shows that the dynamics of the collision are dominated by a single constraint. Find out what this constraint is and explore what features of the dynamics would lead to such a constraint. Proceed to Section 1.5.6 as an option.

(d) *Surprisal plot smooth but not linear*? You have the right variable but the wrong functionality. If no obvious changes [e.g., replacing $A(\Gamma)$

by $A^2(\Gamma)$] seem to help, you have more than one constraint, each of which depends on $A(\Gamma)$ (e.g., translational energy disposal). This will typically be the case if there is some reagent state selection. The condition of detailed balance plus the imposing of $A(\Gamma)$ as a constaint may be the answer. Examine Section 2.5 and then go to (e) where you have a headstart over the others since you already know the variable.

(e) *Surprisal analysis with more than one constraint* means fitting the surprisal to a functional form as in equation (6) with $m > 1$. Choose your variables and try. A least-squares fitting procedure can be used, but the method outlined in Appendix C has some advantages (in particular, a unique minimum[48]). Proceed to Section 1.5.6 as an option.

1.5.6. Optional: Refinements

You now have an acceptable fit of the surprisal in terms of one or more informative constraints. You can judge how good the fit is by computing the quality of fit (Appendix B) between the raw data and your functional form

$$H = \sum P(\Gamma) \ln[P(\Gamma)/P^I(\Gamma)] \tag{45}$$

where

$$P^I(\Gamma) = P^0(\Gamma) \exp\left[-\lambda_0 - \sum_{r=1}^{n} \lambda_r A_r(\Gamma) \right] \tag{46}$$

is the distribution determined by surprisal analysis. For good fit, H should be small compared to the entropy deficiency [cf. equation (B.7)].

The superscript I on the distribution (46) serves as a reminder that exactly the same distribution (with the same magnitude of the λ_r's) would result from the inductive process of maximizing the entropy for given magnitude of the constraints, $\langle 1 \rangle, \langle A_1 \rangle, \ldots, \langle A_n \rangle$ computed as averages over the raw data. To improve upon $P^I(\Gamma)$ one needs to examine the deviance of $P(\Gamma)$ from $P^I(\Gamma)$. Graphical methods are unlikely to be very useful here for the addition of another constraint [i.e., increasing n to $n + 1$ in equation (6)] would change all the λ_r's. The process is therefore best carried out as in nonlinear surprisal analysis (part (e) of Section 1.5.5, and Appendix C) except that now the major contribution to the surprisal has already been identified.

At the risk of endless repetition, we point out again the difference between a prior $P^0(\Gamma)$ and $P^I(\Gamma)$. Like the prior, $P^I(\Gamma)$ can be used as a starting point for analysis. In contrast to the prior which is immutable, the result of the refinement, i.e., the addition of $A_{n+1}(\Gamma)$ as a constraint, will not be just an addition of a term of the form $\lambda_{n+1}A_{n+1}(\Gamma)$ to the exponent in (46). Rather, all the λ_r may well change in the process. The resulting distribution allowing $n + 1$ constraints, say, $P^I_{n+1}(\Gamma)$, is related to the distribution using m constraints, $P^I_n(\Gamma)$, by

$$P^I_{n+1}(\Gamma) = P^I_n(\Gamma)\exp\left[-\delta\lambda_0 - \sum_{r=1}^{n}\delta\lambda_r A_r(\Gamma) - \lambda_{n+1}A_{n+1}(\Gamma)\right]$$

Technical details are provided in Appendixes B and C.

1.5.7. *For Model Builders Only*

Do you have a model (or a pet theory) which predicts a theoretical distribution $Q(\Gamma)$? For you only, we can offer a diversity of options which is not available to those engaged in experimental or computational studies. Like them, you can proceed through Sections 1.5.1–1.5.4 (above) and Section 1.5.8 (below) using your theoretical distribution as the "data," $P(\Gamma)$. This will characterize the specificity, delineate what dynamical constraints are dominant in the model, possibly offer a "short cut" derivation of the main results (using Section 1.5.8 below), etc. In addition, however, you have the option of examining $-\ln[P(\Gamma)/Q(\Gamma)]$ and of computing H [cf. equation (45)]. It must be stressed, however, that $-\ln[P(\Gamma)/Q(\Gamma)]$ is *not* the surprisal *nor* is H the entropy deficiency. What you will find, however, by examining $-\ln[P(\Gamma)/Q(\Gamma)]$ is what important dynamic constraints have been left out of your model. The magnitude of H (Appendix B) will measure the quality of the overall fit. The fit is acceptable, but if you want to do better, proceed to Section 1.5.8.

1.5.8. *Optional: Surprisal Synthesis*

Predicting the distribution by inductive reasoning is different from surprisal analysis only in that you pretend not to look at the data first. Rather, you reason out what should be an informative variable. This provides a constraint and the maximal entropy procedure provides the form of the distribution. If (say, from bulk measurements; see, e.g., References 24 and 25), you also know the magnitude of the constraint you can compute the surprisal parameter. The result is, of course, a distribution with a surprisal linear in the parameter of your choice. Suppose the data do not conform to a linear surprisal. Then it is necessary to maximize the entropy subject to two constraints (see, e.g., References 22, 23, 59, and 60), and so on, adding constraints until an acceptable fit obtains. What assurance do we have that the process will converge? Well, for one thing it will trivially converge whenever there are only a finite number (d) of data points. More fundamentally, one can demonstrate explicit convergence to the exact solution with far fewer constraints than data points.[15,16,61,62]

2. *State-to-State Chemistry*

As long as considerations are limited to the relative frequencies of various groups of final states, life is comparatively simple. The magnitude of the observed *frequencies* may differ for different initial preparation procedures, but the definition of the final-state frequencies is intuitive and obvious. Similarly, for any given theory, the specification of the final-state *probabilities* is clear cut (and here also their magnitude in general will reflect the details of the required averaging over initial states). Given a choice for the reference theory, surprisal analysis and persuasion analysis are comparisons of two intuitively well-understood numbers, the experimental frequency vs. the theoretical probability. Such a comparison is meaningful and useful even when experiment and theory fail to agree. Indeed, it is when they fail to agree that we obtain information (in the technical sense, cf. Appendix B) from the knowledge of the experimental frequencies.

The frequency of a given group of initial states is, however, a more subtle concept. The observed frequency with which a process will initiate from some specific group of states in a given sample depends both upon a dynamically uninteresting quantity (the abundance of that group of states in the sample) and an interesting one (the specific rate coefficient for that group of states). The need to distinguish between the relative abundance of a group of initial states and their relative reactivity (i.e., their relative contribution to the total rate of the process) was recognized a long time ago by Tolman,[63] who coined a special phrase, the "distribution of reactive reactants," to describe the frequency distribution of initial states, which is of interest to us here. This concept has not, however, attained the same acceptance as the more familiar "distribution of nascent states." Yet it is clear that there can be no discussion of state-to-state chemistry without recognizing explicitly the possibility of different specific rate coefficients for different groups of initial states. One purpose of this section is to define the distribution of reactive reactants and to show how this definition enables one to extend the formalism of probability theory (and hence of information theory) to state-to-state processes.

We shall also develop a view that embraces both forward and reverse processes on a more or less equal basis. To avoid the possibility of later confusion, we emphasize that the state label **i** indicates a reagent state for the *forward* process only, and similarly, **f** denotes a product state of the forward process. It is also worthwhile to recall that in our language the term "state" implies a hypothetical situation in which *all* quantum numbers, including those pertaining to the translational degrees of freedom, are discrete and fully specified.

2.1. The Probability Matrix

2.1.1. The Collision Theory Canon

To analyze experiments (or computations) where there is at least a partial selection of reagent states, it is necessary to distinguish between those probabilities that depend on the initial preparation and those that do not. We thus consider a general scheme where the mole fraction of reagents in the state **i** is $X_{\mathbf{i}}$. The fraction of products in the state **f** is determined by the collision theory canon "average over initial states, sum over final states,"

$$P(\rightarrow \mathbf{f}|\mathbf{X}) = \sum_{\mathbf{i}} X_{\mathbf{i}} k(\mathbf{i} \rightarrow \mathbf{f}) \Big/ \sum_{\mathbf{i},\mathbf{f}} X_{\mathbf{i}} k(\mathbf{i} \rightarrow \mathbf{f}) \tag{47}$$

where $k(\mathbf{i} \rightarrow \mathbf{f})$ is the rate coefficient for the process from one detailed quantum state to another. The possible dependence of the product state distribution on the reagent state preparation (**X**) is explicit in the notation. The arrow ($\rightarrow \mathbf{f}$) in the notation for the probability serves to indicate that the probability pertains to the process in which molecules go *into* the state **f**. We shall presently employ ($\mathbf{i} \rightarrow$) to indicate a process *out of* **i**. Although the addition of arrows may appear a cumbersome and superfluous embellishment in view of the identification of **f**, as "products" and **i** as "reagents." We eventually wish to consider both the *forward* ($\rightarrow \mathbf{f}, \mathbf{i} \rightarrow$) as well as the reverse ($\rightarrow \mathbf{i}, \mathbf{f} \rightarrow$) processes, and part of the development will be to identify these conditions under which the arrows are unnecessary and can be dropped.

2.1.2. The Reasonable Person's P Matrix

We have already admitted that individual states as defined here cannot be resolved in practice. Hence we shall focus our attention on groups of reactants or products states where the group label specifies a complete set of *internal* quantum numbers. We shall use $\boldsymbol{\alpha}$ to denote a reactant group and $\boldsymbol{\beta}$ a product group. The labels $\boldsymbol{\alpha}$ and $\boldsymbol{\beta}$ are in boldface type as a reminder that the internal specification is *vectorial* in nature through the "azimuthal" or orientational quantum number for the angular momentum. We shall use the labels α, β (not boldface) to indicate a collection of the g_α internally degenerate groups differing only in this orientational specification. For measurements that do not include polarization or orientation, these larger groups of states suffice. For chemical reactions the set of groups $\{\alpha\}$ will be distinct from the set of groups $\{\beta\}$, while for inelastic collisions the two sets will be congruent.

The previously stated collision theory canon for rate coefficients shows that the "group-to-group" coefficient $k(\alpha \to \beta)$ is

$$k(\alpha \to \beta) = \sum_{\mathbf{i}\in\alpha} \sum_{\mathbf{f}\in\beta} k(\mathbf{i} \to \mathbf{f})/n_\alpha \tag{48}$$

where n_α is the number of reactant states in the group α.

The fraction of products appearing in the group of states β for an initial preparation $\mathbf{X}$, $P(\to\beta|\mathbf{X})$ is given by

$$\begin{aligned} P(\to\beta|\mathbf{X}) &\equiv \sum_{\mathbf{f}\in\beta} P(\to\mathbf{f}|\mathbf{X}) \\ &= \sum_{\alpha} X_\alpha k(\alpha \to \beta) \Big/ \sum_{\alpha,\beta} X_\alpha k(\alpha \to \beta) \end{aligned} \tag{49}$$

Among the possible preparations $\mathbf{X}$ of the system is one in which only one of the initial groups α is populated: $X'_\alpha = \delta_{\alpha\alpha'}$. The product state distribution resulting from this preparation is $P(\to\beta|\alpha\to)$:

$$P(\to\beta|\alpha\to) = k(\alpha \to \beta) \Big/ \sum_{\beta} k(\alpha \to \beta) \tag{50}$$

2.1.3. The Distribution of Reactive Reactants

The algebra of probability theory requires that $P(\to\beta|\mathbf{X})$ be the average of $P(\to\beta|\alpha\to)$ over the initial groups α:

$$P(\to\beta|\mathbf{X}) = \sum_{\alpha} P(\to\beta|\alpha\to)\, P(\alpha\to|\mathbf{X}) \tag{51}$$

It follows from (48)–(51) that the probability $P(\alpha\to|\mathbf{X})$ must be

$$\begin{aligned} P(\alpha\to|\mathbf{X}) &= X_\alpha \sum_{\beta} k(\alpha \to \beta) \Big/ \sum_{\alpha,\beta} X_\alpha k(\alpha \to \beta) \\ &= X_\alpha k(\alpha\to) \Big/ \sum_{\alpha} X_\alpha k(\alpha\to) \end{aligned} \tag{52}$$

The distribution $P(\alpha\to|\mathbf{X})$ is thus seen to depend both on the preparation and the dynamics. It is a weighted distribution of the reactant groups with each group weighed by its reactivity. It is, in fact, none other than Tolman's[63] "distribution of reactive reactants."

Proceeding with exploration of the conventions of probability theory, we can introduce[64,65] a "joint" probability $P(\to\beta, \alpha\to|\mathbf{X})$:

$$P(\to\beta, \alpha\to|\mathbf{X}) = P(\to\beta|\alpha\to)\, P(\alpha\to|\mathbf{X}) \tag{53}$$

It is easy to verify that the column sum of the joint matrix

$$P(\to\beta|\mathbf{X}) = \sum_{\alpha} P(\to\beta, \alpha\to|\mathbf{X}) \tag{54}$$

is the distribution of nascent products and the row sum

$$P(\alpha\rightarrow|\mathbf{X}) = \sum_{\beta} P(\rightarrow\beta, \alpha\rightarrow|\mathbf{X}) \tag{55}$$

is the distribution of reactive reactants as they ought to be.

In terms of the rate coefficients $k(\alpha \rightarrow \beta)$, the joint probability is

$$\begin{aligned} P(\rightarrow\beta, \alpha\rightarrow|\mathbf{X}) &= X_\alpha k(\alpha \rightarrow \beta) \Big/ \sum_{\alpha,\beta} X_\alpha k(\alpha \rightarrow \beta) \\ &= X_\alpha k(\alpha \rightarrow \beta)/k(\rightarrow; \mathbf{X}) \end{aligned} \tag{56}$$

where

$$k(\rightarrow; \mathbf{X}) = \sum_{\alpha,\beta} X_\alpha k(\alpha \rightarrow \beta) \tag{57}$$

is the rate coefficient (for the forward process) for the actual initial preparation $\mathbf{X}$.

2.1.4. *The Poor Person's* P *Matrix*[10]

We have already discussed combining the internally degenerate groups $\boldsymbol{\alpha}$ into larger, orientationally averaged groups α. It is useful to consider further coalescence into even larger, "coarse-grained" groups, which we shall denote Γ for reactants and Γ' for products. Again, group-to-group rate coefficients can be defined by summing over final states and averaging over initial states:

$$k(\Gamma \rightarrow \Gamma') = \sum_{\mathbf{i}\in\Gamma} \sum_{\mathbf{f}\in\Gamma'} k(\mathbf{i} \rightarrow \mathbf{f})/n_\Gamma \tag{58}$$

where n_Γ is the number of initial states in Γ. In terms of the realizable groups α, (58) takes the form

$$k(\Gamma \rightarrow \Gamma') = \sum_{\alpha\in\Gamma} \sum_{\beta\in\Gamma'} \rho(\alpha; E)\, k(\alpha \rightarrow \beta) \Big/ \sum_{\alpha\in\Gamma} \rho(\alpha; E) \tag{59}$$

where $\rho(\alpha; E)$ is the total density of states in the neighborhood of the group α. $\rho(\alpha; E) = g_\alpha \rho_T(\alpha; E)$, where g_α is the internal-state (orientational) degeneracy and ρ_T is the translational-state density.

From the group-to-group rate coefficient, one can define a somewhat less affluent person's $\mathbf{P}$ matrix in this coarse-grained basis for any preparation $\mathbf{X}$:

$$\begin{aligned} P(\rightarrow\Gamma', \Gamma\rightarrow|\mathbf{X}) &= X_\Gamma k(\Gamma \rightarrow \Gamma') \Big/ \sum_{\Gamma,\Gamma'} X_\Gamma k(\Gamma \rightarrow \Gamma') \\ &= X_\Gamma k(\Gamma \rightarrow \Gamma')/k(\mathbf{X}) \end{aligned} \tag{60}$$

with column and row sums that are, respectively, the distribution of nascent products and of reactive reactants:

$$P(\to\Gamma'|\mathbf{X}) = \sum_{\Gamma} P(\to\Gamma', \Gamma\to|\mathbf{X}) \tag{61}$$

$$P(\Gamma\to|\mathbf{X}) = \sum_{\Gamma'} P(\to\Gamma', \Gamma\to|\mathbf{X}) \tag{62}$$

2.1.5. Constructing the **P** *Matrix*

It is important to recognize that the joint initial-state–final-state probability matrix **P** has been introduced primarily to permit certain formal developments within the conventions of probability theory. In actual practice, it will only rarely be the case that data are available for the entire range of possible reactant and product states. If any states are missing, it is, strictly speaking, impossible to form even those parts of the **P** matrix pertaining to states we do know about because of the normalizing denominators in (56) and (60). This does not mean that most of the time we must throw up our hands in despair. Examination of (60) shows that for any rate coefficient known to us we know the corresponding element of the **P** matrix to within a *state-independent* normalization factor. It is thus very much possible to discuss the dependence upon initial and final states for the existing parts of the **P** matrix (cf. Section 2.3.4), even though the full matrix (and hence the normalization constant) is not available. Note, however, that $k(\mathbf{X})$ may well be available from macroscopic measurements.

2.2. Symmetry

2.2.1. Detailed Balance

Detailed balance relates the rates of the forward and reverse processes. When the initial and final quantum states are fully specified, this relation takes its simplest form:

$$k(\mathbf{i} \to \mathbf{f}) = k(\bar{\mathbf{f}} \to \bar{\mathbf{i}}) \tag{63}$$

Here the bars indicate time reversal: the state $\bar{\mathbf{i}}$ differs from $\mathbf{i}$ in that all the momenta are reversed in sign. For our "sharp" group of states, the equivalent relation is

$$\rho(\alpha; E)\, k(\alpha \to \beta) = \rho(\beta; E)\, k(\beta \to \alpha) \tag{64}$$

(We shall drop the explicit indication of time reversal from here on, but it must be borne in mind when the measurements include polarization dependence.) More generally,

$$\rho(\Gamma; E)\, k(\Gamma \to \Gamma') = \rho(\Gamma'; E)\, k(\bar{\Gamma}' \to \bar{\Gamma}) \tag{65}$$

where $\rho(\Gamma; E)$ is the total density of states in the neighborhood of the group Γ. Expressions for state densities for various ways of defining Γ are given in Appendix A. The explicit appearance of E in (64) and (65) is a reminder that all the states in any of these collections have the same total energy.

2.2.2. The **P** *Matrix at a Given Total Energy*

When the preparation procedure $\mathbf{X}$ consists in specifying the total energy with no further state selection and no energy averaging $\mathbf{X} \equiv \mathbf{X}^E$,

$$X_\Gamma^E = \rho(\Gamma; E)/\rho(E) \tag{66}$$

where $\rho(E)$ is the total density of reactant states:

$$\rho(E) = \sum_\Gamma \rho(\Gamma; E) = \sum_\alpha \rho(\alpha; E) \tag{67}$$

It follows from (64) and (65) that the **P** matrices (56) and (60) have the following symmetries:

$$P(\rightarrow\beta, \alpha\rightarrow|\mathbf{X}^E) = P(\rightarrow\alpha, \beta\rightarrow|\mathbf{X}^E) \tag{68}$$

and

$$P(\rightarrow\Gamma', \Gamma\rightarrow|\mathbf{X}^E) = P(\rightarrow\Gamma, \Gamma'\rightarrow|\mathbf{X}^E) \tag{69}$$

2.2.3. The Reverse Reaction

The condition $\mathbf{X} = \mathbf{X}^E$ in (68) and (69) refers to preparation of the "reactants" α and Γ. It is thus a condition on the initial preparation for the forward process. One is moved to inquire what is the corresponding condition on the initial preparation $\mathbf{Y}$ of Γ's in the reverse process. A few lines of algebra suffice to show that the answer is $\mathbf{Y} = \mathbf{Y}^E$:

$$\mathbf{Y}_{\Gamma'} = \rho(\Gamma'; E)/\bar{\rho}(E) \tag{70}$$

where $\bar{\rho}(E)$ is the density of state for the reversed process

$$\bar{\rho}(E) = \sum_{\Gamma'} \rho(\Gamma'; E) \tag{71}$$

In other words, the **P** matrix at constant energy is symmetric for a uniform preparation of the initial states in either direction of the process. Hence (68) and (69) can be rewritten in a more appealing way as

$$P(\beta, \alpha|E) = P(\alpha, \beta|E) \tag{72}$$

and

$$P(\Gamma', \Gamma|E) = P(\Gamma, \Gamma'|E) \tag{73}$$

the E indicating a uniform preparation of initial states irrespective of the direction of the process. Note that the arrows have become superfluous and hence were dropped.

Any preparation **X** at a fixed total energy in which the fraction of any α is either zero or proportional to its density:

$$X_\alpha = \rho(\alpha; E) \Big/ \sum_{\alpha \in \mathbf{X}} \rho(\alpha; E) \tag{74}$$

will possess an inverse preparation **Y** for which the **P** matrix is symmetric:

$$P(\alpha, \beta | \mathbf{X}) = P(\beta, \alpha | \mathbf{Y}) \tag{75}$$

In fact, this inverse preparation will be

$$Y_\beta = \rho(\beta; E) \Big/ \sum_{\beta \in \mathbf{Y}} \rho(\beta; E) \tag{76}$$

where the range of $\beta \in \mathbf{Y}$ is determined by **X**. For any other preparation **X**, in which X_α is *not* proportional to $\rho(\alpha; E)$, an inverse preparation will, in general, not exist.

2.2.4. *The* **P** *Matrix at a Given Temperature*

When the reagents are prepared at a given temperature T, without any further state selection $\mathbf{X} = \mathbf{X}^T$,

$$\mathbf{X}_\mathbf{i}^T = \exp(-E_\mathbf{i}/\mathbf{k}T)/Q(T) \tag{77}$$

where Q is the partition function

$$Q(T) = \sum_\mathbf{i} \exp(-E_\mathbf{i}/\mathbf{k}T) \tag{78}$$

In terms of the groups of states α

$$X_\alpha^T = \rho(\alpha; E) \exp(-E_\alpha/\mathbf{k}T)/Q(T) \tag{79}$$

In a thermal preparation or any other in which there may be a non-infinitesimal energy range, it is convenient to expand the notion of coarse groups of states to include the possibility of containing sharp groups of different total energies. For example, one might wish to think of the group of quantum states all having the same vibrational quantum number but a thermal distribution in all other degrees of freedom. We propose to denote such a group with the symbol V (which is to be regarded as a particular value of Γ). Obviously, there are many possibilities: (V, J), J, etc. In our subsequent development the notation V will be used to stand for any of these. In terms of these coarse groups a thermal preparation would be:

$$X_V^T = Q(V; T)/Q(T) \tag{80}$$

where $Q(V; T)$ is the "partition function" for the designation V

$$Q(V; T) = \int dE \exp(-E/kT)\,\rho(V; E) \tag{81}$$

and $Q(T)$ is the total partition function for the reactants.

The only interesting **P** matrix for a thermal preparation is in terms of such groups V, V':

$$P(\to V', V\to|\mathbf{X}^T) = X_V^T k(V\to V'; T)/k(T) \tag{82}$$

where

$$k(T) = \sum_{V,V'} X_V^T k(V\to V', T) \tag{83}$$

The detailed balance relation for a thermal preparation is

$$Q(V; T)\,k(V\to V'; T) = \bar{Q}(V'; T)\,k(V'\to V; T) \tag{84}$$

(Time reversal is, for simplicity, omitted in the notation for groups of states.) It is relatively easy to show that the symmetry of the **P** matrix at constant T is

$$P(\to V', V\to|\mathbf{X}^T) = P(\to V, V'\to|\mathbf{Y}^T) \tag{85}$$

where $\mathbf{Y}^T$ is a thermal preparation of "products" in the reverse process. As in the constant energy case, a simpler expression is to be preferred:

$$P(V', V|T) = P(V, V'|T) \tag{85'}$$

2.2.5. *An Application of Detailed Balance*

A simple application of detailed balance follows on summing both sides of (85) over V:

$$P(\to V'|\mathbf{X}^T) = P(V'\to|\mathbf{Y}^T) \tag{86}$$

The distribution of nascent products of the forward reaction is identical to the distribution of reactive reactive reactants of the reversed products of the reversed reaction.[7,66] What this means in practice is that if, say starting with thermal reagents, the products' nascent vibrational state distribution has been measured, e.g., for the

$$\mathrm{Cl} + \mathrm{HI} \to \mathrm{I} + \mathrm{HCl}(V)$$

reaction, then the distribution of reactive reagents for the reaction

$$\mathrm{I} + \mathrm{HCl}(V) \to \mathrm{Cl} + \mathrm{HI}$$

(with thermal preparation) is known. Figure 7 shows the construction. If,

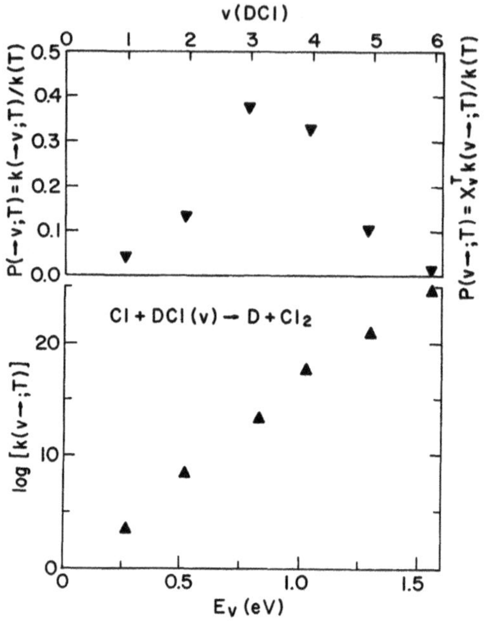

Figure 7. Determination of the distribution of reactive reactants, $P(v \to |T)$, and of the relative rated $k(v \to |T)$ for the Cl + HCl(v) → H + Cl_2 reaction from the measured [adapted from K. G. Anlauf, D. S. Horne, R. G. Macdonald, J. C. Polanyi, and K. B. Woodall, *J. Chem. Phys.* **57**, 1561 (1972)] relative rates, $k(\to v|T)$ for the H + Cl_2 → Cl + HCl(v) reaction [adapted from H. Kaplan, R.D. Levine, and J. Manz[66]]. Top panel, left scale: the normalized products' vibrational state distribution $P(\to v|T)$ for the H + Cl_2 reaction. By detailed balance, equation (86), $P(\to v|T) = P(v \to |T)$, shown on top panel, right scale. Up to an overall normalization $[\bar{k}(T)]$, $P(v \to |T)$ determines the v dependence of $k(v\to; T)$ shown in the bottom panel.

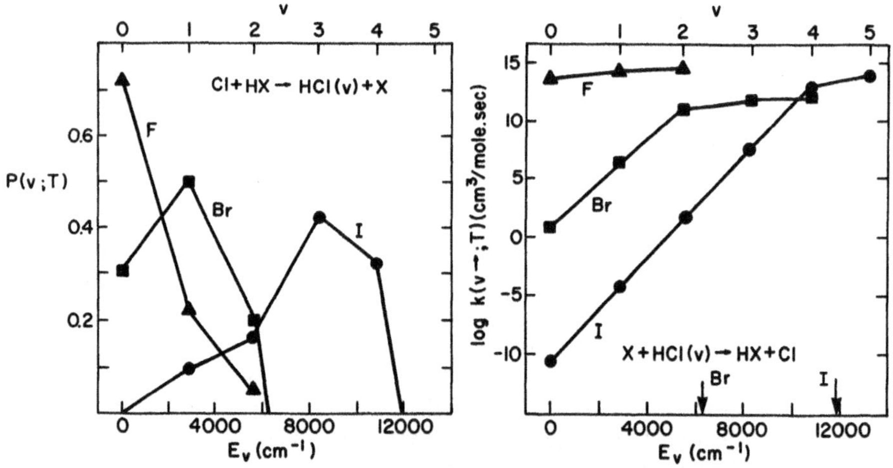

Figure 8. Combining measurements on vibrational energy disposal, vibrational energy consumption, and thermal rate constants (adapted from R. D. Levine[7] where the many sources of experimental data that were used as input are identified). Left: $P(v|T)$ for the Cl + HX → HCl(v) + X [or for the X + HCl(v) → Cl + HX reaction]. The arrows indicate the magnitude of $-\Delta E_0$. Note that the endothermic Cl + HF reaction does not lead to vibrational population inversion ($\lambda_v \approx +4$), and that the higher the exothermicity the more extensive is the inversion. Right: when the thermal constants $k(T)$ are known, the results in the left panel can be used to determine the absolute rates $k(v\to; T)$ for the X + HCl(v) → HX + Cl reactions. Arrows in the right-hand panel indicate the "nominal" energy $\langle E \rangle_{\text{thermal}} + E_a - \Delta H_0^0$ for X = I and Br.

in addition, the thermal rate constant $\bar{k}(T)$ is known, then, since

$$P(V\rightarrow|T) = X_V k(V\rightarrow; T)/\bar{k}(T)$$

one can actually determine the absolute detailed rates. Such a determination is shown in Figure 7, and the results for several

$$\mathrm{X} + \mathrm{HCl}(V) \rightarrow \mathrm{Cl} + \mathrm{HX}$$

reactions are collected in Figure 8.

2.3. Prior and Deviant **P** *Matrices*

The discussion of the deviance of the observed from the prior (or the theoretical) probabilities in the state-to-state case needs to bring out the following points: To begin with, any theory should build in the symmetry of time-reversal invariance. Hence any theoretical probability matrix **Q** must have the same symmetry as the **P** matrix constructed from observed rate constants. Thus, for example, the surprisal at a fixed energy, $-\ln[P(\beta, \alpha|E)/P^0(\beta, \alpha|E)]$, will be a symmetric function of the initial and final groups of states. An identical surprisal (or entropy deficiency) for the forward and reversed process holds not only for $\mathbf{X} = \mathbf{X}^E$ but also for the thermal preparation $\mathbf{X} = \mathbf{X}^T$.[65,66]

Next, one should note that any observed deviance of **P** from $\mathbf{P}^0$ (or **Q**) can arise from two different sources of dynamic bias[25,60,64]:

$$P(V', V|\mathbf{X})/P^0(V', V|\mathbf{X}) = [k(V\rightarrow V'; \mathbf{X})/k^0(V\rightarrow V'; \mathbf{X})]\,[k^0(\mathbf{X})/k(\mathbf{X})] \tag{87}$$

Either the observed rate constant can differ from the prior one (first factor) or the overall normalizing rate $k(\mathbf{X})$ can differ from its prior $k^0(\mathbf{X})$ (second factor). In general, both terms will contribute. Thus far, however, one has only very limited experience with analyzing the deviance of the overall constants, $-\ln[k(\mathbf{X})/k^0(\mathbf{X})]$. More data as well as more theory are clearly warranted.

2.3.1. The Prior Distribution

Finally, a few technical points need to be made about the prior distributions. At any given value of the energy, we have the prior state-to-state rate coefficient given by[60]

$$k^0(\Gamma \rightarrow \Gamma'; E) = R^0\bar{\rho}(\Gamma'; E) \tag{88}$$

where R^0 is an energy-independent constant. Insertion of this into (60) produces $P^0(\Gamma', \Gamma|E)$:

$$P^0(\Gamma', \Gamma|E) = \bar{\rho}(\Gamma'; E)\,\rho(\Gamma; E)/\bar{\rho}(E)\,\rho(E) \tag{89}$$

where $\bar{\rho}(E)$ and $\rho(E)$ are, respectively, the total densities of states for products and reactants. Thus at a fixed energy the prior joint probability factors into the product of the reactant prior and the product prior:

$$P^0(\Gamma', \Gamma; E) = P^0(\Gamma'; E)\, P^0(\Gamma; E) \tag{90}$$

We have mentioned earlier the possibility of explicit incorporation of the conservation of total angular momentum into a theoretical distribution which might take the place of the prior. To do so would, in general, produce correlations between initial and final states so that the factored form (90) would no longer be valid. Fortunately, this correlation is often (but not always) of minor significance.

For thermal preparations or other situations with multiple energies, a much more serious correlation between reactants and products operates, owing to the law of conservation of energy. Since the processes being observed are binary collisions and not, for example, interactions with a heat bath, the products' energy for any given collision must exactly equal the reactants' energy for the same collision.

Given the prior rate coefficient k^0 of (88), this produces a prior $P^0(V', V|T)$ as given below[25,60,67]:

$$P^0(V', V|T) = \frac{\int dE \rho(V; E)\, \bar{\rho}(V'; E) \exp(-E/\mathbf{k}T)}{\int dE \rho(E)\, \bar{\rho}(E) \exp(-E/\mathbf{k}T)} \tag{91}$$

$P^0(\rightarrow V'|T)$ is given by

$$P^0(\rightarrow V'|T) = \frac{\int dE \rho(E)\, \bar{\rho}(V'; E) \exp(-E/\mathbf{k}T)}{\int dE \rho(E)\, \bar{\rho}(E) \exp(-E/\mathbf{k}T)} \tag{92}$$

and $P^0(V\rightarrow|T)$ is

$$P^0(V\rightarrow|T) = \frac{\int dE \rho(V; E)\, \bar{\rho}(E) \exp(-E/\mathbf{k}T)}{\int dE \rho(E)\, \bar{\rho}(E) \exp(-E/\mathbf{k}T)} \tag{93}$$

The form (91) clearly will not factor as did (90), even though the two are based on the same $k^0(\Gamma \rightarrow \Gamma'; E)$. At a given *energy*, the prior distribution of reactive reactants is the same as the mole fractions of reactants. However, since the reactivity varies with total energy, the same cannot in general hold true for preparations involving more than one value of the energy. Priors of the form (91) are given in Appendix A for several choices of V and V'.

For a nonthermal preparation $\mathbf{X}$ involving more than one energy, the prior rate constant (88) gives $P^0(\rightarrow V', V\rightarrow|\mathbf{X})$ as follows:

$$P^0(\rightarrow V', V\rightarrow|\mathbf{X}) = \int dE \bar{\rho}(V'; E)\, f(V; E|\mathbf{X}) \Big/ \int dE \bar{\rho}(E)\, f(E|\mathbf{X}) \tag{94}$$

where $f(E|\mathbf{X})\, dE$ is the fraction of molecules with total energies between

E and $E + dE$ and $f(V; E|\mathbf{X})$ is the fraction in group V in the same energy interval

$$\int f(V; E|\mathbf{X})\, dE = \mathbf{X}_V \tag{95}$$

It was previously noted that the joint **P** was not symmetric for such a preparation and in (94) we see explicitly that it is not for P^0 unless $f(V; E|\mathbf{X}) = \rho(V; E)\, f(E|\mathbf{X})$. The column sum of (94) is

$$P^0(\rightarrow V'|\mathbf{X}) = \int dE\bar{\rho}(V'; E)\, f(E|\mathbf{X}) \Big/ \int dE\bar{\rho}(E)\, f(E|\mathbf{X}) \tag{96}$$

and the row sum is

$$P^0(V\rightarrow|\mathbf{X}) = \int dE\bar{\rho}(E)\, f(E; V|\mathbf{X}) \Big/ \int dE\bar{\rho}(E)\, f(E|\mathbf{X}) \tag{97}$$

2.3.2. Surprisal Analysis at a Given Total Energy

The impossibility of forming the **P** matrix for comparison with $\mathbf{P}^0$ when the data are insufficient for normalization has already been noted. When this common situation obtains, a simpler measure of deviance from the most entropic state-to-state distribution in the $\boldsymbol{\omega}$ matrix[10,38,64] is defined by

$$\begin{aligned} \omega(\Gamma', \Gamma; E) &= k(\Gamma \rightarrow \Gamma'; E)/\rho(\Gamma'; E) \\ &= R^0[k(\Gamma \rightarrow \Gamma'; E)/k^0(\Gamma \rightarrow \Gamma'; E)] \end{aligned} \tag{98}$$

$\omega(E)$ is defined as $\omega(\Gamma', \Gamma; E)$ when there is no internal resolution of reactant or product states:

$$\omega(E) = k(E)/\rho(E) = R^0 k(E)/k^0(E) \tag{99}$$

With these definitions, the surprisal separates as previously indicated into the sum of a contribution from the departure of $k(\Gamma \rightarrow \Gamma'; E)$ from $k^0(\Gamma \rightarrow \Gamma'; E)$ and a contribution from overall normalization:

$$\begin{aligned} -\ln[P(\Gamma', \Gamma'; E)/P^0(\Gamma', \Gamma; E)] &= -\ln[\omega(\Gamma', \Gamma; E)] + \ln[\omega(E)] \\ &= -\ln[k(\Gamma \rightarrow \Gamma'; E)/k^0(\Gamma \rightarrow \Gamma'; E)] \\ &\quad + \ln[k(E)/k^0(E)] \\ &= -\ln \omega(\Gamma', \Gamma|E) \end{aligned} \tag{100}$$

Since we are necessarily being negligent of overall normalization, $\omega(\Gamma', \Gamma; E)$ or $R^0\omega(\Gamma', \Gamma; E)$ is equally good as a measure of the state dependence of the departure from the prior. The following equations indicate alternative expressions which may be useful in some circumstances[27]:

$$k(\Gamma \to \Gamma'; E) = \omega(\Gamma', \Gamma; E)\,\rho(\Gamma'; E) = \omega(E)\,\omega(\Gamma', \Gamma|E)\,\rho(\Gamma'; E)$$
$$= k(E)\,\rho(\Gamma'|E)\,\omega(\Gamma', \Gamma|E)$$
$$= k^0(\Gamma \to \Gamma'; E)\,\omega(\Gamma', \Gamma|E)\,k(E)/k^0(E) \tag{101}$$

Similarly,

$$k(\Gamma\to; E) = k^0(\Gamma \to 'E)\,\omega(\Gamma|E)\,k(E)/k^0(E) = k(E)\,\omega(\Gamma|E) \tag{102}$$

where

$$\omega(\Gamma|E) = P(\Gamma|E)/P^0(\Gamma|E) = k(\Gamma\to; E)/k^0(E) \tag{103}$$

and

$$k(\to\Gamma'; E) = k^0(\to\Gamma'; E)\,\omega(\Gamma'|E)\,k(E)/k^0(E)$$
$$= k(E)\,\omega(\Gamma'|E)\,\bar{\rho}(\Gamma'|E) \tag{104}$$

where

$$\omega(\Gamma'|E) = P(\Gamma'|E)/P^0(\Gamma'|E) \tag{105}$$

On prior grounds any group of reactant states is as reactive as any other:

$$k^0(\Gamma\to; E) = k^0(E) = R^0\bar{\rho}(E) \tag{106}$$

and

$$k^0(\Gamma \to \Gamma'; E) = R^0\bar{\rho}(\Gamma'; E) \tag{107}$$

Hence any dependence of $k(\Gamma\to; E)$ or $k(\Gamma \to \Gamma'; E)$ on the initial group Γ indicates dynamic bias in the energy consumption. On the other hand, dynamic bias in energy release would be exhibited by deviance of $k(\to\Gamma'; E)$ from $R^0\rho(\Gamma'|E)$.

2.3.3. *Surprisal Analysis at a Given Temperature*

The same remarks regarding the difficulty of proper normalization that were made in the previous section also hold for observations at a given temperature. Indeed, since the range of energies spanned in a thermal distribution is infinite, *all* vibrational (or whatever) states will be populated to some degree even if only very slightly. Hence it is even more imperative in this case to accept absence of the normalization contribution to the surprisal. The analogous relation to (100) is

$$-\ln[P(V', V; T)/P^0(V', V; T) = -\ln[k(V\to V'; T)/k^0(V\to V'; T)]$$
$$+ \ln[k(T)/k^0(T)] \tag{108}$$

At a fixed temperature it is no longer true that all groups of reactant states are regarded as equally reactive on prior grounds. Since the reactivity varies with energy, some descriptions V will preferentially sample higher energies and hence typically higher reactivities than others. The prior for energy consumption is given by (93), and the prior for energy release by (92).

Appendix A gives explicit results for $k^0(V \to V'; T)$ and $P^0(V', V; T)$ for a variety of possible V, V'. Some examples are plotted in Figure 9. Several interesting features of $k^0(V \to V'; T)$ emerge, depending on the relative values of $T(V)$ and $T(V')$, and the internal-energy term values for V and V', respectively: (a) For a highly endoergic process, i.e., one for which $T(V') \gg T(V)$, the temperature dependence of $k^0(V \to V'; T)$ is dominated by an explicit factor of $\exp\{-[T(V') - T(V)]/\mathbf{k}T\}$. Roughly speaking, the entire vibrational energy of reactants goes, on prior grounds, to reduction of the activation energy, while the entire vibrational excitation of products adds to it in this limit. (b) At the opposite extreme (a highly exoergic process), the prior expectation is for only weak temperature dependence. (c) In the intermediate region, the exponential temperature dependence gives way to a more moderate enhancement with temperature. (d) At any temperature the prior thermal rate coefficient is a function only of the reduced "energy gap," $[T(V') - T(V)]/\mathbf{k}T$.

2.3.4. Summary: Surprisal Analysis

The more common situation is that insufficient data are available for construction of a $\mathbf{P}$ matrix. It is then convenient to examine the surprisal

$$\begin{aligned} I(\Gamma', \Gamma; \mathbf{X}) &= -\ln R^0 \omega(\Gamma', \Gamma; \mathbf{X}) \\ &= -\ln [k(\Gamma \to \Gamma'; \mathbf{X})/k^0(\Gamma \to \Gamma'; \mathbf{X}) \end{aligned} \tag{109}$$

which [cf. (100)] is the sum of $I(\mathbf{X})$:

$$I(\mathbf{X}) = -\ln \omega(\mathbf{X}) = -n[k(\mathbf{X})/k^0(\mathbf{X})] \tag{110}$$

and of $I(\Gamma', \Gamma | \mathbf{X})$:

$$I(\Gamma', \Gamma | \mathbf{X}) = -\ln [P(\Gamma', \Gamma | \mathbf{X})/P^0(\Gamma', \Gamma | \mathbf{X})] \tag{111}$$

Note, however, that $k(\mathbf{X})$ may be independently available. When this is the case $I(\Gamma', \Gamma | \mathbf{X})$ can be determined:

$$I(\Gamma', \Gamma | \mathbf{X}) = I(\Gamma', \Gamma; \mathbf{X}) - I(\mathbf{X}) \tag{112}$$

for any pair Γ', Γ for which $k(\Gamma \to \Gamma'; \mathbf{X})$ is available. Hence the surprisal, including its overall location factor $I(\mathbf{X})$, may be determined even when many entries of the $\mathbf{P}$ matrix are missing.

Special cases apply when no initial-state selection is attempted so that Γ is the set of all reagent states and $I(\Gamma' | \mathbf{X})$ is then simply the surprisal

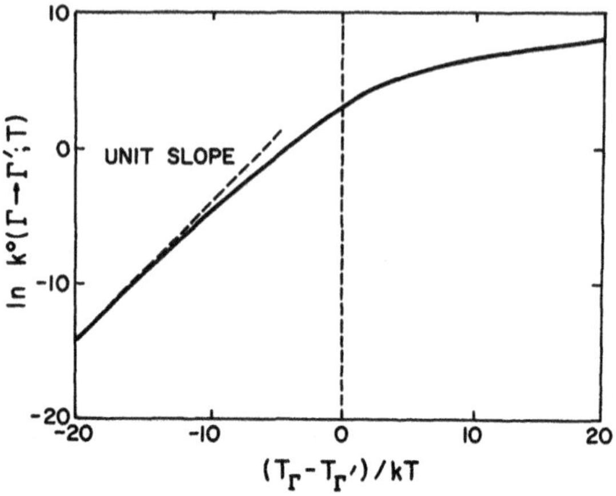

Figure 9. The dependence of the prior rate $k^0(\Gamma \to \Gamma'; T)$ on the reduced energy gap $(T_\Gamma - T_{\Gamma'})/\mathbf{k}T$ (adapted from R. D. Levine, and J. Manz[60]). The term energies refer to a common origin. If energies are measured from the zero point, the energy gap is $(E_\Gamma - \Delta E_0 - E_{\Gamma'})/\mathbf{k}T$). Note the Arrhenius-like dependence for endothermic transitions.

of the energy disposal. Similarly, if there is no products state resolution, $I(\Gamma|\mathbf{X})$ is the surprisal of the reagents' energy consumption.

2.4. *Examples*

2.4.1. *Reagent Energy Consumption*

For reagents at a given temperature (or a given total energy) the distribution of products' energy states in the forward reaction has been shown identical to the distribution of reactive reactants of the reversed reaction (Section 2.2). The examples of analysis of energy disposal discussed in Section 1 are thus equally valid here. The role of reagent vibrational excitation has received particular attention.[60,68] Analysis of state-to-state rate constants (Sections 2.4.3 and 2.5) has served to interpret the observation[7,68] that reagent internal excitation will typically have opposite effects in the endothermic and exothermic regimes.

2.4.2. *Rotational Energy Transfer*

Rotational energy transfer (at a given total energy) is a prime example of a situation where complete **P** matrices are available. Figure 2 from Section 1.2, while not so identified in that section, is an analysis of state-to-state data. Figure 10 shows corresponding results at a given temperature.[69] As is evident from the figure, $I(T)$ is practically T independent.

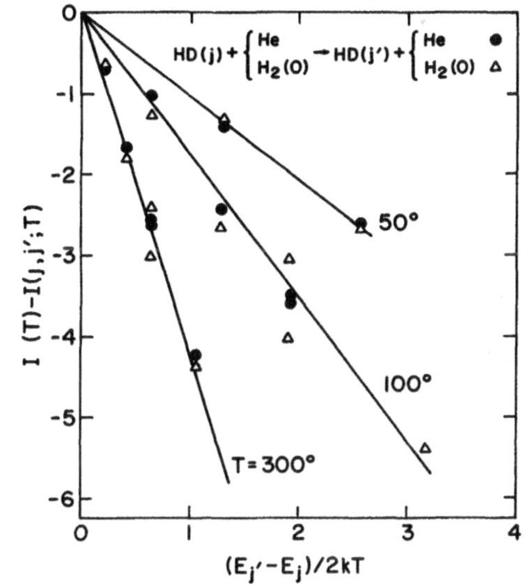

Figure 10. Surprisal analysis of the state-to-state rate constants $k(j \rightarrow j'; T)$ for rotational energy transfer in $HD(j) + M \rightarrow HD(j') + M$ collisions: M = He, dots; $M = H_2$, triangles [close-coupling computations of S. I. Chu, *J. Chem. Phys.* **62**, 4089 (1975) and of S. Green, *Physica* **76**, 609 (1974)]. The detailed rates have been determined by quantal close-coupling computations for $k(j \rightarrow j'; E)$ followed by thermal averaging. The prior rates are from Table A of Appendix A.

2.4.3. Vibrational Energy

Vibrational state-to-state rate constants are largely available only through classical trajectory computational studies. Figures 11 and 12 show the results for energy transfer and for reactive collisions, respectively. The

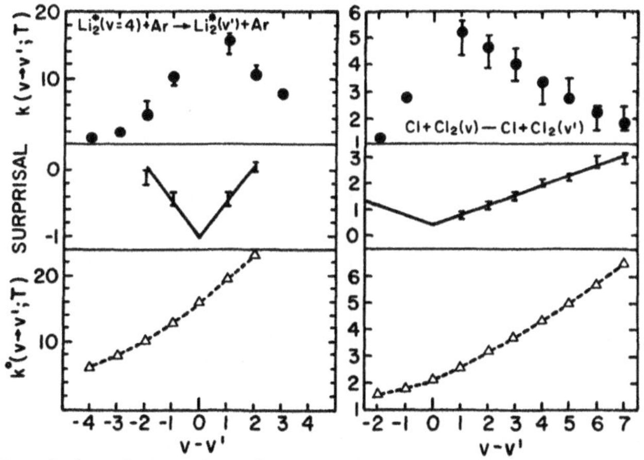

Figure 11. Surprisal analysis and synthesis of rate constants $k(v \rightarrow v'; T)$ for energy-transfer collisions (adapted from I. Procaccia, and R. D. Levine[25]). Left: experimental results [bars, G. Ennen and Ch. Ottinger, *Chem. Phys.* **3**, 404 (1974)]. Right: trajectory computations [bars, D. L. Thompson, *J. Chem. Phys.* **60**, 455 (1974)]. Bottom panels: the prior rates. Middle panels: the surprisal analysis of the data (bars) and the induced surprisal (smooth line). Top panels: the data (bars) and the induced relative rates (dots). Note that the relative rates are not symmetric about the (elastic) $v' = v$ point. Yet, because of the role of the prior rate, the surprisal is essentially a symmetric function of $v - v'$.

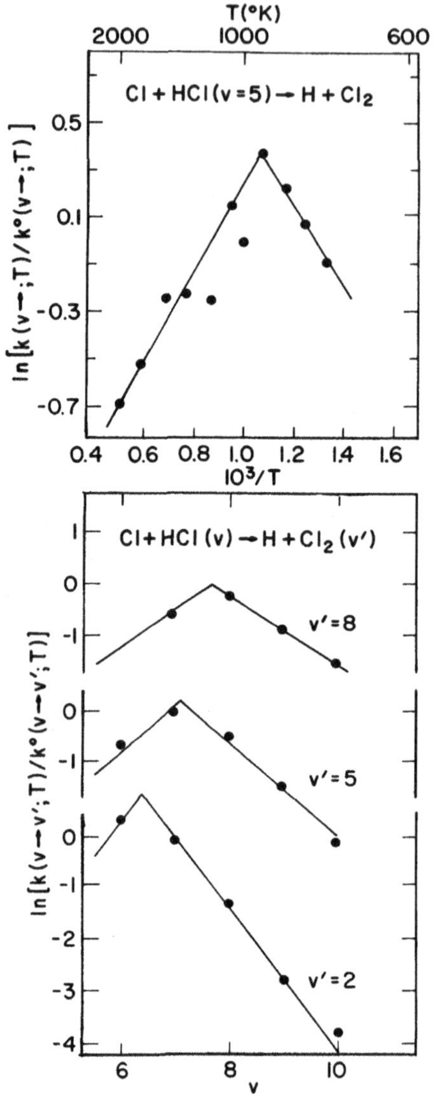

Figure 12. Surprisal analysis of state-to-state rate constants $k(v \rightarrow v'; T)$ for a reactive $Cl + HCl(v) \rightarrow H + Cl_2(v')$ collision (adapted from E. Pollak and R. D. Levine[68]). Data from classical trajectory computations. (Prior rates from Table A of Appendix A.) As v varies, so does the energy gap. The gap is effectively zero (for given v') at the break in each line. The "exponential-gap" dependence of $k(v \rightarrow v'; T)$ accounts for the "threshold" in $k(v \rightarrow; T)$ (as seen in Figure 8).

functional form of the surprisal is that of an "exponential-gap"[7,25,67,68] type. In particular, this implies that for energy consumption (or energy disposal) the surprisal will have an opposite dependence on reagent (or product) vibrational energy, depending on the sign of the energy gap, $\Delta E = E_V - \Delta E_0 - E_{V'}$.[7,25,68] As long as ΔE is negative, $k(v \rightarrow v'; T)$ is enhanced by increasing E_V (which diminishes ΔE). Once $E_V > -\Delta E_0$, reagent vibrational excitation enhances some rate coefficients (for which $E_{V'} > E_V - \Delta E_0$), but diminishes the rest.

2.4.4. Temperature Dependence

The deviance $I(T)$ of the overall rate coefficient from its prior value is found, for atom–diatom collisions, to be often only weakly temperature dependent. The "Arrhenius-like" functionality

$$I(T) \equiv -\ln[k(T)/k^0(T)] = \Delta E_a/\mathbf{k}T \tag{113}$$

is typically sufficient,[25,60] with ΔE_a often quite small.[60] In computing $k^0(T)$ one must take proper account of the minimal energy required for reaction on thermochemical grounds. Indeed, for highly endoergic processes[60] (Figure 9),

$$k^0(T) \simeq R^0 \exp(-\Delta E_0/\mathbf{k}T) \tag{114}$$

where $-\Delta E_0$ is the exoergicity. Hence ΔE_a, equation (113), can be regarded as an additional contribution to the activation energy over and above that required by the thermochemistry. The fact that the observed values ΔE_a are typically small may reflect nothing more than that detailed dynamical results are mostly available for reactions with low intrinsic barriers.

2.4.5. Energy Dependence

The deviance of the total energy dependence of the rate coefficient $I(E) = -\ln[k(E)/R^0\bar{\rho}(E)]$ is less well characterized at present. The available results are mainly on the translational-energy dependence of the reaction cross section for given internal reagent states. Hence E varies simultaneously as E_T is varied. Results for the E_T dependence of $-\ln[\sigma(E_T\rightarrow; E)/\bar{\rho}(E)]$ have indeed been reported. However, as previously noted, $I(E_T; E)$,

$$\begin{aligned} I(E_T; E) &= -\ln[k(E_T\rightarrow; E)/k^0(E_T\rightarrow; E)] \\ &= -\ln[\sigma(E_T\rightarrow; E)/\sigma^0(E_T\rightarrow; E)] \end{aligned} \tag{115}$$

is really the sum of two contributions:

$$I(E_T; E) = I(E_T|E) + I(E) \tag{116}$$

Without a separate evaluation of $I(E_T|E)$ one cannot conclude from $I(E_T; E)$ how $I(E)$ varies with E. There are at present two independent pieces of evidence to suggest that, as for $I(T)$, $I(E)$ may be only weakly E dependent. One is a theoretical attempt at evaluating $I(E_T|E)$ for the alkali-metals–CH_3I reaction.[27] It was noted there that $I(E_T|E)$ itself was sufficient to account for the E_T dependence of $I(E_T; E)$ [i.e., $I(E) \approx 0$]. The other is the experimental observation that for the La (or Sc) + oxidizers reactions[70] $I(E_T; X)$ was itself essentially zero over a fairly wide E_T range. The same was not the case for the quenching collision of Hg* by different diatomics, however.[71]

2.4.6. *Branching Ratios*

A branching ratio is defined as the ratio $k(\Gamma \to \Gamma'|\mathbf{X})/k(\Gamma \to \Gamma''|\mathbf{X})$ of two rate coefficients, and hence can be regarded as a ratio of two elements in a **P** matrix, $\mathbf{P}(\Gamma \to \Gamma'|\mathbf{X})/P(\Gamma \to \Gamma''|\mathbf{X})$. Strictly speaking, therefore, branching ratios have no "special" niche in the approach. Typically, however, one uses the term with special reference to chemically distinct product groups, such that the set of groups $\{a\}$ included in Γ or Γ' is quite large. Under such circumstances and when no initial-state selection is carried out, the branching ratio $P(\to\Gamma'|\mathbf{X})/P(\to\Gamma''|\mathbf{X})$ is often well approximated by its prior value. If $\mathbf{X} = \mathbf{X}^T$ (thermal initial state) it is very much recommended, however, to employ a prior distribution properly averaged over the initial thermal energy distribution (Appendix A).

Inducing improved expressions for branching ratios[17-23,27,65] must be done with some care. The general case is discussed in Section 2.5.3, where dynamic constraints are employed. Alternatively, one can try to employ the product state distribution within the final groups [i.e., $P(\beta'|\Gamma', \mathbf{X})$ and $P(\beta''|\Gamma'', \mathbf{X})$] as known input in a surprisal synthesis.[17-23] The result is[17-23]

$$P(\to\Gamma'|\mathbf{X})/P(\Gamma''|\mathbf{X}) = P^0(\to\Gamma'|\mathbf{X})/P^0(\to\Gamma''|\mathbf{X}) \exp[-I(\beta'|\Gamma', \mathbf{X}) + I(\beta''|\Gamma'', \mathbf{X})] \tag{117}$$

Equation (117) is the best induction for the branching ratio given the final-state distribution (separately normalized for the groups Γ' and Γ''). It is known, however, that, in the presence of interbranch constraints,[17] there are informative observables which are not accounted for by equation (117). Failure of (117) is thus an indication that the more general procedure of Section 2.5.3 is called for (or in loose language, you do know more than you have thus far admitted).

2.5. *Surprisal Synthesis for* **P** *Matrices*

Surprisal synthesis for the joint P matrix is the recommended procedure whenever some initial-state selection is carried out. Even if one is only interested in the products' state distribution, i.e., in $P(\beta|\Gamma)$, it is still advisable to proceed via the joint P matrix (or those unnormalized parts of it that can be constructed). The reason is that such a procedure necessarily allows for the role of the initial state and thereby systematizes what may otherwise appear to be aberrant behavior. In other words, the synthesis includes not only the dynamical constraint but also imposes the condition of detailed balance (see Section 2.2).

Here we shall only quote the final results for a synthesis using a single constraint, say $A(\beta)$, and refer to the literature for the derivation[65] and applications.[20,27] For any initial state the magnitude of the constraint is $\Sigma(\alpha)$:

$$\Sigma(\alpha) = \sum_{\beta} A(\beta)\, P(\beta, \alpha|\mathbf{X}) \tag{118}$$

$\Sigma(\alpha)$ may be a function of α and hence so will its surprisal parameter, $\gamma(\alpha)$. If now $\mathbf{X}$ ix such that the $\mathbf{P}$ matrix is symmetric, the result of the synthesis is[65]

$$P(\beta, \alpha|\mathbf{X}) = P^0(\beta, \alpha|\mathbf{X}) \exp\{-\lambda_0 - [\gamma(\beta) - \gamma(\alpha)][A(\beta) - A(\alpha)]\} \tag{119}$$

Most of the products' final-state distributions known to us can be accounted for using equation (119). In particular, surprisal plots for $P(\beta|\alpha, X)$ which appear nonlinear when plotted vs. $A(\beta)$ can be linearized by the form (119), the reason being that (119) is the distribution induced not only by imposing the dynamical constraint $\langle A(\beta)\rangle = \Sigma(\alpha)$ but also by imposing the symmetry of detailed balance.

Specific guidelines for the application of (119) include the following.

2.5.1. *Rotational Distributions*

Imposing the rotational angular momentum as a constraint, $A(j') = j' + 1/2$ (or, more precisely $= [j'(j' + 1)]^{1/2}$) and using

$$\begin{aligned} \gamma(j') &= \gamma_R j' \\ \gamma(j) &= \gamma_R(2\beta - j) \end{aligned} \tag{120}$$

in (119) leads to[20]

$$P(j', j|E) = P^0(j', j|E) \exp\{-\gamma_0 - \gamma_R[(j' - \beta)^2 - (j - \beta)^2]\} \tag{121}$$

Here, by construction, β may depend on j but not on j'. The functional form (121) accounts for both the j and the j' dependence of (trajectory-generated) $\mathbf{P}$ matrices for reactive collisions[20,23] where β is well approximated by $\beta = aj + i$. It is possible to interpret i as a measure of the impulse release of angular momentum during the reactive collision. For inelastic processes, $a = 1$ and $i = 0$, or $\gamma(j) = \gamma_R j$, leading to [cf. (121)] a Gaussian dependence. For atom–rigid-rotor collisions (Section 1.2) the linear momentum constraint [References 26 and 35, equation (10), and Section 2.5.2 below] provides a better description.

2.5.2. *Translational Distributions*

Imposing the linear momentum as a constraint, $A(k') \equiv k'$ does account for both inelastic and reactive[27,72] poor person's translational $\mathbf{P}(k', k|X)$

$$\begin{aligned}\gamma(k') &= \gamma_T k' \\ \gamma(k) &= \gamma_T(2\kappa - k)\end{aligned} \tag{122}$$

where [cf. Reference (10)] $\kappa = \alpha k + \rho$, with ρ a measure of the impulsive translational energy release. [$\alpha = 1$ and $\rho = 0$ or $\gamma(k) = \gamma_T k$ for inelastic collisions.] Using (121) in (119) leads to the form (41) as previously discussed. This is best seen by using (121) to write (119) as

$$-\ln[P(E_T', E_T|E)/P^0(E_T', E_T|E)] = \lambda_0 + \gamma_T[(k' - \kappa)^2 - (k - \kappa)^2]$$

As $\kappa \to 0$ this surprisal reduces to a form linear in f_T', as previously discussed [equation (43)]. For inelastic-atom–rigid-rotor collisions, (121), with $\gamma(k) = \gamma_T k$, yields

$$-\ln[P(j', j|E)/P^0(j, j'|E)] = \lambda_0 + \theta(j', j)|\Delta g_R|,$$

where

$$\Delta g_R = (E_j - E_{j'})/E$$

and

$$\theta(j', j) = (2\mu\gamma_T/\hbar^2)E/[(1 - E_j/E)^{1/2} + (1 - E_{j'}/E)^{1/2}]$$

is weakly j and j' dependent and increases linearly with E, particularly in the high-energy limit.[30]

2.5.3. *Branching Ratios, Reactive Reactants, and Other Marginals*

From the general form (119) we obtain $P(\beta|\alpha) = P(\beta, \alpha)/\sum_\beta P(\beta, \alpha)$ as

$$P(\beta|\alpha) = P^0(\beta|\alpha)\exp\{-\lambda(\alpha) - [\gamma(\beta) - \gamma(\alpha)][A(\beta) - A(\alpha)]\} \tag{123}$$

Here $\exp[\lambda(\alpha)]$ is the "sum over states" given by

$$\exp[\lambda(\alpha)] = \sum_\beta P^0(\beta|\alpha)\exp\{-[\gamma(\beta) - \gamma(\alpha)][A(\beta) - A(\alpha)]\} \tag{124}$$

For all practical and formal purposes $\exp[\lambda(\alpha)]$ is a "partition function." The conclusion

$$P(\alpha\rightarrow|\mathbf{X}) = P^0(\alpha\rightarrow|\mathbf{X})\exp[-\lambda_0 + \lambda(\alpha)] \tag{125}$$

is thus quite reminiscent of thermodynamics. The fraction of reactive reactants in any given state is expressed as a partition function ratio just as in equilibrium statistical mechanics.

Branching ratios can similarly be expressed as a ratio of two partition functions:

$$P(\alpha \to \Gamma'|\mathbf{X})/P(\alpha \to \Gamma''|\mathbf{X}) = P^0(\alpha \to \Gamma'|\mathbf{X})/P^0(\alpha \to \Gamma''|\mathbf{X}) \times \exp[-\lambda(\Gamma') + \lambda(\Gamma'')] \tag{126}$$

where

$$\exp[(\lambda(\Gamma')] = \sum_{\beta\in\Gamma'} P^0(\beta|\alpha)\exp\{-[\gamma(\beta) - \gamma(\alpha)][A(\beta) - A(\alpha)]\} \quad (127)$$

Note, incidentally, that to determine $P(\Gamma \to \Gamma'|\mathbf{X})/P(\Gamma \to \Gamma''|\mathbf{X})$ one should first average (over the initial states) $P(\alpha \to \Gamma'|\mathbf{X})$ and $P(\alpha \to \Gamma''|\mathbf{X})$ and only then take their ratio. The result that the rigorous expression for any marginal distribution is a partition function ratio has been the first conclusion of surprisal synthesis[17] and has since been rederived from the dynamical approach.[15]

2.6. *Mutual Distributions*

Interest in the analysis of products' state distributions in more than one degree of freedom, say, in both E'_T and E'_R, is just beginning.[73-75] The simplest representation of such a distribution obtains if one takes a separate constraint for each degree of freedom, say, $A_1(k') = k'$ for the translation and $A_2(j') = j'$ for the rotation, and synthesizes the **P** matrix subject to both constraints and to detailed balance. One then finds that

$$-\ln[P(E'_T, E'_R|E_T, E_R)/P^0(E'_T, E'_R|E_T, E_R)] = \text{const} + \mathbf{A}^T\cdot\boldsymbol{\sigma}^{-1}\cdot\mathbf{A} \quad (128)$$

Here A is a column vector with components $k' - \kappa$ and $j' - \beta$, and $\boldsymbol{\sigma}^{-1}$ is a 2×2 matrix,

$$\boldsymbol{\sigma}^{-1} = \begin{pmatrix} \gamma_T & \sigma_{RT}^{-1} \\ \sigma_{RT}^{-1} & \gamma_R \end{pmatrix} \quad (129)$$

such that κ, β, γ_T, and γ_R are the parameters previously introduced [cf. (120) and (121)]. The result (128) appears to account well[75] for all the mutual distributions (in E_V, E_R, $E_V = E - E_T - E_R$) derived from chemiluminescence data. In the absence of significant cross correlation (σ_{RT} small) it also accounts for the success of the reduced variable g_R,

$$g_R = E_R/(E - E_V) \quad (130)$$

in the analysis of the rotational state distribution, $P(E_R|E_V, E)$, within a given vibrational manifold.[8,10,76] In the limits $\kappa = k$ and $\beta = j$ it reduces to the exponential gap characterization of inelastic transitions.[7,25,29,36,67] In particular, at low k's it yields the very useful exponential gap expression

$$-\ln[P(v', v|T)/P^0(v', v|T)] = \lambda_0 + l_V|E_V - E_{V'}|/\mathbf{k}T \quad (131)$$

which has been extensively employed (see, e.g., Figure 11). (We have used l_V to denote the Lagrange parameter when the energy gap is reduced in terms of $\mathbf{k}T$, and retained the symbol λ_V for when E is used in the denominator of the energy gap.)

2.7. Summary

Section 2 examined the new issues that arise when reagent state selection is carried out. Central to our discussion of "state-to-state chemistry" was the principle of detailed balance. Taking cognizance of the implications of this principle one not only can extract more out of the data (as discussed in Section 2.2.5; see, e.g., Figure 7), but can also include it as a dynamic constraint to obtain superior expressions for the surprisal. We did not conclude Section 2 by a guideposts section because our earlier summary (Section 1.5) is still very much valid. All one need do is: (1) Replace everywhere the marginal $P(\beta)$ distribution by the "reasonable person's" joint $P(\beta, \alpha)$ distribution. (2) In analyzing the experimental setup, consider if detailed balance can be applied. If so, then during the process described in Sections 1.5.1 and 1.5.4 do not fail to make use of this fact. In particular, choose your constraints with some reference to the comments of Section 2.5. In attempting Section 1.5.7, do go to the original literature.[65] See also References 20, 22, 23, 25, 27, 30, and 37 on surprisal synthesis for **P** matrices. You still think that being a theorist is easy? Then please note that some guideposts are still missing. We have little experience with mutual distributions, e.g., distributions in both final translational energy and final scattering angle[74] or in both final vibrational and rotational energy.[8,10,75] Similarly, we hardly gave you enough guidance on the energy (or the temperature) dependence of the overall rate coefficients $k(E)$ and $k(T)$, respectively. Let us know what you come up with.

Appendix A. Handbook of Prior Distributions

The present compilation cannot replace the extensive appendixes[10,25,38,59,60,72] which it summarizes. Nor does it discuss inclusion of angular momentum constraints[77,78] or the realistic accounting of the internal energy levels.[19,78] The product molecules are thus regarded throughout as having a rigid-rotor–harmonic-oscillator (RRHO) level scheme. Furthermore, any internal degree of freedom not explicitly selected is regarded as having a continuous energy and is integrated, rather than summed, over. On an absolute scale these are drastic approximations.[19,38,60] However: (a) There is no reason, except for simplicity, to invoke the RRHO approximation. One can always start from the most detailed ($\alpha \to \beta$) result, and correctly average over initial levels and sum over the final ones. For benchmark work, that is the only way. (b) On a relative scale, the RRHO approximation is more accurate than the above description suggests.[60] Hence, unless one is analyzing $k(\mathbf{X})$ (or branching ratios for different chemical arrangements, e.g., References 19 and 79), the RRHO approximation offers a very convenient starting point. Even for branching ratios it is quite reasonable except when the different arrangements have markedly different threshold energies. Even then (see, e.g., Reference 79) the effects of refinements are not dramatic.

A.1. Density of States

The translational density of states $\rho_T(E_T)$ is

$$\rho_T(E_T) = (\pi^{-1}k^2/\mathbf{v})/h = A_T E_T^{1/2} \tag{A.1}$$

where $E_T = \hbar^2 k^2/2\mu = (1/2)\,\mu\mathbf{v}^2$ and $A_T = \mu^{3/2}/2^{1/2}\pi^2 h^3$. The density of states for any group a of degenerate internal states is $\rho(a; E)$:

$$\rho(a; E) = g_a \rho_T(E_T) \tag{A.2}$$

where $E_T = E - E_a$ and g_a is the number (the degeneracy) of internal quantum states of energy E_a. For the A + BC (v, J) case,

$$\rho(v, J; E) = (2J + 1)\,\rho_T(E_T) = (2J + 1)\,A_T(E - E_{VJ})^{1/2} \tag{A.3}$$

For the AB (v, J) + BC (n, L) case,

$$\rho(v, J, n, L; E) = A_T(2J + 1)(2L + 1)(E - E_{VJ} - E_{NL})^{1/2} \tag{A.4}$$

The density of states for any coarser specification is rigorously obtained by summation:

$$\rho(\Gamma; E) = \sum_{a\in\Gamma} \rho(a; E) \tag{A.5}$$

A.1.1. Atom–Diatom

In the RR (rigid-rotor) approximation

$$\rho(E_V; E) = \tfrac{2}{3} A_T A_R (E - E_V)^{3/2} \tag{A.6}$$

Here

$$A_R = (2J + 1)\,dJ/dE_R = (hcB_v)^{-1} \tag{A.7}$$

with B_v as the rotational constant (in cm^{-1}) for the vibrational level v. In the RRHO approximation, $B_v = B_e$ and

$$\rho(E_T; E) = A_T E_T^{1/2}(E - E_T) \tag{A.8}$$

and

$$\rho(E) = \sum_{\alpha} \rho(\alpha, E) = \tfrac{4}{15} A_T A_I E^{5/2} \tag{A.9}$$

where

$$A_I = A_R A_v = (h\omega_e \cdot hcB_e)^{-1}$$

A.1.2. Diatom–Diatom

In the RRHO approximation

$$\rho(E_V, E_N; E) = \tfrac{4}{15}A_T A_R(E - E_V - E_N)^{5/2} \tag{A.10}$$

Here A_R allows for the integration over the rotational states of both diatomics, $A_R = (hc)^{-2} B_v B_n$. If only the vibrational energy of one diatomic, say E_V, is specified, one needs to integrate over E_N:

$$\rho(E_V; E) = \tfrac{8}{95}A_T A_R(E - E_V)^{7/2}/hc\omega_e \tag{A.11}$$

where ω_e is the vibrational frequency of the other diatomic. Finally, $\rho(E) \propto E^{9/2}$.

A.2. Prior Distributions at a Given Total Energy

Directly from the definitions (Sections 1.3 and 2.3)

$$P^0(\alpha|E) = \rho(\alpha|E) \equiv \rho(\alpha; E)/\rho(E) \tag{A.12}$$

or, for a coarser specification,

$$P^0(\Gamma|E) = \rho(\Gamma|E) = \rho(\Gamma; E)/\rho(E) \tag{A.13}$$

Similarly,

$$P^0(\beta, \alpha|E) = P^0(\alpha|E)\, P^0(\beta|E) \tag{A.14}$$

$$P^0(\Gamma', \Gamma|E) = P^0(\Gamma|E)\, P^0(\Gamma'|E) \tag{A.15}$$

Explicit results are obtained using the density of states as given. Special cases of common interest include atom–diatom collisions where at the RHHO level

$$P^0(E_V|E) = \tfrac{5}{2}(E - E_V)^{3/2}/E^{5/2} \tag{A.16}$$

$$P^0(E_T|E) = \tfrac{15}{4}E_T^{1/2}(E - E_T)/E^{5/2} \tag{A.17}$$

$$\begin{aligned} P^0(E_R|E_V, E) &= P^0(E_R, E_V|E)/P^0(E_V|E) \\ &= \tfrac{3}{2}(E - E_V - E_R)^{1/2}/(E - E_V)^{3/2} \\ &= \tfrac{3}{2}(1 - g_R)^{1/2}/(E - E_V) \end{aligned} \tag{A.18}$$

Otherwise,

$$P^0(E_T|E) = C_n E_T^{1/2}(E - E_T)^n/E^{n+3/2} \tag{A.19}$$

where

$$C_n = \Gamma(n + \tfrac{5}{2})/n!\Gamma\tfrac{3}{2} \tag{A.20}$$

$n = 1, 3, 4, 6$ for A + BC, AB + CD, A + BCD, and AB + CDE type of products, respectively.

A.3. Prior Distributions at a Given Temperature

The basic definition is equation (91):

$$k^0(\alpha \to \beta; T) = Q_T^{-1} \int_0^\infty dE_T \rho(\alpha; E)\, \rho(\beta; E) \exp(-E_T/\mathbf{k}T) \quad \text{(A.21)}$$

where E_T is the initial kinetic energy and $\rho(\beta; E)$ is defined to be zero for $E_T < E_\beta$. The integral can be performed analytically[(25,60,67)] and leads to

$$k^0(\alpha \to \beta; T) = N(\beta, T)\, \Delta \exp(\Delta)\, K_1(\Delta) \quad \text{(A.22)}$$

Here $\Delta = \Delta(\alpha, \beta)$ is the reduced energy gap,

$$\Delta(\alpha, \beta) = (-\Delta E_0 + E_\alpha - E_\beta)/2\mathbf{k}T \quad \text{(A.23)}$$

ΔE_0 is the zero-point energy of the products minus that of the reactants ($-\Delta E_0$ is the exoergicity), and E_α and E_β are measured from their respective zero points. $K_1(\Delta)$ is the modified Bessel function (of the second kind of order 1), defined for negative Δ by

$$(-\Delta)^n K_n(-\Delta) = \Delta^n K_n(\Delta) \quad \text{(A.24)}$$

so that (A.22) satisfies detailed balance. $N(\beta, T)$ is a temperature-dependent ($T^{1/2}$) factor,

$$N(\beta, T) = (2\pi\mu\mathbf{k}T/h^2)^{1/2}\, 8R^0/h^2 \quad \text{(A.25)}$$

where the argument β in $N(\beta, T)$ is a reminder that it is the mass and the spectroscopic parameters of the final state that appear in N. To determine $k(\alpha \to \beta; T)$ one need merely multiply (A.22) by g_β.

Coarser descriptions are obtained via the canon

$$k^0(\Gamma \to \Gamma'; T) = \sum_{\alpha\in\Gamma} X_\alpha \sum_{\beta\in\Gamma'} k^0(\alpha \to \beta; T) \quad \text{(A.26)}$$

In the RRHO approximation it is generally possible to obtain closed-form expressions of the general type[(60)]

$$k^0(\Gamma \to \Gamma'; T) = N(\Gamma'; T) \exp(\Delta)\, \Delta \vartheta_{a,b}(\Delta) \quad \text{(A.27)}$$

which are summarized, for atom–diatom collisions, in Table A.

$$\vartheta_{a,a}(\Delta) = (2a + 1)!!\Delta^{a+1} K_{a+1}(\Delta) \quad \text{(A.28)}$$

$$\vartheta_{0,1}(\Delta) = \Delta^2 [K_2(\Delta) - K_1(\Delta)] \quad \text{(A.29)}$$

$$\vartheta_{0,2}(\Delta) = \Delta^2 [2\Delta K_3(\Delta) - (2\Delta + 5) K_2(\Delta)] \quad \text{(A.30)}$$

$$\vartheta_{1,2}(\Delta) = 3\Delta^3 [K_3(\Delta) - K_2(\Delta)] \quad \text{(A.31)}$$

Table A. Prior Rate Constants $k^0(\Gamma_i \to \Gamma_f; T)$ for Atom–Diatom Collisions in the RRHO Level Scheme[a]

	Γ_i				
Γ_f	C_i, v_i, J_i	C_i, v_i	$C_i, E_T(i)$	C_f	$N_f = N(\Gamma_f; T)$
C_f, v_f, J_f	$N_f \Delta e^{\Delta} K_1(\Delta)$ $E_{VJ}(i) - E_{VJ}(f)$	$(2/3)\, N_f e^{\Delta} \vartheta_{0.1}(\Delta)$ $E_V(i) - E_{VJ}(f)$		$N_f e^{\Delta} \vartheta_{0.2}(\Delta) - \dot{E}_{VJ}(f)$	$(2J_f + 1)n$
C_f, v_f	$(2/3)\, N_f e^{\Delta} \vartheta_{1.0}(\Delta)$ $E_{VJ}(i) - E_V(f)$	$(4/3)\, N_f \Delta^2 e^{\Delta} K_2(\Delta)$ $E_V(i) - E_V(f)$		$(8/45)\, N_f e^{\Delta} \vartheta_{1.2}(\Delta)$ $- E_V(f)$	$(\mathbf{k}T/B_e hc)\, n$
$C_f, E_T(f)$			$N_f \Delta^{3/2} K_{3/2}(\Delta)$ $E_T(i) - E_T(f)$		$[(\mathbf{k}T/hc)^2/B_e \omega_e]\, n$ $\times [E_T(f)/\mathbf{k}T]^{1/2}$
C_f	$(4/15)\, N_f e^{\Delta} \vartheta_{2.0}(\Delta)$	$(8/45)\, N_f e^{\Delta} \vartheta_{2.1}(\Delta)$ $E_V(i)$		$(16/15)\, N_f \Delta^3 e^{\Delta} K_3(\Delta)$ 0	$[(\mathbf{k}T/hc)^2/B_e \omega_e]\, n$

[a] $n = (2\pi\mu \mathbf{k}T/h^2)(8\mu R^0/h^2)$; C denotes a chemical species; $E_{VJ}(i) = E(v_i, J_i)$, etc.; the second entry is $\Delta E_0 + 2\mathbf{k}T\Delta$; Δ is defined by (A.23): $\Delta = [-\Delta E_0 + E(\Gamma_i) - E(\Gamma_f)]/2\mathbf{k}T$.

and

$$\vartheta_{a,b}(\Delta) = \vartheta_{b,a}(-\Delta) \tag{A.32}$$

These expressions reduce to the results of Section A.1, i.e.,

$$k^0(\Gamma \to \Gamma'; E) = R^0 \rho(\Gamma'; E) \tag{A.33}$$

upon using the asymptotic expansion

$$\exp(\Delta)\,\Delta^{1/2} K_a(\Delta) \to (\pi/2)^{1/2}\,[1 + (\mu - 1)/8\Delta + \cdots] \tag{A.34}$$

where $\mu = 4a^2$.

Prior distributions obtain as usual from the prior rates, e.g.,

$$P^0(\to\Gamma'; T) = \sum_{\Gamma} Q(\Gamma|T)\,k^0(\Gamma \to \Gamma'; T)/k^0(T) \tag{A.35}$$

where

$$k^0(T) = \sum_{\Gamma} Q(\Gamma|T) \sum_{\Gamma'} k^0(\Gamma \to \Gamma'; T) \tag{A.36}$$

and

$$P^0(\Gamma', \Gamma; T) = Q(\Gamma|T)\,k^0(\Gamma \to \Gamma'; T)/k^0(T) \tag{A.37}$$

Appendix B. The Persuasion

The quality of fit function[12,48,56,61]

$$H = \sum_{\Gamma} P(\Gamma) \ln[P(\Gamma)/Q(\Gamma)] \tag{B.1}$$

is a measure of how well the theoretical probabilities $Q(\Gamma)$ fit the observed frequencies $P(\Gamma)$. It is nonnegative and equals zero for, and only for, a perfect fit, i.e., when $P(\Gamma) = Q(\Gamma)$ for all groups Γ. In information theoretic terms H is the amount of information provided by the frequencies $P(\Gamma)$ when the theoretical probabilities $Q(\Gamma)$ are known beforehand. When the theoretical probabilities account well for the observations no additional information (in the technical sense) is provided by the measured frequencies; the theory is quite persuasive. Not so for a poor fit. It is then necessary to refine the theory that led to the $Q(\Gamma)$'s, until the experimental results present no surprises.

When the theoretical results are quite good, so that the deviance of $P(\Gamma)$ from $Q(\Gamma)$ is, on the average, small, the persuasion reduces to the traditional χ^2 "goodness-of-fit" measure,

$$H \simeq \tfrac{1}{2} \sum_{\Gamma} [P(\Gamma) - Q(\Gamma)]^2/P(\Gamma) \tag{B.2}$$

In practice, an acceptable fit obtains when H is zero to within experimental error δH (determined by the squared uncertainties $[\delta P(\Gamma)]^2$ in the measured frequencies):

$$\delta H = \tfrac{1}{2}\sum_{\Gamma} P(\Gamma)\,[\delta P(\Gamma)/P(\Gamma)]^2 \tag{B.3}$$

Equation (B.3) shows that δH is independent of the theoretical distribution $Q(\Gamma)$ and is 1/2 the average squared fractional error in the P's: $\delta H = (1/2)\langle[\delta P(\Gamma)/P(\Gamma)]^2\rangle$. Note that in (B.3) the uncertainties must conform to the constraints on the distribution, in particular, $\sum \delta P(\Gamma) = 0$. Conversely, if the P's are given, i.e., regarded as exact, then an error $\delta Q(\Gamma)$ in the theoretical probabilities translates into H being nonzero by δH: $\delta H = (1/2)\langle[\delta Q(\Gamma)/Q(\Gamma)]^2\rangle$. Hence, to achieve an overall $n\%$ accord between the theoretical (the Q's) and the experimental probabilities (the P's) requires that $H \leq (n/100)^2/2$. Two competing theories, yielding the probabilities $Q^a(\Gamma)$ and $Q^b(\Gamma)$, cannot be distinguished by the experiment when the difference in their H values,

$$\Delta H = \sum_{\Gamma} P(\Gamma)\ln[Q^a(\Gamma)/Q^b(\Gamma)] \tag{B.4}$$

is comparable to δH. The best theory is, of course, one which is formulated after the results are known. Taking $Q^a(\Gamma) = P(\Gamma)$ in (B.4) one sees that H can be viewed as a measure of how much better can the best (the so-called, "Monday morning") theory do as compared to the theory at hand. The worst possible contender to the best theory is formulated when no information on the system, beyond conservation laws, is available. Then $Q(\Gamma) = P^0(\Gamma)$, the prior distribution, and H is the information content [or entropy deficiency, ΔS; cf. equation (24)] of the experiment. A solid theory should account for much of the measured information content so that its H value should be quite small compared to $\Delta S/R$. If ΔS is already quite small, the prior distribution itself accounts for the data.

Any set of (normalized) probabilities can be compared with experiment using (B.1). An important special case is when the theoretical distribution $Q(\alpha)$ is the induced distribution determined by the procedure of maximal entropy, $Q(\alpha) \equiv P^I(\alpha)$:

$$P^I(\alpha) = P^0(\alpha)\exp\Big[-\lambda_0 - \sum_{r=1}^{n} \lambda_r A_r(\alpha)\Big] \tag{B.5}$$

Here the Lagrange parameters have been chosen so that for $r = 1, 2, \ldots, n$

$$\sum_{r} P(\alpha)\,A_r(\alpha) = \sum_{r} P^I(\alpha)\,A_r(\alpha) \tag{B.6}$$

and so that $P^I(\alpha)$ is normalized. In this case[56,61]

$$H = \Delta S[P] - \Delta S[P^I] \geq 0 \tag{B.7}$$

$\Delta S[P]$ is the entropy deficiency computed using the experimental frequencies but $\Delta S[P^I]$, the entropy deficiency of the induced distribution, can be increased by adding more terms in the exponent of (B.5), until $H = \Delta S[P] - \Delta S[P^I] \ll \Delta S[P]$. The dynamical approach[15] has demonstrated that $H \equiv 0$ can indeed be reached in this fashion.

Appendix C. The Determination of the Lagrange Parameters

A practical procedure for determining the magnitudes of the Lagrange parameters is provided by the use of the procedure of Appendix B.[48] A computer program has been described and is available upon request. To determine a distribution of maximal entropy subject to n constraints A_1 to A_n one introduces a trial distribution $P^T(\alpha)$ of the form

$$P^T(\alpha) = P^0(\alpha)\exp\left[-\lambda_0^T - \sum_{r=1}^{n} \lambda_r^T A_r(\alpha)\right] \tag{C.1}$$

The distribution is taken to be normalized

$$\exp(\lambda_0^T) = \sum_{\alpha} P^0(\alpha)\exp\left[-\sum_{r=1}^{n} \lambda_r A_r(\alpha)\right] \tag{C.2}$$

so that λ_0^T is a function [defined via (C.2)] of the n "trial" Lagrange parameters $\lambda_1^T, \ldots, \lambda_n^T$.

An optimal set of Lagrange parameters is obtained by seeking to minimize the function H [cf. (B.1)]:

$$\begin{aligned} 0 \le H &= \sum_{\alpha} P(\alpha)\ln\left[P(\alpha)/P^T(\alpha)\right] \\ &= \Delta S[P] - \sum_{\alpha} P(\alpha)\ln\left[P^T(\alpha)/P^0(\alpha)\right] \\ &= \Delta S[P] + \lambda_0^T + \sum_{r=1}^{n} \lambda_r^T\left[\sum_{\alpha} P(\alpha) A_r(\alpha)\right] \\ &= \Delta S[P] + \lambda_0^T + \sum_{r=1}^{n} \lambda_r^T \langle A_r\rangle \end{aligned} \tag{C.3}$$

The quality-of-fit function H in (C.3) is now a function of n-independent trial parameters $\lambda_1^T, \ldots, \lambda_n^T$ (recall that λ_0^T is a function of $\lambda_1^T, \ldots, \lambda_n^T$). These are varied until a minimum is reached. There is only one unique minimum[48] provided that the constraints are linearly independent, i.e., that there are no $n + 1$ real numbers C_r such that

$$\sum_{r}^{n} C_r A_r(\alpha) + C_0 = 0 \tag{C.4}$$

H is a concave function of the λ_r^T's so that a very efficient iteration procedure is possible. Practical details have been discussed at length elsewhere.

Here we shall only prove that the resulting set of Lagrange parameters is identical to the set found by the solution of the usual implicit equations [cf. (C.7) below]. From (C.3)

$$\partial H / \partial \lambda_r^T = \partial \lambda_0^T / \partial \lambda_r^T + \langle A_r \rangle \tag{C.5}$$

but from (C.2)

$$-\partial \lambda_0^T / \partial \lambda_r^T = \sum_\alpha A_r(\alpha) P^0(\alpha) \exp\left[-\lambda_0^T - \sum_{r=1}^{n} \lambda_r^T A_r(\alpha) \right] = \sum_\alpha A_r(\alpha) P^T(\alpha) \tag{C.6}$$

At the minimum $\partial H / \partial \lambda_r^T = 0$ or, using (C.5) and (C.6), the optimal set of Lagrange parameters is the solution of

$$\sum_i A_r(\alpha) P^T(\alpha) = \langle A_r \rangle, \qquad r = 1, \ldots, n \tag{C.7}$$

Here $\langle A_r \rangle$ is the correct average value of A_r computed for the "exact" distribution $P(\alpha)$. The inclusion of additional averages, i.e., increasing the number n of trial Lagrange parameters, can only serve to improve the fit. At worst these additional Lagrange parameters will, at the minimum of H, be equal to, or be very near to, zero. This implies (cf. Reference 9) that no additional information is provided by these additional average values. They are "noninformative" in the same way that, say, the magnitude of the specific heat of an ideal gas at thermal equilibrium is noninformative.[8] [Given the magnitude of the mean energy one computes the Lagrange parameter $(1/\mathbf{k}T)$ and hence the specific heat.]

It is important to note that the procedure does not really require that the experimental frequencies, the $P(\alpha)$'s, be known. All that is required are the magnitudes of the n averages $\langle A_r \rangle$ determined for the distribution $P(\alpha)$. [The $\Delta S[P]$ term in (C.3) is not dependent on the λ_r^T's and hence plays no role during the minimization precedure.]

ACKNOWLEDGMENTS

Work supported by the Air Force Office of Scientific Research, USAF, under Grant AFOSR-77-3135 and by grants from the National Science Foundation and the Office of Naval Research.

References

1. N. Abramson, *Information Theory and Coding*, McGraw-Hill, New York (1963).
2. S. Goldman, *Information Theory*, Prentice-Hall, New York (1953).
3. C.E. Shannon, A mathematical theory of communications, *Bell Syst. Tech. J.* **27**, 379–423 (1948).
4. N. Wiener, *Cybernetics*, MIT Press, Cambridge, Massachusetts (1948).

5. R.D. Levine and R.B. Bernstein, Energy disposal and energy consumption in elementary chemical reactions: The information theoretic approach, *Acct. Chem. Res.* **7**, 393–400 (1974).
6. R.D. Levine and R.B. Bernstein, Thermodynamic approach to collision processes, in *Modern Theoretical Chemistry*, Vol. III, W.H. Miller, editor, Plenum Press, New York (1976), pp. 323–364.
7. R.D. Levine, Energy consumption and energy disposal in elementary chemical reactions, in *The New World of Quantum Chemistry*, B. Pullman and R. Parr, editors, D. Reidel, Dordrecht (1976), pp. 103–129.
8. R.D. Levine and A. Ben-Shaul, Thermodynamics of molecular disequilibrium, in *Chemical and Biochemical Applications of Lasers*, Vol. II, C.B. Moore, editor, Academic Press, New York (1977), pp. 145–197.
9. R.D. Levine, Information theory approach to molecular reaction dynamics, *Ann. Rev. Phys. Chem.* **29**, 59–91 (1978).
10. R.B. Bernstein and R.D. Levine, Role of energy in reactive molecular scattering: An information theoretic approach, *Adv. At. Mol. Phys.* **11**, 215–297 (1975).
11. E.T. Jaynes, in *Statistical Physics*, K.E. Ford, editor, Benjamin, Reading, Mass. (1963).
12. M. Tribus, *Thermostatistics and Thermodynamics*, Van Nostrand, Princeton (1961).
13. A. Katz, *Principles of Statistical Mechanics: The Information Theory Approach*, Freeman San Francisco (1967).
14. A. Hobson, *Concepts in Statistical Mechanics*, Gordon and Breach, New York (1972).
15. Y. Alhassid and R.D. Levine, Entropy and chemical change III. The maximal entropy procedure as a dynamical theory, *J. Chem. Phys.* **67**, 4321–4339 (1977).
16. Y. Alhassid and R.D. Levine, Connection between the maximal entropy and the scattering theoretic analyses of collision processes, *Phys. Rev.* A **18**, 89–116 (1978).
17. R.D. Levine and R. Kosloff, The well reasoned choice: An information theoretic approach to branching ratios in molecular rate processes, *Chem. Phys. Lett.* **28**, 300–304 (1974).
18. R.D. Levine and R.B. Bernstein, Branching ratios in reactive molecular collisions, *Chem. Phys. Lett.* **29**, 1–6 (1974).
19. U. Dinur, R. Kosloff, R.D. Levine, and M.J. Berry, Analysis of electronically nonadiabatic chemical reactions—information theory approach, *Chem. Phys. Lett.* **34**, 199–205 (1975).
20. H. Kaplan and R.D. Levine, Dependence of branching ratio in F and HD reaction on initial rotational state of HD, *Chem. Phys.* **13**, 161–171 (1976).
21. R.B. Bernstein, Branching ratios in reactive collisions—theory versus experiment, *Int. J. Quantum Chem. Symp.* **9**, 385–395 (1975).
22. C.L. Vilá and J.L. Kinsey, Information theory for experiments with known conditional probabilities: Application to the prediction of branching ratios, *Chem. Phys.* **26**, 17–22 (1977).
23. J.C. Polanyi and J.L. Schreiber, Test of the information theory of branching ratios by classical trajectory computation, *Chem. Phys.* **31**, 113–136 (1978).
24. I. Procaccia and R.D. Levine, From bulk vibrational relaxation data to the detailed (microscopic) rate constant, *J. Chem. Phys.* **62**, 2496 (1975).
25. I. Procaccia and R.D. Levine, Vibrational energy-transfer in molecular collisions—information theoretic analysis and synthesis, *J. Chem. Phys.* **63**, 4261–4279 (1975).
26. A. Kafri, E. Pollak, R. Kosloff, and R.D. Levine, Translational energy disposal in molecular collisions—transfer of momentum constraint, *Chem. Phys. Lett.* **33**, 201–206 (1975).
27. H. Kaplan and R.D. Levine, Translational energy requirements and disposal in molecular collisions I. Cross sections and probability matrices for alkali atom–alkyl iodide reactions, *Chem. Phys.* **18**, 103–123 (1976).
28. L.A. Collins and N.F. Lane, Quantum-mechanical calculations of rotational-excitation cross sections for HCl, DCl, DF, and HF in collisions with He, *Phys. Rev.* A **14**, 1358–1367 (1976).

29. R.D. Levine, R.B. Bernstein, P. Kahana, I. Procaccia, and E.T. Upchurch, Surprisal analysis and probability matrices for rotational energy-transfer, *J. Chem. Phys.* **64**, 796–807 (1976).
30. I. Procaccia and R.D. Levine, Cross sections for rotational energy-transfer—an information-theoretic synthesis, *J. Chem. Phys.* **64**, 808–817 (1976).
31. I. Procaccia and R.D. Levine, Comment on the rotational excitation in hydrogen-halide–rare-gas collisions, *Phys. Rev.* A **14**, 1569 (1976).
32. B.E. Wilcomb and P.J. Dagdigian, Determination of state resolved rotationally inelastic cross sections: LiH$(j = 1)$–Ar, *J. Chem. Phys.* **67**, 3829 (1977).
33. M.D. Pattengill and R.B. Bernstein, Surprisal analysis of classical trajectory calculations of rotationally inelastic cross sections for the Ar–N_2 system; influence of the potential energy surface, *J. Chem. Phys.* **65**, 4007–4017 (1976).
34. J.C. Polanyi, N. Sathyamurthy, and J.L. Schreiber, Rotational energy transfer theory I. Comparison of quasiclassical and quantum-mechanical results for elastic and rotationally inelastic HCl–Ar collisions, *Chem. Phys.* **24**, 105–110 (1977).
35. D.F. Heller, Exponential gap law for atom–diatom scatting, *Chem. Phys. Lett.* **45**, 64–70 (1977); E. Pollak, Rotational surprisals for H + H_2, a simple theory, *Chem. Phys. Lett.* **47**, 513–516 (1977); A.E. DePristo and H. Rabitz, Quantum number and energy scaling of rotationally inelastic scattering cross sections, *Chem. Phys.* **24**, 209–210 (1977).
36. R.D. Driver, T.A. Brunner, N. Smith, and D.E. Pritchard, Rotational energy transfer in $Na_2(A^1\Sigma)$–Xe collisions: Level to level dynamics, *J. Chem. Phys.*, in press.
37. A. Kafri and R. Kosloff, V–RT transfer—study of two opposing dynamic constraints, *Chem. Phys.* **23**, 257–271 (1977).
38. J.L. Kinsey, Microscopic reversibility for rates of chemical reactions carried out with partial resolution of the product and reactant states, *J. Chem. Phys.* **54**, 1206–1217 (1971).
39. J.C. Light, Chapter 19 in this volume.
40. E.E. Nikitin, *Theory of Elementary Atomic and Molecular Processes in Gases*, Clarendon Press, Oxford (1974).
41. R.A. Marcus, On the theory of chemical reaction cross sections I. Statistical-dynamical model, *J. Chem. Phys.* **45**, 2630–2638 (1966).
42. M. Quack and J. Troe, Information, memory and statistical theories of elementary chemical reactions, *Ber. Bunsenges. Phys. Chem.* **80**, 1140–1149 (1976).
43. W.H. Miller, Unified statistical model for "complex" and "direct" reaction mechanisms, *J. Chem. Phys.* **65**, 2216–2223 (1976).
44. M. Quack and J. Troe, Unimolecular processes 4. Product state distributions after dissociation, *Ber. Bunsenges. Phys. Chem.* **79**, 469–475 (1975).
45. A. Ben-Shaul, R.D. Levine, and R.B. Bernstein, Prior-expectation distribution functions for energy disposal and energy consumption in reactive molecular collisions, *J. Chem. Phys.* **61**, 4937–4938 (1974); A. Ben-Shaul, Prior distributions in molecular collision theory, *Chem. Phys.* **22**, 341–366 (1977); E. Pollak, Prior statistical distributions for the collision of an atom with a diatom, *J. Chem. Phys.* **68**, 547–563 (1978).
46. R.V.L. Hartley, Transmission of information, *Bell Syst. Tech. J.* **7**, 535 (1928).
47. W.H. Miller, Classical *S*-matrix in molecular collisions, *Adv. Chem. Phys.* **30**, 77–136 (1975).
48. Y. Alhassid, N. Agmon, and R.D. Levine, An algorithm for finding the distribution of maximal entropy, *Chem. Phys. Lett.* **53**, 22–26 (1978).
49. I. Procaccia and R.D. Levine, Potential work—statistical-mechanical approach for systems in disequilibrium, *J. Chem. Phys.* **65**, 3357–3364 (1976).
50. P. Pechukas, Statistical approximation in collision theory, in *Modern Theoretical Chemistry*, Vol. III, W.H. Miller, editor, Plenum Press, New York (1976), pp. 269–322.
51. R.D. Levine and R.B. Bernstein, *Molecular Reaction Dynamics*, Clarendon Press, Oxford (1974).
52. J.L. Kinsey, Laser induced fluorescence, *Ann. Rev. Phys. Chem.* **28**, 349–372 (1977).

53. J.L. Kinsey, Molecular beam reaction, in *Chemical Kinetics,* J.C. Polanyi, editor, Butterworth-Med. Tech. Press, Oxford (1972), pp. 173–212.
54. D.R. Herschbach, Molecular dynamics of chemical reactions, *Pure Appl. Chem.* **47,** 61–73 (1976).
55. M.B. Faist, R.D. Levine, and R.B. Bernstein, Angular entropy: Information content of molecular scattering angular distributions, *J. Chem. Phys.* **66,** 511–523 (1977).
56. E. Pollak and R.D. Levine, Surprisal analysis of products translational energy distribution in molecular collisions, *Chem. Phys.* **21,** 61–80 (1977).
57. R.B. Bernstein and B.E. Wilcomb, On the products' translational distribution for the reactions of CH_3I and CF_3I with K and Ba, *J. Chem. Phys.* **67,** 5809–5815 (1977).
58. T.P. Schafer, P.E. Siska, J.M. Parson, F.P. Tully, Y.C. Wong, and Y.T. Lee, Crossed molecular beam study of $F + D_2$, *J. Chem. Phys.* **53,** 3385–3387 (1970); K.G. Anlauf, P.E. Charters, D.S. Horne, R.G. McDonald, J.C. Polanyi, W.J. Skrlac, D.C. Tardy, and K.B. Woodall, Translational energy-distribution in the products of some exothermic reactions. *J. Chem. Phys.* **53,** 4091–4092 (1970).
59. A. Ben-Shaul, R.D. Levine, and R.B. Bernstein, Entropy and chemical change II. Temperature and entropy deficiency of products' state distribution, *J. Chem. Phys.* **57,** 5427–5447 (1972).
60. R.D. Levine and J. Manz, Effect of reagent energy on chemical reaction rates: Information theoretic analysis, *J. Chem. Phys.* **63,** 4280–4303 (1975).
61. U. Halavee and R.D. Levine, Quantal transition probabilities from classical trajectories: Application to collinear reactive H and Cl_2 system, *Chem. Phys. Lett.* **46,** 35–41 (1977).
62. R. Kosloff, A. Kafri, and R.D. Levine, Rotational excitation of interstellar OH molecules, *Astrophys. J.* **215,** 497–502 (1977).
63. R.C. Tolman, *Statistical Mechanics with Applications to Physics and Chemistry,* Chemical Catalog Co., New York (1927).
64. C. Rebick, R.D. Levine, and R.B. Bernstein, Energy requirements and energy disposal: Reaction probability matrices and a computational study of a model system, *J. Chem. Phys.* **60,** 4977–4989 (1974).
65. H. Kaplan, R.D. Levine, and J. Manz, Microscopic reversibility and probability matrices for molecular collisions: Information theoretic synthesis *Mol. Phys.* **31,** 1765–1782 (1976).
66. H. Kaplan, R.D. Levine, and J. Manz, Dependence of reaction-rate constant on reagent excitation: Implications of detailed balance, *Chem. Phys.* **12,** 447–461 (1976).
67. M. Rubinson and J.I. Steinfeld, Entropy analysis of product energy-distributions in nonreactive inelastic collisions, *Chem. Phys.* **4,** 467–475 (1974).
68. E. Pollak and R.D. Levine, Different roles of reagent vibrational excitation for endothermic and exothermic reactions, *Chem. Phys. Lett.* **39,** 199–204 (1976).
69. I. Procaccia and R.D. Levine, Rotational excitation of HD by collisions with He, *Physica A* **82,** 623–630 (1975).
70. D.M. Manos and J.M. Parson, Crossed molecular beam study of chemiluminescent reactions of Group IIIB atoms with O_2, *J. Chem. Phys.* **63,** 3575–3585 (1975).
71. K. Liu and J.M. Parson, Deactivation of $Hg(^3P_2)$ in crossed molecular beams, *J. Chem. Phys.* **65,** 815–822 (1976).
72. M. Tamir and R.D. Levine, Translational energy requirements and disposal in molecular collisions 2. Branching ratios and products state distributions for exchange reactions of alkali atoms with alkali-halides, *Chem. Phys.* **18,** 125–144 (1976).
73. R.D. Levine and R.B. Bernstein, Energy disposal and energy requirements for elementary chemical reactions, *Faraday Discuss. Chem. Soc.* **55,** 100–112 (1973).
74. R.B. Bernstein and M.B. Faist, Information content of detailed differential reaction cross sections (molecular-beam scattering velocity angle maps) vs. resolved product state distributions, *J. Chem. Phys.* **65,** 5436–5444 (1976).
75. H.J. Korsch and R.D. Levine, Joint products state distributions in molecular collisions, *Chem. Phys.* **30,** 333–341 (1978).

76. R.B. Bernstein, Information theoretic analysis of rotational distributions from quantal and quasiclassical computations of reactive and nonreactive scattering, *Int. J. Quantum Chem. Symp.* **10**, 267–276 (1976).
77. A. Ben-Shaul, Statistical models and prior distributions in theory of chemical reactions, *Chem. Phys.* **22**, 341–366 (1977).
78. A. Kafri, Computations and application of classical phase space integrals in reactive molecular collisions, *Chem. Phys.* **13**, 309–321 (1976).
79. M.B. Faist and R.D. Levine, On the product electronic state distribution in reactions of alkali dimers with halogen atoms, *Chem. Phys. Lett.* **47**, 5–10 (1977).

Author Index

Page numbers in italics refer to the reference sections at the ends of chapters.

Subject Index

GPSR Compliance

The European Union's (EU) General Product Safety Regulation (GPSR) is a set of rules that requires consumer products to be safe and our obligations to ensure this.

If you have any concerns about our products, you can contact us on ProductSafety@springernature.com

In case Publisher is established outside the EU, the EU authorized representative is:

Springer Nature Customer Service Center GmbH
Europaplatz 3
69115 Heidelberg, Germany

Batch number: 10761593

Printed by Printforce, the Netherlands